Environmental Politics and Theory

Series Editor
Joel Jay Kassiola, Department of Political Science, San Francisco State University, San Francisco, CA, USA

The premise of this series is that the current environmental crisis cannot be solved by technological innovation alone. The environmental challenges we face today are, at their root, political crises involving political values, institutions and struggles for power. Therefore, environmental politics and theory are of the utmost social significance.

Growing public consciousness of the environmental crisis and its human and more-than-human impacts, exemplified by the worldwide urgency and political activity associated with the problem and consequences of climate and earth system change make it imperative to design and achieve a sustainable and socially just society.

The series publishes inter- and multi-disciplinary scholarship that extends the theoretical dimensions of green political theory, international relations, philosophy, and earth system governance. It addresses the need for social change away from the hegemonic consumer capitalist society to realize environmental sustainability and social justice.

Joel Jay Kassiola · Timothy W. Luke
Editors

The Palgrave Handbook of Environmental Politics and Theory

Editors
Joel Jay Kassiola
Department of Political Science
San Francisco State University
San Francisco, CA, USA

Timothy W. Luke
Department of Political Science
Virginia Tech
Blacksburg, VA, USA

ISSN 2731-670X   ISSN 2731-6718 (electronic)
Environmental Politics and Theory
ISBN 978-3-031-14345-8   ISBN 978-3-031-14346-5 (eBook)
https://doi.org/10.1007/978-3-031-14346-5

© The Editor(s) (if applicable) and The Author(s), under exclusive license to Springer Nature Switzerland AG 2023, corrected publication 2023
Chapter "Eco-Anxiety and the Responses of Ecological Citizenship and Mindfulness" is licensed under the terms of the Creative Commons Attribution 4.0 International License (http://creativecommons.org/licenses/by/4.0/). For further details see license information in the chapter.
This work is subject to copyright. All rights are solely and exclusively licensed by the Publisher, whether the whole or part of the material is concerned, specifically the rights of translation, reprinting, reuse of illustrations, recitation, broadcasting, reproduction on microfilms or in any other physical way, and transmission or information storage and retrieval, electronic adaptation, computer software, or by similar or dissimilar methodology now known or hereafter developed.
The use of general descriptive names, registered names, trademarks, service marks, etc. in this publication does not imply, even in the absence of a specific statement, that such names are exempt from the relevant protective laws and regulations and therefore free for general use.
The publisher, the authors, and the editors are safe to assume that the advice and information in this book are believed to be true and accurate at the date of publication. Neither the publisher nor the authors or the editors give a warranty, expressed or implied, with respect to the material contained herein or for any errors or omissions that may have been made. The publisher remains neutral with regard to jurisdictional claims in published maps and institutional affiliations.

Cover credit: Oleh_Slobodeniuk/E+/Getty Images

This Palgrave Macmillan imprint is published by the registered company Springer Nature Switzerland AG
The registered company address is: Gewerbestrasse 11, 6330 Cham, Switzerland

About This Book

The Palgrave Handbook of Environmental Politics and Theory is a comprehensive and updated intellectual resource for students new to the compelling field of environmental politics and theory and for scholars of this important subject. It contains introductory discussions as well as cutting-edge original works that contribute to the advancement of the field.

There are twenty-six chapters and an Editors' Introduction by a diverse group of international contributors. With longer and more substantive essays than the usual Handbook, the Palgrave Handbook seeks to be the reference work for many disciplines studying the environment. It is divided into five sections: (1) Environmental Political Theory; and Environmental Politics and Theory: (2) in the Policy-making Process, (3) in the City, (4) in Specific International Regions, and (5) in the Anthropocene Age. Thus, the *Palgrave Handbook* aims to consolidate theory and action addressing the dire environmental problems confronting contemporary society.

PRAISE FOR *The Palgrave Handbook of Environmental Politics and Theory*

"Environmental emergency is the theme running through this timely handbook. The editors and a diverse set of international authors move beyond mainstream environmental understandings to challenge their problematic, often latent assumptions that impede a deeper grasp of the ecological crisis and the fundmental challenges it poses. Toward this end, these first-rate essays examine an extensive range of issues—from the Anthropocene and degrowth, environmental justice and democracy, sustainable agriculture and animal rights, eco-feminism and biopower, to mention just a few. The authors look beyond eco-pessimism to shape a hopeful vision for stimulating and guiding a socio-ecological transformation."

—Frank Fischer, *Professor Department of Agricultural Policy and Politics, University of Humboldt in Berlin Humboldt University of Berlin, Germany*

"We have known for decades, but the IPCC 6th Assessment report made it crystal clear. Without immediate and deep decarbonization across all sectors of society, it will be impossible to stop global mean warming at 1.5 degrees C. The time for transformative climate action is now. *The Palgrave Handbook on Environmental Politics and Theory* takes this dire and inescapable environmental reality as the starting point for a new era of green political theorizing. The volume seeks to reclaim a space for critically engaged scholarship that encourages dissent and speaks up for the many communities that are mobilizing for a radical change of political course. The environmental context within which this Handbook appears is urgent and troubling. However, by confronting systems of denial and inviting conceptual and ethical innovation, the chapters form a powerful collective response."

—Eva Lövbrand, *Associate Professor in Environmental Change, Linköping University, Sweden*

"The volume delivers on its promise of providing pioneering and provocative analyses of foundational and societally transformative issues for our turbulent times. So much so that the editors might have just as easily entitled the book, *A Field Guide for Understanding our Age of Perpetual Crises*. The volume, made up of 26 chapters, covers 'high level' topics ranging for the urgent need to move beyond perpetual economic growth, and how this means moving beyond both carbon energy and capitalism (and the legacy of extractivist colonialism). Other important and emerging areas such as eco-anxiety, ethical dimensions of human-animal-nature relations, gender and deep climate adaptation are also represented It also has 'applied theoretical' considerations of urbanism and cities, housing, property rights, the law, the role of the state, and youth environmental activism. Written by an impressive array of both established scholars and new voices, the editors (Kassiola and Luke) are to be congratulated for assembling such a variety of topics, scholars and approaches, which has produced an excellent addition to the Palgrave Handbook series."

—John Barry, *Professor Green Political Economy and Co-Director of the Centre for Sustainability, Equality and Climate Change, Co-Chair, Belfast Climate Commission, School of History, Anthropology, Philosophy and Politics, Queens University Belfast, Belfast, United Kingdom*

Contents

Introduction: The Time for Social and Political Transformation Based on the Environment Is Now   1
Joel Jay Kassiola and Timothy W. Luke

Environmental Politics and Theory

Environmentalism and Political Ideologies   27
Fabrice Flipo

Democracy, Citizenship and Nationalism in Environmental Political Theory   43
Andy Scerri

Eco-Anxiety and the Responses of Ecological Citizenship and Mindfulness   65
Michel Bourban

Earth Breaking Bad: The Politics of Deep Adaptation   89
Timothy W. Luke

Animal Citizens: Do We Need to Rethink the Status of Animals or Human Citizenship Itself?   105
Turquoise Samantha Simon

Degrowth: A State of Expenditure   127
Giacomo D'Alisa and Onofrio Romano

The Nature of the State: A Deep History of Agrarian Environmentalism   147
Jake Greear

The Environmental Political Role of Counter-Hegemonic
Environmental Ethics: Replacing Human Supremacist Ethics
and Connecting Environmental Politics, Environmental
Political Theory and Environmental Sciences 173
Joel Jay Kassiola

Critical Ecofeminism: A Feminist Environmental Research
Network (FERN) for Collaborative and Relational Praxis 195
Jennifer L. Lawrence, Emily Ray, and Sarah Marie Wiebe

Property and the Anthropocene: Why Power on Things Is
Central to Our Ecological Predicament 223
Benoît Schmaltz

Environmental Politics and Theory in the Policy-Making Process

Ecosystem Policy and Law: A Philosophical Argument
for the Anticipatory Regulation of Environmental Risk 259
John Martin Gillroy

Tracing Instrumented Expert Knowledge: Toward a New
Research Agenda for Environmental Policy Analysis 299
Magalie Bourblanc

The Rise of Environmental Health as a Recognized Connection
and Academic Field 323
Corinne Delmas

Sustainable Housing: International Relations Between Housing
and the Environment Revisited 345
Sophie Nemoz

The Mobilization of the Philanthropic Sector for the Climate:
A New Engagement? 367
Anne Monier

Environmental Politics and Theory in the City

Cities and Nature: Conceptualizations, Normativity
and Political Analysis 385
Nir Barak

Henri Lefebvre's "Right to the City:" Key Elements
and Objections 407
Joshua Mousie

Going Beyond the Great Divide Between Nature and Culture: The Concept of "Relocalized Society" to Account for the Local Agri-Food Networks   427
Clémence Nasr

Environmental Politics and Theory in Specific International Regions

Environmental Justice and the Global Rights of Nature Movement   467
Chris Crews

Theorizing with the Earth Spirits: African Eco-Humanism in a World of Becoming   503
Anatoli Ignatov

The Social Construction of International Environmental Policies in the Caribbean: The Case of Sargassum   539
Andrea Parra-Leylavergne

Contemporary Youth Environmental Activism: Lessons from France and Italy   567
Paolo Stuppia

Environmental Politics and Theory in the Anthropocene

The Anthropocene New Stage: The Era of Boundaries   599
Florian Vidal

The Anthropocene and Global Environmental Politics   627
Philipp Pattberg and Michael Davies-Venn

Foucault's Biopolitics and the Anthropocene: Making Sense of Ecopower   649
Pierre-Yves Cadalen

Technonaturalism: A Postphenomenological Environ-Mentality   667
Alexander Stubberfield

Correction to: Contemporary Youth Environmental Activism: Lessons from France and Italy   C1
Paolo Stuppia

Bibliography   701

Index   705

Notes on Contributors

Barak Nir is a Lecturer (Assistant Professor) in the Department of Politics and Government at Ben Gurion University of the Negev, Israel. He received his Ph.D. from the Hebrew University for his research on social and political aspects of urban sustainability and was afterward a Fulbright Postdoctoral Fellow at Columbia University, and at the Technion-Israel Institute of Technology. His current research focuses on the relationship between national citizenship and urban citizenship (city-zenship) in light of cities' rising power in national and global politics and growing demands for more political autonomy vis-a-vis the state.

Bourban Michel is an Assistant Professor in Environmental Ethics at the University of Twente, the Netherlands. He completed his Ph.D. at the University Lausanne, Switzerland, and the University of Paris-Sorbonne Paris IV, Paris, France, and carried out his postdoctoral research at Kiel University, Kiel, Germany, and the University of Warwick, Coventry, United Kingdom. His current research focuses on ecological citizenship and cosmopolitan citizenship, climate ethics and climate engineering, eco-anxiety, and the ethics of technology.

Bourblanc Magalie is a political scientist at CIRAD, the French Agricultural Research Centre for Development, Joint Research Unit on "Water Management, Actors and Uses" of the University of Montpellier (MUSE), Montpellier, France. She is also an Extraordinary Lecturer at the Faculty of Natural and Agricultural Sciences, University of Pretoria, Pretoria, South Africa. She recently published, "Expert Reporting as a Framing Exercise: The Controversy Over Green Macroalgal Blooms' Proliferation in France," in *Science and Public Policy*. And co-authored, "The Role of Expert Reporting in Binding Together Policy Problem and Solution Definition Processes" in the edited volume, *The Political Formulation of Policy Solutions*. Her research interest focuses on the role of expert reporting and technical policy instruments within environmental policy-making.

Cadalen Pierre-Yves is currently a Post-Doctoral Researcher at the CRBC (the Center for Celtic and Breton Research)—the University of Western Brittany (*Universite de Bretagne Occidentale*), Brest, France, and Associate to the CERI (Center for International Studies), Sciences Po, Paris, France. He recently obtained his Ph.D. in Political Science and International Relations. His works are mainly related to the relations of power around the Environmental Commons. His former studies included a Bachelor in Philosophy and led him to theorize the new forms of power related to the Anthropocene era. A recent publication is: "Republican Populism and Marxist Populism: Perspectives from Ecuador and Bolivia," in *Discursive Approaches to Populism Across Disciplines: The Return of Populists and the People* (Palgrave Macmillan).

Crews Chris is a Visiting Assistant Professor of International Studies at Dennison University, Granville, Ohio, U.S.A. He earned his Ph.D. in Politics from the New School for Social Research in New York City. His dissertation research focused on the intersections of social and environmental justice and Indigenous rights through the lens of the Anthropocene, with special attention to issues of land and agrarian struggles (Nepal in particular). He worked for six years at the India China Institute in New York City where he coordinated international research programs with scholars, activists, and practitioners from India, China, Nepal, Tibet, and the U.S.A. He also led a digital mapping project on sacred landscapes in Western Tibet and helped with the publication of a series of folk story collections from the Himalayas.

D'Alisa Giacomo is a Post-Doctoral Researcher at the Center for Social Studies at the University of Coimbra, Portugal. He is a political ecologist whose research is interdisciplinary, action-oriented, and based on a community service motivation. It is aimed to contribute to the broader field of sustainability scholarship and is invested in substantial efforts to influence policies. Recently, he has co-edited a special issue of *Ecological Economics* on "Commons and Social Movements." He is the co-author of the monograph: *The Case for Degrowth* (Polity Press).

Davies-Venn Michael researches global environmental governance. A policy analyst, his work addresses international frameworks and measures for environmental protection and sustainable development. Presently, his research focuses on implementation of climate mitigation and climate adaptation within the Paris Agreement framework and between developed and developing regions. He was named Junior Fellow in the Ethics of the Anthropocene Programme at the Vrije Universiteit, Amsterdam, the Netherlands, where he is also a member of the Institute for Environmental Studies. As a communication professional, he worked in politics advising international organizations such as the United Nations Development Program (UNDP) and Canadian public officials. His commentaries on climate change and sustainability are regularly published by the Heinrich Boll Foundation in *Energy Transition* and in similar other outlets.

Delmas Corinne is a Professor of Sociology at the *Universite Gustave Eiffel* (UGE) at Marne La Vallee in Champs-sur-Marne, France, and a member of the *Laboratoire Techniques, Territoires et Societes* (Technologies, Territories and Societies Laboratory) (LATTS, UGE/CNRS, National Center for Scientific Research). Her research focuses on expertise, work, and professional groups which she approaches from the perspective of knowledge and techniques, gender, health, and the environment.

Flipo Fabrice is a Philosopher at the *Institut Mines-Telecom Business School* and Researcher at the Political and Social Change Laboratory, *Paris-Cite* University, Paris, France. He authored several books on political ecology in a political theory perspective such as *Nature et Politique* and *The Coming Authoritarian Ecology* (Wiley). He is also a specialist of the ecology of digital infrastructure.

Gillroy John Martin is Professor of Philosophy, Law and Public Policy and Founding Director of Environmental Policy Design Graduate Programs at Lehigh University, Bethlehem, Pennsylvania, U.S.A. His research involves using the use of pre-positivist Enlightenment philosophical arguments as paradigms for application to contemporary law and policy. His current work is a three-book project: *Philosophical Method, Policy Design and the International Legal System* (Palgrave Macmillan). This project utilizes deciphered paradigms from Hume, Hegel, and Kant to illuminate, respectively, the origin, current dilemmas, and future imperatives of transnational legal authority. The first book, *An Evolutionary Paradigm for International Law: Philosophical Method, David Hume and the Essence of Sovereignty*, was published in 2013.

Greear Jake is an Instructor in the Department of Political Science and Public Affairs at Western Carolina University in Cullowhee, North Carolina, U.S.A. He teaches courses in political theory, American politics, and the politics of race and the environment. His research approaches environmental politics from philosophical, historical, and scientific perspectives. His recent publications include "Decentralized Production and Affective Economies: Theorizing the Ecological Implications of Localism" in *Environmental Humanism*.

Ignatov Anatoli is an Associate Professor of Sustainable Development at Appalachian State University, Boone, North Carolina, U.S.A. His current research interests include environmental political theory, African political thought, African land politics and policy, post-colonial theory, endogenous development, and law and development. His research has been focused on traditions of sustainability and ecological political thought in Africa that remain largely invisible within the dominant Euro-American orientations in political theory and development studies. He is completing a monograph, *Decolonizing Land Politics: Legal Pluralism, Conflict, and the Reinvention of Political*

Authority in Ghana. His scholarly work has appeared in: *Africa, Political Theory, GeoHumanities, Contemporary Political Theory, Theory & Event* as well as edited volumes.

Kassiola Joel Jay is a Professor of Political Science at San Francisco State University, San Francisco, California, U.S.A. He is the author of one of the first books in the emerging field of environmental political theory, *The Death of Industrial Civilization* and, more recently, editor of *Explorations in Environmental Political Theory*. Lately, he has turned to Chinese Confucianism which he terms "Confucian Green Theory" as an alternative to the unsustainable and unequal Western worldview and the neoliberal consumer society. His most recent publication is: "Zhang Zai's Cosmology of Qi/qi and the Refutation of Arrogant Anthropocentrism: Confucian Green Theory Illustrated" in *Environmental Values*.

Lawrence Jennifer L. is an Assistant Professor in the Urban and Environmental Planning Program at the University of Virginia, Charlottesville, Virginia, U.S.A. Combining critical and creative approaches, her scholarship examines contradictions within environmental governance, particularly with reference to petrochemical industries and geographies. Her research interrogates the systemic production of crises to understand how extractive logics of governance impact lived experiences directly in frontline communities as well as through the protracted disaster of climate change. Demonstrating spatio-temporal tensions between shaping understanding of environmental degradation, this work highlights contradictions at the intersection of economic system, resource extraction, and socio-environmental (in)justice. Her recent publications include: *Biopolitical Disaster* (Routledge), and the *Resilience Machine* (Routledge) and articles in: *New Political Science, Geoforum, Political Geography,* and *The Oxford Handbook of Complex Risks and Resilience*.

Luke Timothy W. is University Distinguished Professor of Political Science, Chair of the Department of Political Science, and Interim Director of the School of Public and International Affairs at Virginia Polytechnic Institute and State University, Blacksburg, Virginia, U.S.A. Much of his research focuses on the intersections of environmental political theory with global governance, political economy, and cultural politics. He is an Associate Editor of *New Political Science* and a Founding Editor of *Fast Capitalism* with the Center for Theory at the University of Texas. His books in the field of environmental politics and theory include: *Anthropocene Alerts: Critical Theory of the Contemporary as Ecocritique* (Telos), *Museum Politics: Powerplays at the Exhibition* (Minnesota), *Capitalism, Democracy, and Ecology: Departing from Marx* (Illinois), and *Ecocritique: Contesting the Politics of Nature, Economy, and Culture* (Minnesota).

Monier Anne holds a Ph.D. in Social Sciences (ENS/EHESS) (Higher Education/the School of Advanced Studies in the Social Sciences) Paris,

France, and is a Researcher at the ESSEC (Higher Education in Economic Sciences and Business) Business School Philanthropy Chair, Paris, France.

A specialist in philanthropy, she has published two books on this topic: *Nos chers Amis Americains* (PUF), and, with Sylvain Lefevre, *Philanthropes en democratie* (PUF/VDI), as well as several articles in academic journals like the *Socio-Economic Review, Geneses* and *Politix*. Her research interests are: philanthropy, environmental issues, climate movement, sociology of elites, transnational relations, inequalities, social movements, and cultural policies. She is now working on a new research project on climate philanthropy. She also teaches at the School of Public Affairs at Sciences Po, Paris, France.

Mousie Joshua is an Associate Professor of Philosophy at Oxford College of Emory University, Atlanta, Georgia, U.S.A. His work focuses on the intersection of socio-political and environmental philosophy, and he is currently developing concepts that help us understand the political dimensions of built environments. His publications have appeared in: *Environmental Philosophy, Ethics and the Environment*, and *Journal of Social Philosophy*.

Nasr Clémence is a Ph.D. in Political and Social Sciences, a Research Associate at the Centre for Political Research at Sciences Po CEVIPOF (Paris Institute of Political Studies, Sciences Po), Paris, France, and temporary Teaching and Research Associate at the Jean Monnet Faculty, Paris-Saclay University, Gif-sur-Yvette, France. She works on socialist thought and political ecology. At the intersection of social philosophy and environmental theory, she focuses on cooperation and its role in the agri-food sector. Among her works, an article entitled: "De-concentrating Megacities: Political Theory and Material Normativity," will be published in *Political Theory* in January, 2023.

Nemoz Sophie is an Associate Professor at the University Bourgogne Franche-Comte (UBFC), Besancon, France, and is senior member of the Laboratory of Sociology and Anthropology (LaSA, UBFC) and responsible for Bachelor degrees. After her double Master's degrees in Sociology and Anthropology, she completed a doctoral thesis on the international and intercultural history of political ecology in residential settings and the process of eco-innovation in Europe. Thereafter, she worked on a postdoctorate at the Institute for Environmental Management and Land-Use Planning (IGEAT-Brussels, Belgium). She conducted research on rebound effects linked to the implementation of energy-efficiency policies. She also published on climate justice in the European Community. Recently, she has also co-authored "Beyond Technology: A Research Agenda for Social Sciences and Humanities Research on Renewable Energy in Europe," in *Energy Research & Social Science*. Her research revolves around the field of environmental political theory focusing on the social and spatial dimensions and the temporalities. She leads the scientific programme, RIFTS, at the *Maison des sciences de l'homme et de l'environment* Claude Nicolas Ledoux, a research unit under the supervision of the CNRS (National Center for Scientific Research). She is the coordinator of the Research Network "Sociology of Environment and Risks" within the French Sociological Association. At the international level, she is a member of the European Board "Environment and Society" of the European Sociological Association.

Parra-Leylavergne Andrea is a Researcher in Political Science for the Caribbean Laboratory of Social Sciences of the French National Center of Scientific Research, CNRS. Her research is mainly focused on the international aspects of the environmental governance in the Americas and the Caribbean. Her publications include: "EU-Latin America and Caribbean Diplomacy: From Interregional to Intercultural Dialogue" published in the journal, *Negotiations, Conflict, Decision and Deliberation*. She has also co-authored, "Constitutional Aspects of FTAs: a Columbian Perspective" in the international journal, *European Journal of International Law*, among others.

Pattberg Philipp is Professor of Global Environmental Governance and Policy, Department of Environmental Policy Analysis, Institute for Environmental Studies, Vrije Universiteit, Amsterdam, the Netherlands.

Ray Emily is an Associate Professor of Political Science at Sonoma State University, Rohnert Park, California, U.S.A. Her research is in environmental political theory and politics with particular interest in the intersections of doomsday prepping, the politics of outer space, and climate change. Her recent work includes: "Neoliberalism and Prepping for Disaster," in *New Political Science*.

Romano Onofrio is Associate Professor of Post-Development Sociology in the Department of Humanities of the University of Rome 3, Rome, Italy. His research ranges over social theory, post-modernity, degrowth, political participation, word and consumption, Balkan countries, Mediterranean and Adriatic societies. Among his recent works are: *Towards a Society of Degrowth* (Routledge) and *The Sociology of Knowledge in a Time of Crisis* (Routledge). He has contributed to the following collective works: *Degrowth: A Vocabulary for a New Era* (Routledge) and *Pluriverse: A Post-Development Dictionary* (Tuilka Books).

Scerri Andy is Associate Professor of Political Science and International Studies at the Virginia Polytechnic Institute and State University, Blacksburg, Virginia, U.S.A. His research applies an interpretivist approach to environmental politics in the post-industrial Settler Societies which draws insights from the political realist project in political theory. He is the author of *Greening Citizenship* (Palgrave) and *Postpolitics and the Limits of Nature* (SUNY), as well as articles in: *Environmental Politics, Social Movement Studies, Citizenship Studies, Ecological Economics, Critical Review of International Social and Political Philosophy, Philosophy and Social Criticism*, and elsewhere.

Schmaltz Benoît holds a Master's degree of Public Law from the *Universite Jean Monnet Saint-Etienne*, University of Lyon, Lyon, France, and the

CERCRID (Center on Critical Research on the Law) *Centre de Recherches Critiques sure le Droit*. He holds a doctoral degree in Public Law from Lyon 3 University in Lyon, France. He did his Ph.D. about the public/private law divide from the perspective of property. He published *Les personnes publiques proprietaires*. He won a prize for his research from the French association for research in administrative law (AFDA). The book developed a new comprehensive theory of subjective rights to demonstrate the specificities of public owners compared to their private counterparts. He has written mainly about legal topics, but his interest in ecology grew progressively to become his only field of research. He is now working on two main projects. First, the writing of a new book about the bridging of Law and the Anthropocene. Second, a long-term multidisciplinary research program about the decarbonization of the Lyon-Saint-Etienne area in France.

Simon Turquoise Samantha is a Ph.D. student in Political Science (Institute for Nation and State Research), Universite de Lorraine, Nancy, France. After a master's degree in philosophy of law and a master's degree in political science, she is currently writing a thesis on animal advocacy parties in Europe. Her research fields are social movements (especially the fight for animal rights and ecology), political parties, and institutions. Her works are marked by a strong interdisciplinarity, mixing political science, philosophy, political theory, sociology, and law.

Stubberfield Alexander is Visiting Adjunct Assistant Professor in the Political Science Department and the International Studies program at the Virginia Polytechnic Institute and State University, Blacksburg, Virginia, U.S.A. He received his doctorate in Virginia Tech's "Alliance for Social, Political, Ethical and Cultural Thought," a cutting-edge interdisciplinary Ph.D. program in the Humanities and Social Sciences. He has published previously on American political development that appeared in: *Re: Reflections and Explorations: A Forum for Deliberative Dialogue* and *SPECTRA: the ASPECT Journal*. His interests fall within environmental political theory, environmental social science, philosophy, environmental history, social theory, environmental politics, and ecological modernization.

Stuppia Paolo holds a Ph.D. in Political Science (University Paris 1 Pantheon-Sorbonne, Paris, France). He is an Associate Researcher at CESSP (*Centre Europeen de Sociologie and de Science Politique-Centre National de la Recherce Scientifique*) (The European Center of Sociology and Political Science), University of Paris 1, Paris, France, and in the Department of Sociology of Cal Poly Humboldt, Arcata, California, U.S.A. His research is focused on social movements, especially student and environmental protests, counterculture, radical commitments, and youth politics. He is currently co-directing a research project retracing the origins and evolutions of the Back-to-the-Land movement in California and Oregon since the 1960s. His latest publications include, co-authored: *Geopolitique de la jeunesse: Engagement et* *(de)mobilisations* and "Activation of Student Protest: Reactions, Repression, and Memory at Nanterre University, Paris, 1968-2018," in *When Student Protests: III- Universities in Global North*.

Vidal Florian is an Associate Researcher at the LIED (CNRS) (Laboratory on Interdisciplinary Studies on Energy, French National Center on Research in the Sciences), Paris Cite University, Paris, France. He is also an Associate Fellow at the French Institute of International Relations (IFRI). He holds a Ph.D. in Political Science and specializes on the resources-energy nexus in light of the Anthropocene concept. He works on mining issues in remote areas (polar region, deep sea areas, outer space) in light of the ecological transition. He is currently involved in a research project funded by the French National Research Agency (ANR) on strategic metals in the context of the transition policy.

Wiebe Sarah Marie is an Assistant Professor in the School of Public Administration at the University of Victoria, Victoria, British Columbia, Canada, where she teaches in the Community Development Program and is an Adjunct Professor at the University of Hawai'i, Manoa, Hawai'i, U.S.A. She is a Co-Founder of the FERN (Feminist Environmental Research Network) Collaborative with Jennifer Lawrence and Emily Ray and has published in journals, including: *New Political Science, Citizenship Studies and Studies in Social Justice*. Her book, *Everyday Exposure: Indigenous Mobilization and Environmental Justice in Canada's Chemical Valley* (UBC Press), won the Charles Taylor Book Award and examines policy responses to the impact of pollution on the Aamjiwnaang First Nation's environmental health. With Jennifer Lawrence, she is the co-editor of *Biopolitical Disaster* and co-editor of *Creating Spaces of Engagement Justice and the Practical Craft of Deliberative Democracy*. As a collaborative researcher and filmmaker, she worked with Indigenous communities on sustainability-themed films and co-directs the *Seascape Indigenous Storytelling Studio* with research partners from the University of Victoria, University of British Columbia, and coastal Indigenous communities.

List of Figures

Property and the Anthropocene: Why Power on Things Is Central to Our Ecological Predicament

Fig. 1	General diagram showing the metabolic processes and the relation between society and nature	236
Fig. 2	Diagram showing the appropriation act in current societies, which is performed by a series of appropriation units (P). The appropriated materials and energy are later circulated (Cir) by different routes, in some cases transformed (Tr), and finally consumed by industry and cities. All these processes, in turn, generate a flow of wastes towards nature, or excretion (Exc). Along with domestic relations with nature (input and output), societies also import and export commodities to and from other societies	236

Ecosystem Policy and Law: A Philosophical Argument for the Anticipatory Regulation of Environmental Risk

Fig. 1	Matrix of value: Humanity and nature	265
Fig. 2	Types of risk for policy choice	268
Fig. 3	Comparative paradigms: Market vs. Kant's PPLD	294

The Anthropocene New Stage: The Era of Boundaries

Fig. 1	Resources and human entanglement	613

Introduction: The Time for Social and Political Transformation Based on the Environment Is Now

Joel Jay Kassiola and Timothy W. Luke

> The cumulative scientific evidence is unequivocal. Climate change is a threat to human well-being and planetary health. Any further delay in concerted global action in adaptation and mitigation will miss a brief and rapidly closing window of opportunity to secure a livable and sustainable future for all. (very high confidence)
>
> —Working Group II, IPCC, 2022[1]
>
> It's now or never, if we want to limit global warming to 1.5C (2.7F); without immediate and deep emissions reductions across all sectors, it will be impossible. This assessment [of the IPCC Working Group III, *Climate Change 2022: Mitigation of Climate Change*, released on 4 April 2022] shows that limiting

[1] Working Group II Contribution to the Sixth Assessment Report of the Intergovernmental Panel on Climate Change (IPCC), *Climate Change 2022: Impacts, Adaptation and Vulnerability, Summary for Policymakers*, edited by Hans-Otto Portner and Debra C. Roberts, Working Group II Co-chairs, 27 February 2022, D.5.3. 35. https://www.ipcc.ch/working-group-wg-2/ accessed: 20 April 2022.

J. Jay Kassiola (✉)
San Francisco State University, San Francisco, CA, USA
e-mail: kassiola@sfsu.edu

T. W. Luke
Virginia Polytechnic Institute and State University, Blacksburg, VA, USA
e-mail: twluke@vt.edu

© The Author(s), under exclusive license to Springer Nature Switzerland AG 2023
J. J. Kassiola and T. W. Luke (eds.), *The Palgrave Handbook of Environmental Politics and Theory*, Environmental Politics and Theory, https://doi.org/10.1007/978-3-031-14346-5_1

warming to around 2C (3.6F) still requires global greenhouse gas emissions to peak before 2025 at the latest, and be reduced by a quarter by 2030.
—Jim Skea
Co-chair, Working Group III, IPCC, 2022[2]

Our Present Environmental Emergency

As we write our Introduction to this *Handbook on Environmental Politics and Theory* on the 52nd anniversary of the first Earth Day protests/celebrations, the news from the most recent reports from the Intergovernmental Panel on Climate Change (IPCC) (as well as other parallel scientific reports regarding other environmental threats like biodiversity loss, water and air pollution, wildfires, etc.) is dire. Environmental degradation is close to the limits beyond which human actions to prevent irreversible ecological disaster will be ineffective. We are long past the 100-year projections of "collapse" conditions made in 1972 innovative *Report on The Limits to Growth*, which inspired so many contemporary environmental movements and environmental political thinkers.[3] The latest IPCC Report gives humanity until 2025 to begin to radically reduce greenhouse gases and then decrease these pollutants by 25% in 2030, if we are to avoid the catastrophic consequences of climate change. This is alarmingly short notice to humans who have not been inclined to reduce overall greenhouse gas emissions through household decisions, national policies or international agreements (except temporarily during economic recessions and the current COVID-19 pandemic for brief respites until the increases return as economic growth restarts.[4]

The *Palgrave Handbook of Environmental Politics and Theory* comes together then from daunting and discouraging contemporary conditions. The themes addressed by the *Handbook*'s contributors have never been timelier during the brief history of formalized academic work on environmental politics and theory. We present the *Handbook* with a sense of utmost humility in the face of staggering historical complexities, ecological tensions and deadly hazards that confront humanity as well as other living beings and our shared ecosystems today. Nonetheless, we are confident that the viewpoints included and insights presented, in the *Handbook* should make a significant impact

[2] "UN Climate Report: 'It's now or never' to limit global warming to 1.5 degrees." United Nations, *UN News: Global Perspectives, Human Stories*, 4 April 2022. https://news.un.org/en/story/2022/1115452 accessed: 20 April 2022.

[3] Donella H. Meadows, Dennis L. Meadows, Jorgen Randers and William W. Behrens III, *The Limits to Growth: A Report to the Club of Rome's Project on the Predicament of Mankind*, Second Edition, New York: New American Library, 1975.

[4] See, for example, greenhouse emissions for the first pandemic year in the United States of 2020 declined by 10%, only to see such emissions rise in 2021 in the United States by 6.2%. See, Brad Plumer, "U.S. Greenhous Gas Emissions Bounced Back Sharply in 2021." *New York Times*, January 10, 2022, Section A. 11. www.nytimes.com accessed: 25 April 2022.

upon the unfolding project of environmental politics and theory. We hope it will influence the choices made by individual households, local governments, national decision-makers and international policy deliberations in the crucial next few years as we near the IPCC's pressing deadlines for making substantial greenhouse gas emission reductions.

We hope the *Handbook* will be used both by introductory students and engaged scholars of the immensely wide-ranging multidisciplinary and interdisciplinary subject matter of environmental politics and theory as well as by corporate managers, third sector organizations and state policy-makers trusted with the awesome responsibility of protecting the environment and averting disaster in the future.

Objectives of the Handbook

How can such profound and difficult objectives be attained in one volume—even a comprehensive one like this? One part of the answer is diversity: a diversity of authors; analytical perspectives; topical subjects addressed; substantive questions posed; new methods utilized; and types of evidence provided and fresh discourses included in this project. Consequently, the *Handbook* represents many different disciplines within the social sciences (with political science and political theory as the most prevalent), working in quite varied scholarly networks and divergent national traditions.

It also contains chapters intended for novice students new to systematic analyses of salient environmental topics by introducing the nature of the key issues and their development within various academic fields. For this important group of readers, we aspire to deepen their interests and curiosity about the environment and humanity's role. Additionally, it can stimulate more specialized and advanced study of environmental politics, including the innovative theoretical contributions outlined in the *Handbook*'s chapters. Hence, this volume can be both the initial intellectual resource for the new student *and* an indispensable instrument for them to advance their studies beyond the introductory level.

For the scholar and researcher in environmental politics, we aim to provide cutting-edge discussions of groundbreaking subjects and theories plus original treatments of established subjects. The special nature of the handbook-genre in size and heterogeneity of contributions often realizes these pedagogic and theoretical goals that are seldom joined otherwise in one publication.

As for the policy-makers who normally would not have the time nor interest to read mainstream publications written to benefit students or academics, particular sections of the *Handbook* can be useful. The contributions devoted to environmental politics and theory in the policy-making process, and those dedicated to environmental politics and theory in specific international regions will be of highly relevant whether they are focused on domestic political issues and decision-making or concerned with the realms international environmental politics and policies. All academics, especially political scientists, hope that their intellectual endeavors and writings will influence those responsible for political decisions on behalf of their constituents. We hope that the *Handbook*'s chapters on policy-making can influence the thinking and decisions of environmental policy-makers (as well as all of the other sections of the Handbook), even if that impact is less direct.

How can the field of environmental politics and theory contribute to present-day efforts to avoid environmental catastrophe as well as learning to live more sustainably and justly?

Furthermore, how will this particular handbook assist accomplishing these goals? We think the main contribution of this young field established formally only in the 1990s revolves around the conception of what the environmental crisis is.

Going back to the work of biologist, Rachel Carson, and her classic 1962 book, *Silent Spring*, and the computer and systems scientists who produced the path-breaking volume, *The Limits to Growth*, the modern environmental movement and the subjects examined in environmental studies started out as a set of problems that largely were regarded as scientific in character. They were addressed by natural scientists from different intellectual disciplines, like biologists, chemists, geologists, meteorologists, oceanographers, physicists, systems theorists, etc. seeking to describe, explain and predict various environmental conditions as Carson and the Meadows et al. group did. We want to redefine and expand this apolitical scientific approach to the nature and study of the environmental crisis with new sets of research problems and methods. We would not reject the natural scientific contributions altogether, but rather we wish to add the important role of politics, especially the significance of political values and political theoretical discourse to environmental studies. This approach thereby highlights the new field of environmental political studies, specifically, the normative field of environmental political theory. We have made this critical normative political expansion of the natural scientific conception of environmental studies the main thesis of our respective books and subsequent writings that became part of the creation of the new study of environmental political theory.[5]

Simply put, this new approach emphasized that the environmental crisis is a product of the way humanity lives in industrialized networks linking all of the world's nations. The way we all now live is the result of industrial urban values that generate the social goals first of capitalist and colonizing West and the behaviors used to achieve them by spreading them unevenly around the planet by agents and structures working inside and outside of all nations encountered over several centuries of political development, national expansion and cultural change; most importantly, these changes have been

---

[5] See, Joel Jay Kassiola, *The Death of Industrial Civilization: The Limits to Economic Growth and the Repoliticization of Advanced Industrial Society*, Albany: The State University of New York Press, 1990; Timothy W. Luke, *Ecocritique: Contesting the Politics of Nature, Economy and Culture*, Minneapolis: The University of Minnesota Press, 1997.

linked by fundamental capitalist assumptions about the presumptive merits of endless or limitless economic growth in urban industrial society. Once this insight was recognized and accepted by scholars in one form or another, the subfields of critical, interpretative and normative environmental political theory as well as empirical environmental political study evolved as shown in the diverse contents of this volume.

Natural scientific environmental studies, such as the research and reports produced by the IPCC, remain essential to inform the scientific community and the general public with regard to current environmental conditions and threats, their hypothetical causes and actions for remediation. However, if we really want to discover the ultimate cause of environmental threats such as global warming and climate change, environmental political theory tells us to examine our social values, to be specific: the taken-for-granted, industrial values and the culture founded upon them, the most influential of which is limitless economic growth.

Our main objective in the long process of producing this Handbook (see below for details) is to create a work that will become the reference book for the entire academic field's members to be consulted by environmental decision-makers. We hope that the readers of this *Handbook* will learn much from its authors' chapters and will refer to it often in their contemplation and actions regarding the environment. We hope that the collection will inform you about how to live and how our society structures the lives of its citizens. Ideally, how we change. As the IPCC scientists tell us, we absolutely have no time to waste before we must transform our social and political beliefs and actions to become environmentally sustainable and socially just.

If one is skeptical about accomplishing these unprecedented imperative goals, we think the words of a Jewish sage quoted by Kassiola in his book 32 years ago, are even more relevant and instructive now that the environmental crisis is even more severe and pressing: "The day [life] is short; the task is great; the workmen [human beings] are lazy; the reward is great, and the Master is insistent."[6]

> Life is, indeed, short, and the task facing postindustrial citizens is, in fact, great: human beings are, indeed, lazy: we succumb easily to inertia, we fear change, and the pain and insecurity it brings. Furthermore, we are reluctant to challenge the dominant worldview and social structure of our existing social order. Our goals are difficult to formulate and intimidating to achieve. However, if enlightened postindustrial citizens can overcome these obstacles and meet the demanding tasks, the reward of a sustainable, just and humane world is also

---

[6] Rabbi Tarfon, *Ethics of the Fathers*, Philip Birnbaum, Translator and Annotator, New York: Hebrew Publishing, 1949, Chapter Two, Verse 20, 16. The words in brackets were provided in the text of this edition. The quotation may be found in Kassiola, *The Death of Industrial Civilization*, 217.

great. For Rabbi Tarfon, the Master was insistent. For us, the urgent [environmental] limits-to-growth Issues confronting contemporary humanity are no less compelling.[7]

This collection does address the fear, pain and insecurity that societal transformation brings by proposing the formulation of new goals and the means to achieve them. The authors do provide conceptual, factual and value analyses to acknowledge the wisdom of the following conviction: "Ironically, the pain of a collapsing culture is also an opportunity: to change is hard, but not to change is impossible."[8]

THE LONG GENESIS AND CREATION SAGA OF THE HANDBOOK

Most academic books have a lengthy, multi-stage production process: initial conception to further research, writing and revising several drafts, preparation for publication, with some extra (-long) time added for reviews, responses and further revisions. In addition to these components, this *Handbook* has an extraordinarily challenge in its creation process owing to the COVID-19 pandemic.

Many globally significant events have occurred since the *Handbook* was conceived in 2017. Greta Thunberg's first strike for the climate occurred; the COVID-19 virus was identified as it spread throughout the world and other three years later its variants are still wreaking havoc on humanity (Americans alone have already surpassed 1 million deaths); the global economy declined precipitously because of the pandemic in 2020 and has not yet fully rebounded; the election of a climate change-prioritizing American president defeating a climate change-denying president; a Russian invasion and resulting war with Ukraine; and, most importantly for the topic of the *Handbook*, the IPCC has issued its most alarming forecasts in February and April of 2022.

This *Handbook* began in conception with a telephone conversation on June 29, 2017, between Michelle Chen, the then Editor of Palgrave Macmillan's list of books on American politics, political theory and public policy and Joel Kassiola, the longtime Series Editor for the Palgrave Macmillan Series on Environmental Politics and Theory (EPT).

Kassiola reflected on the EPT Handbook proposal and decided he would accept the challenge but only with a co-editor. The first and only person he thought suited for this role was his longtime EPT colleague and fellow member of the first EPT professional group associated with the Western Political Science Association, Professor Tim Luke of Virginia Polytechnic Institute and State University who accepted the invitation.

---

[7] Kassiola, *The Death of Industrial Civilization*, 217.

[8] George Lakey, *Strategy for a Living Revolution*, San Francisco: W.H. Freeman, 1973, 28. Quoted in: Kassiola, *The Death of Industrial Civilization*, 199.

We exchanged ideas for several months about how to structure the Handbook and finally settled on a detailed framework for the Handbook that was comprehensive in scope, while utilizing the generous amount of space a Handbook format permits. We discovered during this formative period that there were so many policy issues, concepts, movements, historical developments, theories, theorists, activists, organizations and problems to address within the tremendously wide subject of environmental politics and theory that we would need to be open-minded about topics to be included in the volume. Although we had our own conception about the structure and content of the *Handbook*, we concluded that it would be advisable to receive feedback from contributors about their own intellectual interests and questions to address in their contributions which we encouraged.

Of course, we could never have foreseen the onslaught of the global pandemic affecting authors' ability to conduct research and write their chapters due to their personal and professional lives being thrown into turmoil by campus closures and lockdowns. During this time, Ms. Madison Allums took over responsibility for the Editorship of our *Handbook*. We would like to express our deepest gratitude to Ms. Allums as well as to Palgrave Macmillan management staff for their extraordinary patience, understanding and support for a handbook project that was beset by many pandemic-caused delays and sustaining this endeavor beyond the usual time allowed for manuscripts.

COVID-19 AND THE HANDBOOK: TRANSFORMING THE UNAVOIDABLE INTO A WINDOW OF OPPORTUNITY

Early on, we decided to make a virtue out of necessity in this unprecedented production saga of the Handbook caused by COVID-19. The extra time taken up during the pandemic could be used to work with contributors in honing their chapter topics as well as revising and refining their initial drafts. In this way, surprisingly, we think the final product is a superior version to the one that would have been produced for its initial delivery date, a distinct silver lining to the dreadfully dark cloud of COVID. Five years in production is a long time for such a book, but we believe *The Palgrave Handbook of Environmental Politics and Theory* is unique. Moreover, it will prove valuable to environmental students, scholars and policy-makers in our current stage of environmental emergency. We hope you agree.

One distinguishing characteristic of the Handbook made possible by the longer interval from conception to completion is the large number of European-based contributors. Given the international and global nature of EPT subject matter and Palgrave Macmillan's parent company (Spring Nature Group, headquartered in Berlin), it was understandable that its Review Board requested inclusion of a good number of European-based scholars. The field of EPT transcends any one nation's boundaries. In addition, an international perspective on the environmental crisis and its resulting politics is likely to expand and improve upon an exclusively American environmental political worldview. Therefore, we embraced the opportunity to make the Handbook more comprehensive in the topics discussed and literatures examined. It is now useful for its readers in ways that an exclusively American group of contributors would not be.

Regarding the recruitment of European-based scholars to be contributors, a hero in the long gestation process of the *Handbook* was a colleague in France, Professor Fabrice Flipo, who was instrumental in communicating with many of these researchers. Through his networks, Professor Flipo reached environmental scholars in Europe and connected us with many contributors. These outreach efforts went well beyond our initial goals, and this *Handbook* is probably the most internationally diverse volume on the politics of the environment.

We want to express our thanks to Professor Flipo and to all the European-based contributors and of course our American EPT colleagues as well as scholars from Canada, Israel and the Caribbean, who have been extremely patient awaiting the publication of the Handbook in addition to their extra work in revising and improving their chapters, sometimes several times. We also want to express our gratitude to the Palgrave Macmillan Review Board's suggestion about recruiting European-based researchers which we think enhances the *Handbook's* value to its readers.

THE IPCC THREATS, THE NEED FOR POSITIVE VISIONS FOR SOCIAL TRANSFORMATION AND EPT

In closing this review of the gestation process of the Handbook and its long journey from conception to completion through the global pandemic, we wish to highlight the most recent and perhaps most important development regarding the environment during the 5-year period of the *Handbook's* creation: the 2022 Reports from the IPCC cited in the two epigraphic quotations. These Reports' dire findings by hundreds of climate scientists from around the world reinforce the message from this U.N. scientific body repeated since its inception in 1988 that our climate is deteriorating in ways dangerous to humans and nonhuman species as well as the planet itself. Now there is an urgent warning that we have extremely little time to stop increasing greenhouse gas emissions (in 3 years from publication of the Reports in 2022) and, that we must reduce them substantially by 25% (in 8 years from publication) to avert calamitous climate-based consequences. These conditions come as a result of human economic activities and industrial values, especially endless economic growth that increasingly produces greenhouse gases that trap heat and produce global warming. Climate change, the IPCC scientists are telling the world today, is upon us right now (2022) and not 100 years or even 50 years from now as the earlier *Limits to Growth*-era publications projected.

It is necessary to mention that there are critics of the IPCC and its alarming forecasts like the latest Reports. These include ideological climate change-deniers who consider the IPCC's alarm over climate change to be misplaced, or a hoax because it is not actually occurring.

Others might argue that it is real but is "natural" and not caused by humanity, or, that it is an attack on our "free market" system and purely ideological.[9]

However, the IPCC is also criticized by climate change advocates who contend that the new and urgent reports in 2022 merely perpetuate "doom and gloom" threats. They claim that these reports are extreme, predicting the least amount of time allegedly available before prevention or mitigation of a climate disaster is possible. Furthermore, some argue that such grim reports have characterized the entire modern environmental movement since it began with the "collapse" projection of *The Limits to Growth* in 1972.[10]

These critics of extreme environmental negativity, while sympathetic, point out that the catastrophic threat stance, espoused by the IPCC during its entire existence and characteristic to the entire modern environmental movement has not produced its desired effect of fundamental social change throughout its 50-year life. What this failure to achieve the required social and political transformation demonstrates is that such fundamental change will not result from frightening projections alone. Instead some positive vision is needed for generating a desirable alternative sustainable social order that citizens and elected leaders can champion.

We must have a worthy goal to attain rather than solely a catastrophe to be avoided, while keeping the values of the social and political order intact.

Here is where environmental political theory (and hopefully this *Handbook*) can make a decisive contribution by proposing, defending and demonstrating how to implement a sustainable and just social order that is persuasive, in proving positive reasons for accepting it to replace current normatively undesirable and environmentally unsustainable society. The future of our planet and its living inhabitants is at stake. A number of the contributors directly address this profound issue, particularly the chapters by Bourban, Cadalen, D'Alisa and Romano, Greear, Ignatov, Kassiola, Luke, Lawrence, Ray and Wiebe, Simon and Stubberfield. It is our goal that the entire *Handbook* will guide and stimulate the creative imaginations of its readers to envision positive social change.

The subtitle for this Introduction to the Handbook captures the extreme urgency of the recent IPCC Reports: "The Time for Social and Political

---

[9] For a profile of such a climate change denier organization, the Heartland Institute, see, Naomi Klein, "Capitalism vs. the Climate" in Klein's *On Fire: The (Burning) Case for a Green New Deal*, New York: Simon and Shuster, 2019, 70–103.

[10] For a presentation of the history and critique of limits environmentalism, see, Damian F. White, Alan P. Rudy and Brian J. Gareau, *Environments, Natures and Social Theory: Towards a Critical Hybridity*, London: Palgrave, 2016, Chapter 3: "Limits/No Limits? Neo-Malthusians, Promethians and Beyond," 52–70.

Transformation Based on the Environment is Now." We humbly submit this Handbook to the global public to advance global awareness of the political thinking about the environment and to suggest methods for change that readers can support in various ways: by voting for candidates and policy initiatives that support the prescriptions contained in this Handbook; by joining and supporting environmental organizations and advocacy groups for social transformation to achieve environmental sustainability and social justice.

It has been an exciting challenge producing the *Handbook*, and we are delighted to have persisted to see its eventual publication. We want to acknowledge and deeply thank all of the contributors and the publisher for their immense patience and continued support for this project, but such an intellectual challenge as this *Handbook* is dwarfed by the immediate threats and the resulting imperative for change the world faces today regarding the environment and its and our future. As Rabbi Tarfon noted, our "Master," the subject matter of the environmental crisis and the politically transformative response to it, is insistent and, therefore, we have no choice but to think and act fundamentally differently as well as much more effectively to save and protect the environment and all of the planet's life dependent upon it. May the *Palgrave Handbook of Environmental Politics and Theory* contribute to our species' successful efforts to prevent disaster for all species and the planet. We fervently trust that the Handbook's contributions will inspire further research and innovative transformative ideas and actions on the part of its readers.

THE HANDBOOK ITSELF

As befits EPT subject matter, we have attempted to be as inclusive and wide-ranging as possible in selecting chapter subjects and discussions. We have permitted contributors ample space to convey and explain their main theses, and, most importantly, to defend them with reasoned and/or factual evidence. Thus, the resulting Handbook is extremely heterogeneous and includes case studies and conceptual analyses; normative arguments (with prescriptions and value assessments) and empirical data-based presentations; historical studies and contemporary discussions; activist-focused ethnographic and theoretically oriented chapters, etc. Taken collectively, all the chapters reflect the multi-dimensional and interdisciplinary nature of the domain of environmental political discourse. By being so diverse in subject matter within environmental politics and theory, we hope that the reader of the *Handbook* (no matter their level of experience or advancement in learning) will find essays germane to their interests and discover new interests in the immense realm of political studies of the environment.

The Structure and Summary Content of the Handbook

The Handbook is divided into what we consider some of the main categories and themes in contemporary environmental politics and theory. They capture the chapters' chief subjects and aims by a sectional structure and the essays' placement within it.

We use this framework of Environmental Political Theory; Environmental Politics and Theory in the Policy-making Process; Environmental Politics and Theory in the City; Environmental Politics and Theory in Specific International Regions; and, Environmental Politics and Theory in the Anthropocene, to present the insights of our contributors in a clear and comprehensive manner.

Section I: Environmental Political Theory

The first theoretical section in the Handbook is the largest and most diverse, reflecting the huge scope of environmental political theory. Environmental political theory encompasses the convergence of normative and analytical political theory revolving around the environment. It has developed into a *sui generis* subfield within environmental politics, combining components of the history of Western political thought (although Non-Western political theory is becoming increasingly studied) with contemporary political analysis applied to thinking about the environment.[11]

This wide-ranging subfield is eclectic, yet analytical connections between the diverse chapters in this first section of the Handbook are instructive. An example might be to compare Flipo's survey of political ideologies and how they relate to environmentalism with Scerri's discussion of democratic theory and green citizenship. Environmental citizenship is discussed not only in Scerri's chapter but also in chapters by Bourban and Simon. Kassiola's analysis of the political impact of environmental ethics aligns well with Simon's discussion of the ethical and political status of nonhuman animals and their possible eco-citizenship, as well as with the concept of "critical ecofeminism" introduced by the chapter written by Lawrence, Ray and Wiebe.

In his chapter, Flipo addresses the subject of: "environmentalism and political ideologies" and how liberalism, socialism and conservativism relate to environmentalism as a social movement and political theory. He identifies and compares these three modern political worldviews with ecologism and

---

[11] For an example of the former, book-length application of the Western political theory canon (with one exception) to the environment see, Peter F. Cannavo and Joseph H. Lane, Jr. eds. *Engaging Nature: Environmentalism and the Political Theory Canon*, Cambridge: The MIT Press, 2014. For the latter, analytical and normative political theory applied to the environment see, Steve Vanderheiden, *Environmental Political Theory*, Cambridge, UK: Polity Press, 2020. On the increasing literature *on* Comparative Political Theory or Non-Western Political Theory, one example would be: Fred Dallmayr, ed. *Comparative Political Theory: An Introduction*, New York: Palgrave Macmillan, 2010.

discusses how they relate. The author concludes that environmental thought is "distinct" from these standard Western ideologies.

Scerri's chapter contains analyses of one political ideology, democracy and the concept of "democratic citizenship" within the context of environmental political theory. The author emphasizes the importance of "moralizing about politics" and green citizenship for environmental political theory, especially regarding the value of equity. Equity is vital, the author claims, to ensure "equal subjection of all to constraint by democratically legitimate laws and regulatory institutions"; therefore, green democrats must work toward political institutional changes that advance greater economic and political equality.

Like Scerri, Bourban addresses the subject of ecological citizenship and ecological mindfulness as responses to our current "eco-anxiety." As a result, Bourban's chapter joins with several others in the *Handbook* proposing an alternative to the current precarious state of humanity and the environment. The author says that "[a]nxiety has become a defining feature of our time. It is both true of the most fundamental human emotions and one of the most common forms of psychological disorder."

Bourban defines "eco-anxiety" with three characteristics: "uncertainty about the future but also fear and insecurity in confronting ecological threats and, finally… paralysis of action," identifying three eco-anxiety disorders wherein the eco-anxiety becomes pathological. The chapter concludes with ways of coping with eco-anxiety, especially "mindfulness" (paying attention) which is a "green virtue" supporting ecological thinking.

In light of the pervasive "eco-anxiety" discussed by Bourban, Luke's chapter on "the politics of deep adaptation" explores the work of those who hold these fears inevitable due to the impending collapse of many earth environments, global atmospheric conditions and existing societies because of unchecked rapid climate change. Focusing on the work of Jem Bendell, the founder of the Deep Adaptation Forum, forsakes the existing academic-third sector-international organization sustainability complex as a failed project which mystifies the inevitable societal collapse coming from now unattainable greenhouse gas emission reductions. Organizing sustainability efforts must happen in a few years in order to succeed in time.

Like the ideas of Paul Kingsnorth and Dougald Hine at the Dark Mountain Project, Bendell concedes that industrial civilization's ideology of endless economic growth is a type of "uncivilization" requiring new approaches to survive the climate chaos that is already manifesting in the twenty-first century. The roots of denying this collapse can be found in climate science, academic careerism, positivistic Panglossianism and a collective inability to recognize normal sustainability science and are charged with postponing the decline and doom at the heart of unending endless growth. Hence, "deep adaptation" offers a program for "postsustainability" tied to getting past eco-modernist visions of resilience to embrace different resilient behaviors for human relinquishing, restoring and reconciling with the predicaments of collapse that already present. The politics of deep adaptation then resituate sustainable life as how to embrace climate chaos and survive in these forever changed conditions with valued behaviors and norms suited to the end of fossil fuels and their endless growth mania.

Simon takes up the concept and issue of citizenship. In this chapter, it is citizenship of nonhuman animals, their status and whether there such animal citizens can exist. The status of animals is also a component of the field of environmental ethics discussed in the chapter by Kassiola.

Simon discusses the history of canonical Western political theory wherein citizenship is restricted to humans alone: anthropocentric, human-centered citizenship. She proceeds to criticize a theory of animal citizenship offered by two philosophers (Kymlicka and Donaldson) in their path-breaking work, *Zoopolis*. Simon concludes that "this approach [in *Zoopolis*] is as refreshing as it is disorienting, but presents several major problems." She recommends that this book be interpreted as a utopia and in this way its readers can make the argument for animal citizens as "a source of inspiration for a new approach to political theory." Simon concludes that humans can reconsider how they treat animals by following the author's utopian approach.

The author concludes by examining three proposals of how to integrate ecological and animal issues into politics that lead to transforming free citizens into "*eco*citizens" (emphasis in original), a middle way between anthropocentrism and a sole focus on animals to the neglect of the environment.

Expanding upon the important need for alternatives to the current hegemonic, unsustainable and unjust social order, D'Alisa and Romano's chapter introduces the value of the "degrowth" society.

For several years, a "degrowth" body of thought and social movement have arisen to criticize the modern, industrial neoliberal ultimate value of endless or unlimited economic growth. In their chapter, D'Alisa and Romano propose to the reader a counter-hegemonic value and social order that is not obsessed with unlimited economic growth. These authors present the intellectual roots of the degrowth society, primarily through the works of twentieth-century French social theorist, Georges Bataille. His central concept of "*depense*" (or "expense," "consumption") is divided into necessary expenses for life versus excess luxury expenses not vital for life.[12]

The authors end with the social change perspective of Antonio Gramsci and the need to change the "common sense" of a society.

The nature of the State and agrarian environmentalism are explored by Greear in his chapter. He claims these ideas help us to understand the historical development of agrarian environmentalism by examining the thought of writer and environmental activist, Wendell Berry. Greear's discussion can provide

---

[12] For a detailed analysis of this key concept of Bataille's and the degrowth body of thought see, one of the co-author's works, Romano, in his "*Depense*," in Giacomo D'Alisa, Federico Demaria and Giorgos Kallis, eds. *Degrowth: A Vocabulary for a New Era*, New York: Routledge, 2015, 86–89.

insight into how to structure a sustainable and just social order as an alternative to the failing one that dominates now.

This chapter presents an example of a "post-colonial environmental political" conflict in Bougainville Island of Papua New Guinea between global extractive capitalism (Rio Tino Zinc) and the island's indigenous people who want to protect their land from damages caused by mining.

Agrarian environmentalism should be built upon the value and actions to advance environmental justice. The apolitical nature of Berry's thought is noted by Greear as a handicap to the new agrarian environmental movement. He elaborates upon ancient and modern forms of agrarianism throughout Western thought.

Kassiola's chapter makes the argument for unifying the field of environmentalism in its normative and empirical branches through environmental ethics. It serves a political role by challenging the dominant anthropocentric— or human-centered, supremacist—worldview that underlies the Western social order, especially in its modern form. Kassiola explains the rise of the field of environmental ethics with Lynn White, Jr.'s pioneering challenge to anthropocentric Christianity as responsible for our environmental crisis. He explores subsequent development and ethical expansion of the field with the environmental ethical theories of Peter Singer, Paul Taylor and Aldo Leopold. This chapter illustrates how environmental ethics can inform moral and political progress by helping to build an alternative sustainable and just global society.

Lawrence, Ray and Wiebe introduce the idea and theory of "critical ecofeminism" in their chapter. This mode of analysis aids in understanding the root causes of the current environmental crisis, particularly the climate change emergency. This critical ecofeminist theory helps to envision a superior replacement social order. The authors consider this theory to be the fourth wave of ecofeminism that is "intersectional" in nature, combining issues that involve: gender, class and race. They present earlier ecofeminist scholars' thought and how it integrates intersectional issues about how power over nature by humans (naturism) parallels the power over women by men (ecofeminism).

The intersectional critique of traditional feminism for ignoring women of color, lower-class women and indigenous women is applied by the authors to ecofeminism as well. Thus, a fourth wave or iteration of ecofeminism is needed. Therefore, ecofeminism must become "critical" of oppression by colonialism and acknowledge the plight of indigenous women around the world in order to form a "decolonial ecofeminism." The chapter concludes with critical ecofeminism praxis involving specific projects, including the Feminist Environmental Research Network [FERN] advocating "participatory and arts-based methods of community-engaged scholarship."

Section I closes with a diagnosis of our current ecological predicament provided by Schmaltz who discusses the concept of "property" and social power. Despite being a core concept of neoliberal and earlier forms of capitalism and the social order founded upon them, the idea of "property" is seldom explicitly analyzed as it is in Schmaltz's chapter, and even less so within environmental politics. The author's main thesis is that property is: "central to our ecological predicament"; provided in the chapter's epigraph by Richard Heinberg: "Climate change is a problem of power."

The definition of "property" offered by the author is: "The right to exercise some power on things we legally own... *property is social power*" (emphasis in original). Schmaltz's reasoning is as follows: since property inescapably involves the law and the law's object is power, property is about power as well.

Schmaltz then introduces the concept of "social metabolism" ("the organized exchange of energy, materials and information between nature and society....") and uses this concept to discuss property in social metabolism as a central issue in the Anthropocene.

Section II: Environmental Politics and Theory in the Policy-Making Process

Section II contains chapters addressing the impact of environmental politics and theory on the policy-making process. Many students of politics contend that policy-making is the essence of their subject matter. No matter the level of government, whether local, state (provincial), national, international (or global, such as the World Trade Organization), on the legislative or executive decisions, policy-making is the way any state acts to allocate social goods and bads among its constituents thereby serving its population. The environment and environmental politics and theory can impact the policy-making process in several ways, for example, small island states addressing territory flooded by the sea as a result of global warming; the health and safety of its citizens threatened by water and air pollution; many land use issues involving preservation of national resources and wildlife; or the protection and management of food supply by preserving crop-growing areas and food sources. Of course, policies regarding energy sources, especially fossil fuels or non-carbon emitting renewables, are among the most important environmental policy questions facing humanity today. The accelerating climate emergency affects all decisions made by state jurisdictions and at the international level, given the global nature of climate change.

Section II's chapters encompass discussions of the general environmental policy-making and regulatory processes from a theoretical point of view (Bourblanc and Gillroy) and specific public policy issue areas: environmental health (Delmas) and sustainable housing (Nemoz), with Monier's chapter creatively addressing often overlooked possible roles for philanthropic organizations in protecting the climatic environment by influencing the policy-making process.

Clearly, humanity will only be able to meet myriad environmental challenges by relying upon effective state action on all levels of government with formal legislative action and administrative executive regulatory decision-making. This requires sound scientific expertise about current environmental conditions in combination with broad understanding of factors that support a good environmental policy-making process.

Bourblanc and Gillroy provide innovative theoretical accounts for making environmental policy. Bourblanc examines reliance upon Science and Technology Studies (STS), especially what is termed "instrumented expert knowledge," to analyze the environmental policy-making process. With scientific knowledge crucial to environmental public policy-making, Bourblanc's chapter raises the important preliminary question of the politics of the scientific knowledge-generating method itself. How so different scientific epistemologies, primarily scientific "equipped" or "devices-based" knowledge, affect how we analyze environmental policy-making? Gillroy's chapter centers on the key component to environmental policy, environmental risk. His argument emphasizes the need for morality, specifically Kantian morality, to generate an original conceptual framework and theory of environmental risk within the policy-making process.

Delmas highlights the under-theorized topic of environmental health within environmental politics and theory. As we learn that millions of people (plus billions of other creatures) die each year from environmental causes (e.g., air and water pollution), the relationship between the environment and our health is theoretically and practically urgent.

After years of the COVID-19 pandemic and millions of deaths, we have experienced the harsh lessons of the tenuous links between public health, scientific knowledge, policy-making and politics on the national and global levels. Environmental health will assume greater importance, perhaps prime importance, in the years ahead as we discover even more threats to human and animal health (see, e.g., possible dangers to fertility from chemical pollution) caused by our industrial way of life, as discussed by Delmas. The general environmental policy-making process analyzed and prescribed by Bourblanc and Gillroy might be applied fruitfully to health as well.

Nemoz chronicles the world's efforts over the past 45 years to provide sustainable housing on the international level through United Nations' agencies' activities and conferences. She examines the politics involved in this policy issue's international historical development. As global population and urbanization growth accelerate, the challenge of how to build environmentally sustainable housing for the increasing human population becomes increasingly important on the international level. Nemoz poses a crucial question: What can we learn for current sustainable housing policy applications from past international relations and dynamics regarding the creation of sustainable housing and the environment?

Monier's chapter explores fieldwork in three case studies of national associations of climate-based philanthropic foundations that are climate-based in order to describe the developing landscape of climate philanthropy. The author examines two additional organizations of broader scope, one European-based and the other international. The chapter presents the results of inductive methods of ethnography and archival analyses including interviews conducted in the five, climate change-focused administrative entities in different nations in order to illuminate what is termed: "the understudied topic of philanthropy for climate." With climate change widely recognized today as an emergency, Monier's research and findings about the possibility of philanthropic foundations as agents of societal transformation to prevent climate catastrophe is significant. This chapter opens up a new area for research into a vital source for assistance provided by philanthropic foundations in the struggle to protect our climate.

Section III: Environmental Politics and Theory in the City

This section's theme of the relation between environmental politics and theory and the city is important as the world becomes ever-increasingly urbanized with more than half of the global population now living in cities (since 2007) with hundreds of millions of additional urban residents projected in the future (in China alone).

Barak discusses the ideas underlying the concept of "sustainable urbanism" and the dangers of the mistaken city-nature monism and dualism. As an alternative, the author explores the advantages of city-nature nexuses producing city-nature hybrids and the integration of both elements in this complex interrelationship. Barak explains the empirical consequences of the monistic and dualistic dichotomy and the normatively undesirable results of such erroneous thinking.

Mousie examines the impact of philosopher Henri Lefebvre's concept of "the right to the city" upon environmental urbanists and activists and strives to clarify how this concept was originally understood by Lefebvre and subsequently developed by later students of the city.

Mousie prescribes ways in which the "right to the city" could be changed in meaning in order to be effective in contemporary urban conditions that are much different from Lefebvre's in 1967.

Through Mousie's analyses of Lefebvre's rich conceptual work and the expansion of its original meaning, the author provides an enhanced theory of urban environments and how it can assist us in constructing a sustainable and just political life for all.

Nasr also addresses the historically important dualism in the West between humanity (culture, society) and nature. The chapter advocates we "go beyond" this dualism by acknowledging the erroneous nature of this millennia-long "great divide" and rethinking the concept of "society." The author prescribes a "relocalized society" inspired by agri-food networks that emphasize recent relocalization of food production and sales. This reduces "food miles" between food production and consumption and also creates new, local social networks of cooperation connecting food and agriculture. The author suggests the need to reexamine the composition of "society" conceptually and spatially, using Bruno Latour's theory of society. To Nasr, agri-food networks can be a model for a necessary transition toward the transformation of our current unsustainable and competitive society, envisioning a cooperative model for the future with relocalized social bonds forming a sustainable and more satisfying society.

Section IV: Environmental Politics and Theory in Specific International Regions

The Handbook's next section moves from the national level of analysis to the international level with chapters examining how environments and their surrounding politics impact life and policy-making in specific international regions. The environment around us varies and so do the politics regarding environmental problems and our response to them.

Crews opens up this section with a discussion of the "Rights of Nature" international movement based on the concept of "ecological kinship" between humans, nonhumans and natural objects. This transcends the usual anthropocentric hierarchical division between different types of entities with humanity as superior, entitled to exploit nonhumans and the components of nature. Newly proposed ecological kinship has become the foundation for social change movements reflected in the politics of several nations around the world including Bolivia, Ecuador and the United States. Such environmental movements advocate for the expansion of the legal rights of nature and its components, (e.g., mountains and rivers), based on kinship relations between humans, nonhuman species and nonsentient ecological entities termed: "kincentric ecologies." According to Crews' account, this ecological kinship model endows nature with more care and respect from humans so that our politics and laws produce an enhancement of environmental justice.

These Rights of Nature movements already includes international efforts, ranging from U.N. actions such as the United Nations Declaration on the Rights of Indigenous Peoples to the International Labor Organizations' Indigenous and Tribal People's Convention. The key purpose of these international movements is for national legal systems to recognize that nature and humans are inescapably interconnected. Furthermore, ideas of rights and justice must expand beyond exclusively human needs and incorporate the existence of a "multispecies world" into a "multispecies justice" concept in their policy- and law-making. This prescribed expansion of justice and rights is specifically examined by Crews in Bolivia under Evo Morales and in the United States with the Lake Erie Bill of Rights describing the City of Toledo, Ohio's legal efforts to protect its ecosystem through a Rights of Nature perspective.

The section's discussions move to Africa with a chapter by Ignatov, "Theorizing with the Earth Spirits: African Eco-Humanism in a World of Becoming." This chapter is the author's effort to fill the *lacuna* of African political thought within Environmental Politics and Theory. We are excited to include his report from fieldwork in Africa and hope it will inspire other researchers to follow Ignatov's lead and investigate the thought and practices of the African continent as yet another source of ideas, values and actions that are alternatives to currently unsustainable industrial hegemony around the world.

Ignatov describes the role of the Chief of Medicine in Ghana with responsibility for the environment of the entire community including the nonhuman world, granting it "respect and moral worth." One conversation with such a Chief of Medicine includes the following statement: "Our feet are the gods. The earth is the gods. Trees are gods, rivers are gods, and stones are gods. We owe our very lives to these mentioned elements of nature."

Other forms of African environmental thought and values support this chapter's thesis that the African worldview can be a rich intellectual resource for a non-anthropocentric, non-dualist perspective on the nature-human relationship to counter the predominant anthropocentric one that has contributed to the dangerous situation we are in today.

The next chapter in this section by Parra-Leylavergne pertains to international politics surrounding the case of sargassum (a form of seaweed) that endangers local ecosystems in the Caribbean. Like African thought and practices described by Ignatov, the Caribbean is under-analyzed in environmental politics and theory literature along with the whole Global South.

Thus, we are pleased to include Parra-Leylavergne's innovative discussion of the international environmental politics of the Caribbean-island countries in response to the sargassum problem.

The author utilizes the theory of social constructivism to frame and comprehend the responses by Caribbean political systems regarding sargassum and describes environmental policies in the Caribbean addressing this environmental threat on the local, national, regional and international scale.

This chapter then reviews the important international environmental treaties and actors in the Caribbean and illustrates these connections with the management of sargassum as a case study as well as marine resources management throughout the Caribbean region. Parra-Leylavernge poses a series of important questions about how local, national, regional and international actors think and act with regard to the sargassum issue for tourist-dependent island states of the Caribbean. The author perceptively inquires about *when* the response to the sargassum challenge leaves the realm of natural science and becomes a political problem. This is an important question to raise for all issues, if the domain of environmental politics and theory is to have a clear conception to guide its students, researchers and policy-makers.

The fourth section concludes with a chapter on contemporary youth environmental activism. Case studies by Stuppia focus on three, climate change, youth, activist international organizations in France and Italy. With the global celebrity of Greta Thunberg and her immense media presence, plus environmental protests over climate change (highlighting next generational impacts), environmental activism has shifted to youth in a large and more significant way. Stuppia's fieldwork and survey-based chapter is a very timely and important case study of this age group that could add energy and creativity to the challenging effort for social and political transformation. Such a transformational movement needs to lead to policies that will prevent environmental dangers as well as change our social values and practices to sustainable and more desirable ones. Given the lack of progress in the long history of the U.N.'s Framework Conference on Climate Change and looming climate disaster, the introduction of new youthful forces advocating for a new green way of life should be welcome.

Stuppia's socio-demographic data and analyses of selected organizations with regard to "ages, classes, genders, and colors of dissent" and levels of radicalness can provide insights into the nature of political active youth across the world. One particularly important idea is "DIO Politics" or "Do-It-Ourselves" politics whereby the youth involved in such environmental organizations seek to act politically themselves around issues of their own choice and are not driven by political ideologies nor existing political parties (like the Greens) nor mainstream environmental organizations (like environmental NGOs, like Greenpeace).

The author's interviews with participants and leaders of the three organizations provide a window into what the environmental activist youth are thinking and why they have acted the way they do in these climate justice movements. As environmental problems worsen and more people are adversely affected, international protests will become increasingly important as well. The author significantly concludes that environmental protests' nature and participants have changed in the last 25 years from conventional political expressions by different groups to the current youth-dominated unconventional and unprecedented actions driven by projections of disaster for their adulthood.

Section V: Environmental Politics and Theory in the Anthropocene

The final section of the Handbook begins with a chapter by Vidal expositing the meaning and consequences of the now well-known geological hypothesis of the Anthropocene (see Paul Crutzen and his colleagues) whereby humanity is claimed to be the dominant force on the planet. The author explores this scientific conjecture from an astrobiological perspective focusing on the habitability of life on Earth. He goes on to state the transformative conclusion that "the status quo for humankind can no longer be sustained over the long term." Therefore, the chapter describes the paradigmatic transformation required regarding the relationships between humans, nonhumans and the planetary environment that have been "complexified" by modern humans.

What stands out in this discussion is the astrobiological emphasis upon the ontological question (the nature of being or the nature of the real) of the human condition as it is composed of what is termed the "thermo-industrial civilization" during the current environmental crisis of the Anthropocene Era. Like many of the Handbook's contributors, the chapter contends, based on the scientific evidence presented, that a civilizational transformation will be required if a species-wide environmental catastrophe is to be avoided.

The beginning of the chapter is a thorough and updated review of the scientific literature detailing the evidence for the scientific claim that various threats exist that threaten to transgress existing planetary boundaries. These findings generate, according to the author, a fraught open question of whether or not the human species (or other living species) can survive. Yet, the encouraging news, the author notes, is that the planet has shown great resilience when existentially challenged in the past. However, success in averting disaster will require, according to the chapter, recognition of the existence of these planetary boundaries and a required civilizational transformation to a post-thermo-industrial civilization (which must be described and given substance to be valuable, as we have noted earlier) in order to avoid crossing the planetary boundaries of the "Anthropocene's new [and dangerous] stage of boundaries."

The latter part of the chapter commendably begins the crucial positive substantive discussion of an alternative to the unsustainable thermo-industrial civilization with the illustration of a nation-state, Japan, as a possible policy model of how to respond to the ecological threats of this new and precious stage of the Anthropocene Age. The author borrows the concept of "adaptation learning" as an important first step to the creation of this needed sustainable social order.

The second chapter in this section by Pattberg and Davies-Venn on environmental politics and theory in the Anthropocene provides a detailed examination of the specific geological meaning as well as a rare insight into the political processes involved in accepting the supposed apolitical scientific geological name and concept for our contemporary time. What Vidal in the first chapter within this section and mostly all scientists and environmentalists assume about the qualities of this newly proposed Anthropocene Age, including the alleged dominance by humanity over the planet and all of its living nonhuman inhabitants is explicitly examined for its political nature by the authors, for example, the background and starting date of the hypothesized new age.

What is clear to the authors is that the issues surrounding the prescribed geological concept of the Anthropocene are surely more than geo-scientific in nature because social and political factors are inescapable even if various scientific certifying organizations publicly ignore them.

Pattberg and Davies-Venn illustrate the politics involved in one such scientific organization, the Anthropocene Working Group, by emphasizing how leadership, gender and discipline-membership issues carry their own political biases. Thus, the authors find considerable politics involved within the geological process of declaring a new Age. For many readers who have taken the 20-year-old hypothesis of the Anthropocene as a closed scientific matter will find the authors' data about the certification process revelatory and dismaying. This produces the upshot that the Anthropocene hypothesis is more than a scientific hypothesis but a political one as well.

The authors conclude that the Anthropocene hypothesis must transcend narrow technical geological considerations and focus upon political values, such as fairness and equality in a new governance perspective.

They also have attached a helpful Addendum Key to clarify for the novice reader a technical glossary of the concepts and definitions involved in the Anthropocene thesis and resulting scientific debate.

With the Vidal and Pattberg/Davies-Venn chapters, we have two different approaches to the Anthropocene question: one starts out with the scientific findings demonstrating the extreme and dangerous ecological situation humanity and other-than-human-species are in with respect to planetary boundaries with the resulting urgency of transforming our social order (Vidal). The other approach creatively looks behind the scientific curtain to reveal the scientific and political issues involved in the nomenclature change to a new Geological Era of the Anthropocene. The authors of this second approach also prescribe a transformed way of assessing the Anthropocene hypothesis, one that is explicitly political (Pattberg/Davies-Venn).

The third chapter in this section written by Cadalen relies upon Michel Foucault's famous theoretical concept of "biopolitics" and presents it within the context of the Anthropocene to inspire a new proposed concept of "ecopower."

Cadalen begins his essay with a profound point about the Anthropocene concept: "Politics is thus the key to the Anthropocene problems and solutions." This position is one that the two previous chapters' authors would strongly agree upon. The Anthropocene concept, hypothesis and Age are all political in essence. This remains true even if the geological (and other) scientists who dominate the debate about different criteria for the geological period name and status, along with the geological discipline-based standards of acceptability of the proposal of a human-dominated era overlook the decisive facts of politics for our future which must include a transformed social order, if we are to achieve sustainability.

Cadalen introduces the reader to Foucault's influential concept of "biopolitics" driven by a new form of "biopower" or "the power over life and death." He emphasizes Foucault's important insight for environmental political theory regarding the intersection of political power of a sovereign and physical nature. As Foucault says: "... the sovereign will be someone who will have to exercise power at the point of connection where nature, in the sense of physical elements, interferes with nature in the sense of the human species, at the point of articulation where the *milieu* [the environment] becomes the determining factor of nature."

Cadalen discusses the social science history of the idea of the Anthropocene, including the claim that biopolitics, or politics focusing on life, has failed because of *milieux* [environmental] destruction as "humans have acquired a centrality while their *milieux's* transformations threaten the long-term conditions of their reproduction." To remedy this dire environmental problem, Cadalen offers a new concept of "ecopower," that he argues is necessary if all species reliant upon nature are to be preserved. This is a new concept by Cadalen offered to reflect the new circumstances of sovereign power and ecopower, where "ecopower" is defined as "the power to perpetuate life conditions of human species and the vertebrate living or to produce their destruction… as Biopolitics seems to have considerable failed in the move to preserve life, a radically new political concept appears to be needed as the degradations of our environment accelerate abruptly…."

Cadalen refers to the thought of twentieth-century political theorist, Nikos Poulantzas to explain the contemporary politics of life and death of species in order to extend and improve upon Foucault's "biopower" concept in an effort to "foster life" with his concept of "ecopower."

The fourth and concluding chapter to this section and the entire Handbook turns away from the idea of the Anthropocene and human planetary supremacy as well as its basis in dualism (nature vs. humanity, or culture) to propose "technonaturalism." This proposed concept is clarified and supported throughout the chapter by Stubberfield, continuing the common theme of the Handbook chapters rejecting the prevailing dualism between nature and society in favor of a more egalitarian and holistic understanding of the hybrid of: "humans, nonhumans and extra-humans," as Stubberfield puts it.

He also finds deficiencies in the Anthropocene idea in his chapter, mainly its narrowness and thus limited range of application that omits the crucial determinant facts for environmental degradation by displacing the responsibility for the deterioration of the environment to all humans equally when we know because of global inequality this is not true.

To correct this fault, the author prescribes a "new naturalism," negating the Anthropocene concept by assisting in the comprehension of the environmental crisis and is a "rule of artifice." This is accomplished with the concept of "technonaturalism" highlighting the political aspects of "artificial persons," combining humans and nonhumans into functional machinery. Stubberfield describes this new approach that accepts death of the (nonhuman) nature thesis yet rejects dualism between nature and society as well as focusing on the hybrid combination of humans, nonhumans and extra-humans by relying upon a postphenomenological and critical urban geography of Don Ihde and the environmental political theoretical writings of Tim Luke.

Stubberfield builds upon Ihde's theory of hybrid "interrelational ontology" a rejection of modern false dichotomies such as, "mind/body, human/technology, technology/culture and technology/nature" for a hybrid environmental subjectivity connected to the theories of several social theorists including: Mills, Foucault, Luke, McKay and Mumford.

The upshot of this chapter is the centrality of moving beyond the humanity/nature misleading division and replacing it with a socio-ecological perspective—an idea that permeates several chapters within this Handbook and that may be considered foundational within environmental politics and theory. The chapter is an investigation into ways these integrative social theorists' ideas can be used for a new hybrid technonatural worldview utilizing

Ihde's thought emphasizing human–machine relations. How might the environment be envisioned in a new interrelational ontology that the author terms "postphenomenological environmentality?" is posed by the author.

For Stubblefield, the significant conclusion from Ihde's critical geography and the other society theories is that the Anthropocene Era is not created by humanity alone but by a combination of artificial technical hybrid persons. The resulting innovative theory produces a technonatural political theory of the environment by applying the paradigm of Ihde and the insights of Luke to "de-naturalize" the environmental by adopting "anti-anthropocentric technical analyses."

Conclusion

The chapter by Stubberfield with its prescription for a more politically sensitive technonatural theory to rival the current natural scientific paradigm of the depoliticized Anthropocene, and its optimistic ending of a "new planetary becoming" with an "ever widening vision," is an excellent position upon which to conclude the volume. This collection analyzes the political structure for the environmental threats confronting humanity, nonhuman nature and the planet at a time of dire environmental emergency, yet the contributors regard the future with an "ever widening vision" of moral and political progress for the reader to consider. What transformations in our thinking and acting are necessary and desirable?

We hope that the *Palgrave Handbook of Environmental Politics and Theory* will be valuable to the its readers in diagnosing our ills in practice and errors in perception, while providing new evaluations that will prove positively fruitful in providing alternatives to the prevailing paradigms for environmental analysis to ponder and assess new departures, and, thereby, learn from confronting these challenging perspectives. Our critical learning process must continue and deepen because, like Rabbi Tarfon's process of enlightenment and action through studying sacred text, while we are unable to finish the task, neither are we free to desist from it.[13] As students of environmental politics and theory, we must keep at our vital task of greater understanding to envision a more just and sustainable social order even if we cannot finish this essential work in our lifetime.

We hope the *Palgrave Handbook of Environmental Politics and Theory* guides this grand endeavor and that its diverse readership of students, scholars and policy-makers will rely upon the *Handbook* and its multitude of environmental observations, judgments and recommendations to inspire their thoughts and actions.

---

[13] See, Kassiola, *The Death of Industrial Civilization*, 217.

Environmental Politics and Theory

Environmentalism and Political Ideologies

Fabrice Flipo

WHAT KIND OF ECOLOGISM?

In a well-quoted book, Andrew Dobson proposed differentiating between "light green" and "dark green"; the former acknowledges the ecological issues such as climate change and resources depletion but positions itself in favor of pursuing growth, finding cleaner technologies, while the latter challenges the first with three arguments: the insufficiency of technological solutions, the exponential effects of growth (which are counter-intuitive), and the interdependency of the issues, which invalidate all partial solutions.[1] We can add that time and space are key parameters: cleaner technologies, if they are ever available, will not be in time or of the required magnitude.[2] Growth or degrowth: Only two possibilities? No, some movements advocate for an organizational change in production and consumption modes, trying to preserve certain gains

---

[1] Dobson, Andrew, *Green Political Thought*, London, Routledge, 2003, p. 62.

[2] Flipo, Fabrice, «Les trois conceptions du développement durable», *Développement Durable et Territoire*, vol. 5, no. 3, décembre 2014.

This article is inspired by Fabrice Flipo, Authoritarian ecology, Wiley, 2018.

F. Flipo (✉)
Institut Mines – Telecom BS, Paris, France
e-mail: fabrice.flipo@imt-bs.eu

from modern technologies while criticizing the argument of dematerialization. In the case of France, the *négaWatt* association in the field of energy or the *Confédération paysanne* in the field of agriculture can be considered as representatives of this third current. *NégaWatt* played a very important role in placing an energy scenario on the agenda that is not only 100% renewable but also very realistic[3] in so far as the scenario does not foresee a massive reduction in travel or inhabited areas. Considering more than two scenarios is also the tendency of the available scenario exercises such as the Millennium Ecosystem Assessment, which proposes four[4]: the TechnoGarden, which corresponds to "light green" or dematerialization of growth; the Adapting Mosaic, a radical version of which would correspond to a chosen degrowth, through justice; Order from Strength, which would be an endured degrowth, through conflict and partial collapse; and finally, Global Orchestration, which designates a successful global cooperation, both in social justice and harmony with nature.

Another important distinction has to be made, at least in a French context, between the "protection of nature" or "environmentalism" that reflects a sectoral, narrow activity and "ecologism," which refers to the more general issue of the quest for "ecological societies," a "new paradigm" that sometimes has the same emphasis in the prophetic discourse as the heralds of the socialist movement that was born in the nineteenth century.[5] For French ecologists, "environmentalism" is above all a refusal to take a position on the "distal" causes and in particular the differentiated responsibilities within the social order, which leads them not to be very precise about the solutions either because they are necessarily the solutions of one party of humans against another party that opposes them. For Yves Frémion, one former French green leader, protection of the environment and political ecology are different movements that do not have the same history or the same founders[6]; it is not the same movement with more or less "radicality." The sociologist Alphandéry rightly notes that the most frequently cited authors in the movement (Illich, Moscovici, Gorz, Dumont, Passet, etc.) have all challenged the reductionism of environmental ecology,[7] pointing that looking only at nature degradation isn't enough to foster change.

---

[3] https://negawatt.org.

[4] Millennium Assessment Report, "Living Beyond Our Means: Natural Assets and Human Well-Being, 2004", 2004.

[5] As a reminder, in French, the term "ecologist" is reserved to designate scientists, while "ecologism" is the name of the activist movement ("écologues" as scientists, "écologistes" as activists); in English, "ecologist" designates both. And "environmentalists" are "environnementalistes."

[6] Frémion, Yves, *Histoire de la révolution écologiste*, Hoëbecke, 2007.

[7] Alphandéry, Pierre, Bitoun, Pierre et Dupont, Yves, *L'équivoque écologiste*, Paris, La Découverte, 1991.

Having said this, what is environmentalism? If the political ideas are essentially contested,[8] it can be defined by the conflict that emerges when it appears in the face of other political ideologies. Therefore we confront ecologism to liberalism, socialism and conservatism.

THE SKEPTICAL LIBERAL

Clarifying the issue requires first defining "liberalism." In French political theory, then, we come upon a problem: liberalism is often reduced to its political and legal dimension, as in *Philosophie politique* by Luc Ferry and Alain Renaut.[9] Inversely, in Anglo-Saxon countries, economics is often the obvious element to consider in philosophy.[10] However, an American consideration only poses this problem in reverse: How to align with continental philosophy, in these conditions? The book by Catherine Audard,[11] the French transaltor of Rawl's *Theory of Justice*, seems to propose a satisfactory compromise, although it mostly subscribes to the social liberalism of J. S. Mill. Audard characterizes liberalism by three elements founded on the idea of the liberty and sovereign power of the individual, who is considered not to be subject to any indefinite subordination.[12] The concrete realization of this liberty occurs through the implementation of three distinct spheres of action, each of which is essential to the achievement and stability of the two others: the self-regulating order of the market, or "invisible hand," the Rule of law, and the representative government. This modern form of liberty no longer needs religious virtue, nor an absolute State, nor community pressure ("mechanical" solidarity or solidarity based on similarities, according to Durkheim[13]), three elements that are known to be at the core of pre-modern or non-liberal societies. The site of power must remain "empty" in the sense that those who occupy it are only temporary tenants. This is what democracy means, for liberalism: the institution of liberty—the only possible way to the institution of liberty.

How does this "invisible hand" work? David Ricardo (1772–1823) illustrated the process with a famous example. Both England and Portugal each produce fabric and wine. The cost of producing wine in England is higher than in Portugal, due to the lack of sunshine; the cost of producing fabric is higher in Portugal than in England, due to the absence of expertise in matters of machines. If each country produces wine and fabric in their own territory, they find themselves less wealthy than if they trade fabric for wine, thus making

---

[8] Gallie, W. B., «Essentially Contested Concepts», *Proceedings of the Aristotelian Society*, vol. 56, 1955, pp. 167–198.

[9] Ferry, Luc et Renaut, Alain, *Philosophie politique*, Paris, PUF, 2007.

[10] Dupuy, Jean-Pierre, *Libéralisme et justice sociale*, Paris, Hachette, 1992.

[11] Audard, Catherine, *Qu'est-ce que le libéralisme ?*, Paris, Gallimard, 2009.

[12] Ibid.

[13] Durkheim, Emile, *De la division du travail social*, Paris, Félix Alcan, 1893.

good use of their respective comparative advantages.[14] This assumes that they are free to trade and not restricted by the State. This logic of exchange has remained the same up to the present day, and we find it essentially unchanged in the works of Paul Krugman.[15] Trade is indifferent to borders: it is exactly the same to trade with one's neighbor as with to trade with the other side of the world if the economic interest is there. Given that trading implies almost always specializing, the developments of trade and of the division of labor are two sides of the same coin. Trade also defined a very specific conception of labor, which now designates an activity that generates added value.[16] As Marx pointed out, the search for added value leads to the extension of trade to the global market, and to the development of "relative surplus value" in the form of machines, which, with equal human labor, allow more production.[17] Hence, a linearly ordered sense of history. The discourse that President Sarkozy used at the University of Dakar in 2007 is a particularly caricatured and brutal testimony to the extreme diffusion of this modernizing common sense that sometimes remains underground; he said that if Africans "had not entered into history" it was because they "lived with the seasons," that they only knew the

> eternal cycle of time paced by the endless repetition of the same gestures and the same words [...] In this universe where nature commands everything, Man escapes the anxiety of History that tortures modern Man but remains frozen in the middle of an unchanging order where everything seems written in advance. Man never looks toward the future. It never occurs to him to break away from the repetition to invent his own destiny.[18]

Friedrich Hayek (1899–1992) adds that elevating production in a Benthamian utilitarianism *is not* the only objective of liberalism. And, for him, not the main: What he calls *catallaxy* is a common order founded on the only economic interest understood as free confrontation of supply and demand[19] with the idea that the State cannot make a better choice than the individual himself and must therefore remain limited. At the extreme, this position argues that all overarching points of view are conflated with totalitarianism because they must necessarily have an exorbitant amount of information about people to be relevant.

---

[14] Ricardo, David, *Des principes de l'économie politique et de l'impôt (1817)*, Paris, Flammarion, 1999.

[15] Krugman, Paul, *La mondialisation n'est pas coupable*, Paris, La Découverte, 2000.

[16] Postone, Moishe, *Temps, travail et domination sociale (1986)*, Paris, Mille et Une Nuits, 2009.

[17] Marx, Karl, *Le Capital (1867)*, Paris, PUF, 1993.

[18] The speech can be found on Wikipedia. https://fr.wikipedia.org/wiki/Discours_de_Dakar.

[19] Hayek, Friedrich A., *La route de la servitude (1945)*, Paris, PUF, 2013.

When confronted to ecologists, liberals are initially skeptical about the existence of an ecological crisis, their view being rooted in the last 150 years of linear progress of man's domination over nature. For them, past experience tells us that technology has mastered nature, and it will work again in the case of the so-called "ecological crisis," which is only an adjustment of the existing system, through smart cities, autonomous vehicles, hydrogen, thermonuclear fusion and biodiversity creation. Bjorn Lomborg is highly representative of this current. To the extent that ecologism calls for collective control of our destiny, giving life plans a salvational and collective dimension, liberalism tends to equate it with a veiled socialism: it refuses the pluralism of the conceptions of Good and the freedom of the market. Moreover, ecologists aspire to such changes that it seems to require a revolution or a rejection of the formal institutions of democracy. This is something Dobson also noted. At the same time, ecologism seems anti-modern and tinged with conservatism due to its appeal to re-inserting man into nature and its criticism of technology; yet liberalism conceives itself as emerging from the "closed" society of the Ancien Régime, in which progress was absent and the "natural order" of religion dominated every aspect of life. This is also why one of the core ideas of ecologism and environmentalism seem scandalous: the "rights of nature" or intrinsic value of nature. This claim has four justifications: biosphere isn't only a means toward human ends, but also an end in itself, following Kant's distinction; nature isn't a product of work, and therefore cannot be fully appropriated, only used in a sustainable manner ("fructus" is permitted, "abusus" is prohibited); nature is a precondition for any form of government, culture and civilization to come;[20] nature is a living being, which can suffer and has rights.

Being both socialist and conservative, ecologists appeared "red-brown," wanting to socialize the economy by politicizing consumption (against the atomistic and decentralized order of the market) or by practicing "direct action," extra-parliamentary, but also defending the Earth and the "natural order," in the same way as the Ancien Régime, which was based on a hierarchical social order seen as natural (nobility, clergy et third estate); which led them to be regularly qualified as "Khmers verts," notably in the media. This is because red khmers in Cambodia value rural life, a collective control of common destiny and an authoritarian regime.[21]

It is true that the two types of liberty liberals value (Benthamite-Ricardian utilitarianism and Hayekian libertarian freedom of choice) aren't compatible with a ecologism, for different reasons: utilitarian liberty implies infinite (ricardian) growth and libertarian choice implies a rejection of any global coordination. Ecologism revives the question of virtue, that liberalism thought belonged to the "liberty of the Ancients" following the distinction proposed

[20] Jonas, Hans, *Le principe responsabilité: une éthique pour la civilisation technologique*, Paris, Le Cerf, 1990.
[21] Ben Kiernan, *How Pol Pot Came to Power: Colonialism, Nationalism, and Communism in Cambodia, 1930–1975*, Yale University Press, 1985. Rééd. en 2004.

by Benjamin Constant in 1819:[22] a relation of the individual to the Whole that is required for the web of life, due to the practical interdependencies that it entails. The virtue of every person is a condition for the good of all. This is one of the reasons why liberalism accuses ecologism of rejecting democracy and modernity.

But ecological arguments are rooted in a science: ecology, not in a religion. They are of secular nature. This is a key difference with Ancien Régime. For ecologists, liberals consider themselves independent of nature, but this is an illusion, it is a denial of (ecological) scientific conclusions. Liberalism considers nature in a Cartesian way, constituted of mental substance (*res cogitans*) and corporeal substance (*res extensa*) that are displaceable and modifiable at will locally without notable consequences globally. Confidence in technology prevailed; and technology is based on an instrumental value of nature, taken as a pure resource. This order, usually called "anthropocentrism," opens on a massive use of mineral and non-living resources, and this is indeed the case of industrial civilization; but it obscures the ecological dimensions of the world in an attitude contrary to the scientific spirit that it deems characteristic of the modernity that it claims. For ecologism, anthropocentrism puts human beings as the only ends-in-itself and, therefore, is self-destructive, among other consequences, given that there's no known possibility of living outside biosphere, with an exception for very short trips. And many features of nature aren't replaceable: this justifies the "dark" green position. Against anthropocentrism, human beings are of nature, in nature. They rely on nature. Human beings cultures and civilizations are not out of nature but determining different states of nature.[23]

Ecologism can then be reinterpreted: its holism is less a "return" to a previous understanding of the social order than the desire to reveal and insert into the public debate what the liberal order would like to mask. Ecologism advocates for science-based (free) choices. It not only demands the rights that liberalism only promises, it also challenges the liberal paradigm, its vision of the world, in the past and in the future, and in particular in its propensity to sow the seeds of an eco-fascism. This eco-fascism would correspond to what the Marxist political thinker Nikos Poulantzas theorized about liberalism in the 1930s: an alliance of capitalists and strong power to defend the regime rather than find solutions to the ongoing crisis. In the 1930s, the crisis was mainly financial, and industrial, with capitalists refusing to pay workers more, this leading to an overproduction crisis.[24] Adolf Hitler took the power with a support of the capitalist class, who sees the *Fürher* as the one who was capable to restore order, especially among the working class; but the creature escapes

---

[22] Constant, Benjamin, *Ecrits politiques (1819)*, Paris, Gallimard, 1997.

[23] Moscovici, Serge, *De la nature. Pour penser l'écologie*, Paris, Métaillé, 2002.

[24] Poulantzas, Nicos, *Pouvoir politique et classes sociales de l'état capitaliste*, Paris, Maspéro, 1968; Poulantzas, Nicos, *Fascisme et dictature*, Paris, Maspéro, 1970.

its creators. Nowadays, similarly, the capitalist order is strongly reluctant to change its industrial and financial order, to face the ecological crises.

To summarize, ecologism denaturalizes liberalism and modernity, opening up to other possible worlds; it secularizes the faith of the moderns in technology, opening up into a technical democracy. In doing so, it draws from nature a principle that is outside of the artificial order that it critiques in the name of another that it deems more consistent with emancipation. On its side, "direct action" refers most of the time to a simple exercise of (real) freedom, correctly informed of its long-term consequences. This is what Hans Jonas calls "responsibility."[25] A step above, direct action refers to civil disobedience, not to terrorism or war, which are coherently seen has incompatible with the inherent or intrinsic value of nature, especially living beings, including humans. According to Dobson, "direct action" consists of contesting infrastructure in the field rather than at the polls.[26] Theoretically, liberalism could be ecologist, according to the arguments of Marcel Wissenburg:[27] it would be enough for individuals to convince others in the context of an open debate. But the political reality is different, dominated by giant firms that control socio-technical networks, drawing a very different landscape from the one theorized by neoclassical economists, and taken for granted by liberal ideologists.

The Difficult Way to Eco-Socialism

What about socialism? At first glance, it shares several characteristics with ecologism. It contests the liberal order, notably on the basis that liberties it promises are only formal, not real: influence on laws, markets and government is granted for capitalists, not for proletariat. It takes the form of social movements that have been accused of wanting to impose their conception of the Good, versus an atomistic and individualistic liberal system that is supposedly "neutral" and "democratic." It has also been accused of wanting to provoke disorder and the arrival of strong powers by waving the threat of future catastrophes. These three traits and others lead liberals to see ecologism as a kind of ideological alternative that allows leftists to reframe themselves, given the failure of Marxism-Leninism. Jean-Marie Le Pen, the long-time leader of the extreme-right-wing French National Front party, is remembered for having famously qualified ecologists as "pastèques" (watermelons): green on the outside, but red on the inside. American liberals and libertarians consider also ecologists "watermelons".[28] Socialism ought to have welcomed ecologism with open arms, then. This has not been the case. Andrew Dobson gives us

---

[25] Jonas, *Le principe responsabilité: une éthique pour la civilisation technologique*.

[26] Dobson, *Green Political Thought*, p. 142.

[27] Wissenburg, Marcel, *Green Liberalism—The Free and the Green Society*, London, University College London, 1998.

[28] Nouhalat, Laure et Guérin, Franck, *Climatosceptiques – la guerre du climat*, 2014. Film.

some direction: for the socialists, ecologism is conservative and reactionary, because they want to spin the wheel of history backwards; the ecologists are considered to be the petits-bourgeois incapable of challenging capitalism, which is the true nature of liberalism. On their side, the ecologists ask what the concrete content of socialism is, and to what degree it breaks with accumulation and faith in technology. A better distribution of wealth would not be enough to put an end to the unsustainable nature of industrial practices.

Why has socialism turned out to be so little receptive to ecologism, in reality, even though it would affirm clearly that it is a vehicle for global equality? Why does it broadly take up Promethean and Cartesian anthropology, which are also claimed by liberalism? We must come back to socialism. The *Histoire générale du socialisme* by Jacques Droz is a good basis,[29] the only one with the aim of producing a general typology like the one proposed by René Rémond in the 1950s for the French right,[30] and still used.

Jacques Droz distinguishes three periods. The first takes up only a few dozen pages: these are pre-industrial socialisms. They are characterized by their critique of "social injustice"[31] and do not bear the name of "socialism"; in this regard, they can be found all over the world, or almost, in all eras: China, India, Ancient Greece, etc. Then follow the authors who are considered to be the precursors of modern socialisms and communisms: the *Code de la nature* by Morelly (1755), the *Discours sur l'inégalité* by Rousseau (1754), the *Télémaque* by Fénelon, Diderot, Voltaire, Baron Holbach, etc. in short, it is in large part these authors who have recently been qualified as "radical Enlightenment thinkers."[32] Droz adds the French revolutionary experience in 1789, of which he recognizes the bourgeois dimension, regretting that Robespierre and Saint Just only come slowly to social democracy. The classical socialist reading of the revolution effectively considers it be bourgeois, only allowing the conquest of political rights. The first properly socialist attempts emerge later from the pauperism provoked by liberalism and industrialization, then subsequently the convergence with the workers' movements as of 1875 and socialisms properly speaking. For the socialists, the liberal perspective is doubly flawed: not only is it incorrect, when it affirms that liberties are respected and that wealth is produced, but it is also ideological in the sense of a veil covering the reality of the bourgeoisie, this social class that only overthrew the nobility and the clergy to better take power. Liberal discourse has a legitimizing function for the new authorities. The years between the French Revolution and the July Revolution (27, 28 and 29 of July, 1830) saw the workers' movement grow and France become wealthier— but pauperism spread. This phenomenon, unexpected by the elites, prompted

---

[29] Droz, Jacques, *Histoire du socialisme. Tome 1*, Paris, PUF, 1974.

[30] Rémond, René, *Les droites aujourd'hui*, Paris, Louis Aubibert, 2005.

[31] Droz, *Histoire du socialisme. Tome 1*, p. 10.

[32] Israel, Jonathan, *Les Lumières radicales. La philosophie, Spinoza et la naissance de la modernité 1650–1750*, Paris, Amsterdam, 2005.

inquiries like those of Eugène Buret or Louis Villermé.[33] Pauperism is distinguished from poverty: poverty is isolated and transient, pauperism, on the other hand, is part of an epidemic, it strikes an ever-growing population, which is developing through economic progress. Very quickly, however, socialism was anchored in the critique of inequality with the goal of "establishing a society of harmony"[34] in the new conditions of modernity: the market, industrialization, States with vast territories. Droz follows Marx, with differentiating between early "utopian" socialists, who could be revolutionary or communist but didn't rely on the propulsive force of the proletariat, and "scientific" socialism, rooted in historical materialism, which acknowledges both the centrality of the labor movement and a role for an enlighted Avant-garde.

Marx and Engels saw four types of socialism, in the *Communist Manifesto*, aside from the socialism or a revolutionary ("scientific") communism that they aligned with: a "reactionary" socialism, a "corporatist" or "petit-bourgeois" socialism, a "conservative" socialism and "critical-utopian" socialisms. The first ("reactionary") includes various trends that we can roughly group under the idea of seeking to re-establish medieval corporations in the modern mode of production, which for Marx is both reactionary and utopian: reactionary, because it comes down to trying to re-establish what was abolished by the bourgeois revolution, and utopian because nothing like this is possible any longer due to the emergence of industry, which expends free labor. "Corporatis" ou "petit-bourgeois socialism" is embodied in the figure of Sismondi who essentially called for a State intervention program with the goals of protecting the working class, struggling against the excesses of competition and regulating progress in order to prevent unemployment and economic crises. This program is not about emancipation but only securing different stakeholders (including the working class) immediate and corporate revendications. "Conservative" socialism accepts the new situation and looks toward the future but all the while refuses to take the revolutionary path; here, Marx targeted Proudhon in particular. And "critical-utopian" communism (Fourier, Saint-Simon, Owen, etc.) designate the individuals who substitute their own ingenuity for social activity: "Future history resolves itself, in their eyes, into the propaganda and the practical carrying out of their social plans."[35]

For Droz, classical Marxism, which follows the utopians socialists, constitutes the "moral unity" of socialism.[36] The *Communist Manifesto* went relatively unnoticed when it was released but it provided what has been subsequently interpreted as a synthesis of the thinking of Marx and Engels,

---

[33] Buret, Eugène, *De la misère des classes laborieuses en Angleterre et en France*, 1840; Villermé, Louis, *Tableau de l'état physique et moral des ouvriers employés dans les manufactures de coton, de laine et de soie*, 1840.

[34] Considerant, Victor, *Le socialisme devant le vieux monde*, Paris, Librairie Phalanstérienne, 1848.

[35] Marx, Karl, *Le Manifeste du Parti Communiste*, 1962ᵉ édition, Paris, 10/18, 1847.

[36] Droz, *Histoire du socialisme. Tome 1*, p. 5.

themselves expressing the truth of socialism. The text sees the bourgeoisie as a revolutionary force:

> Each step in the development of the bourgeoisie was accompanied by a corresponding political advance [but wherever it had passed, it] has drowned out the most heavenly ecstasies of religious fervor, of philistine sentimentalism, in the icy water of egotistical calculation. It has resolved personal worth into exchange value, and in the place of the numberless indefeasible chartered freedoms, has set up that single, unconscionable freedom – Free Trade.
>
> The proletariat alone is a genuinely revolutionary class. The other classes decay and finally disappear in the face of Modern Industry; the proletariat is its special and essential product.[37]

Built on Hegel's vision of history,[38] a typical process is described. First, there was the primitive commune, where property has no meaning. Then Greek Antiquity invented democracy, without economic progress. They are followed by Romans who invented property, through slavery, without democracy. During Middle Age work was infeoded to landlords and war. Then bourgeoisie went on the rise, freeing work, with free markets and other liberal democracy institutions, described earlier in this chapter: representative democracy, civil and political liberties, rule of law and free market. Capitalism took place against landlords and noblesse. Regarding the entire story, the French Revolution was a bourgeois revolution, not a social one. Capitalism arise, and is regulated by laws, which resulted of bourgeois attitude and norms, not of nature. Class struggle is now an explicit process. Proletariat is exploited by the capitalists, who are the owners of production means. Each capitalist fights others, trying to make profit. This system has many revolutionary consequences: cities slowly arise; peasantry decrease, being replaced by labor class; landlords power decrease also; profit seeking leads to labor exploitation and innovation, especially in the field of machinery; trade and labor division are destroying ancient social orders and abolishing frontiers; capitalists are less and less numerous, given they buy each other out and create ever larger companies with ever more capital; at the end of the process the labor class is immense, and takes the power, allowing productives forces running for its own interest. This is why revolution was supposed to happen in the most advanced countries, England and Germany. The fall of the Russian monarchy was a surprise, which forced Lenin and Trotsky to theorize socialism in what was considered as a backwarded country.

We understand now why that differences with ecologism are numerous, leading to a difficult eco-socialism, even if brilliant theoretic synthesis are

---

[37] Marx, *Le Manifeste du Parti Communiste*.
[38] Hegel, G. W. F., *La raison dans l'histoire (1822)*, Paris, 1993.

available.[39] First socialism tends to see ecologism as a "secondary front": the same goal, abolition of capitalism, should pave the way to the protection of nature and the workers. But neither the socialist revolution nor the reform was comfortable with the ecological issue, in history—globally speaking. For ecologists, taking ecological issues as a secondary front means not taking the issue seriously: not supporting ecological struggles; prioritizing class struggle, when the two are in conflict such as in the case of employment or revenue; talking about "ecological planification" but for supporting nuclear power, such as the communist party in France. Even nowadays China has a grand discourse on ecocivilization which is not fairly in line with practices. And labor organization are not very offensive on the issue. On the socialist side, many are confident with technology. "Degrowth" is hardly an audible slogan, given that it could be understood, on the capitalist side, as a decrease of proletariat's share of revenue. Capitalism itself is susceptible to at least two definitions that inform the debate: either the private ownership of the means of production (which is compatible with degrowth, if freedom is based on an ecological virtue, not on profit) or the endless accumulation of capital (which could be publicly owned and is not a guarantee of degrowth). The sociological ideal–typical ecologist is also different from the proletariat or popular classes, although both can be linked and articulated: he is highly graduate, on average, even if not well-paid, and not a member of the bourgeoisie. The difference between profiles could easily be seen during 2018 and 2019 in France, with Marches for the Climate walking on the streets without merging with the Yellow Jackets. In light of classical Marxism, the ecologism is made of local alternatives such as ZADs ("zone à défendre," "areas to be defended," that is, places occupied by activists, on a long period) and cooperatives, which seems pre-Marxist and at best falls under either a "critical-utopian" or "conservative" socialism that Jacques Droz considered to have disappeared with the start of the twentieth century. Some ecologist current idealizes often primitive societies, such as *Deep Green Resistance*.[40] Other are more compatible with "reactionary" socialism, based on the (reformist) socialization of the market. The European Green New Deal could be seen as close to this idea, even if very liberal. A third form of eco-socialism would be "critical-utopian." This eco-socialist current is particularly sustained, through local alternatives such as ZADs ("zone à défendre," "areas to be defended," that is, places occupied by activists, on a long period), similar to those theorized by Hakim Bey.[41]

Experiences of real socialism and postmodernism have pluralized and complicated perspectives. Even if the "socialisms of the 21st century" in Latin America (Venezuela, Ecuador, Bolivia) are still part of the dominant tradition of a state development of productive forces, they began to include some

---

[39] O'Connor, James, «Capitalism, Nature, Socialism: A Theoretical Introduction», vol. 1, no. 1, 1988, pp. 11–38.

[40] https://deepgreenresistance.fr/.

[41] Bey, Hakim, *TAZ, zone autonome temporaire*, Paris, Editions de l'Eclat, 1997.

ecological issues, especially under the aspirations of indigenous people. Yet socialist parties have incorporated the issue, but a working synthesis is still difficult to obtain, even if some experiences seem to succeed, such as the Zapatista revolution or the Kurdish region of Rojava, lead by Abdullah Öcalan, himself inspired by the social ecology of Murray Bookchin.[42] Some socialist leaders are also advocating an alliance between the graduate fraction of the working class, to which ecologists often belong, and working classes.

Ecologism vs Conservatism

What is "conservatism"? Is "conservative revolution" an oxymoron? Then how can we explain its widespread use? Addressing this question, Dobson cites Malthus, mentioning the *Principle of Population*, underscoring that this is the reference that is most often taken up by the Marxists as a symbol of conservatism because it embodies the idea of the supposedly natural limits to human progress, both technological and organizational.[43] This frames adequately the issue. Dobson cites John Gray, a supporter of Thatcher inviting the Greens to cease thinking of themselves as a protest movement and side with them, since the integrity of the environment seems a conservative idea, in his mind. Gray highlighted the alignment with the theories of Burke, according to which society is not an assortment of ephemeral contracts between individuals but a continuity between the dead and the unborn. He identifies other points of convergence: a skepticism about the possible advances of humanity, a condemnation of large-scale social experiments (which probably relates more to Popper[44]), the idea that an individual can only flourish in the context of community life, and finally risk aversion, with the precautionary principle.

Is Gray right? Is "dark green" ecologism conservative? No, responds Dobson, who points out the divergences: the ecologists are not trying to save a cultural tradition or work of history; quite on the contrary, they want to invent a new one; conservatism is as anthropocentric as the other ideologies, because if there is a defense of the environment and the non-human, it is from a romantic perspective that is nostalgic for past lifestyles, not for the sake of attributing dignity to nature. Burke is more interested in the respect for ancestors than in future generations.[45] Finally, a distinction exists "between the malleability of the human *condition* and the malleability of human *nature*. It is perfectly possible to believe that the human condition is fixed [because it is ecological] while human nature is not, and this is indeed what political

---

[42] Bookchin, Murray, *L'écologie sociale: penser la liberté au-delà de l'humain*, Marseille, Wildproject, 2020.

[43] Dobson, *Green Political Thought*, p. 173.

[44] Popper, Karl R., *Misère de l'historicisme*, Paris, Plon, 1956.

[45] Dobson, *Green Political Thought*, p. 176.

ecologists believe."[46] Citing John Gray again, Dobson affirms the conservatives who claim to be ecologists remain attached to the productivist *status quo* for the sake of not disrupting modern society, in reality.

At first glance, these objections echo French ecologism: What is intended by "nature" is partly the biosphere and partly a call for authenticity and emancipation against the dominant, technocratic and productivist culture; the ecologists easily accept a strong malleability of human nature (but not of the human condition); third-worldism is part of a concern for social justice ("think global and act local") and therefore of an ecology that cannot be reduced to local pollution nor nostalgia for an idealized past. When French ecologists mention the flow of generations, it is to imagine the future rather than to recreate the past, or claim their attachment to past social structures. Precaution concerns ecosystems more than societies; the most radical currents do not hesitate to envision perspectives such as the collapse of Western civilization or enormous power outages; some of them even hope for it and anticipate it, thus the movement of cities in transition, in a certain way.

French thought distinguishes three type of conservatism, in two major families: liberal (Orléanist) and illiberal.[47] The first is broad, ranging from centrist positions, always conservative on economic aspects but liberal or not on societal issues such as race or gender, and always in favor of democratic institutions. The analysis of the second is more complex because its existence and its scope are not agreed upon by observers. In France the major features of illiberal conservative theories are Bonald, De Maistre, Maurras, Vichy and the "radical" right-wing in contemporary France. We effectively found some common traits with ecologism, such as the critique of industry and modernity. But again differences are strong, and even if conservatives show some interest toward ecological issue, it is mainly in an instrumental and opportunistic ("populist" and "electoralist") manner. The key difference is given by Carl Schmitt, the famous legal theoretician of the Nazi regime:[48] "politics," for conservatives, means the conservation of the political unit, and its homogeneity, against all other units, which are potential enemies. Being strong in order to live. And accumulating power does not come without an ecological cost. We should remember here that the GDP as indicator has been framed during the Second World War to measure the military power of a nation-state:[49] it wasn't a peaceful invention. This explains also why ecologists are pacifists, very often—such as the name "Greenpeace" indicates, an organization born in demonstrations against nuclear weapons.

---

[46] Ibid., p. 178.

[47] Nay, Olivier, *Histoire des idées politiques*, Paris, Armand Colin, 2016.

[48] Schmitt, Carl, *Théorie de la constitution*, 2013ᵉ édition, Paris, PUF, 1928; Schmitt, Carl, *La notion de politique; Théorie du partisan*, Paris, Flammarion, 2009.

[49] Fourquet, François, *Les comptes de la puissance. Histoire de la comptabilité nationale et du plan*, Paris, Encres, 1980.

Some conservatives are genuinely attracted to ecology, however, because of its anti-modern character. The former French Green leader, Antoine Waechter, recently wrote a book with the historian Fabien Niezgoda[50] who is part of the circles of GRECE (*Groupement de recherche et d'études pour la civilisation européenne*), one of the iconic institutions of the *New Right*, a school of thought that has been the subject of much discussion since its emergence at the end of the 1960s and whose main theorist, Alain de Benoist, published a book in favor of degrowth in 2007.[51] Many sociologists have noticed this "ecologist ambivalence".[52] But these are very small political currents, which represent almost nothing compared to the main currents. The politician classifies them as romantic currents, in the sense that their aspirations are unrealistic in relation to their own convictions. For other currents, what ecology means is very different from what it means for ecologists.[53]

Conclusion

Three conclusions arose from this brief overview of ecologism. First, taken through the confrontation with other ideologies, ecologism seems to be distinct from what Karl Mannheim consider to be the main ideologies of modern times: socialism, liberalism and conservatism.[54] It favors a type of society which does not merge easily with the closest other one, socialism. Second, ecologism is hardly conservative. Some ecologist thinkers claim to be conservative and some conservatives join them in this sense,[55] but it is poorly convincing, as we saw. Right and extreme-right leaders do support ecologist struggles only by accident, when an issue has also a national or patrimonial dimension, or when they cannot follow another path. From Trump to Bolsonaro or François Fillon in France, conservative leaders, even if moderately conservatives, are almost always acting against ecologists. The key difference relies in the cosmopolitic nature of ecologism taken as a biospheric issue. It is less clear when struggles are NIMBY ("not in my backyard"). Third conclusion: the threat of an eco-fascism is real. It could arise in the future, in a situation of collapse or quasi-collapse. The mistake is to believe that such a regime would be "ecologist" or environmentally friendly. It will rather be close

---

[50] Waechter, Antoine et Niezgoda, Fabien, *Le sens de l'écologie politique – une vision par-delà droite et gauche*, Paris, Sang de la Terre, 2017.

[51] De Benoist, Alain, *Demain la décroissance*, Paris, Edite, 2007.

[52] Alphandéry, Bitoun et Dupont, *L'équivoque écologiste*.

[53] E.g., see https://www.ericzemmour.org/eric-zemmour-la-vraie-ecologie-est-dessence-conservatrice/.

[54] Mannheim, Karl, *Idéologie et utopie*, Paris, Editions de la MSH, 2006; Mannheim, Karl, *La pensée conservatrice*, Editions de la revue Conférence, 2009.

[55] Scruton, Roger, *De l'urgence d'être conservateur: territoires, coutumes, esthétique, un héritage pour l'avenir*, Paris, L'Artilleur, 2016.

to the Millennium Ecosystem Assessment scenario "Order from strength." This underlines again why ecologism carries also a democratic issue.

References

Alphandéry, Pierre, Bitoun, Pierre et Dupont, Yves, *L'équivoque écologiste*, Paris, La Découverte, 1991.
Audard, Catherine, *Qu'est-ce que le libéralisme*, Paris, Gallimard, 2009.
Bey, Hakim, *TAZ, zone autonome temporaire*, Paris, Editions de l'Eclat, 1997.
Bookchin, Murray, *L'écologie sociale: penser la liberté au-delà de l'humain*, Marseille, Wildproject, 2020.
Buret, Eugène, *De la misère des classes laborieuses en Angleterre et en France*, 1840.
Considerant, Victor, *Le socialisme devant le vieux monde*, Paris, Librairie Phalanstérienne, 1848.
Constant, Benjamin, *Ecrits politiques (1819)*, Paris, Gallimard, 1997.
De Benoist, Alain, *Demain la décroissance*, Paris, Edite, 2007.
Dobson, Andrew, *Green political thought*, London, Routledge, 2003.
Droz, Jacques, *Histoire du socialisme. Tome 1*, Paris, PUF, 1974.
Dupuy, Jean-Pierre, *Libéralisme et justice sociale*, Paris, Hachette, 1992.
Durkheim, Emile, *De la division du travail social*, Paris, Félix Alcan, 1893.
Ferry, Luc et Renaut, Alain, *Philosophie politique*, Paris, PUF, 2007.
Flipo, Fabrice, «Les trois conceptions du développement durable», *Développement Durable et Territoire*, vol. 5, no. 3, décembre 2014.
Fourquet, François, *Les comptes de la puissance. Histoire de la comptabilité nationale et du plan*, Paris, Encres, 1980.
Frémion, Yves, *Histoire de la révolution écologiste*, Hoëbecke, 2007.
Gallie, W. B., «Essentially Contested Concepts», *Proceedings of the Aristotelian Society*, vol. 56, 1955, pp. 167–198.
Hayek, Friedrich A., *La route de la servitude (1945)*, Paris, PUF, 2013.
Hegel, G. W. F., *La raison dans l'histoire (1822)*, Paris, 1993.
Israel, Jonathan, *Les Lumières radicales. La philosophie, Spinoza et la naissance de la modernité 1650–1750*, Paris, Amsterdam, 2005.
Jonas, Hans, *Le principe responsabilité : une éthique pour la civilisation technologique*, Paris, Le Cerf, 1990.
Krugman, Paul, *La mondialisation n'est pas coupable*, Paris, La Découverte, 2000.
Mannheim, Karl, *La pensée conservatrice*, Editions de la revue Conférence, 2009.
Mannheim, Karl, *Idéologie et utopie*, Paris, Editions de la MSH, 2006.
Marx, Karl, *Le Capital (1867)*, Paris, PUF, 1993.
Marx, Karl, *Le Manifeste du Parti Communiste*, 1962[e] édition, Paris, 10/18, 1847.
Millennium Assessment Report, «Living Beyond Our Means: Natural Assets and Human Well-Being, 2004», 2004.
Moscovici, Serge, *De la nature. Pour penser l'écologie*, Paris, Métaillé, 2002.
Nay, Olivier, *Histoire des idées politiques*, Paris, Armand Colin, 2016.
O'Connor, James, «Capitalism, Nature, Socialism: A Theoretical Introduction», vol. 1, no. 1, 1988, pp. 11–38.
Popper, Karl R., *Misère de l'historicisme*, Paris, Plon, 1956.
Postone, Moishe, *Temps, travail et domination sociale (1986)*, Paris, Mille et Une Nuits, 2009.

Poulantzas, Nicos, *Fascisme et dictature*, Paris, Maspéro, 1970.
Poulantzas, Nicos, *Pouvoir politique et classes sociales de l'état capitaliste*, Paris, Maspéro, 1968.
Rémond, René, *Les droites aujourd'hui*, Paris, Louis Aubibert, 2005.
Ricardo, David, *Des principes de l'économie politique et de l'impôt (1817)*, Paris, Flammarion, 1999.
Schmitt, Carl, *La notion de politique ; Théorie du partisan*, Paris, Flammarion, 2009.
Schmitt, Carl, *Théorie de la constitution*, 2013ᵉ édition, Paris, PUF, 1928.
Scruton, Roger, *De l'urgence d'être conservateur : territoires, coutumes, esthétique, un héritage pour l'avenir*, Paris, L'Artilleur, 2016.
Villermé, Louis, *Tableau de l'état physique et moral des ouvriers employés dans les manufactures de coton, de laine et de soie*, 1840.
Waechter, Antoine et Niezgoda, Fabien, *Le sens de l'écologie politique – une vision par-delà droite et gauche*, Paris, Sang de la Terre, 2017.
Wissenburg, Marcel, *Green liberalism – The free and the green society*, London, University College London, 1998.

Democracy, Citizenship and Nationalism in Environmental Political Theory

Andy Scerri

INTRODUCTION

The concept "democracy" arose in Ancient Athens. It describes a constitutionally governed polity that sanctions the political power of the organized poorer majority of citizens. Athenians contrasted democracy with positive and negative iterations of two regime types; aristocracy and oligarchy, rule by wealthy or high-born citizens, and monarchy and tyranny, rule by a king or prince.[1] In Ancient Rome and Medieval and early-modern Europe, enthusiasm for democracy waned further. During this period, however, advocates for political empowerment of the poorer majority were active. Most well-known is the republican realist, Niccolò Machiavelli. Machiavelli revised accounts of the Roman republic's anti-monarchical system in just this direction. He sought to reduce the influence exerted over government by wealthy, well-connected citizens. For him, government always tends away from sanctioning "people

---

[1] The Ancients had no positive definition of democracy, yet defined oligarchy positively as aristocracy and tyranny as monarchy. John Dunn, *Setting the People Free: The Story of Democracy* (Princeton: Princeton University Press, 2019), 4. Also, Josiah Ober, "The Original Meaning of 'Democracy': Capacity to Do Things, Not Majority Rule," *Constellations* 15, no. 1 (2008): 8.

A. Scerri (✉)
Department of Political Science and International Studies, Virginia Tech, Blacksburg, VA, USA
e-mail: ajscerri@vt.edu

power" and toward oligarchy or tyranny. The principle goal of any constitution must be to temper the inevitability of such a tendency. This is because elite citizens inevitably place their own interests in maximizing wealth and preserving power above the public interest in maintaining a resilient polity. For Machiavelli, this means that the many poorer citizens "should be made the guardians of liberty."[2]

For the contemporary "democratic republicans," these considerations make the central problem of politics one of maintaining civic liberty or equity, which has two dimensions. First, ensuring equal conditions of participation in law- and policy-making for all citizens and, second, ensuring the equal subjection of all to constraint by laws and regulatory institutions. A democratic republican theory of citizenship, therefore, considers the relationship between equity, democracy and the public interest, and a normative orientation to defending the latter against the malign influence exerted by elites over public policy.[3]

In this chapter, I expand on these democratic republican themes and survey oft-cited accounts of the relationship between citizenship, democracy and the public interest in "national" resilience in recent Environmental Political Theory (EPT). As will become clear, I contrast *democratic* readings of republican tradition over the two normative approaches long dominant in EPT, *civic humanist republicanism* and *progressive liberalism*.[4] My objective is to reveal how, even though efforts to "green" citizenship in civic humanist and progressive liberal lenses and the feminist and agonistic democratic critiques that such efforts have prompted continue to provide EPT with a bountiful harvest, these each ultimately place too strong an emphasis on the first dimension of equity. The now significant body of EPT dedicated to interrogating the relationship between citizenship, democracy and nationalism is concerned almost entirely with the nonetheless laudable task of supporting the equal participation of all in law- and policy-making.

To justify this claim, I trace this one-dimensionality to a now well-documented shift in how political philosophers and theorists in general came to understand their topic in the second half of the twentieth century. Critics of this shift uncover a move away from Machiavellian "republican realism"[5]

---

[2] Niccolò Machiavelli, *The Discourses*, trans. Leslie J. Walker (London: Penguin 2003), 116 (D.I.5).

[3] See, Bruno Leipold, Karma Nabulsi and Stuart White, eds., *Radical Republicanism: Recovering the Tradition's Popular Heritage* (Oxford: Oxford University Press, 2020). Also, Yiftah Elazar and Geneviève Rousselière, eds., *Republicanism and the Future of Democracy* (Cambridge: Cambridge University Press, 2019).

[4] In general, republicans reject civic humanist readings of the tradition as unsatisfactory because grounded in what is "essentially a form of perfectionism," Frank Lovett, "Republicanism and Democracy Revisited," in *Republicanism and the Future of Democracy*, ed. Yiftah Elazar and Geneviève Rousselière (Cambridge: Cambridge University Press, 2019), 119, and regard "liberalism as … impoverished or incoherent" Maurizio Viroli, *Republicanism*, trans. Anthony Shugaar (New York: Hill & Wang, 2002), 61.

[5] Luca Baccelli, "Political Imagination, Conflict, and Democracy: Machiavelli's Republican Realism," in *Machiavelli on Liberty & Conflict*, ed. David Johnston, Nadia Urbinati

and toward forms of "moralising" about politics or "political moralism."[6] In conclusion, I demonstrate the value of a methodologically realistic democratic republicanism to EPT by showing how a focus on the second dimension of equity—ensuring the equal subjection of all to constraint by democratically legitimate laws and regulatory institutions—can direct our attention to the kinds of structural reforms that are needed to confront the well-organized and well-funded movements that, largely based in and orchestrated from within the Anglophone liberal-democracies, have since the 1970s sought actively to ensure national and global institutional "lock-in" of political-economic reliance on burning fossil fuels.

Awkward Siblings: Democracy and "Liberal Morality"

Democracy remained anathema to most political theorists, republicans included, until well into the nineteenth century.[7] Prior to this shift, however, a historically novel corpus of political thought, liberalism, emerged in the wake of the "Atlantic revolutions."[8] Advocates of liberalism, such as Benjamin Constant, argued that classical republicanism was ill-equipped for dealing with modern conditions.

These were, specifically, the rise of the "free-thinking" or "rational" individual, a mercantile capitalist economy and the nation-state. At risk of parody, liberalism can be thought of as having emerged out of the *Lebensphilosophie* which represents the moral worldview of scholars such as Constant, who felt themselves hemmed in by monarchical tyranny on one side and demagogic oligarchy, "populism," on the other.[9] At the core of the liberal morality stands freedom as noninterference in the exercise of private will (rationality), the universal right to equal treatment by authority (emancipation) and justice (progress).[10] Liberal theorists can, therefore, be said to have modified classical

---

and Camila Vergara (Chicago: University of Chicago Press, 2017); Richard Bellamy, "The Paradox of the Democratic Prince: Machiavelli and the Neo-Machiavellians on Ideal Theory, Realism, and Democratic Leadership," in *Politics Recovered: Realist Thought in Theory and Practice*, ed. Matt Sleat (New York: Columbia University Press, 2019); Alison McQueen, *Political Realism in Apocalyptic Times* (Cambridge: Cambridge University Press, 2018).

[6] Raymond Geuss, *Philosophy and Real Politics* (Princeton: Princeton University Press, 2008), 101; Bernard Williams, *In the Beginning Was the Deed: Realism and Moralism in Political Argument* (Princeton: Princeton University Press, 2005), 2.

[7] Dunn, xx.

[8] Annelien de Dijn, *Freedom: An Unruly History* (Cambridge, MA: Harvard University Press, 2020), 215.

[9] David Runciman, *Politics* (London: Profile Books, 2014).

[10] Duncan Bell, "What Is Liberalism?," *Political Theory* 42, no. 6 (2014): 684, citing Gary Gerstle, "The Protean Character of American Liberalism," *American Historical Review* 99, no. 4 (1994): 1046.

republican concerns with civic liberty, transformed into noninterference by state authority; constitutionalism, transformed into formal equal rights under the rule of law and a democratized form of equal representation; and, the public interest, transformed into the "natural" outcome of the value-neutral harmonization of competing interests over time.

Given the liberal transformation, republicans grew more sympathetic to democracy, while generally, although not universally, agreeing that any viable democratic state would need to be at least "tendentially liberal."[11] This is not merely a normative assertion but an analytic, sociological claim. Observe any modern polity—at least, a liberal-democracy[12]—and you will soon realize that it is made up of "liberal individuals," persons who as did Constant, take for granted the moral value of individual equal rights, free will and neutral legal protections.[13] However, those sympathetic to democratic norms will also observe something else. "Actually existing" liberal-democratic institutions prioritize one peculiar right, the right to private property, one peculiar form of equality, formal legal equality and one peculiar form of democratic power, electoral representation.[14] Hence, it is no surprise that normative support for democracy began to grow among republicans in the early nineteenth century, as the market economy reached more deeply into persons' lives and more extensively across the planet.[15] Modern democratic republicans, therefore, seek to account normatively for "social" protest waged by majoritarian, early-on predominantly labor, movements against untrammeled private property rights and the economic inequality that such rights entail.[16]

From around the 1860s onward until sometime in the 1970s, labor-based struggles did somewhat successfully—if, as we will see, ephemerally—prompt many to question liberal claims that rights, formal equality and representativeness constitute the legitimate boundaries of a modern polity. What the democratic movements brought into view was the possibility that modern government could and should support economic conditions adequate to ensuring a dignified life for the poorer majority. From this period onward, republicans began to interpret democracy not simply as indicating the power of the many against the few, nor as being tendentially constrained by liberal

---

[11] Bellamy, 172.

[12] Note that the institutional products of this synthesis are known as liberal-democracies rather than liberal-republics. See, Josiah Ober, *Demopolis: Democracy Before Liberalism, in Theory and Practice* (Cambridge: Cambridge University Press, 2017), 160.

[13] Williams, 9; also, C.B. Macpherson, *The Political Theory of Possessive Individualism* (Oxford: Oxford University Press, 1962).

[14] Bernard Manin, *The Principles of Representative Government* (Cambridge: Cambridge University Press, 1997).

[15] Alex Gourevitch, *From Slavery to the Cooperative Commonwealth: Labor and Republican Liberty in the Nineteenth Century* (Cambridge: Cambridge University Press, 2014).

[16] Leipold et al.

morality but, simultaneously, as an impulse or aspiration oriented to realizing civic liberty or equity.[17]

Yet, over this period, republicans and liberals alike had noticed that successive waves of Civil Rights and women's activism were building on, contesting, and extending key "social" achievements of the labor movement: not only adult suffrage, but also industrial safety regulation, urban sanitation and public health regulation and, in some jurisdictions, minimum wage laws and social welfare schemes.[18] And, by the early twentieth century, in light of the empirical observation that many citizens suffered under the burdens of industrial poisoning both in work and at home, from tainted food and beverages, and from inadequate access to clean air and water, a nascent body of explicitly *environmental* political theory emerged.[19] A key insight of early contributions to EPT was that all extant political regimes had been and remained absolutely dependent on industrial production practices and distribution techniques that were fueled by the combustion of fossil fuel resources.[20] A finding that would be elaborated subsequently in the understanding that concentrated ownership and control of these and other resources posed a significant problem for political order in general and democratic movements in particular.[21]

While democratic republicans tended to see the civil, political, social and environmental struggles that extended from the nineteenth and into the mid-twentieth century as evidence of ongoing class conflict and redistributionary struggle, theorists sympathetic to liberal morality saw something very different. In the broadest of senses, liberals interpreted these achievements as evidence of the progressive march, or regressive defeat, of human's innate desire for moral improvement through increased rationality. Those prioritizing liberal morality over democratic norms have understood political order to be the product of individuals reaching rational consensus, or potential consensus, on the question of how best to govern, which may or may not be achieved through

---

[17] Ibid., also, Dunn, Gourevitch and John Medearis, *Why Democracy Is Oppositional* (Cambridge, MA: Harvard University Press, 2015).

[18] Daniel T. Rodgers, *Atlantic Crossings: Social Politics in a Progressive Age* (Cambridge, MA: Harvard University Press, 1998).

[19] Robert Gottlieb, *Forcing the Spring: The Transformation of the American Environmental Movement*, 2nd ed. (Washington, DC: Island Press, 2005).

[20] Murray Bookchin, *The Ecology of Freedom: The Emergence and Dissolution of Hierarchy* (Oakland: AK Press, 2005); Herbert Marcuse, *Eros and Civilization* (London: Sphere Books, 1970); Arne Naess, *Ecology, Community, Lifestyle: Outline of an Ecosophy*, trans. D. Rothenburg (Cambridge: Cambridge University Press, 1993).

[21] Joachim Radkau, *Nature and Power: A Global History of the Environment* (Cambridge: Cambridge University Press, 2008); *The Age of Ecology* (Cambridge: Polity, 2014); Andreas Malm, *Fossil Capital: The Rise of Steam Power and the Roots of Global Warming* (London: Verso, 2016); John Bellamy Foster, *Socialism and Ecology: The Return of Nature* (New York: Monthly Review Press, 2020); Jason W. Moore, *Capitalism in the Web of Life: Ecology and the Accumulation of Capital* (London: Verso, 2015). Contemporaneously, Galbraith; Engler.

the use of propaganda to disseminate rationality-undermining ideology. Alternately, and the position that democratic republicans take is that political order should be understood as the product of fear, force and the threat of force, and luck, fortune or timeliness among different classes holding empirically definable material and so, ideological, interests.[22]

One way to interpret this tension at the core of normative political theory is to focus on the questions that are raised by liberal-democratic polities' tendency to prioritize private property over other conceptions of rights. That is, in practice, to prioritize economic over civic liberty. A focus on such questions reveals shared concern with upholding, curtailing—or, in the outlier cases posed by Marxism and Anarchism, eliminating—private property rights. On the surface, recognizing this tension might be taken to imply that the central problem of politics is best conceived in terms of either class conflict or individual consensus-building in relation to questions of economic distribution. Yet, in a republican lens, economic distributive conflict presupposes something else: the establishment of order that, by definition, serves the interests of some and not others. There can be no functioning economy, let alone private property rights, electoral representation, or economic or civic liberty in the absence of an at least minimally functional political order.[23] Essentially political things—the artifices of constitution, law and regulatory institutions—stand behind distributive conflict, because citizens "essentially inhabit two different worlds" simultaneously.[24] The "natural" world of physical forces and the "human" or "artificial" world of politics.

To unpack this point, consider that, as democratic activists such as labor unionists constantly remind us, it is laws and regulatory institutions that prioritize private property rights and, as such, economic liberty. It is no secret that those able to most effectively maximize economic liberty can build massive concentrations of wealth, and that this brings a corresponding degree of political power and influence. Consider too, the historical burdens carried by those subject to the power of those able to exploit such rights, and by those denied such rights; Peasants, unskilled, semi-skilled and service workers, enslaved and post-slavery Black Americans, women, Indigenous peoples from Australia and Argentina to Canada and Finland, irregular migrants, the majority of refugees and so on. The ostensibly value-neutral laws and institutions of liberal-democracy deny individual members of these groups access to the full

---

[22] See, Dennis Wrong, *The Problem of Order* (Cambridge, MA: Harvard University Press, 1994).

[23] Both civic or liberal and democratic republicans agree on this point. Compare, Philip Pettit, *On the People's Terms: A Republican Theory and Model of Democracy* (Cambridge: Cambridge University Press, 2012), 36 with John P. McCormick, *Reading Machiavelli: Scandalous Books, Suspect Engagements, and the Virtue of Populist Politics* (Princeton, NJ: Princeton University Press, 2018), 46.

[24] Quentin Skinner, *Hobbes and Republican Liberty* (Cambridge: Cambridge University Press, 2008), 126.

benefits that accompany maximal leverage of economic liberty, such as accrue to those who inherit wealth, either individually or collectively, for example.

Ironically, however, this means that what is important about liberal-democratic politics is not in the first instance distributional conflict or consensus-building, or the simple ascription of private property rights-bearing citizenship to everyone. Rather, what is important is that the power which concentrated economic wealth brings allows those possessing it to act with a degree of relative impunity in relation to the laws and institutions that constrain or exclude others. The influence and discretion that wealth delivers to those possessing it—the capacity to, say, "buy" citizenship in "tax havens" such as Panama or "offshore" assets to jurisdictions such as Barbados; to directly and in many cases individually influence industrial safety, environmental, or financial regulatory efforts; to escape military service obligations; or, meritocratic entry criteria applied by educational institutions, for example—are practices to which poorer, that is, politically weaker, citizens and noncitizens do not have access. The historical inevitability of a wealthy minority able to flout or deeply influence laws, and a poorer majority bound not only by laws but also, due to poverty or exclusion, an inability to influence laws or alter citizenship status, while simultaneously subjected to arbitrary willfulness on the part of the domestic and global wealthy in daily economic life[25] is what anchors politics in the problem of equity.

CITIZENSHIP AND THE CONVERGENCE ON MORALITY

Unsurprisingly, citizenship is the site around which debates over equity are conducted, and provides common ground for both consensus- and conflict-oriented normative theorizing.[26] Cast in terms of citizenship, the historical political transformations alluded to in the section above can be said to reflect the development of the *civil* rights to equal treatment and noninterference under law, as demanded against monarchs by classical liberal revolutionaries such as Constant in the seventh and eighteenth centuries; the reform of *political* rights to electoral representation demanded by progressive liberals over the nineteenth and early twentieth centuries; the revolutionary *social* welfare rights demanded by organized labor against bureaucratized industrial states from the 1930s until the 1980s; and, the *stakeholder* rights and duties demanded by "neoliberal" counterrevolutionaries from the 1980s onward.[27]

In a lens shaped by liberal moral concerns with rational consensus, however, these transformations of citizenship mark progress toward or regress from the

---

[25] Gourevitch, 2014; also, Elizabeth Anderson, with et al., *Private Government: How Employers Rule Our Lives (and Why We Don't Talk About It)* (Princeton, NJ: Princeton University Press, 2017).

[26] Andy Scerri, *Greening Citizenship: Sustainable Development, the State and Ideology* (Basingstoke: Palgrave, 2012), x.

[27] Ibid., 113.

ideal, which is taken to be constitutive of justice. For example, depending on classical or progressive sympathies, the achievements of the welfare state can be said to augur in the direction of progress away from or regress toward the injustices of serfdom.[28] Conversely, the neoliberal counterrevolution ongoing since the 1980s could be said to represent a step toward the restoration of justice based on economic liberty, or a regressive disfigurement of social justice.[29]

In a lens shaped by democratic republican norms, in contrast, historical transformations of citizenship are regarded as the products of contingent and

> interrelated processes of state building, the emergence of commercial and industrial society, and the construction of a national consciousness, with all three driven forward in various ways by class struggle and war. The net effect [being] to create a "people" ... entitled to be treated as equals before the law and possess[ing] equal rights to buy and sell goods, services and labor; whose interests [are] overseen by a sovereign political authority; and who share ... a national identity. [As such] the development of legal rights stems from a subordinate group employing formal and informal political strategies to win concessions from those with power in [a] fight to be treated with equal concern and respect.[30]

In this lens, the problem of inequity is not in the first instance to be thought of in terms of redistributive justice but of the democratic legitimacy of the laws and institutions within or between nation-states.

With this distinction in mind, I now argue that something very interesting happens in EPT from the mid-1990s onward. In short, the concept of "citizenship" was not only re-thought in the shadow of growing scientific knowledge and public awareness of the social and environmental costs of economic growth but, at the same time, was increasingly subjected to interpretation through a lens fashioned by a wide-reaching transformation that took place among political philosophers and theorists in general. A growing body of "realist" scholarship in political philosophy, political theory and the

---

[28] Compare, J.K. Galbraith, *The New Industrial State*, 2nd ed. (London: Pelican, 1967) with Friedrich Hayek, *The Constitution of Liberty* (Chicago: Chicago University Press, 2011).

[29] Compare, Fareed Zakaria, *The Future of Freedom: Illiberal Democracy at Home and Abroad* (New York: W. W. Norton, 2007) with Nadia Urbinati, *Me the People: How Populism Transforms Democracy* (Cambridge, MA: Harvard University Press, 2019).

[30] Richard Bellamy, "Introduction: The Theories and Practices of Citizenship," in *What Is Citizenship? Theories of Citizenship: Classic and Contemporary Debates* (Vol. 1 of *Citizenship: Critical Concepts in Political Science*), ed. Richard Bellamy and Madeleine Kennedy-MacFoy (London: Routledge, 2014), 14.

history of political thought recognizes that a kind of convergence on a peculiar way of moralizing about politics took place in the post-1945 period.[31] This convergence was given further impetus by the collapse of the Soviet Bloc in the late 1980s, and has persisted until very recently. In the view of critics of this convergence, liberal-democracy had come to be idealized as a regime type or model of order that, absent some kind of endogenous or exogenous shock, was grounded by real or potential rational consensus on laws and institutions, a "social contract." This, the value-neutrality of liberal-democratic institutions meant that over time, order was the product of a kind of "dynamic self-reinforcing equilibrium."[32] Abandoned were views of politics as a distinct realm of activity characterized by formalized conflict, the contingency of order and constraint. In place of such realist considerations, scholars took to viewing politics as an extension of the realm morality and ethics and, so, as characterized by rational consensus, equilibrated stability and justice. Such a conception of politics is said to be inherently moralizing because operationalized on the basis of an "simpleminded Binary";[33] good citizens promote equilibrium, and evil citizens detract from it. From the mid-twentieth century onward, the commanding heights of political philosophy and theory were captured by scholars who conceived their task as being to "pass ethical judgments on the world by appealing to general moral principles designed to help us make sense of what justice requires of our politics and institutions."[34]

Green Citizenship and the Convergence on Morality

This is all well and good, and critiques of the realist critique of moralizing about politics abound.[35] However, I believe that EPT can beneficially take a page from the critics' book, and that democratic republicanism supplies one narrative that might help respond to the aporias which the critics expose. In the next few paragraphs, I offer what I hope is an accurate if overgeneralized summary of how this convergence affected the study of citizenship and democracy in EPT.

In the 1990s, EPT began to incorporate insights from empirical research that demonstrated a significant increase in citizenly awareness of environmental problems in general and climate change in particular. As I point out above, a certain "greening" of liberal-democratic societies had in fact begun in the late

---

[31] The literature now escapes citation. See, Geuss, Williams, most recently, Katrina Forrester, *In the Shadow of Justice: Postwar Liberalism and the Re-making of Political Philosophy* (Princeton, NJ: Princeton University Press, 2019).

[32] Ober, 2017, 7.

[33] Raymond Geuss, *Politics and the Imagination* (Princeton: Princeton University Press, 2010), 33.

[34] Forrester, ix.

[35] For example, see contributions by Michael Freeden, Alison McQueen and William Scheuerman, in Matt Sleat, ed. *Politics Recovered: Realist Thought in Theory and Practice* (New York: Columbia University Press, 2018).

1960s and early 1970s. Alongside the progressive wing of the labor movement, a crop of "new" social movements sought to universalize expectations that work should be motivated by creativity rather than avarice, and that all individuals' lives should be meaningful, grounded in subjective self-reflection rather than objective status-group expectations, and conducted in harmony with nature.[36]

Very soon, it became clear that civic humanist and liberal efforts to green citizenship in fact "share the same architecture."[37] There "seems to be a consensus ... that the very enlisting of the idea [of citizenship for environmentalist ends] implies a recognition that sustainable development requires shifts in attitudes at a deep level."[38] On this view, green citizenship was said to provide "the possibility of checking self-interest against the common good in systematic ways, because this is part of what citizenship—as concept and practice—is about."[39] Elaborated by scholars working in both the civic humanist republican and progressive liberal traditions, the new "green" theories of citizenship "shared five central claims: 1. the need to challenge nature/culture dualism; 2. to dissolve the divide between the public and private spheres; 3. to undermine state-territorialism; 4. to eschew social contractualism; and, 5. to ground justice in awareness of the finiteness and maldistribution of planetary ecological space."[40]

Regarding 1., advocates of green citizenship assert that such citizenly growing awareness of social and environmental harms marks the collapse of the long-prevalent, ontologically *dualistic* understanding of humanity, as rightfully engaged in an effort to dominate nature, and its displacement by a *holistic* understanding grounded in the more ontologically justifiable belief that humans are but one participant in nature: "Everything is connected to everything else."[41] Contrary to some of their critics, advocates of both civic humanist republican "ecological" and liberal "environmental" citizenship recognize that

> few concepts are as deeply embedded in the dualisms of Western political thought as that of citizenship ... written into the very concept ... is a privileging of the epistemic that constructs political space through the reinforcing dualisms

---

[36] Scerri, 82.

[37] Tim Hayward, "Ecological Citizenship: Justice, Rights and the Virtue of Resourcefulness," *Environmental Politics* 15, no. 3 (2006): 441.

[38] Andrew Dobson and Ángel Valencia Sáiz, "Introduction," in *Citizenship, Environment, Economy* (London: Routledge, 2005), 157.

[39] Ibid.

[40] Andy Scerri, "Green Citizenship and the Political Critique of Injustice," *Citizenship Studies* 17, no. 3–4 (2013): 294.

[41] Originally articulated by Bart van Steenbergen, "Towards a Global Ecological Citizen," in *The Condition of Citizenship*, ed. B. van Steenbergen (London: Sage, 1994).

of mind/matter, nature/culture, reason/emotion, men/women, public/private and so on.[42]

Regarding 2., combined with recognition of the affront to liberal claims for institutional neutrality afforded by the increasingly dire predictions of climate science, growing public awareness of the science have forced "green" liberals to concede that, indeed, it is desirable to dissolve the dualistic distinction between the private sphere of individual right and "free" will, and the public sphere of political authority. Liberal environmentalists thus reconceive action in the private sphere—such as purchasing "sustainable" or "ethical" goods and services, riding a bicycle or taking public transport, vegetarian-izing one's diet or recycling—insofar as these are intentionally oriented to achieving "green" ends, as environmentally rational and therefore "always already" political acts. For green liberals, environmental justice is of paramount importance and should be grounded in individuals' environmentally rationalized responses to knowledge that planetary ecological space is inter-connected, finite and maldistributed.[43]

For civic humanist republicans, the argument against dualism and the public/private divide is more circuitous but similarly holistic and individualistic. Individual choices can be sites for green citizenship because these complement the ethico-moral formation of the virtuous self, which in turn, leads to deeper more ethically sincere political engagement.[44] Private ecological virtue is, therefore, paramount. Civic humanists, therefore, problematize liberalism for prioritizing an unrealistically rational, individualistic conception of environmental rights. What is rather demanded is an active notion of citizenly virtue and concomitant responsibilities for bringing about sustainability.[45] Hence, the task at hand is one of connecting solidarity, commitment and democracy to citizenship. This is to be achieved by challenging injustice, not as the product of self-interest, but as a flaw in institutions that perpetuate capitalistic competitiveness, nationalistic chauvinism and political statism. Civic humanist republican, ecological and green citizenship is in this view committed to a "weakly perfectionist"[46] and so idealist understanding of human nature. Such theories are oriented to the gradual achievement of the global good life

[42] Teena Gabrielson and Katelyn Parady, "Corporeal Citizenship: Rethinking Green Citizenship Through the Body," *Environmental Politics* 19, no. 3 (2010): 374–5. Compare with John Barry, *Rethinking Green Politics* (London: Sage, 1999); Andrew Dobson, *Citizenship and the Environment* (Oxford: Oxford University Press, 2003), 99.

[43] Ibid.

[44] John Barry, *The Politics of Actually Existing Unsustainability* (Oxford: Oxford University Press, 2012).

[45] "Resistance Is Fertile: From Environmental to Sustainability Citizenship," in *Environmental Citizenship*, ed. A. Dobson and D. Bell (Cambridge: MIT Press, 2006), 23.

[46] Anne Fremaux, *After the Anthropocene: Green Republicanism in a Post-capitalist World* (Cham, Switzerland: Palgrave Macmillan, 2019), 225, 39.

by active citizens, for whom freedom is defined in a positive moral sense, and made practical through active participation in political decision-making at all scales from the local to the global.

Regarding 3., some advocates of liberal "environmental" citizenship seek to extend the reach of environmental justice-defined rights while simultaneously extending the definition of who and what such rights cover. If pollution is global, then justice and citizenship rights should be sensitive to global rather than national considerations, and if nonhuman animals feel pain and communicate, then rights should also cover them, thus transcending state-territorialism. Accompanying this transcendence is abandonment of the liberal ideal of freedom as noninterference, which can and should no longer be regarded as the supreme criterion of morality.[47] Altering liberalism along green lines requires modifications "too deep to preserve the 'liberal' label in all its force."[48] Indeed, liberals abandon the norm of representativeness that, as I describe it above, was granted by mercantile states as a concession to classical liberal demands for civil and progressive liberal demands for political rights of citizenship from the seventeenth and eighteenth centuries onward. The formal political equality that stands behind representativeness is reconceived "not as a device for the aggregation of [green rational] preferences and interests but as [grounds for] a dialogue within discursive communities."[49]

Regarding 4., green liberals, therefore, retool representation, and its theoretical justification in the "social contract," in favor of active participation in the greening of democracy. With this move, progressive liberal prioritization of moral suasion aimed at convincing individuals to embrace "green" reason is rendered all but indistinguishable from the civic humanist ideal of perfecting citizenly virtue. And, finally, regarding 5., what seems to be going on here is that emphasis on the ontological backdrop for the rights-bearing environmental liberal and the obligated ecological civic humanist leads to a misconstrual of political conflict, the exercise of power under formal constraint, as merely a matter of persuading other citizens that green reason or virtue is superior to other values. By grounding justice in citizens' awareness of the finiteness and maldistribution of planetary ecological space, political hopes seem to be pinned to beliefs that indisposed political actors will somehow one day simply recognize the folly of their ways and embrace green rationality or virtue. It is in this sense that green citizenship theory takes on the moralizing qualities that realist critics deride as an unhelpful turn to "politics as applied ethics" and merely "passes ethical judgments on the world."[50] As it

[47] Marcel Wissenburg, *Green Liberalism: The Free and the Green Society* (London: UCL Press, 1998).

[48] Manuel Arias-Maldonado, "The Democratisation of Sustainability: The Search for a Green Democratic Model," *Environmental Politics* 9, no. 4 (2000): 56.

[49] Ibid.

[50] Williams, 1; Forrester, ix.

stands, green citizenship punts into the long grass, the essentially political task of engaging in conflict to achieve equity.

CRITIQUES OF GREEN CITIZENSHIP

Shortcomings of this turn to citizenship did not go unnoticed. By the middle of the first decade of the twenty-first century, three incisive critiques had emerged.

In one critique, feminists see, even in intentionally holistic efforts to green citizenship, a problematic dualistic "assumption of a generic model of citizenship that masks realities of gender (and other forms of) inequality."[51] Notably, Teena Gabrielson and Katelyn Parady regard advocates of civic humanist ecological and progressive liberal environmental citizenship as having failed to adequately transcend the "dualisms" of Western political thought. And, for having embraced an unhelpful "narrative of declension" in which liberalism and civic republicanism sustain an "elitist epistemology that harkens back to some imagined golden era of full equal rights and transparently virtuous democratic participation."[52] Moreover,

> green citizenship empower[s] those positioned to know or imagine a particular conception of what a green 'good life' would entail. Those who are not so positioned are not fully recognized as citizens and are largely excluded from participation.[53]

The feminists counter that "our notions of the good life necessarily draw from our [corporeal] positioning within both natural and social contexts."[54] A genuinely holistic "corporeal [green] citizenship recognizes the centrality of the environment to human subjectivity by acknowledging the variety of places that bodies inhabit and the diversity of human relations with the natural world."[55]

This said, however, the feminists might just be pushing back a step, the moralizing tendencies of green citizenship. What Gabrielson and Parady exclude is the essential prerequisite for political order: *Formalized* conflict over equity. By embracing holism (everything is connected to everything else) and analytic, if not normative, individualism (corporeal subjectivity), normative argument is left with nowhere to turn but the embrace of *informal* conflict, the outcome of which will be determined by the strongest, or perhaps

---

[51] Sherilyn MacGregor, "No Sustainability Without Justice: A Feminist Critique of Environmental Citizenship," in *Environmental Citizenship*, ed. A. Dobson and D. Bell (Cambridge: MIT Press, 2006), 102.

[52] Gabrielson and Parady, 374, 80.

[53] Ibid., 375.

[54] Ibid., 381.

[55] Ibid., 386.

most charismatic or immoral individual. In a democratic lens, this cannot do. It is something of a truism to state that bodies inhabit a variety of places and a diversity of relations with nature. Politics—formal laws and regulatory institutions—is a wholly artificial concession that, albeit incompletely and ephemerally, weak citizens obtain from the strong to waylay the abject domination that would be produced in politics' absence. The dualistic "public" sphere of laws and regulatory institutions is neither an epistemological feature nor an ontological category but rather, a contingent "artificial" product of historical struggle which produces significant but not satisfactory elite concessions. The feminist critiques perhaps blur the line between politics, formalized conflict, and raw power, the informal product of "natural" physical or biological processes.

The second critique of green citizenship builds on Ernesto Laclau and Chantal Mouffe's highly popular agonistic theory of democracy. Green agonists such as Amanda Machin recognize that, while the feminist critiques are laudably "sensitive to the existence of conflict, they [like civic humanists and liberals] stop short of considering the necessary political decisions that must be made, regardless of political discord."[56] In an agonistic democratic lens, green citizenship theories "share a common flaw; a presupposition of rational consensus and an underplaying of the importance and difficulty of decision."[57] Agonists want to resolve this "quieting of democratic difference"[58] by recognizing that "the drawing of the frontier between the legitimate and the illegitimate is always a political decision, and... it should therefore always remain open to contestation."[59] In Mouffe's oft-cited words, "Consensus is needed on the institutions constitutive of democracy and on the ethico-political values informing *the political* association—liberty and equality for all—but there will always be disagreement concerning their meaning and the way they should be implemented."[60]

The focus for analysis in an agonistic lens is the normatively desirable disagreement said to inhere in "the political." The political is defined as that realm of collective experience in which anti-essentialist, antagonistic forms of agency, agonistic respect for difference, disequilibrium and, to use the favored jargon, Heideggerian being-as-becoming take precedence. In contrast, "politics" is viewed in normatively undesirable terms, as the realm of the essential, of necessity, order, control, conventions, institutions and constraints on human agency and interactions, which is to say, again employing the jargon, the realm of Schmittian decisionism. For agonistic democrats, the political is the axiological principle that always and everywhere opposes politics. The

---

[56] Amanda Machin, "Decision, Disagreement and Responsibility: Towards an Agonistc Green Citizenship," *Environmental Politics* 21, no. 6 (2012): 858.

[57] Ibid., 848.

[58] Ibid., 856.

[59] Ibid., 858.

[60] Chantal Mouffe, *On the Political* (London: Routledge, 2005), 31.

political confronts the order of politics with its antitheses; disorder, contingency, unconventional ways and means, spontaneous actions, unconstrained creativity, the jouissance of the moment and so on.[61]

In the agonistic democratic view, green citizenship makes the task of normative theorizing too easy. Advocates rely on a presumption of perfectibility—the appeal of the environmentally rational choice, or the desire to exercise ecological virtue—to bracket-off the essential imperfectability of "the political." For Mouffe, democratic movements cannot replicate politics. Rather, movements should "engage with the state" by deploying "counter-hegemonic practices" that

> do not eliminate [but are] an 'ensemble of differences,' all coming together, only at a given moment, against a common adversary. Such as when ... environmentalists, feminists, anti-racists and others come together to challenge dominant models of development and progress.[62]

Yet, such formulations beg two questions. First is what movements are meant to do once the political "ensemble of difference" has disrupted politics. Second is the empirical fact that, perhaps setting aside the very small cohort of the "educated precariat"[63] who read agonistic democratic theory, the lives of many if not most of the citizens who constitute the poorer majority have been encumbered by laws and regulatory institutions that, increasingly since the 1970s, have been designed to realize precisely what Mouffe defines as the contents of the political. Consider, for example, the normalization of disorder (dissolution of the regulatory state and encouragement of the entrepreneurial "disruptor"), contingency ("independent contracting" in the gig economy), unconventional ways and means (the predatory business model central to gig economy firms such as Uber, Lyft and, so on), unconstrained creativity (the removal of government support for deontological oversight of labor relations), the jouissance of the moment (consumerism and social media). As argued by Lois McNay, the agonistic turn is essentially socially and politically weightless and, at best, rests on foundational claims that are antithetical to the formal democratic ends the theory otherwise seeks.[64]

The third critique of green citizenship is my own, and I do not labor it here. Suffice to say that in my view, liberal and civic republican green citizenship as well as its corporeal feminist and agonistic democratic critiques are theories for

---

[61] See, *The Return of the Political* (London: Verso, 1993); *The Challenge of Carl Schmitt* (London: Verso, 1999); *The Democratic Paradox* (London: Verso, 2000); *On the Political*; "The Importance of Engaging the State," in *What Is Radical Politics Today?*, ed. J. Pugh (Basingstoke: Palgrave, 2009).

[62] "The Importance of Engaging the State," 236–7.

[63] Hauke Brunkhorst, *Critical Theory of Legal Revolutions: Evolutionary Perspectives* (London: Bloomsbury, 2014), 308.

[64] Lois McNay, *The Misguided Search for the Political: Social Weightlessness in Radical Democratic Theory* (Cambridge: Polity, 2014), Urbinati, *Me the People*.

another time past.[65] As does much of the high theory associated with the great convergence, green citizenship responds atavistically to a political world that "oligarchic" anti-democratic activists, many with direct links to fossil fuel and related industries, were already in the 1970s taking active steps to dismantle. The economic model sustained by liberal-democracy between the 1930s, in the United States and Oceania and the 1940s in Western Europe, and the 1970s was a product of sustained democratic opposition to oligarchic power, as the actions of well-organized, membership-based and funded political, civil, labor, socialistic, feminist and environmentalist movements show.[66]

In addition, and against the agonistic democratic critics in particular, insofar as "democracy functions when people agree to go along with policies that they do not substantively agree with," it seems fair to say that sometime in the 1970s, oligarchic actors and their political allies simply began to disagree with democracy, with the aspiration to acheivement of state sanctioned "people power." Oligarchic actors and their political allies embarked upon a "secession from democracy,"[67] an "organizational counteroffensive [that] was swift"[68] and which remains deeply effective. The historical record is clear. An increasingly well-funded and globally organized collective, bearing albeit divergent nationalities, methods and philosophies, nonetheless convened to defend a remarkably homogenous belief: that economic liberty should be defended from encroachment by advocates for democracy.[69] The singular achievement of the "profound infrastructure of political organization"[70] that grew out of the anti-democratic reaction of the 1970s was in fact to "depoliticize," "moralize" or "erode" citizenship.[71] That is, to exclude from political challenge the private property rights upon which economic liberty and so, to a large degree, institutional fossil fuel "lock-in" depend.

---

[65] Scerri, 2013, 294; also, *Postpolitics and the Limits of Nature: Critical Theory, Moral Authority, and Radicalism in the Anthropocene* (Albany, NY: SUNY Press, 2019), 162.

[66] Philip Mirowski and Dieter Plehwe, eds., *The Road from Mont Pèlerin: The Making of the Neoliberal Thought Collective* (Cambridge: Harvard University Press, 2009); Jeffrey A. Winters, *Oligarchy* (Cambridge: Cambridge University Press, 2011); Quinn Slobodian, *The Globalists: The End of Empire and the Birth of Neoliberalism* (Cambridge, MA: Harvard University Press, 2018).

[67] Urbinati, 123.

[68] J. Hacker and P. Pierson, "Winner-Take-All Politics: Public Policy, Political Organization, and the Precipitous Rise of Top Incomes in the United States," *Politics & Society* 38, no. 2 (2010): 176.

[69] Lars Cornelissen, "How Can the People Be Restricted? The Mont Pèlerin Society and the Problem of Democracy, 1947–1988," *History of European Ideas* 43, no. 5 (2017).

[70] Philip Mirowski, "The Eighteenth Brumaire of James Buchanan: Review of Nancy Maclean, Democracy in Chains," *boundary 2* 46, no. 1 (2019).

[71] Margaret Somers, *Genealogies of Citizenship: Markets, Statelessness, and the Right to Have Rights* (Cambridge: Cambridge University Press, 2008); Bryan S. Turner, "The Erosion of Citizenship," *British Journal of Sociology* 52, no. 2 (2001); Andy Scerri, "Moralizing About the White Working-Class 'Problem' in Appalachia and Beyond," *Appalachian Studies* 25, no. 2 (2019).

CONCLUSION

In light of the above discussion, it seems fair to say that a certain moralization in the work of theorists of green citizenship and their critics has led its advocates to occlude the central problem of politics, that of ensuring that the very wealthy and their political allies are constrained to participate in the devising of laws and institutions on terms that parallel everyone else, while also ensuring that all are constrained by those laws and institutions. Understood in this light, one of the key tasks accruing to green citizens would be to tame what Machiavelli's democratic republican followers define pungently as oligarchic power, especially that derived ultimately from fossil fuel sources. What we witness since the 1970s—which is to say, coincidentally, over a period that coincides with the moralistic "convergence" in high theory—is the reassertion of oligarchic power.

In fact, it seems fair to say that today, the biggest danger to green democratic aspirations is the growing influence of North–South alliances between the very wealthy, especially those with deep-seated interests in fossil fuels. Such alliances involve actors in the "dirty three" liberal-democracies, Australia, Canada and the United States, as well as the Middle Eastern, African and Latin American fossil fuel autocracies and the Russian Federation. No amount of individual agency, virtuous local activism, awareness of corporeality, or agonistic celebration of ensembles of difference is going to alter this fact. To date, some theorists of green citizenship have thought and acted as if politics were exhausted by moral suasion, by seeking fulfillment of the first dimension of the politics of equity. The mistaken assumption at the heart of these approaches has been that "democracy" means "rule by the people" and not state sanction of the poorer majority's "collective capacity to do things in the public realm."[72]

A first step in constructing a green democratic response to this situation may be to (re)place at the front and center of analysis what, in a democratic republican lens, has historically constituted the central problem of politics, formalized conflict over equity—that is, over the inherently democratic desire for equal terms of participation in the creation of *and* subjection to the laws and institutions of a given polity. Practically, the key implication of this rethinking of green citizenship is to conceive of green democratic movement activism as an exercise in holding the very wealthy and their political allies to account for breaches of equity, initially within their own nation-states, and in alliance with democratic movements elsewhere, globally. Citizenship is, after all, the product of national democratic struggle against oligarchic power, whether such power originates domestically or from some far-flung imperium.

Understood in this light, this chapter has sought to contribute to the development of programmatic demands for constitutional and institutional reforms, the value of which are measured in terms of the two dimensions of equity.

---

[72] Ober, 2008, 6.

Not perhaps as immediate or morally fulfilling as shopping for local organic produce, going vegetarian, or petitioning local government for density in planning, or as exciting as disruptive protest within an ensemble of difference, but I would argue, essential nonetheless. Such reforms might include building a critical mass of citizen support for the establishment of a third chamber or peoples' tribunate or oversight committees at local, national and international levels of government, as recommended by "Machiavellian" democratic republicans.[73] One way to ensure that such an addition to the bicameral system of government remains sensitive to democratic interests would be to implement class-sensitive sortition (randomized selection) for election to such a third chamber. This could be achieved by setting the criteria for eligibility to sit for election according to some combination of social indicators, primarily annual real income and, not inconceivably, complemented by gender or racial-ethnic indicators. Given the depredations that the mass of poorer citizens have been subjected to since the 1970s, efforts to build support to establish such a chamber—empowered to act in relation to distributive questions; to levy penalties on miscreants in, for example, a carbon taxation regime; or to veto legislation developed by elected representatives on grounds that it adversely affects majoritarian interests in sustaining an equitable and resilient polity—would I feel be greatly appealing. Green democratic efforts might also be oriented to building support for constitutional amendments in support of similarly elected citizen-science policy monitoring programs, or citizen review panels for trade treaties between nation-states, or even in relation to the granting of favored nation status by rich Northern to poor Southern nations, thus ensuring that labor and environmental standards are acceptable to poorer citizens in both. Given the plausible belief that the wealthy and their political allies will continue to support environmentally corrosive and maldistributed economic growth, even in the face of mounting evidence that this is disastrous for people and the planet, more reflection is needed on how green democrats might support institutional reforms that allow poorer citizens to subject the wealthy and their political allies to effective and so genuinely democratic oversight and constraint. That is, to increase political equity.

References

Anderson, Elizabeth. *Private Government: How Employers Rule Our Lives (and Why We Don't Talk About It)*. Princeton, NJ: Princeton University Press, 2017.

Arias-Maldonado, Manuel. "The Democratisation of Sustainability: The Search for a Green Democratic Model." *Environmental Politics* 9, no. 4 (2000): 43–58.

Baccelli, Luca. "Political Imagination, Conflict, and Democracy: Machiavelli's Republican Realism." In *Machiavelli on Liberty & Conflict*, edited by David Johnston, Nadia Urbinati and Camila Vergara, 352–71. Chicago: University of Chicago Press, 2017.

---

[73] As outlined in, John P. McCormick, *Machiavellian Democracy* (Cambridge: Cambridge University Press, 2011).

Barry, John. *The Politics of Actually Existing Unsustainability*. Oxford: Oxford University Press, 2012.
———. "Resistance Is Fertile: From Environmental to Sustainability Citizenship." In *Environmental Citizenship*, edited by A. Dobson and D. Bell, 21–48. Cambridge: MIT Press, 2006.
———. *Rethinking Green Politics*. London: Sage, 1999.
Bell, Duncan. "What Is Liberalism?" *Political Theory* 42, no. 6 (2014): 682–715.
Bellamy, Richard. "Introduction: The Theories and Practices of Citizenship." In *What Is Citizenship? Theories of Citizenship: Classic and Contemporary Debates (Vol. 1 of Citizenship: Critical Concepts in Political Science)*, edited by Richard Bellamy and Madeleine Kennedy-MacFoy, 1020. London: Routledge, 2014.
———. "The Paradox of the Democratic Prince: Machiavelli and the Neo-Machiavellians on Ideal Theory, Realism, and Democratic Leadership." In *Politics Recovered: Realist Thought in Theory and Practice*, edited by Matt Sleat, 166–93. New York: Columbia University Press, 2019.
Bookchin, Murray. *The Ecology of Freedom: The Emergence and Dissolution of Hierarchy*. Oakland: AK Press, 2005 [1982].
Brunkhorst, Hauke. *Critical Theory of Legal Revolutions: Evolutionary Perspectives*. London: Bloomsbury, 2014.
Cornelissen, Lars. "How Can the People Be Restricted? The Mont Pèlerin Society and the Problem of Democracy, 1947–1988." *History of European Ideas* 43, no. 5 (2017): 507–24.
de Dijn, Annelien. *Freedom: An Unruly History*. Cambridge, MA: Harvard University Press, 2020.
Dobson, Andrew. *Citizenship and the Environment*. Oxford: Oxford University Press, 2003.
Dobson, Andrew, and Ángel Valencia Sáiz. "Introduction." In *Citizenship, Environment, Economy*, 157–62. London: Routledge, 2005.
Dunn, John. *Setting the People Free: The Story of Democracy*. Princeton: Princeton University Press, 2019 [2005].
Elazar, Yiftah, and Geneviève Rousselière, eds., *Republicanism and the Future of Democracy*, Cambridge: Cambridge University Press, 2019.
Forrester, Katrina. *In the Shadow of Justice: Postwar Liberalism and the Re-making of Political Philosophy*. Princeton, NJ: Princeton University Press, 2019.
Foster, John Bellamy. *Socialism and Ecology: The Return of Nature*. New York: Monthly Review Press, 2020.
Fremaux, Anne. *After the Anthropocene: Green Republicanism in a Post-capitalist World*. Cham, Switzerland: Palgrave Macmillan, 2019.
Gabrielson, Teena, and Katelyn Parady. "Corporeal Citizenship: Rethinking Green Citizenship Through the Body." *Environmental Politics* 19, no. 3 (2010): 374–91.
Galbraith, J.K. *The New Industrial State*. 2nd ed. London: Andre Deutsch Ltd/Pelican Books, 1967.
Gerstle, Gary. "The Protean Character of American Liberalism." *American Historical Review* 99, no. 4 (1994): 1043–73.
Geuss, Raymond. *Philosophy and Real Politics*. Princeton: Princeton University Press, 2008.
———. *Politics and the Imagination*. Princeton: Princeton University Press, 2010.
Gottlieb, Robert. *Forcing the Spring: The Transformation of the American Environmental Movement*. 2nd ed. Washington, DC: Island Press, 2005.

Gourevitch, Alex. *From Slavery to the Cooperative Commonwealth: Labor and Republican Liberty in the Nineteenth Century.* Cambridge: Cambridge University Press, 2014.
Hacker, J., and P. Pierson. "Winner-Take-All Politics: Public Policy, Political Organization, and the Precipitous Rise of Top Incomes in the United States." *Politics & Society* 38, no. 2 (2010): 152–204.
Hayek, Friedrich. *The Constitution of Liberty.* Chicago: Chicago University Press, 2011 [1960].
Hayward, Tim. "Ecological Citizenship: Justice, Rights and the Virtue of Resourcefulness." *Environmental Politics* 15, no. 3 (2006): 435–46.
Leipold, Bruno, Karma Nabulsi, and Stuart White, eds. *Radical Republicanism: Recovering the Tradition's Popular Heritage.* Oxford: Oxford University Press, 2020.
Lovett, Frank. "Republicanism and Democracy Revisited." In *Republicanism and the Future of Democracy*, edited by Yiftah Elazar and Geneviève Rousselière, 117–29. Cambridge: Cambridge University Press, 2019.
MacGregor, Sherilyn. "No Sustainability Without Justice: A Feminist Critique of Environmental Citizenship." In *Environmental Citizenship*, edited by A. Dobson and D. Bell, 101–26. Cambridge: MIT Press, 2006.
Machiavelli, Niccolò. *The Discourses.* Translated by Leslie J. Walker. London: Penguin 2003.
Machin, Amanda. "Decision, Disagreement and Responsibility: Towards an Agonistc Green Citizenship." *Environmental Politics* 21, no. 6 (2012): 847–63.
Macpherson, C.B. *The Political Theory of Possessive Individualism.* Oxford: Oxford University Press, 1962.
Malm, Andreas. *Fossil Capital: The Rise of Steam Power and the Roots of Global Warming.* London: Verso, 2016.
Manin, Bernard. *The Principles of Representative Government.* Cambridge: Cambridge University Press, 1997.
Marcuse, Herbert. *Eros and Civilization.* London: Sphere Books, 1970. 1955.
McCormick, John P. *Machiavellian Democracy.* Cambridge: Cambridge University Press, 2011.
———. *Reading Machiavelli: Scandalous Books, Suspect Engagements, and the Virtue of Populist Politics.* Princeton, NJ: Princeton University Press, 2018.
McNay, Lois. *The Misguided Search for the Political: Social Weightlessness in Radical Democratic Theory.* Cambridge: Polity Press, 2014.
McQueen, Alison. *Political Realism in Apocalyptic Times.* Cambridge: Cambridge University Press, 2018.
Medearis, John. *Why Democracy Is Oppositional.* Cambridge, MA: Harvard University Press, 2015.
Mirowski, Philip. "The Eighteenth Brumaire of James Buchanan: Review of Nancy Maclean, Democracy in Chains." *boundary 2* 46, no. 1 (2019): 197–219.
Mirowski, Philip, and Dieter Plehwe, eds. *The Road from Mont Pèlerin: The Making of the Neoliberal Thought Collective.* Cambridge: Harvard University Press, 2009.
Moore, Jason W. *Capitalism in the Web of Life: Ecology and the Accumulation of Capital.* London: Verso, 2015.
Mouffe, Chantal, ed. *The Challenge of Carl Schmitt.* London: Verso, 1999.
———. *The Democratic Paradox.* London: Verso, 2000.
———. "The Importance of Engaging the State." In *What Is Radical Politics Today?*, edited by J. Pugh, 230–37. Basingstoke: Palgrave Macmillan, 2009.

———. *On the Political*. London: Routledge, 2005.
———. *The Return of the Political*. London: Verso, 1993.
Naess, Arne. *Ecology, Community, Lifestyle: Outline of an Ecosophy*. Translated by D. Rothenburg. Cambridge: Cambridge University Press, 1993 [1980].
Ober, Josiah. *Demopolis: Democracy Before Liberalism, in Theory and Practice*. Cambridge: Cambridge University Press, 2017.
———. "The Original Meaning of 'Democracy': Capacity to Do Things, Not Majority Rule." *Constellations* 15, no. 1 (2008): 3–9.
Pettit, Philip. *On the People's Terms: A Republican Theory and Model of Democracy*. Cambridge: Cambridge University Press, 2012.
Radkau, Joachim. *The Age of Ecology*. Cambridge: Polity, 2014 [2011].
———. *Nature and Power: A Global History of the Environment*. Cambridge: Cambridge University Press, 2008 [2002].
Rodgers, Daniel T. *Atlantic Crossings: Social Politics in a Progressive Age*. Cambridge, MA: Harvard University Press, 1998.
Runciman, David. *Politics*. London: Profile Books, 2014.
Scerri, A. "Green Citizenship and the Political Critique of Injustice." *Citizenship Studies* 17, no. 3–4 (2013): 293–307.
Scerri, Andy. *Greening Citizenship: Sustainable Development, the State and Ideology*. Basingstoke: Palgrave, 2012.
———. "Moralizing About the White Working-Class 'Problem' in Appalachia and Beyond." *Appalachian Studies* 25, no. 2 (2019): 202–21.
———. *Postpolitics and the Limits of Nature: Critical Theory, Moral Authority, and Radicalism in the Anthropocene*. Albany, NY: SUNY Press, 2019.
Skinner, Quentin. *Hobbes and Republican Liberty*. Cambridge: Cambridge University Press, 2008.
Sleat, Matt, ed. *Politics Recovered: Realist Thought in Theory and Practice*. New York: Columbia University Press, 2018.
Slobodian, Quinn. *The Globalists: The End of Empire and the Birth of Neoliberalism*. Cambridge, MA: Harvard University Press, 2018.
Somers, Margaret. *Genealogies of Citizenship: Markets, Statelessness, and the Right to Have Rights*. Cambridge: Cambridge University Press, 2008.
Turner, Bryan S. "The Erosion of Citizenship." *British Journal of Sociology* 52, no. 2 (2001): 189–209.
Urbinati, Nadia. *Me the People: How Populism Transforms Democracy*. Cambridge, MA: Harvard University Press, 2019.
van Steenbergen, Bart. "Towards a Global Ecological Citizen." In *The Condition of Citizenship*, edited by B. van Steenbergen, 141–52. London: Sage, 1994.
Viroli, Maurizio. *Republicanism*. Translated by Anthony Shugaar. New York: Hill & Wang, 2002.
Williams, Bernard. *In the Beginning Was the Deed: Realism and Moralism in Political Argument*. Princeton: Princeton University Press, 2005.
Winters, Jeffrey A. *Oligarchy*. Cambridge: Cambridge University Press, 2011.
Wissenburg, Marcel. *Green Liberalism: The Free and the Green Society*. London: UCL Press, 1998.
Wrong, Dennis. *The Problem of Order*. Cambridge, MA: Harvard University Press, 1994.
Zakaria, Fareed. *The Future of Freedom: Illiberal Democracy at Home and Abroad*. New York: W.W. Norton, 2007.

Eco-Anxiety and the Responses of Ecological Citizenship and Mindfulness

Michel Bourban

> The health, economic, political, and environmental implications of climate change affect all of us. The tolls on our mental health are far reaching. They induce stress, depression, and anxiety; strain social and community relationships; and have been linked to increases in aggression, violence, and crime. Children and communities with few resources to deal with the impacts of climate change are those most impacted.
>
> Susan Clayton et al. (2017) *Mental Health and Our Changing Climate*

> Concern about climate change coupled with worry about the future can lead to fear, anger, feelings of powerlessness, exhaustion, stress, and sadness, referred to as ecoanxiety and climate anxiety. Studies indicate this anxiety is more prevalent among young people.
>
> Susan Clayton et al. (2021) *Mental Health and Our Changing Climate – 2021 Edition*

INTRODUCTION

Anxiety has become a defining feature of our time. It is both one of the most fundamental human emotions and one of the most common forms of psychological disorder. About one-third of the adult population reports

---

Present Address:
M. Bourban (✉)
University of Twente, Enschede, The Netherlands
e-mail: m.bourban@utwente.nl

having anxiety problems, with almost one-fifth meeting the criteria for clinical disorder.[1] We live in an era of anxiety culture: climate change, pandemics, mental illnesses, rapid technological changes, migration flows and many other contemporary challenges feed into a growing feeling of uncertainty, insecurity and powerlessness.[2] Although some forms of anxiety can be non-clinical, many can lead to disorders, such as phobia, generalized anxiety disorder and post-traumatic stress disorder.[3]

A new but rapidly spreading form of anxiety is ecological anxiety or eco-anxiety. Climate change and other global environmental changes such as biodiversity loss, ocean acidification and deforestation affect people psychologically on an ever wider scale. So far, eco-anxiety and its close cousin, climate anxiety,[4] have been extensively discussed in the media through newspaper articles, documentaries and interviews. However, the amount of academic literature on the topic is still relatively slight: mental health is often overlooked in the discourse on climate change and other current ecological issues.[5] Even though there is a small but growing quantity of scientific literature on the topic in various disciplines, there is still a lack of research on the definition of "eco-anxiety."[6]

This chapter explores the key features of eco-anxiety and proposes a possible way to cope with its undesirable side effects and its pathological consequences. The first section highlights three major features of eco-anxiety, suggests three possible eco-anxiety disorders and identifies three categories of people that seem to be more vulnerable to experiencing eco-anxiety and its disorders. The

---

[1] Daniel Freeman and Jason Freeman, *Anxiety: A Very Short Introduction* (Oxford: Oxford University Press, 2012), 111.

[2] Michael Schapira, Ulrich Hoinkes, and John Allegrante, "Anxiety Culture: The New Global State of Human Affairs?," *EuropeNow: A Journal of Research & Art* (2018); Nicole Shea and Emmanuel Kattan, "Anxiety Culture," Ibid. (2018).

[3] Freeman and Freeman, *Anxiety: A Very Short Introduction*, xiv.

[4] Climate anxiety is a form of eco-anxiety related to climate change. Because of the salience of climate change in political, economic, scientific, and media discourses, climate anxiety has become the dominant form of eco-anxiety, so that the two expressions can almost be used synonymously. It is, however, important to keep in mind that other forms of eco-anxiety are possible and do exist.

[5] Two recent books on climate anxiety contribute to closing this research gap: Sarah Jaquette Ray, *A Field Guide to Climate Anxiety. How to Keep Your Cool on a Warming Planet* (Oakland: University of California Press, 2020); Megan Kennedy-Woodard and Patrick Kennedy-Williams, *Turn the Tide on Climate Anxiety. Sustainable Action for Your Mental Health and the Planet* (London and Philadelphia: Jessica Kingsley Publishers, 2022). See also the two important reports on climate anxiety by the American Psychological Association: Susan Clayton et al., "Mental Health and Our Changing Climate: Impacts, Implications, and Guidance" (Washington: American Psychological Association and ecoAmerica, 2017); Susan Clayton et al., "Mental Health and Our Changing Climate: Impacts, Inequities, Responses" (Washington: American Psychological Association and ecoAmerica, 2021).

[6] Pihkala Panu, "Anxiety and the Ecological Crisis: An Analysis of Eco-Anxiety and Climate Anxiety," *Sustainability* 12, no. 19 (2020).

second section adopts the normative framework of ecological citizenship and a virtue ethics approach to explore a promising response to eco-anxiety. The goal of this essay is to better understand eco-anxiety and its possible disorders and to find possible ways to live with it.

Defining "Eco-Anxiety"

Just as with anxiety, there is no single definition of eco-anxiety. My objective is to identify key features of eco-anxiety without claiming to give a definitive definition and without trying to excessively restrict our understanding of anxiety.

Three Features of Eco-Anxiety

Let us start with the following working definition[7]:

> Eco-anxiety is a subjective trait, state or disposition turned toward a possible objective state of the planet in the near or distant future. Its object is severe ecological risks and dangers that are not yet here (in space) or present (in time), but which might happen at some point in a nearer or more distant future. It is a fear resulting from an acute awareness of the risks raised by global ecological issues. Eco-anxiety can lead to a generalized feeling of discouragement; taken to too high a degree, it can also become pathological, a fear of a fear or an exaggeration of the probabilities of environmental dangers.

Let us now highlight the building blocks of eco-anxiety. First, like any form of anxiety, eco-anxiety is *future-oriented*. The envisioned future can be more or less determinate and nearer or further away, but it is both (a) threatening and (b) uncertain. Regarding (a), the main object of eco-anxiety is *ecological risks*, such as sea-level rise, megafires and extreme climate events such as hurricanes, droughts, heatwaves and floods. Its object can also be the consequences of these phenomena on human and/or non-human beings, such as human and animal suffering, ecosystem degradation, species extinction, the disruption of agricultural systems and food supply chains, economic shocks, socio-political instability along with starvation, mass migration, conflict and even societal collapse.[8] Regarding (b), risks and uncertainty are intrinsically linked. Risks are the product of magnitude and probability: the magnitude is a

---

[7] In this sub-section, I draw on Michel Bourban, "Eco-Anxiety: A Philosophical Approach," in *Anxiety Culture: The New Global State of Human Affairs*, ed. John Allegrante et al. (Baltimore: Johns Hopkins University Press, 2023). My definition of eco-anxiety is an adaptation of the definition of anxiety proposed by André Comte-Sponville in André Comte-Sponville, *Dictionnaire Philosophique* (Paris: PUF, 2013), 79.

[8] For a recent study of ecological threats to human societies, including the risk of collapse events, see C. E. Richards, R. C. Lupton, and J. M. Allwood, "Re-framing the Threat of Global Warming: An Empirical Causal Loop Diagram of Climate Change, Food Insecurity and Societal Collapse," *Climatic Change* 164, no. 3 (2021). Literature on the topic

measure of the seriousness of the loss and damage at stake; the probability is a measure of the likelihood of this loss and damage occurring.[9] That something is uncertain means that it currently has no measurable probability; it does not mean that its objective probability, if known, would be small.[10]

Questions of probability can, however, become less relevant if the magnitude of the possible loss and damage is massive. Take the two core planetary boundaries through which all the other boundaries operate: the climate system and biosphere integrity.[11] A cascade of tipping points in the climate system, such as rapid permafrost thawing, weakening of terrestrial and oceanic carbon sinks, and Amazon forest dieback could lead the entire Earth system into a "Hothouse Earth" trajectory, in which global warming may be substantially accelerated.[12] Likewise, the sixth mass species extinction leads to the degradation of ecosystem services and jeopardizes the integrity of ecosystems, posing an existential threat to life-support systems that are essential for both human and non-human beings.[13] Due to the magnitude of the loss and damage at hand, accelerated climate disruption and accelerated biodiversity loss are more

---

of societal collapse caused by ecological problems is growing. The emerging transdisciplinary field of "collapsology" addresses the possible causes of the collapse of civilization as we know it, along with the possible ways to live in a post-collapse world: see, e.g., Pablo Servigne and Raphaël Stevens, *Comment Tout Peut S'éffondrer. Petit Manuel De Collapsologie À L'usage Des Générations Présentes* (Paris: Seuil, 2015). Collapsologists are quite influential in the media and contribute to the spread of eco-anxiety. Some scholars have criticized this new field because of its survivalist tone and its apolitical discourse: see, e.g., Pierre Charbonnier, "Splendeurs Et Misères De La Collapsologie. Les Impensés Du Survivalisme De Gauche" [The Splendor and Squalor of Collapsology], *Revue du Crieur* 13, no. 2 (2019).

[9] Henry Shue, "Deadly Delays, Saving Opportunities: Creating a More Dangerous World?," in *Climate Ethics: Essential Readings*, ed. Stephen Gardiner et al. (Oxford: Oxford University Press, 2010), 147.

[10] Ibid., 148.

[11] Will Steffen et al., "Planetary Boundaries: Guiding Human Development on a Changing Planet," *Science* 347, no. 6223 (2015). The nine boundaries are the following: the big three—the climate system, the ozone layer, and the ocean; the four biosphere boundaries—biodiversity, land, fresh water, and nutrients; and the two aliens—novel entities and aerosols. Five planetary systems have already been pushed beyond their critical limits: climate, biosphere integrity (biodiversity), the nitrogen and phosphorus cycles (nutrients), land use, and novel entities (especially plastics): see Linn Persson et al., "Outside the Safe Operating Space of the Planetary Boundary for Novel Entities," *Environmental Science & Technology* (2022). If these transgressions persist, the entire planet may be pushed into a new state that would be much less hospitable for human societies, not to mention other species.

[12] Will Steffen et al., "Trajectories of the Earth System in the Anthropocene," *Proceedings of the National Academy of Sciences* 115, no. 33 (2018).

[13] Gerardo Ceballos, Paul R. Ehrlich, and Peter H. Raven, "Vertebrates on the Brink as Indicators of Biological Annihilation and the Sixth Mass Extinction," ibid. 117, no. 24 (2020).

than mere risks; they represent "transcendental damages,"[14] that is, dangers that threaten the very condition of human existence on the planet, or at least the conditions of a flourishing human life.[15]

Second, eco-anxiety is not only a state of uncertainty, but also of *fear* and *insecurity* in the face of ecological risks and transcendental dangers. Eco-anxiety is an emotion that reaches deep, a gut feeling that the future is insecure. It is a constant or structural fear that we are contributing, with our individual and collective lifestyles, to creating a more dangerous world for young people and future generations. Anxiety and fear are often used interchangeably. It is, however, possible to distinguish them according to the object of our emotion. While the object of fear is usually clear and close in space and time, the object of anxiety is vague and can be distant in space and time. This is why it is so difficult to get rid of anxiety: if we do not know precisely what is making us anxious, it is difficult to deal with the threat.[16] As Nicole Shea and Emmanuel Kattan put it, "Fear of immediate danger has been replaced by anxiety over an uncertain future and shapeless though imminent catastrophes."[17] Despite this distinction, it is still possible to consider anxiety and fear as two very close emotions. Our understanding of anxiety is composed of a tangle of concepts and experiences, such as uncertainty, risk, threats, dangers and fear.[18]

Since the objects of eco-anxiety are ecological risks and transcendental dangers, there is indeed an inescapable vagueness in our experience of it. However, eco-anxious people do not necessarily perceive the future as shapeless, vague or amorphous; the futures they have in mind can be all too clear, with specific extreme climate events (heatwaves, hurricanes, droughts and so on) and their consequences (human and animal suffering, economic shocks, pandemics and so on). The feeling of fear and insecurity that comes with eco-anxiety does not require the experience or direct observation of a dangerous anthropogenic ecological event to exist; imagining a risk or danger is sufficient to make us eco-anxious. There are enough details and information in scientific

---

[14] Dominique Bourg, "Dommages Transcendantaux," in *Du Risque À La Menace: Penser La Catastrophe*, ed. Dominique Bourg, Pierre-Benoît Joly, and Alain Kaufmann (Paris: PUF, 2013).

[15] Societal collapse and transcendental damage link eco-anxiety with existential anxiety, a form of anxiety focusing on the threats to the very existence of human beings and human societies. Global environmental changes can raise deep feelings of ontological insecurity: see Panu, "Anxiety and the Ecological Crisis: An Analysis of Eco-Anxiety and Climate Anxiety," 6–7.

[16] Freeman and Freeman, Anxiety: A Very Short Introduction, 11–12.

[17] Shea and Kattan, "Anxiety Culture," 1

[18] John Allegrante et al., "Anxiety as a New Global Narrative," in *Anxiety Culture: The New Global State of Human Affairs*, ed. John Allegrante et al. (Baltimore: Johns Hopkins University Press, 2023).

articles and reports to feed our imagination. There are also multiple documentaries, movies and novels portraying possible ecological disasters in the nearer or more distant future that can trigger eco-anxiety.[19]

One of the earliest meanings of "scenario," a word now ubiquitous in scientific reports and articles on climate change and other global environmental changes, is a reasoned effort of anticipation in a film or a novel. More and more Anthropocene fictions explore apocalyptic and/or post-apocalyptic scenarios in which anthropogenic ecological disasters completely change the world as we know it.[20] Even though these fictions draw more or less accurately on scientific knowledge, they influence popular perceptions of the near and distant future. Anthropocene fictions link numbers, graphs and figures with our daily lives. They bring the notion of the Anthropocene to life and make it tangible. They play a part in making eco-anxiety a pervasive feature of our time.

Although eco-anxiety is oriented toward an uncertain future, it is based on solid facts about the current state of the planet. For instance, it is an unequivocal fact that human economic activities have warmed the atmosphere, ocean and land, and that this has already led to widespread and severe impacts, such as heatwaves, heavy precipitation, droughts and cyclones.[21] Eco-anxiety is not an exaggerated anticipation of highly unlikely future ecological phenomena; it is first and foremost a lucid reaction to an accurate empirical description of global environmental changes, some of which are already occurring. It is not an irrational thinking circle that needs to be broken or something delusional; it is a "scientifically accepted danger to the way of life we know on earth."[22] The feeling of anxiety starts to emerge from the moment one realizes the existence

---

[19] They include documentaries such as Jeff Orlowski's *Chasing Coral*, David Attenborough and Johan Rockström's *Breaking Boundaries*, and Fisher Stevens and Leonardo DiCaprio's *Before the Flood*, eco-fiction movies such as Kevin Reynolds' *Waterworld*, Roland Emmerich's *The Day After Tomorrow*, and Bong Joon-ho's *Snowpiercer*, and climate fiction novels such as Ian McEwan's *Solar*, Ronald Wright's *A Scientific Romance*, and Kim Stanley Robinson's *Science in the Capital* trilogy. For the first climate fiction anthology, see John J. Adams, *Loosed Upon the World: The Saga Anthology of Climate Fiction* (New York: Saga Press, 2015).

[20] Adam Trexler and Adeline Johns-Putra, "Climate Change in Literature and Literary Criticism," *WIREs Climate Change* 2, no. 2 (2011); Adam Trexler, *Anthropocene Fictions: The Novel in a Time of Climate Change* (Charlottesville and London: University of Virginia Press, 2015); Adeline Johns-Putra, "Climate Change in Literature and Literary Studies: From Cli-Fi, Climate Change Theater and Ecopoetry to Ecocriticism and Climate Change Criticism," *WIREs Climate Change* 7, no. 2 (2016); Michael Svoboda, "Cli-Fi on the Screen(s): Patterns in the Representations of Climate Change in Fictional Films," ibid., no. 1.

[21] IPCC, "Summary for Policymakers," in *Climate Change 2021: The Physical Science Basis. Contribution of Working Group I to the Sixth Assessment Report of the Intergovernmental Panel on Climate Change*, ed. V. Masson-Delmotte et al. (Cambridge: Cambridge University Press, 2021).

[22] Kennedy-Woodard and Kennedy-Williams, *Turn the Tide on Climate Anxiety. Sustainable Action for Your Mental Health and the Planet*, 28.

of two discrepancies: the discrepancy between the magnitude and severity of ecological problems and one's capacity to deal with these problems; and the discrepancy between one's knowledge of the state of the planet and willingness to do something about it and other people's knowledge and (un)willingness to change.[23] Eco-anxiety is an almost unavoidable reaction to awareness of the state of the planet and of these two discrepancies.

Third, eco-anxiety can lead to a form of *paralysis* or suspension of action. Fear can sometimes be motivating, for instance when it pushes us to avoid its object or its source, but it can also be paralyzing.[24] Eco-fictions can incite us to avoid the disaster they depict by changing our behaviors and policies, but they can just as easily block our individual actions by showing the futility of our efforts. Likewise, the avalanche of data on the deteriorating state of the world in both scientific and newspaper articles can lead to "infowhelm," which prompts passivity and inaction. As Sarah Jaquette Ray stresses, "Doomsayers can be as much a problem for the climate movement as deniers, because they spark guilt, fear, apathy, nihilism and ultimately inertia. Who wants to join that movement?"[25] Eco-anxiety can lead to a generalized feeling of discouragement regarding the future and what we can do about it, both individually and collectively.

Global environmental changes can lead to two different forms of inaction. The first typically appears when self-regulatory mechanisms of moral disengagement are activated. These psychological mechanisms allow people to "rationalise their reprehensible behaviour, and thus [permit] weakness of will and/or self-interested desires to thwart their moral motivation to abide by their moral judgement."[26] They include discrediting evidence of harm, advantageous comparison, diffusion of responsibility, displacement of responsibility,

---

[23] Eco-anxiety is partly caused by what other people think and do, and especially by what they are not prepared to do. The feeling of the environmentally aware person who sees other people driving gas-guzzling SUVs is comparable to that of the vegetarian or the vegan who sees other people eating a barbecue: the discrepancy between what they know (about climate change and animal suffering) and are willing to do about it and what other people know and do is so stark that they cannot help but become anxious about the future.

[24] Sabine Roeser, "Risk Communication, Public Engagement, and Climate Change: A Role for Emotions," *Risk Analysis* 32, no. 6 (2012).

[25] Ray, *A Field Guide to Climate Anxiety. How to Keep Your Cool on a Warming Planet*, 35.

[26] Wouter Peeters, Lisa Diependaele, and Sigrid Sterckx, "Moral Disengagement and the Motivational Gap in Climate Change," *Ethical Theory and Moral Practice* 22, no. 2 (2019): 430.

unreasonable doubt, selective attention and delusion.[27] Let us call this *inaction as psychological defense*, inaction as a means to avoid the discomfort that comes with questioning our beliefs, convictions and lifestyles.

The second form of inaction usually emerges after one has become aware of the state of the planet because of a direct experience of ecological impact, or because of an indirect source such as a book, a report or a documentary. Here, the facts are not denied, misrepresented or underestimated: the gravity and the emergency of the situation are acknowledged. However, this knowledge can lead to helplessness, powerlessness and resignation. As the American Psychiatric Association (APA) stresses in a report on mental health and climate change:

> the psychological responses to climate change, such as conflict avoidance, fatalism, fear, helplessness and resignation are growing. These responses are keeping us, and our nation, from properly addressing the core causes of and solutions for our changing climate, and from building and supporting psychological resiliency.[28]

In this case, we move from inaction as defense to *inaction as paralysis*: it is no longer about self-protection, but about discouragement, about the feeling that the issues at hand are just too massive and out of control for us to be able to address them.

Three Eco-Anxiety Disorders

In addition to leading to paralysis, eco-anxiety can also become pathological and be connected to mental disorders. Given the scale and the gravity of the problems discussed in scientific articles on global environmental changes[29] and the content of the scenarios explored in Anthropocene fictions, feeling overwhelmed is understandable. "Doom and gloom" narratives can cause despair, which in turn can demotivate people in ways that could exacerbate ecological problems.[30] When eco-anxiety gives way to despair, it becomes a fear of a fear,

---

[27] Stephen M. Gardiner, "A Perfect Moral Storm: Climate Change, Intergenerational Ethics and the Problem of Moral Corruption," *Environmental Values* 15, no. 3 (2006); Peeters, Diependaele, and Sterckx, "Moral Disengagement and the Motivational Gap in Climate Change."

[28] Clayton et al., "Mental Health and Our Changing Climate: Impacts, Implications, and Guidance," 4.

[29] For instance, Earth system scientists warn us in a recent study that "the evidence from tipping points alone suggests that we are in a state of planetary emergency: both the risk and urgency of the situation are acute." Indeed, "the intervention time left to prevent tipping could already have shrunk towards zero, whereas the reaction time to achieve net zero emissions is 30 years at best": see Timothy M. Lenton et al., "Climate Tipping Points—Too Risky to Bet Against," *Nature* 575 (2019): 595.

[30] Catriona McKinnon, "Climate Change: Against Despair," *Ethics and the Environment* 19, no. 1 (2014): 33.

an exaggeration of the probabilities of transcendental dangers. At that point, there is little chance that eco-anxiety will leave space for possible solutions to avoid the realization of the darkest scenarios.

The limited research on the topic suggests that although most forms of eco-anxiety appear to be non-clinical, some pathological cases that come with anxiety disorders also exist.[31] A recent study on climate anxiety stresses that the most severe cases of the mental health effects of climate change are related to the direct experience of severe climate impacts and include post-traumatic stress disorder (PTSD), depression, the exacerbation of psychotic symptoms, suicidal ideation and suicide completion.[32] Although it remains unclear whether or not eco-anxiety should be categorized as a mental health condition, it is clear that the direct and indirect effects of ecological problems can have detrimental effects on our mental health.[33] Eco-anxiety can for instance lead to "climate depression,"[34] which can in turn lead in its extreme forms to suicidal thinking—Ray proposes the notion of "eco-nihilism" to refer to this idea that we should simply erase ourselves because we are so bad for the planet.[35]

The fact that eco-anxiety can become pathological is not that surprising, as traditional forms of anxiety disorders can be related to ecological factors. Phobias, which are characterized by an excessive and unreasonable fear of a specific object or situation, include natural environment phobias.[36] Excessive and unreasonable fear of storms, water and fire can be exacerbated by climate impacts such as hurricanes, sea-level rise and megafires. People with phobias tend to overestimate the likelihood of being exposed to harm and underestimate their capacity to cope with the situation they fear. The same can be said about people suffering from generalized anxiety disorder (GAD). In this case, worry becomes uncontrollable in that people start to worry about worry. This makes them intolerant of uncertainty and pushes them to believe that they are poor at solving problems.[37] One can easily imagine how constant exposure to the flow of information about the deteriorating state of the world can lead people to become persistent worriers. A third possible eco-anxiety disorder is PTSD. The APA defines a traumatic event as one in which "the

---

[31] Panu, "Anxiety and the Ecological Crisis: An Analysis of Eco-Anxiety and Climate Anxiety."

[32] Ashlee Cunsolo et al., "Ecological Grief and Anxiety: The Start of a Healthy Response to Climate Change?," *The Lancet Planetary Health* 4, no. 7 (2020).

[33] Kennedy-Woodard and Kennedy-Williams, *Turn the Tide on Climate Anxiety. Sustainable Action for Your Mental Health and the Planet*, 42–45.

[34] Ibid., 71–72.

[35] Ray, *A Field Guide to Climate Anxiety. How to Keep Your Cool on a Warming Planet*, 40.

[36] Freeman and Freeman, *Anxiety: A Very Short Introduction*, 58–60.

[37] Ibid., 88–90. A closely related emotion here is stress, a feeling that we are not able to deal with the problems at hand, a belief that we cannot cope with the demands facing us (ibid., 11).

person experienced, witnessed, or was confronted with an event or events that involved actual or threatened death or serious injury, or a threat to the physical integrity of self or others."[38] Natural disasters are identified as a possible cause of PTSD. People with PTSD believe they are still seriously threatened by the trauma they have experienced. They feel like nowhere is safe. The APA sums up the impacts of a changing climate on mental health the following way: "Climate change-fueled disaster events impact individual mental health and include trauma and shock, PTSD, anxiety and depression that can lead to *suicidal ideation* and *risky behavior*, feelings of abandonment, and physical health impacts."[39]

Importantly, some categories of people are more vulnerable to anxiety disorders. Phobia is more common in children and young people; women are twice as likely to suffer from forms of phobia than men. Women are also twice as likely to be exposed to GAD and PTSD as men.[40] This leads to the question: who is more exposed to eco-anxiety?

Three Categories of Exposed Population

The limited available data indicate that three categories of people stand out as more vulnerable to experiencing eco-anxiety. The first category is *people directly exposed to ecological disasters*, especially those who rely closely on the land and land-based activities, such as Indigenous peoples and farmers.[41] Physical ecological losses and their catastrophic effects on traditional ways of life and cultures, disruption of environmental knowledge systems and the resulting feelings of identity loss, and anticipated future losses of place, land, species and culture can all lead to eco-anxiety.[42]

The second category is *environmental scientists* working in the field and collecting data, especially climate scientists. A recent survey found that many IPCC authors are suffering from climate anxiety, with more than 60% of the respondents saying that they experience anxiety, grief or other distress

---

[38] cited in Ibid., 103.

[39] Clayton et al., "Mental Health and Our Changing Climate: Impacts, Inequities, Responses," 6—emphasis original.

[40] Freeman and Freeman, *Anxiety: A Very Short Introduction*, 59–61, 87, 105; Kennedy-Woodard and Kennedy-Williams, *Turn the Tide on Climate Anxiety. Sustainable Action for Your Mental Health and the Planet*, 44.

[41] Cunsolo et al., "Ecological Grief and Anxiety: The Start of a Healthy Response to Climate Change?"

[42] Ashlee Cunsolo and Neville R. Ellis, "Ecological Grief as a Mental Health Response to Climate Change-Related Loss," *Nature Climate Change* 8, no. 4 (2018): 275. The authors of this study highlight an emotion closely related to that of eco-anxiety: ecological grief, the "grief felt in relation to experienced or anticipated ecological losses, including the loss of species, ecosystems and meaningful landscapes due to acute or chronic environmental change."

because of concerns over climate change. Eighty-two percent of the respondents report that they think they will see catastrophic impacts of climate change in their lifetime, and six in ten expect the world to warm by at least 3 °C above pre-industrial levels by 2100.[43] Environmental scientists suffer from both ecological grief and eco-anxiety. They directly witness not only the loss of biodiversity, but also the loss of their life's work: this includes biologists studying threatened and endangered species, earth scientists recording the destruction of coral reefs and glaciers, and all the other specialists who watch the shrinking or disappearance of complex ecosystems.[44] This second category also includes scholars from other fields who draw on the work of environmental scientists in their research and teaching activities, as well as students who carry out research and the millions who take courses in environmental studies across the world. This is important to note for college instructors of environmental studies courses like environmental politics and environmental political theory.[45]

The third and probably the most important category is *children and young people*. In their review of different surveys at the national level, Megan Kennedy-Woodard and Patrick Kennedy-Williams highlight that while 40% of young British people aged 16–24 described feeling overwhelmed because of the environmental emergency, 57% of American teenagers said that climate change made them feel scared.[46] They add that in Australia, 96% of young people aged 7–25 consider climate change to be a serious problem, with 89% saying they are worried about the effects of climate change, and 70% being concerned that adults do not or will not take their opinion on climate change seriously. The largest study to date highlights that a huge and growing number of children and young people from a diverse range of countries with a diverse range of income and exposure to climate impacts are suffering from

---

[43] Jeff Tollefson, "Top Climate Scientists Are Sceptical That Nations Will Rein in Global Warming," *Nature* 599, no. 4 November (2021).

[44] Ray, *A Field Guide to Climate Anxiety. How to Keep Your Cool on a Warming Planet*, 32; Kennedy-Woodard and Kennedy-Williams, *Turn the Tide on Climate Anxiety. Sustainable Action for Your Mental Health and the Planet*, 46–47.

[45] Just as an example, Ray gives the following account of her students' reaction to her environmental study courses: "their despair about the state of the planet and their feelings of guilt (leading to powerlessness) could threaten their ability to show up to class, stay motivated to graduate, and then go into the world with the resolve required to tackle all these problems. Much to my chagrin, my office hours and classrooms became group therapy sessions, and I found myself totally ill-equipped to manage the demand... The course material was not just a rite of passage, challenging my students to mature into critically thinking adults. It, combined with myriad other stresses of college, was sending my students off the rails, and they were taking me with them" (Ray, *A Field Guide to Climate Anxiety. How to Keep Your Cool on a Warming Planet*, 12–13).

[46] Kennedy-Woodard and Kennedy-Williams, *Turn the Tide on Climate Anxiety. Sustainable Action for Your Mental Health and the Planet*, 40–41.

eco-anxiety.[47] While 59% say that they feel very or extremely worried about climate change, more than 45% say that their feelings about climate change negatively affect their daily lives. Eighty-three percent think that people have failed to take care of the planet, 75% find the future frightening, 56% believe that humanity is doomed, 55% think that the things they most value will be destroyed, 55% say that they will have fewer opportunities than their parents, 52% think that their family security will be threatened, and 39% are hesitant to have children. More than 50% say that they feel afraid, sad, anxious, angry, powerless, helpless and/or guilty. The psychological state of children and young people is affected not only by climate impacts, but also by their perception of governments' failure to respond to climate change, which leads to feelings of betrayal and abandonment by adults. The study highlights that all these stressors will have "considerable, long-lasting, and incremental negative implications for the mental health of children and young people." It also frames government failure as a "failure of ethical responsibility to care" and a source of "moral injury," understood as "the distressing psychological aftermath experienced when one perpetrates or witnesses actions that violate moral or core beliefs."[48]

Coping with Eco-Anxiety

Resilience

Since global ecological changes are here to stay, in one form or another, we need to develop ways to cope with eco-anxiety, to learn how to live with it by mitigating its undesirable side effects such as worry, stress and paralysis as well as its pathological consequences such as phobia, GAD and PTSD. Anxiety, in both its clinical and non-clinical forms, finds its primary articulation in psychology. Possible means to reduce levels of anxiety include psychological therapy (principally cognitive behavior therapy) and medication (through anxiolytics and antidepressants), but also lifestyle changes, especially increased physical exercise, healthy diets and yoga.[49]

There are different adaptive strategies to face eco-anxiety beyond professional assistance from mental health specialists, for instance by cultivating hope, confidence and resilience. I focus here on resilience.[50] Resilience is a notion widely used in environmental theory, especially in sustainability studies.

---

[47] Caroline Hickman et al., "Climate Anxiety in Children and Young People and Their Beliefs About Government Responses to Climate Change: A Global Survey," *The Lancet Planetary Health* 5, no. 12 (2021).

[48] Ibid., 864.

[49] G. M. Manzoni et al., "Relaxation Training for Anxiety: A Ten-Years Systematic Review with Meta-Analysis," *BMC Psychiatry* 8 (2008); Freeman and Freeman, *Anxiety: A Very Short Introduction*, 111–23.

[50] For an analysis of hope and confidence in the context of eco-anxiety, see Bourban, "Eco-Anxiety: A Philosophical Approach."

In a narrow sense, resilience is a measure of the ability of an ecosystem to recover from a very stressful event. As Paul Thompson and Patricia Norris explain, "Ecosystems are said to be resilient when they possess a nexus of stocks, flows and feedbacks that return to an equilibrium after perturbation."[51] In a broader sense, resilience can also be used for other types of recovery or rebound: a community or a whole country can express resilience when it recovers from an economic or social crisis; likewise, an individual can be resilient when they recover from major stressors such as traumas. I am interested here in psychological resilience as a response to eco-anxiety.

Ecological Citizenship

Ecological citizenship is a possible way to develop resilience in our time of rapid global environmental changes. The term "citizenship" confers a sense of action, of participation in the moral and/or political community to which one belongs. The expression "ecological citizenship" was coined in the mid-1990s as a renewed and expanded notion of citizenship that would help humanity deal with global environmental problems, such as anthropogenic mass extinction, climate change and ozone depletion.[52] In contrast to other forms of citizenship, ecological citizenship is primarily interested not in political participation in the decision-making process that determines the terms of social cooperation, but rather in changes in behaviors and underlying attitudes. The main reason for this is that the most sustainable changes in behavior do not come from social, economic and political measures put in place by local or national governments, but from voluntary changes in the attitudes that underlie our behaviors. Policies to promote sustainability such as fiscal incentives and disincentives can indeed change people's behaviors, but most of these changes usually do not last longer than the policies, because they often do not change people's underlying attitudes. Changing people's attitude

---

[51] Paul B. Thompson and Patricia A. Norris, *Sustainability: What Everyone Needs to Know* (New York: Oxford University Press, 2021), 58.

[52] See, e.g., Bart van Steenbergen, "Towards a Global Ecological Citizen," in *The Condition of Citizenship*, ed. Bart van Steenbergen (London: Sage, 1994); Peter Christoff, "Ecological Citizens and Ecologically Guided Democracy," in *Democracy and Green Political Thought: Sustainability, Rights and Citizenship*, ed. Brian Doherty and Marius de Geus (London: Routledge, 1996); Mark J. Smith, *Ecologism: Towards Ecological Citizenship* (Buckingham: Open University Press, 1998); John Barry, *Rethinking Green Politics: Nature, Virtue and Progress* (London: Sage, 1999); Andrew Dobson, *Citizenship and the Environment* (Oxford: Oxford University Press, 2003); Andrew Dobson and Angel Valencia Sáiz, *Citizenship, Environment, Economy* (London and New York: Routledge, 2005); Tim Hayward, "Ecological Citizenship: Justice, Rights and the Virtue of Resourcefulness," *Environmental Politics* 15, no. 3 (2006); Andrew Dobson and Derek Bell, *Environmental Citizenship* (Cambridge: MIT Press, 2006). Other expressions such as "environmental citizenship," "green citizenship," "sustainability citizenship" and "Earth citizenship" are possible, but "ecological citizenship" seems to be the most generic one and has become the dominant one because of the influence of Andrew Dobson's writings in green political theory.

first, in contrast, can lead to more secure and long-lasting changes in behaviors. Truly sustainable behaviors and societies cannot be based exclusively on self-interested actions that are themselves based on economic incentives and disincentives, such as carbon taxes, rubbish taxes and road-pricing schemes. Sustainability requires voluntary shifts in attitudes at a deep level—deeper than those reached by fiscal measures.[53]

One key objective of ecological citizenship is to develop a broader picture of human motivation than the one provided by an approach focused on self-interested behaviors aligned with incentives and disincentives, as is typical of the economics worldview and approach to environmental studies. To the external or extrinsic motivation to protect the environment procured by legal and economic instruments, ecological citizenship adds internal or intrinsic motivations based on the duties, responsibilities and especially the *virtues* of environmentally aware citizens. Virtuous citizens internalize the purpose and value of good environmental practice, thus basing their obedience not only on mere external motivations of price, punishment or prohibition, but on self-imposed duties and autonomous virtuous activities.[54] Individual attitudes and behaviors are indeed influenced by regulation, education and incentives set by governments, but citizens can develop an individuality that is relatively or partially independent from the economic and political structures that inform their attitudes and behaviors.[55]

Ecological citizenship is a possible way to rebound from eco-anxiety and its psychological effects. Ecological citizens are more likely to develop eco-anxiety, since their civic commitment is based on an environmental awareness that is itself based on knowledge of the state of the planet and/or on experienced ecological impacts. At the same time, they are also more likely to cope with eco-anxiety, since they perceive citizenship as intrinsically linked with environmental action, both at the individual level of lifestyle choices and

---

[53] Andrew Dobson and Ángel Valencia Sáiz, "Introduction," *Environmental Politics* 14, no. 2 (2005); Andrew Dobson and Derek Bell, "Introduction," in *Environmental Citizenship*, ed. Andrew Dobson and Derek Bell (Cambridge and London: The MIT Press, 2006); Andrew Dobson, "Environmental Citizenship: Towards Sustainable Development," *Sustainable Development* 15, no. 5 (2007). For an investigation into the requirements of environmental sustainability, see Michel Bourban, "Strong Sustainability Ethics," *Environmental Ethics* 43, no. 4 (2022).

[54] James Connelly, "The Virtues of Environmental Citizenship," in *Environmental Citizenship*, ed. Andrew Dobson and Derek Bell (Cambridge and London: The MIT Press, 2006), 49, 63.

[55] Dobson, "Environmental Citizenship: Towards Sustainable Development," 276–77. As one of the greatest advocates of individuality wrote, "Human nature is not a machine to be built after a model, and set to do exactly the work prescribed for it, but a tree, which requires to grow and develop itself on all sides, according to the tendency of the inward forces which make it a living thing" (John Stuart Mill, *'On Liberty' and Other Writings*, ed. Stefan Collini (Cambridge: Cambridge University Press, 2019), 60). This contrast between the "inward forces" and the external factors that push individuals to act is precisely what is stressed by ecological citizen theorists when they distinguish between intrinsic and extrinsic motivations.

at the collective level of policy-making. Not only do they develop a sound knowledge of environmental issues; they also act on it, both at the individual and at the collective level. Even though their knowledge and/or experiences can initially make them more vulnerable to eco-anxiety and its disorders, they do not let themselves be paralyzed by their psychological burden. They draw their motivation and energy from a very powerful ally: mindfulness.

Mindfulness

What are the virtues characteristic of the ecological citizen? Possible candidates include justice,[56] temperance,[57] simplicity,[58] and energy sobriety.[59] Mindfulness represents a major virtue that can aid in building resilience and reduce levels of eco-anxiety. In his list of green virtues that are useful in facing global environmental changes, Dale Jamieson explains that mindfulness can help us "improve our behavior" by helping us to "appreciate the consequences of our actions that are remote in time and space."[60] Sarah Jaquette Ray adds that mindfulness is a "practice of staying in the moment" that was originally used by Brahmans and Buddhist monks and that is now increasingly appreciated by neuroscientists and psychologists for its ability to "enhance self-regulation and foster a sense of agency."[61] Mindfulness is about paying attention, being present by taking a step back from the flow of information outside and the flow of emotions inside. According to David Treleaven, "By virtue of paying close, nonjudgmental attention to their inner world, people who practice mindfulness are more self-responsive to their own emotion and can even have less emotion exhaustion. [Mindfulness] also increases their capacity to be present with challenging emotions and thoughts without overreacting."[62] Daniel Freeman and Jason Freeman complete the picture by stressing that mindfulness is a possible treatment for anxiety and anxiety disorders: they conceive it as a "synthesis of modern Western psychological thinking and ancient Buddhist beliefs and practices, particularly meditation, that emphasizes

---

[56] Dobson, *Citizenship and the Environment*, 132–35.

[57] Dale Jamieson, "When Utilitarians Should Be Virtue Theorists," *Utilitas* 19, no. 2 (2007): 181.

[58] Joshua Colt Gambrel and Philip Cafaro, "The Virtue of Simplicity," *Journal of Agricultural and Environmental Ethics* 23, no. 1 (2009).

[59] Michel Bourban, "Ethics, Energy Transition, and Ecological Citizenship," in *Comprehensive Renewable Energy (Second Edition)*, ed. Trevor M. Letcher (Oxford: Elsevier, 2022).

[60] Dale Jamieson, *Reason in a Dark Time: Why the Struggle Against Climate Change Failed—And What It Means for Our Future* (Oxford: Oxford University Press, 2014), 187.

[61] Ray, *A Field Guide to Climate Anxiety. How to Keep Your Cool on a Warming Planet*, 42.

[62] David Treleaven, *Trauma-Sensitive Mindfulness: Practices of Safe and Transformative Healing* (New York: W. W. Norton, 2018), 36.

learning to live in the moment, and understanding that your thoughts and feelings are temporary, transient, and not necessarily a reflection of reality."[63]

Mindfulness cannot allow us to entirely escape negative emotions such as eco-anxiety; rather, it helps us to manage them, first by facing them, then by tolerating them, and finally by accepting them. The first step toward mindfulness is awareness, not of the external world, but this time of our inner world: we stop fighting our negative emotions and instead recognize them as such. Mindfulness starts with recognizing our feelings of vulnerability in the face of ecological risks and transcendental dangers: "We can open ourselves up and deal with feeling vulnerable."[64] Through mindfulness, "Our feelings become entities that we can talk about and look straight in the eye, rather than run away from."[65] Mindfulness also helps to reassure us by giving us the tools to trace the source of our eco-anxiety: this emotion is not absurd or exaggerated; originally, as we saw above, eco-anxiety is just a rational reaction to what we know about the state of the planet. Whatever other people say about this reaction, it is first and foremost a justified response to a real danger.

In what sense is mindfulness a *virtue*?[66] Jamieson defines virtues as "non-calculative generators of behaviors."[67] They are character traits that motivate us to act regardless of our calculative abilities and of the behavior of others. This is crucial in our current context, because "when it comes to large-scale collective action problems, calculation invites madness or cynicism":[68] madness, because the sums are too complicated and sometimes impossible to do; cynicism, because nothing seems to change if I fail to cooperate. Driving a SUV will not in itself change the climate, nor will my refraining from driving it stabilize the climate.[69] While utilitarian calculations seem to lead us to a downward spiral of non-cooperation, virtues sustain patterns of behavior whatever others do or do not do: they "give us the resiliency to live meaningful lives even when our actions are not reciprocated."[70] Mindfulness is about being present in the moment, paying attention to our direct and more distant natural environment, no matter how others behave.

---

[63] Freeman and Freeman, *Anxiety: A Very Short Introduction*, 123.

[64] Kennedy-Woodard and Kennedy-Williams, *Turn the Tide on Climate Anxiety. Sustainable Action for Your Mental Health and the Planet*, 18.

[65] Ray, *A Field Guide to Climate Anxiety. How to Keep Your Cool on a Warming Planet*, 49.

[66] I do not rely here on a religious or theological conception of virtue, but on a secular approach to virtue ethics focused on the dispositions that make us "good" ecological citizens.

[67] Jamieson, "When Utilitarians Should Be Virtue Theorists," 167.

[68] Ibid.

[69] Walter Sinnott-Armstrong, "It's Not My Fault: Global Warming and Individual Moral Obligations," in *Perspectives on Climate Change*, ed. Walter Sinnott-Armstrong and Richard Howarth (Amsterdam: Elsevier, 2005).

[70] Jamieson, *Reason in a Dark Time: Why the Struggle Against Climate Change Failed—And What It Means for Our Future*, 186.

Now, in what sense can we conceive mindfulness as a *green* virtue? Being mindful is being more engaged here and now. Mindfulness contributes to promoting human and non-human flourishing, individually and collectively. It makes us recognize that we do not and cannot flourish in an ecological vacuum: our individual and collective well-being depends on ecological resources and services provided by a flourishing natural world. As Philip Cafaro highlights:

> our flourishing and nature's flourishing are intertwined. It is no accident that the same actions and the same personality traits typically help us to be good neighbors and citizens and good environmentalists. The same ecosystems, in many cases, facilitate the flourishing of human and non-human beings; pollution and declining ecosystem health harm both people and other organisms.[71]

Mindfulness is a green virtue because it contributes to both human and ecological flourishing. But becoming aware of the fact that individual, social and ecological flourishing is interrelated is only half of the story; the other half consists in living according to this awareness. Mindfulness is also an acquired and stable disposition to live a lifestyle that is more conducive to ecological flourishing. It leads to a change in the way we conceive of our relationship to ourselves, to others and to the natural world.

For this reason, mindfulness, as well as other green virtues such as carbon sobriety, is a solution to the problem of causal inefficacy: the belief that our individual actions, behaviors and lifestyles have no significant or observable effect on global environmental problems such as climate change.[72] Green virtues make us question the influential consequentialist belief that our actions need to be measurably impactful and yield immediate results. This assumption is a barrier to individual resilience.[73] Mindfulness motivates us to reduce our environmental impact, in particular our carbon footprint, *not because of the observable, measurable or calculable impact of our actions on global environmental changes, but because we feel that it is the right thing to do*. Mindful ecological citizens want to act virtuously, whether or not this allows them to maximize collective utility. Virtues have an intrinsic value; they matter in themselves, whatever their consequences on other's people actions. They can have an instrumental value when virtuous people encourage other people to act more virtuously (for instance by leading by example), but their primary value is to make us become better persons.[74]

---

[71] Philip Cafaro, "Environmental Virtue Ethics," in *The Routledge Companion to Virtue Ethics*, ed. Lorraine Besser-Jones and Michael Slote (New York: Routledge, 2015), 432.

[72] Augustin Fragnière, "Climate Change and Individual Duties," *WIRE Climate Change* 7, no. 6 (2016): 800.

[73] Ray, *A Field Guide to Climate Anxiety. How to Keep Your Cool on a Warming Planet*, 53.

[74] Michel Bourban and Lisa Broussois, "The Most Good We Can Do or the Best Person We Can Be?," *Ethics, Policy & Environment* 23, no. 2 (2020).

Mindful ecological citizens are therefore not only more aware of the state of the planet (the *environmental awareness component*); they also act on that knowledge by contributing to individual and collective efforts to address global environmental changes (the *environmental action component*). Mindfulness implies both gathering sound knowledge and acting on it. This means that mindful ecological citizens help to reduce two persistent gaps between theory and practice that have been contributing to environmental degradation for a long time.

The first gap is that between the scientific knowledge of the state of the world and the individual and collective ecological awareness. The first Assessment Report by the Intergovernmental Panel on Climate Change (IPCC) was published more than three decades ago, in 1990. During the Earth Summit in Rio two years later, policy-makers from all over the world launched a call for multilateral action on the most urgent ecological issues by writing and signing the United Nations Framework Convention on Climate Change (UNFCCC) and the Convention on Biological Diversity (CBD). Many scientists had actually called for urgent political action before that, especially Donella Meadows et al. in the landmark *Limits to Growth* report, published 50 years ago.[75] The same year, in 1972, the United Nations Conference on the Human Environment took place in Stockholm. In the Stockholm Declaration that resulted from this international summit, policy-makers had already recognized that the environment should be placed at the top of the political agenda of all countries. This gap between scientific knowledge and ecological awareness is now slowly closing, but it took almost half a century to do so, and some recent events, such as the election of Trump in the United States and of Bolsonaro in Brazil, show that it remains non-negligible.

The second gap is the one between ecological awareness and individual and collective action. Despite a growing body of conventions, declarations, protocols and agreements at the international level over the last few decades, policies that adequately address global environmental changes are still too slow to emerge. Climate change is a striking example. Half of cumulative anthropogenic CO_2 emissions between 1750 and 2010 have occurred since 1970, with larger increases toward the end of this period.[76] Greenhouse gas emissions grew on average by 1.3% per year from 1970 to 2000, and by 2.2%

---

[75] H. Donella Meadows et al., *The Limits to Growth* (New York: Universe Books, 1972). Dobson stresses that this report played a central role in green political theory: "*The Limits to Growth* report of 1972 is hard to beat as a symbol for the birth of ecologism in its full contemporary guise" (Andrew Dobson, *Green Political Thought*, 4th ed. (London and New York: Routledge, 2007), 25).

[76] IPCC, "Summary for Policymakers," in *Climate Change 2014: Mitigation of Climate Change. Contribution of Working Group II to the Fifth Assessment Report of the Intergovernmental Panel on Climate Change*, ed. O. Edenhofer, R. Pichs-Madruga, Y. Sokona, E. Farahani, S. Kadner, K. Seyboth, A. Adler, I. Baum, S. Brunner, P. Eickemeier, B. Kriemann, J. Savolainen, S. Schlömer, C. von Stechow, T. Zwickel and J. C. Minx (Cambridge: Cambridge University Press, 2014), 6.

per year from 2000 to 2010.[77] After a three-year period in which emissions remained largely steady, global fossil CO_2 emissions rose by approximately 1.6% in 2017 and 2.7% in 2018.[78] In 2020, global fossil CO_2 emissions fell by about 7% below 2019 levels because of the measures implemented to slow the spread of the COVID-19 pandemic.[79] Although these reductions in global CO_2 emissions are unprecedented, they are likely to be temporary, because they do not reflect structural changes in the economic, transport or energy systems. For instance, investments in response to the 2008–2009 global financial crisis led to a rebound of global emissions to their pre-crisis trajectory by 2010.[80] Preliminary data for 2021 already suggest a rebound in global fossil CO_2 emissions of 4.9% compared to 2020.[81]

Why is this second gap still persisting despite the progressive reduction of the first gap due to the spread of environmental awareness? Two psychological phenomena can help to explain this paradoxical situation: *cognitive dissonance* between what we know and what we do, between the results of scientific research and collective and individual actions; and *moral dissociation* between what we should do and what we actually do, between the duties of ecological citizenship and our individual and collective behavior.[82] Mindful ecological citizens contribute to reducing these two persisting gaps by fighting against cognitive dissonance and moral dissociation. They are not only more environmentally aware than other citizens; they are also adapting their everyday behaviors and political choices accordingly. They feel eco-anxiety on a daily basis, but they have learnt not to let it lead to inaction as psychological defense or as paralysis.

---

[77] Ibid.

[78] C. Le Quéré et al., "Global Carbon Budget 2018," *Earth System Science Data* 10, no. 4 (2018).

[79] Corinne Le Quéré et al., "Fossil CO_2 Emissions in the Post-Covid-19 Era," *Nature Climate Change* 11, no. 3 (2021).

[80] Ibid.

[81] P. Friedlingstein et al., "Global Carbon Budget 2021," *Earth System Science Data Discussions* 2021 (2021).

[82] In the field of animal ethics, Gary Francione points toward a similar psychological phenomenon: that of "moral schizophrenia," which arises from the gap between our moral belief that it is wrong to impose unnecessary suffering on animals and our behavior which contributes to inflicting an overwhelming amount of suffering on animals (Gary L. Francione, *Introduction to Animal Rights: Your Child or the Dog?* [Philadelphia: Temple University Press, 2007], Chap. 1). I first introduced the notion of "moral dissociation" in a paper on animal ethics and environmental ethics with Lisa Broussois: Michel Bourban and Lisa Broussois, "Nouvelles Convergences Entre Éthique Environnementale Et Éthique Animale: Vers Une Éthique Climatique Non Anthropocentriste," *VertigO - la revue électronique en sciences de l'environnement* 32 (2020).

CONCLUSION

Because of the continuous flow of information on the rapidly deteriorating state of the planet in scientific articles, reports, documentaries, and eco-fiction novels and movies, it has become difficult, if not impossible, not to feel eco-anxious. Global environmental changes such as climate change are omnipresent, both directly through their more and more numerous, severe, and frequent impacts, and indirectly through the stream of more and more accurate and available information on their current and future effects.

The building blocks of eco-anxiety are future orientation and uncertainty, fear and insecurity, and paralysis and inaction. Eco-anxiety is related to a range of other psychological states, such as ecological grief, climate depression, eco-nihilism and existential anxiety. Degrees of eco-anxiety vary according to factors such as age and location, but three categories of people seem to be more vulnerable to eco-anxiety: people directly exposed to ecological disasters, environmental scientists and the researchers and students who draw on their research, and children and young people. Possible eco-anxiety disorders include phobias, GAD and PTSD.

Different ways to deal with eco-anxiety are being explored in the emergent literature on the topic. This chapter proposed the normative framework of ecological citizenship as a way to cultivate mindfulness, a green virtue that allows us to become more resilient in the face of eco-anxiety. Mindfulness makes us aware of the fact that individual, social and ecological flourishing are interrelated, but by connecting environmental awareness with environmental action, it also pushes us to act on that awareness. It is also a solution to the problem of causal inefficacy: whether or not our actions have an observable, measurable or calculable impact, whether or not others reciprocate our actions, behaviors and lifestyle, we ought to reduce our environmental footprint, because this allows us to live more virtuously, to become better persons. Finally, mindfulness is a way to reduce cognitive dissonance (between what we know and what we do) and moral dissociation (between what we should do and what we actually do), two persistent psychological phenomena that have been contributing to environmental degradation for too long. The need for mindful ecological citizens is more important than ever.

REFERENCES

Adams, John J. *Loosed Upon the World: The Saga Anthology of Climate Fiction*. New York: Saga Press, 2015.

Allegrante, John, Ulrich Hoinkes, Michael Schapira, and Karen Struve. "Anxiety as a New Global Narrative." In *Anxiety Culture: The New Global State of Human Affairs*, edited by John Allegrante, Ulrich Hoinkes, Michael Schapira and Karen Struve. Baltimore: Johns Hopkins University Press, 2023.

Barry, John. *Rethinking Green Politics: Nature, Virtue and Progress*. London: Sage, 1999.

Bourban, Michel. "Eco-Anxiety: A Philosophical Approach." In *Anxiety Culture: The New Global State of Human Affairs*, edited by John Allegrante, Ulrich Hoinkes, Michael Schapira and Karen Struve. Baltimore: Johns Hopkins University Press, 2023.

———. "Ethics, Energy Transition, and Ecological Citizenship." In *Comprehensive Renewable Energy (Second Edition)*, edited by Trevor M. Letcher, 204–20. Oxford: Elsevier, 2022.

———. "Strong Sustainability Ethics." *Environmental Ethics* 43, no. 4 (2022): 291–314.

Bourban, Michel, and Lisa Broussois. "The Most Good We Can Do or the Best Person We Can Be?" *Ethics, Policy & Environment* 23, no. 2 (2020): 159–179.

———. "Nouvelles Convergences Entre Éthique Environnementale Et Éthique Animale: Vers Une Éthique Climatique Non Anthropocentriste." *VertigO - la revue électronique en sciences de l'environnement* 32 (2020): 1–29.

Bourg, Dominique. "Dommages Transcendantaux." In *Du Risque À La Menace: Penser La Catastrophe*, edited by Dominique Bourg, Pierre-Benoît Joly and Alain Kaufmann, 109–26. Paris: PUF, 2013.

Cafaro, Philip. "Environmental Virtue Ethics." In *The Routledge Companion to Virtue Ethics*, edited by Lorraine Besser-Jones and Michael Slote, 427–44. New York: Routledge, 2015.

Ceballos, Gerardo, Paul R. Ehrlich, and Peter H. Raven. "Vertebrates on the Brink as Indicators of Biological Annihilation and the Sixth Mass Extinction." *Proceedings of the National Academy of Sciences* 117, no. 24 (2020): 13596–602.

Charbonnier, Pierre. "Splendeurs Et Misères De La Collapsologie. Les Impensés Du Survivalisme De Gauche." *Revue du Crieur* 13, no. 2 (2019): 88–95.

Christoff, Peter. "Ecological Citizens and Ecologically Guided Democracy." In *Democracy and Green Political Thought: Sustainability, Rights and Citizenship*, edited by Brian Doherty and Marius de Geus, 151–69. London: Routledge, 1996.

Clayton, Susan, Christie Manning, Kirra Krygsman, and Meighen Speiser. "Mental Health and Our Changing Climate: Impacts, Implications, and Guidance." Washington: American Psychological Association and ecoAmerica, 2017.

Clayton, Susan, Christie Manning, Meighen Speiser, and Alison N. Hill. "Mental Health and Our Changing Climate: Impacts, Inequities, Responses." Washington: American Psychological Association and ecoAmerica, 2021.

Comte-Sponville, André. *Dictionnaire Philosophique*. Paris: PUF, 2013.

Connelly, James. "The Virtues of Environmental Citizenship." In *Environmental Citizenship*, edited by Andrew Dobson and Derek Bell, 49–73. Cambridge and London: The MIT Press, 2006.

Cunsolo, Ashlee, and Neville R. Ellis. "Ecological Grief as a Mental Health Response to Climate Change-Related Loss." *Nature Climate Change* 8, no. 4 (2018): 275–281.

Cunsolo, Ashlee, Sherilee L. Harper, Kelton Minor, Katie Hayes, Kimberly G. Williams, and Courtney Howard. "Ecological Grief and Anxiety: The Start of a Healthy Response to Climate Change?". *The Lancet Planetary Health* 4, no. 7 (2020): e261–e63.

Dobson, Andrew. *Citizenship and the Environment*. Oxford: Oxford University Press, 2003.

———. "Environmental Citizenship: Towards Sustainable Development." *Sustainable Development* 15, no. 5 (2007): 276–85.

———. *Green Political Thought*. 4th ed. London and New York: Routledge, 2007.
Dobson, Andrew, and Derek Bell. *Environmental Citizenship*. Cambridge: MIT Press, 2006.
———. "Introduction." In *Environmental Citizenship*, edited by Andrew Dobson and Derek Bell, 1–17. Cambridge and London: The MIT Press, 2006.
Dobson, Andrew, and Angel Valencia Sáiz. *Citizenship, Environment, Economy*. London and New York: Routledge, 2005.
Dobson, Andrew, and Ángel Valencia Sáiz. "Introduction." *Environmental Politics* 14, no. 2 (2005): 157–162.
Fragnière, Augustin. "Climate Change and Individual Duties." *WIREs Climate Change* 7, no. 6 (2016): 798–814.
Francione, Gary L. *Introduction to Animal Rights: Your Child or the Dog?* Philadelphia: Temple University Press, 2007.
Freeman, Daniel, and Jason Freeman. *Anxiety: A Very Short Introduction*. Oxford: Oxford University Press, 2012.
Friedlingstein, P., M. W. Jones, M. O'Sullivan, R. M. Andrew, D. C. E. Bakker, J. Hauck, C. Le Quéré, et al. "Global Carbon Budget 2021." *Earth System Science Data Discussions* 2021 (2021): 1–191.
Gambrel, Joshua Colt, and Philip Cafaro. "The Virtue of Simplicity." *Journal of Agricultural and Environmental Ethics* 23, no. 1 (2009): 85–108.
Gardiner, Stephen M. "A Perfect Moral Storm: Climate Change, Intergenerational Ethics and the Problem of Moral Corruption." *Environmental Values* 15, no. 3 (2006): 397–413.
Hayward, Tim. "Ecological Citizenship: Justice, Rights and the Virtue of Resourcefulness." *Environmental Politics* 15, no. 3 (2006): 435–446.
Hickman, Caroline, Elizabeth Marks, Panu Pihkala, Susan Clayton, R. Eric Lewandowski, Elouise E. Mayall, Britt Wray, Catriona Mellor, and Lise van Susteren. "Climate Anxiety in Children and Young People and Their Beliefs About Government Responses to Climate Change: A Global Survey." *The Lancet Planetary Health* 5, no. 12 (2021): e863–e873.
IPCC. "Summary for Policymakers." In *Climate Change 2014: Mitigation of Climate Change. Contribution of Working Group III to the Fifth Assessment Report of the Intergovernmental Panel on Climate Change*, edited by O. Edenhofer, R. Pichs-Madruga, Y. Sokona, E. Farahani, S. Kadner, K. Seyboth, A. Adler, I. Baum, S. Brunner, P. Eickemeier, B. Kriemann, J. Savolainen, S. Schlömer, C. von Stechow, T. Zwickel and J. C. Minx, 1–30. Cambridge: Cambridge University Press, 2014.
———. "Summary for Policymakers." In *Climate Change 2021: The Physical Science Basis. Contribution of Working Group I to the Sixth Assessment Report of the Intergovernmental Panel on Climate Change*, edited by V. Masson-Delmotte, P. Zhai, A. Pirani, S. L. Connors, C. Péan, S. Berger and N. Caud, et al., 3–31. Cambridge: Cambridge University Press, 2021.
Jamieson, Dale. *Reason in a Dark Time: Why the Struggle Against Climate Change Failed—And What It Means for Our Future*. Oxford: Oxford University Press, 2014.
———. "When Utilitarians Should Be Virtue Theorists." *Utilitas* 19, no. 2 (2007): 160–83.
Johns-Putra, Adeline. "Climate Change in Literature and Literary Studies: From Cli-Fi, Climate Change Theater and Ecopoetry to Ecocriticism and Climate Change Criticism." *WIREs Climate Change* 7, no. 2 (2016): 266–82.

Kennedy-Woodard, Megan, and Patrick Kennedy-Williams. *Turn the Tide on Climate Anxiety. Sustainable Action for Your Mental Health and the Planet.* London and Philadelphia: Jessica Kingsley Publishers, 2022.

Le Quéré, Corinne, R. M. Andrew, P. Friedlingstein, S. Sitch, J. Hauck, J. Pongratz, P. A. Pickers, et al. "Global Carbon Budget 2018." *Earth System Science Data* 10, no. 4 (2018): 2141–94.

Le Quéré, Corinne, Glen P. Peters, Pierre Friedlingstein, Robbie M. Andrew, Josep G. Canadell, Steven J. Davis, Robert B. Jackson, and Matthew W. Jones. "Fossil CO_2 Emissions in the Post-Covid-19 Era." *Nature Climate Change* 11, no. 3 (2021): 197–199.

Lenton, Timothy M., Johan Rockström, Owen Gaffney, Stefan Rahmstorf, Katherine Richardson, Will Steffen, and Hans Joachim Schellnhuber. "Climate Tipping Points—Too Risky to Bet Against." *Nature* 575 (2019): 592–95.

Manzoni, G. M., F. Pagnini, G. Castelnuovo, and E. Molinari. "Relaxation Training for Anxiety: A Ten-Years Systematic Review with Meta-Analysis." *BMC Psychiatry* 8 (2008): 41.

McKinnon, Catriona. "Climate Change: Against Despair." *Ethics and the Environment* 19, no. 1 (2014): 31–48.

Meadows, Donella, H., Dennis L. Meadows, Jergen Randers, and William Behrens III. *The Limits to Growth.* New York: Universe Books, 1972.

Mill, John Stuart. *'On Liberty' and Other Writings.* Edited by Stefan Collini. Cambridge: Cambridge University Press, 2019.

Panu, Pihkala. "Anxiety and the Ecological Crisis: An Analysis of Eco-Anxiety and Climate Anxiety." *Sustainability* 12, no. 19 (2020): 7836.

Peeters, Wouter, Lisa Diependaele, and Sigrid Sterckx. "Moral Disengagement and the Motivational Gap in Climate Change." *Ethical Theory and Moral Practice* 22, no. 2 (2019): 425–447.

Persson, Linn, Bethanie M. Carney Almroth, Christopher D. Collins, Sarah Cornell, Cynthia A. de Wit, Miriam L. Diamond, Peter Fantke, et al. "Outside the Safe Operating Space of the Planetary Boundary for Novel Entities." *Environmental Science & Technology* (2022).

Ray, Sarah Jaquette. *A Field Guide to Climate Anxiety. How to Keep Your Cool on a Warming Planet.* Oakland: University of California Press, 2020.

Richards, C. E., R. C. Lupton, and J. M. Allwood. "Re-framing the Threat of Global Warming: An Empirical Causal Loop Diagram of Climate Change, Food Insecurity and Societal Collapse." *Climatic Change* 164, no. 3 (2021): 49.

Roeser, Sabine. "Risk Communication, Public Engagement, and Climate Change: A Role for Emotions." *Risk Analysis* 32, no. 6 (2012): 1033–40.

Schapira, Michael, Ulrich Hoinkes, and John Allegrante. "Anxiety Culture: The New Global State of Human Affairs?" *EuropeNow: A Journal of Research & Art* (2018).

Servigne, Pablo, and Raphaël Stevens. *Comment Tout Peut S'éffondrer. Petit Manuel De Collapsologie À L'usage Des Générations Présentes.* Paris: Seuil, 2015.

Shea, Nicole, and Emmanuel Kattan. "Anxiety Culture." *EuropeNow: A Journal of Research & Art* (2018).

Shue, Henry. "Deadly Delays, Saving Opportunities: Creating a More Dangerous World?" In *Climate Ethics: Essential Readings*, edited by Stephen Gardiner, Simon Caney, Dale Jamieson and Henry Shue, 146–62. Oxford: Oxford University Press, 2010.

Sinnott-Armstrong, Walter. "It's Not My Fault: Global Warming and Individual Moral Obligations." In *Perspectives on Climate Change*, edited by Walter Sinnott-Armstrong and Richard Howarth, 221–53. Amsterdam: Elsevier, 2005.

Smith, Mark J. *Ecologism: Towards Ecological Citizenship*. Buckingham: Open University Press, 1998.

Steffen, Will, Katherine Richardson, Johan Rockström, Sarah E. Cornell, Ingo Fetzer, Elena M. Bennett, Reinette Biggs, et al. "Planetary Boundaries: Guiding Human Development on a Changing Planet." *Science* 347, no. 6223 (2015): 1259855.

Steffen, Will, Johan Rockström, Katherine Richardson, Timothy M. Lenton, Carl Folke, Diana Liverman, Colin P. Summerhayes, et al. "Trajectories of the Earth System in the Anthropocene." *Proceedings of the National Academy of Sciences* 115, no. 33 (2018): 8252–59.

Svoboda, Michael. "Cli-Fi on the Screen(s): Patterns in the Representations of Climate Change in Fictional Films." *WIREs Climate Change* 7, no. 1 (2016): 43–64.

Thompson, Paul B., and Patricia A. Norris. *Sustainability: What Everyone Needs to Know*. New York: Oxford University Press, 2021.

Tollefson, Jeff. "Top Climate Scientists Are Sceptical That Nations Will Rein in Global Warming." *Nature* 599, no. 4 (2021): 22–24.

Treleaven, David. *Trauma-Sensitive Mindfulness: Practices of Safe and Transformative Healing*. New York: W. W. Norton, 2018.

Trexler, Adam. *Anthropocene Fictions: The Novel in a Time of Climate Change*. Charlottesville and London: University of Virginia Press, 2015.

Trexler, Adam, and Adeline Johns-Putra. "Climate Change in Literature and Literary Criticism." *WIREs Climate Change* 2, no. 2 (2011): 185–200.

van Steenbergen, Bart. "Towards a Global Ecological Citizen." In *The Condition of Citizenship*, edited by Bart van Steenbergen, 141–52. London: Sage, 1994.

Open Access This chapter is licensed under the terms of the Creative Commons Attribution 4.0 International License (http://creativecommons.org/licenses/by/4.0/), which permits use, sharing, adaptation, distribution and reproduction in any medium or format, as long as you give appropriate credit to the original author(s) and the source, provide a link to the Creative Commons license and indicate if changes were made.

The images or other third party material in this chapter are included in the chapter's Creative Commons license, unless indicated otherwise in a credit line to the material. If material is not included in the chapter's Creative Commons license and your intended use is not permitted by statutory regulation or exceeds the permitted use, you will need to obtain permission directly from the copyright holder.

Earth Breaking Bad: The Politics of Deep Adaptation

Timothy W. Luke

Introduction

After decades of racing to higher levels of economic growth thanks to higher rates of fossil fuel consumption, and despite yellow caution flags waved by dozens of environmentalist critics and conservation organizations about the threat of rapid climate change intrinsic to these two entangled trends, the 2020s increasingly are marked by ecological catastrophes. Instead of maintaining the pretense that ecological degradation always can be managed, manipulated or mitigated, Jem Bendell and other proponents of "deep adaptation" in policy studies circles, like Paul Kingsnorth, Dougald Hine and their associates in the Dark Mountain Project among arts and humanities groups, openly entertain the proposition that ecological overshoot is real. In turn, as influential "thought leaders," they advance daunting concepts, like "collapsology" or "uncivilization," to illustrate how to live amid permanently worsening environmental disruption as well as close to complete societal collapse.

Devastation on this scale no longer can be dismissed in tame affluent regions of the world as merely extraordinary ephemeral events in evening news clips, like those depicting extreme desertification in arid African areas, the thinning or disappearing ice in the Arctic Ocean or massive cyclones

---

T. W. Luke (✉)
Virginia Polytechnic Institute and State University, Blacksburg, Virginia, USA
e-mail: twluke@vt.edu

© The Author(s), under exclusive license to Springer Nature Switzerland AG 2023
J. J. Kassiola and T. W. Luke (eds.), *The Palgrave Handbook of Environmental Politics and Theory*, Environmental Politics and Theory,
https://doi.org/10.1007/978-3-031-14346-5_5

tearing through poor farming villages in Bangladesh, in the wilder impoverished zones of the planet. Such destruction the past two or three decades repeatedly is befalling nearly all human settlements, including millions with more comfortable lives in once secure sites in the affluent regions of the world.

From massive bushfires in New South Wales and Victoria that scorched millions of hectares, displaced thousands of people and killed millions of animals, flash flooding across New York City and Zhengzhou that inundated subways, tunnels and roads or severe heatwaves in Northern Europe and India that killed hundreds unable to endure hot days that never cooled overnight to recurrent sea rise events in Venice, Miami, Shanghai and Alexandria, extraordinary tundra fires in Alaska and Siberia or the total collapse of the Texas electrical grid during the huge but not unprecedented 2021 ice storm, the plausibility and persistence of human-made rapid climate change have gained widespread credence in rich nations that had for too long been living in total denial of its dangers.

Once more robust stable systems—artificial and natural—are becoming brittle, wobbly, less predictable. Seemingly one-off events are intensifying, proliferating and widening. With these catastrophic clusters, resilience may hold only so many times until collapse happens. Slowly or rapidly, it comes.

The linkage of such recurrent widespread catastrophes has been tied scientifically to the Earth breaking bad due to carbon dioxide, methane, nitrous oxide and industrial fluorocarbons, which are the main greenhouse gases spewing into the atmosphere, from increasingly reckless fossil fuel consumption that has accelerated tremendously since World War II ended. Yet these technical findings have been tended, on the one hand, to be neglected or ridiculed by industrial planners, major oil companies and rival scientific networks, while, on the other hand, being ignored or suppressed by local developers, partisan activists and national law-makers who cynically allowed their clients, voters and citizens to be complacent for decades.

Nonetheless, voters all too often have been willingly distracted by the technological seductions of affluent modernity to change their energy-intensive lifestyles in light of these troubling trends, even as they have degraded food sources, water quality, public utilities and governmental integrity to the point that widespread societal collapse is likely or, in fact, already unfolding. Crucial decades—from the early 1960s to late 1990s—when meaningful material changes could have been made were frittered away by most corporate and government decision-makers. Now, however, the ill-effects of rapid climate change are regarded as quite real, particularly by commercial, financial and managerial interests that hitherto dismissed such topics as little more than new hooks for their corporate social responsibility campaigns to stimulate investor awareness in their firms.

A Politics for Deep Adaptation

These rapid changes are dramatic, disruptive and destabilizing; now management thinkers and business groups have begun considering the most plausible of practices suitable for a "politics of deep adaptation."[1] This task, of course, is both attributed and assigned to those considered to be sensible, pragmatic thinkers, or "sustainability professionals" from the fields of leadership studies, management and sustainability studies, who are willing to assert "the scale of the sustainability challenge means it is best characterized as a 'transition' from unsustainable ways of living and working," which requires "'sustainable leadership'—ways of relating that promote change that is mutually beneficial for the person, organization, stakeholders and world at large."[2]

This chapter then will explore the tenets of "deep adaptation," which purport to provide new paths for dealing with today's ecological crises by accepting that the still-contested predicates of "the Anthropocene" lead

---

[1] See Jem Bendell, "Deep Adaptation: A Map for Navigating Climate Tragedy," Institute of Leadership and Sustainability (IFLAS) Occasional Paper 2, University of Cumbria (July 27, 2020). https://www.cumbria.ac.uk/research/centres/iflas/ Plainly, the organized responses of mainstream educational, governmental and scientific institutions in the 1970s to such troubling environmental news were focused on the research imperatives of defining, and then actually discovering "limits to growth" by developing large-scale, data models to predict how and when environmental crises would become most threatening (see, for example, Donella H. Meadows, Dennis L. Meadows, Jorgen Randers and William W. Behrens III, The *Limits to Growth: A Report to the Club of Rome's Project on the Predicament of Mankind*. Second Edition (New York: New American Library, 1975). Bendell and others concede merits of the earlier research by the Club of Rome. Yet, to respond to contemporary challenges, they also advance new recommendations for "deep adaptation" as a pragmatic approach for building a broad-based consensus about how to live today. The comprehensive study of the limits of growth today. The studies from the 1970s saw the bottlenecks becoming more severe decades out. Yet, those decades now are here. We are living through those times, and they are a mix of better and worse conditions. Decisions had been postponed then "to make later," no longer can be avoided. Before, they were still distant predictions about possible conditions of ecological and social collapse. In the 1970s, their plausible ill-effects were downplayed. Today, these choices and their potential downsides no longer are distant abstractions. Instead, they have become concretely threatening material realities. See Pablo Severigne and Raphael Stevens, *How Everything Can Collapse: A Manual for Our Times* (Cambridge: Polity Press, 2020); David Wallace-Wells, *The Uninhabitable Earth: Life After Warming* (New York: Tim Duggan Books/Random House, 2020); and, Brian T. Wilson, *Headed into the Abyss: The Story of Our Time, and the Future We'll Face* (Swampscott, MA: Anvilside Press, 2019).

[2] Bendell, "Deep Adaptation," https://www.cumbria.ac.uk/research/centres/iflas/ For additional insight, see C. N. Waters, et al."The Anthropocene is functionally and stratigraphically distinct from the Holocene," *Science* 351, aad2622 (2016); and, Roy Scranton, *Learning to Die in the Anthropocene: Reflections on the End of a Civilization* (San Francisco, CA: City Lights Publishers, 2015); and, Dipesh Chakrabarty, (2009) "The climate of history: Four theses," *Critical Inquiry*, 35 (2009), 197–222 as well as Paul Kingsnorth and Dougald Hine, *Uncivilization: The Dark Mountain Manifesto* (2009) https://dark-mountain.net/about/manifesto/

directly to confronting social collapse and uncivilization in the eclipse of "the Holocene."[3]

This yet to be fully defined geological epoch has still not been officially endorsed by many stratigraphers and other earth scientists. Essentially, it is polyvalent conceptual tangle of many divergent trends pointing toward differing outcomes, multiplying alternative futures and running at varying rates, given alternate modeling assumptions. In 2018, however, Bendell operationally assumed the Anthropocene could soon compound itself as widespread ecological catastrophes, which would inevitably entail societal collapse and human extinction as modern fossil-fueled lifestyles began to fully fragment under the pressures of rapid climate change. While this willingness to embrace life in the Anthropocene as leading to "INTHE," or "Inevitable Near Term Human Extinction," the theory and practice for deep adaptation sketch "conceptual maps" for "synthesizing the main things" people could be "doing differently in light of such "inevitable collapse and probable catastrophe" as "the 'deep adaptation agenda'.".[4] How novel, innovative and effective this agenda is, and the "who, whom" behind how it might be implemented, then, is well-worth reconsidering as superstorms, megadroughts and hyperheatwaves, once regarded as 100-, 500- or 1000-year events recur year after year.

Deep adaptation strategies also seem imperative to their advocates, because concerned climatologists and environmental activists today are deeply distraught over the prospects of limiting global warming to 1.5 over preindustrial temperature levels. Attaining that goal was quite possible in the 1970s, as Exxon, Shell and Chevron technicians then reported to their senior managers. These firms, however, suppressed these findings in order to produce more coal, gas and oil, but hardened their industrial infrastructure to adapt to the eventual sea rise and more intense storms their products were accelerating. The

---

[3] See Michael Shellenberger, *Apocalypse Never: Why Environmental Alarmism Hurts Us All* (New York: Harper, 2020); Timothy W. Luke, *Anthropocene Alerts: Critical Theory of the Contemporary as Ecocritique* (Candor, NY: Telos Press Publishing, 2019); Clive Hamilton, *Defiant Earth: The Fate of Humans in the Anthropocene* (Cambridge: Polity Press, 2017); and, J.R. McNeill and Peter Engelke, *The Great Acceleration: An Environmental History of the Anthropocene since 1945* (Cambridge, MA: Belknap Press, 2016).

[4] Bendell, "Deep Adaptation," https://www.cumbria.ac.uk/research/centres/iflas/ Such full and frank discursive claims about the unanticipated and unintended consequences of such complex coupled social/biological systems too often are deflected by Panglossian taxonomists, creating neat little boxes to mystify the variation and intensity in the development of environmental debates, by presuming to be the most highly sophisticated observers of Humanity and Nature as they slog toward the deliberative democracy they are certain lies beyond the expert ministrations of ecomanagerialist sustainability studies. Yet, their pragmatic textbook vignettes of democratic theory all too often occlude dramatic everyday episodes of antidemocratic practice in what so many regard as the workings of "Earth System Governance." See, for example, Frank Biermann, *Earth System Governance: World Politics in the Anthropocene* (Cambridge, MA: MIT Press, 2014) or John Dryzek, *The Politics of the Earth: Environmental Discourses*, fourth edition (Oxford: Oxford University Press, 2021).

destructive impact of this speeding degradation of the Holocene era's overall environmental stability, in turn, has been accurately tracked for a generation. When James Hansen presented his initial findings in about these unintended consequences of greenhouse gas emissions in 1988 at US Senate hearings, there arguably was a considerable measure of time remaining for humanity to gradually adapt.

If shifting to more renewable energy sources, reducing energy consumption, as well as increasing energy efficiencies had been fast-tracked collectively by 1990, like the decisions behind 1987 Montreal Accord to ban chlorofluorocarbon emissions protect the planet's ozone layer, hitting safer stable increments over preindustrial CO_2 levels by 2020 was entirely possible. Those choices, however, were generally not made outside of a small handful of more developed nations, like the efforts in Denmark, Norway, Iceland or Costa Rica. Instead, most nations from the Kyoto to the Glasgow climate conferences increased their greenhouse gas emissions, but promised to meet their targets in the far-off years of 2040, 2050 or 2060. In so doing, Bendell maintains the prospects for greater societal collapse and severe ecological degradation already are baked into 2030.

For example, when estimated in gigatons (Gt) per annum, "cumulative net CO_2 emissions from 1850 to 2019 were 2400 ± 240 $GtCO_2$ (high confidence). Of these, more than half (58%) occurred between 1850 and 1989 [1400 ± 195 $GtCO_2$]," but, more tragically, with the explosive economic growth after the end of the Cold War in 1991, greenhouse gas emissions wildly accelerated with "about 42% between 1990 and 2019 [1000 ± 90 $GtCO_2$]," and nearly "17% of historical cumulative net CO_2 emissions since 1850 occurred between 2010 and 2019 [410 ± 30 $GtCO_2$]."[5] If the IPCC's models are valid, Bendell's operational assumptions appear more and more plausible.

Overshooting the limits on the planet's tolerable carbon budget only was a looming threat in 1990 when Senator Al Gore, Jr. in the United States was working on his 1992 book, *Earth in the Balance: Ecology and the Human Experience*. As part of his run-up for the 1992 presidential elections, he asserted that the United States and world should work toward climate change avoidance rather than adaptation.[6] For Gore, the environmental crisis was a moral crisis. At that juncture, it could, and would, be solved by individuals— alone and together—deciding to make the best moral choices as citizens and

---

[5] Working Group III (AR6), *Climate Change 2022: Mitigation of Climate Change: Working Group III contribution to the WG III Sixth Assessment Report of the Intergovernmental Panel on Climate Change* (New York: United Nations Environmental Program, 2022), 10–11. https://report.ipcc.ch/ar6wg3/pdf/IPCC_AR6_WGIII_FinalDraft_FullReport.pdf

[6] Return to review, for example, Al Gore, Jr, *Earth in the Balance: Ecology and the Human Spirit* (New York; Houghton Mifflin, 1992) as well as his two follow-up documentary films, "An Inconvenient Truth" (2007) and "An Inconvenient Sequel: Truth to Power" (2017).

consumers to purchase fewer energy-intensive products, live more simply in green communities and thereby realize the higher goods, like community, family, friends or spiritual tied, needed by the Earth and humanity.

Over a generation later, however, it is quite evident Gore's 1992 ambitions for humanity have been ignored, given that around 40% of all historical greenhouse gas emissions have happened since 1990. Consequently, "the inconvenient truth" is that Earth's delicate ecological equilibria all are unraveling irreversibly—over and against the transnational interests of all the planet's peoples, who increasingly are beset by the accelerating ill-effects of state failure, rapid climate change, stagnating standards of living and environmental degradation.[7] Today, the time horizon for limiting global warming to 1.5 centigrade over preindustrial temperatures is no longer several decades; it is less than eight years with about 80% of the allowable carbon budget emissions already having been spent by 2020, as the 2022 IPCC report notes.

With the deep disruptions in energy use caused by NATO bloc sanctions against Russian fossil fuel exports after Moscow invaded Ukraine in 2022, global waves of "revenge travel" after nearly three years of lockdowns during the COVID-19 pandemic, and energy shortages in Europe caused by several countries poorly planned transitions to renewable energy sources have forced some nations to return to burning coal, like Germany and the Netherlands, to endure unprecedented heat waves, ice storms or recurrent floods. As a result, catastrophic ecological collapse appears increasingly inevitable. Even when set apart from his this perspective, these complex developments seem to be changing everything:

> It is a truism that we do not know what the future will be. But we can see trends. We do not know if the power of human ingenuity will help sufficiently to change the environmental trajectory we are on. Unfortunately, the recent years of innovation, investment and patenting indicate how human ingenuity has increasingly been channeled into consumerism and financial engineering. We might pray for time. But the evidence before us suggests that we are set for disruptive and probably uncontrollable levels of climate change, bringing starvation, destruction, migration, disease and war.[8]

Here, the "deep adaptation agenda" means facing the hard reality lying within ecological, economic, political and/or societal collapse, not pushing how to adapt to more smog in the city, wind turbines on the horizon or solar panels on the neighbor's house down the hill.

---

[7] See Working Group III (AR6), *Climate Change 2022: Mitigation of Climate Change: Working Group III contribution to the WG III Sixth Assessment Report of the Intergovernmental Panel on Climate Change* (New York: United Nations Environmental Program, 2022). https://report.ipcc.ch/ar6wg3/pdf/IPCC_AR6_WGIII_FinalDraft_Full Report.pdf

[8] Bendell, "Deep Adaptation," https://www.cumbria.ac.uk/research/centres/iflas/

For Bendell, the time for those polite public policy parlor games have past. It simply is too late. Without saying as much, sustainability now must become a lifestyle for basic survival—material, physical, spiritual and technical—amid the industrial civilized world not as it has worked in the twentieth century, but rather as how it will unfold in the quick, nonlinear shifts of de-Holocenation as this geological epoch morphs into something far more inhospitable to life on Earth.

Disrupting the Systems of Denial

Bendell also realizes there is an ethos of denial in environmentalist circles about inescapable ecological disaster. After all, since the earliest years of the conservation movement, initiatives for conserving natural resources somewhere did happen, managing overused range ecologies did meet with some success in places, a few wildlife species were protected, and the professional identity of environmental activists, sustainability experts and ecological scientists rests upon trusting such successes can continue. Bendell admits,

> many people have said to me that 'it can't be too late to stop climate change, because if it was, how would we find the energy to keep on striving for change?' With such views, a possible reality is denied because people want to continue their striving. What does that tell us? The "striving" is based in a rationale of maintaining self-identities related to espoused values. It is understandable why that happens. If one has always thought of oneself as having self-worth through promoting the public good, then information that initially appears to take away that self-image is difficult to assimilate.[9]

Nonetheless, his aspirations for the "Deep Adaptation Agenda" are to disrupt this systemic evasion, collective denial or personal repression fostered in mainstream schools of sustainability studies as they skirt around imminent societal collapse.[10]

Ultimately, the theory and practice of mainstream sustainability programs and policies have not effectively shielded the environment from permanent harm. For personal and institutional reasons, however, sustainability experts must ignore these facts and maintain their stance of hope. In this personal vein, such environmental workers are highly-trained people "who have invested time and money in progressing to a higher status within existing social structures," which makes them "more naturally inclined to imagine reform of those systems than their upending. This situation is accentuated if we assume our livelihood, identity and self-worth is dependent on the perspective that progress on sustainability is possible and that we are part of that progressive process."[11]

---

[9] Ibid.
[10] Ibid.
[11] Ibid.

Likewise, on the institutional side, sustainability professionals are embedded in networks of financial, ideological and political support tied to,

> non-profit, private and governmental sectors. In none of these sectors is there an obvious institutional self-interest in articulating the probability or inevitability of societal collapse. Not to members of your charity, not to consumers of your product, not to voters for your party. There are a few niche companies that benefit from a collapse discourse leading some people to seek to prepare by buying their products... But the internal culture of environmental groups remains strongly in favor of appearing effective, even when decades of investment and campaigning have not produced a net positive outcome on climate, ecosystems or many specific species.[12]

Sustainability professionals, environmental activists and the general public seeing these patterns are not easy, and accepting them can be extremely hard even when they are identified.

At the same time, the swift changes in the polar ice caps, increasing ocean temperatures and acidity, rising atmospheric temperatures, the aridity and nutrient loss in soils, frequent and intense extreme weather events, and the decline of many nonhuman animal species, all put an exclamation point on Bendell's warnings about the turning point emerging from the chaos of the earth system. Furthermore, this is the moment to get a strong hold upon any emotional, moral, psychological or spiritual "difficulties with realizing the tragedy that is coming, and that is in many ways upon us already," because they are true challenges for individuals and societies. Still, each one of "these difficulties need to be overcome so we can explore what the implications may be for our work, lives and communities," after endless growth stalls, fragmentation wracks society and the collapse of the world, as many have known it for decades or centuries, happens.[13]

In some sense, Bendell admits the mainstream currents of theory and practice, where many sustainable development experts and organizations are still paddling around in circles and have been struggling to identify the vague outlines of "INTHE" their ordinary everyday bureaucratic processes over the past 15 years. As a response to "the Anthropocene," one can see the IPCC explicitly urging societies and economies to cut deep and wide adaptation

---

[12] Ibid. Bendell does not put in these terms, but "deep adaptation" involves what others would see forsaking most underpinnings of fossil capitalism, modern everyday life, the basis of existing civil society as well as the state of nature that are all amalgamated together in urban industrial civilization and its once reliable environmental foundations. See Fabian Scheidler, *The End of the Megamachine: A Brief History of a Failing Civilization* (Atresford, Hampshire: Zero Books, 2020). As Kingsnorth and Hine define the outlines of "uncivilization," they proclaim, "we believe that the roots of these crises lie in the stories we have been telling ourselves...the myth of progress, the myth of human centrality and the myth of our separation from "nature," because they are now altogether "more dangerous for the fact that we have forgotten they are myths,"" *Uncivilization*. https://dark-mountain.net/about/manifesto/

[13] Bendell, Ibid.

paths into the thickets of rapid climate change. A new United Nations Global Adaptation Network has been established to facilitate greater "knowledge sharing" and "joint collaboration." The Paris Climate Accord's member states developed their own "Global Goals on Adaptation" (GGA). All of its signatory countries agreed to draft National Adaptation Plans (NAPs) and then report their aspirations in detail as they filed them at the UN. Similarly, the International Fund for Agricultural Development (IFAD), African Development Bank (AfDB), Asian Development Bank (ADB), Global Facility for Disaster Reduction and Recovery (GFDRR) and the World Bank also made available fresh financing for governments to boost the "resilience" of their citizens and their communities.[14]

Thus, projects like the Green Climate Fund and Climate Action Programme were launched to tap international funding agencies, global bond markets and green philanthropy to trigger new thinking and preparations for adapting directly to climate change risks. Such energies also were channeled into vision statements for "Disaster Risk Reduction," which were placed under the aegis of other international agency, namely the United Nations International Strategy for Disaster Reaction (UNISDR). Working to curtail the plausible damages that are caused by typical environmental hazards, like earthquakes, floods, droughts and cyclones, these capacity-building measures prepared communities to respond when such disasters strike them. These innovations are not trivial. Indeed, they have been helpful. Yet, the intellectual framing around such adaptation strategy is faulty inasmuch as it assumes countries and their citizens will continue to roll with the punches ordinary natural hazards deliver. They do not confront the looming hyper-hazards of "nature" itself is losing its resilience in rapid climate change. Thus, all of these measures will prove not to be irrelevant but rather utterly insufficient for the deep adaptation challenge either directly right ahead and/or abruptly rising now as still not well-known Holocene slips away, and the known unknowns of Anthropocene arrive with radically greater dangers and deeper risks.

Such "business as usual" sustainability policies and practice only are repurposing, according to Bendell's analysis, existing integral, but also outmoded, pieces of an obsolete administrative assemblage as their strategy for postponing environmental disaster, which only are profiled as "known knowns." They occur, but their quantitative frequency, intensity and scope already are constituting a qualitative change as greater heat waves kill more people, more forest fires induce greater numbers of annual deaths from fire particulates, more massive flooding drowns people who believed they would never be flooded, and larger crop losses to prolonged droughts starve millions unable to flee or receive food aid.

Global accords for slowing rapid climate change have developed elaborate new international monitoring, mitigation and management regimens for greenhouse gas surveillance, but they do little more than measure and

---

[14] Ibid.

report the accelerating failure to attain anything more. National reports made, however, to toothless international governance regimes are counterproductive. This green linkage-politics identifies this systemic insufficiency as greater "resilience," even though the rustling sounds of complete collapse are rising in volume just over horizon. Time is growing shorter and shorter, and calamities closing faster and faster around the planet.

At the end of the day, these institutional moves mark the corrupt complicity of too many contemporary professional pursuits, like sustainable developments, corporate social responsibility, sustainability studies, technocratic ecomanagerialism or earth governance projects, in accelerating the societal collapse.

This also is the core of the Dark Mountain Project's critique of these debased delusions. As Kingsnorth and Hine observe, the conceptual coins circulating as shiny ecological solutions are counterfeit currencies. On heads, there are stamped shiny ecological solutions; but, on tails, they only are tarnished cultural delusions. Indeed, they all too often usually involve the necessity of urgent political agreement and a judicious application of human technological genius. Things may be changing, runs the narrative, but there is nothing we cannot deal with here, folks. We perhaps need to move faster, more urgently. Certainly, we need to accelerate the pace of research and development. We accept that we must become more "sustainable." But everything will be fine. There will still be growth, and there will still be progress: these things will continue, because they have to continue, so they cannot do anything but continue. There is nothing to see here. Everything will be fine.[15]

The "Deep Adaptation Agenda" flatly repudiates such mythic incantations, but one must realize they also echo through many halls occupied by the doyens of environmental politics and theory. Whether they present their political programs, decade after decade, as smug deliberative democrats, cheerful Future Earth affiliates, earnest green statists or dutiful Earth Governance Project cadres, such political players' rhetoric confirms what "rapid climate change" continually makes very plain, "more effectively than any carefully constructed argument or optimistically defiant protest, how the machine's need for permanent growth will require us to destroy ourselves in its name."[16]

When the president of Exxon proudly can predict during 2022, even if all of the world's automobiles are only running completely on lithium ion battery packs tied to electric motors by 2040, his transnational oil company will produce and sell as much fossil fuel as it did in 2013, and his frank disclosure makes many things crystal clear. In particular, it is "climate change, which brings home at last our ultimate powerlessness. These are the facts, or some of them. Yet facts never tell the whole story. ("Facts," Conrad wrote, in *Lord*

---

[15] Kingsnorth and Hine, *Uncivilization*.
[16] Ibid.

Jim, "as if facts could prove anything.") The facts of environmental crisis we hear so much about often conceal as much as they expose."[17]

IMPLEMENTING THE "DEEP ADAPTATION AGENDA"

Given these twists and turns in a world rapidly being refabricated as greenhouse gas capture and containment construct, Bendell's political project for "Deep Adaptation" accepts that the cultural, economic and societal collapse unfolding all around the world as one more facet of the Anthropocene. It is not necessarily apocalyptic as a mode of everyday life. On the contrary, it appears more likely to be many connected clusters of continuous catastrophes, which will weave other "artifactual ecologies" through the already "natural environments" created by the existing world system of nation-states and multinational markets.

In this respect, Bendell follows the lead set by Severigne and Stevens, namely the coming "collapse" becomes another enveloping, surrounding or milieu-making environment rather than the demise of humanity, end of history or final hours of the planet. While it is neither the end of the world nor the Apocalypse, it also is not, "a simple crisis from which we can emerge unscathed or a one-off disaster that we can forget in a few months, like a tsunami or a terrorist attack. A collapse is 'the process at the end of which basic needs (water, food, housing, clothing, energy, etc.) can no longer be provided [at a reasonable cost] to a majority of the population under legal supervision. So, it's a large-scale, irreversible process—just, like the end of the world, except it is not the end! It looks as if the consequences will last for a long time, and we'll need to live through them. And one thing is certain: we don't have the means to know what they will consist of."[18]

This claim opens the door for Bendell to provisionally present "deep adaptation" as one important intellectual guide that leads to the means for living through this near universal collapse—practically, psychologically and politically.

Basically, Bendell imagines there are four major behavioral domains essential for developing deeply adaptive individuals and societies. The first is "*resilience*," or "the capacity to adapt to changing circumstances so as to

---

[17] Ibid. In fact, the managers of carbon-intensive, endless industrial growth, like the highest corporate officers in today's major fossil fuel companies, largely have signed on to advance the decarbonization of global economy. Through carbon taxation or CO_2 sequestration incentives, the research and development divisions of these firms are now producing plant-based bio-fuels, scrubbing greenhouse gas emissions off their books by pulling CO_2 out of their refinery operations to pipe it out to depleted gas fields to stabilize extracted cavities, and/or capturing more methane, once flared at well-heads, to mix with other contained methane stocks to sell as another "natural gas." See Timothy W. Luke, "Caring for the Low Carbon Self: The Government of Self and Others in the World as a Gas Greenhouse." *Towards a Cultural Politics of Climate Change, Devices, Desires, and Dissent*, eds. Harriett Bulkeley, Matthew Patterson, and Johannes Stripple (Cambridge: Cambridge University Press, 2016), 66–80.

[18] Severigne and Raphael Stevens, *How Everything Can Collapse*, 3.

survive with valued norms and behaviors."[19] The second domain is recognizing nothing can survive a general collapse. Consequently, it is crucial that the second change is to accept the necessity of *"'relinquishment.'* It involves people and communities letting go of certain assets, behaviors and beliefs where retaining them could make matters worse. Examples include withdrawing from coastlines, shutting down vulnerable industrial facilities, or giving up expectations for certain types of consumption."[20] Many of these assets, behaviors and beliefs became possible, more universal or widely practices thanks to fossil fuels. Without easy access to such energy by most people, these trade-offs are unavoidable and inevitable.

In their place, a third shift is toward the work of *"restoration"* may promise greater satisfaction, granted that it "involves people and communities rediscovering attitudes and approaches to life and organization that our hydrocarbon-fueled civilization eroded. Examples include re-wilding landscapes, so they provide more ecological benefits and require less management, changing diets back to match the seasons, rediscovering non-electronically powered forms of play, and increased community-level productivity and support."[21] Fossil fuels enabled expanded opportunities for more people to enjoy lifestyles centered on "having," while restrictions on energy usage instead point another direction toward life courses centered on ways of "being," like better diet, more leisure, access to wild nature, rich communal interaction or localized modes of production. Finally, the fourth change anchoring the vessel of "deep adaptation" is *"reconciliation"* to be pursued "in recognition of how we do not know whether our efforts will make a difference, while we also know that our situations will become more stressful and disruptive, ahead of the ultimate destination for us all. How we reconcile with each other and with the predicament we must now live with will be key to how we avoid creating more harm by acting from suppressed panic."[22] This shift toward a more virtuous balance of other-regarding and self-regarding activity by all members of a deeply adaptive society seems apropos. Learning to inhabit and connect these four closely linked value domains should prove mutually supportive for those navigating through the hazards of general societal collapse.

At the same time, in this 2018 manifesto, Bendell also concedes it is not for him to rise as some omniscient law-giving founder, who must "map out more specific implications of a deep adaptation agenda. Indeed, it is impossible to do so, and to attempt it would assume we are in a situation for calculated attempts at management, when what we face is a complex predicament beyond our control." In being out of control, it is his "hope the deep adaptation agenda of resilience, relinquishment and restoration can be a useful framework for

---

[19] Bendell, "Deep Adaptation.".
[20] Ibid.
[21] Ibid.
[22] Ibid.

community dialogue in the face of climate change."[23] Bendell seems aware the adoption and implementation of these values will vary in terms of personal wealth, status, power and capability.

Many environmental political programs and theories, however, are relatively inattentive to class, since authors often believe the imperative of "saving the Earth" will motivate almost everyone to change. In practice, however, the privileged and powerful few, or those often tagged as "Davos Man" and "Davos Woman," already have planned how to cope with rapid climate change and its impending societal collapse. Those plans of these few are surviving at the top together, while leaving "the many" to relinquish many of their life-chances. In navigating through any societal collapse, very few will consume gasoline. Similarly, not many will be able to resiliently accumulate wealth, but they might be free to restore themselves by living very simply out in the woods or makeshift urban encampments. And, they will likely be compelled to reconcile such conditions of life-long material precarity in a climate-changed world, because they number among "the many," who are the majority of the population under legal supervision of "the few" throughout years or decades of the collapse. Of course, "the few" constituting powerful minorities will not agree that "small is beautiful," at least for them. They also are neither willing to embrace the asceticism of "voluntary simplicity" nor likely to trod slowly down "soft energy paths."[24] Ironically, however, some network in these great and good few has enthusiastically feted Jem Bendell as a "Young Global Leader" for the World Economic Forum, because they regard his ideas as being well-worth propagating among the less powerful, privileged and prominent members of society they "legally supervise."

Very similar ethico-political options were debated in great detail during the oil crises in the 1970s by various counter-cultures as well as poorer mainstream communities well outside of major metropolitan centers. In the struggles to maintain their chosen alternative lifestyles for coping with necessity against aggressive neoliberal economics, however, they became millions of left-behind bystanders who got less and less as the years rushed along after the end of the Cold War. Getting ahead in the 1970s and 1980s required energy-intensive consumption to propagate the production of waste, which was, and still is, "the sign of success" for those willing to sacrifice their time and energy daily in richer countries devoted its "the winner to takes all" ethic.

---

[23] Ibid.

[24] For studies that arguably anticipated how to answer Bendell's call to make hard choices about resilience, relinquishment, restoration and reconciliation 50 years ago, see E.F. Schumacher's path-breaking 1973 study, *Small is Beautiful: A Study of Economics as if People Mattered* (New York: Harper Collins, 2010); Ivan Illich, *Tools for Conviviality* (New York: Harper and Row, 1973); Vernard Eller, *The Simple Life* (Nashville, TN: Abingdon Press, 1971); Duane Elgin, *Voluntary Simplicity* (New York: Harper, 1980); Amory B. Lovins, *Soft Energy Paths: Towards a Durable Peace* (Harmondsworth: Penguin Books, 1977).

That said, Bendell and his associates still believe that their provisional agenda for deep adaptation raises the most important questions for everyone to answer in their everyday lives as they begin to experience the current "long emergency."[25] Even though this crisis involves more than rapid climate change, its demands should prompt more citizens to address the four existential questions, which truly demand new answers for today, amid or before the coming collapse:

> Resilience asks us "how do we keep what we really want to keep?" Relinquishment asks us "what do we need to let go of in order to not make matters worse?" Restoration asks us "what can we bring back to help us with the coming difficulties and tragedies?" Reconciliation asks "with what and whom can we make peace with as we face our mutual mortality?[26]

These questions by Bendell's lights must frame both larger policy issues and more personal life choices.

When all is said and done, today's "deep adaptation agenda" is the work of maverick academics tired of the greenwashing corporate social responsibility curricula boosted by most business schools. Bendell's mission with followers from the Deep Adaptation Forum, Extinction Rebellion, the Green Party as well as other allies elsewhere, like Citizen Climate Lobbyists, Green New Dealers and the Sunrise Movement, is to work toward a joint realization of "shifts in being" and "shifts in doing." Like the green counterculture of the 1970s, Bendell's Deep Adaptation Forum maintains these cultural shifts will motivate people in the 2020s or 2030s, as currents of collapse rise, to explore climate psychology, alternative economics, relocalizing urban settlements on smaller scales, reimagining what schools teach students to do, and training morally-minded leaders to recruit more people to navigate through the collapse of national economies and global ecologies.[27] On the one hand, their political project is admirable; but, on the other hand, it also is a ready-made alternative package to persuade, or comfort millions of consumers that high industrial modernity's fossil fuel lifestyles will continue only for the elites.

Meanwhile, these millions must be convinced, educated or led to accept these green, lean or clean justifications for living in the "deeply adaptive" conditions of industrial stagnation, economic degrowth, intentional frugality and moral reconciliation with far fewer personal opportunities. Some ecological artists and climate refugees from "uncivilization" apparently are ready

---

[25] These dilemmas were discussed in considerable detail by James Howard Kunstler 15 years ago, see his The Long Emergency: *Surviving the End of Oil, Climate Change, and Other Converging Catastrophes of the Twenty-First Century* (New York: Grove Press, 2007).

[26] Bendell, "Deep Adaptation.".

[27] See Jem Bendell and Rupert Read, *Deep Adaptation: Navigating the Realities of Climate Chaos* (Cambridge: Polity Press, 2021) to follow more of the aspirations of this new environmental politics and theory circle as it works to put its programs into motion.

to wind their way to their own private "Dark Mountains" as the Earth keeps breaking bad. Others, however, will demand to stay more engaged in mainstream pursuits.[28] The "constructive collapsological" creeds of the Deep Adaptation Forum then appear to some citizens, clients and consumers of urban industrial civilization among the better alternatives for somehow surviving the Anthropocene, while avoiding "INTHE" moments with a greater sense of collective purpose and personal agency.

---

[28] Remarkably, Bendell maintains this acceptance of precarity during a general collapse will cheerfully be welcomed by many with whom he has crossed paths: "In my work with mature students, I have found that inviting them to consider collapse as inevitable, catastrophe as probable and extinction as possible, has not led to apathy or depression. Instead, in a supportive environment, where we have enjoyed community with each other, celebrating ancestors and enjoying nature before then looking at this information and possible framings for it, something positive happens. I have witnessed a shedding of concern for conforming to the status quo, and a new creativity about what to focus on going forward. Despite that, a certain discombobulation occurs and remains over time as one tries to find a way forward in a society where such perspectives are uncommon. Continued sharing about the implications as we transition our work and lives is valuable." See Bendell, "Deep Adaptation.".

Animal Citizens: Do We Need to Rethink the Status of Animals or Human Citizenship Itself?

Turquoise Samantha Simon

For a utopian interpretation of "animal citizenship" and an ecological regeneration of human citizenship.

Introduction

This chapter explores the unsettling proposal of "animal citizenship," an apparent contradiction in terms that is nevertheless a subject of study in political philosophy, such as the one presented by Will Kymlicka and Sue Donaldson in their work, *Zoopolis: A Political Theory of Animal Rights* (Kymlicka & Donaldson, 2011). Here, in order to assess the appropriateness of this idea through the lens of both dominant classical political thought and the societal—now consensual—objective of improving animal welfare, I take a dialectical approach to a number of avenues of reflection. First, I describe the way in which the concepts of citizenship, and of politics more broadly, have been constructed as being proper to humans, and thus as excluding

---

Translated by Hayley Wood

T. S. Simon (✉)
IRENEE (Institute for Nation and State Research), Université de Lorraine, Nancy, France
e-mail: samantha.simon@univ-lorraine.fr

animals. Second, I undertake a critical analysis of the argument that certain animals should be recognized as citizens, arguing that this is an impracticable yet fruitful proposal when considered for what it is, namely as an activist-led doctrine with a utopian vision. Finally, I propose a middle way between the absolute restriction of citizenship to humans and the need to change our relationship to living things: the idea of ecocitizenship, led by the concepts of deep ecology and responsibility.

The ways in which animal issues are represented or included in politics is an almost entirely foreign area of enquiry to French academia, but well-chartered territory in the Anglophone literature, to the extent that some authors describe a "political turn in animal ethics" (Garner & O'Sullivan, 2016) as having taken place since the 1990s. This shift in the field of animal studies over the last decade or so consists not only of asking questions about the moral status of animals, but also of considering how animals can be better taken into consideration in the political sphere (Cochrane, 2020, pp. 4–5). This new field thus aims to establish the conditions of possibility for a collective, institutional—rather than simply individual and moral—transformation of our relationship with animals. Here I consider one particularly startling proposal: the idea of animal citizenship.

The term itself has an immediately provocative—if not frankly oxymoronic—dimension, insofar as citizenship is an artificial status, constructed by humans, and giving rise to a privilege and status reserved not only solely for humans, but solely for some of them, as the battle over undocumented immigration rages on in the United States and Western Europe. What is more, from the point of view of moral and political philosophy, living in a political community or "*polis*" is itself one of the very qualities—moral, political, legal, and even geographic qualities—that separate humans from animals, conversely seen as natural. Citizenship and animality are thus two concepts that appear on the face of it to be incompatible as two mutually exclusive categories as part of the hegemonic Western anthropocentric worldview. Animals are not therefore excluded from citizenship, as may be the case for some human beings, but wholly unconcerned by it. And the existence of the systems of legal protection and political representation for animals that have been conceived of and implemented in several states does not disprove this idea, but rather confirms it. Belgium may have dedicated animal welfare ministers,[1] French law now recognizes animals as "sensitive living beings,"[2] an Argentinian court in Mendoza recently granted a female

---

[1] As Belgium is a federal state, and animal welfare is devolved at the regional level, the country has three animal welfare ministers: Bernard Clerfayt in Brussels (https://clerfayt.brussels/fr/competences/ministre-bien-etre-animal), Céline Tellier in Wallonia (https://tellier.wallonie.be/home/competences/bien-etre-animal.html) and Ben Weyts in Flanders (https://www.benweyts.be/wie-is-wie/ben-weyts).

[2] French Civil Code, Art. 515–14 (https://www.legifrance.gouv.fr/codes/article_lc/LEGIARTI000030250342/).

chimpanzee the status of a nonhuman legal person,[3] Germany's Basic Law states that animals must be protected[4] as our "co-creatures,"[5] and the Swiss Federal Constitution protects their "dignity"[6]—yet no country anywhere in the world has given animals citizen status. Should we therefore dismiss the idea out of hand as simply ludicrous? Perhaps not.

Issue and Framework

I believe that the proposal in fact deserves consideration for at least two reasons. First, in recent decades it has generated undeniable—albeit primarily academic—interest. Will Kymlicka, a Canadian political philosopher known internationally for his research on multiculturalism, was the first to defend the position, proposing not a form of near-citizenship, nor a merely symbolic form of citizenship, but to fully extend human citizenship and its prerogatives to certain animals. Kymlicka's academic credentials, the quality of his argument, but also the debates provoked by this idea (Pelluchon, 2019, pp. 122–136), invite us to take it seriously. But beyond its being a stimulus for intellectual debate, the idea of animal citizenship also has a value of a wholly different kind: even if the proposal is an initially disconcerting one, it deserves to be investigated because current animal welfare conditions raise questions that must be answered. The state of our relations with the world around us, but with animals in particular, is extremely concerning because it has resulted in both the gradual mechanization of livestock—to the point of making the Cartesian theory of the "animal machine"[7] (Descartes, 1991, p. 302 and p. 360) look

[3] Tercer Juzgado de Guarantias, Mendoza, November 3, 2016, file no. P-72.254/15, "Presented by A.F.A.D.A. about the chimpanzee Cecilia—Non-human individual": "JUDGMENT: I.-Grant the Habeas Corpus action (…) II.- Declare chimpanzee Cecilia, who lives in the Province of Mendoza zoo, a non-human legal person" (https://www.animallaw.info/sites/default/files/PRESENTED%20BY%20AFADA%20ABOUT%20THE%20CHIMPANZEE%20CECILIA%202016.pdf).

[4] Act amending the Basic Law (Gesetz zur Änderung des Grundgesetzes (Staatsziel Tierschutz), no. 53, July 31, 2002 (https://www.bgbl.de/xaver/bgbl/start.xav#__bgbl__%2F%2F*%5B%40attr_id%3D%27bgbl102s2862.pdf%27%5D__1642871940618);
and Grundgesetz, Art. 20a: "The state, also with responsibility for future generations, protects the natural foundations of life and animals within the framework of the constitutional order through legislation and, in accordance with the law and the law, through executive power and jurisdiction" (https://dejure.org/gesetze/GG/20a.html).

[5] Tierschutzgesetz (Animal Welfare Act), First Section §1: "The purpose of this law is to protect the life and well-being of the animal as a co-creature from the responsibility of man for the animal. No one may cause pain, suffering or harm to an animal without reasonable cause" (https://www.gesetze-im-internet.de/tierschg/BJNR012770972.html).

[6] Federal Constitution of the Swiss Confederation, Art. 120–2: "The Confederation shall legislate on the use of reproductive and genetic material from animals, plants and other organisms. In doing so, it shall take account of the dignity of living beings as well as the safety of human beings, animals and the environment, and shall protect the genetic diversity of animal and plant species" (https://www.fedlex.admin.ch/eli/cc/1999/404/en).

[7] Descartes (1991, p. 302), Letter to the Marquess of Newcastle, November 23, 1646: "I know that animals do many things better than we do, but this does not surprise me.

like a self-fulfilling prophecy—and the sixth mass extinction of wild animals. Were this not concerning enough, in the Anthropocene epoch, it is not just biodiversity that is under threat, but the survival of humanity itself. These issues have become embodied in the development of animal rights and ecology movements, which call for a transformation not only of our practices, but of the very foundations of our individual and collective relationship with animals. This disturbing state of affairs may provide an opportunity to interrogate the classical approach, which sees humans as the sole moral and political being, not only in opposition to animals, but also at their expense.

Tracing the background to, and reasons for, the germination of this idea may be straightforward, but the question that directly follows—how it might potentially be implemented—is much more problematic, and yet paradoxically rarely the subject of non-philosophical investigation. From a more legal standpoint, the status envisaged, of citizen, denotes a differential status (Denquin, 2003) that enables the individual who holds it to play an active role in political life via specific procedures. But recognition of a status means little if one cannot exercise the rights and duties associated with it. Citizenship thus involves the possession of certain qualities, such as the capacity to express one's opinions or a certain level of awareness of the general interest, and its acquisition is conditioned by criteria such as age and nationality. If even some humans are excluded from or deprived of citizenship, how can animals be granted such status?

Based on these initial considerations and questions, my approach will be guided by two principles: first, the resolution to *take this idea seriously*, no matter how startling it may be, and second, to *remain serious* by removing

---

It can even be used to prove they act naturally and mechanically, like a clock which tells the time better than our judgment does. Doubtless when the swallows come in spring, they operate like clocks. The actions of the honeybees are of the same nature, and the discipline of cranes in flight, and apes in fighting, if it is true that they keep discipline. Their instinct to bury their dead is no stranger than that of dogs and cats who scratch the earth for the purpose of burying their excrement; they hardly ever actually bury it, which shows that they act only by instinct and without thinking." See also Descartes (1991, p. 360), Letter to More, February 5, 1649: "But there is no prejudice to which we are all more accustomed from our earliest years than the belief that dumb animals think. Our only reason for this belief is the fact that we see that many of the organs of animals are not very different from ours in shape and movement. Since we believe that there is a single principle within us which causes these motions—namely the soul, which both moves the body and thinks—we do not doubt that some such soul is to be found in animals also. I came to realize, however, that there are two different principles causing our motions: one is purely mechanical and corporeal and depends solely on the force of the spirits and the construction of our organs and can be called the corporeal soul; the other is the incorporeal mind, the soul which I have defined as a thinking substance. Thereupon I investigated more carefully whether the motions of animals originated from both these principles or from one only. I soon saw clearly that they could all originate from the corporeal and mechanical principle, and I thenceforward regarded it as certain and established that we cannot at all prove the presence of a thinking soul in animals. I am not disturbed by the astuteness and cunning of dogs and foxes, or all the things which animals do for the sake of food, sex, and fear; I claim that I can easily explain the origin of all of them from the constitution of their organs."

my analysis from the activist-led—and sometimes, it must be agreed, unrealistic—enthusiasm of its proponents. The problem as it is presented here is thus the appropriateness of granting such a status to animals which, despite what recent advances in ethology have shown, still do not appear to possess the articulated language and capacity for abstraction required for true participation in political life. This leads us to the underlying question: *is citizen status the most effective way of giving animals greater political consideration?* My argument takes a three-stage dialectical approach. First, I reveal how the concept of citizenship has been constructed as being proper to humans, and thus to all intents and purposes as excluding animals. Second, I undertake a serious exploration of the proposed alternative of animal citizenship and identify its limitations. And finally, I sketch out the foundations for discussing a middle way that might resolve the contradiction between the classical approach and animal citizenship: the idea of ecocitizenship.

I. Citizenship as Proper to Humans

The concept and status of citizen were conceived of and constructed as being consubstantial with human nature (A). But the qualities proper to humans that were claimed to provide access to a specific moral, political and legal status have gradually been called into question (B).

A. *Citizenship as a Construction Consubstantial with Human Nature*

Western moral and political philosophy have been shaped by and for humans, being first founded in an ontological debate about the qualities proper to humans, and second taking a normative approach to deduce moral and political principles from them (to be discussed below). The theories of Aristotle, Kant, Rousseau and Hobbes may be very different, but all share this logic. As it would be impossible and unnecessary to go into the details of this process here, I will instead focus on the most telling and relevant elements with regard to my investigation into the construction of the relationship between humans, animals and politics.

While Antiquity did produce some dissident voices, including Pythagoras,[8] Plutarch[9] and Porphyry (2014), a consensus developed around the approach described and inspired by Aristotle, whose political thought took both an ontological and normative approach. In his treatise *On the Soul*, Aristotle does not entirely deprive animals of a soul, but makes a distinction between three different kinds: the nutritive soul of plants, the perceptive soul—which is combined with the first kind in animals, and primarily senses—and finally the rational or intellectual soul, which is combined with the first two in

---

[8] See the analysis of Empedocles and Pythagoras by Elisabeth de Fontenay (1998, pp. 84–85).

[9] "On Eating Meat" (Plutarch, 2021, pp. 128–168).

humans and enables knowledge (Aristotle, 2018, p. 26). Taking a teleological approach, all beings must be able to realize their nature. From this, Aristotle naturally deduces the famous statement in the *Politics* that "man is more of a political animal than any [...] gregarious animals" (Aristotle, 1996, p. 13), insofar as nature has given humans rationality which enables them to tell the just from the unjust and language, or, the means to express it. Linguists do not always agree on whether Aristotle was referring to a difference in degree or nature (Jaulin & Güremen, 2019, p. 8), but this has no impact on the fact that the *polis*—and its *agora*, a political gathering place in ancient Greece—remains the place where human beings realize themselves. In parallel, it becomes absurd to consider animals as being "political" or as citizens, since nature itself has deprived them of the necessary qualities to fulfill this role, like voting.

Later, and with very few exceptions, Christian and humanist thought came together in reaffirming this position. It was Kant who developed a substantive concept of dignity founded on the characteristics of the human species, proposing criteria modeled on the qualities proper to humans, namely reason,[10] autonomy[11] and freedom.[12] These elements enable humans to achieve their dignity by divesting themselves of nature and becoming its *end*,[13] thus elevating themselves above animality, which is confined to the status of a *means*.[14] From this stems the categorical imperative: "So act that you use humanity, whether in your own person or in the person of any other, always at the same time as an end, never merely as a means" (Kant, 1997, p. 38).

This moral idea, developed in the *Groundwork of the Metaphysics of Morals*, can be transposed to the political sphere through the idea of citizenship. Just as a human must always be treated as an end, and never merely as a means, the citizen both decides and obeys.

---

[10] Kant (1997, p. 37): "[The] human being and in general every rational being exists as an end in itself, not merely as a means to be used by this or that will at its discretion."

[11] According to Kant (1997, p. 42), thanks to reason, the human being "obeys no law other than that which he himself at the same time gives."

[12] Kant (1997, p. 57): "independence from the determining causes of the world of sense [...] is freedom. With the idea of freedom the concept of autonomy is now inseparably combined, and with the concept of autonomy the universal principle of morality, which in idea is the ground of all actions of rational beings."

[13] Kant (2009, p. 12): "Nature has willed that the human being should produce everything that goes beyond the mechanical arrangement of his animal existence entirely out of himself [...] For nature does nothing superfluous and is not wasteful in the use of means to its ends. Since it gave the human being reason, and the freedom of the will grounded on it, that was already a clear indication of its aim in regard to that endowment."

[14] Kant (2007, p. 167): "The fourth and last step that reason took in elevating the human being entirely above the society with animals was that he comprehended (however obscurely) that he was the genuine end of nature [...] he became aware of a prerogative that he had by nature over all animals, which he now no longer regarded as his fellow creatures, but rather as means and instruments given over to his will for the attainment of his discretionary aims."

A century after Hobbes (1991, p. 34), who believed that despite having many shared qualities, humans surpassed animals in their capacity for anticipation and potential for abstraction related to language, two conditions for the social contract, Rousseau (2013, pp. 36–37), in his *Discourse on the Origin of Inequality,* did not deny the sensitivity of animals, but considered only humans to possess natural characteristics such as rationality, freedom and perfectibility. Humans are thus the only living beings that can be a party to the social contract, the objective of which is to "find a form of association which will defend and protect, with the whole of its joint strength, the person and property of each associate, and under which each of them, uniting himself to all, will obey himself alone, and remain as free as before" (Rousseau, 2008, p. 55). There is thus a conceptual contemporaneity to the dual dimension of the citizen, since the citizen governs at the same time as being governed. From this point of view, citizenship is thus meant to enable humans to preserve their natural freedom within the *polis* by exercising their sovereignty with other citizens.

From a more legal standpoint, the cardinal characteristics of citizenship, as outlined by Massimo La Torre (2012) in Volume III of the *Traité International de Droit Constitutionnel,* establish conditions of exercise with which animals are clearly unable to comply. Citizenship is, in this view, participatory and libertarian. It supposes the ability to govern, and freely consenting to being governed. In each community, citizenship thus requires the ability to make decisions about the common good, through specific procedures such as elections. But citizenship also supposes the ability to present oneself at these elections, with suffrage determining its scope. While citizenship has had a tendency to expand in its scope (Denquin, 2003), it does not encompass the entire population of a state, and its effects remain conditional upon the possession of certain qualities.[15] Citizenship thus remains a differential status, even within humanity itself, and it goes without saying that animals who do not meet any of the conditions required to exercise it are excluded from it. Furthermore, citizenship can only be recognized between equals and makes people legally and politically equal. Yet even the most radical animal rights thinkers have not dared espouse, but merely brushed up against, what I consider to be the unlikely and dangerous idea of equality—whether real or notional—between humans and animals.[16]

---

[15] It should be noted, however, that many recent studies have revealed the limitations of such a system—notably for people with disabilities—and explored possible alternatives, for example by linking citizenship with vulnerability. See in particular Durand (2018).

[16] Peter Singer (1975), for example, opens his most famous work with the strikingly and provocatively titled chapter "All Animals are Equal," but in fact calls for neither equality in fact, nor in law, but an "equal consideration of interests": a very different proposition, since it admits that animals and humans do not have equivalent interests. Tom Regan (1983, p. 187), meanwhile, argues that humans and animals are both "subjects-of-a-life," which gives them "inherent value" and justifies their deserving of "rights." These are not, however, the same "rights." And finally, in his provocatively titled *Introduction to Animal Rights: Your Child or the Dog?* even Gary Francione (2000), perhaps the most radical

Looking back at this overview of how the idea of an exclusively interhuman moral and political community developed, we can see how the incompatibility between citizenship and animality is located in its very foundations, but also present in each of its characteristics, since neither equality, nor participation, nor conventionality, are possible with animals.

B. The Challenge to the Qualities Proper to Humans

These approaches have created a clear distinction between humans—as moral, political beings—and animals—as natural, gregarious beings. They have however come in for criticism of two kinds.

First, critics have challenged the very principle of seeking to identify the qualities proper to humans. They point out the porous, evolving aspect of these qualities, since the criteria of distinction proposed are constantly called into question, notably following new discoveries in ethology, such as animal consciousness[17] and the political behavior of great apes (De Waal, 1998). Each time an animal is found to possess a quality that was thought to be the sole preserve of human beings, the question is revisited in order to identify the one that will this time steadfastly endure. Some authors, such as Elisabeth de Fontenay, have even highlighted the rather laughable, or amusing, aspect of this quest. Without herself questioning either the existence, or the necessity, of a distinction between humanity and animality, she writes, in the deliciously titled work *Sans offenser le genre humain* [*Without Offending Humans*], "when one thinks of the succession of age-old, irrefutable signs of anthropological difference, and observes the retreat from the attack by advances in life sciences on the sacrosanct human difference, one cannot help but laugh" (De Fontenay, 2008, p. 48). The humanist approach, based on the distinction between humanity and animality and the superiority of the former over the latter, has also been challenged on the basis of its shortcomings, since it has not only in practice failed to prevent the very worst crimes against humanity but may even partly spring from a similar impulse of disregard and supremacism. Critics of this kind include Jacques Derrida (2008, p. 104), who developed the idea of carnophallogocentrism, i.e., the similarity between the domination of men over women and of humans over animals, and Claude Lévi-Strauss (1973, p. 53), for whom humanism was corrupted by the fact

---

animal rights thinker of all, nevertheless firmly states the need to privilege humans over animals.

[17] See the Cambridge Declaration on Consciousness (July 7, 2012): "The absence of a neocortex does not appear to preclude an organism from experiencing affective states. Convergent evidence indicates that non-human animals have the neuroanatomical, neurochemical, and neurophysiological substrates of conscious states along with the capacity to exhibit intentional behaviors. Consequently, the weight of evidence indicates that humans are not unique in possessing the neurological substrates that generate consciousness. Nonhuman animals, including all mammals and birds, and many other creatures, including octopuses, also possess these neurological substrates." (http://fcmconference.org/img/CambridgeDeclarationOnConsciousness.pdf).

that the boundary used to separate men from animals had been transposed within humanity itself. Put simply, he argues that the distinction and hierarchy between humans and animals, men and women, and whites and blacks originate in the same impulse of false claims of domination or superiority.

Finally, the arbitrary, performative and self-proclaimed nature of this quest was criticized or even mocked by Nietzsche (1913, p. 194), who parodied Kant with his provocative suggestion that "possibly the ant in the forest is quite as firmly convinced that it is the aim and purpose of the existence of the forest." The criteria of moral, legal, political and even religious consideration are decreed by human beings, for human beings, and on the basis of what human beings are. In this view, an entire branch of moral and political philosophy that considers humanity, citizenship and politics is in fact based on a kind of solipsism, i.e., the attitude of the thinking subject for whom his or her own consciousness is the sole reality and other consciousnesses and the external world are merely representations of the subjects' brain.

Some critics have also focused on the conception of animality proposed which some see as the origin of its abuse. They include Derrida, who takes aim at the term "animal," which is now used to denote all animals except humans, and thus imposes, perhaps artificially, a mask of uniformity across an astounding diversity of over 1,500,000 species, putting great apes on a par with sponges.[18] Furthermore, such an—artificial—distinction cannot be made without hierarchization or inequality. The distinction between human beings and animals is made in favor of the former, and it is because humans and animals are seen as being of a different nature that the former are able to justify the right to exploit the latter. This is one of the key problems for anti-speciesists,[19] who argue that all major philosophical enterprises aiming to identify the qualities proper to humans have been based on the quest for an

---

[18] Derrida (2008, p. 34): "Confined within this catch-all concept, within this vast encampment of the animal, in this general singular, within the strict enclosure of this definite article ('the Animal' and not animals'), as in a virgin forest, a zoo, a hunting or fishing ground, a paddock or an abattoir, a space of domestication, are *all the living things* that man does not recognize as his fellows, his neighbors, or his brothers. And that is so in spite of the infinite space that separates the lizard from the dog, the protozoon from the dolphin, the shark from the lamb, the parrot from the chimpanzee, the camel from the eagle, the squirrel from the tiger (...)."

[19] Richard Ryder coined the term "speciesism," but it was constructed, defined and popularized as a concept by Peter Singer (1975) in *Animal Liberation*. According to Singer, speciesism is "a prejudice or attitude of bias in favour of the interests of members of one's own species and against those of members of other species" (p. 6). He adds: "Racists violate the principle of equality by giving greater weight to the interests of members of their own race when there is a clash between their interests and the interests of those of another race. Sexists violate the principle of equality by favouring the interests of their own sex. Similarly, speciesists allow the interests of their own species to override the greater interests of members of other species. The pattern is identical in each case" (p. 9). Singer considers sentience to be the only relevant criterion of moral consideration (pp. 7–8) and proposes applying the principle of equal consideration of interests to animals (p. 5).

uncontestable distinction between humanity and animality and more specifically on an identifiable trait that human beings, unlike animals, might possess. *As such, animals have been thought of not in relation to what they are, but in relation to what they are not.* In addition, animals have been seen not only as failed prototypes of humans but also as a biological and symbolic reserve for use by humans. The principle of seeking to identify the qualities proper to humans as a basis for political theory has also been criticized on the grounds that its result, but perhaps also its objective, has been the potential to make animals, and other living things, an available resource (Waldau, 2013, p. 99).

These traditional moral and political constructions have therefore been identified as obstacles to greater consideration of the interests of animals by our moral, political, economic and legal systems, and have led some authors to develop alternative proposals, including the idea of animal citizenship. Such authors fall into two groups. Those in the first group want to extend the human moral and political community to animals, taking a particular characteristic shared by these two categories of living beings as a relevant criterion for doing so. The most successful of the proposed criteria has been the concept of sentience, i.e., the "capacity to suffer and/or experience enjoyment" (Singer, 1975, p. 8). This approach thus involves separating animals (and their protection) from the idea of the environment (and its protection) and integrating them into the political, legal and moral sphere as individuals with rights. It does not therefore deal with the domination of humans over nature, but simply takes animals out of the category of exploitable living beings. The second group consists of ecocentrists. In line with the deep ecology philosophy developed by Arne Naess, they aim to move beyond solely thinking in terms of the qualities proper to humans and take a holistic approach to the interconnections and links between different living entities within an ecosystem.[20] In this approach, all living beings should be respected not because of a shared quality with humanity, such as sentience, but simply because of what they are, and because of what they bring to the global harmony of species.

---

[20] I refer the reader to the deep ecology philosophy conceived of by Norwegian philosopher Arne Naess. See in particular Naess (1989). According to his seminal early article "The shallow and the deep, long-range ecology movement," the shallow ecology movement is concerned with the "[fight] against pollution and resource depletion. [Its] central objective [is] the health and affluence of people in the developed countries" (Naess, 1973, p. 95). The deep ecology movement, on the other hand, "has deeper concerns which touch upon principles of diversity, complexity, autonomy, decentralization, symbiosis, egalitarianism and classlessness" (p. 95). To put it another way, shallow ecology consists of anthropocentric care for the environment solely as a resource, while deep ecology is holistic and ecocentric and seeks to connect with the nonhuman living world and to link human, animal and environmental issues in an ecological theory and a movement.

II. Animal Citizenship as a Utopia

The theory of animal citizenship developed by Will Kymlicka is based on the first, anti-speciesist, of these logics. Here I present a summary (A) and a critique (B) of his work.

A. Overview

In *Zoopolis,* published in 2011, Canadian philosophers Will Kymlicka and Sue Donaldson consider the political status of animals. Since Kymlicka himself has noted and explored the link between the two areas (Kymlicka & Donaldson, 2014), I will take his lead in observing that his background and interest in multicultural citizenship provide the foundations for his work on animal rights. In his earlier work, Kymlicka challenges the fact that, in multicultural societies, the view of the dominant ethnic group is imposed on others, and argues that growing ethnic diversity in Western societies requires new and more inclusive forms of citizenship and national identity (Banting & Kymlicka, 2005; Kymlicka & Banting, 2006). In particular, he advocates compensation mechanisms and various forms of protective status based on the specific characteristics and needs of the minorities in question.

His later theory of the political status of animals, developed with Sue Donaldson, seems in places to echo these ideas, which lead to the notion of animal citizenship as follows. First, the two authors challenge the conventional view developed by political theory, criticizing it for being expressed in rationalist, intellectual terms, and arguing that it has failed to succeed in establishing the acquisition of a protective status, not only for animals, but also for vulnerable humans (Kymlicka & Donaldson, 2011, p. 32). They then replace the foundation of rationality with that of selfhood (p. 32)[21] and pose the protection of vulnerability as the objective of a new political theory (p. 33).[22] The question is thus no longer how to establish a *polis* as close as possible to human reason, but how to establish a *polis* that recognizes and protects the most vulnerable beings.

Second, and naturally drawing on the anti-speciesist theories developed by Peter Singer and Tom Regan, Kymlicka and Donaldson propose sentience and the fact of leading a subjective existence as the sole but necessary criterion for attributing what they call "inviolable rights" to animals (p. 19). To put it clearly, in the view of Will Kymlicka and Sue Donaldson, all animals (including humans), but only animals (as Singer also restricts his theory to one that omits

---

[21] "This then is our basic starting point: like many other AR theorists, we defend inviolable rights for animals as a response to the vulnerability of selfhood or individual consciousness."

[22] "Our theory is not one based on any account of the essence of being human, any more than it is based on the essence of, say, being a dog. Our theory is instead based on an account of one of the key purposes of justice, which is the protection of vulnerable individuals."

living non-sentient beings, namely plants) should enjoy these inviolable rights. The two authors demand nothing short of a prohibition on harming, killing, imprisoning, possessing and enslaving the living beings in question (p. 49), clearly amounting to widespread adoption of a vegan lifestyle. They also criticize the classical approach for denying animals agency, i.e., the capacity to demonstrate their preferences (pp. 57–58).

But these "universal negative rights" are not enough. Kymlicka and Donaldson want to combine them with "positive relational rights" (p. 12). They advocate a society that can learn to decipher and take into account these preferences, but also use them as the basis for structuring their own political theory by setting out three forms of status that are dependent not on belonging to a particular species nor on the interest of their exploitation by humans, but on the animal's behavior in relation to humans. The rights and duties associated with each status are added to (rather than contradict) the inviolable rights. First, they consider wild animals, i.e., those that live in a natural environment without demonstrating a desire to interact with humans. In Kymlicka and Donaldson's view, these animals should be granted a form of *sovereignty*. In addition to their "inviolable rights," their territory and autonomy should also be respected. This status creates prohibitions for human beings, but not obligations to those nonhuman animals. It does not require an end to all interaction with human beings, but limits this interaction to individual acts of assistance to injured animals, esthetic appreciation of their territory and reasonable specimens of natural produce (fruits, mushrooms) (pp. 156–209).

The two authors then consider the case of "liminal" animals—wild animals that still live in contact with human beings, such as urban pigeons and rats—which they argue should be given *denizenship*. They should not receive specific protection, beyond their inviolable rights, but their population size may however be controlled non-violently using contraceptive grain (pp. 210–251).

Finally, Kymlicka and Donaldson consider domestic animals, i.e., those that show a desire to live in relation to humans, or dependent on them. They assert that these animals have the ability to express preferences, and that we need to listen to and take these into consideration and to take seriously the capacity of animals to understand the rules of living in what they call "human-animal society." Based on this principle, they propose granting these animals *citizenship* which establishes, in addition to the inviolable rights discussed above, several obligations on the part of humans, but also several "obligations"—the term used by Kymlicka and Donaldson—on the part of animals. This form of citizenship encompasses nine measures: ensuring basic socialization, freedom of movement and sharing the public space, a duty of protection, the possibility of using animal products and labor without exploitation, a duty of care, ensuring a sexual and reproductive life, an obligation to provide a healthy diet and finally, most relevant to us, ensuring political representation across all institutions, through the translation, interpretation and expression of their interests by competent, skilled persons (p. 153).

B. Assessment

This approach is as refreshing as it is disorienting but presents several major problems. First, it overlooks the fact that citizenship is useful only to those able to take up the rights and obligations that stem from it. Paradoxically, focusing on the agency of animals tends to overlook their animality. It is by no means a sign of disrespect to point out that while animals can demonstrate agency, they cannot demonstrate *political* agency. Supposing the contrary amounts, in fact, to anthropomorphism since politics involves an ability for abstraction and the capacity to express it that animals do not seem to possess, at least not sufficiently. While I do not believe that animals are devoid of autonomy, consciousness or the capacity to communicate their preferences, political participation is more than simply communicating one's individual interests—and even our collective interests. As Rousseau (2008, p. 66) describes, it consists of expressing the general will, which is different from "the sum total of individual wants"—something animals are unable to do. And even if they did express a "political will," it would need to be translated, interpreted and represented as is the case with children. In my view, this process is a major practical obstacle that prevents animal citizenship from being a credible proposal. Furthermore, the fact that any such political will would need to be translated into a human medium in order to be expressed ultimately merely confirms the practical impossibility of recognizing citizen status for any entities other than human beings.

Second, stretching the bounds of citizenship to such an extent arguably strips this differential status of all effectiveness. Citizenship must remain a privilege, separate from human, and even more so animal, rights. A differential status loses its effectiveness and its value once it is applied in too broad a manner. In France and Europe, citizenship is often closely linked to nationality: so much so that although the idea of European citizenship—and nationality—has been mooted, residency alone usually does not produce a right to vote.[23] While the concept of multicultural citizenship has been proposed (Kymlicka, 1995) and critiqued (Joppke, 2001), the European nations, unlike certain states in the United States and Canada, generally restrict voting rights to national citizens, and in some cases to citizens of the European Union. Furthermore, citizenship provides more than simply the right to vote as it also enables individuals to stand for election. Could an animal or child stand for election? Clearly not.

Third and finally, by challenging concepts as fundamental as citizenship and sovereignty, in a world in which human rights are still, and increasingly so, under threat, the authors of *Zoopolis* are in danger of undermining the credibility of their objectives, no matter how commendable these may be. If the objective is to offer animals a "political voice," is it essential to propose an idea as controversial as citizenship? Let us not forget that, legally speaking,

---

[23] See https://www.vie-publique.fr/fiches/23928-les-etrangers-ont-ils-le-droit-de-vote.

animals are still goods in most countries. While animal welfare and rights have become a legitimate political issue in recent years, we are a long way off support for an attempted societal and political transformation that could lead to widespread adoption of a vegan lifestyle. Let me be clear here: this perspective is surely an enviable future for animals, for the environment, and for human beings. We could and should certainly hope for a moral, legal and political transformation of our relationship with animals. But we live in democratic systems in which these desirable transformations and their consequences are subject to consensus approval, or at least majority approval, and while animal welfare appears to be an increasingly important political issue, it is not seen as a priority. Therefore, the idea of "animal citizenship" is as challenging as it is interesting, but its implementation and consequences raise important and unresolved questions.

While Kymlicka and Donaldson present their argument in an extremely methodical and rigorous way, it is perhaps not from a practical and purely realistic and rational angle that we should read *Zoopolis*, at least not at present. I would argue instead that a great deal can be gained from a utopian reading of the book. The polysemic word "utopia" resists any oversimplifying definition (Riot-Sarcey, 2006), but analysis of the etymology of this neologism coined by Thomas More (2012) in the sixteenth century reveals an essential dimension of utopia as a place that is both ideal and unreal.[24] The definition proposed by Alberto Tenenti (1996, p. 832) appears to be consistent with this idea, since he simply defines the word as "a set of intellectual products that, in forms that seek to be exemplary, express societal aspirations or dreams that are either realized and perfect, or yet to come but radiantly depicted." The relationship between utopia and reality is not only one of distancing. Utopias are an "evasion of reality with a desire to influence it" (Tenenti, 1996, p. 833). How can we interpret Kymlicka and Donaldson's utopia in a way that influences reality? The authors of *Zoopolis* have laid a groundwork that, while not necessarily practicable, is still inspiring. The relational approach and the ideas of sovereignty, denizenship and citizenship that it produces may provide not only the symbolic reserve of a moral change with regard to animals, but also a source of inspiration for a new approach to political theory. Perhaps the human exception and the moral and political status that stems from it are in truth not only the *sole*, but the *best* possible frameworks for imagining better consideration of animals in politics. Thus, even if the idea of animal citizenship currently leads to a dead end, it may inspire us not to rethink the status of animals, but to rethink the role of human citizens in relation to them.

---

[24] As the prefix "u" comes from the Greek "eu" meaning "good" or "ou," which is a negation, utopia can either mean the "good place," an ideal place, or an unreal place that does not exist.

III. Ecocitizenship as a Middle Way

A number of stimulating proposals have recently emerged for considering how to integrate animal issues, and ecological issues more broadly, into politics. Here I will analyze three complementary strands—or levels—that I believe to be of particular relevance: collectively rethinking the current understanding of the social contract as exclusively between humans (A), envisaging the renewed role of citizens within it (B) and investigating how true animal representation could work (C).

A. Rethinking the Social Contract

As the survival of humanity depends on its surroundings, and since animals and ecosystems are worthy of protection not solely for the utility they represent, but also for their own sake, some authors have proposed broadening out the social contract into a contract with living things (Serres, 1995; Pelluchon, 2019). The fact that nonhuman living things, children or people with disabilities cannot enter into a contract, or express themselves regarding the general will, in no way means they should be excluded from the sphere of this contract and cannot benefit from it.

Proposals of this kind include the work of two French authors, which I will present only in brief for reasons of concision and relevance. The first is Michel Serres (1995), whose book *The Natural Contract* was originally published in 1990. This rather provocative work highlights the fatal dimension of a social contract solely between humans and calls on us to listen to the language expressed by the Earth, which owes its survival to humans just as humans owe their survival to it in turn. Calling humans "parasites" and rejecting the highly anthropocentric term "environment,"[25] Serres invites humans to enter into a true contract of symbiosis and reciprocity (p. 38) with the natural world, replacing politics with what he calls "physiopolitics" (p. 44). More recently, in her book *Nourishment: A Philosophy of the Political Body*, Corine Pelluchon (2019) sets out the foundations for a renewed social contract that takes living things into account without making them parties to it, and is not based on the requirement of reciprocity. This new contract does not overturn previous theories but is inspired by, and includes, future generations and living things in its definition of the common good.

These two thought-provoking proposals stand out amid an extensive literature in this area, but remain no less symbolic, since their translation into the legal and political spheres has not been considered in any depth. Before

---

[25] Serres (1995, p. 33): "So forget the word *environment*, commonly used in this context. It assumes that we humans are at the center of a system of nature. This idea recalls a bygone era, when the Earth [...], placed in the center of the world, reflected our narcissism, the humanism that makes of us the exact midpoint of excellent culmination of all things. No. [...] Thus we must indeed place things in the center and us at the periphery, or better still, things all around and us within them like parasites."

looking briefly at how such a social contract might be applied and introduced in practice, let us first look at how a new approach to citizenship might involve this inclusion of living things in the common good.

B. *From Free Citizens to Responsible Ecocitizens*

The decision to adopt an ecocentric position—moving away from a sole focus on animal issues—is an important one. In my view, a new approach to citizenship must be holistic, systemic and ecological in the full sense of the word. As such, a half-hearted "environmental" form of ecology is inadequate as it aims only to protect the environment of humans for their benefit and assigns no inherent value to each living entity. But sentientist anti-speciesism, which calls for the extension of peculiarly human rights and duties to animals alone, is also inadequate because, following a logic close to the one that it criticizes, it is based on a re-evaluation of the status of animality through a disregard for plant life because of the latter not being sentient. This is why I believe the most appropriate perspective here is an ecocentric one, as it is not only compatible with, but complementary to, a true individual and systemic consideration of animals. At the beginning of this chapter, we saw how the foundation, and objective, of the theory of citizenship is to organize the *polis* in a way that respects the freedom and autonomy of humans. As these qualities cannot be denied, politics is conceived of by humans, and while not all humans are citizens, only humans can be citizens as presently conceived of. But this does not mean that we cannot broaden out this idea of citizenship, and in particular the duties associated with it. The development of zoonotic diseases and the public health impact of pollution are both alarm bells warning us that the survival of human beings depends on ecosystems being in good health. While humans can only realize their nature in the *polis*, they can only survive if the *polis* has a healthy relationship with the surroundings from which it draws its subsistence. The freedom and autonomy that set humans apart from other animals do not, therefore, take away one of the natural characteristics that connects them to all other living creatures: their vulnerability or dependence on the environment.

As a result, the fact that humans are the sole beings in possession of a reason that is able to conceive of the general will and express it, can be used not to establish a right to a relationship of domination, exploitation and plundering, but to establish a relationship of responsibility—understood here in the sense of Hans Jonas—toward living things. According to Jonas, human power over nature does not bring humanity a right to exploit it but rather a responsibility toward it:

> Power conjoined with reason carries responsibility with it. This was always self-understood in regard to the intrahuman sphere. What is not yet fully understood is the novel expansion of responsibility to the condition of the biosphere and the future survival of mankind, which follows simply from the extension of power over these things—and from its being eminently a power of destruction. Power

and peril reveal a duty which, through the commanding solidarity with the rest of the animate world, extends from our being to that of the whole, regardless of our consent. (Jonas, 1984, pp. 138–139)

The concept of responsibility is also useful because it concerns both responsibility toward the past and toward the future (Jonas, 1984, pp. 90–93).

Following Jonas' idea, ecocitizens, who recognize their vulnerability and responsibility in addition to their freedom and autonomy, will therefore need to both literally respond to the past actions of humanity and bear the burden of preserving the future of life on Earth. The concept of "ecocitizenship" could therefore indicate a reconnection between humanity and animality: indeed, the prefix "eco" comes from the Greek "oikos," meaning house, inviting us to think of humans, the political community and living things as one whole living together on one planet as their home. This idea enables us not only to preserve and reaffirm human privilege, but also to make the human exception the foundation not of a right to subjugate nature, but of obligations that, even if they are solely between human beings, form a framework for the protection of living things for their own sake. Jonas (1984, p. 137) points to this in his claim that "one [...] may say that the common destiny of man and nature, newly discovered in the common danger, makes us rediscover nature's own dignity and commands us to care for her integrity over and above the utilitarian aspect."

C. Representing Nonhuman Living Things in Politics

How can nonhuman living things be represented politically? Bruno Latour already incorporates this idea into his definition of politics, by making a distinction between the usual meaning of the word as the "struggle and compromises between interests and human passions, in a realm separate from the preoccupations of nonhumans" (Latour, 2004, p. 247), which he calls the "politics of the Cave," and its "proper sense," i.e., the "progressive composition of the common world" (p. 47). More specifically, Latour, who asks how we can introduce "concern for nature" into political life" (p. 2) in many ways, also seems to offer an important perspective on the substantive approach to take in relation to animals. He begins from the observation that in the geological era of the Anthropocene, many of the enduring dualisms between human and nonhuman, subject and object, and culture and nature have become obsolete: "Where we were dealing earlier with a 'natural' phenomenon, at every point now we meet the 'Anthropos' [...] and, wherever we follow human footprints, we discover modes of relating to things that had formerly been located in the field of nature" (Latour, 2017, p. 120). According to Latour, this *interconnection* between "nature" and "culture," or "human" and "nonhuman" (Latour, 2004, p. 61), means that we must no longer think of these notions in terms of opposition or reconciliation, but as a "pairing" (p. 61), by uniting them under the concept of "world," or "worlding" (Latour, 2017,

p. 35). Latour argues that these outdated dualisms have poisoned democratic debate insofar as they have arbitrarily separated the human world, endowed with a soul and intentionality, from the nonhuman world, considered to be inert (Latour, 2017, p. 67), and that by "defending the rights of the human subject to speak and to be the sole speaker, one does not establish democracy; one makes it increasingly more impracticable every day" (Latour, 2004, p. 69).

Drawing on the work of Michel Serres, James Lovelock and Donna Haraway, Latour rejects this idea by arguing that nonhuman elements are the opposite of an inert force, and may, like the Mississippi, possess an extremely powerful "agency" (Latour, 2017, p. 52). He thus seeks to establish the concept of *political agency*, in order to "stop taking nonhumans as objects" and make them into social actors (Latour, 2004, p. 76), doing so not only without challenging—but in fact enshrining—*logos* as the foundation and *sine qua non* condition of politics, by enabling living nonhumans to speak through a "spokesperson."

In concrete terms, what form might this representation of living things in politics take? Looking beyond the rapid development—particularly in Europe—of animal rights parties (Lucardie, 2020), i.e., political parties that are exclusively or primarily dedicated to protecting animal welfare,[26] a more enduring and certain form of representation might be considered. While Alasdair Cochrane (2020, pp. 101–103) proposes requiring existing decision-making bodies to include dedicated animal rights representatives, others have argued for a "third chamber" entirely dedicated to the protection of animals and ecosystems. Aymeric Caron (2016, pp. 423–450) has suggested a "Committee of the Living,"[27] while Corine Pelluchon has built on—while departing from—the work of Dominique Bourg (2017, p. 75) and Pierre Rosanvallon[28] to propose a body called the "Assembly of Nature and Living Beings," which would have a right to veto and whose representatives would be randomly selected or elected from people with proven competence and commitment to the environment and animals (Pelluchon, 2017, p. 274), a condition also proposed by Cochrane.[29] It is thus through the institutionalization and representation of the interrelationship[30] that we might see the emergence of what

---

[26] For example, the Dutch Party for the Animals (Partij voor de Dieren), the French Animal Rights Party (Parti Animaliste), the German Animal Protection Party (Tierschutzpartei), the Italian Animal Rights Party (Partido Animalista Italiano), the Spanish Party Against Mistreatment of Animals (Partido Animalista Contra el Maltrato Animal) and the Animal Welfare Party in the UK.

[27] Caron sees biodemocracy as involving entrusting power to experts. See also Caron (2017, p. 207).

[28] Pelluchon supports the idea of a third chamber but calls it something different: Bourg and Rosanvallon use the term "Long-Term Assembly," while she proposes an "Assembly of Nature and Living Beings" (Pelluchon 2019, p. 273).

[29] Cochrane (2020, p. 108).

[30] A term used by legal scholar Judith Rochfeld (2019, p. 163 and p. 170).

Serge Audier (2020) calls "ecorepublicanism." And this does not require us to legally recognize certain animals as having citizenship, denizenship or sovereignty.

Conclusion

Looking back on the foundations of citizenship, I have thus shown that it is consubstantially linked to humanity, giving proposals of granting citizenship to animals, or even living things in general,[31] a primarily symbolic or utopian value. Yet the fact remains that the Anthropocene, declining animal welfare, and the destruction of ecosystems more broadly call for an urgent and profound transformation of our relationship with nonhuman living things. In my view, drawing on the work of Hans Jonas (1984, p. 138), who reminds us that "power conjoined with reason carries responsibility with it," the most achievable and most fruitful path lies through a reinterpretation—rather than a negation—of the particular role of humans. The theoretical and practical transformation of citizens into true, responsible ecocitizens aims to restore living things their rightful place, from which they have been unfairly and illegitimately rejected and removed. Deep ecology, which provides a way to resolve the contradiction between an ecosystemic approach and an individualistic approach, reconciles the idea of the interrelationship between living things with the inherent value of each of its entities. As such, it enables us to break away from the dualist, binary, anthropocentric, human supremacist thinking that for too long has, in both word and deed, created a separation between things that are in fact inextricably linked and interdependent: humans and nonhumans, and by extension the "natural" and the "political." This symbiosis is within reach, dependent on humans now owning their particular qualities and translating this deep interrelationship into politics. May the ecocitizen be the change they want to see! We have a world to win by doing so.

References

Audier, S. (2020). *La cité écologique. Pour un éco-républicanisme*. Editions La Découverte.
Aristotle. (2018). *On the soul and other psychological works* (F. D. Miller, Jr., Trans.). Oxford University Press (Original work 4th century BCE).
Aristotle. (1996). "The Politics" (B. Jowett, Trans.). In S. Everson (Ed.), *The Politics, and the Constitution of Athens*. Cambridge University Press (Original work 4th century BCE).
Banting, K., & Kymlicka, W. (2005). "Les politiques de multiculturalisme nuisent-elles à l'État-providence?" *Lien social et Politiques*, 53, 119–127. https://doi.org/10.7202/011650ar
Bourg, D. (2017). *Inventer la démocratie du XXIème siècle. L'Assemblée citoyenne du futur*. Les Liens qui Libèrent.

---

[31] For a critique of this idea, see Ost (2003, pp. 164–172).

Caron, A. (2016). *Antispéciste*. Editions Don Quichotte.
Caron, A. (2017). *Utopia XXI*. Flammarion.
Cochrane, A. (2020). *Should Animals have Political Rights?* Polity Press.
De Fontenay, E. (1998). *Le Silence des bêtes: La philosophie à l'épreuve de l'animalité*. Fayard.
De Fontenay, E. (2008). *Sans offenser le genre humain. Réflexions sur la cause animale*. Albin Michel.
De Waal, F. (1998). *Chimpanzee Politics: Power and Sex among Apes*. Johns Hopkins University Press (Original work published 1982).
Denquin, J.-M. (2003). "Citoyenneté." In S. Rials & D. Alland (Eds.), *Dictionnaire de culture juridique* (p. 200). PUF.
Derrida, J. (2008). *The Animal that Therefore I Am* (D. Wills, Trans.). Fordham University Press (Original work published 2006).
Descartes, R. (1991). *The philosophical writings of Descartes, Volume III: The correspondence* (J. Cottingham, A. Kenny, D. Murdoch, & R. Stoothoff, Eds.). Cambridge University Press.
Durand, B. (2018). "Citoyenneté, vulnérabilité et handicap." *Pratiques en santé mentale*, 64, 25–30. https://doi.org/10.3917/psm.182.0025
Francione, G. (2000). *Introduction to Animal Rights: Your Child or the Dog?* Temple University Press.
Garner, R., & O'Sullivan, S. (2016). *The Political Turn in Animal Ethics*. Rowman & Littlefield.
Hobbes, T. (1991). *Leviathan* (R. Tucker, Ed.). Cambridge University Press (Original work published 1651).
Jaulin A., & Güremen R. (2019). *Aristote. L'animal politique*.
Jonas, H. (1984). *The Imperative of Responsibility* (D. Herr, Trans.). The University of Chicago Press (Original work published 1979).
Joppke, C. (2001). "Multicultural Citizenship: A Critique." *European Journal of Sociology / Archives Européennes de Sociologie / Europäisches Archiv Für Soziologie*, 42(2), 431–447.
Kant, E. (2007). "Conjectural Beginning of Human History." In G. Zöller & R. B. Louden (Eds.), *The Cambridge edition of the works of Emmanuel Kant: Anthropology, history, and education*. Cambridge University Press (Original work published 1755).
Kant, E. (1997). *Groundwork of the metaphysics of morals* (M. Gregor, Trans.). Oxford University Press (Original work published 1785).
Kant, E (2009). "Idea for a Universal History with a Cosmopolitan Aim" (A. Wood, Trans.). In A. O. Rorty & J. Schmidt (Eds.), *Kant's idea for a universal history with a cosmopolitan aim* (pp. 9–23). Cambridge University Press (Original work published 1784).
Kymlicka, W. (1995). *Multicultural citizenship: A liberal theory of minority rights*. Oxford University Press.
Kymlicka, W., & Banting, K. (2006). "Immigration, Multiculturalism, and the Welfare State." *Ethics & International Affairs*, 20(3), 281–304.
Kymlicka, W., & Donaldson, S. (2011). *Zoopolis: A political theory of animal rights*. Oxford University Press.
Kymlicka, W., & Donaldson, S. (2014). "Animal Rights, Multiculturalism, and the Left." *Journal of Social Philosophy*, 45(1), 116–135. https://doi.org/10.1111/josp.12047

La Torre, M. (2012). "Citoyenneté." In M. Troper & D. Chagnollaud (Eds.), *Traité international de droit constitutionnel. Suprématie de la Constitution*, Vol. 3 (pp. 258–374). Dalloz.

Latour, B. (2004). *Politics of nature: How to bring the sciences into democracy* (C. Porter, Trans.). Harvard University Press (Original work published 1999).

Latour, B. (2017). *Facing gaia: Eight lectures on the new climactic regime* (C. Porter, Trans.). Polity Press (Original work published 2015).

Lévi Strauss, C. (1973). *Anthropologie structurale*, II. Plon.

Lucardie, P. (2020). "Animalism: A Nascent Ideology? Exploring the Ideas of Animal Advocacy Parties." *Journal of Political Ideologies*, 25(2), 212–227.

More, T. (2012). *Utopia* (D. Baker-Smith, Trans.). Penguin (Original work published 1516).

Naess, A. (1973). "The Shallow and the Deep, Long-Range Ecology Movement." *Inquiry*, 16, 95–100.

Naess, A. (1989). *Ecology, Community and Lifestyle: Outline of an Ecosophy* (D. Rothenberg, Trans. and Rev.). Cambridge University Press (Original work published 1976).

Nietzsche, F. (1913). *Human All-Too-Human: A Book for Free Spirits, Part II* (P. V. Cohn, Trans.). Macmillan Company (Original work published 1878).

Ost, F. (2003). *La nature hors la loi. L'écologie à l'épreuve du droit*. Editions La Découverte (Original work published 1995).

Pelluchon, C. (2019). *Nourishment: A Philosophy of the Political Body* (J. E. H. Smith, Trans.). Bloomsbury Publishing (Original work published 2015).

Pelluchon, C. (2017). *Manifeste animaliste. Politiser la cause animale*. Alma Editeur.

Plutarch. (2021). *Plutarch's Three Treatises on Animals: A Translation with Introductions and Commentary* (S. T. Newmyer, Trans.) Routledge.

Porphyry. (2014). *On Abstinence from Killing Animals* (G. Clark Trans.). Bloomsbury.

Regan, T. (1983). *The Case for Animal Rights*. University of California Press.

Riot-Sarcey, M. (2006). "Introduction." In M. Riot-Sarcey (Ed.), *Dictionnaire des Utopies* (pp. IX–XII). Larousse.

Rochfeld, J. (2019). *Justice pour le climat ! Les nouvelles formes de mobilisation citoyenne*. Odile Jacob.

Rousseau, J.-J. (2008). *Discourse on Political Economy and the Social Contract* (C. Betts, Trans.). Oxford University Press (Original work published 1762).

Rousseau, J.-J. (2013). *Discours sur l'origine et les fondements de l'inégalité parmi les hommes*. Editions J'ai Lu (Original work published in 1755).

Serres, M. (1995). *The Natural Contract* (E. MacArthur & W. Paulson, Trans.). The University of Michigan Press (Original work published 1990).

Singer, P. (1975). *Animal Liberation*. Random House.

Tenenti, A. (1996). "Utopie." In P. Raynaud & S. Rials, *Dictionnaire de Philosophie Politique* (pp. 832–833). PUF.

Waldau, P. (2013). *Animal Studies. An Introduction*. Oxford University Press.

Degrowth: A State of Expenditure

Giacomo D'Alisa and Onofrio Romano

A long time ago, I once lived a whole week luxuriating in all the goods of this world: we slept without a roof, on a beach, I lived on fruit, and spent half my days alone in the water. I learned something then that has always made me react to the signs of comfort or of a well-appointed house with irony, impatience, and sometimes anger. (Albert Camus)

Growth and the Depoliticization of Ecological Concern

What is growth? It is a cumulative process through which a society increases the use of materials and energy extensively. It is the final output of a specific imaginary according to which, in order to live well, a person needs stuff, and in order to live better she needs more and more (stuff). In a liberal society, each person can legitimately aspire to meet her wants and wishes, extracting, controlling, using, consuming, wasting and disposing of all the resources she

---

G. D'Alisa
FCT (SFRH/ BPD/ 116505/ 2016)-CES University of Coimbra, Coimbra, Portugal

O. Romano (✉)
University of Rome 3, Rome, Italy
e-mail: onofrio.romano@uniroma3.it

needs. Growth is the aggregate outcome of these legitimate attitudes implemented by the individuals circulating in our Westernized society. It is the core common sense of liberal individualism, its cultural bedrock.

Each cultural imaginary has its provisioning system, or if the reader prefers, its system of production. Pursuing a particular imaginary generates a material necessity of production and consumption; capital forces seem to provide the most effective societal provisioning system which allows the liberal individuals overcoming environmental limits and societal obstacles that prevent their culturally legitimate self-realization. Moreover, capital effectiveness in pursuing multiple individual aspirations and personal self-realization lays the foundation for expanding on the extended scale of capitalist economic forces even if those forces are entrenched with patriarchal (Salleh, 2017) and colonial values (Mignolo, 2018).

In mainstream economics, personal self-realization comes via maximizing personal utility. However, neoliberal economists do not acknowledge that it also comes via the socialization of the costs that the activities put in place to maximize utility generate. As an illustration, the reader can think about a person who wishes to possess a Ferrari in her life because for the latter driving a Ferrari is vital to enjoying a life worth living. Once she has got the luxury car because she can pay for it, she can also socialize the cost of the CO_2 equivalent the Ferrari emits for the kilometers she drives the car. Of course, this is true not only at the individual level but also at the corporate level. According to William Kapp (1963), the ability to socialize the cost is a measure of the businesses' success. Consequently, in liberal economies, if one wants to make real her dreams, she needs to run successful enterprises.

The cowboy logic pinpoints this kind of economy, as the economist Kennet Boulding suggested in the 60s (Boulding, 1966). The cowboy has in front of him an illimitable plain to colonize and exploit wildly. This is the logic of growth that does not match with sustainability. Boulding proposed moving from an unlimited growth economy to a sustainable one. Sustainbility can be achieved, Boulding continued, if Westernized[1]] humans leave behind the logic of cowboy and embrace the logic of an astronaut. Astronauts think in terms of a spaceship economy, i.e., an economy whose resource is finite and scarce thus, that is, focuses on cyclical process of recycling. However, his recommendations were not even able to scratch the growth imaginary.

Nowadays, two of the most potent capitalists are investing in spaceships to move to Mars, founding their spatial enterprises precisely on the values, beliefs and principles of the growth imaginary. Of course, they are not the culprit of the growth imaginary shared primarily in Westernized society. However, the two persons, whose sentences and imaginary we will use to offer concrete examples of the growth logic, are specimens of growth. Nonetheless, their decisions have crucial material and immaterial consequences on pursuing the unsustainable and unfair growth path. It is well known that Jeff Bezos, the

---

[1] Our added to Boulding arguments.

founder of the Amazon corporate, is a fan of Star Trek. Since he was young he dreamed to travel in a spaceship.[2] However, he does not want to follow Boulding's suggestion, he aspires to overcome the limit to growth for protecting the earth, as the following sentence clarifies:

> Earth is finite. To protect our gem of a planet and enable a future of abundance and growth, Earth needs space. Space gives Earth room to grow, new resources, more frontiers to explore, and a way forward that unites us all. We just need future generations to help visualize this future...[3]

Bezos is still a cowboy. Even if he does not ride horses or bulls, he travels in spaceships. It is not by chance that in the first photo appearing on the website that promotes "the Club for the Future" he wears the cowboy hat and invites everybody, but particularly the members of the young generation, to send a postcard to the space with their dream for the future. Blue Origin gave birth to the foundation "Club for the Future", i.e., the enterprise that will allow Bezos, and all other people that can win the auction,[4] to travel to the space in July 2021. Of course, as it is announced on the website this will also help to preserve "Earth" finding new energy and material resources. The subtle argument of this enterprise is the following: human community should not care about sustainability, it is not a political issue, it is a Bezos' concern. Sustainability is his business. In this profitable business, of course, Bezos is not alone. Elon Musk, the Tesla founder, is on it too. He shares the same idea: if the earth alone is not sufficient, why humanity does not go multiplanetary.[5] After offering solutions to climate change with his Tesla electric car, Musk has joined the spaceship enterprises to help overcoming the limiting earth. Again, this growth imaginary depoliticizes sustainability (Swyngedouw, 2015). As a consequence, society should not care about ecological disruption, or socio-environmental injustices, the two growth moguls do it for us. "Make you thoughtless,"[6] and depoliticize your socio-ecological concern: the successful growthers will do it. This is the main, even if subtle, problematic message of the growth culture.

DEGROWTH AND DEGROWTHERS

Degrowth is a challenge to the growth imaginary. It is a social and academic movement that has emerged from the belly of the beast; indeed, it is a reflexive

---

[2] Read this article on *The Guardian* https://www.theguardian.com/technology/2021/feb/03/jeff-bezos-and-the-world-amazon-made or its declaration: https://www.blueorigin.com/news/jeff-bezos-first-human-flight

[3] Read here: https://clubforfuture.org/news/earth-needs-space/.

[4] Read here: https://www.blueorigin.com/news/jeff-bezos-first-human-flight.

[5] Read here: https://www.spacex.com/human-spaceflight/.

[6] This is the refrain of the Camorra's boss to his protégé in the series of Gomorra: see https://www.youtube.com/watch?v=-4QORgagblU.

byproduct of the European thoughts and socio-ecological struggles. However, it bridges with and searches inspiration from sister movements that across the world struggle against the developmentalist regime and argue in favor of decolonial paths.[7] The developmentalist regime originated in the geographies of the West in the middle of the twentieth century and since then expanded worldwide where it has found several oppositional forces that enact a pluriverse of alternatives (Khotari et al., 2019). More recently, degrowth has forged alliances with radical and eco-feminist scholars and activists. However, Gregoratti and Raphael (2019) have convincingly argued that unfortunately feminists such as Maria Mies and Marilyn Waring among many others have been passed unnoticed in degrowth scholarship for too long, as their analytical critic of growth dated back to the 80s.

Degrowth was used for the first time as a term in French (*la décroissance*) in the 70s, where groups of green leftist intellectual and scholars radicalize further the discussion around the limit to growth and the scarcity of natural resources. Main authors were André Gorz, Jacques Grinevald and Ivo Rens. Georgescu-Roegen's discussion about the relation between the economic process and the entropy law influenced very much all of them. Degrowth re-emerged in a French environmental activist milieu that campaigned against cars and advertisements. Michel Bernard, Bruno Clémentin and Vincent Cheynet leaded the group and promoted at tend of the 90s the idea of sustainable degrowth mainly through two magazines: *Silence* and *Casseurs de pub*.

At the beginning of the twenty-first century, Serge Latouche became the main proponent of degrowth. In the 80s, he had joined the French movement MAUSS—Mouvement Anti-utilitariste dans les sciences sociales (Anti-utilitarian Movement in the Social Sciences), whose leading figure was Ailan Caillé that strenuously criticized the utilitarianism as the ultimate justification of all human actions. Latouche in the 90s also collaborated with scholars critical of development such as Wolfang Sachs, Arturo Escobar and Gustavo Esteve. Latouche's works influenced very much the diffusion of the degrowth movement in Italy and in Spain.

Degrowth calls for the decolonization of public debate from the idiom of economism and for the abolishment of economic growth as a social objective. Beyond that, degrowth signifies also a desired direction, one in which people use fewer natural resources and organizes and lives differently than the Westernized individuals. "Sharing", "simplicity", "conviviality", "care," and the "commons" are primary significations of what the degrowth society looks like.

Degrowthers aspire to live simply so the others, human and nonhuman being, can simply live; they conceive an imaginary founded on the interdependency and eco-dependency of existence and the immanent vulnerability

---

[7] The great part of this section has been published in the introduction of D'Alisa et al. (2015) and D'Alisa (2019)

of the beings (D'Alisa, 2019). As matter of principle, degrowth scholarship critically documents the troublesome connection between economic expansion, social inequality and ecological disruption driven by individualism, patriarchal values and colonial imaginary (for a short compendium see D'Alisa et al., 2015).

Usually, degrowth is associated with the idea that smaller can be beautiful. However, the emphasis is on different, not only less. Degrowth signifies a society with a smaller metabolism, but more importantly, a society with a metabolism which has a different structure and serves new functions. It envisions a society with different gender relations and roles, different distribution of paid and unpaid work, different cultural interactions, and different co-evolutive paths between the human and nonhuman species. The degrowth imaginary focuses on the reproductive economy of caring for humans and the environment, and the reclaiming of commons. Caring in common is embodied in new forms of living and producing, such as eco-communities and cooperatives, and can be supported by new government institutions, such as work-sharing and working-time reduction, a universal care income coupled with a maximum income. These institutions can liberate time from paid work and make it available for unpaid communal and caring activities.

So degrowth is not the same as negative GDP growth. Still, a reduction of GDP, as currently counted, is a likely outcome of actions promoted in the name of degrowth. A green, caring and communal economy is likely to secure the good life, but unlikely to increase gross domestic activity two or three percent per year. Advocates of degrowth ask how the inevitable and desirable decrease of GDP can become socially sustainable, given that, under capitalism, economies tend to either grow or collapse. A leaner societal metabolism—less energy and material throughput (Georgescu-Roegen, 1971) is necessary but not sufficient for degrowthers. It needs to be coupled with equitable conditions for supporting lives that merit the joy to be lived (Orozco, 2015) free from racism, colonialism and patriarchal values.

As degrowth imaginary implies a deep political and economic re-organization of Westernized society, even if degrowthers promote smooth and voluntary transitions (Kallis, 2018), a degrowth future will not come without environmental justice struggles (Anguelovski and Martinez Alier, 2014), class struggles (Barca 2019; Leonardi 2019), feminist struggles (Salleh, 2017) and decolonial struggle (Nirmal and Rocheleau, 2019). As degrowth scholars have highlighted many degrowthers' actions are oppositional and reject the main features of contemporary growth economies (e.g., capital accumulation, property, colonial and racial activities, patriarchal values, productivity, socialization of businesses' costs, etc....), from the point of view of the champions of the current growth civilization, degrowthers are the uncivil society that need to be fought against (D'Alisa et al., 2013).

In the last decade, degrowth scholarship has bloomed, as well as its media and political reach, and more and more people are convinced that degrowth principles, visions and claims are relevant in a context of continuous economic and environmental crises, that the current COVID-19 pandemic cannot but exacerbate them (Paulson et al., 2020).

THE ROAD OF DÉPENSE TOWARD DEGROWTH

In degrowth scholarship, the debate is open on how to implement a degrowth society. As follows, we try to suggest a specific path toward degrowth—one among hundreds of others equally legitimate. This path is particularly inspired to Georges Bataille's thought about *dépense* and (what he called) the "general economy".

The French author is certainly a surprisingly precursor of degrowth:

> My purpose is to show [...] on the one hand the deep deformations of the general balance that the annual development of industry has brought about, on the other the prospects of an economy not centered on growth [...] It will be necessary to introduce new theoretical considerations and to found the general representation of the economic game on the description of the systems in place before the capitalist accumulation.
>
> (Bataille, 1998, 279)

Bataille starts from a matter of fact, in some way opposed to the preliminary assumptions of current degrowth advocates: i.e., the abundance of energy that invests the planet and its inhabitants. This abundance is essentially ensured by the sunlight radiation on earth. Energy released by the sun profusely pours on living beings. Plants, animals and men are able to catch only an infinitesimal part of the circulating energy for the satisfaction of their vital needs, that is for biological reproduction. But these processes are not able to incorporate all the available energy circulating on the planet. So living beings are spontaneously solicited to catch and spend an additional share of energy to "grow," thus going beyond the threshold of the accomplishment of mere survival processes. But their growing capacity is also limited. The living system is not able to dispose of all the available energy. Thus, the surplus energy, which cannot be used by living beings due to their physiological limits, builds up. It starts circulating and pressing on earth. Finding no utilization, the surplus undertakes a process of gradual dissipation, until it burns out. If we consider only the "biological" domain, we can define *dépense* as the dissipation of the share of energy that exceeds the absorption capacity of living beings. As Bataille puts it:

> The living organism [...] receives in theory more energy than it is necessary to sustain life: the surplus energy (wealth) can be employed for the growth of a system (such as an organism); if the system cannot anymore grow or if the surplus cannot be entirely absorbed by its growth, then we assist at a loss without return; surplus is spent, willingly or not, gloriously or in a catastrophic way.
>
> (2003, p. 73)

The process, simple and almost mechanical, has important implications when we move from the biological sphere to the anthropological domain: from this point of observation, the energy could be redefined as the fuel of the action, more specifically, the fuel that "calls us to act," the mere presence of which urges men to choose for it (and, if necessary, to justify) a destination.

Like all other living beings, humans are able to spend for its sustenance and growth only a tiny portion of the available energy. At this stage, the use of energy takes a distinctly "servile" character. Man is driven by a natural impulse, by his basic, reproductive needs, like every other living organism. Intentionality is not needed. There is no need to activate those distinctive dimensions of the human being that consist of reflection, the elaboration of a sense and political mediation. Man is on this side of consciousness. He is a thing (although animated) among things.

Problems arise when we stand before the residual energy exceeding the amount needed for servile use, that is the most conspicuous share. In fact, surplus energy claims a "sovereign" use, i.e., emancipated from the instrumental relation with life reproduction (Hegel, 1976). The meeting with excess energy is a crucial moment, as it tests the very consistency of the human, after the satisfaction of those natural needs that assimilate man with all other living organisms, animals and plants. This moment marks the transition from unconsciousness to consciousness, from the animal to the human. Here, the state of urgency that in the human being suspends all questions on the ultimate meaning of life, the world, and of the general biological system in which he instinctively and unthinkingly works for the procurement of vital resources, finally ends.

The surplus immediately appears as an "accursed share." It places humankind before the question of freedom, of choice. There are no more "natural" indications on how to use energy. Instinct does not decide for us anymore. Human kind must now elaborate a meaning, an end, in the name of which to draw the fuel of action and channel it into canons of value autonomously designed. It is necessary to decide about the destination for the fuel, on the basis of a philosophical intention, of a project, which can no longer be drawn from the mere automatism of natural processes. It is a curse but at the same time a great chance and a way out from the growth society:

> In the sphere of human activity the dilemma takes this form: either most of the available resources (i.e., work) are used to fabricate new means of production – and we have the capitalist economy (accumulation, the growth of wealth) – or the surplus is wasted without trying to increase production potential – and we have the festive economy. In the first case, human value is a function of productivity; in the second, it is linked to the most beautiful outcomes of art, poetry, i.e.,: the full growth of human life. In the first case, we only care about the time to come, subordinating the present time to future; in the second, only the present instant becomes relevant, and life, at least from time to time and as much as possible, is freed from the servile considerations that dominate a world devoted to the growth of production.

(Bataille, 1998, p. 277)

It is the different ways of using the surplus that determines, in fact, the specific characters and differences of human consortia in time and space. The surplus can be spent in sacrifices or in glory, in religious asceticism or in the festive *re-ligare*, and in war or in peace (as, for example, the teachings of Tibetan society, assigning it almost entirely to the maintenance of an important monastic caste).

How is it used in our modern civilization? In light of the general economy, it is possible to clearly identify the underlying problem of the "growth society" born with modernity and on this basis to suggest a reformulation of the degrowth project, amending some inconsistencies of its dominant form today.

The birth of the growth society is marked by the implosion of the communitarian totality. The individual begins to wander in search of the means of survival. This becomes its dominant occupation (according to the "cowboy logic" above mentioned). It loses contact with the totality. Out of the community, everything is reabsorbed by the survival enterprise. All of this reinforces a wrong perception of the state of energy in the living system and anyway it doesn't affect the mechanics of dissipation at all:

> The centrality of necessity arises for the particular living being, or for the limited groups of living beings. But man is not only the separated being who disputes his share of resources to the living world or to other men [...]
>
> If he denies it, as he continually is obliged to do by the conscience of a necessity, of the indigence characterizing the separated being (who incessantly lacks resources, who is only an eternal needy), his denial doesn't change anything in the dynamic of global energy: this builds up without limits in the productive forces; in the end like a river in the sea, it must escape us and get lost for us.
>
> (Bataille, 2003, p. 75)

The development of the individualization process reduces the collective ability to manage energy and, in particular, to work off the surplus by ritual forms of *dépense* (the collective body becomes a sort of functional, soulless hub, as is well described in the classical works by Elias, Durkheim, Simmel). Surplus management is no longer a collective task; the individual is now the only and lonely holder of each sovereign act. She/he autonomously takes their determinations about the use of the over-servile share of the circulating energy. This is way in Westernized society the spaceship logic cannot be an obstacle to the cowboy logic, but on the contrary it comes to stimulate the latter to develop further (on Mars, in the space).

On a more properly philosophical or, better, ideological level, modern narrative prescribes that the accursed share, although individually managed, must be spent in moral, intellectual, and civil growth. No more *dépense*, no vulgar waste, but an active search for a moral sense to be attached to one's earthly path. The modern subject, already charged with the unbearable weight of the surplus, is then invited to use extra-servile time for his moral completion.

Needless to say, this is a completely inadequate reply to the issued posed by excess energy, given its relevance. Both on the political–institutional side (the burden uploaded on the individual) and on the ideological–moral level (the surplus must be employed in the search for meaning), modernity does not provide credible ways out of the anguish generated by the surplus. This "*manque*," this lack of response generates multiple consequences in Western societies, of tactical adjustments or, if one prefers, a series of "real answers," against the merely ideological ones made available. We highlight, in particular, three answers: one concerning the shared imaginary, another on the institutional side, and the last one aimed at the functional relocation of the "removed" surplus. All these answers are interpretable, following Durkheim (1960), as forms of solidification of the effervescences occurred in the *statu nascenti* of modernity.

The first answer coincides with the exasperation of the original servile moment, that is to say the tension toward unlimited economic "growth."

Our civilization was formed under the primacy of accumulation, of the consecration of wealth to the increase of production potential. Our moral and political conceptions are still dominated by a principle: the excellence of the development of productive forces.

(Bataille, 1998, p. 278)

The emphasis placed on servilism, in fact, is a strategy for removing the impending surplus. Making the original emergency situation eternal, giving absolute primacy to the activities necessary for survival (and for growth, that represents its euphoric declination) and transforming them in a real collective obsession allow us to forget the issue of the action meaning.

> At first this prolific movement halted the war activity by absorbing the essential of the surplus: the development of modern industry determined the relative period of peace from 1815 to 1914. The development of productive forces, the growth of resources, made it possible at the same time the rapid demographic multiplication of advanced countries (it is the carnal aspect of the bony structure represented by the proliferation of factories). But in the long run the growth that technical innovations made possible became difficult. In its turn it became the generator of an increased surplus.

(Bataille, 2003, p. 76)

The perpetuation of the survival engagement frees us from that state of paralysis that raises before the call to "be" pronounced by the impending surplus. Staying an animal, human is liberated from the fatigue of the human.

Here returns the "ideological genesis of need" as Baudrillard (1972) called it, that is the erasing of any symbolic function of the object and its reduction to a mere commodity. This process goes hand-in-hand with the reduction of the individual to his supposed innate needs. Since individual desires, motivations and sense are no longer actualizations of the coercion exercised by the community, the actor recognizes his pure essence, beyond the social ties, in his own needs: they are a new source of determination, rationally identifiable not vitiated by symbolic rubbish. At the molecular level, the mechanism produces a long-term toxic dependence from commodities. At the molar level, instead, we see the transformation of the potential energy surplus into a situation of "real" scarcity of the available resources and of danger for the environmental balances.

Second, institutions adopt more and more the regulatory principle of life for life's sake. The modern "democratic" institutions, against their legitimating theory, are no longer an arena for collective discussion on the "meaning" of life, on the sense of being together, on what is "good," and on how to achieve it, but they become a mere neutral machine, passive before each person's determinations, only aimed at ensuring exclusively the protection, the reproduction, the promotion and the valorization of the species biological life. Life itself expels the collective construction of its meaning from the public arena. As already underlined, it will then be the individuals, with their free will, to decide on how to use the surplus and therefore which sense to add to their existential path. This is the meaning of the "biopolitical" turn considered, albeit from a different and very questionable perspective, by Foucault (1976, pp. 119–142).

The third answer concerns the *dépense* destiny. It becomes more and more a real self-dissipation practice. In spite of the pretense to spend surplus energy for civil and moral growth, *dépense* is not at all buried in the darkness of tradition. It is simply kicked off the "official" public arena, "privatized" and hidden. Individuals take charge of the expenditure function (once acted out by collectivity in the eminent moments of its rituals), by coasting trade practices: from perverse sexuality to alcoholism, from gambling to conspicuous consumption, etc. Bataille called them the "shameful belching" of the petty bourgeoisie. In the bourgeois era, there is no longer the eminent and sumptuous collective *dépense*, but the private dissolution informally consumed in the secret rooms, under the hood of shame, far from public visibility.

> Man's glorious expenditure was brought back to the extent of commercial exploitation. The form of individual expenditure, which excludes the true splendor, is the only admitted by capitalist production. It is oriented by the mass-produced objects; the simulacrum replaces luxury [...]
>
> In fact, the individual without social ties cannot aspire to splendor. If he succumbs to the fascination of luxury, he does so without tact: he destroys the sense of splendor. Comfort and boredom are the result of this ever-greater poverty of wealth.
>
> (Bataille, 2000, p. 82)

Bataille has not had time to see that these forms of *dépense*, especially since the second postwar period, with the advent of electronic media, mass culture and consumer society, have actually ceased to be experienced in the secret rooms and now permanently connote the Western regime out in the open: this is what elsewhere we have called "demodernization" (Romano, 2008), evoked by a long series of authors (from Baudrillard to Bell, from Maffesoli to Magatti) and consisting of a sort of labor division between the economic system—which continues to function according to the usual competitive mechanisms of the capitalist market—and the cultural system, which instead derails toward an ethic of immediate enjoyment, unbridled consumption and value reversibility. Private *dépense* no longer needs to hide and it is increasingly becoming a pillar of "techno-nihilist" capitalism (Magatti, 2009, 2012).

Despite this form of acknowledgment, *dépense* practices continue to be staged within a degraded frame, by no means appropriate to the importance of surplus employment. The anxiety generated by the burden of surplus energy doesn't find, in this context, a suitable chance of disposal but only palliative responses, thus it inexorably spreads.

> [W]e use surplus to multiply "services" that facilitate life, and we are led to reabsorb some of it in the increase of hours of free time. But these diversions have always been insufficient: their existence in *excess* has always voted multitudes of human beings and large amounts of useful goods to the destruction of wars.
>
> (Bataille, 2003, 75)

The "Depensing" Degrowth and Its Institutional Case: Claiming Verticalism

Considering the approach delineated above, a degrowth society cannot rely on the generalization of an individual effort of restraint. If we aim at escaping from the growth society, it is necessary to give back the managing of surplus at the collectivity. Otherwise we will leave it in the hand of few specimens of the growth logic. It is necessary to re-build a form of totality, defusing the particularization (individualization) at the basis of the bad infinity of growth.

A shift toward degrowth is unlikely if we do not radically rethink the formal–institutional dimension of the project. If degrowth scholarship rests on the same institutional "form" that frames the growth regime, only suggesting to overcome it by competing on "values" (a different idea of happiness and human completion), it is doomed to failure. This same form is "horizontalism" (Romano, 2014). Here immanence becomes the privileged dimension. In general, it is believed that we can find the true meaning of a social organism by looking at its single players and the networks they interweave (Adorno, Canetti, & Gehlen, 1996). At a political level, horizontalism considers a social order much more desirable insofar as it leaves out the subject "as it is," promoting a process of self-revelation. The more social players are free to act and interact based upon their own preferences, the more society as a whole will be happy. It is the horizontalism of the growth logic that legitimates Bezos and Musk to invest the accumulated surplus in their dream to go to the space, instead of let's say to prevent ecological catastrophes on the earth.

Horizontality appears to be the "natural" order, more harmonious and suited to individual moods (Benedict, 1952; Scheler, 1960; Mannheim, 1991). It is precisely this social casing that gives rise to the "particularization" denounced by Bataille, by which the human being starts to consider as his main problem the lack of resources and consequently to address any effort toward growth:

> as a rule, particular existence always risks succumbing for lack of resources. It contrasts with general existence whose resources are in excess and for which death has no meaning. From the particular point of view, the problems are posed in the first instance by a deficiency of resources. They are posed in the first instance by an excess of resources if one starts from the general point of view.
>
> (Bataille 1988, p. 39)

So, it is necessary to socially restore the view on the "general existence" if we aim at stopping the growth dynamic and the inequality across geographies and humans, as well as the ecological predicaments it produces. It is necessary to regain a form of verticality.

In some way, degrowthers' claims recall Polanyi's typical arguments against the commodification of labor, money and nature that boost the blind growth in the nineteenth century (i.e., against horizontalism). In the final analysis, they stigmatize the effects of horizontal deregulation, claiming for restoring human autonomy (Castoriadis, 1975; Asara et al., 2013), i.e., the primacy of people and planet needs (Muraca, 2013). When societies lose their sovereignty over the factors of production, then social, economic and ecological disruption follows. But, contrary to Polanyi, most of them do not go so far as to promote the restoration of a new vertical regime. The hegemony of horizontalism prevents this logical and natural outcome. They mostly displace the focus of their diagnosis in the domain of "values." See, for example, the emphasis by Serge Latouche on the decolonization of the imaginary (Latouche, 2015).

Ecological and social disruption are not seen as the effects of the "form" of the dominant regime but more often of the prevailing myth of growth for growth's sake that rages in the socially shared imagination (Latouche, 2011). So, the fight is relocated in the sphere of the imaginary: it is necessary to abandon the value of growth, completely rethinking the set of values that frame our lives. The current key themes are more or less selfishness, competition, working hard, consuming hard, globalism, heteronomy, efficiency and rationalism. Degrowth supporters oppose this list with a completely reversed one: altruism, cooperation, promoting good living (work–life balance), sociability, localism, autonomy, beauty and reasonableness. By following these key themes, we would be led to another world, where the compulsion to growth is completely erased (Latouche, 2007). Change is imagined as the promotion here and now of anti-growth "values" by staging social alternatives from the grassroots and radicalizing their horizontal form (Alexander & Gleeson, 2018). So, in the voluntary simplicity strategy, individual "agency" is the privileged dimension (Boonstra & Joosse, 2013): activists secede from the public arena where the majority of people lie, in order to build a small world together with those who only share the same values and visions.

> A degrowth or steady-state economy will depend for its realization on the emergence of a post-consumerist culture, one that understands and embraces 'sufficiency' in consumption.
>
> (Alexander, 2013, p. 300)

We contend that this strategy risks being ineffective, because growth is not a value in itself of our society but it derives directly from the liberation of the elementary particles decreed by horizontalism: once "disembedded" from society, individuals are naturally led to undertake the path of growth, due to the feeling of precariousness increased by isolation, as Bataille teaches us.

In a society framed by horizontalism, the individualized being (or the isolated small community) is bound by the precarious nature of his existence and therefore obsessed with the problem of his survival. When isolated, he embraces a fundamentally servile position and reverts to the status of an animal, in which obtaining resources is central (from the medieval peasant expelled by his community to Jeff Bezos' space adventures, it's always like that). The individual point of view that emphasizes the insufficiency of resources gets applied to the general collective.

The reader could guess: why does real social change require the changing of the institutional "forms" and not only a values change? Why do we have to assume that the effects of horizontal neoliberalism are similar to the alleged effects of the (horizontal) degrowth regime, although they are founded on very different values? This is mainly due to a structural feature of horizontalism. As Magatti (2009) asserts, this is based on a clear separation between "functions" and "meanings." A horizontal social system—regardless of the intentions of their promoters—does not fit into a particular idea of justice, into a "form of life" (Pellizzoni, 2021). It doesn't obey any "value." Horizontality prepares the conditions for the erasing of the political dimension, i.e., for depoliticization. It is indifferent to any principle, aiming only at ensuring that each singularity (the citizen and his networks) can freely play his game, on the basis of his specific values. If horizontality is the elected form, then it is impossible to legitimize any authority that establishes specific values that everyone has to compulsorily follow. Horizontalism necessarily engenders neutralitarian institutions.

Obviously, we are talking about the ideological case of the regime, not of its reality: horizontalism is first of all a narrative that hides a "real verticalism", becoming visible mainly in the crisis situations, like the pandemic shows. Anyway, it is inconsistent to hope in a reform of values resting on a horizontal institutional frame and narrative, like Latouche (2007), Fotopoulos (1997), and Asara et al. (2013) do. Maybe it can work in the early stages of the system but, in the absence of a central intentionality, nothing assures that in the medium–long term all the society members will follow the same values: when the elementary particles of society are free to run, everyone runs where it wants. Horizontalism, itself, is indifferent to values (not the same for verticalism). This "passivity" of the system toward values engenders, ultimately, ecological, social and economic deregulation. Moreover, it is the real origin of the emphasis on "growth."

Growth is nothing other than the translation of the modern principle of neutrality: it is "rightly" indifferent to any goal, if not to that of increasing everybody's material chances to choose and implement one's goals. It is even indifferent, for example, to the most simple goal that vertical form had in the past i.e., the redistribution of resources. As the system growth will benefit everyone, why keep insisting on the necessity of redistribution to stop the increasing inequality? In these structural conditions, the ethical dimension is revealed to be totally harmless for the horizontal regime, which rather promotes the unlimited proliferation of values and meanings, even reciprocally antithetical. So, it is incongruous to challenge it only with new values. It requires that the whole citizenship adopt a certain set of values (namely those linked to the "degrowth" society): an impossible precondition.

So, we have to merge the fight for new values with the fight for a new "form." It is the only way to attain a sovereign regime that could assure the reproduction of renewable resources and the preservation of nonrenewable resources, replacing the obsession for growth with the restoring of the eminent function of *dépense*. Sure, the values fight remains crucial. Verticalism (the form) itself is not enough. As we well know, twentieth-century Western verticalist regimes have all promoted growth with great efficacy. In this sense, the shift from verticalism to horizontalism was accomplished under the same growth mark. But in the twentieth century, environmental awareness was not so developed. More generally, there was no consciousness of the negative aftermaths of growth. Together with a shift toward verticality, we have to sustain the values and the common sense of degrowth more and more (D'Alisa & Kallis, 2020), at the horizontal level (i.e., from the grassroots).

The vertical form without degrowth values can produce more damages than horizontalism (hyper-exploitation of nature, inequalities in favor of the elites and so on). But, contrary to horizontalism, verticalism is not indifferent to values. If we feed the vertical form with degrowth values, we can hope to obtain a degrowth society. The opposite is not true: degrowth values inside a horizontal frame do not produce a degrowth society, for the above mentioned reasons, i.e., because horizontalism is indifferent to any value. So we must aim at a verticalist regime framed by degrowth values. In this way, we restore the collective management of surplus and individuals are finally freed from the ghost of scarcity, being their existence fully protected by the community. This is the precondition for democracy, which—as Arendt (1998) taught us—can be exercised only by humans being freed from the survival needs.

Reloading the State: A Gramscian Perspective

If a form of verticalism has to be envisioned, the main issue becomes: how the fundamental contribution of the degrowth practices and discourses can help to implement a form of verticalism that promotes the common good for all? How degrowth vision can transform state apparatuses and agencies toward a sustainable a fair future? Despite increasing attention to social "transformations", the related scholarship, including that on degrowth transformations (Asara et al., 2015) has not questioned sufficiently power asymmetries or shed light on structural obstacles (Brand, 2016a, b). The word "transformation" is used mostly to indicate a desired direction, but is analytically weak (Brand, 2016b).

On the one hand, we maintain, the analytical lens of Eric Olin-Wright's model of transformation (Wright, 2009) is useful to classify different transformational strategies. Wright identifies three logics and visions of systemic transformations: ruptural, interstitial and symbiotic. Each vision is associated with a political tradition, and it is developed around a pivotal political actor, and has a particular strategic logic with respect to the state structures. The ruptural strategy consists of a frontal attack on the state, aiming at the construction of new emancipatory institutions after existing state institutions have been dismantled. This is the logic of revolutionaries. Interstitial metamorphosis instead is focused on the promotion of horizontalist alternatives within the crack of the capitalist system, building the new in the interstices of the capital markets and outside the *longa manus* (long arms) of state apparatuses. This strategy is in line with anarchist visions of building parallel self-governmental systems in the civil society arena. The strategy of symbiotic metamorphosis envisions a coevolving trajectory of transformation based on compromises with the dominant political-economic forces. It aspires "to use" existing institutions and state agencies, and to mobilize popular power to transform them. This approach is in line with a more reformist, social, democratic political tradition.

On the other hand, Gramsci's lexicon is also a vocabulary for thinking how a transition could evolve, overcoming a division between grassroots and policy, or bottom-up and top-down action. The grassroots is not an alternative to the state—civil society is the one half of the integral state transformation. Following a Gramscian approach to the state composition, we can argue that what grassroots practices often do is to construct a counter-hegemony that reorders common senses—the fate of this counter-hegemony depends on its ability to occupy the political sphere and to use the collective force of the state apparatuses and agencies to spread the new common senses in different spheres of the society, well beyond the niches of activists and practitioners of alternatives to capitalism and growth logic, transforming the current institutions and (re)producing new ones in line with the new common senses that the horizontal practices performed in the civil society make possible to emerge. For example, alternative economies, say food cooperatives or community currencies, are new civil society institutions that nurture new common senses. As they expand, they undo the common sense of growth and make degrowth ideas potentially hegemonic, creating conditions for a social and political force to change political institutions in the same direction. Grassroots actions alone are insufficient from a Gramscian perspective. Greater forces or institutions often limit them. For example, alternative food networks are limited by access to land or high land prices, by legislative rules that prioritize corporate agriculture and the dominant practices of agribusiness, by price dumping in "liberalized" food markets or by the rising costs of public health or education that make it hard for small scale farmers to secure their living. Young, back-to-the-land farmers who want to produce and distribute food differently end up exploiting themselves, overworking in order to sell at a fair and affordable price. Likewise, co-housing or cooperative housing initiatives are swamped by private capital and gentrification in liberalized housing markets without rent controls.

So it is necessary to provide policies in order to open space and to release resources in support of the new practices, values and common senses alternative economies enact and convey, such as food sovereignty and co-habitation. A reduction of working hours and a basic income make it more possible for people to devote time to alternative food networks and gain collective sovereignty on their nutrition. Policies for rent control, price controls or subsidies for alternative housing projects directly benefit new economy initiatives and foster new forms of co-housing and make them visible to common people without time to commit themselves to communitarian project. The demand though for such political/institutional changes would not come without a critical mass of people involved in—and making a living from—alternative grassroots economies. Alternative food networks, open software communities or solidarity practices, such as popular health clinics change the common sense of participants and allow them to imagine different knowledge, health, care or education systems. Participants and those who experience these projects become then a potential base for articulating social demands and for changing political institutions (e.g., intellectual copyright or welfare provision) to support their projects.

In this sense, organizing for change in civil society and political organizing for occupying the sphere of political society are the two sides of the same coin. To reverse the process that consolidates the values and beliefs of the growth society, those who want change cannot simply take power, say through elections and then implement the policies they want.

Imagine for the sake of illustration a wild scenario that a revolution or an election puts Herman Daly (the recognized ecological economist that has developed the idea of steady-state economy) as the leader of the U.S., or Kate Raworth (the famous Oxford professor who developed the doughnut economics) of the U.K. Even in those extreme cases, little would change, unless there was a common sense cultivated in society that steady state or postgrowth is the way to go. Without such a culture, even the most enlightened leaders would find quickly their policies undermined by un-cooperative administrations, conservative populations, and elite struggling for their liberally legitimate privilege. It will not be easy for Herman Daly to implement a maximum income or to cut the wealth of Jeff Bezos and Elon Musk for guaranteeing full access to house and healthy food, for promoting low-carbon production and slow mobility. Who can legitimately put before the desires and wishes of these self-made tycoons that are putting their intelligence and entrepreneurship to overcome the limits of the Earth and go to find those resources on Mars and further in the space?

Of course, the readers could object that to end up with such leaders, the society itself must have changed. Transformation then involves a coevolutionary change between civil and political society. A degrowth transformation requires first social relationships and activities that provide viable livelihoods and produce in the ground, and not in the abstract, common senses that prioritize "degrowth-oriented" values and objectives. In parallel, those who believe and live by these values have to organize politically for the implementation of policies that reflect these common senses. For Gramsci then cultural change is fundamental. He rejects the idea that those who want change can somehow take control of political society and force their decisions on others. This does not mean that his model allows only for slow, long-term change of beliefs. A social and political rupture is possible and can force many people to change dramatically their everyday lives in a new direction. Change can take place fast, from one day to another (as a result of a crisis or a revolution) or simply because a tipping point in social awareness about a topic has been overcome. Think about for a moment how slow for a revolutionary could appear the strategy adopted by Greta Thunberg, every Friday striking alone in front of the Swedish Parliament, and on the contrary how fast the movement Friday For Future has developed. What is a core argument here is that people will accept significant changes only if they find them resonating with their everyday needs and prevalent beliefs—otherwise, they will attempt to re-establish the prior condition.

The state is not a monolithic entity, and it's not simply the ensemble of politicians and bureaucrats, but an integral, dialectical process between civil and political society with a constant interplay of the battle for ideas with the battle for institutions of enforcement. The state changes as new ideas and common senses emerge and get reordered in civil society, and as social groups struggle for new institutions that embody, facilitate and enforce the new common senses. A transition beyond growth would require an end to the ideological and institutional hegemony of growth—this means an abolition of growth institutions and a demise of taken-for-granted growth values.

Ecological economists have paid little attention to politics and strategy, and have considered their job done in coming up with a policy proposal that makes sense. Gramsci's theory shows that change passes through a reordering of common senses at the level of civil society—and this involves rooted practices that demonstrably work, and within which new, post or degrowth ideas and values start making sense. From this perspective, degrowth's emphasis on connecting activism with science is important. It is not enough for us as scholars to come up with good ideas. If we want to see change, we need to act as action researchers to see these changes happen. And there is no better place to start than our own home or workplace. We have to show that it is possible to rediscover the "abundance" of resources if we only entrust the survival issue to the collective consortium, then devoting ourselves to discuss on how to enjoy the surplus and to invent always new collective rituals for *dépense*.

References

Adorno, W. T., Canetti, E., & Gehlen, A. (1996). *Desiderio di vita: Conversazioni sulle metamorfosi dell'umano* (U. Fadini (Ed.)). Milan: Mimesis.

Alexander, S. (2013). "Voluntary Simplicity and the Social Reconstruction of Law." *Degrowth from the grassroots up. Environmental values*, Vol. 22: 287–308.

Alexander S., & Gleeson, B. (2018). *Degrowth in the Suburbs: A Radical Urban Imaginary*. New York: Palgrave Macmillan.

Anguelovski, I., & Martínez Alier, J. (2014). "The 'Environmentalism of the Poor' Revisited: Territory and Place in Disconnected Glocal Struggles." *Ecological Economics*, Vol. 102: 167–176.

Arendt, H. (1998). *The Human Condition*. Chicago: University of Chicago Press.

Asara, V., Kallis, G., & Profumi, E. (2013). "Degrowth, Democracy and Autonomy." *Environmental Values*, Vol. 22: 217–239.

Asara, V., Otero, I., Demaria, F., et al. (2015). "Socially Sustainable Degrowth as a Social–Ecological Transformation: Repoliticizing Sustainability." *Sustain Science*, Vol. 10: 375–384. https://doi.org/10.1007/s11625-015-0321-9

Barca, S. (2019). "The Labor(s) of Degrowth." *Capitalism, Nature, Socialism*, Vol. 30, 2: 207–216.

Bataille, G. (1988). *The Accursed Share: An Essay on General Economy. Vol. I: Consumption*. New York: Zone Books.

Brand, U. (2016a). "How to Get Out of the Multiple Crisis? Towards a Critical Theory of Social-Ecological Transformation." *Environmental Values*, Vol. 25, 5: 503–525.

Brand, U. (2016b). "'Transformation' as a New Critical Orthodoxy. The Strategic use of the Term 'Transformation' does not Prevent Multiple Crises." *Gaia*, Vol. 25, 1: 23–27.

Bataille, G. (1998). *Choix de lettres 1917–1962*. Paris: Gallimard.

Bataille, G. (2000). *Il limite dell'utile*. Milan: Adelphi.

Bataille, G. (2003). *La parte maledetta*. Turin: Bollati Boringhieri.

Baudrillard, J. (1972). *Pour une critique de l'économie politique du signe*. Paris: Gallimard.

Beck, U. (1992). *Risk Society: Towards a New Modernity*. London: Sage.

Benedict, R. (1952). *Patterns of Culture*. London: Routledge & Kegan Paul.

Boonstra, W. J., & Joosse, S. (2013). "The Social Dynamics of Degrowth." *Environmental values*, Vol. 22: 171–189.

Boulding, K. E. (1966). *The Economics of the Coming Spaceship Earth. Environmental Quality in a Growing Economy: Essays from the Sixth RFF Forum*. H. Jarrett. Baltimore, John Hopkins University Press: 3–14.

Castoriadis, C. (1975). *L'institutiton imaginaire de la société*. Paris: Seuil.

D'Alisa G., Demaria F., & Cattaneo C. (2013). "Civil and Uncivil Actors for a Degrowth Society." *Journal of Civil society*, Vol. 9. 212–224.

D'Alisa, G., Demaria, F., & Kallis, G. (eds.). (2015). *Degrowth: A Vocabulary for a New Era*. Routledge.

D'Alisa, G., & Kallis, G. (2016). "A Political Ecology of Maladaptation: Insights from a Gramscian Theory of the State." *Global Environmental Change*, Vol. 38: 230–242.

D'Alisa, G. (2019), "Degrowth", *Dicionário Alice*. Accessed: https://alice.ces.uc.pt/dictionary/?id=23838&pag=23918&id_lingua=1&entry=24248.%20ISBN%20978-989-8847-08-9. the July 12, 2021.

Donati, P. (2012). *Relational Society: A New Paradigm for Social Sciences*. New York: Routledge.

Durkheim, É. (1960). *Les formes élémentaire de la vie religieuse*. Paris: Presses Universitaires de France.

Fotopoulos, T. (1997). *Towards an Inclusive Democracy*. London: Cassell Continuum.

Foucault, M. (1976). *La volonté de savoir*. Paris: Gallimard

Georgescu-Roegen, N. (1971). *The Entropy Law and the Economic Process*. Cambridge, MA: Harvard University Press.

Gregoratti, C., & Raphael R. (2019). "The Historical Rots of a Feminist 'Degrowth': Maria Mies and Marilyn Waring's Critique of Growth." In Chertkovskaya, E., Paulsson A., & Barca, S. (eds.) *Towards a Political Economy of Degrowth*. Rowman & Littlefield.

Hegel, G. W. F. (1976). *The Phenomenology of Spirit*. New York: Oxford University Press.

Kallis, G. 2018. *Degrowth. The Economy | Key Ideas*, Newcastle upon Tyne: Agenda Publishing.

Kapp, K. W. (1963), *The Social Costs of Business Enterprise*. Nottingham: Spokesman.

Kothari, A., Salleh, A., Escobar, A., Demaria, F., & Acostam A. (2019). *Pluriverse. A Postdevelopment Dictionary*. Tulika Books. New Delhi.

Latouche, S. (2007). *La scommessa della decrescita*, Milan: Feltrinelli.

Latouche, S. (2011). *Décoloniser l'imaginaire*. Paris: Parangon. 2011.

Latouche, S. (2015). "Imaginary (Decolonization of)." In D'Alisa, G., Demaria, F., & Kallis, G. (eds.) *Degrowth: A Vocabulary for a New Era*. New York and London: Routledge, 117–120.

Leonardi E. (2019). "Bringing Class Analysis Back in: Assessing the Transformation of the Value Nature Nexus to Strengthen the Connection between Degrowth and Environmental Justice." *Ecological Economics*, Vol. 156, 83–90.

Magatti, M. (2009). *Libertà immaginaria: Le illusioni del capitalismo tecno-nichilista*. Milan: Feltrinelli.

Magatti, M. (2012). *La grande contrazione*. Milan: Feltrinelli.

Mannheim, K. (1991). *Ideology and Utopia: An Introduction to the Sociology of Knowledge*. London: Routledge.

Mignolo W. D. (2018). "Pensamiento Decolonial." In D'Alisa G., Demaria F., & Kallis G. (eds.) *Decrecimiento: un vocabulario por una nueva era*. Icaria editorial y Fundación Heinrich Boell, México. 343–347.

Muraca, B. (2013). "*Décroissance*. A Project for a Radical Transformation of Society." *Environmental Values*, Vol 22: 147–169.

Nirmal, P., & Rocheleau, D. (2019). "Decolonizing Degrowth in the Post-Development Convergence: Questions, Experiences, and Proposals from Two Indigenous Territories." *Environment and Planning: Nature and Space*, Vol. 2(3): 465–492.

Orozco Perez, A. (2015). "Prólogo: Palabras vivas ante un sistema biocida." In D'Alisa G., Demaria F., & Kallis G. (eds.) *Decrecimiento: un vocabulario por una nueva era*. Barcelona: Icaria Editorial. 27–33.

Paulson S., D'alisa G., Demaria F., Kallis G. with Feminisms and Degrowth Alliance. (2020). "From Pandemic Toward Care-Full Degrowth, Interface," https://www.interfacejournal.net/wp-content/uploads/2020/05/Paulson-etal.pdf

Pellizzoni, L. (2021). "Nature, Limits and Form-of-Life," *Environmental Politics*, https://doi.org/10.1080/09644016.2020.1868864

Romano, O. (2008). *La comunione reversiva*. Rome: Carocci.

Romano, O. (2014). *The Sociology of Knowledge in a Time of Crisis. Challenging the Phantom of Liberty*. New York & London: Routledge.

Romano, O. (2020). *Towards a Society of Degrowth*. New York & London: Routledge.

Salleh A. (2017) *Ecofeminism as Politics: Nature, Marx and the Postmodern*. London. Zed Books, Second edition.

Scheler, M. (1960). *Die Wissensformen und die Gesellschaft*. Bern: Francke Verlag.

Swyngedouw, E. (2015). "Depoliticized Environments and the Promises of the Anthropocene." In R. L. Bryant (ed.), *The International Handbook of Political Ecology*. Cheltenham: Edward Elgar, pp. 131–146.

Wright, E.O. (2009). *Envisioning Real Utopias*. Verso: London.

The Nature of the State: A Deep History of Agrarian Environmentalism

Jake Greear

INTRODUCTION

"I wish to be as peaceable as my land."—Wendell Berry

The most straightforward case of a post-colonial eco-revolution in the world is probably the three-decade-old conflict on the Island of Bougainville, some 900 miles northeast of Australia. Dubbed the "Coconut Revolution," this conflict began in the late 1980s when dispossessed native Bougainvilleans over-ran and shut down Panguna Mine, which at the time was one of the world's largest copper mines, operated by Rio Tinto Zinc.[1] Since opening in 1972, the mine's significant profits had enriched Western stockholders and had been the main source of funding for the government of Papua New Guinea, of which Bougainville Island was a part. A small proportion of the profits were

[1] Kenneth W. Grundy, *River of Tears: The Rise of the RioTinto Zinc Mining Corporation* (London: Earth Island Press 1974): 101–104.

I am grateful to Veronica Sotolongo, Matthew Tuten, and Liam Currie for research and editing help.

J. Greear (✉)
Western Carolina University, Cullowhee, NC, USA
e-mail: jpgreear@wcu.edu

diverted to local development and landowner compensation. However, while some Bougainvilleans found a way to profit from the mine, the seismic shifts in the Bougainvillean economy and society precipitated class stratifications and other disruptions to the society and culture. The meager and unevenly accruing benefits came to be seen as an insult added to the broader injuries of colonization and the ecological devastation of the island traditionally called "Mekamui."[2] After their demand for $10 billion in reparations was denied, the indigenous rebel faction expelled the mine operators and defeated the Papua New Guinea Defense Forces, despite their being aided by the Australian military and industry-backed, private military contractors.[3]

At the time of this writing, three decades later, the fighting has stopped, but the struggle is still unresolved. The mine remains under a moratorium,[4] but even some Bougainvilleans who took part in the struggle want to see the mine re-opened, albeit under more favorable terms. In a long-promised referendum, Bougainvilleans recently voted overwhelmingly in favor of independence from Papua New Guinea.[5] The mine represents the most plausible fiscal path to true sovereignty for the island.[6] And yet, while the advantages of "development" are surely clear to Bougainvilleans, there has been continued opposition to various offers made to reopen the mine.[7]

The situation on Bougainville is an interesting case study in post-colonial environmental politics and one in which the ideological battle line between global extractive capitalism and indigenous land claims seems to be starkly drawn. The viewpoint opposed to mining is summed up by a Bougainvillean farmer who is featured in one of several short documentaries posted in recent years on YouTube and other media platforms. The man holds up a small corn plant, which he seems to regard as a crucial visual aid, as he speaks:

---

[2] Terence Wesley-Smith and Eugene Ogan, "Copper, Class, and Crisis: Changing Relations of Production in Bougainville," *The Contemporary Pacific* 4, no. 2 (Fall 1992): 245–267.

[3] Anthony J. Reagan, "Causes and Course of the Bougainville Conflict," *The Journal of Pacific History* 33, no. 3 (November 1998): 269–285.

[4] Liam Fox, "Bougainville President Says Panguna Mine Moratorium Remains in Place," *Radio Australia, Australian Broadcasting Company*, accessed February 9, 2021, https://www.abc.net.au/radio-australia/programs/pacificbeat/bougainville-president-says-panguna-mine-off-limits/13134904.

[5] Rod McGuirk, "Bougainville Votes for Independence from Papua New Guinea," *The Diplomat*, accessed December 14, 2019, https://thediplomat.com/2019/12/bougainville-votes-for-independence-from-papua-new-guinea/.

[6] See also Joshua McDonald, "Will Bougainville Reopen the Panguna Mine?" *The Diplomat*, accessed November 22, 2019, https://thediplomat.com/2019/11/will-bougainville-reopen-the-panguna-mine/.

[7] Darby Ingram, "Bougainville President Rejects Panguna Mine Claims," *National Indigenous Times*, accessed February 1, 2021, https://nit.com.au/bougainville-president-rejects-panguna-mine-claims/.

> I am not happy to hear BCL [the subsidiary of Rio Tinto Zinc] is coming back. BCL has destroyed our land and taken the minerals from the ground. In the law of nature God set […], we survive from food that we get from plant life, plants get food from minerals in the ground. BCL returning means they will destroy more of our land, and our children will suffer. If government wants to develop Bougainville, focus on agriculture.[8]

The speaker here stakes out a quintessentially agrarian position on questions of environmental ethics and political economy. It is agriculture, from this perspective, that obeys God's law, and upholds and enacts the order of nature. The mine, on the other hand, contravenes that law and quite literally undermines the natural order. Historian of technology, Lewis Mumford, pointed out that mining shares with capitalism a certain abstract, quantitative logic, which sets both apart from the organic economy.

> The miner's notion of value, like the financier's, tends to be a purely abstract and quantitative one. Every other type of primitive environment contains food, something that may be immediately translated into life—game, berries, mushrooms, maple-sap, nuts, sheep, corn, fish—while the miner's environment alone is—salt and saccharine aside—not only completely inorganic, but completely inedible. The miner works not for love of nourishment, but to "make his pile."[9]

Some strains of environmentalism—especially wilderness-oriented strains—mark the advent of farming, or some crucial tipping point within agricultural history, as the moment of humanity's fall from ecological grace.[10] And yet, agrarianism is itself a mode of eco-critique that has undeniable popularity and intuitive appeal. Both capitalism, with its presumption of infinite exponential economic growth, and the very sites of resource extraction it depends upon (i.e., mines), bear the hallmarks of unsustainability. Infinite growth is impossible and mines are eventually depleted. In juxtaposition to both, it is at least possible to imagine a truly ecological agrarian economy—one that works within, not outside of or against the trophic cycles of a solar-powered biosphere.

This chapter explores agrarianism as a mode of environmental political critique through several sections. Section 1 is a brief exposition of the "new agrarian" environmental movement. Section 2 engages the environmental writings of Wendell Berry as an influential Western proponent of agrarian environmentalism, engaging the criticism that this movement is handicapped by a fundamentally apolitical focus on private virtue. Sections 3 and 4 draw upon

---

[8] Daniel Jones, "Panguna Mine Dilemma," filmed 2008, *Eel Films*, https://www.youtube.com/watch?v=Sv8Q5hH0cys.

[9] Lewis Mumford, *Technics and Civilization* (New York: Harcourt Brace and Co., 1934): 77.

[10] Specific arguments along these lines are discussed in the final section.

Leo Marx and others to trace the problems and the promise of contemporary Western agrarianism to their literary roots, focusing primarily on Virgil and Hesiod. Section 5 draws on recent archaeological scholarship to argue that agrarian political critique as well as the agrarian environmental aesthetic is rooted in the production of "rural space" that attends the formation of the earliest states. The final section compares and contrasts two visions of nature within contemporary environmental thought: agrarian nature and nature-as-wilderness. I conclude with the suggestion that agrarian nature is, paradoxically perhaps, "the nature of the state," but is not to be disvalued on that basis.

The Promise of Environmental Agrarianism

Insofar as movements such as the one on Bougainville Island are "environmental" movements, they have often been understood as fundamentally different from the environmental movements of the Global North. Joan Martinez-Alier, for example, has contrasted the "environmentalism of the poor" with the "cult of wilderness" and the "doctrine of eco-efficiency" prevalent in Europe and North America.[11] Timothy Doyle makes similar distinctions between the environmentalisms of "majority and minority worlds," and suggests a posture of respect for the wide diversity of environmental movements globally.[12]

The differences illuminated in these analyses are real; however, agrarian environmentalism is a global common thread that such categorizations overlook. Environmental justice struggles across the Global South are often fundamentally agrarian, as they frequently seek to defend traditional agricultural ways of life in the face of large-scale industrial and extractive development projects orchestrated by alliances of post-colonial governments and multinational corporations.[13] In the Global North, meanwhile, agrarian issues have become at least as prevalent as wilderness preservation in environmental activism and discourse over the last half century, as seen in the growth of

---

[11] Joan Martinez-Alier, *The Environmentalism of the Poor: A Study of Ecological Conflicts and Valuation* (Northampton: Edward Elgar Publishing, 2003).

[12] Timothy Doyle, *Environmental Movements in Minority and Majority Worlds: A Global Perspective* (New Brunswick: Rutgers University Press, 2005).

[13] Some examples of movements that marry agrarianism with environmental justice are the *Movimiento Sin Tierra* in Brazil, the Chipko and rivers movements in India, as well as various anti-mining movements from the Philippines to Peru. Such movements often mobilize the interests of small agriculturalists and foreground agrarian political themes. See, for example: Alexander Dunlap, "'Agro si, mina NO!' The Tia Maria Copper Mine, State Terrorism, and Social War by Every means in the Tambo Valley, Peru," *Political Geography* 77, no. 1 (2019): 10–25. See also Mariana Walter and Lucrecia Wagner, "Mining Struggles in Argentina. The Keys of a Successful Story of Mobilization," *The Extractive Industries and Society* 8, no. 4 (2021). The international agrarian reform movement Via Campesina has attempted to forge solidarity among such movements across the post-colonial world, due to the similarities among them. For a general discussion of environmental movements in the Global South, see Martinez-Alier, *Environmentalism of the Poor*.

organic and regenerative agriculture, the local food and slow food movements, and the "new agrarianism" that offers a broad sustainability-focused critique of consumerism and the industrial growth economy.[14] Agrarian environmentalism holds promise, therefore, in its potential to forge connections and alliances between environmental movements of the North and South.[15]

Agrarianism is also timely as a mode of eco-critique because the ways of life and systems of values traditionally criticized by agrarians are no longer just purportedly corrupting forces within the human world; they are now, with the onset of destructive anthropogenic climate change, forces threatening civilizational collapse, ecological catastrophe and global destruction. James Montmarquet offers the following definition of agrarianism: "In its weakest sense we may take 'agrarianism' to assert merely that agriculture is an honorable (and virtuous) way of life; much more contentiously, the agrarian may assert that this is the most honorable way of life, or at least that it is morally superior to ways of life dedicated to typical urban pursuits."[16] To put a finer point on it, agrarianism has often favorably contrasted the agricultural way of life with a series of disfavored alternatives. For pre-modern agrarians, these included commerce, warfare and power politics.[17] In the modern context, we see these classical concerns extended to include a skepticism toward industrial manufacturing, technological change and extractive industry. The ecological appeal of agrarianism is plain. In contrast to ever-expanding industrial and extractive economies, the mode of life that agrarians celebrate seems to bear the hallmark of ecological salvation: sustainability. To be sure, the actual history of agriculture is a history replete with class domination and environmental degradation, but at least in principle agriculture works with and within the renewable energy cycles of the biosphere.

---

[14] See Eric T. Freyfogle, ed., *The New Agrarianism: Land, Culture, and the Community of Life* (Washington D.C.: Island Press, 2001).

[15] For example, see Wendell Berry, "Foreword," in *The Vandana Shiva Reader*, Vandana Shiva (Lexington: The University Press of Kentucky, 2014).

[16] J.A. Montmarquet, "Philosophical Foundations for Agrarianism," *Agriculture and Human Values* 2, no. 2 (1985): 5.

[17] Skepticisms toward commerce are clear themes in classical Western agrarian texts, including those of Hesiod, Virgil and Cato the Elder. Beyond an aesthetic pacifism that characterizes pastoral imagery, the works of Hesiod and Virgil in particular are notable for providing a counterpoint to the martial ethos that was otherwise so prevalent in the classical world. Hesiod provides the clearest example of agrarian antipathy toward power politics. Hesiod and Virgil will be explored in more detail below. Anti-mercantile agrarianism is also present in classical Chinese thought. See Roel Sterckx, "Ideologies of Peasant and Merchant in Warring States China," in *Ideologies of Power and Power of Ideologies in Ancient China*, ed. Yuri Pines, Paul R. Goldin, and Martin Kern (Boston: Brill, 2015): 211–248.

Wendell Berry's Eco-Agrarianism

If agrarian environmentalism holds significant promise, Wendell Berry deserves particular attention as one of its foremost literary proponents in the Anglo-American context. Since he began writing on agricultural topics over a half century ago, Berry has argued for a revival of an agrarian society centered on farming communities and particularly the family farm, which he regards as a basis not only for social and spiritual health, but also for an ecologically viable economy. For Berry, the ecological importance of the family farm lies in its functioning as a crucial point of affective contact and ethical praxis between human civilization and the ecosystems that support it.[18] The replacement of the family farm by large agribusiness corporations in the twentieth century, Berry contends, severs this crucial affective, ethical and practical linkage between people and the land, fundamentally altering the functional interface between humanity and nature. Corporations driven almost solely by the short-term profit motive have become humanity's proxies, replacing the much more complex sets of motivations and imperatives that had structured the interactions between traditional farmers and the land for thousands of years.[19]

And "the land" here must be understood quite literally. Over half of the solid surface of the earth is dedicated to some form of agriculture.[20] When systematic global changes, even subtle ones, occur within the agricultural sector, therefore, the potential geophysical impact is enormous. As Berry argues at length in numerous essays, the changes that have taken place over the last century in the agricultural sector have not been subtle, and they can be characterized fundamentally as a shift from farming as cultivation to farming as mining.[21] Soils are mined for their nutrients, their water conserving and carbon sequestering capabilities are diminished and the potential losses to the corporations are prevented by mining fertilizers, mining fossil fuels to synthesize fertilizers and mining water from rapidly depleting aquifers. The results are diminished long-term food production capacity and pollution of water, soil and air. Contemporary industrial agriculture in this sense has more in common with commercial mineral extraction than it does with what used to be called "farming."[22]

Critics of Berry and of other agrarian environmentalists have pointed out that in a rapidly urbanizing world of 8 billion people, it is fatuous to pretend

---

[18] Wendell Berry, *The Long-Legged House* (New York: Harcourt, 1969): 79.

[19] Wendell Berry, *Sex, Economy, Freedom, & Community* (New York: Pantheon, 1992): 36–8.

[20] See Paul B. Thompson, "Chapter 1: Sustainability and Environmental Philosophy," in *The Agrarian Vision: Sustainability and Environmental Ethics* (Lexington: University Press of Kentucky, 2010): 18–41.

[21] See especially Berry's essays in *The Art of the Commonplace: The Agrarian Essays of Wendell Berry*, ed. Norman Wirzba (Berkeley, CA: Counterpoint, 2002).

[22] Wendell Berry, *The Art of the Commonplace*, 287.

that everyone can become self-sufficient farmers. Over half of the people on the planet live in cities and that percentage is growing.[23] Arguments can be made, moreover, that urban dwellers use fewer resources and contribute less overall to negative ecological impacts, at least in industrialized countries.[24] Agrarian environmentalism is, however, not rendered obsolete by an urbanizing world. A transformation of the food economy and the kind of "reinhabitation of the landscape" that Berry has continually advocated would entail changed patterns of production, consumption and dwelling practices in both urban and non-urban environments. Suburbs could be replaced by productive, urban agricultural geographies, more densely populated walkable "mini-cities," or re-ruralized communities of small farms. Larger cities could become more densely populated and more starkly divided from nearby rural hinterlands on which they may more closely depend for food. These trends are already occurring in some places and are associated with economic, social and environmental benefits.[25]

Significant among the benefits of such an urban-friendly, neo-agrarian transition is the potential for mitigating global warming. Organic and sustainable farming practices such as poly-cropping, green manuring and no-till planting are now seen as a significant tool in reversing the loss of soil carbon that has resulted from conventional industrial farming practices.[26] If these soil-regenerating techniques were to be applied to all of the world's cultivated and grazed land, a recent study indicates that the building of topsoil and soil organic matter alone could store as much carbon each year as what is emitted by the global transportation sector.[27]

The economic roadblock to these agricultural practices is the cost of labor. Farming practices that build carbon-rich soils, keep waterways clean and contribute to domesticated and wild species diversity are probably only around 20% less efficient (in the near-term economic sense of efficiency) in terms of production per acre.[28] The more significant difference lies in how much labor is required per unit of agricultural output when the best ecological farming

[23] Hannah Ritchie and Max Roser, "Urbanization" OurWorldInData.org (2018), accessed February 1, 2022, https://ourworldindata.org/urbanization.

[24] For a synopsis of data for the United States bearing on this question, see Russell McLendon, "Urban or Rural: Which Is More Energy-Efficient?" *Treehugger*, accessed February 1, 2022, https://www.mnn.com/earth-matters/translating-uncle-sam/stories/urban-or-rural-which-is-more-energy-efficient.

[25] F.K. Benfield, *People Habitat: 25 Ways to Think about Greener, Healthier Cities* (Washington, D.C.: Island Press, 2014).

[26] Judith D. Schwartz, "Soil as Carbon Storehouse: New Weapon in Climate Fight?" *Yale Environment 360*, accessed March 13, 2019, https://e360.yale.edu/features/soil_as_carbon_storehouse_new_weapon_in_climate_fight.

[27] R.J. Zomer, Deborah A. Bossio, Rolf Sommer, and Louis V. Verchot, "Global Sequestration Potential of Increased Organic Carbon in Cropland Soils," *Scientific Reports* 7, no. 1 (November 2017): 1–8.

[28] Claire Lesur-Dumoulin, Eric Malézieux, Tamara Ben-Ari, Christian Langlais, and David Makowski, "Lower Average Yields But Similar Yield Variability in Organic Versus

practices are employed.[29] This "drawback" is, however, likely to become a benefit as global unemployment trends resulting from industrial automation begin to impose increasing social and economic costs on advanced industrial economies.[30] As with green energy, the labor-intensiveness of sustainable agriculture is likely to soon be seen as a mark in its favor.

Ultimately, such a new agrarian transition will, according to Berry, rely upon a transformation of consciousness at the individual level. Berry has often expressed a "distrust of movements."[31] For Berry, the needed changes must come from the ground up, not from experts, not from leaders and not from some "political bunch."[32] "One must begin in one's own life the private solutions," Berry writes, "that can only *in turn* become public solutions."[33] Berry has little to say about what such public solutions might look like, even if the private changes were to come about en masse. Politics, and even political activism, for Berry seem to intrude as a polluting substance on the ethical moorings provided by the fundamentally private, place-centered, agrarian form of life he extols. As one commentator recently wrote, "It is not simply that the personal is political; for Berry, there is no other political."[34] That the contemporary sustainable food movement largely shares in this individualistic, moralistic and arguably apolitical approach is evident in the way this movement has largely degenerated in popular culture to mere green consumerism.[35]

This apoliticality is a significant handicap to the new agrarian environmentalist movement. While Berry and others have helped inspire a global movement, the movement remains marginal and limited. As Nora Hanagan has argued in a convincing Deweyan critique of Berry, truly moving the now globalized system of industrial agriculture toward a neo-agrarian vision of sustainability will likely require organized resistance, civil disobedience, political coalition building and the implementation of economic and environmental

Conventional Horticulture: A Meta-analysis," *Agronomy for Sustainable Development* 37, no. 5 (2017): 45.

[29] James Morison, Rachel Hine, and Jules Pretty, "Survey and Analysis of Labour on Organic Farms in the UK and Republic of Ireland," *International Journal of Agricultural Sustainability* 3, no. 1 (2005): 24–43.

[30] Craig Pearson discusses the benefits of "community well-being and rural social capital" associated with the labor density of regenerative farming systems in Craig J. Pearson, "Regenerative, Semi-closed Systems: A Priority for Twenty-first-century Agriculture," *Bioscience* 57, no. 5 (2007): 409–418.

[31] See Wendell Berry, "In distrust of movements," *The Land Report* 65 (1999): 3-7.

[32] Wendell Berry, *A Continuous Harmony: Essays Cultural and Agricultural* (New York: Harcourt Brace Jovanovich, 1970): 87.

[33] Wendell Berry, *The Unsettling of America: Culture and Agriculture*(San Francisco: Sierra Club Books, 1977): 23.

[34] William Major, "Other Kinds of Violence: Wendell Berry, Industrialism, and Agrarian Pacifism," *Environmental Humanities* 3, no. 1 (2013): 31.

[35] I discuss the limitations of eco-labeling and sustainability as a consumer choice in Jake P. Greear, "Decentralized Production and Affective Economies: Theorizing the Ecological Implications of Localism," *Environmental Humanities* 7, no. 1 (2016): 107–127.

policies at the local, regional, national and perhaps even global level.[36] Given the political aporia within the thought of one of the most influential agrarian environmentalist writers, it is worthwhile to explore the larger tradition of agrarian thought and how its aesthetic and ethical naturalism is related to its engagement with politics and to the state itself. It is to this deeper history of agrarian thought that I turn in the following sections.

Agrarianism: Georgic and Pastoral

It may seem a safe assumption that prior to the modern age and the comforts it supplies, farming was such a toilsome reality and such a universal imperative that no one would ever think to extoll its merits. But in fact, agrarianism is an ancient mode of thought discernable, if only as a minor tradition, in the oldest known literatures, religions and philosophies. If we understand agrarianism as a broad category of political and social critique, it has two recognized aspects in the Western literary tradition: pastoral and georgic. Pastoral agrarianism can be traced back to the Greek bucolic poetry of Theocritus and celebrates agrarian settings as a site of leisure, innocence, peace and harmony.[37] The georgic mode can be traced back to Hesiod's *Works and Days*.[38] Georgic poetry and prose also celebrate rural landscapes and lifeways, but do so by thematizing work instead of leisure, and by offering ostensibly practical advice on farming. While both modes are often thought to have Greek origins, it was the Roman poet, Virgil, who drew upon these precursors to inaugurate these genres with his *Eclogues* and *Georgics*, respectively.

Exactly how agrarianism, pastoral and georgic are related and differentiated is a matter on which scholars differ. Leo Marx[39] and Lawrence Buell[40] are two literary theorists who essentially make agrarian political thought a species or instance of pastoral. Timothy Sweet,[41] however, proposes an expansive sense of georgic—as all literature that is dedicated primarily to negotiating the human interaction with nature—that would encompass agrarian political thought and probably pastoral literature as well.

---

[36] Nora Hanagan, "From Agrarian Dreams to Democratic Realities: A Deweyan Alternative to Jeffersonian Food Politics," *Political Research Quarterly* 68, no. 1 (2015): 34–45.

[37] See Part One of Paul Alpers, *What is Pastoral* (Chicago: Chicago University Press, 1996).

[38] See Stephanie Nelson, "Hesiod, Virgil, and the Georgic Tradition," in *The Oxford Handbook of Hesiod* (Oxford: Oxford University Press, 2018): 368.

[39] Leo Marx, *The Machine in the Garden* (Oxford: Oxford University Press, 1964): 5.

[40] Lawrence Buell, *Environmental Imagination* (Cambridge: Harvard University Press, 1995): 439.

[41] Timothy Sweet, *American Georgics* (Philadelphia: University of Pennsylvania Press, 2022): 2.

These frameworks are illuminating for these authors' purposes, but not authoritative. For present purposes, I provisionally adopt a different conceptual framework, considering pastoral and georgic as the two key tributaries in the history of Western agrarian thought.[42]

In Leo Marx's influential reading, the Virgilian pastoral is centered on an ideal subject position represented by the shepherd inhabiting a "middle landscape"—neither city nor wilderness. He writes,

> To arrive at this haven it is necessary to move away from Rome [the City] in the direction of nature. But the centrifugal motion stops short of unimproved raw nature. [...] This ideal pasture has two vulnerable borders: one separates it from Rome, the other from the encroaching marshland. It is a place where Tityrus is spared the deprivations and anxieties associated with both the city and the wilderness. [...] Virgil quickly itemizes the solid satisfactions of the pastoral retreat: peace, leisure, and economic sufficiency. The key to all of these felicities is the harmonious relation between Tityrus and the natural environment. It is a serene partnership.[43]

The significance of pastoral to contemporary environmental consciousness is acknowledged in the distinction made by Donald Worster between "arcadian ecology" and "imperial ecology" as the two principal and opposed approaches to nature in Western modernity.[44] And indeed, environmentalists may plausibly see the georgic tradition as antithetical to ecological thought, as it portends the exploitative, anthropocentric ethos of industrial modernity, bending and harnessing nature to the human will. Perhaps it is not surprising, though, that some ecologically oriented literary theorists have recently resurrected an "ecological georgic," arguing that the arcadian, pastoral alternative trades upon an overly sentimental, romantic and fantastical aestheticization of nature, packaged for urban consumption and, therefore, provide little of value to contemporary ecological thought. Georgic, on the other hand, may, according to these scholars, ground an environmental ethic in a more pragmatic, engaged and authentic appreciation for the natural world.[45] One might likewise suspect that a georgic perspective would offer resources for a more hard-headed mode of political engagement. These juxtapositions, however, gloss over the commonalities between these two literary modes, which become

---

[42] This use of terms is implied in Montmarquet's broad definition of "agrarianism" quoted above. (Montmarquet, 1985).

[43] Marx, *The Machine in the Garden*, 22–3.

[44] Donald Worster, *Nature's Economy* (Cambridge: Cambridge University Press, 1994): 3–55.

[45] Sweet, *American Georgic*. Michael G. Ziser, "Walden and the Georgic mode," *Nineteenth-Century Prose* 31, no. 2 (2004): 186 – 208. David Fairer "'Where Fuming Trees Refresh the Thirsty Air': The World of Eco-Georgic," *Studies in Eighteenth-Century Culture* 40, no. 1 (2011): 201–218.

apparent if we read Hesiod's *Works and Days* as an early instance of ecological agrarianism.

Hesiod's Ecological Agrarianism

The poetry of Hesiod's *Works and Days* is often taken as the earliest surviving instance of georgic verse, but in spite of the distinction often drawn, it shares many commonalities with Virgil's *Eclogues*. In Hesiod, we find a convergence (or perhaps a common origin) of democratic political critique, practical agricultural wisdom and a certain naturalistic ethic and aesthetic. Although a significant portion of the poem is framed as a compendium of practical wisdom for the yeoman farmer—which also functions as backhanded moral advice to Hesiod's estranged brother, Perses—the primary importance of Hesiod for most classicists lies not in any advancements in the agricultural arts, nor in the exhortation dished up for Perses, but rather in the radical break with the Homeric, warrior-aristocratic ethos that the poem represents.[46] Hesiod celebrates work over war, he exalts the small farmer over sea-faring merchants and "bribe-swallowing lords," and he sides with humble victims against the predatory military class.[47] (In these ways, Hesiod's poetry is ultimately a critique of the State, such as it existed in early Iron Age Greece). Like Virgil's *Eclogues*, Hesiod's narrative functions to place a deliberate distance between the central sympathetic figure—the yeoman farmer—and the centers of political power: city, court or *polis*.

A naturalistic aesthetic is conveyed along with the poem's political critique through the didactic form and the agrarian content of the latter half of the poem. The subject position of innocence that Hesiod's pious yeoman occupies is established by his reverent and attentive participation in the natural order. As Stephanie Nelson points out, Hesiod's text notably marks no normative distinction between wild and domesticated nature.[48] In a series of proverbial prescriptions mainly concerning husbandry, tillage and viticulture, Hesiod artfully depicts his own pastoral ideal of the prosperous farmstead situated within a natural environment that is neither fully wild nor fully controlled or artificial.

---

[46] See Levi Bryant, "Military Technology and Socio-political Change in the Ancient Greek City," *The Sociological Review* 38, no. 3 (1990): 484–516.

[47] See lines 225–237 of Hesiod's *Works and Days*. References to Hesiod's work here and below are to line number rather than page number, and are taken from Glen W. Most, ed. and trans., *Theogony, Works and Days, Testimonia*, (New Haven, CT: Harvard, Loeb Classical Library, 2006).

[48] Stephanie Nelson, *God and the Land, The Metaphysics of Farming in Hesiod and Vergil* (Oxford: Oxford University Press 2008): 152.

"When the Atlas-born Pleides arise, start the harvest..."[49]

"Take notice when you hear the voice of the crane every year, calling from above the clouds, she brings the sign for plowing..."[50]

"After this, Pandion's daughter, the dawn-lamenting swallow, rises into the light for human beings, and the spring begins anew. Forestall her, prune the vines first..."[51]

Harmony between the agriculturalist and nature is not depicted through musical metaphors and themes of leisure and interpersonal harmony, as in Virgil's *Eclogues*, but it is nonetheless a central theme. For Hesiod, as for Berry, this harmony is depicted in terms of work, which, when properly carried out, is portrayed as justly rewarded by a theistically-ordered cosmos.

> Peace, the nurse of the young, is on the earth, and far-seeing Zeus never marks out painful war; nor does famine attend straight-judging men, nor calamity, but they share out in festivities the fruits of the labors they care for. For these the earth bears the means of life in abundance, and on the mountains the oak tree bears acorns on its surface, and bees in its center; their woolly sheep are weighed down by their fleeces; and their wives give birth to children who resemble their parents. They bloom with good things continuously. And they do not go into ships, for the grain giving field bears them crops.[52]

Both the naturalistic aesthetic and the political critique in Hesiod's agrarianism are part and parcel of a valuation of private over public life and a vision of the virtuous individual next to which any vision of a virtuous community is secondary, if not altogether neglected. Human community is not absent from Hesiod's agrarian imagery, but the agrarian society depicted is remarkably individualistic. The nuclear family (understood as potentially including enslaved laborers) is the basic unit of society, and Hesiod famously portends *laissez-faire* economic theory in celebrating the industriousness that results from a kind of "harmony of discord"—when farmer competes with farmer, blacksmith with blacksmith and so on.[53] Like Berry's agarianism and Virgil's pastoral, Hesiod's work bears an arguably democratic political critique as well as a resonant vision of harmony with nature, but also expresses a disvaluation of politics itself. "Perses, do store this up in your spirit, lest gloating Strife keep your spirit away from work, while you gawk at quarrels and listen to the assembly. ... When you can take your fill of [the fruits of your labor] then you might foster quarrels and conflict for the sake of another man's wealth."[54]

---

[49] Hesiod, *Works and Days*, 383.
[50] Ibid., 450.
[51] Ibid., 570.
[52] Ibid., 230–237.
[53] Ibid., 11–26.
[54] Ibid., 121–130.

In light of these common themes in the long tradition of Western agrarian thought, it may be helpful to reflect on the relationship between agriculture, nature and the state at a yet more fundamental level, and in a yet more ancient historical context. Toward this end, the next section briefly surveys current research on the relationship between agriculture and political power in the emergence of the earliest states.

Ruralization in Early States

Agriculture, and particularly grain agriculture, arose independently in a few regions around the world over the last 10,000 years—relatively recently if we consider the roughly 200,000-year history of anatomically modern humans. The first states arise even more recently—only about 5,000 years ago. The classical tradition of Western political thought, from Machiavelli to Hobbes and Locke, suggested that fear of violence and the desire for security were key to the logic of state formation. States, on this account, arise as defensive pacts. However, most modern theories about the origins of the state have considered the advent of agriculture to be decisive. These two lines of reasoning are not mutually exclusive. The intensive cultivation of cereal grains could lead to state formation precisely because settled, grain-rich communities needed to provide for the common defense against raiders from the outside who sought to enjoy the fruits without the labor. Agriculture could also initiate state formation for non-security-related reasons. It could be simply that grain production led to population increases in sedentary societies, which then leads to state formation as population density rises and various mounting collective action problems demand bureaucratic solutions. One theory connecting agriculture to state formation asserts that the states arose to organize complex irrigation systems and distribute water rights.[55]

Archeological, historical and ethnographic evidence indicates, however, that agriculture does not lead inexorably to state formation any more than patterns of intergroup violence do. James C. Scott makes a point of this in his recent book, *Against the Grain: A Deep History of the Earliest States*.[56] A key point for Scott is that there were, in the ancient world at least, no yam or taro states. Root crops did not lend themselves to taxation, centralized storage and large-scale redistribution and, therefore, wherever root crops became the dominant agricultural staple states did not tend to arise.[57] Cereal agriculture may therefore be a necessary antecedent to the earliest states, but even if necessary, cereal agriculture is no more an efficient cause of the earliest states than warfare as

---

[55] One prominent proponent of the plausible but now generally discredited "hydraulic hypothesis," centering on irrigation agriculture as a key antecedent to the state, is Karl S. Wittfogel, *Oriental Despotism: A Comparative Study of Total Power* (New Haven: Yale, 1957).

[56] James C. Scott, *Against the Grain: A Deep History of the Earliest States* (New Haven, CT: Yale University Press, 2017).

[57] Ibid., 214.

settled cereal-growing communities predate the earliest states by at least a few millennia.

This is evidenced in the fertile crescent, where there appears to have been a flowering and then a decline of what some archaeologists call "mega-villages" during the three or four millennia after the advent of agriculture but prior to the appearance of the first states in Kemet and Mesopotamia.[58] The largest of these mega-villages yet discovered is at the site of Çatalhöyük in modern-day Turkey. We could call the Çatalhöyük settlement a "city" based on population size and density. At its height, about 9,000 years ago, it housed as many as 8,000 people in close proximity.[59] And yet, in spite of what was clearly a settled society with some degree of agricultural surplus, there exists in the archaeological record, which spans over two thousand years of near-continuous occupation, no evidence of any social stratification, kingship or statehood. There are no monumental architecture, no ceremonial center and no centralized grain storage. In fact, there is no centralized anything.[60] Cities of comparable population size in the Classic Mayan or early Mesopotamian civilization could, by contrast, make claims to full-fledged city-statehood, with ruling monarchs, a hierarchy of social classes and bureaucratic organizations, all evidenced in the archaeological record.[61]

That such late-neolithic mega-villages arose, prospered, expanded, developed technologically and persisted for many centuries before dispersing, indicates not only that a certain kind of "urbanism" can persist without bureaucratic organization, but also that densely populated, settled societies that are at least partially agricultural can arise and persist for long periods without becoming states. However, while the earliest states did not invent agriculture, and while agriculture does not produce states in any deterministic sense, the state form does have a special relationship with farming and in almost every case states transformed and expanded agriculture in sociologically significant ways.[62]

According to Norman Yoffee, a leading scholar of early state formation, one of the dramatic effects of the appearance of the fist city-states is what he calls the "ruralization" of the surrounding landscape.

---

[58] See Nerissa Russell, "Spirit Birds at Neolithic Çatalhöyük," *Environmental Archaeology* 24, no. 4 (2019): 377–386.

[59] For an introduction to Çatalhöyük, see Ian Hodder, *The Leopard's Tale* (London: Thames and Hudson, 2006).

[60] Ibid., 7.

[61] See Robert J. Sharer and Loa P. Traxler, *The Classic Maya* (Stanford. CA: Stanford University Press, 2006): 688.

[62] According to Bruce Trigger, between 70 and 90 percent of available labor had to be devoted to food production, leaving precious little surplus to be exploited by ruling classes. Bruce Trigger, *Understanding Early Civilizations* (Cambridge, Cambridge University Press, 2003): 313.

For many of the earliest cities, the urban demographic implosion was accompanied by an equally important creation of the countryside. This process of ruralization can be observed in two dimensions. First, existing towns and villages became networked to urban places. The social and economic roles of non-urban dwellers were tied to decisions made in the cities; specialized institutions of production and consumption in the countryside [...] were altered by the demands of urban rulers and elites, and ranks of urban officials were conceived precisely to carry out new activities. Second, countrysides became relatively depopulated as many people became incorporated in the new cities, [...] Subsequently new villages, towns and hamlets arose in the backdraft of urbanization. This condition also led to the intensification of specialized activities, such as pastoralism and nomadism, which flourished not only to supply goods and services to cities but also served as refuges for urban flight.[63]

Ruralization would have taken a variety of forms in different locations, and prehistoric rural sociology is necessarily a speculative business. Nonetheless, a general trend is discernible, in which early territorial states lead to more land being brought under cultivation by agricultural specialists living in relatively dispersed settlement patterns outside of early cities. Archaeological evidence from the North American proto-state of Cahokia (circa 1000 CE) indicates the dynamics that likely characterized the transformation from pre-state agricultural villages to a ruralized landscape under the dominion of early states. Shortly after the building and settlement of Cahokia, the patterns of settlement in the surrounding landscape changed significantly as violence subsided and agriculture intensified.[64] Fortified villages, which entail dense nucleated clusters of often communal dwellings, were replaced by settlements without defensive palisades, consisting of what seem to be single-family "spatially discrete farmsteads situated on bottomland ridge crests and slopes."[65] This relatively pacified, "ruralized" Mississippian space was coeval with an intensification of agriculture. It appears to have persisted only for a short time—a period of one or two hundred years—which some have referred to as the *Pax Cahokiana*, after which Cahokia declined as a political power.[66] As it did so, rival chiefdoms arose throughout the greater Mississippi valley and a period

---

[63] Norman Yoffee, *Myths of the Archaic State* (Cambridge: Cambridge University Press, 2005): 60.

[64] Erik Browne writes, "In the decades before 900 AD, many people in the region had been forced to live behind stockade walls due to endemic violence." Erik E. Browne, *Mound Sites of the Ancient South: A Guide to the Mississippian Chiefdoms* (Athens, GA: University of Georgia Press, 2013): 63.

[65] G.R. Milner, "Mississippi period population density in a segment of the central Mississippi River valley," *American Antiquity* 51, no. 2 (1986): 228.

[66] Timothy Pauketat writes of the demise of the "Cahokian peace," "Without the mantle of Cahokian peace covering the Mississippi, village based tensions and ethnic level tensions re-emerging, with squadrons of warriors prowling the landscape and with one village's warriors fighting those of other villages." T.R. Pauketat, *Cahokia: Ancient America's Great City on the Mississippi* (New York: Penguin Random House, 2009): 168.

of generalized hostility resumed characterized by inter-polity, inter-ethnic and inter-village conflict.[67]

According to Bruce Trigger, in his influential *Understanding Early Civilizations* (2005), a similar dynamic is indicated in the evolution of the Incan Empire which had enough of a pacifying influence in the Andean valleys to allow farmers to expand into what had been unoccupied hinterlands at different elevations. "The Inka state," he writes, "greatly benefitted highland farmers, by abolishing local warfare, thus permitting them to leave their hillforts and resettle near the thirty-five-hundred-meter-line where they could grow the widest possible array of crops."[68] These cases suggest that pacification is one of the mechanisms at work in the "ruralization" that Yoffee points to and which he shows is widespread in the archaeological record, including Wari (a pre-Inca Andean state), and the pre-Aztec metropolis of Teotihuacan.[69]

Trigger's study of seven early civilizations suggests that the transformation of the countryside was most pronounced in early territorial states such as the Inka Empire and Kemet, as opposed to city-state systems such as the Mayan and Mesopotamian civilizations. In these territorial states,

> the populations [of capital cities] tended to be of modest size. Even national capitals, with perhaps fifty thousand inhabitants, were no more populous than the capitals of substantial city-states, although the populations that sustained them were many times larger. This was because these centers were inhabited almost exclusively by the ruling class and by administrators, craft specialists, and manual labourers [...] Because the government protected the state as a whole from armed attack, farmers preferred for practical reasons to live in dispersed homesteads and villages near their fields rather than in urban centers.[70]

The process of "ruralization" in city-state systems such as in Mesopotamia seems to have looked quite different, sometimes reverting to a scaled-up version of the pre-Cahokian, nucleated farming villages of the Mississippi Valley. Soon after Uruk is urbanized (around 3000 BCE), rival city-states emerged, and rather than rural dispersal and a simple de-nucleation of settlements, we see significant rural depopulation. Farmers retreat inside the city walls, venturing out daily to work nearby fields. "To a considerable degree, the ambitions of city rulers and the fears of rural peasants would have converged to persuade or compel the latter into a piecemeal flight from their dispersed agricultural enclaves to the larger centers."[71] In this insecure environment,

---

[67] Pauketat, *Cahokia: Ancient America's Great City on the Mississippi*, ??.

[68] Bruce Trigger, *Understanding Early Civilizations* (Cambridge, Cambridge University Press, 2003): 300.

[69] Yoffee, *Myths of the Archaic State*, 52.

[70] Trigger, *Understanding Early Civilizations*, 131–132.

[71] Robert McCormick Adams and Hans J. Nissen, *The Uruk countryside: The Natural Setting of Urban Societies* (Chicago: University of Chicago Press, 1972): 21.

the lower Mesopotamian hinterland "had become a dangerously unsettled district" in which the "only inhabitants were small groups of wary nomads or semi-sedentary folk, moving with their flocks."[72]

Over time, many city-state systems do, of course, become consolidated into territorial states, as happened in Mesopotamia under Sargon the Great and later in the Mediterranean under the *Pax Romana*. But even unconsolidated city-state systems may produce a relatively pacified rural space. The *poleis* of Classical Greece resisted unification for over five centuries, and yet the Hellenic countryside, although politically fractured, was secure enough by the late eighth century BCE to support what is arguably the first agrarian society based on numerous, dispersed, privately owned, small farmsteads in an intensively cultivated landscape.[73]

The advent of the state itself, according to Yoffee, is best seen not as a process of increasing complexity, but rather as one of simplification, or at least *attempted* simplification. Drawing upon James C. Scott's *Seeing Like a State,* Yoffee contends that ancient states "tried no less than modern ones to refashion and simplify social arrangements and make them 'legible.'"[74] Processes of social, political and economic simplification are part of the story of the agrarian transformation of the countryside. Many craft specialists moved out of agricultural villages into the urban centers while urban ruling elites attempted to control out-lying territories by simplifying and hierarchizing social and political relations among those who remain, mostly agricultural specialists, whether of the peasant or yeoman variety.[75] Centralized administrative politics, moreover, sought to supplant the complex politics of pre-state and non-state societies within administered territories.[76]

---

[72] Ibid., 32.

[73] Victor David Hanson, *The Other Greeks: The Family Farm and the Agrarian Roots of Western Civilization* (New York: Free Press, 1995): 91–126.

[74] Yoffee, *Myths of the Archaic State,* 94. See also Gilles Deleuze and Felix Guattari, *A Thousand Plateaus: Capitalism and Schizophrenia,* trans. Brian Massumi (Minneapolis: University of Minnesota Press, 1987), 474–500. Deleuze and Guattari refer to state space as "striated space." Conversely, non-state space—particularly the space of nomadic societies—they call "smooth space." The sense of this conceptual distinction is most apparent in the physical architecture of the earliest cities-states. Modern cities are in fact far more organic—more village-like—than the most ancient urban spaces, which are highly planned, often carefully oriented with sophisticated astrological considerations and elaborated with rectilinear and symmetrical architectural motifs. On the cosmological architecture of early cities, see also Kevin Lynch, *Good City Form* (Boston, MA: MIT Press, 1981). It is the agricultural landscapes of early state territories, however, which afford the principal example for Deleuze and Guattari, as the land is physically striated with the plow but also orchestrated and taxed according to grid-like or tree-like conceptual schemas overlaid on agricultural populations and territories by governments.

[75] See Trigger, *Understanding Early Civilizations,* 120–121. See also Yoffee, *Myths of the Archaic State,* 60 – 61.

[76] Trigger, *Understanding Early Civilizations,* 142.

As Yoffee notes, simplification was an elite aspiration that often failed, even when the state itself persisted.[77] Rural hinterlands, in any case, are always hotbeds of potential resistance,[78] and the political criticism inherent in early agrarianism attests to this. The first agrarians and peasant utopians begin to imagine a world apart from war, commerce and power politics, in which the people might peaceably supply their own means of life through a working harmony with nature—a world where "the worthy ruler feeds himself by ploughing side-by-side with his people, and rules while cooking his own meal,"[79] which means, of course, a world without rulers per se. What I am suggesting in this section and the next is that these agrarian ideas should be understood as emerging very early, but emerging from a landscape and a world that has been transformed by the state.

With these observations, I do not intend to adopt an uncritical attitude to the state. If anything, I am sympathetic with the normative thrust of Scott's book, *Against the Grain*, which is to say, against the state. However, my aim in this section is not to adjudicate the merits or demerits of early states or even the state form of society in general. At least not directly. Rather, in this chapter, I wish to elucidate is what we might call the "environmental hermeneutics" of the early state. That is, how might the experience of the landscape and the perception of the landscape change under these changing conditions, as the landscape itself is changed and occupied differently? *What I suggest is that there is a particular vision and experience of "nature" that is peculiar to agrarianism in both its ancient and modern versions, and in both its literary and political aspects, and that this agrarian "nature," while antagonistic to the state, may be fundamentally also produced by the state.* The following section seeks to further explicate and explore this agrarian "nature of the state," with reference to recent eco-criticism.

Nature Alienation: Agrarian and Wilderness Perspectives

Western environmental theorists have often postulated a significant break in human consciousness of the natural world at the transition from hunting and gathering to farming. As Max Oelshlager writes,

> The onset of Neolithic culture forever altered both intellectual and material culture [...] Rather than attempting to live in harmony with wild nature, as hunter-gatherers had done since time immemorial, farmers literally rose up and

---

[77] Yoffee, *Myths of the Archaic State*, 112. Here Yoffee is again following Scott, who makes this argument with respect to modern states.

[78] See James C. Scott, *The Art of Not Being Governed* (New Haven: Yale University Press, 2009).

[79] Quote attributed to Ch'en Hsiang around 350 BCE by A.C. Graham "The 'Nung-chia' 農家 'School of the Tillers' and the Origins of Peasant Utopianism in China," *Bulletin of the School of Oriental and African Studies* 42, no. 1 (1979): 66.

attempted to dominate the wilderness. Boundaries were drawn between the natural and the cultural and conceptual restructuring was inevitable.[80]

Deep ecology's "back to the Pleistocene" rallying cry is based on this view of things. As Earthfirst founder Dave Foreman writes,

> Before agriculture was midwifed in the Middle East, humans were in the wilderness. We had no concept of "wilderness" because everything was wilderness and *we were part of it*. But with irrigation ditches, crop surpluses, and permanent villages we became *apart from* the natural world. ...Between the wilderness that created us and the civilization created by us grew an ever-widening rift.[81]

This view of a Neolithic fall from ecological grace is not confined to twentieth-century deep ecology. Timothy Morton, who professes a post-natural approach to eco-criticism that is ostensibly at odds with Oelschlager's and Foreman's defense of wilderness, nevertheless comes to much the same conclusion, arguing that the advent of agriculture erected a "massive firewall between humans and nonhumans."[82] For Morton, the Neolithic moment of our separation from "non-human otherness" is marked by the emergence of the very concept of "nature." Therefore, in Morton's view, it seems the deep ecologists' project of reconnecting with wilderness is only a late-coming symptom of this rudimentary alienation, not a cure for it. On this point, though, the only real difference between the deep ecology viewpoint and Morton's own lies in the characterization of that from which humans become alienated after they start farming.

Agrarian environmentalism offers a clearly distinct account of nature alienation which perhaps entails a distinct conceptualization of nature itself. Here, it is not the transition *to* farming that marks a crucial turning point in environmental subjectivity, but the later transition *away from* an agrarian mode of life and into industrial modernity. We may return here to Wendell Berry for a prominent articulation of this view. "The word 'agriculture,'" he writes,

> means "cultivation of land." And *cultivation* is at the root of the sense both of *culture* and of *cult*. The ideas of tillage and worship are thus joined in culture. And these words all come from an Indo-European root meaning both 'to revolve' and 'to dwell.' To live, to survive on the earth, to care for the soil, and to worship, all are bound at the root to the idea of a cycle.[83]

---

[80] Max Oelschlaeger, *The Idea of Wilderness: From Prehistory to the Age of Ecology* (New Haven: Yale University Press, 1991): 28.

[81] Dave Forman, *Confessions of an Eco-Warrior* (New York: Harmony Books, 1991): 69 (original emphasis).

[82] Timothy Morton, *Being Ecological* (Cambridge: The MIT Press, 2018): 59 (original emphasis).

[83] Berry, *The Unsettling of America*, 87.

At some later point, though, he writes,

> The Wheel of Life became an industrial metaphor; rather than turning in place, revolving in order to dwell, it began to roll on the "highway of progress" toward an ever-receding horizon. The idea, the responsibility, of return weakened and disappeared from agricultural discipline. Henceforth, *any* resource would be regarded as an ore.[84]

Martin Heidegger came to similar conclusions about the environmental merits of what he would have thought of as "peasant life." To live on the land and carry on the quiet, cyclical labors of subsistence farming, for Heidegger, is the very essence of being-in-the-world, or existence within an immediate environment.[85] Modern science and technology, urban existence, and mass society, on the other hand, represented for him an existential separation from nature, environment or lifeworld. Lynn White Jr. also offered a variant of the agrarian account of nature alienation, arguing that the introduction of the heavy plow in Northern Europe around 900 CE marks the crucial turning point in the Western relationship to nature.[86]

Contemporary eco-Marxists rely upon yet another articulation of the agrarian environmentalist version of the nature alienation narrative which is based on the Marxian concept of the "metabolic rift" inherent in capitalist modes of production. According to John Bellamy Foster et al., the pioneering work of chemist Justus von Liebig on the cycling of nitrogen and phosphorous had drawn the attention of Marx who came to understand that capitalist production interrupted these cycles; the agricultural products of increasingly industrialized farms were transported over ever longer distances into cities, to be consumed by growing populations of urban workers, resulting in twin problems of urban pollution and rural soil depletion.[87] Other eco-Marxists inspired by Foster et al. have expanded the logic of the metabolic rift to consciously attend to not only the rift in material flows, but also the concomitant rift or rupture in the flows of "sensuous knowledge" that had connected small-scale agriculturalists to land and environment.[88] In this view, the metabolic rift in matter/energy flows under capitalist production was also

[84] Berry, *The Art of the Commonplace*, 287.

[85] This basic narrative is arguably discernible thought Heidegger's corpus, but see especially Martin Heidegger, "Why Do I Stay in the Provinces?" in *Heidegger, Philosophical and Political Writings*, ed. Manfred Stassen (New York: Continuum International, 2006).

[86] Lynn White Jr., "The Historical Roots of Our Ecologic Crisis," *Science* 155, no. 3767 (1967): 1203–1207.

[87] John Bellamy Foster, Brett Clark, and Richard York, *The Ecological Rift: Capitalism's War on the Earth* (New York: New York University Press, 2010).

[88] See Mindi Schneider and Philip McMichael, "Deepening, and Repairing, the Metabolic Rift," *The Journal of Peasant Studies* 37, no. 3 (2010): 461–484. See also Ivan Scales, "Green Consumption, Ecolabelling and Capitalism's Environmental Limits," *Geography Compass* 8, no. 7 (2014): 477–489.

and at the same time a "knowledge rift" and an "epistemic rift" at the root of the general ecological blindness of bourgeois modernity. Similar to the accounts of Berry and Heidegger, here it is the pre-capitalist farmer—whether peasant, yeoman or communard—whose relationship to nature manifests the basic ingredients of ecological harmony and integrity.

Both the agrarian and wilderness versions of the nature alienation narrative may be questioned. Here, I venture to speak a word for agrarian nature. Although arguable, it is not obvious that stabbing animals to death with stone-tipped spears or burning forests to create game openings is a more harmonious, attuned or ecologically conscious mode of relation to the non-human world than slashing and burning and planting manioc or tending an established farmstead. Furthermore, there is much overlap between mobile foraging societies and farming societies, both historically and among contemporary non-state societies. The earliest cultivators and domesticators certainly combined farming with hunting and foraging. In Neolithic farming villages in the fertile crescent, artifacts arguably suggest more cultural continuity with the Paleolithic hunters who painted the walls of Lascaux cave than with their fellow farmers in the early city-states of Mesopotamia. Indeed, the rise of the state itself may be a more significant and decisive moment in the history of environmental consciousness than the advent of sedentism and agriculture.

But, rather than marking a break with nature, *the advent of the state may be seen as producing nature—in the sense of bringing about new modes of relation to the non-human world that comprise an important part of what we today mean by the term "nature" and even by the term "wilderness."* The ruralization of space by early states—which I have suggested can be seen as a relative pacification of a space of dispersed subsistence—would seem to make possible new, and in some ways more intimate, even if also more geographically limited, modes of interaction with the biotic landscape.

One of the classical senses of the Greek term *phusis* (and the Latin *natura*) is the simple contrast with artifice. Nature is that which is not made, willed or caused by humans. Ruralized space bears the physical stamp of the human will, of course, but in an equally valid sense it may open a space for escape from the human world. In other words, it may be easy for environmental theorists such as Oelshlager or Morton to over-estimate the extent to which, in an eco-phenomenological sense, the pre-agricultural "wilderness" is an *inhuman* space. The "nature" in which hunter-gatherers and pre-state agriculturalists were immersed was, it seems, not often experienced in peaceful reverie or solitude. It was likely very often a socialized and politicized space, if not a space characterized by the ever-present threat posed by potentially hostile neighboring bands of humans, not to mention deadly predators. As one historical anthropologist points out, many hunter-gatherers in the distant past occupied densely populated prime habitats, which would systematically differ from the marginal and sparsely populated territories occupied by the few remaining mobile forager bands of today.[89] The idea of nature, as contemporary environmentalism conceives it, is a space of the non-human, but also importantly, and I contend relatedly, a space of peace from the human perspective. And it is a space of peace and freedom precisely because it can become a space of escape from the human world and from the strife inherent in the interdependence of the human worlds of commerce, politics and war.

Conclusion

Bill McKibben, in *The End of Nature*, tells a story about how his restorative walks in the Vermont woods are occasionally ruined by the sound of a nearby chainsaw—an obtrusive, even violent sound of human artifice. In fact, the sound does not even have to be actually heard to ruin the experience of nature in this sense. Just the anticipation of the sound, the awareness of its possibility, has much the same "denaturalizing" effect.[90]

For McKibben, the point of this story is to illustrate the sense in which the global systemic alteration of the environment by humans, from global warming to stratospheric pollution, marks the death of nature in precisely this ecophenomenological sense. Even if you don't see the difference, you know it's there. And, of course, it *is* there, physically, materially. So sometimes you *do*, in fact, see it or feel it—when the jonquils bloom in January or when you recognize the effects of acid rain on a stand of spruce. It is now, McKibben writes, as if "the chainsaw is always in the woods…"[91] But how much, really, does this post-natural environmental hermeneutic of the Anthropocene environmentalist differ from the pre-natural one of the Pleistocene hunter-gatherer? Perhaps what is lost here is not a primordial nature, but an "unnatural nature"; the agrarian environmentalist's nature; the Arcadian nature that was produced first by the State.

I do not suggest that this "unnatural" agrarian nature should be disvalued on the basis of its entanglement with the state. Rather, what I propose on the basis of this analysis is that agrarian environmentalism is fundamentally utopian and inextricably political in the same way that political anarchism, Marxism, and radical democracy are. If radical democracy can be understood as an attempt to preserve many advantageous aspects of the state without the rigid hierarchies that have otherwise been its nearly universal and arguably essential feature, if Marxism wishes to eventually enjoy industrial plenitude without the state's control over the means of production, and if political anarchism wishes to attain a civil society without coercive force, then agrarian environmentalism similarly wishes to preserve a space of nature and an ecological way of life that is threatened by the very thing that heretofore made it possible: the state. Like these more overtly political projects, agrarian environmentalism ought to

---

[89] Azar Gat, *War in Human Civilization* (Oxford: Oxford University Press, 2008): 18.
[90] Bill McKibben, *The End of Nature* (New York: Random House, 1991): 40.
[91] Ibid.

understand itself as a radical political project engaged in the remaking, if not the transcendence, of the state.

References

Adams, Robert McCormick and Hans J. Nissen. *The Uruk Countryside: The Natural Setting of Urban Societies.* Chicago: University of Chicago Press, 1972.
Alpers, Paul. *What is Pastoral.* Chicago: Chicago University Press, 1996.
Azar Gat. *War in Human Civilization.* Oxford: Oxford University Press, 2008.
Benfield, F.K. *People Habitat: 25 Ways to Think About Greener, Healthier Cities.* Washington, D.C.: Island Press, 2014.
Berry, Wendell. *The Long-Legged House.* New York: Harcourt, 1969.
Berry, Wendell. *A Continuous Harmony: Essays Cultural and Agricultural.* New York: Harcourt Brace Jovanovich, 1970.
Berry, Wendell. *The Unsettling of America: Culture and Agriculture.* San Francisco: Sierra Club Books, 1977.
Berry, Wendell. *Sex, Economy, Freedom, & Community.* New York: Pantheon, 1992.
Berry, Wendell. "In Distrust of Movements." *The Land Report* 65 (1999): 3–7.
Berry, Wendell. *The Art of the Commonplace: The Agrarian Essays of Wendell Berry*, edited by Norman Wirzba. Berkeley: Counterpoint, 2002.
Berry, Wendell. "Foreword," In *The Vandana Shiva Reader*, Vandana Shiva. Lexington: The University Press of Kentucky, 2014.
Browne, Erik E. *Mound Sites of the Ancient South: A Guide to the Mississippian Chiefdoms.* Athens, GA: University of Georgia Press, 2013.
Bryant, Levi. "Military Technology and Socio-political Change in the Ancient Greek City." *The Sociological Review* 38, no. 3 (1990): 484–516.
Buell, Lawrence. *Environmental Imagination.* Cambridge: Harvard University Press, 1995.
Deleuze, Gilles and Felix Guattari. *A Thousand Plateaus: Capitalism and Schizophrenia.* Translated Brian Massumi. Minneapolis: University of Minnesota Press, 1987.
Dunlap, Alexander. "'Agro si, mina NO!' The Tia Maria Copper Mine, State Terrorism, and Social War by Every Means in the Tambo Valley, Peru." *Political Geography* (2019): 10–25.
Fairer, David. "'Where Fuming Trees Refresh the Thirsty Air': The World of Eco-Georgic." *Studies in Eighteenth-Century Culture* 40, no. 1 (2011): 201–218.
Forman, Dave. *Confessions of an Eco-Warrior.* New York: Harmony Books, 1991.
Foster, John Bellamy, Brett Clark, and Richard York. *The Ecological Rift: Capitalism's War on the Earth.* New York: New York University Press, 2010.
Fox, Liam. "Bougainville President Says Panguna Mine Moratorium Remains in Place." *Radio Australia, Australian Broadcasting Company.* Accessed February 9, 2021, https://www.abc.net.au/radio-australia/programs/pacificbeat/bougainville-president-says-panguna-mine-off-limits/13134904.
Freyfogle, Eric T., Ed. *The New Agrarianism: Land, Culture, and the Community of Life.* Washington D.C.: Island Press, 2001.
Graham, A.C. "The 'Nung-chia' 農家 'School of the Tillers' and the Origins of Peasant Utopianism in China." *Bulletin of the School of Oriental and African Studies* 42, no. 1 (1979): 66.

Greear, Jake P. "Decentralized Production and Affective Economies: Theorizing the Ecological Implications of Localism." *Environmental Humanities* 7, no. 1 (2016): 107–127.
Grundy, Kenneth W. *River of Tears: The Rise of the RioTinto Zinc Mining Corporation*. London: Earth Island Press, 1974.
Hanagan, Nora. "From Agrarian Dreams to Democratic Realities: A Deweyan Alternative to Jeffersonian Food Politics." *Political Research Quarterly* 68, no. 1 (2015): 34–45.
Hanson, Victor David. *The Other Greeks: The Family Farm and the Agrarian Roots of Western Civilization*. New York: Free Press, 1995.
Heidegger, Martin. "Why Do I Stay in the Provinces?" In *Heidegger, Philosophical and Political Writings*, edited by Manfred Stassen. New York: Continuum International, 2006.
Hodder, Ian. *The Leopard's Tale*. London: Thames and Hudson, 2006.
Ingram, Darby. "Bougainville President Rejects Panguna Mine Claims." *National Indigenous Times*. Accessed February 1, 2021, https://nit.com.au/bougainville-president-rejects-panguna-mine-claims/.
Jones, Daniel. "Panguna Mine Dilemma." Filmed 2008, *Eel Films*. https://www.youtube.com/watch?v=Sv8Q5hH0cys.
Lesur-Dumoulin, Claire, Eric Malézieux, Tamara Ben-Ari, Christian Langlais, and David Makowski. "Lower Average Yields But Similar Yield Variability in Organic Versus Conventional Horticulture: A Meta-analysis." *Agronomy for Sustainable Development* 37, no. 5 (2017): 45.
Lynch, Kevin. *Good City Form*. Boston, MA: MIT Press, 1981.
Major, William. "Other Kinds of Violence: Wendell Berry, Industrialism, and Agrarian Pacifism." *Environmental Humanities* 3, no. 1 (2013): 31.
Martinez-Alier, Joan. *The Environmentalism of the Poor: A Study of Ecological Conflicts and Valuation*. Northampton: Edward Elgar Publishing, 2003.
Marx, Leo. *The Machine in the Garden*. Oxford: Oxford University Press, 1964.
Mcdonald, Joshua. "Will Bougainville Reopen the Panguna Mine?" *The Diplomat*. Accessed November 22, 2019, https://thediplomat.com/2019/11/will-bougainville-reopen-the-panguna-mine/.
McGuirk, Rod. "Bougainville Votes for Independence from Papua New Guinea." *The Diplomat*. Accessed December 14, 2019, https://thediplomat.com/2019/12/bougainville-votes-for-independence-from-papua-new-guinea/.
McKibben, Bill. *The End of Nature*. New York: Random House, 1991.
Milner, G.R. "Mississippian Period Population Density in a Segment of the Central Mississippi River Valley." *American Antiquity* 51, no. 2 (1986): 228.
Montmarquet, J.A. "Philosophical Foundations for Agrarianism." *Agriculture and Human Values* 2, no. 2 (1985): 5.
Morison, James, Rachel Hine, and Jules Pretty. "Survey and Analysis of Labour on Organic Farms in the UK and Republic of Ireland." *International Journal of Agricultural Sustainability* 3, no. 1 (2005): 24–43
Morton, Timothy. *Being Ecological*. Cambridge: The MIT Press, 2018.
Most, Glen W., Ed. and trans. *Theogony, Works and Days, Testimonia*. New Haven, CT: Harvard, Loeb Classical Library, 2006.
Mumford, Lewis. *Technics and Civilization*. New York: Harcourt Brace and Co., 1934.

Nelson, Stephanie. *God and the Land, The Metaphysics of Farming in Hesiod and Vergil.* Oxford: Oxford University Press, 2008.
Nelson, Stephanie. "Hesiod, Virgil, and the Georgic Tradition." *The Oxford Handbook of Hesiod.* Oxford: Oxford University Press, 2018: 368.
Oelschlaeger, Max. *The Idea of Wilderness: From Prehistory to the Age of Ecology.* New Haven: Yale University Press, 1991.
Pauketat, T.R. *Cahokia: Ancient America's Great City on the Mississippi.* New York: Penguin Random House, 2009.
Pearson, Craig J. "Regenerative, Semi-closed Systems: A Priority for Twenty-first-century Agriculture." *Bioscience* 57, no. 5 (2007): 409–418.
Reagan, Anthony J. "Causes and Course of the Bougainville Conflict." *The Journal of Pacific History* 33, no. 3 (November 1998): 269–285. https://www.jstor.org/stable/25169410.
Ritchie, Hannah and Max Roser. "Urbanization." OurWorldInData.org (2018). Accessed February 1, 2022, https://ourworldindata.org/urbanization.
Russell, Nerissa. "Spirit Birds at Neolithic Çatalhöyük." *Environmental Archaeology* 24, no. 4 (2019): 377–386.
Scales, Ivan. "Green Consumption, Ecolabelling and Capitalism's Environmental Limits." *Geography Compass* 8, no. 7 (2014): 477–489.
Schneider, Mindi and Philip McMichael. "Deepening, and Repairing, the Metabolic Rift." *The Journal of Peasant Studies* 37, no. 3 (2010): 461–484.
Schwartz, Judith D. "Soil as Carbon Storehouse: New Weapon in Climate Fight?" *Yale Environment 360.* Accessed March 13, 2019, https://e360.yale.edu/features/soil_as_carbon_storehouse_new_weapon_in_climate_fight.
Scott, James C. *The Art of Not Being Governed.* New Haven: Yale University Press, 2009.
Scott, James C. *Against the Grain: A Deep History of the Earliest States.* New Haven, CT: Yale University Press, 2017.
Sharer, Robert J., and Loa P. Traxler. *The Classic Maya.* Stanford. CA: Stanford University Press, 2006.
Sterckx, Roel. "Ideologies of Peasant and Merchant in Warring States China." In *Ideologies of Power and Power of Ideologies in Ancient China,* edited by Yuri Pines, Paul R. Goldin, and Martin Kern, 211–248. Boston: Brill, 2015.
Sweet, Timothy. *American Georgics.* Philadelphia: University of Pennsylvania Press, 2022.
Thompson, Paul B. "Chapter 1: Sustainability and Environmental Philosophy." In *The Agrarian Vision: Sustainability and Environmental Ethics,* 18–41. Lexington: University Press of Kentucky, 2010.
Trigger, Bruce. *Understanding Early Civilizations.* Cambridge, Cambridge University Press, 2003.
Walter, Mariana and Lucrecia Wagner. "Mining Struggles in Argentina. The Keys of a Successful Story of Mobilization." *The Extractive Industries and Society* 8, no. 4 (2021).
Wesley-Smith, Terence and Eugene Ogan. "Copper, Class, and Crisis: Changing Relations of Production in Bougainville." *The Contemporary Pacific* 4, no. 2 (Fall 1992): 245–267. https://www.jstor.org/stable/23699898.
White, Lynn Jr. "The Historical Roots of Our Ecologic Crisis." *Science* 155, no. 3767 (1967): 1203–1207.

Wittfogel, Karl A. *Oriental Despotism: A Comparative Study of Total Power.* New Haven: Yale, 1957.

Worster, Donald. *Nature's Economy.* Cambridge: Cambridge University Press, 1994.

Yoffee, Norman. *Myths of the Archaic State.* Cambridge: Cambridge University Press, 2005.

Ziser, Michael G. "Walden and the Georgic Mode." *Nineteenth-Century Prose* 31, no. 2 (2004): 186–208.

Zomer, R.J., Deborah A. Bossio, Rolf Sommer, and Louis V. Verchot. "Global Sequestration Potential of Increased Organic Carbon in Cropland Soils." *Scientific Reports* 7, no. 1 (November 2017): 1–8.

The Environmental Political Role of Counter-Hegemonic Environmental Ethics: Replacing Human Supremacist Ethics and Connecting Environmental Politics, Environmental Political Theory and Environmental Sciences

Joel Jay Kassiola

It's all a question of story. We are in trouble just now because we do not have a good story. We are in between stories. The old story, the account of how the world came to be and how we fit into it, is no longer effective. Yet we have not learned a new story... A radical reassessment of the human situation is needed, especially concerning those basic values that give to life some satisfying meaning. We need something that will supply in our times what was supplied formerly by our traditional religious story. If we are to achieve this purpose, we must begin where everything begins in human affairs—with the basic story, our narrative how things came to be, how they came to be as they are and how the future can be given some satisfying direction. We need a story that will educate us, a story that will heal, guide and discipline us.

–Thomas Berry

The Dream of the Earth, 123–124.[1]

Philosophy ought to question the basic assumptions of the age. Thinking through critically and carefully, what most people take for granted is, I believe,

[1] Thomas Berry, *The Dream of the Earth*, San Francisco: Sierra Club Books, 1988, 123–124.

J. J. Kassiola (✉)
Department of Political Science, San Francisco State University, San Francisco, CA, USA
e-mail: kassiola@sfsu.edu

the chief task of philosophy, and it is this task that makes philosophy a worthwhile activity.

–Peter Singer

"All Animals Are Equal," 181.[2]

INTRODUCTION: THE ACADEMIC COMPARTMENTALIZATION OF ENVIRONMENTAL ETHICS, POLITICAL SCIENCE AND NATURAL SCIENCES

I have been teaching a course in the Philosophy Department on environmental ethics for 15 years and a course in a Political Science Department on environmental political theory ("Politics, the Environment and Social Change") for 35. This is a distinctive combination because generally my colleagues in both departments rarely venture outside of their respective academic disciplines in their teaching or research and publications. This separation in the academy between philosophy and political science in pedagogy is also reflected in the research agendas of both disciplines. Although mention of ethical issues and principles may be made in environmental political studies and vice versa in environmental ethics research, including some references to politics, especially policy-making and political change, the interrelationship of these two environmental disciplinary analyses with specific arguments is rarely discussed.

My practice in fully integrating both fields in teaching, research and writing is the result of my having studied in a multi-disciplinary graduate degree program that combined both philosophy and politics.[3] This preceded my interest in the environmental crisis as a political philosophical subject and was also prior to the establishment of environmental political theory in the 1990s with what is now well recognized as a multi- and interdisciplinary subfield within political science examining the environment in all of its forms from a normative political theoretical perspective.[4]

---

[2] Peter Singer, "All Animals Are Equal," *Philosophical Exchange*, Vol. 5, No. 1, 103–116. All paginations to sources found in my collected edition of foundational readings within environmental ethics will be made to their reprinted versions in: Joel Jay Kassiola, ed. *Environmental Ethics: Foundational Readings, Critical Responses*, San Diego: Cognella Academic Publishing, 2021.

[3] The program is offered at Princeton University. The other disciplines participating in this unique academic program are: History, Classics, and Religion Departments.

[4] I envisioned and advocated for the creation of this subfield in one of the first books about this subject: *The Death of Industrial Civilization: The Limits to Economic Growth and the Repoliticization of Advanced Industrial Society* (Albany: State University of New York Press, 1990). For more recent publications in this subfield that combines environmental political theory and the history of political thought, see, Peter F. Cannavo and Joseph H. Lane, Jr. eds. *Engaging Nature: Environmentalism and the Political Theory Canon*. (Cambridge: The MIT Press, 2014). For a more analytical, as opposed to historical, approach see, Steve Vanderheiden, *Environmental Political Theory*, Cambridge, UK: Polity, 2020.

The lack of mutual engagement between political science studies and environmental ethics is caused by the division between the empirical social sciences, in general, and political science, in particular, and philosophy, especially normative environmental ethics. One exception is normative political theory wherein ethical questions and theories are prominent (for example, in the historical canon and secondary literature of Western political theory). Typically, the social sciences, as well as the natural sciences, including environmental sciences, do not pose ethical questions. If they do, they do not pursue them in a manner that includes ethical discourse and reasoning as well as ethical theories. Consider this recent statement on a proposal for scientists to respond "to the normative demands of nature:" "Scientists don't normally respond to normative demands of any kind. The normative is not part of the domain of objects that science studies, or a property of the objects it studies."[5]

Both social and natural sciences remain focused on empirical descriptions and causal explanations of their chosen subjects while eschewing ethical issues and discourse and are largely uninformed about ethical theories. Similarly, environmental ethics' discussions usually omit systematic scientific consideration of specific environmental problems or political institutional responses to them resulting in the fragmentation and isolation of the academic fields addressing the environment, as indicated in the subtitle to this chapter.

The goal of this chapter of the Handbook is to argue that cross-disciplinary and interdisciplinary engagement between the academic fields of political science, especially environmental politics and environmental political theory, along with environmental sciences, with environmental ethics is imperative, if progress is to be achieved in understanding and ameliorating the many environmental challenges facing humans (and nonhumans and the natural environment itself) as the first quarter of the twenty-first century comes to a close. I maintain that in its critique and expansion of the hegemonic, human-centered and human supremacist ethics—anthropocentrism[6]—environmental ethics can be the uniting factor between the environmental disciplines. Such a combined integrative multi-disciplinary or interdisciplinary approach of environmental ethics can play an important political role in our urgently needed response to the current environmental crisis.

---

[5] Nikolas Kompridis, "Disciplinary Variations on the Anthropocene: Temporality and Epistemic Authority: Response to Kyle Nichols and Bina Gogineni," in: Akeel Bilgrami, ed. *Nature and Value*. (New York: Columbia University Press, 2020), p. 67.

[6] For an introductory discussion of the nature of anthropocentrism, see, L. Goralnik and M.P. Nelson, "Anthropocentrism," in: Ruth Chadwick, ed. *Encyclopedia of Applied Ethics*, Second Edition, Vol. 1, San Diego: Academic Press, 2012, 145–155. For my own interpretation of anthropocentrism, see, Joel Jay Kassiola, "Zhang Zai's Cosmology of Qi/qi and the Refutation of Arrogant Anthropocentrism: Confucian Green Theory Illustrated," *Environmental Values*, Vol. 31, October 2022, 533–554.

The Birth and Development of Environmental Ethics

With the Introduction above about the academic compartmentalization of environmental ethics, political science and environmental science as background, I would like to offer an explanation for this deleterious academic separation between fields. The relatively young subfield of environmental ethics arose in the late 1960s and early 1970s as a challenge to the millennia-long and still dominant worldview in the West (now worldwide as a result of economic and cultural globalization): "anthropocentrism." This human-centered and human supremacist ethic originated in Western thought with Socrates.[7]

This long period of anthropocentric thought, coextensive with Western thought as a whole, began with Socrates and Plato and continues right up to today as current publications below indicate (see footnote 19 below). When morals and moral philosophy were considered, it meant relations between human beings exclusively. Nonhumans and environmental ecosystems were not included among those entities granted moral standing or moral status and, thereby, were subordinate to the interests and morals of human beings.

It is not until the mid-twentieth century with the pioneering thought of Aldo Leopold, Lynn White, Jr. and Peter Singer, who were among the first advocates of an expanded environmental ethics as opposed to the longstanding exclusive human-centered ethics, that the subfield of environmental ethics arose.[8]

---

[7] For a brief introduction to ethics, see, Lewis Vaughn, "Ethics and the Moral Life," in his *Beginning Ethics: An Introduction to Moral Philosophy*. (New York: W.W. Norton, 2015), Chapter 1, 13–27; partially reprinted in *Environmental Ethics*, 21–27.

For the reference to Socrates, Vaughn writes: "These [moral] beliefs help guide our [human] actions, define our [human] values, and give us [humans] reasons for being the persons we are. Ethics [or moral philosophy, or the philosophical study of morality] addresses the powerful question that Socrates formulated twenty-four hundred years ago: How ought we [humans] live?" See Vaughn, reprint, 21.

For another discussion of the anthropocentric nature of traditional Western ethics, see, Louis P. Pojman, ed. *Environmental Ethics: Readings in Theory and Application*, Third Edition, (Belmont, CA: Wadsworth/Thomson Learning, 2001), "What Is Ethics?" pp. 3–8, where the editor lists 5 human purposes of morality, including: 1) "to ameliorate human suffering;" and, 2) "to promote human flourishing;" 6.

[8] See path-breaking works by: Aldo Leopold, "The Land Ethic," in his *A Sand County Almanac: And Sketches Here and There*. (New York: Oxford University Press, 1989 (1948)), pp. 201–226, reprint, 133–146. Lynn White, Jr. "The Historical Roots of Our Ecological Crisis." *Science*, Vol. 155, No. 3767, 1967, 1203–1207, reprint, 85–93. Peter Singer, "All Animals are Equal." *Philosophical Exchange*, Vol. 5, No. 1, 1974, 103–116, reprint, 173–186.

Because of the massive response to White's famous and influential essay (more on this unprecedented academic response later in the discussion), it is often considered to be the founding article for the subfield of environmental ethics. In addition, it was White's essay and his claim of the historical detrimental influence of anthropocentric Christianity that also began the offshoot, dual disciplinary subfield of ecotheology. However, all three authors with their seminal essays have generated huge individual literatures and have been reprinted many times right up to today. They have inspired many followers including a

All three thinkers proposed creative enlargements of the subject matter of anthropocentric ethics to include: the "land," or what became eco-centric ethics, understood in Leopold's broad way: "soils, waters, plants and animals," (Leopold); the egalitarian ideas of Saint Francis of Assisi (White); and the moral equality of human and nonhuman animals based on their sentience, or the capacity to feel pain and joy, (Singer), or what became zoocentric ethics.[9]

Environmental ethics was developed later by the bioethical philosopher, Paul W. Taylor. He prescribed the further expansion of the range of equal moral standing to all life forms, including plants along with animals, compared to Singer's more limited zoocentrism that excluded non-sentient plants. Taylor argued for a biocentric ethics that maintained the expansion of morality including equal moral rights to all living creatures: humans, non-human animals and plants, explicitly denying human moral superiority and rightful dominance in his biocentric theory.[10]

Logically, if not chronologically, environmental ethics developed from traditional, Western human-centered, human supremacist, anthropocentric ethics to non-human animals (White and Singer); to the inclusion of all living beings as members of the planetary "community of life,"[11] (Taylor); to the "eco-centrism" of Aldo Leopold and his "land ethic" which included: "soils, waters, plants, animals, or collectively: the land."[12] For Leopold, ecosystems such as bodies of water and land masses are included in addition to non-human animals and plants as subjects with moral standing and possessing moral rights in his eco-centric ethics.[13]

Environmental Ethics' Revolutionary Denial of Human Moral Superiority and Its Various Human and Non-Human Environmental Partners

The reasons for the rejection and expansion of hegemonic anthropocentric ethics since the beginning of Western thought (White emphasizes the Biblical book of *Genesis* and the Judeo-Christian account of creation including the role of the divine justification of human superiority over nature through the story

---

large social movement (animal rights movement for Singer's position of the moral equality between nonhuman animals and humans), and deserve recognition for their innovation and creativity. Perhaps all three thinkers should be thought of as co-founders of the field with their renowned works?

[9] See the above works by these three authors. The specific quotation from Leopold is taken from reprint, 134.

[10] See Paul W. Taylor, "The Ethics of Respect for Nature." *Environmental Ethics*, Vol. 3, 1981, 197–218, reprint, 199–215. See section on "The Denial of Human Superiority," reprint, 208–214.

[11] Taylor, "The Ethics of Respect for Nature," reprint, 204–206.

[12] Leopold, "The Land Ethic," reprint, 134.

[13] Leopold, "The Land Ethic," reprint, 134–136.

of Adam and Eve in the Garden of Eden)[14] differ in each case of the revolutionary environmental ethical thinkers: Leopold, White, Singer and Taylor. However, the details of their different arguments are not germane to the main point I wish to make in this chapter about the birth and development of environmental ethics and its collective political role: the refutation of anthropocentric ethics and worldview as well as the rejection of exploitative and harmful environmental policies engendered by this ethical doctrine's taken-for-granted, worldwide acceptance. The refutation of anthropocentric ethics will demonstrate the need for a replacement for this harmful ethical theory with an expanded conception of the humanity-nature ethical relation. Greater protectionist environmental policies should follow from this broader environmental ethical conception of ethics and its encompassing various non-human and ecological ethical partners.

White used the legend of Saint Francis of Assisi's compassion for non-human animals to support his rejection of (Western) Christianity as "the most anthropocentric religion the world has seen."[15] Singer relied upon the concept of "sentience," or the ability to feel pain and joy shared by humans and non-human animals to make his case for zoocentric moral equality.[16] Taylor used his conception of a universal "community of life" wherein all living entities share the common traits of deserving survival and seeking their own good in their own way; all species deserve equal respect, according to his egalitarian biocentric outlook.[17] Finally, Leopold emphasized the idea of an interdependent ecological community of ecosystems through which energy is transferred, including non-sentient soils and waters.[18]

What is important to recognize for the purposes of this chapter regarding all four, transformative environmental ethical theories is that humans are not ethically exceptional or unique. We are not deemed to be of superior ethical status compared to non-humans and natural objects, in marked contrast to the anthropocentric tradition of Western Civilization right up to the present.[19]

---

[14] See White, "The Historical Roots of Our Ecological Crisis," reprint, 89–90. For a detailed Biblical exegesis of the specific chapters in *Genesis*, 1–3 that detail this story involving Adam and Eve and the historically important verse, *Genesis* 1:28 and how it should be interpreted in assessing White's textual interpretation, see, Peter Harrison, "Subduing the Earth: *Genesis* 1, Early Modern Science, and the Exploitation of Nature," *Journal of Religion*, Vol. 79, No. 1, 86–109, reprint, 95–114.

[15] See, White, "The Historical Roots of Our Ecological Crisis," reprint, 90.

[16] See, Singer, "All Animals are Equal," reprint, 177. He borrowed this crucial idea for his zoocentric position from the nineteenth century British philosopher, Jeremy Bentham, see, reprint, 177.

[17] See, Taylor, "The Ethics of Respect for Nature," reprint, 204.

[18] See, Leopold, "The Land Ethic," reprint, 134–136, 140–143, for his concepts of: "community," and "the land pyramid.".

[19] On the Western intellectual tradition of human ethical superiority and resulting rightful domination of nature, see, William Leiss, *The Domination of Nature*, Boston, Beacon Press, 1974. Also, see, Paul Wapner, *Living through the End of Nature: The*

Leopold's imagery transformed humans from the "conquerors" of nature to a "plain members and citizens of it," conferring equal membership in the land community as envisioned by Leopold. This understanding of our role in nature: not as dominant superior but respectful equal reveals a cultural, ethical and political revolution with profound political import. Leopold writes:

> All ethics so far evolved rest upon a single premise: that the individual is a member of a community on interdependent parts... the land ethic simply enlarges the boundaries of the community... In short, a land ethic changes the role of *Homo sapiens* from conqueror of the land community to plain member and citizen of it. It implied respect for his [sic] fellow-members, and also respect for the community as such.[20]

White's prescription of Saint Francis' view over anthropocentrism is noteworthy for its explicit political metaphor: after calling Saint Francis the "greatest radical in Christian history since Christ," White says:

> The key to understanding of Francis is his belief in the virtue of humility—not merely for the individual but for man [sic] as a species. Francis tried to depose man [sic] from his monarchy over creation and set up a democracy of all God's creatures.[21]

Taylor raises this fundamental point of the environmental ethical rejection of human supremacy in a philosophically rigorous manner by posing the essential challenge to anthropocentrism: "In what sense are humans alleged to be superior to other animals?" and answers in the following manner:

> We [humans] are different from them [animals] in having certain capacities that they lack. But why should these capacities be a mark of superiority? From

---

Future of American Environmentalism, Cambridge: The MIT Press, 2010, Chapter 4, "The Dream of Mastery," 79–105.
I shall discuss current philosophical defenses of hegemonic anthropocentrism later in the essay, but for now, I shall merely cite notable exceptions to this taken-for-granted anthropocentric cultural assumption that is rarely explicitly articulated, much less argued for with supporting evidence. We will not make environmental or moral progress until this assumed doubtless position is made explicit and the subject of reflection and debate, as I hope to stimulate with this chapter.
For the few examples of explicit contemporary defenders of anthropocentrism, I was able to find the following: Andrew Light, "Taking Environmental Ethics Public," in David Schmidtz and Elizabeth Willott, eds. *Environmental Ethics: What Really Matters, What Really Works*. Second Edition. (New York: Oxford University Press, 2012, 654–664. Anthony Weston, "Before Environmental Ethics." *Environmental Ethics*, Vol. 14, No. 4, 1992, 321–338. William F. Baxter, "People or Penguins," in Lewis Vaughn, ed. *Doing Ethics: Moral Reasoning, Theory, and Contemporary Issues*. Fifth Edition. (New York: W.W. Norton, 2019), 442–446. And, Carl Cohen, "The Case for the Use of Animals in Biomedical Research," in Vaughn, *Doing Ethics*, 407–413.

[20] Leopold, "The Land Ethic," reprint, 134.

[21] White, "The Historical Roots of Our Ecological Crisis," reprint, 92.

what point of view are they judged to be signs of superiority and what sense of superiority is meant? After all, various nonhuman species have capacities that humans lack. There is the speed of a cheetah, the vision of an eagle, the agility of a monkey. Why should not these be taken as signs of *their* superiority over humans?... *Humans are claiming human superiority from a strictly human point of view, that is, from a point of view in which the good of humans is taken as the standard of judgment.* All we need to do is to look at the capacities of nonhuman animals (or plants, for that matter) from the standpoint of *their* good to find a contrary judgment of superiority.[22]

Taylor's important insight about how anthropocentrism wrongfully assumes the human perspective as justification is a fatal begging of the question fallacy by the anthropocentric position that reveals its faulty reasoning.[23]

This logical point alone inspired me to seek a replacement for the anthropocentric worldview. I have sought to do this in the research I have termed: Confucian Green Theory. In it, I argue for expanding morality and rejecting anthropocentrism as introduced by: Leopold, White, Singer and Taylor with the birth and development of environmental ethics. I perceive a logical progression (including Chinese Confucian and Neo-Confucian philosophy) from anthropocentrism (human-centered and human supremacist), zoocentrism (sentience equality between humans and non-humans), biocentrism (life-centered equality) and, finally, ecocentrism (ecosystem-centered equality) leading to the ultimate cosmicentrism (cosmos-centered equality) of the Confucian tradition. This is a metaphysical, ontological, cosmological vision that encompasses non-sentient, non-living things as well as living beings. It is drawn from the Neo-Confucian tradition of thought, mainly from the eleventh century in China, in particular the cosmological theory of *Ch'i* (or *Qi* or *qi*) of Zhang Zai.[24]

Zhang's cosmology, or theory of the universe, centers on a ceaseless cyclical cosmological process of concentration and diffuseness, change, and transformation. It is comprised of the renowned Chinese concept of "*ch'i*" (or "*Qi*" or "*qi*"), "energy-matter," and driven by the celebrated Chinese traditional

---

[22] Taylor, "The Ethics of Respect for Nature," reprint, 208 (the first emphasis is mine, the second emphasis is in original). Of course, plants can photosynthesize which humans cannot.

[23] Taylor, "The Ethics of Respect for Nature," reprint, 208.

[24] See my, "Zhang Zai's Cosmology of *Qi/qi* and the Refutation of Arrogant Anthropocentrism.".

concepts of *"yin/yang"* or complementary forces that keep the cosmological process going endlessly.[25] A Confucian scholar summarizes Zhang Zai's cosmology in the following manner:

> In his [Zhang Zai] vision nature is the result of the fusion and intermingling of the vital forces that assume tangible forms. Mountains, rivers, rocks, trees, animals, and human beings are all modalities of energy matter [*ch'i*], symbolizing that the creative transformation of the Tao [the Way] is forever present... It [Zhang's cosmological theory] is impartial to all modalities of being and not merely anthropocentric.[26]

For the purposes of this essay, what is of most importance regarding the Neo-Confucian non-anthropocentric cosmology of Zhang Zai is the statement by Tu about Zhang's vision of the cosmos being "impartial to all modalities of being and not merely anthropocentric." Zhang Zai's *ch'i* cosmology rejects anthropocentrism or human-centered and human supremacist ethics over all of nature: animals, plants, ecosystems and inanimate objects.

This is the defining characteristic and main contribution of environmental ethics as a subfield of philosophy: the opposition to anthropocentrism that has dominated Western thought and life since its beginning and continues to do so. The current defenders of this worldview proclaim the near-universal acceptance as a ground to justify human supremacist environmental policy to the public without examination. Consider the following statements from such defenders of anthropocentrism:

> ... whether we like it or not, most humans think about nature in human-centered terms... Since many people are motivated by anthropocentric reasons environmental pragmatists... will of necessity have to endorse anthropocentrism in a public policy context... Always first pursue the anthropocentric justification for the environmental policy in order to persuade a broader array of people to embrace the view, because... anthropocentric justifications can most plausibly speak to people's ordinary moral intuitions more persuasively than nonanthropocentric justifications.[27]

---

[25] See, Wing-Tsit Chan's classic work which he translated and compiled the sources: *A Source Book in Chinese Philosophy* (Princeton: Princeton University Press, 1963), and the chapter on Zhang Zai (Chang Tsai) that includes translated works by Zhang as well as commentary by Chan. See Chapter 30, "Chang Tsai's Philosophy of Material Force," 495–517. On the various meanings of *"ch'i,"* see, Appendix "On Translating Certain Chinese Philosophical Terms," 784.
On the traditional concepts of *"yin/yang,"* see, Robin R. Wang, *Yinyang: The Way of Heaven and Earth in Chinese Thought and Culture*, Cambridge, UK: Cambridge University Press, 2012.

[26] Tu Wei-Ming, "The Continuity of Being: Chinese Visions of Nature," in, Tu Wei-Ming, *Confucian Thought: Selfhood as Creative Transformation*. (Albany: State University of New York Press, 1996), 41–42.

[27] Light, "Taking Environmental Ethics Public," 657, 660.

Another environmental philosopher prescribing anthropocentrism writes about the "contemporary anthropocentrized world:" "The contemporary anthropocentrized world, which is, in fact, the product of an immense project of world reconstruction that has reached a frenzy in the modern age has become simply the taken-for-granted reference point for what is 'real,' for what must be accepted by any responsible criticism."[28]

Or, this bold statement of anthropocentrism: "My criteria are oriented to people, not penguins. Damage to penguins, or sugar pines, or geological marvels, is, without more, simply irrelevant... my observations about environmental problems will be people-oriented as are my criteria. I have no interest in preserving penguins for their own sake."[29]

Another anthropocentric philosopher declares Peter Singer's argument for zoocentric ethics: "worse than unsound, it is atrocious... Human beings do have rights; there is a moral status very different from that of cats or rats... Humans owe to other humans a degree of moral regard that cannot be owed to animals."[30]

What distinguishes these statements in support of anthropocentrism is how specific they are. Most of the time human exceptionalism and supremacy are merely assumed or presupposed with no explicit argument provided for these sweeping and morally questionable assumptions. Factual claims about the hegemony of anthropocentrism in modern society are no doubt true: as one of the above philosophers put it: "... we have sound empirical evidence that humans think about the value of nature in human terms."[31] Anthropocentric philosophers are empirically correct about the factual support of anthropocentrism in contemporary society created by the overwhelming majority of its citizens adopting uncritically the anthropocentric paradigm which is instilled into children at a very young age and reinforced throughout their adult lives by the mass media.

Yet such arguments fail to recognize the presence of the naturalistic fallacy in their arguments: the inferring of a normative conclusion from a factual premise in their defense of anthropocentrism. This is the logical mistake of taking the fact of popular anthropocentric belief as decisive evidence for accepting the morality of anthropocentrism: the masses of people could be wrong, which would not be the first time. The evidence that a large majority of people support human supremacy over nature used to support the exploitation of nature ("people over penguins or sugar pines") does not justify that it is how we morally *should* believe or act. *Factual anthropocentrism does not support moral anthropocentrism*; it only indicates the challenging modern social context that non-anthropocentrism must overcome to be socially accepted and implemented in policy.

---

[28] Weston, "Before Environmental Ethics," 324.

[29] Baxter, "People or Penguins," 443.

[30] Cohen, "The Case for the Use of Animals in Biomedical Research," 410.

[31] Light, "Taking Environmental Ethics Public," 659.

As Weston stated, anthropocentrism has, indeed, reached an extreme degree (or "frenzy") in the modern age and is at the foundation of modern society; however, a discussion of the inherent relationship between modernity and anthropocentrism and its reigning hegemony in contemporary global society would take us too far afield in this essay.[32] However, I do want to emphasize that the currently dominant status of anthropocentrism gives profound political significance to non-anthropocentric environmental ethics. It could possess revolutionary impact upon the existing anthropocentric foundation for virtually the entire current world with globalization of the Western anthropocentric perspective on the humanity-nature relation. Our global anthropocentric outlook conceives of humans as separate from, superior to and, dominant over, nature. Such a hegemonic worldview rationalizes the human exploitation of nature and, thereby, is at the heart of the current environmental crisis brought on by humans exceeding the boundaries of our planet whether it is regarding greenhouse gases, withdrawing water faster than it be can be recharged, hunting animals to extinction and so on.

My aim in this chapter is to bridge the isolation between different discourses, fields and theories among: environmental politics, environmental political theory, environmental sciences and environmental ethics. This is done in order to demonstrate the relevance and significance for environmental studies as a whole, mainly bearing on the central issue of the rightful place of humanity within nature and how non-anthropocentric environmental ethics should and can influence environmental policy-making on the many environmental problems confronting humans today.

This discussion proposes a much different non-anthropocentric environmental politics in contrast to contemporary anthropocentric environmental politics with its view of humanity as dominant over a subordinate nature that has brought humanity, non-humans and natural ecosystems to the brink of dire environmental conditions.[33] We are presently in urgent need of a new set of conditions and policies associated with a non-anthropocentric social order and values that are both environmentally sustainable and socially just.

The transformative impact of environmental ethics should now be clear. The ending of hegemonic Western anthropocentrism, since *Genesis* , is long overdue and is desperately needed before an environmental catastrophe occurs

---

[32] See my inclusion of the thought of Francis Bacon, the seventeenth century, British philosopher and founder of the modern scientific method as evidence for this claim about modernity and anthropocentrism. See, Kassiola, *Environmental Ethics*, Chapter 3A, "Francis Bacon: Quotations from Various Works," 59–61.

Also on this topic of anthropocentrism's relation to modernity, see, Carolyn Merchant, *The Death of Nature: Women, Ecology, and the Scientific Revolution*. (New York: Harper-Collins, 1983), especially, Chapter 7, "Dominion Over Nature," 164–191; partially reprinted, 63–71.

[33] For a succinct discussion of the current environmental crisis, including the crossing of several planetary boundaries, see, Kate Raworth, *Doughnut Economics: 7 Ways to Think Like a 21st Century Economist*. (White River Junction, VT: Chelsea Green Publishing, 2017), Appendix: "The Doughnut and Its Data," 254–258.

as the current environmental conditions worsen. This essay argues that environmental ethics as a subfield of philosophy and its paradigmatic denial of human-centered, human supremacist ethics (no matter which version of non-anthropocentrism is adopted and why) can help bring about this necessary, planet-saving, political change. Moreover, this new philosophical subfield should be considered by all students of the environment, be they: political scientists, political theorists or natural scientists.

Following this far-reaching outlook, (a new "basic story" in Berry's terms), the next section will present an argument on the ultimate, ontological cosmological cause of the environmental crisis: anthropocentrism. It was in a revolutionary article generally considered to be the breakthrough essay that launched the field of environmental ethics.

White's "Roots" of the Environmental Crisis, the "Greatest Psychic Revolution" of Anthropocentrism and the Need for a New Non-Anthropocentric Revolution

In what is commonly considered the foundational work in the subfield of environmental ethics by Lynn White, Jr. in 1967, the author shockingly contended that Christianity's "victory over paganism was the greatest psychic revolution in the history of our culture." He then added that this victory of Christianity "made it possible to exploit nature in a mood of indifference to the feelings of natural objects," and, finally, White concluded that "Especially in its Western form, Christianity was the most anthropocentric religion the world has seen."[34]

This article is one of the most influential and commented-upon articles published in the last 50 years, according to a report on a data base of over 700 articles written in response to this landmark essay by White.[35]

White argued that Christian anthropocentrism "in absolute contrast to ancient paganism and Asia's religions... not only established a dualism of man and nature but also insisted that it is God's will that man exploit nature for his proper ends."[36] Thus, this seminal essay on environmental ethics' creation maintained that anthropocentrism was introduced in the West by Christianity and was responsible for our environmental crisis. Anthropocentrism rejected

---

[34] White, "The Historical Roots of Our Ecological Crisis," reprint, 89–90 for the quotations in this paragraph.

[35] A trio of researchers report on the immense significance of the White essay as follows: "It [White's "Roots" article] became one of the journal's [*Science*] most cited articles (by 2016, 924 citations in the Web of Science's core collection and 4600 citations in Google's Scholar's collection)." See, Bron Taylor, Gretel Van Wieren, and Bernard Daley Zaleha, "Lynn White, Jr. and the Greening-of-Religion Hypothesis." *Conservation Biology*. Vol. 30, No. 5, 2016, 1001. The data base reference is from 1002 of this work.

[36] White, "The Historical Roots of Our Ecological Crisis," reprint, 90.

the animism of pagan thought that was respectful of non-human nature treated as a moral equal to humans in favor of the current anthropocentric-dominating worldview that grounded the exploitative relations to nature and its "resources" for humans' use and benefit.

In 1967, White perceived that what we needed (and what we need even more today, some 50 years and billions of pounds of C02 and other greenhouse gases later) is a second such psychic revolution in values and worldview, this time *rejecting* anthropocentrism and human supremacy in favor of a non-anthropocentric worldview (Berry's "new story" or new cosmology). It is a paradigmatic Copernican Revolution to a non-anthropocentrism ceasing to conceive of humanity as the center and dominator of the universe. Instead, we should envision humanity to be as Leopold described: "a plain member of the biotic community;" as Taylor suggests: a species morally equal to all the other living species; as Singer recommends: morally equal with all other non-human animals; or, as I propose, non-exceptional participants in Zhang Zai's cosmological *ch'i* process driven by the forces of *yin/yang*, condensing during the concentration phase and diffusing during that phase only to begin the concentrating stage again, and so on, endlessly. No superior status is granted to humans by any of the non-anthropocentric theorists and our environmental policies should reflect this crucial moral fact.

An Example to Support Non-Anthropocentric Environmental Ethics' Significance for Environmental Politics, Political Theory and Natural Sciences: The Anthropocene Age

If environmental politics can reflect a true non-supremacist relation between humanity and nature as a replacement for a false supremacist and exploitative anthropocentrism, environmental policy-making and treatment of nature will be enhanced not only ethically but also with regard to policies to achieve ecological sustainability. By continuing harmful anthropocentrism, we court environmental disaster as well as commit moral wrongs. *In these two respects, there cannot be a more urgent topic than to align and inform empirical environmental politics, normative, environmental political theory, and physical and biological environmental sciences with a non-anthropocentric environmental ethics.* This chapter is dedicated to that critical aim. However, for this chapter within a Handbook devoted to "Environmental Politics and Theory" to be successful, it must expose the consequences of academic disconnection and isolation into self-contained silos which currently separate the subfields of environmental research and scholarship from environmental ethics and its signature characteristic of nonanthropocentrism: embracing the rejection of human-centered and human-supremacist ethics.

One example of this faulty and damaging academic compartmentalization will illustrate the effectiveness of environmental ethics in improving the other subfields of environmental studies. This is the recent proposal by environmental natural scientists who assert that we have left the Geological period that the Earth had been in for the past 12,000 years or so, the Holocene ("recent whole"), and have entered a new Geological Age, the "Anthropocene," where humans are hypothesized to be dominant in their planetary role and, moreover, where the future of the planet rests entirely on human behavior, hence the human-centered label for this purported new Geological age.[37] Two environmental scientists (one of whom was a Nobel Prize winner in Chemistry, Crutzen) conclude their inaugural statement proclaiming the new Geological Age of The Anthropocene after a list of humanity's huge impacts on the environment with the following statement: "Considering these and many other major and still growing impacts of human activities on earth and atmosphere, and at all, including global, scales, it seems to us more than appropriate to emphasize the central role of mankind [sic]...."[38]

Nothing but empirically measurable human impacts on Earth are considered by these natural scientists who proposed this new epoch. One of these examples is: human population over the last 300 years:

> increased tenfold to 6000 million [8000 million today], accompanied e.g. by a growth in cattle population to 1400 million (about one cow per average family). Urbanization has increased tenfold in the past century . . . 30-50% of the land surface has been transformed by human action . . . more than half of accessible fresh water is used by mankind [sic]; human activity has increased the species extinction rate by thousand to ten thousand fold in the tropical rain forests.[39]

An environmental ethics-based critic of this Anthropocene recommendation by Crutzen and Stoermer observes that the data reflecting dramatic changes on our planet initiated by humans moved the scientists to merely conceive a new Geological epoch with its own name instead of compelling them to exercise ethical judgment and "understand them [the data of humanity's extraordinary impacts on our planet] as dire warning signs, demonstrating beyond a doubt the perilous legacies of highly invasive industries, and signaling the unprecedentedly urgent need to terminate and transform harmful practices, and to move our cultures and economies in Earth-friendly directions."[40]

---

[37] See, Paul J. Crutzen and E. F. Stoermer, "The 'Anthropocene'," *International Geosphere Biosphere Program Global Change Newsletter*, Vol. 41, 2000, 16–18, reprint, 317–319. For a discussion of what this proposed concept of the "Anthropocene" means for our understanding of modernity and our contemporary, environmental political circumstances see, Clive Hamilton, Christophe Bonneuil and Francois Gemenne, eds. *The Anthropocene and the Global Environmental Crisis: Rethinking Modernity in a New Epoch*, Oxford: Routledge, 2015.

[38] Crutzen and Stoermer, "The 'Anthropocene'," reprint, 318.

[39] Crutzen and Stoermer, "The 'Anthropocene'," reprint, 317.

[40] See, Christine J. Cuomo, "Against the Idea of an Anthropocene Epoch: Ethical, Political, and Scientific Concerns". *Biogeosystem Technique*, Vol. 4, No. 1, 2017, pp. 4–8, reprint, 321–326. This quotation is found on 322.

Christine Cuomo opposes the Anthropocene Geological Age initiative and in its place prescribes that what we need today is not a new name for our current Geological condition but a new social order with new ethics and economics that result in an end to the "harmful practices" and values that underlie them; *social change is needed not linguistic change* with the latter's implicit result of maintaining the social and value status quo of modern, consumer capitalist society given the natural scientists' new Geological age prescription. This critique reflecting "ethical, political and scientific concerns" (see the essay's subtitle) is consistent with the major claim of this chapter that the normative ethical perspective of environmental ethics, if adopted by social and natural scientists, would have desirable and needed consequences for environmental policy-making and behavior.

This moralist opponent of the naming and new age proposal goes further and terms the Crutzen and Stoermer conjecture regarding the Anthropocene a "moral atrocity" because: "Instead of normalizing moral atrocities [like climate change and biodiversity loss through species extinction] by proclaiming the birth of a new epoch, *ethical interventions are required to address the serious and systematic harms of the last century and restore ecological health.*"[41] Here I would add "moral health" to Cuomo's final thought since she mentions "ethical interventions." We need to achieve social justice along with ecological sustainability and health implied in Cuomo's critique of the Anthropocene idea and more consistent with the main thrust of environmental ethics and this chapter's argument.

Can there be any clearer illustration of the academic disconnection between physical scientists who prescribe the "Anthropocene" as the label for a new Geological Age because of humanity's unique influence and capacities and those who focus on environmental ethics? The natural scientists are literally not seeing the same world through the same paradigm (especially since the natural scientists are trained not to raise or engage in normative ethical inquiry and reasoning). Therefore, they reach very different conclusions. Yet, the end of this debate between natural scientists and environmental ethicists is not a stalemate created by cognitive relativism with each position standing firm within their respective disciplinary and paradigmatic realms.

As students of ethics know, but not likely those thinkers who ignore ethical considerations or subject matter: the subject matter of ethics overrides conflicting views stemming from positions in nonmoral realms.[42] Whenever there is a conflict between moral principles and the principles or facts from nonmoral fields, like law or natural sciences, the moral ideas and prescriptions or obligations outweigh the others. In the word of one moral philosopher:

---

[41] Cuomo, "Against the Idea of an Anthropocene Epoch," reprint, 324 (my emphasis).

[42] See Vaughn's discussion of "the dominance of moral norms" in his description of ethics; Vaughn, reprint, 27. He defines "ethics" "as the philosophical study of morality," and "morality" as the "beliefs concerning right and wrong, good and bad—beliefs that can include judgments, values, rules, principles, and theories," reprint, 21.

"Moral norms seem to stand out from all of these [nonmoral norms] in an interesting way: they dominate. Whenever moral principles or values conflict in some way with nonmoral principles or values, the moral considerations usually override the others. Moral considerations seem more important, more critical, or more weighty."[43]

The pre-eminence of moral thinking is crucial when it is in conflict with nonmoral claims like empirical scientific ones. Consequently, the moral philosopher's criticism of the Anthropocene Geological epoch and name change is essential given its grounding in moral judgments about the modern social order that generated the factual changes bolstering the scientists' argument for their hypothesis of a new Geological age. As Cuomo's rejection of the Anthropocene Age prescription implies, merely noting the many environmental changes caused by modernity without morally assessing the very social order that caused them, (such as: modern capitalism, with its values of: individualism, materialism and the focus of this chapter, anthropocentrism), as Crutzen and Stoermer do, is not sufficient. The re-naming is misleading, if environmental ethical inquiry is not pursued.

I hope this illustration of the prescription of the Anthropocene era clarifies and supports my claim about the need for the political and scientific subfields of environmental studies to engage with nonanthropocentric environmental ethics. Bridging such epistemological gaps between fields, paradigms, and methods will be a steep challenge, but our efforts are imperative for the environmental health of the planet and all of its inhabitants and, I would add, realizing the advancement of global justice.

Environmental politics, environmental political theory, environmental sciences, and environmental ethics currently all distinctively, but largely unrelatedly, address issues associated with the environment from their respective disciplinary and subfield viewpoints and with their own methods. I have argued that these subfields need to interconnect and interact productively with each other. This is especially important for environmental ethics, given the overriding relevance and decisive nature of this philosophical subfield because of the necessity to expand reductionist and arrogant anthropocentrist supremacist morality. In addition, a turn toward environmental ethics by environmental researchers is necessary because of the profound political consequences that follow from the refutation and replacement of the structure upon which our modern, neoliberal, capitalist, consumer society and its dependency upon endless economic growth culture is founded as well as on its domination and exploitation of nature based on its anthropocentrism.

I realize that combining these environmental subfields, subjects and methods seems daunting and goes against the current tide of academic specialization. Nonetheless, the wide-ranging and cross-disciplinary nature of the subject matter of the environment requires it to be studied comprehensively in an interdisciplinary manner, not exclusively within the sciences or

---

[43] Vaughn, "Beginning Ethics," reprint, 27.

social sciences. Teaching, research and publications must include philosophy's environmental ethics, and the other normative discourses of political theory, aesthetics and theology, where the latter's relevance is demonstrated by White's eco-theological classic article and the dominant role of anthropocentric Christianity in Western (now globalized) cosmology and society.

Works of environmental political theory, such as the ones included in this Handbook,[44] and hundreds of articles in eco-theology in the wake of White's path-breaking and subfield-founding essay on Christian anthropocentrism and the environmental crisis are excellent hybrid examples of what is needed. This chapter advocates a further step in our "moral progress" of expanding the moral self beyond the reductionist, limited and unjust anthropocentrism in order to extend our moral horizons concerning the environment.[45]

Singer emphasizes the expansion of morality, moral progress, as a result of ethical reasoning. He analogizes ethical reasoning to an escalator where one might only want to go a short distance but once on the escalator (reasoning), we must go all the way to the end. He writes: "... once reasoning has got started it is hard to tell where it will stop... Ethical reasoning, once begun, pushes against our initially limited ethical horizons [instances of partiality] leading us always toward a more universal point of view."[46]

Following these insights of Singer about the nature of moral progress, we could say that environmental ethics has fostered an environmental libertarian movement achieving moral progress from anthrocentrism to zoocentrism, biocentrism, eco-centrism and, finally, cosmicentrism, creating a continuous moral progress—in theory—with much practical work to do in order to overcome dominant anthropocentrism in practice. This is where environmental ethics meets practical politics and should inform the structure of the latter.

And as part of the liberating process of the progressive expansion of anthropocentric ethics, this essay prescribes the prioritization of the shift in focus from our ordinary environmental thinking and policy-making presuming the faulty anthropocentrism to a general turn toward environmental ethics in combination with the environmental studies' subfields in an urgent search for a replacement of anthropocentrism in the ethical realm and which grounds environmental policy as components of a new postmodern nonanthropocentric cosmology and social order (Berry's needed "new basic story").

---

[44] For other examples of such environmental political theory works beside the previously cited *Death of Industrial Civilization*, see my edited volume, *Explorations in Environmental Political Theory: Thinking About What We Value*, Oxford: Rouledge, 2015; Cannavo and Lane, *Engaging Nature*, and Vanderheiden, *Environmental Political Theory*.

[45] On the concept of "moral progress" see, Peter Singer, *The Expanding Circle: Ethics, Evolution, and Moral Progress*, Princeton: Princeton University Press, 2011; 1981 edition with a new Preface and Afterword.

[46] Singer, *The Expanding Circle*, 114, 119. For Singer, a libertarian movement (like ending discrimination against African-Americans, women and other minorities) "demands an expansion of our moral horizons;" see his, "All Animals are Equal," reprint, 173.

An Environmental Ethical Exploration for a Replacement Ethic for Anthropocentrism

The issue of the replacement for factually dominant but misguided anthropocentrism takes on great significance for two reasons: (1) the harm caused to the environment by an anthropocentric human supremacy belief and the resulting instrumentalizing and exploitation of nature solely for human interests threatens to bring about an environmental catastrophe for humans, nonhumans and our Earth's ecosystems; (2) it is difficult to defeat a hegemonic cosmology or worldview with no alternative; as the saying goes: "You cannot beat something with nothing." This adage is especially true when that "something" has been dominant in the West for millennia and structures every aspect of our contemporary society, especially if we consider a divine justification for human supremacy over nature in the Biblical book of *Genesis* as White suggests in his founding essay for environmental ethics in his "Historical Roots of Our Ecological Crisis."

He prescribes not giving up Christianity despite its anthropocentrism and its responsibility for the environmental crisis, instead reforming it with the nonanthropocentric ideas of Saint Francis of Assisi, whom he calls "the greatest radical in Christian history since Christ."[47] White's logic for this hybrid ecotheological prescription is insightful and persuasive: "Since the roots of our [human, environmental] trouble are so largely religious, the remedy must also be essentially religious, whether we call it that or not."[48] By parallel reasoning, I would contend that the roots of the environmental crisis are ethical: the disastrous consequences of anthropocentric ethics. Hence, the remedy must also be ethical: a nonanthropocentric ethics that denies human ethical separatism and supremacy. White ends his essay creating the subfield of environmental ethics with the proposal of Saint Francis as the "patron saint for ecologists."[49] This Franciscan belief in a universal species democracy is reminiscent of Leopold's humanity being a "plain member and citizen" of the land-based community and not its conqueror (see earlier discussion of Leopold's land ethic theory) and Taylor's species egalitarianism where all living species are morally equal because of their shared membership in the "community of life" and universal traits such as seeking survival and their own good. What motivates these environmental ethical thinkers here (and it is noteworthy to observe the specific political metaphors both White and Leopold rely upon to express their conception of a nonanthropocentric worldview) is to reject and replace the anthropocentric hierarchy with humanity at the top as the exceptional, autocratic, self-interested ruler of all the other species.

And so, the search for replacing the anthropocentric cosmology that began with White continues. Here again, I believe environmental ethics can make a

---

[47] White, "The Historical Roots of Our Ecological Crisis," reprint, 92.
[48] White, "The Historical Roots of Our Ecological Crisis," reprint, 93.
[49] White, "The Historical Roots of Our Ecological Crisis," reprint, 93.

unique contribution to our need to find an acceptable replacement for human-centered and human supremacist anthropocentric viewpoint.

This is merely a sample of nonanthropocentric replacement theories and worldviews discussed within environmental ethics that began with the thought of Aldo Leopold and Lynn White, Jr. and developed further by Singer and Taylor, and I would add Zhang Zai's Neo-Confucian cosmicenteric cosmology, as illustrations of what the field of environmental ethics can contribute for environmental politics, environmental political theory and environmental sciences.

Conclusion

No matter what your academic environmental interests and training may be, I hope you will consider and learn from this brief survey of environmental ethics and its continuous theoretical development and moral progress from the starting dominant but misguided human supremacist and harmful anthropocentrism. As Singer perceptively noted, we can achieve moral progress, as has been realized in the instance of environmental ethics through its brief existence as a subfield within philosophy; moral theoretical progress has also been attained through its expansion of the moral self and moral horizons beyond the human-centered and supremacy view of deeply entrenched anthropocentrism. In addition, by accepting the prescription of this chapter of learning from the field of environmental ethics and its relevance to the three subfields of environmental political studies and sciences would be fulfilling what makes "philosophy a worthwhile activity," according to Singer, "thinking through, critically and carefully, what makes people take for granted" (see epigraph): hegemonic anthropocentrism. Difficult as it might be to oppose this socially dominant worldview, the courageous example of the philosopher Socrates must continue to both challenge and inspire us.

When all life on Earth is seriously threatened by our own anthropocentric policies and behavior approaching or actually crossing planetary limits, it is imperative that we learn from environmental ethics and its rejection of anthropocentrism. We need to expand the moral self beyond our own species, what Singer terms the moral wrong of "speciesism", akin to racism and sexism, etc., oppressing the rest of nature.[50] The environmental ethics literature is an intellectual resource to aid us in recognizing the need to expand our moral perspective. *Not to adhere to the overall message of counter-hegemonic environmental ethics and not supersede the false anthropocentric paradigm and argument of human superiority and its resulting exploitation of nature may prove as fatal to the human species as it is proving to be for other nonhuman species, in addition to being morally wrong.*

---

[50] For Singer's definition of "speciesism," see his, "All Animals Are Equal," reprint, 178, where he says: "... the speciesist allows the interests of his own species to override the greater interests of members of other species... Most human beings are speciesists".

I hope this chapter persuades the reader to consider the pursuit of this subfield of philosophy a necessary and worthwhile activity as well as to relate environmental ethics to the other components of environmental studies in order to unify these vital environmental fields and achieve greater social justice and efficacious environmental policy. Such a pursuit, I believe, is required for both moral progress and environmental sustainability. And, at this perilous environmental moment, no more important task confronts humanity.

REFERENCES

Baxter, William, F. "People or Penguins," in: Lewis Vaughn, ed. *Doing Ethics: Moral Reasoning, Theory, and Contemporary Issues.* Fifth Edition. New York: W.W. Norton, 2019, 442–446.

Cannavo, Peter F. and Joseph H. Lane, Jr. eds. *Engaging Nature: Environmentalism and the Political Theory Canon.* Cambridge: The MIT Press, 2014.

Chan, Wing-Tsit. Translated and Compiled. *A Source Book in Chinese Philosophy.* Princeton: Princeton University Press, 1963.

Cohen, Carl. "The Case for the Use of Animals in Biomedical Research," in: Lewis Vaughn, ed. *Doing Ethics: Moral Reasoning, Theory, and Contemporary Issues.* Fifth Edition. New York: W.W. Norton, 2019, 407–413.

Crutzen, Paul J. and E.F. Stoermer. "The Anthropocene," *International Geosphere Biosphere Program Global Change Newsletter*, Vol. 41, 2000, 16–18.

Cuomo, Christine J. "Against the Idea of an Anthropocene Epoch: Ethical, Political, and Scientific Concerns," *Biogeosystem Technique*, Vol. 4, No. 1, 2017, 4–8.

Goralnick, L. and M.P. Nelson. "Anthropocentrism," in: Ruth Chadwick, ed. *Encyclopedia of Applied Ethics*, Second Edition. Volume 1. San Diego: Academic Press, 2012, 145–155.

Harrison, Peter. "Subduing the Earth: Genesis 1, Early Modern Science, and the Exploitation of Nature," *Journal of Religion*, Vol. 79, No. 1, 86–109.

Kassiola, Joel Jay. "Zhang Zai's Cosmology of *Qi/qi* and the Refutation of Arrogant Anthropocentrism: Confucian Green Theory Illustrated." *Environmental Values*, Vol. 31, October 2022, 533–554.

Kassiola, Joel Jay. ed. *Environmental Ethics: Foundational Readings, Critical Responses.* San Diego: Cognella Academic Publishing, 2021.

Kassiola, Joel Jay. ed. *Explorations in Environmental Political Theory: Thinking About What We Value.* Oxford: Routledge, 2015.

Kassiola, Joel Jay. *The Death of Industrial Civilization: The Limits to Economic Growth and the Repoliticization of Advanced Industrial Society.* Albany: State University of New York Press, 1990.

Kompridis, Nikolas. "Disciplinary Variations on the Anthropocene: Temporality and Epistemic Authority: Response to Kyle Nichols and Bina Gogineni," in: Akeel Bilgrami, ed. *Nature and Value.* New York: Columbia University Press, 2020, 63–67.

Leiss, William. *The Domination of Nature.* Boston: Beacon Press, 1974.

Leopold, Aldo. "The Land Ethic," in: Aldo Leopold, *A Sand County Almanac: And Sketches Here and There.* New York: Oxford University Press, 1989 (1948), 201–226.

Light, Andrew. "Taking Environmental Ethics Public," in: David Schmidtz and Elizabeth Willott, eds. *Environmental Ethics: What Really Matters, What Really Works.* Second Edition. New York: Oxford University Press, 2012, 654–664.

Merchant, Carolyn. *The Death of Nature: Women, Ecology, and the Scientific Revolution.* New York: HarperCollins, 1983.

Pojman, Louis P. "What is Ethics?" in: Louis P. Pojman, ed. *Environmental Ethics: Readings in Theory and Application.* Third Edition. Belmont, CA: Wadsworth/Thomson Learning, 2001, 3–8.

Raworth, Kate. *Doughnut Economics: 7 Ways to Think Like a 21st Century Economist.* White River Junction, VT: Chelsea Green Publishing, 2017.

Singer, Peter. *The Expanding Circle: Ethics, Evolution, and Moral Progress.* Princeton: Princeton University Press, 2011 (1981).

Singer, Peter. "All Animals Are Equal." *Philosophical Exchange*, Vol. 5, No. 1, 1974, 103–116.

Taylor, Bron, Gretel Van Wieren, and Bernard Daley Zaleha. "Lynn White, Jr. and the Greening-of-Religion Hypothesis," *Conservation Biology*, Vol. 30, No. 5, 2016, 1000–1009.

Taylor, Paul W. "The Ethics of Respect for Nature," *Environmental Ethics*, Vol. 3, 1981, 197–218.

Tu, Wei-Ming. "The Continuity of Being: Chinese Visions of Nature," in: Tu Wei-Ming, *Confucian Thought: Selfhood as Creative Transformation.* Albany: State University of New York Press, 1996, 35–50.

Vaughn, Lewis. "Ethics and the Moral Life," in: Lewis Vaughn, *Beginning Ethics: An Introduction to Moral Philosophy.* New York: W.W. Norton, 2015, 13–28.

Wang, Robin R. *Yinyang: The Way of Heaven and Earth in Chinese Thought and Culture.* Cambridge: Cambridge University Press, 2012.

Wapner, Paul. *Living Through the End of Nature: The Future of American Environmentalism.* Cambridge: The MIT Press, 2010.

Warren, Karen J. "The Power and Promiseof Ecological Feminism," *Environmental Ethics*, Vol. 12, 1990, 125–146.

White, Lynn Jr. "The Historical Roots of Our Ecological Crisis," *Science*, Vol. 155, No. 3767, 1967, 1203–1207.

Zhang Zai. Complete Works. 6.108, in and translated by, Sui-Chi Haung, "Chang Tsai's [Zhang Zai] Concept of Ch'I," *Philosophy East and West*, Vol. 18, October, 1968, 247–260.

Critical Ecofeminism: A Feminist Environmental Research Network (FERN) for Collaborative and Relational Praxis

Jennifer L. Lawrence, Emily Ray, and Sarah Marie Wiebe

Introduction

The world is in the middle of a climate emergency. To explain this emergency and to make collective plans to survive it, there is considerable need to caution against reproducing the same patriarchal and capitalist-driven inequalities that produced this emergency. We argue in this chapter that critical, intersectional ecofeminist analysis helps us understand what conditions led us to this emergency and can serve as a guiding framework for imagining better futures.[1] We are inspired by the motivations that Greta Gaard lays out in her 2017 monograph *Critical Ecofeminism* in which she advances contemporary conversations about sustainability using a feminist lens to ask:

J. L. Lawrence
University of Virginia, Charlottesville, VA, USA
e-mail: jlawrence@virginia.edu

E. Ray
Sonoma State University, Rohnert Park, CA, USA
e-mail: emily.ray@sonoma.edu

S. M. Wiebe (✉)
University of Victoria, Victoria, BC, Canada
e-mail: swiebe@uvic.ca

How can we understand the entanglement of alienation, hierarchy and domination in terms that are simultaneously social, economic, ecological and political? And how can this understanding be used to leverage stronger and more joyful alliances for climate justice, reflecting insights, and commitments that are simultaneously feminist, queer, anticolonial, and trans-species? (xxii)

Toward this end, we first discursively frame climate emergencies; explore some of the ideologies and systems responsible for climate change; and finally look to collaborative and intersectional ecofeminism as a useful framework for resistance, and for praxis, in the effort to build a better world beyond a climate of everyday emergency conditions. Specifically, we contend that a revitalized critical ecofeminist lens can offer important ways of experiencing, feeling, seeing and understanding the pressing and urgent problems related to climate change while calling attention to the harmful politics of extraction and seeking to uphold non-extractive relationships of solidarity. A more reciprocal, relational and emancipatory approach to the pressing environmental climate emergency we face presents a vital space for drawing together diverse voices who are imagining and fighting for alternative futures[1] with their bodies on the frontlines.

This chapter has two main objectives: first, to explain how this critical and revitalized relational ecofeminism can help us to notice our current climate emergency and to imagine worlds beyond it; and second to elaborate what we mean by "critical ecofeminism" and articulate its value to political science. By relational, we mean ecofeminism that does not simply observe injustice and crisis from a distance, and does not extract data, stories and experiences from impacted communities, but rather works in conversation with those impacted by environmental and gendered injustice and provides an avenue between theory and practice. This imaginative praxis requires engagement beyond the academy, including a critical approach to extractive practices and politics, creative articulations of alternative possibilities to the colonial status quo, and collaborations across communities including artistic forms of resistance to systems of oppression. First, we put forward ways of detecting climate emergencies through discursive framing of fast- and slow-moving disasters and through community activism. Then, we discuss ecofeminist frameworks for engaging the climate emergency. Finally, we introduce the design and implementation of a collaborative feminist environmental research network, called FERN, as resource for community-engaged scholarship and organization, co-founded by the authors.

---

[1] The plural is deployed here intentionally, in the spirit of divesting from colonial thinking about temporality, and embracing the manifold of open futures "beyond the strictures of colonial modes of thinking" (Moulin, 2016), as well as Gesturing Towards Decolonial Futures (GTDF), an arts and research collective that curates "artistic, pedagogical, cartographic, and relational experiments that aim to identify and de-activate colonial habits of being, and gesture towards the possibility of decolonial futures." To read more about GTDF, see: www.decolonialfutures.net.

Convergence Zones

From the emergent Extinction Rebellion movement with roots in the colonial heart of the UK to the widespread social mobilization against resource extraction in unceded Wet'suwet'en territory in settler-colonial Canada reveal, the changes we are seeing, feeling and witnessing are urgent and command our attention (Extinction Rebellion 2021). This chapter articulates a sense of urgency about the current climate emergency while also highlighting avenues for intervention, disruption, alternatives and collaboration. As we will discuss in the final section of this chapter, by convening the Feminist Environmental Research Network (FERN), we aim to curate these critical conversations, support collaborative research and grow community in order to envision and enact alternative futures to the oppressive, colonial status quo through our grounded normative approach to environmental political thought.

What spaces of convergence might we envision between radical ecofeminism, critical Indigenous studies and environmental justice not just in theory, but in practice? This paper is an attempt to put these polyphonic voices in conversation, highlighting possible avenues for shifting narratives about climate change to create space for visions of emancipatory, sustainable and decolonial futures. In contention with narratives of climate denial, or the impending doom of our drastically changing climate, many communities are rising up, speaking out and actively creating radical change: social, political, economic and ecological.

Indigenous political theorists offer a rich foundation for place-based solidarity and grounded approaches to the pressing colonial problems of our time, which are deeply entwined with our climate emergency (Aikau and Gonzales 2019; Coulthard and Simpson 2016). These scholars remind us that colonization is ongoing and capitalist resource extraction exacerbates the colonial present as well as fuels the no-longer slow-moving crisis of climate change, and perhaps more importantly, demonstrates that decolonization is not confined to mere moments of denial or resistance. Rather that it is sustained effort to dismantle "relations of power and conceptions of knowledge that foment the reproduction of racial, gender and geopolitical hierarchies that came into being or found new and more powerful forms of expression in the modern/colonial world" (Maldonado-Torres 2006: 117).

Climate Emergency and the Ecofeminist Lens

With erratic weather shifts, rising seas poised to consume coastlines and forests burning with voracious force, it is evident that we are in climate emergency. Numerous island communities across the Pacific and around the world, from Hawai'i to the Caribean, find themselves in compromised environments with limited resources to respond to the systemic changes affecting their livelihoods. Many communities are actively organizing to articulate their local responses to this global phenomenon (KUA 2020a). On February 26,

2020, Hawai'i-based environmental grassroots organization KUA (Kua'aina Ulu 'Auamo) signed the Climate Strong Declaration as a call to action to create more sustainable and resilient island communities, calling for island-based leadership in responding to the UN Sustainable Development goals to complement the Local 2030 Islands Network launched at the UN General Assembly as a network of island leaders. The declaration noted island communities' lived experience with climate emergency: dealing with compromised critical infrastructures, overburdened and under-resourced health care, food, education and housing systems, changes in marine environments devastating fisheries and degrading important ecosystems on which island communities rely for their livelihoods, and geographical isolation in conjunction with limited political power (KUA 2020b). These vulnerabilities become particularly apparent during emergency and disaster response scenarios (Lawrence and Wiebe 2017). Examples of these (among many others) can be found around the globe, from Attawapiskat to Aamjiwnaang to Puerto Rico, TO Haiti and New Orleans.

Further signaling the severity of the climate emergency, scientists around the world are chiming in about our unprecedented climate crisis. For example, the United Nations Intergovernmental Panel on Climate Change (IPCC) noted that 2019 marked a year of exceptional global heat, retreating ice and rising sea levels. In 2021, the intergovernmental body continued to sound the alarm that climate change is widespread, rapid and intensifying, noting that (Tollefson 2021). With this swell of urgency in relation to shifting the debate about climate change, Oxford Dictionary declared "climate emergency" the word of the year in 2019 (UNEP, 2021; Zhou 2019). The discourse, science and widespread global mobilization all point to a drastically changing climate that commands political attention.

Jurisdictions around the world are declaring climate emergencies in their governing assemblies. From the forest fires in Australia and across the Western Provinces and States of North America to the hurricanes and their aftermath in archipelagos like Puerto Rico, humans have much to learn from these dramatic shifts about the radical elemental forces in our surrounding more-than-human environments. These environments are enduring the impacts of climate stress. Scholars are challenged today with finding innovative ways to hear these concerns and collaborate with those speaking up and out for environmental change—youth activists, Indigenous elders, students, mothers and fathers, to name but a few—and find ways to listen to those advocating for fully emancipated futures where all can live free from insecurity and harm.

A gendered lens is critical to these conversations; yet, it is not enough to uproot hierarchical structures of power. For academics trying to engage in radical political critique through applied academic praxis, this requires humility and decentering the egoism of the human as a master over nature in our anthropocentric culture, something feminists have been calling our attention to for decades (Einion and Rinaldi 2018; O'Reilly 2007; Plumwood 1993; Mies and Shiva 2014; MacGregor 2021; Merchant 1990). Women have long organized around pressing concerns related to contaminated environments, health and reproductive justice, while also calling for attention to treating climate change as more than a technological problem requiring a scientific response (Gaard 2015). Feminist scholarship broadly, and ecofeminist analysis in particular, is well-situated to expand the horizons of engagement to address structural inequalities as well as discursive contexts apparent in responses to climate change. As Greta Gaard has pointed out, the overconsumption of climate change is deeply gendered (Gaard 2015). Her analysis highlights how there can be no climate justice without gender justice. This requires widening out the scope of ecofeminist thought to build upon radical, intersectional ecofeminist critique to account for a diversity of voices.

Enhancing ecofeminist praxis includes the foregrounding of black and indigenous voices who have historically been excluded from spaces of knowledge production, as well as queer and posthumanist feminist ecological citizenship (MacGregor 2014; Alaimo 2016; Sandilands 1999). Ecofeminist political thought, posthumanism and critical Indigenous studies press the boundaries of how to understand the problem of climate change and the threats it poses to human and more-than-human lives. We take our cue from Gaard, who coined the concept of "critical ecofeminism," to advocate for an intersectional ecofeminist lens to examine climate change as a structural, material, discursive and colonial problem while also calling attention to the alternative practices and ways of thinking. In this way, we understand critical ecofeminism as intersectional feminism as it helps to "unite and empower different groups of oppressed people without collapsing or dismissing the meaningful differences between these groups" (Johnson et al. 2020).

Theories of feminism are often articulated in "waves," or general trends in feminist thought and practice across history and bound up with different theories of gendered injustice, demands and different political actors. For example, the "first wave" is associated with the suffragette movement of the nineteenth century, the second wave with the civil rights movement and equal opportunity demands, the third wave with cultural and identity concerns, and the fourth wave is still unfolding. Ecofeminism emerged during the second wave, with "critical ecofeminism" reflecting the fourth wave and benefitting from "past lessons about gender and racial essentialisms, as well as from more contemporary critical dimensions of economic, posthumanist, and postcolonial analysis" (Gaard 2017: xxiii). The theory of intersectionality is often attributed to Kimberlé Crenshaw, civil rights advocate, philosopher and leading scholar of critical race theory. "Intersectionality" can be understood as a set of theoretical and organizing principles that highlight entangled relations of power along gender, class and race (Crenshaw 1989). It offers an honest look at the structured realities of oppression that are embedded into institutions and ultimately shape privilege and discrimination.

An intersectional approach to the climate emergency can reveal how particular bodies experience the impacts of extraction differently. The experiences of Indigenous and black women draw this into sharp focus. Deborah King notes that black women in particular have long been aware of the multiple pressures they face based on both race and gender. In her words: "For us, the notion of double jeopardy is not a new one" (King 1988: 42). In Frances Beale's *Double Jeopardy: To be Black and Female,* she calls racism the "afterbirth of capitalism" and argues that black women suffer under capitalism and racism differently than their black male counterparts. Beale argues that the black liberation movement has, at times, called for black women to return to their natural place as mothers and homemakers, while outside of the home, black women experienced more economic exploitation than black men or white women. Beale recognized the possibilities for solidarity between white feminist movements and black liberation, but on certain conditions:

> The white women's movement is far from being monolithic. Any white group that does not have an anti-imperialist and anti-racist ideology has absolutely nothing in common with the Black women's struggle. In fact, some groups come to the incorrect assumption that their oppression is due simply to male chauvinism... While it is true that male chauvinism has become institutionalized in American society, one must always look for the main enemy – the fundamental cause of the female condition...if the white groups do not realize that they are in fact fighting capitalism and racism, we do not have common bonds. (Beale 1970: 121)

Here, Beale identifies capitalism as the primary antagonist of women's liberation.

In alignment with Beale's assertion that capitalism and racism are deeply intertwined, the main antagonist we write against is extractive capitalism. King is not convinced that each jeopardizing condition can be collapsed into one or another, and argues that class, race and gender should be taken as independent but interconnected issues (King 1988). She is critical of ideological monism found in many liberation movements, which asserts that all forms of oppression can be reduced to one primary issue. For example, Marxists have argued that concern for race and gender misdirect attention away from the central issue of class antagonism.

Marxism has long struggled with "the woman problem." As Heidi Hartmann describes, it tends to reduce inequalities between men and women into a set of problems addressed to capitalism, and not to patriarchy. Not only can Marxist analyses of the conditions of labor and living be "sex blind," as Hartmann puts it, but they can also be dismissive of categories of identity beyond class (Hartmann 1979). Ecofeminism, with its long-standing interest in lived experiences and activism, already has strong affinities with analyses of power that directly engage race, class and gender, among other categories of identity that are often under-represented in ecofeminism and Marxist feminism.

Intersectional Ecofeminism and Environmental Justice

Ecofeminism has, in some demonstrable ways, started to connect with intersectionality. Even in complicated texts like that of Mies and Shiva, they made clear their connections between colonialism, racism, environmental degradation, and violence toward women (Mies and Shiva 2014). Others, like Carol Adams and Lori Gruen (2014) made connections between the oppression of non-human animals and women, while also challenging the perceived divisions between these categories of human and other (Haraway 1998). Ecofeminist work should be diligent about addressing the multiple pressures of capitalism, race and gender, and address the different ways that people are affected by climate change, the multitude of strategies they use to fight the systems that promote climate change, and the ways they organize to protect and enrich their communities.

Activists like Vandana Shiva and the late Wangari Maathai have dedicated themselves to drawing connections between the vulnerability of women in the Global South to climate change and patriarchy and to the resilience of these same women and their organizing efforts to rehabilitate forests and watersheds, and to protect biodiversity through seed saving and sharing (Sturgeon 1999). These movements, however, are typically analyzed through an ecological feminism and international development lens, rather than explicitly as an intersectional effort. This presents an opportunity for critical ecofeminist organizing, which is essential at a historical time of planetary emergency, pandemic and rising inequalities (Bernacchi 2020).

Attention to race and environment in the United States is mostly reserved for scholarship on environmental justice. While environmental justice accounts for race and class, scholarship often limits a gendered analysis to acknowledging the special contributions of women as organizers and grassroots activists. For example, literature overviewing environmental movements tend to highlight activists like Lois Gibbs of Love Canal, Rachel Carson, author of the seminal environmentalist text, *Silent Spring*, and more recently, the efforts of Greta Thunberg without more deeply interrogating the social, economic and political contexts of their activism (Gottlieb 2005). Moreover, the role of women's leadership in environmental justice movements is evident in decolonial scholarship (Aikau and Gonzales 2019; Goodyear-Ka'ōpua 2013; Simpson 2011). Despite these critical works, all too often, gender is not addressed as its own sphere of experiences, and instead is often subordinated to analyses based on race and class. Attending to race, class and gender, explicitly and acknowledging the complicated relationships between each of these categories and how they come to bear on the relationship between people and the environment, aids critical ecofeminist scholarship to address oppression with more accuracy and care to specificity of place and experience. Ecofeminism can learn from intersectionality to address some of the criticisms of traditional Marxist feminism, which sometimes excludes race from consideration, and overextends a Western, white, middle-class view of womanhood to apply to all women. Audre Lorde writes about some of the consequences of feminism centered on a universal woman with white, middle-class characteristics: "As white women ignore their built-in privilege of whiteness and define *woman* in terms of their own experience alone, then women of Color become 'other,' the outsider whose experience and tradition are too 'alien' to comprehend" (Lorde 1984: 117). Ecofeminism has similarly struggled with identifying who women are and how women of color and women in the Global South are included and depicted in academic and activist discussions. Women of color are often identified as victims of patriarchy, climate change and capitalism. Such limited depictions fail to portray more nuanced relational forms of agency and capacities for radical change and transformation. Moving toward more dynamic and whole articulations of agency and power, "intersectionality also highlights new linkages and positions that can facilitate alliances between voices that are usually marginalized in the dominant climate agenda" (Kaijser and Kronsell 2014: 419). The women who are often the subject of ecofeminist analysis and activism are sometimes viewed as women in peril.

Ecofeminism, following long-standing patterns in feminism, tends to focus exclusively on the "category of 'woman'," as though it stands freely apart from race, class, ability and other categories that makes it difficult to talk about being a woman as a homogeneous group. Kimberlé Crenshaw (1989) draws attention to the problem of single-issue analysis:

> Unable to grasp the importance of Black women's intersectional experiences, not only courts, but feminist and civil rights thinkers as well have treated Black women in ways that deny both the unique compoundedness of their situation and the centrality of their experiences to the larger classes of women and Blacks. (1989: 150)

The experiences of Black women are often reduced to race or gender, and if feminist and ecofeminist theory continue to use a universal concept of "woman," then many women will continue to be excluded from scholarship and activism. bell hooks evokes the possibilities of feminism for everybody and the power of feminist thought and politics to create beloved community where "dreams of freedom and justice" are realized and also acknowledges that the "feminist revolution alone will not create such a world" (hooks 2000, x). While hooks believes in the potential of feminism to change "all our lives," it is also essential to "end racism, class elitism, [and] imperialism" (Ibid. 2000). Crenshaw similarly unveils the mythology of a universal feminism:

> The value of feminist theory to Black women is diminished because it evolves from a white racial context that is seldom acknowledged. Not only are women of color in fact overlooked, but their exclusion is reinforced when *white* women speak for and as *women*. The authoritative universal voice – usually white male

subjectivity masquerading as non-racial, non- gendered objectivity – is merely transferred to those who, but for gender, share many of the same cultural, economic, and social characteristics. (Crenshaw 1989: 154, [emphasis in the original])

The critique of feminism holds true for ecofeminism and thus carries the same challenges of speaking from a white, middle class, Western point of view. Departing from Crenshaw's argument that feminist theory is less valuable when it centers on white women, ecofeminism similarly needs to come to terms with the problems inherent in its relationship to feminist theory. In doing so, ecofeminism can move forward a critique of exploitative conditions for labor and living that includes race and class in a more nuanced analysis, particularly in the context of environmental harms and climate change. For example, Traci Voyles' book, *Wastelanding*, examines the histories of colonialism, racism and sexism in the environmentally catastrophic period of uranium mining on Diné land and in their communities with particular attention on the impact of women and their activism (Voyles 2015). See also further discussion of Indigenous community members' lived experiences in *Everyday Exposure: Indigenous Mobilization and Environmental Justice in Canada's Chemical Valley* (Wiebe 2016).

CONTENDING WITH THE THIRD WAVE: ECOFEMINISM AND MARXIST CRITIQUE

One of the critiques of third-wave feminism, of which intersectionality is a prominent part, is the overemphasis on the importance of identity and recognition, at the expense of focusing on neoliberalism as an ideology in which the exploitation of environment, labor, gender and race is written off as externalities or the just deserts of those suffering a lack of ambition and resourcefulness, resulting in the lack of development of human capital, according to the neoliberal worldview. Despite the shortcomings of second-wave feminism between the 1960s and 1990s, this wave put forth a strong Marxist feminist critique of capitalism and patriarchy. Writing in the LIES Journal, the collective authors who go by FLOC (2015) states:

> Marxist feminism has given us some conceptual tools to understand how and why patriarchal gender relations, and the relational categories 'man' and 'woman,' continue to be reproduced in capitalism. As a body of inquiry it demonstrates that men and women exist and are materially real; not in a biological sense, but as produced through a matrix of social relationships and institutions sustained by the needs of capital and of men as a group. (60)

Marxism emphasizes the primacy of class relations as the driver of exploitation and recognizes the state as the means by which the bourgeoisie control the means of production and the subsequent social and cultural systems that emerge from a particular mode of production. What Marxism does not do very well is acknowledge race and gender as experiences and conditions integral to understanding the relations and mechanisms of capitalism.

One of the most important critiques of second-wave feminism, and Marxist feminism in particular, came from women of color and women from the Global South asserting the universal "woman" at the center of Marxist feminist was conceived as Western, white and middle class. "As feminists increasingly took account of differences that exist among women, many feminists also moved from the tenets of modernism with its notion of a unified subject, that is, a universal (female) nature, to several postmodern tenets, especially the notion of a multiple and socially construed subject" (Mack-Canty 2014: 158). Third-wave and intersectional feminism spent significant energy on dealing with the question of identity in ways that benefit the essentialism problem in ecofeminism. Ecofeminism has, in many ways, reproduced gender essentialism in its theorizing (Norgaard 1998; Mies and Shiva 2014). A particularly tenacious line of reasoning in ecofeminism is that there is a natural connection between women and the earth. This perspective holds that women, as mothers, share with the earth the capacity for reproduction and the creation of new life. This particular perspective is fraught with unquestioned assumptions about who women are, which activities make a person a real woman, and what makes women "more natural" than men and, presumably, other women who do not reproduce or do not identify as nurturers. This version of ecofeminism is preoccupied with protecting femininity, vulnerability and nature—from the true nature of women to the nature in which they live and labor. Gender essentializing is evident in the work Mies and Shiva in their writing about reproduction (Mies and Shiva 2014) and Starhawk in her advocacy for paganism and goddess cults as part of ecofeminist practice (Gallixsee, n.d.).

Drifting from the universal woman also became adrift from Marxian analysis of oppression. However, Marxian analysis with intersectionality offers a way to deal with the essentialism debates by addressing identity in socio-economic context. As Nancy Fraser argued, identity does not need to be pitted against socio-economic justice (Fraser 2013). In fact, identity discussions are critical to understanding the subjects for which feminist, intersectional and environmental movements claim to advocate. Linking intersectionality to ecofeminism provides a set of tools for helping ecofeminist scholarship work through the debates that have hamstrung the field (Gaard 2011), even as economic and environmental pressures compound for women in the age of climate change.

To do so, we turn our attention to "decoloniality" and to Indigenous political theorists to engage in solidarity-building for alternative anticolonial political systems, and Indigenous leaders mobilizing at the frontlines of resource extraction initiatives who demonstrate with their bodies on the line that radical action is necessary for transformative politics. Recent work from Macarena Gomez-Barris (2017), Thea Riofrancos (2020) and Dana

Powell (2018) all illustrate the power of women's movements against extractive projects to slow or halt the steady march of continual accumulation by extraction, while also cultivating future imaginaries of what a decolonial and post-extractive world can be. From their efforts, we have much to learn about the intersections of settler-colonialism, gender and the environment.

Capitalism Will not Save Us: The Climate Emergency is a Colonial Crisis

Colonialism and its drive to extract labor, resources and land rights are part of the matrix of economic, political and administrative systems that advances climate change. One of the positions from which to address colonialism is decoloniality. Decoloniality has multiple meanings and is nearly inseparable from on the ground practices of resisting colonialism and building new relations that are rooted in worldviews that reject colonialism.

We take our starting point from Aníbal Quijano, Peruvian sociologist who introduced "coloniality" and the dimensions of power often hidden behind the metanarrative of modernity, of which "coloniality" exists as the "darker and hidden side of modernity" (Mignolo 2011: 140). Quijano argues that the beginning of America—not the beginning of organized society on the continent—ushered in the beginning of a fundamentally new modern global capitalism with labor and extraction organized around capital, and labor newly organized around the emerging category of race as "biologically structural and hierarchical differences between the dominant and dominated" (Quijano 2000: 216).

With modernity grounded in a certain form of racialized capitalism, coloniality as praxis takes shape as historically and geographically contingent approaches to deWesternization. This is not to imply the rejection of Western thought or scientific modes of knowledge production generally; rather, here we seek to emphasize more multidimensional modes of knowledge generation: "Decoloniality…is not a new paradigm or mode of critical thought. It is a way, option, standpoint, analytic, project, practice, and praxis" (Mignolo and Walsh 2018: 5). Decoloniality can be found in global resistance movements and in global efforts to create ways to transmit knowledge and socially organize around different organizing principles than capital and exploitation. Decoloniality reveals what the myth of modern progress requires, including creating racialized and gendered hierarchies that naturalize difference to justify subordination and genocide. This myth promotes the Western story of modernization as a universal benefit to all civilization, which serves to split the "modern" from "premodern" in such a way that the Western imposition of imperial power is considered a mission for the betterment of all "developing" peoples and nations.

According to Mignolo and Walsh (2018), decoloniality cannot be the better universal option, but is broken into "two movements"; "Decoloniality as conceived here therefore consists of two movements: one, its affirmation as an option among options (diverse and heterogeneous but grounded as any co-existing options, from Christianity, to neoliberalism, to Marxism, Islamism)" (Ibid.). The first movement places decoloniality in a range of ideological and theological worldviews. One can hold multiple worldviews and be committed to decoloniality. This perspective is not an exclusive one. Decoloniality is a substantive and diverse orientation to the world. The second movement emphasized how conceptions and enactments of decolonization offer many options for organizing and building momentum, with vast and varied references that draw upon ideas from theology and ideology, conflict or collaboration. Arguments for decoloniality must also work to wrestle decoloniality from "the temptations of totalitarian totality."

This second movement insists that decoloniality should not be taken up in the same exclusive and universal terms as Western thought has been since the Enlightenment and that decoloniality is not converted into a competing metanarrative to modernity, but rather refuses universalization as part of the core attributes of decolonial thought and action. Taken in these terms, decoloniality is a relational set of strategies, frameworks, and movements with porous boundaries across disciplines, geography, cosmologies, epistemologies and actions. Given that colonization, capitalism and heteropatriarchy shape-shift, following Aikau and Gonzales, "we need decolonizing practices that are creative, adaptive, innovative and ongoing" (2019: 3).

The fluid, open nature of decoloniality can render it vulnerable to co-optation by States and institutions that wish to rebrand their colonial practices as progressive atonement and "reconciliation" for past and present colonial misdeeds.[2] Tuck and Yang warn that this often takes the form of rendering decoloniality-as-praxis into decoloniality-as-metaphor, which "kills the very possibility of decolonization; it centers whiteness, it resettles theory, it extends innocence to the settler, it entertains a settler future" (2012: 3). Inspired by the works of Mignolo and Walsh, decoloniality then is praxis, it cannot stand alone as a theoretical framework or as action without theory. Moreover, decolonization must correspond with self-determination for Indigenous peoples, communities and lived experiences (Corntassel 2012; Corntassel and Bryce 2012).

One of the valences of this large and multifaceted project is decolonial feminism, which can be understood in similar terms to the pluralistic and "option of options" style of decoloniality. Decolonial feminism, following Mignolo and Walsh (2018), works to unseat the universal "woman" especially as she is typically made in the image of the white Western woman (Mignolo and Walsh 2018: 39). Decoloniality and its inherent pluralism help

---

[2] For more critical analysis and insights on the discourse of reconciliation, see "The Scam of Reconciliation" Red Nation podcast interview with Uahikea Maile and Eva Jewell: https://podcasts.apple.com/ca/podcast/the-scam-of-reconciliation-w-eva-jewell/id1482834485?i = 1,000,505,615,815.

to queer and transgress the boundaries of what it means to be a woman and what it means to practice feminism. Decolonial feminism is attentive to the multiple ways gender is built, imposed, experienced and challenged, including the limitations of feminism itself, although "feminism" is not always named or invoked in struggles that account for gendered power imbalances For example, short-term labor camps (dubbed "man camps" since they are predominately occupied by men) established in oil producing regions of the upper midwest and southern Canada are part of a destructive network of sexual violence toward women, particularly indigenous women, and environmental destruction, but there exists only limited research focusing on these dimensions (Parson and Ray 2020). Ecofeminism, when productively entangled with decoloniality, critique of capitalism and intersectionality, can make important discursive and strategic contributions to theory and practice of resisting domination and rebuilding new modes of organization. We look to just a few examples of these movements in action that, in addition to fighting colonial violence, fight the practices contributing to climate change.

Decolonial Feminism: Encounters with Climate Change

Widespread social mobilization across the Province of British Columbia, Canada and the world articulate and enact emancipatory solidarity movements for alternative, anti-capitalist, anti-extractive and decolonial politics. Hereditary chiefs of the Wetland leaders of the Likhts'amisyu Clan have taken the federal government to court in Canada, demanding that the state take climate change seriously, which they consider to be an existential threat and a violation of their Sect. 7 Charter of Rights and Freedoms protections, including the right to life, liberty and security of the person (Proctor 2020). Their lawsuit challenged the Canadian government's approval of the 670 km Coastal GasLink Pipeline through their territory without the free, prior and informed consent of hereditary leadership and drew attention to ongoing disputes about Indigenous land ownership and jurisdiction (Luxon 2019). This issue is not a standalone event, but a microcosm of larger geopolitical tensions about the state's continuous push for capitalist extraction to the detriment of Indigenous rights and the health of the planet.

The state encroachment on Wet'suwet'en territory without consent of those most directly affected also alerts our attention to the limits of liberal reconciliation politics and the gendered dimensions of extraction. For instance, Indigenous leaders from Wet'suwet'en territories have long expressed concern about the slow moving, often overlooked environmental violence of "man camps" and the ways in which extractive labor affects Indigenous women's lands, bodies and livelihoods (Kojola and Pellow 2021; Nixon 2011; Parson and Ray 2020). While the Provincial Government of British Columbia passed the United National Declaration on the Rights of Indigenous Peoples (UNDRIP) into law on October 24, 2019—twelve years after its introduction to the

United Nations General Assembly—committing itself to a clear plan for reconciliation, the state's efforts to ease extraction on Wet'suwet'en territory call this commitment into question (Larsen 2019).

With "Reconciliation is Dead" flags surrounding them and the sounds of their drums reverberating, staring down the approaching authorities, as helicopters dropped tactical teams from above, Wet'suwet'en women leaders, chiefs and land defenders sang, danced and held ceremony as they were arrested for breaching a court injunction (Reconciliation is Dead 2021). The advancing police extinguished the matriarchs' sacred fire and tore down red dresses placed to hold and honor the spirits of missing and murdered Indigenous women, girls and two-spirit people. Allowing the GasLink pipeline to intersect Wet'suwet'en territory would bring with it a camp of 450 men along the now infamous Highway of Tears (Levin 2016). This stretch of road—many of which is without cell service—is well-known as a site for the disappearance of many Indigenous women and girls. This heartbreaking standoff between the police arresting Wet'suwet'en leaders draws into clear view the politics of slow violence (Nixon 2011), environmental injustice and settler-colonialism (Coulthard 2014; Kojola and Pellow 2021) and the disproportionate harm felt by Indigenous women as a result of extractive processes. Moreover, this extractive violence exposes the discursive rhetoric of reconciliation and continuation of patriarchal colonialism in Canada. The climate emergency is felt across borders, across the globe, and highlights the need for critical ecofeminist approaches to climate emergency.

LISTENING AS CRITICAL ECOFEMINIST PRAXIS: INDIGENOUS LAND MANAGEMENT

In the summer of 2020, California burned. In northern California, not far from the San Francisco Bay, the sky color was determined by the smoke layer, and the only precipitation was a steady rain of ash. The air quality reading on mobile devices was a deep purple warning with the cartoon figure of a man in a gas mask. How do you protect yourself from breathing air? This presented a fundamental contradiction for survival: breathing air would make us sick and shorten our lives, and yet breathing air is essential for human life. Californians appeared divided in their analysis of the record-setting fire season. Most mainstream news outlets connected the fires to climate change-driven drought, poor funding, low public concern for forest management, and a dangerously extractive relationship to fire on public lands (Stelloh 2020; "*The Guardian*" 2020).

This competed with another narrative, one pushed by the 45th President: that Californians simply needed to rake their leaves and take personal responsibility for the state of their public lands (White 2020). In this account, the choking fires were not a reckoning with climate change as an emergency, this was not a call to radically rethink our relationship with the environment and with our socio-economic systems.

The impacts of the fires were felt differently across the region, with Latine farmworkers in low-wage labor positions exposed to toxic smoke and debris while wealthy celebrity vineyard owners could watch from afar in any number of homes located safely away from danger. During the summer of burning, the Forest Service partnered with the Karuk and Yurok tribes, giving tribes permission to conduct traditional controlled burns on their own lands that were stolen from them during the height of the California genocide (Sommer 2020; Madley 2016). Tribal Chairman of the North Fork Mono, Ron Goode, puts it this way: "This is old land... It's been in use for thousands and thousands of years. And so what we're doing out here is restoring life" (Sommer 2020). The fires in California were not just a jarring demonstration of climate emergency, they revealed stark inequalities in climate change vulnerability and the challenging position of native tribes seeking to share their life-saving knowledge while also operating in a colonial context with the state.

Noticing the Climate Emergency

Many people move through the world without making the conscious connection that their bodies, lives, homes and families are affected by the interconnected forces of climate change. We are not just living in a climate emergency, but a global public health emergency. Human and more-than-human bodies are sites of concern. The health of our atmospheres, oceans, forests and waterways relate to the well-being of human bodies. To understand the severity of living in a climate emergency requires a shift in material praxis, discourse and emancipatory possibilities for alternative futures, a waking up to the lived-realities facing our environments and us. This is a critical moment to realize that we can no longer be complacent in our normalized way of living in this world premised upon limitless extraction and thoughtless waste. As we highlight below, there are alternatives, oases of community restoration and regeneration across the globe that call for our attention as we simultaneously critique this world and enact new ones.

Our current climate emergency demands attention to its gendered dimensions. Women and queer voices are at the forefront of the global environmental movement. Young activists from across diverse backgrounds are taking on the intersecting trifecta of capitalism, misogyny and extraction. Greta Thunberg, Takaya Blaney, Kathy Jetnil-Kijiner and Amber Pelletier are but some of the constellating young women leaders and activists animating this conversation, calling out the patriarchal and colonial underpinnings of extractive capitalism and putting decision-makers on alert. Their actions remind us that environmental justice involves multiple dimensions. For instance, an awareness of the ideological forces that enabled an uneven distribution of goods and bads among societies—i.e., petrochemical and polymer refineries situated in close proximity to communities of color—which have asymmetrical consequences for those living with toxins posing harm to their bodies and ecosystems (MacGregor 2021; Kojola and Pellow 2021; Wiebe 2016). Moreover, their resistance to capitalist patriarchy exposes the embedded hierarchical forces apparent in procedural, bureaucratic and administrative responses to experiences of environmental injustice; and further, these women give expression to the need to depart from the extractivist neoliberal status quo that gives primacy to the atomistic, possessive and property-owning individual. Instead, their efforts highlight how we exist in relation to our surrounding more-than-human environments and in fact are dependent on them for survival. This multispecies orientation to environmental justice is gaining momentum in environmental political thought, and our collaborative efforts aim to contribute to this critical conversation (Celermajer et al. 2021).

The climate emergency signals that we need new languages, words and politics for understanding our intersectional, embedded and entangled relationships with/in our hurting ecosystems. As such, critical ecofeminism presents a critique of the liberal status quo way of thinking about, seeing and feeling in relation to the environment as property to be owned, abused, contained and controlled for human interest only. Classical liberalism posits that freedom is founded on property ownership, a stable state to provide security and legal frameworks for addressing justice, and the opportunity for those rational, moderate, motivated individuals to succeed. As such, liberalism and its cousin, neoliberalism, are used to address climate change through policy incrementalism and market-based solutions. They provide no challenge to the economic and political order that produces climate change and environmental injustice.

Since liberalism and neoliberalism contribute to the climate emergency, they cannot be seen as starting points for solutions. Critical ecofeminism goes further to examine and interrogate the historical, structural, colonial, economic, misogynistic and foundational forces that enabled these systemic inequities to persist through categorizations of Western superiority and human dominance over nature and male dominance over women. A critical, relational ecofeminist orientation does not stop at critique, but also centers creation through collaboration. For instance, the ecofeminism we flesh out in this paper challenges rigid dualisms or essentialisms while centering creative forms of resistance and articulations of radical alternatives. We contend that this intersectional, feminist mode of critique necessitates relational awareness of more-than-human environments and co-creation of pathways for sustainable possibilities.

Ecofeminist Knowledge Production as Anti-Oppressive Praxis

Climate emergency conditions require us to confront paradigms of endless economic growth and critically reappraise the meaning and force of everyday discourses and practices that shape material realities through words like "progress," by way of policies that enable continual violence against women and nature, and also through the circulation of mythologies that repress potential for emancipatory futures. The entrenched framing of scholarship and praxis as separate, fixed, domains not only conserves capitalist world systems, but also maintains the roots of violence by erasing and silencing ecofeminist perspectives and actions to confront climate emergency.

For example, the derogatory and persistent adage "those who can't do, teach" indicates that people involved in the production of knowledge do not have enough skill to engage in the things they are teaching about. This concept has been taken up alongside populist and anti-intellectual factions in recent years, ignoring the fact that the overwhelming majority of faculty in the United States are contingent workers with no access to tenure, and who are working for less than a living wage (Childress 2019).

This mythological and divisive frame emboldens the supposed town/gown divide, the idea that academics are set apart from society in an ivory tower, out of touch with the experiences of society as if the exploitative processes of capitalism stop at the border of university campuses, as if universities are not built on and sustained by stolen land and through extracted labor. A recent project by Kalen Goodluck, Tristan Ahtone and Robert Lee entitled "Landgrab University" (Goodluck et al. 2020) helps us to critically evaluate how diversionary tactics enable status quo power relations and capital accumulation to persist through the production of knowledge. This so-called division between research and praxis functionally oppresses possibilities for collective organizing and deeper social understanding of our shared experiences and fates under capitalism whereas the heart of ecofeminist work is the "centrality of praxis which links intellectual, political and activist work" (Gaard 2012: 15). Thus, the false frame of knowledge production and action as separate actions must be confronted on the basis of its illusory power. Instead, knowledge and praxis are interdependent, integrated and imbued with creative possibility.

As it has historically done, ecofeminism in this critical moment offers "its ideas and its political commitments to justice and sustainability" which have "always been materialist and always been rooted in the quotidian labors that reproduce life" (MacGregor 2021: 55). Moreover, it allows for an important interrogation about the framing of absences, disconnections and the emergence of new strands of thought within environmental discourses. Here, the production of knowledge and the "politics of knowledge production in the field of environmental politics" (Ibid.) are confronted. For critical ecofeminism, and critical projects writ large, the research-praxis-nexus is not new.

In *Pedagogy of the Oppressed*, Paulo Freire built on a Marxist philosophy of praxis to articulate the importance of the knowledge-praxis nexus in the struggle for freedom. He insists, "within the word we find two dimensions, reflection and action" and that "if one is sacrificed – even in part – the other immediately suffers" therefore, "there is no true word that is not at the same time a praxis. Thus, to speak a true word is to transform the world" (Freire

2018: 87). In the same spirit of critical pedagogy, which requires accessibility, openness and freedom for all to speak, critical ecofeminism shakes the foundation of oppressive structures.

In her seminal work *Feminism and the Mastery of Nature*, Val Plumwood praises the contributions of ecofeminism to "both activist struggle and to theorising links between women's oppression and the domination of nature" and calls for a critical ecofeminism to be wielded as a "critical and analytical force" and a "powerful political tool" (Plumwood 1993: 1).

Assessing how we move from extractive practices to ones informed by ecofeminism, anti-oppressive knowledge, relationality necessitates a critical appraisal of ecofeminism, not simply as an ideological orientation, but as an anti-oppressive, co-creative process and political tool that bridges thought and practice. While it may seem that efforts to connect "work to concrete experience" can result in frustrated outcomes and unrealized ideals in the quest for emancipation, the creative construction of alternative worlds, even unfinished and imperfect ones, allow for resistance to take hold and imagination to flourish; bell hooks remarks on this experience in her feminist theory classes where she indicates that "student frustration is directed at the inability of the methodology, analysis and abstract writing to make the work connect to their efforts to live more fully, to transform society, to live a politics of feminism," that is a politics of freedom (hooks 2000: 88).

Taking inspiration from the expressed aims of critical theory as an explicitly political project, critical ecofeminist scholarship is not about a value-neutral quest for knowledge. Rather, critical ecofeminism acknowledges the importance of thought, reflection on and engagement with structural violence and dominant worldviews, which have historically exploited women and nature, in order to create anti-oppressive materialities and liberatory futures. In this way, critical ecofeminism provides a way to imagine and build futures through constructive anti-oppressive practice in a process of inquiry that maintains ambivalence and uncertainty, while enabling us to question that which appears "normal" and taken for granted. Research involves an ongoing renegotiation to create spaces for ourselves and others who are commonly excluded from knowledge creation (Potts and Brown 2015). Moreover, it answers the call for critical theory to embrace a more intersectional approach that "promotes emancipatory aims by uncovering injustices, power structures, and problematic assumptions within the ideological status quo" (Hammond 2021: 293). The creation of knowledge and anti-oppressive struggles are/should be domains for all, not just the privileged or powerful. Yet, in a global fossil-empire, language, policies and practices coalesce to maintain exploitative systems of production.

Certainly, in a moment of climate emergency an ecofeminist confrontation of fossil empire and the knowledge structures that sustain it is of critical importance. More than a refusal, ecofeminism is a constructive effort to find a better way, to create paths forward, to cultivate meaningful work and regenerative possibilities. This orientation to knowledge production presents a challenge for some institutions of higher education whose research portfolios and endowments are supported by petrochemical industries and other merchants of doubt whose efforts minimize or silence scholarship that calls into account the role that industry interests play in crafting and legitimizing data and evidence about environmental harm. The fossil-fuel industry has a record of delegitimizing citizens' claims to harm, in particular within sites of extractive violence with environmental and human health impacts (Lawrence 2018: 172), but also plays an active role in funding research to preserve the ability of the industry to continue their life-negating practices (Kamola 2019). Naomi Oreskes and Erik Conway articulate the ways in which powerful industries (Big Tobacco and Big Oil, in particular) have effectively merged knowledge domains to the material realities of our contemporary world through the production of climate emergency and public health crises (2011). Casting doubt on climate and public health researchers is an insidious, yet legal practice whereby powerful industries can utilize the tools of the state to silence, attack and delegitimize.

Prominent climate researcher Michael Mann was the mark of such a campaign which targeted his unpublished climate data and research by utilizing the Freedom of Information Act (FOIA) to discredit his work by accessing his private correspondence (Union of Concerned Scientists 2017). Although his university (University of Virginia) initially supported Mann, the administration ultimately bowed to the pressures of ATI, the American Tradition Institute, now the Energy and Environmental Legal Institute, a Koch funded think tank with deep financial ties to the fossil-fuel industry. After years in court and fighting other industry-funded attacks on his work and personal life, Mann won his case but warned of the deterrent and chilling effects of such harassment on young climate researchers at a moment when they are desperately needed in the fight against climate emergency:

> I fear the chill that could descend. I worry especially that younger scientists might be deterred from going into climate research (or any topic where scientific findings can prove inconvenient to powerful vested interests). As someone who has weathered many attacks, I would urge these scientists to have courage. The fate of the planet hangs in the balance. (Mann 2016)

These attacks compound the efforts of corporations to frame the climate emergency, and women's exposures and erasures, in a way that obviates responsibility. While these targets, omissions and efforts to silence may not be direct evidence of the power of research and knowledge production to shape the material world, to upend status quo power relations, and to imagine, generate and build liberatory futures, such efforts indicate an anxiety about the possibilities of transformative knowledge to democratize power. In her article, "Making Matter Great Again?: Ecofeminism, New Materialism and the Everyday Turn in Environmental Politics," Sherilyn MacGregor speaks truth to power and forcefully reminds us that "ecofeminist theory *has never not* been grounded in materiality" and "everyday material practices" (2021: 41).

It is clear that the creation of knowledge shapes policy, and thus informs and forms our realities. However, we should not be so arrogant to either understand the research-praxis-nexus as a new phenomenon that is particular to ecofeminism nor should we see the production of knowledge as something that occurs solely within the halls of academia. The critical theory project was always about an unbounded approach to thinking about and creating "alternatives to the dominance of technical reason, disciplinary modes of power, and false consciousness that govern our contemporary everyday lives" (Boros 2019: 5). This refrain about the deep entanglements of the material world is further supported by Marcuse in a candid articulation of critical theory's purpose to move beyond the object of theorization:

> It is the conviction of its founders that critical theory of society is essentially linked with materialism... The theory of society is an economic, not a philosophical system. There are two basic elements linking materialism to correct social theory: concern with human happiness, and the conviction that it can be attained only through a transformation of the material conditions of existence. The actual course of the transformation and the fundamental measures to be taken in order to arrive at a rational organization of society are prescribed by analysis of the economic and political conditions in the given historical situation. The subsequent construction of the new society cannot be the object of theory, for it is to occur as the free creation of the liberated individuals. (Marcuse 2009: 135)

Cultivating explicit anti-oppressive, anti-extractive knowledge and material realities is foundational to critical ecofeminism. While there is not, and perhaps should not be, a framework or prescription for how this is enacted, ecofeminist praxis characterized by democratized processes of inquiry and everyday practices that include "negotiation, reciprocity, and empowerment" are essential (Lather 1986: 257). If it is true that "any form of transformation of society must include a discussion of the social organization and political action necessary to achieve this goal" (Akard 1983: 213), then we must respond with a celebration of spaces of resistance and hopes for fully emancipated futures. In this spirit, we highlight several projects that harness the potential of creating with nature, with rich diversity, that cultivates "power with" as opposed to "power over." While these organizations may not attach the explicit label of ecofeminism to their work, their efforts to draw together the concerns of feminists with ecological ones evidence the proliferation of efforts through which ecofeminist praxis is being enacted and inspiring new movements.

CRITICAL ECOFEMINIST PROJECTS

Navdanya, an organization led by Vandana Shiva and whose name means "nine seeds," demonstrates how collective action for policy change can work. Navdanya works to conserve diversity and reclaim the commons through

"Earth-centric, Women-centric, and Farmer-led" movements for the protection of biological and cultural diversity. Just as rich soil allows for growth and sustenance, a critical ecofeminist embrace of diverse ways of knowing "is a gift of life, of heritage, and continuity" (Navdanya). For more than three decades, this organization has resisted the commodification of seeds and food by embracing indigenous knowledge and seed saving practices which pave the way for a democratized earth. Food sovereignty is not only a social struggle, but also a feminist issue. "Promoting food sovereignty can advance women's rights based on gender, but only if organizers and actors involved in this movement open their eyes and take on board feminist analyses and practices" (Leroy 2017). The link between reproduction and agency is made clear in the work of Navdanya to flatten hierarchies within the global food economy, seeing food as a giver of life, not a commodity.

Second, the All We Can Save Project is a solidarity movement and community care initiative to nurture emergent climate feminists of all ages and genders. This is a powerfully compelling and creative project to build feminist leadership on the climate emergency. Co-founded by Drs. Ayanna Elizabeth Johnson and Katherine K. Wilkinson, All We Can Save promotes the work of women who are "already leading boldly and effectively and throwing doors open to welcome people into climate work" (All We Can Save). Amplifying the perspectives of Black, Indigenous and other women of color as forerunners to a "transformational climate-feminist ecosystem," All We Can Save works to address "burdens, systemic barriers, and burnout" through solidarity and community building, a true inspiration for imaginative and critical ecofeminist research/practice. For example, they have developed educational resources for educators and offer mentorship to support those who are still finding their place in climate work.

Third, Women's Voices for the Earth (WVE) is a non-profit organization whose mission is to amplify women's voices to eliminate toxic chemicals that harm health and communities. Bridging work, equity, and environment, WVE imagines a world free of toxins at "every point in the cycle of production and consumption – from extraction, to processing, to use, to disposal" (Women's Voices). This vision has been the basis of more than 25 years of work to cultivate a thriving world through efforts to shift chemical policy and highlight intersections of race, class, gender, health and the environment. Acknowledging the disproportionate impact that toxic chemicals have on women and children is foundational to ecofeminist work. This functions as a counterpoint to representations that have been restricted under colonial forms of governance. Honest and historically accurate representations are not only consistent with calls to action from indigenous leaders, but importantly lay a foundation toward social change.

Finally, WeDo is a global women's advocacy organization for a just world that promotes and protects human rights, gender equality and the integrity of the environment. WeDo is working to build feminist solidarity and resilience in times of crisis, including at this moment of climate emergency and global public health crisis through advocacy of gender responsive climate policy in programs such as the Women's Global Call for Climate Justice Campaign and the Women Climate Justice Advocates program (WeDo). For WeDo the creation of knowledge to empower women is central as is the ability to support families and communities by acknowledging and attending to the "deeply personal and embodied crisis" that each person faces in their own intersecting ways. Centering care, community and compassion are foundational to the work of WeDo in its struggle for global justice, recognition of the violence of borders, and the cultivation of democratic and multilateral spaces to address "critical areas of concern including climate change, biodiversity, macro-economic issues, and health" (WeDo).

There are hundreds of other organizations and movements, and thousands of voices, that we can highlight in relation to the power of critical ecofeminism to create new realities and to push back upon the framing of climate emergency as totalizing and inevitable. There are anti-extractive resistance movements across the world which are primarily led by women—from the mountains of Appalachia in which blockades have been erected to slow the devastation of the ecosystem as Equitrans decimates habitats to make way for the proposed Mountain Valley Pipeline, and the Unist'ot'en Camp, an indigenous reoccupation of the Wet'suwet'en land in northern British Columbia. In these spaces, and so many others, women's bodies are on the frontlines of resistance (Cirefice and Sullivan 2019; Wiebe 2016), even as their lives are threatened through state violence as well as the slow disaster of climate emergency.

Ariel Salleh's concept of "embodied materialism" allows us to see the connected ways that knowledge and these diverse struggles to imagine and create liberatory futures. Engaging in anti-oppressive scholarship is one path to encourage structural shifts. She notes: "Too many political programs rest in ossified and disembodied belief systems, whereas an embodied materialism is a transitional idea, a tool for making change at this moment now. Once attitudes and structures shift, the ecofeminist critique can be discarded" (Salleh 2017: 298–299). Until that time, we continue our work and consider what might be next for FERN as an anti-oppressive endeavor to examine the roots of exploitation and seek to eradicate them through political action.

Concluding Reflections

In this chapter, we suggest that advancing a praxis of critical ecofeminist scholarship requires a movement to address the multiple pressures of capitalism, race and gender, alongside the different ways that people are affected by climate change, the multitude of strategies they use to fight the systems that promote climate change, and the ways they organize to protect and enrich their communities. This requires creativity, critical reflection, imagination, informed action and collaboration. For over a decade, we have sought to foster a sense of community for scholars who bring feminist and gender-based approaches to the pressing environmental challenges of our time.

Through active participation in the Western Political Science Association (WPSA) environmental political theory community for over a decade, we have organized panels, pre-conference workshops and produced publications that weave together critical ecofeminism, environmental justice and political theory. We are excited about convening this intellectual community and look forward to opening up this conversation more broadly to engage with critical, creative and intersectional approaches to climate change and ongoing crises such as the climate emergency. By founding FERN—a collaborative Feminist Environmental Research Network—we aim to amplify the voices reflecting on these pertinent matters, particularly those at the frontlines facing the visceral impacts of climate change in their everyday lives.

FERN is a collaborative research network that aims to grow community and cultivate collaboration through critical scholarship and creative approaches leading to ameliorative action. This includes participatory and arts-based methods of community-engaged scholarship. Through social media, a listserv and website, we aim to elevate conversations about environmental issues such as the climate emergency and promote opportunities for further academic-activist connections. As a research collaborative, we seek to convene a space for critical engagement with ideas and pathways for sustainable futures while promoting just relations between all living things, thus taking seriously a more equitable relational orientation to human/more-than-human relations and contending with neoliberal extractivism. There are three main areas of emphasis for this collaborative effort: *growing community, cultivating collaboration,* and *critique and creativity*. **Growing community:** as an emergent and rhizomatic community, FERN is the connective tissue for a body of scholarship that investigates environmental issues from a gendered lens. This community of practice, virtually and in person, seeks to reach from the academy into community respectfully and reciprocally, while centering care and refusing extractive relations. **Cultivating Collaboration**: as a transdisciplinary collective, FERN engages ideas within the academy and beyond, including activist, artistic and policy researchers from diverse backgrounds and life-experiences to address pressing environmental challenges. This approach celebrates collaboration to address environmental injustices through multiple methods and angles of vision. **Critique and Creativity**: from an imaginative lens, FERN expresses hope for future generations. We are committed to a shared vision that engages creative methods and media to envision sustainable, decolonial and emancipatory futures. FERN is dedicated to supporting research, art, activism and open dialogue about socio-environmental relations with special attention to interconnected injustices, the lived realities of climate change and oppressive systems of power (anthropocentrism, racism, sexism, patriarchy, colonialism and capitalism). We hope you join us to grow this community as we collaboratively interrogate the intersecting forces of extractive capitalism, colonialism and elevate a more radical, loving, caring praxis of relational engagement with more-than-human lives inside the academy and beyond.

Acknowledgements We also acknowledge our relative positions of privilege as we write about this topic as three non-Indigenous, female, able-bodied scholars with relative security within academic institutions who intend to write in solidarity, and amplify the voices of those involved in the ongoing struggle against dominant systems of power tied to crises of the climate emergency, extractive capitalism and ongoing colonialism. See more at www.ferncollaborative.com and contact us at: ferncollaborative@gmail.com.

References

Adams, Carol and Lori Gruen. *Ecofeminism: Feminist Intersections with Animals and the Earth*, 2014. New York: Bloomsbury.

Aikau, Hōkūlani K and Vernadette Vicuña Gonzalez, eds. *Detours: A Decolonial Guide to Hawai*.

Akard, Patrick. "The 'Theory-Praxis Nexus' in Marcuse's Critical Theory." *Dialectical Anthropology*, 1983, 8(3): 207–215.

All We Can Save Project. [Available Online]: https://www.allwecansave.earth/project.

Alaimo, Stacy. *Exposed: Environmental Politics and Pleasures in Posthuman Times*. Minneapolis: University of Minnesota Press, 2016.

Beale, Francis. "Double Jeopardy: To Be Black and Female" in Bambara, Toni Cade (Ed.). *The Black Woman: An Anthology*. (pp. 109–122). New York: Washington Square Press, 1970.

Bernacchi, Erika. "The Ecofeminist Response to Covid 19." *Fuori Luogo. Rivista di Sociologia del Territorio, Turismo, Tecnologia*, 2020, 8(2): 23–30.

Boros, Diana. "Critical Theory and the Challenge of Praxis." *Contemporary Political Theory*. 2019, 18(1): 5–7.

"California's Wildfire Hell: How 2020 Became the State's Worst Ever Fire Season." *The Guardian*. Guardian News and Media. Accessed December 30, 2020. https://www.theguardian.com/us-news/2020/dec/30/california-wildfires-north-complex-record.

Celermajer, Danielle, David Schlosberg, Lauren Rickards, Makere Stewart-Harawira, Mathias Thaler, Petra Tschakert, Blanche Verlie, and Christine Winter. "Multispecies Justice: Theories, Challenges, and a Research Agenda for Environmental Politics." *Environmental Politics*, 2021, 30(1–2): 119–140.

Childress, Herb. *The Adjunct Underclass: How America's Colleges Betrayed Their Faculty, Their Students, and Their Mission*. University of Chicago Press, 2019.

Cirefice, V'cenza and Lynda Sullivan. "Women on the Frontlines of Resistance to Extractivism." *Policy and Practice: A Development Education Review*, 2019 (29).

Corntassel, Jeff. "Re-Envisioning Resurgence: Indigenous Pathways to Decolonization and Sustainable Self-Determination." *Decolonization: Indigeneity, Education & Society*, 1(1): 86–101.

Corntassel, Jeff and Cheryl Bryce. "Practicing Sustainable Self-Determination: Indigenous Approaches to Cultural Restoration and Revitalization." *Brown Journal of World Affairs*, Spring/Summer, 2012, XVIII(11): 151–162.

Coulthard, Glen and Leanne Betasamosake Simpson. "Grounded Normativity/Place—Based Solidarity." *American Quarterly*, 2016, 68(2): 249–255.

Coulthard, Glen. *Red Skin, White Masks*. Minneapolis: University of Minnesota Press, 2014.

Crenshaw, Kimberlé. "Demarginalizing the Intersection of Race and Sex: A Black Feminist Critique of Antidiscrimination Doctrine, Feminist Theory, and Antiracist Politics". *University of Chicago Legal Forum*, 1989 (1): 139–167.

Einion, Alys and Jen Rinaldi, eds. *Bearing the Weight of the World: Exploring Maternal Embodiment*. Bradford: Demeter Press, 2018.

Extinction Rebellion. Accessed online March 19 2021: https://extinctionrebellion.uk/.

Freire, Paulo. *Pedagogy of the Oppressed*. New York. Bloomsbury publishing USA, 2018.

FLOC. "To Make Many Lines, to Form Many Bonds//Thoughts on Autonomous Organizing." *LIES Journal*, 2015, 2(August): 58–70.

Fraser, Nancy. *Fortunes of Feminism*. London. Verso Books, 2013.

Gaard, Greta. "Ecofeminism Revisited: Rejecting Essentialism and Re-Placing Species in a Material Feminist Environmentalism." *Feminist Formations*, 2011, 23(2): 26–53. https://doi.org/10.1353/ff.2011.0017.

Gaard, Greta. "Feminist Animal Studies in the U.S.: Bodies Matter." *DEP Deportate, Esuli E.Profughe* 2012, 20: 14–21.

Gaard, Greta. "Ecofeminism and Climate Change." *Women's Studies International Forum*, 2015, 49: 20–33.

Gaard, Greta. *Critical Ecofeminism*. London. Lexington Books, 2017.

Gallixsee, Alli. "Regenerative Culture, Earth-Based Spirituality, and Permaculture." Starhawk's Website. Accessed September 28, 2021. https://starhawk.org/.

Goodluck, Kalen, Tristan Ahtone and Robert Lee. "The Land Grant Universities Still Profiting Off of Indigenous Homelands." *High Country News*. Accessed August 18, 2020. [Available Online]: https://www.hcn.org/articles/indigenous-affairs-the-land-grant-universities-still-profiting-off-indigenous-homelands.

Goodyear-Ka. *Seeds We Planted*. Minneapolis: University of Minnesota Press, 2013.

Gómez-Barris, Macarena. *The Extractive Zone: Social Ecologies and Decolonial Perspectives*. Duke University Press, 2017.

Gottlieb, R. *Forcing the Spring: The Transformation of the American Environmental Movement*. Washington D.C.: Island Press, 2005.

Hammond, Marit. "Imagination and Critique in Environmental Politics." *Environmental Politics*, 2021, 30 (1–2): 293.

Haraway, Donna. *When Species Meet*. Minneapolis: University of Minnesota Press, 1998.

Hartmann, Heidi. "The Unhappy Marriage of Marxism and Feminism: Towards a More Progressive Union." in J.A. Kournay, J.P. Sterba, and R. Tong (Eds.), *Feminist Philosophies: Problems, Theories, and Applications* (424–433). Upper Saddle River, NJ: Prentice Hall, 1979.

hooks, bell. *Feminism is for Everybody*. Pluto Press, 2000, x.

Johnson, Chelsea, Latoya Council and Carolyn Choi. "Ecofeminism is Intersectional Feminism." *Ms. Magazine*. Accessed April 22, 2020. [Available Online]: https://msmagazine.com/2020/04/22/eco-feminism-is-intersectional-feminism/.

Kaijser, Anna and Kronsell, Annica. "Climate Change Through the Lens of Intersectionality." *Environmental Politics*, 2014, 21(3): 417–433.

Kamola, Isaac A. *Making the World Global: US Universities and the Production of the Global Imaginary*. Duke University Press, 2019.

King, Deborah K. 1988. "Multiple Jeopardy, Multiple Consciousness: The Context of a Black Feminist Ideology." *Signals*, 14(1): 42–72.

Kojola, Erik and David N. Pellow. "New Directions in Environmental Justice Studies: Examining the State and Violence." *Environmental Politics*, 2021, 30 (1–2): 100–118.

KUAa. "KUA Signs Climate Strong Islands Declaration: A Call to Action to Create more Sustainable and Resilient Communities." February 26, 2020 Blog Post. Accessed online March 19, 2021: http://kuahawaii.org/climate-strong-islands-declaration/.

KUAb. "Climate Strong Islands Declaration: Final Version." February 2020. Accessed online March 19, 2021: http://kuahawaii.org/wp-content/uploads/2020/02/Final-Climate-Strong-Islands-Declaration.pdf.

Larsen, Karin. "'We are Moving Forward Together': Premier Urges Feds to Follow B.C.'s Lead Enshrining UNDRIP." *Canadian Broadcasting Corporation*. CBC News. December 3 2019. Accessed online March 19, 2021: https://www.cbc.ca/news/canada/british-columbia/assembly-of-first-nations-recognizes-b-c-s-historic-undrip-legislation-1.5382649.

Lather, Patti. "Research as Praxis." *Harvard Educational Review*, 1986, 56(3): 257–278.

Lawrence, Jennifer L. "Fossil-Fueling Vulnerability." in Bohland, James, Jack Harrald, and Deborah Brosnan. *The Disaster Resiliency Challenge: Transforming Theory to Action*. Springfield. Charles C. Thomas Publisher, 2018: 172.

Lawrence, Jennifer L. and Sarah Marie Wiebe, eds. *Biopolitical Disaster*. Routledge, 2017.

Levin, Dan. "Dozens of Women Vanish on Canada's Highway of Tears, and Most Cases are Unsolved." *New York Times*. May 24 2016. Accessed online March 19, 2021: https://www.nytimes.com/2016/05/25/world/americas/canada-indigenous-women-highway-16.html.

Leroy, Aurélie. "Food Sovereignty: A Feminist Struggle?" *Committee for the Abolition of Illegitimate Debt*. March 17, 2017. Accessed online September 24, 2021: http://www.cadtm.org/Food-sovereignty-A-feminist.

Lorde, Audre. *Sister Outsider*. Freedom, CA: Crossing Press, 1984.

Ludovisi, Stefano Giacchetti. *Critical Theory and the Challenge of Praxis: Beyond Reification*. New York: Routledge, 2016.

Luxon, Micah. "Who Owns Indigenous Land?" *BBC*. February 10, 2019. Accessed October 1, 2021. https://www.bbc.com/news/world-us-canada-47034740.

Maldonado-Torres, N. "Césaire's Gift and the Decolonial Turn." *Radical Philosophy Review*, 2006, 9(2): 111–138.

MacGregor, Sherilyn. "Making Matter Great Again? Ecofeminism, New Materialism and the 'Everyday Turn' in Environmental Politics." *Environmental Politics*, 2021, 30(1–2): 41–60.

MacGregor, Sherilyn. "Only Resist: Feminist Ecological Citizenship and the Post-Politics of Climate Change." *Hypatia*, 2014, 29(3): 617–633.

Mack-Canty, Colleen. "Third-Wave Feminism and the Need to Reweave the Nature/Culture Duality." *NWSA Journal*, 2014, 16(3): 154–179.

Madley, Benjamin. *American Genocide*. New Haven: Yale University Press, 2016.

Mann, Michael E. "I'm a Scientist Who Has Gotten Death Threats. I Fear What May Happen Under Trump." *Washington Post*. December 16, 2016. [Available Online]: https://www.washingtonpost.com/opinions/this-is-what-the-coming-attack-on-climate-science-could-look-like/2016/12/16/e015cc24-bd8c-11e6–94ac-3d324840106c_story.html?utm_term=.a46028c1c026.

Marcuse, Herbert. *Negations: Essays in Critical Theory*. London: MayFlyBooks, 2009.

Merchant, Carolyn. *The Death of Nature: Women, Ecology, and the Scientific Revolution*. New York: First Harper, 1990.

Mies, Maria, and Shiva, Vandana. *Ecofeminism*. 2nd Edition. New York: Zed Books, 2014.

Mignolo, W. *The Darker Side of Western Modernity: Global Futures, Decolonial Options*. Durham: Duke University Press, 2011.

Mignolo, Walter D. and Catherine E. Walsh. *On Decoloniality*. Duke University Press, 2018.

Moulin, Carolina. "Decolonial Temporalities Plural Pasts, Irreducible Presents, Open Futures." *SciELO em Perspectiva: Humanas*, 2015. [Available Online]: https://humanas.blog.scielo.org/blog/2015/11/30/decolonial-temporalities-plural-pasts-irreducible-presents-open-futures/#.YWNFPUZKi3I.

Navdanya. [Available Online]: https://www.navdanya.org/site/.

Norgaard, Kari Marie. "The Essentialism of Ecofeminism and the Real." *Organization & Environment*, 11(4), (1998): 492–497.

Nixon, Rob. *Slow Violence and the Environmentalism of the Poor*. Cambridge: Harvard University Press, 2011.

O'Reilly, Andrea, ed. *Maternal Theory: Essential Readings*. Bradford: Demeter Press, 2007.

Oreskes, Naomi, and Erik M. Conway. *Merchants of Doubt: How a Handful of Scientists Obscured the Truth on Issues from Tobacco Smoke to Global Warming*. Bloomsbury Publishing USA, 2011.

Parson, Sean, and Ray Emily. "Sustainable Colonization: Tar Sands as Resource Colonialism." *Capitalism, Nature, Socialism*, 2020.

Plumwood, Val. *Feminism and the Mastery of Nature*. Routledge, 1993.

Potts, Karen, and Leslie Brown. "Becoming an Anti-Oppressive Researcher." in Strega, Susan and Leslie Brown. *Research as Resistance: Critical, Indigenous and Anti-Oppressive Approaches*. Toronto: Canadian Scholars Press, 2015, 1–16.

Powell, Dana E. *Landscapes of Power: Politics of Energy in the Navajo Nation*. Duke University Press, 2018.

Proctor, Jason. "'Walk the Walk" Wet'suwet'en Chiefs Sue Ottawa to Force Crown to Act on Climate Change", *Canadian Broadcasting Corporation*. CBC News. February 12 2020. Accessed online March 19, 2021: https://www.cbc.ca/news/canada/british-columbia/wet-suwet-en-climate-change-federal-court-1.5461273.

Quijano, Aníbal. "Coloniality of Power and Eurocentrism in Latin America." *International Sociology*, June 2000, 15(2): 215–232.

Parson, Sean, and Emily Ray. "Drill Baby Drill: Labor, Accumulation, and the Sexualization of Resource Extraction." *Theory & Event*, 2020, 23(1): 248–270. https://muse.jhu.edu/article/747105.

"Reconciliation is Dead: Unist'ot'en Women Arrested in Ceremony." Facebook video. February 13 2020. Accessed online March 19, 2021: https://www.facebook.com/watch/?v=194181774992595.

Riofrancos, Thea. *Resource Radicals: From Petro-Nationalism to Post-Extractivism in Ecuador*. Duke University Press, 2020.

Salleh, Ariel. *Eco-Sufficiency & Global Justice: Women Write Political Ecology*. New York: Pluto Press, 2009.

Salleh, Ariel. *Ecofeminism as Politics: Nature, Marx and the Postmodern*. Zed Books Ltd., 2017.

Sandilands, Catriona. *The Good-Natured Feminist: Ecofeminism and the Quest for Democracy*. University of Minnesota Press, 1999.

Shiva, Vandana and Lucy Bradley. "Ecofeminism." PodAcademy. November 2, 2014. [Available Online]: http://podacademy.org/bookpods/ecofeminism/

Simpson, Leanne Betasamosake. *Dancing on Our Turtle's Back*. Winnipeg: ARP Books.

Sommer, Lauren. "To Manage Wildfire, California Looks to What Tribes have Known All Along." *NPR*. Accessed online March 19, 2021: https://www.npr.org/2020/08/24/899422710/to-manage-wildfire-california-looks-to-what-tribes-have-known-all-along.

Stelloh, Tim. "California Exceeds 4 Million Acres Burned by Wildfires in 2020." NBCNews.com. NBCUniversal News Group, October 5, 2020. https://www.nbcnews.com/news/us-news/california-exceeds-4-million-acres-burned-wildfires-2020-n1242078.

Sturgeon, Noel. "Ecofeminist Appropriations and Transnational Environmentalisms." *Identities Global Studies in Culture and Power* 6(2–3), (1999): 255–279.

Tollefson, Jeff. "IPCC Climate Report: Earth is Warmer Than It's Been in 125,000 Years." August 9, 2021. *Nature*. Accessed online September 24, 2021: https://www.nature.com/articles/d41586-021-02179-1.

Tuck, Eve, and K. Wayne Yang. "Decolonization is Not a Metaphor." *Decolonization: Indigeneity, education & society*, 2012, 1(1): 1–40.

Union of Concerned Scientists. "How the Fossil-Fuel Industry Harassed Climate Scientist Michael Mann." October 12, 2017. [Available Online]: https://www.ucsusa.org/resources/how-fossil-fuel-industry-harassed-climate-scientist-michael-mann.

United Nations Environment Programme (UNEP). "Facts About the Climate Emergency." Accessed online March 19, 2021: https://www.unep.org/explore-topics/climate-change/facts-about-climate-emergency.

Voyles, T. B. *Wastelanding: Legacies of Uranium Mining in Navajo County*. Minneapolis: University of Minnesota Press, 2015.

WeDo. "Feminist Solidarity and Resilience in Times of Crisis." [Available Online]: https://wedo.org/feminist-solidarity-resilience-in-times-of-crises/.

White, Jeremy B. "Trump Blames California for Wildfires, Tells State 'You Gotta Clean Your Floors'." Politico PRO, August 21, 2020. https://www.politico.com/states/california/story/2020/08/20/trump-blames-california-for-wildfires-tells-state-you-gotta-clean-your-floors-1311059.

Wiebe, Sarah M. *Everyday Exposure: Indigenous Mobilization and Environmental Justice in Canada's Chemical Valley*. Vancouver: UBC Press, 2016.

Women's Voices for the Earth. [Available Online]: https://www.womensvoices.org/about/who-we-are/.

Zhou, Naaman. "Oxford Dictionaries Declares 'Climate Emergency' the Word of 2019." *The Guardian*. November 21 2019. Accessed online March 19, 2021: https://www.theguardian.com/environment/2019/nov/21/oxford-dictionaries-declares-climate-emergency-the-word-of-2019.

Property and the Anthropocene: Why Power on Things Is Central to Our Ecological Predicament

Benoît Schmaltz

"Climate change is a problem of power".

Richard HEINBERG.[1]

INTRODUCTION

Facing the Anthropocene,[2] the geological epoch in which humans, not nature, are the dominant force disturbing the Earth System,[3] the very foundations of our societies shall be questioned and perhaps reshaped to address the ecological crisis including climate disruption. Property is among the usual suspects, for causing ecological damages in the name of selfish greed or economic development. Modern private property is often intimately associated with the

---

[1] R. HEINBERG, *Power: Limits and Prospects for Human Survival*, Gabriola, New Society Publishers, 2021, p. 12.

[2] P. P. J. CRUTZEN, "Geology of mankind", *Nature*, 2002, vol. 415, No. 6867, pp. 23–23; J. A. ZALASIEWICZ (ed.), *The Anthropocene as a Geological Time Unit: A Guide to the Scientific Evidence and Current Debate*, Cambridge, Cambridge University Press, 2019.

[3] "Anthropocene", *Oxford English Dictionary* (2014).

B. Schmaltz (✉)
Universite Jean Monnet/Universite de Lyon, Saint-Etienne, France
e-mail: benoit.schmaltz@univ-st-etienne.fr

modern worldview in which man dominates nature and only sees environmentalism as a secondary goal, far behind a materialistic idea of progress embodied in (infinite) growth.

Property confronted with the Anthropocene raises a wide set of issues. A main debate is around "commons," a concept with varying definitions depending on the context. Starting in economy, with the famous Hardin's tragedy of the commons,[4] promoted by Ostrom,[5] it has been discussed in many ways such as a comedy,[6] or even a tragedy of anti-commons.[7]

In a legal perspective, property rights can be questioned as a tool for ecological preservation[8] or as they apply to natural resources,[9] or to assess how they will be affected by climate change.[10] John G. Sprankling,[11] for example, analyzes how American property law should respond to the Anthropocene challenge by a shift from a static to a dynamic law system, without abandoning the Takings Clause.[12] Ugo Mattei, with Fritjof Capra[13] and Alessandra Quarta,[14] suggests a more radical paradigm shift, away from exclusive property, toward "commons." In doing so, they join people advocating the controversial idea of a Capitalocene.[15]

In France, legal scholarship has shown an intense activity toward environmental issues from the viewpoint of property, from the need to a new

---

[4] G. HARDIN, "The Tragedy of the Commons", *Science*, vol. 162, No. 3859, American Association for the Advancement of Science, 1968, pp. 1243–48.

[5] E. OSTROM, *Governing the Commons: The Evolution of Institutions for Collective Action*, Canto classics, Cambridge, Cambridge University Press, 2015 (1990).

[6] C. ROSE, "The Comedy of the Commons: Custom, Commerce, and Inherently Public Property", *The University of Chicago Law Review*, 1986, vol. 53, No. 3, p. 711.

[7] M. A. HELLER, "The Tragedy of the Anticommons: Property in the Transition from Marx to Markets", *Harvard Law Review*, 1998, vol. 111, No. 3, p. 621; M. HELLER, "The Tragedy of the Anticommons: A Concise Introduction and Lexicon", *The Modern Law Review*, 2013, vol. 76, No. 1, pp. 6–25.

[8] J. ADLER, "Introduction: Property in Ecology", *Natural Resources Journal*, 2019, vol. 59, No. 1, p. x.

[9] R. BARNES, "The Capacity of Property Rights to Accommodate Social-Ecological Resilience", *Ecology and Society*, 2013, vol. 18, No. 1.

[10] H. DOREMUS, "Climate Change and the Evolution of Property Rights", *UC Irvine Law Review*, 2011, vol. 1, No. 4, p. 1091.

[11] J. J. G. SPRANKLING, "Property Law for the Anthropocene Era", *Arizona Law Review*, 2017, vol. 59, p. 36.

[12] For more, see H. DOREMUS, "Takings and Transitions", *Florida State University Journal of Land Use and Environmental Law*, 2018, vol. 19, No. 1.

[13] F. CAPRA, U. MATTEI, *The Ecology of Law: Toward a Legal System in Tune with Nature and Community*, Oakland, Berrett-Koehler Publishers, 2015.

[14] U. MATTEI et A. QUARTA, *The Turning Point in Private Law*, Cheltenham, Edward Elgar Publishing, 2018.

[15] A. MALM, *Fossil Capital: The Rise of Steam-Power and the Roots of Global Warming*, London, Verso, 2016.

paradigm in our legal relationships with non-human entities,[16] eligible or not to legal personhood,[17] to a more technical approach regarding private and civil law.[18] The concept of "commons" is also widely mobilized, whether as an alternative to exclusive property[19] or as a mere "scale of commonality."[20]

Despite all these works, property is not, to my humble opinion, fully understood in the ecological context. And it has to be if we don't want to act as Don Quixote tilting against windmills. Property is not well understood in law, where its nature of fundamental mechanism involved in every legal relation is not acknowledged, and this explains a lot of misplaced critics against private property.

In this chapter, I will try to specify the place and function of the concept of property in the general theory of environmental politics. Technicalities belonging to legal science and legal innovations will only be mentioned to exemplify a point and I will not investigate the complex issue of implementing "commons" in legal systems. This essay derives from a wider ongoing research project trying to combine different fields of knowledge in a coherent theory of property as the fundamental anthropogenic transformative power, and the main cause of the ecological crises encompassed by the word "Anthropocene."

Property, as I will demonstrate, is the fundamental mechanism that allows to assemble law, sociology, economics, and physics from the viewpoint of the ecological predicament. It is so because it is power: physical power as it always involves energy, physical work done; economic power as property can organize reality so it produces wealth; social power as property is the infrastructure of its sources. Power that law intends to formalize and govern, though many times using words and concepts very different from those of property. Power that is the cause of all our "development," as well as the cause of the ecological impacts and disruptions. Property is the power at the core of the Anthropocene and taming that power is, as in Richard Heinberg's last book, the very point of addressing this ecological predicament.

To put my thesis in one assertion: *property is the fundamental mechanism by which humans exercise power on other humans and non-human realities, and the Anthropocene calls for disciplining that power.* In that perspective, "property and the Anthropocene" refers to the legal formalization of our relationship to the environment and the rules of conduct we should design to act within planetary limits, acceptable ecological impacts and especially climate neutrality.

---

[16] S. VANUXEM, *Des choses de la nature et de leurs droits*, Sciences en questions, Versailles, Quae, 2020.

[17] M.-A. HERMITTE, «La nature, sujet de droit?», *Annales. Histoire, Sciences Sociales*, 2011, No. 1, pp. 173–212.

[18] M. HAUTEREAU-BOUTONNET, *Le Code civil, un code pour l'environnement*, Les sens du droit, Paris, Dalloz, 2021.

[19] P. CRÉTOIS, *La part commune: critique de la propriété privée*, Paris, Éditions Amsterdam, 2020.

[20] J. ROCHFELD et al., *L'échelle de communalité—Propositions de réformes pour intégrer les biens communs en droit*, Paris, GIP justice, 2021.

The work here is only a first step, for which I express a profound gratitude toward Joel Kassiola, hoping he will recognize this work as a humble and partial answer to his call: "*the world desperately requires a 'new' (or perhaps old, but updated) holistic vision as we proceed through the 21st century and beyond.*"[21]

During my PhD (2008–2014) in French administrative law, my interest was toward property in the context of the public law/private law divide.[22] After a few years as an academic, I was able to link property and ecology. This led to reflexions sketched in this contribution, trying to cohere legal, social, economic, and biophysical sciences around property, the mechanism expressing the power of humans, *upon* and *through* things, lying at the core of their socio-ecological metabolism. The final objective is a rational conception of property in line with *scientific materialism*,[23] and more generally with scientific methods and most established results. As Mario Bunge puts it: "*If this be scientism, let it be so.*"[24]

The main idea of scientific materialism is that "*everything that exists really is material*" and should be explained with "*exactness and consistency with contemporary science.*"[25] This approach is non-reductionist and non-eliminative in that it never considers everything could or should be explained by, say, the laws of quantum physics. A key concept here is *emergence*, the fact that complex systems possess properties absent from their components: life emerged from chemicals interactions between basic elements; conscience emerged from neurochemical connections and culture emerged from the sociological and ecological interactions of humans with their environment and between themselves.[26]

This approach is another way to reduce the divide between nature and culture which are intertwined by complex systemic relations. This does not mean we are wrong to distinguish between human cultures and the natural world, nor that we must abandon any kind of anthropocentrism. But this implies that human and non-human realities can be distinguished without ever being separated. From an ecological perspective, this means that human cultures, sciences, and technics are embedded in the natural world and vice

---

[21] J. Kassiola, "Confucianism and Contemporary Environmental Politics", *China Policy Institute Blog*, 4 dec. 2015, https://blogs.nottingham.ac.uk/chinapolicyinstitute/2015/12/04/confucianism-and-contemporary-environmental-politics/.

[22] B. Schmaltz, *Les personnes publiques propriétaires*, Nouvelle bibliothèque de thèses, Paris, Dalloz, 2016.

[23] M. Bunge, *Scientific Materialism*, Dordrecht, Reidel Publishing Company, 1981.

[24] Ibid., p. xiii.

[25] Ibid., p. 17.

[26] In French, the main reference would be M. Silberstein (ed.), *Matériaux philosophiques et scientifiques pour un matérialisme contemporain*, Paris, Éditions Matériologiques, 2013; In English, Papineau, D., *Philosophical Naturalism*, Oxford: Blackwell, 1996; Moser, P. K., Trout, J. D., *Contemporary Materialism: A Reader*, London; New York, Routledge, 1995.

versa. Prudence is, therefore, a necessary ethics and our choices should not rely on a blind faith in technology, but rather be thoroughly guided by environmental evaluation. Maybe it is not "deep ecology," and maybe it is below the level of cultural transformation advocating by our kind editor, but I cannot help thinking that modernized Confucianism is not so different from my ecologically instructed scientific materialism, which I identify with (good) modernity.

As a French legal scholar, I have no choice but to present my ideas in a canonical two parts and two subparts plan. Property will be qualified as the core matrix of the Anthropocene. It is the legal formalization of our socio-ecological metabolism: the appropriation of nature with our anthropogenic ecological footprint as consequence (I). Addressing the Anthropocene thus implies to consider property as its central issue, whether by an ecological exercise of property rights or by an ecological political economy (II).

Property, Core Matrix of the Anthropocene

Our materialistic definition of property is that it is the right to exercise some power on things we legally own. This power is not one among others, but the very fundamental mechanism of law at the basis of all legal relations. In that perspective, property is not only power ON things, but also power THROUGH things, on others: PROPERTY IS SOCIAL POWER (A). Property also appears as the fundamental mechanism of "*social metabolism*" and thus the very basis of our anthropogenic ecological footprint (B).

A. Property as Social Power

Property is not a legal category and the concept cannot decide any but a few cases because it is not meant as a particular legal category, but as the formalization of the very, though often concealed, object of law: power. Property is the basic relationship between a legal subject and the reality, including the social reality of human interrelations. That's why property must be conceived as the fundamental mechanism of law (1), duplicating the fundamental mechanism of social power (2).

(1) Property as the Fundamental Mechanism of Law

If one considers law from the viewpoint of a great library with books dealing with contract law, tort law, corporate law, and so on, property law only appears as one of the many fields of the legislation. But if one considers law from a scientific viewpoint and asks himself what the object of law is and what is the role of property, another picture comes to mind. The object of law is society and the rules of conduct that shall govern individuals and institutions. Formulating those rules implies a legal formalization of reality. Positive law does so with a lot of implicit formalizations, with archaic (mis)understandings of the real world, physical or social. Influence of the past, here, is paramount. Legal theory enjoys more freedom and I feel free to think *legally* as a modern human being educated in *scientific materialism*. That leads me to consider that property is not one among the other fields of law, but the fundamental mechanism by which legal subjects harness material and social realities to serve interests they favor, attain their goals, or fulfill their aspirations.

Common and civil law are not so different as they seem regarding property.[27] They share the same core principles and rules as the same original myths, like the idea of an absolute property[28] such as William Balckstone's famous quote: "*the sole and despotic dominion which one man claims and exercises over the external things of the world, in total exclusion of the right of any other individual in the universe.*"[29] They also share the same never-ending debates on its definition.

On both sides of the Atlantic, the industrial revolution made necessary new conceptions of property, new forms of ownership and new kinds of properties. As K. Vandevelde explains it thoroughly,[30] a more complex economy made obsolete the Blackstonian property. Property rights could no longer be absolute and had to apply to intangible assets like trademarks or business goodwill or even social benefits. Hence, the success of a more flexible conception of property as a "bundle of rights" in Hohfeld's[31] and Honoré's[32] famous works. An evolution not so different of the ones that occurred in French or other civil law jurisdictions.

The result is not the end of controversies, quite the contrary. In common law, despite its wide success in legal thinking and in Courts, the bundle of rights metaphor has a lot of critics.[33] An answer was given by erecting the "right to exclude"[34] as the criterion of property, now incorporated in an

---

[27] Y. EMERICH, *Conceptualising Property Law*, Cheltenham; Northampton, Edward Elgar Publishing, 2018.

[28] J.-P. CHAZAL, « Le propriétaire souverain: archéologie d'une idole doctrinale», *Revue trimestrielle de droit civil*, 2020, vol. 1, pp. 1–33.

[29] W. BLACKSTONE et al., *Commentaries on the laws of England*, vol. 2, Oxford; New York, Oxford University Press, 2016, where Simon Stern remarks, with references: "*Blackstone evidently added these sweeping observations for rhetorical effect, never anticipating that they would be pressed into service as a durable framing device under the label of 'Blackstonian property'*"; Also see T. MERRILL, "Property and the Right to Exclude", *Nebraska Law Review*, 1998, vol. 77, No. 4, p. 753 with references note 61.

[30] K. VANDEVELDE, "The New Property of the Nineteenth Century: The Development of the Modern Concept of Property", *Buffalo Law Review*, 1980, vol. 29, No. 2, p. 325.

[31] W. N. HOHFELD, *Fundamental Legal Conceptions as Applied in Judicial Reasoning*, 26 YALE L.J. 710, 746–47 (1917).

[32] A. M. Honoré, 'Ownership', in A. G. GUEST (ed.), *Oxford Essays in Jurisprudence*, Oxford, Oxford University Press, pp. 107–147 (1961).

[33] J. PENNER, *The Idea of Property in Law*, Oxford, Oxford University Press, 2000.

[34] T. MERRILL, "Property and the Right to Exclude", *Nebraska Law Review*, 1998, vol. 77, No 4; T. MERRILL, "Property and the Right to Exclude II", *Brigham-Kanner Property Rights Conference Journal*, 2014, vol. 3, p. 1.

"architecture" of property.[35] Property is the right to a "thing" (tangible or not), a right that supposes some kind of "right to exclude" to be effective, without denying any kind of "inclusion" or specific "governance" to preserve other legitimate interests. Easements and other regulatory mechanisms are here to serve this purpose, reminding autonomous owners that they belong to a wider society.

In France, the French Revolution is a momentum-separating feudalist property with its intricacy of concurrent rights on a same thing and modern unitary individualist property. The definition of property is thus a matter of interpretation of the article 544 of the "Code civil," which states, since 1804: "*ownership is the right to enjoy and dispose of things in the most absolute way, excepted in a way prohibited by statutes or regulations.*" There are three French schools of thought about property: classicists, gathering most of the prominent professors and to be seen in most Handbooks; modernists, that followed the seminal work of Frédéric Zenati in 1981[36]; and structuralists, as William Dross separated himself from the latter.[37]

Zenati's modern conception of property is rooted in Roman law, against its "romanist" interpretation. Nineteenth-century scholars were educated during the "Ancien Régime" and consequently interpreted the Code with their feudal background. To avoid technicalities and details, let us say that the modern conception of property chooses a simple definition of "property" from the model of the Roman *dominium*. It was the power (a *potestas*, like *imperium*) of the *dominus* on the things (and people, as the master) that were in his dependency (*propietas* being the quality of a thing to belong to someone). In our modern individualistic view, that power is now exercised by the subjective right of property we have on each thing we own.

The right of property is then defined as the right to enjoy and dispose of a thing according to the statutes and regulations. To enjoy a thing has a twofold meaning: (1) a dominion precisely defined and efficiently enforced, and (2) the free and immediate benefit of the services a thing can provide in that secured dominion. To dispose of a thing is simply the right to voluntarily change the legal situation of a thing: to sell it, to lease it, etc. The "right to exclude" is ontologically necessary, whatever the degree of exclusivity that is granted by law to owners, because rights of property are not effective without an official enforcement process.

In that modern conception of property, the right of property is not one among others, but THE subjective right, *i.e.*, the right by which the subject

---

[35] T. W. MERRILL, H. E. SMITH, *The Architecture of Property*, Rochester, Social Science Research Network, 2019.

[36] F. Zenati, *Essai sur la nature juridique de la propriété: contribution à la théorie du droit subjectif*, th. Lyon 3, 1981.

[37] W. DROSS, *Droit Civil, Les Choses,* Paris, LGDJ-Lextenso éd, 2012.

projects itself on the objective reality. Hence the idea of property as a fundamental mechanism of the law. The mechanism by which the subject has power and enter in relation with reality, and with others, by the mediation of things.

This school of thought also adopts a broad definition of things that makes this theory of property a comprehensive way to describe the legal relationships, with property at the core. Besides tangible movable and immovable physical assets, things are also those intangible assets that industrial capitalism made necessary for property law to encompass: for example, patents and intellectual properties, trademarks and tradenames, business goodwill. But in that conception, reified social relations also make relevant properties. The simplest example is debt: when Peter is the creditor of John, Peter is the owner of a right (creance), which he can sell or lease, or use as a collateral ("gage" in French Law), etc. It is probably one of the most controversial, but also essential components of the theory, while many legal scholars try to separate property law and contract law. But it is line with the Code civil that labeled contract law "means by which one acquires property."[38]

In my PhD dissertation, I drew upon Zenati's school of thought, considering that this simple and systemic conception of property, easily applicable to any legal context, made a good formal and neutral concept of "owner" that could be used to compare private and public property law. But to distinguish between a private owner and a public one like the state, I had to integrate to the model a very specific feature: subjective rights of authority, defined has rights to command (by statute, regulation or any direct bounding legal prerogative constituting someone debtor without his consent of an obligation of any kind). The eminent domain and many other prerogatives, among them the most evident the right to tax, manifest the autonomous status of public owners, using authority as an auxiliary of ownership.

Eventually, that led me to establish the model of subjective legal power of the subject. Starting from the *summa divisio iuris* between persons and things, the two fundamental subjective rights are authority rights, a subject exercises on other subjects bound to obey his direct command, and property rights, a subject exercises on things, and indirectly on persons through those special things that are obligations compelling other subjects to give, to do or not do.

As authority rights are mostly commanding *owners* to give up things (takings, taxes) or to use them in the general interest (regulations), property is, indeed, the fundamental mechanism of law. Further inquiries confirmed the intuition that it is so because property, power on things, is a fundamental mechanism of social life, as power on people *through* things.

---

[38] William DROSS, *op. cit.*, *passim*, harshly criticizes the classicist's theory as obsolete, but also harshly rejects the idea of a property of "rights" that are created by contracts. To my point of a view, it is though quite contradictory to state that such rights are "goods" but not "things." I must say I find even more contradictory the same separation in common law where "things" of property law are mere "interests in something".

(2) Property as the Fundamental Mechanism of Social Power

Having formalized in law a general theory of legal subjects and subjective rights of authority and property, I wanted to find its strongest justification in its coherence with fundamental concepts of sociology. If property was a fundamental mechanism of legal relations, it had to be because property was at the core of the general functioning of societies. Of course, this intuition seemed obvious, if we consider the analogy between property and economy, between ownership and "capital."

If property is the power on things, individual power in the form of property rights or, collective power in the form of the aggregate of things a society mobilizes through ownership, then property sits at the heart of the most important social transformations. The Neolithic Revolution and the emergence of states can be viewed as the profound transformation of the collective power on things—by exploiting land through agriculture and developing new economic activities and relations due to the surplus generated—and its distribution among individuals in a more stratified society.[39]

That the control of tangible and intangible assets is a source of power for individuals and states is an omnipresent idea in sociological thinking. Marxist authors, like Immanuel Wallerstein's world system analysis,[40] immediately come to mind but it is also the point of Charles Tilly's *Coercion, capital, and European states, AD 990–1990*.[41] In those works, and many others, I could have found materials for the combination of public authority rights and private or public property rights in a broad explanation of social change and state construction. But that was not satisfying.

The encounter with the work of Michael Mann was decisive. In his 1984 article[42] and the subsequent 4 volumes of *The Sources of Social Power*,[43] he developed a social theory of power he tried to justify from an historical perspective. That theory also distinguishes itself from other sociological traditions such as Marxist, liberal, and functionalist. It is influenced by Max

---

[39] J. A. SABLOFF, P. L. W. SABLOFF (ed.), *The Emergence of Premodern States: New Perspectives on the Development of Complex Societies*, SFI Press seminar series, No 2, Santa Fe, SFI Press, 2018.

[40] I. WALLERSTEIN, *The Modern World-System*, New York; Berkeley, 1974–2011.

[41] C. TILLY, *Coercion, Capital, and European States, AD 990–1990*, Studies in social discontinuity, Cambridge, Blackwell, 1990.

[42] M. MANN, "The Autonomous Power of the State: Its Origins, Mechanisms and Results", *European Journal of Sociology/Archives Européennes de Sociologie*, 1984, vol. 25, No. 2, pp. 185–213.

[43] V. 1. *A History of Power from the Beginning to AD 1760*, Cambridge, Cambridge University Press, 1986; v. 2. *The Rise of Classes and Nation-States, 1760–1914*, Cambridge, Cambridge University Press, 1993; v. 3. *Global Empires and Revolution, 1890–1945*, Cambridge, Cambridge University Press, 2012; v. 4. *Globalizations, 1945–2011*, Cambridge, Cambridge University Press, 2013.

Weber, but goes far beyond Weberian sociology.[44] It follows the master in considering that "societies" never exist as such, but are networks of relations and power networks between individuals.[45]

Mann's theory of power consists of a model combining four sources of social power as overlapping power networks: ideological, economic, military, and political (IEMP). The main feature of this approach is, to me, its coherence with *emergence*. *Collective* power emerges from the complex relations of *individual* power involved in overlapping *networks*. It is a form of methodological individualism that avoids reductionism and is compatible with a naturalistic approach of culture and social relations,[46] which is quite appealing to the scientific materialist.

Mann defines power as "*the capacity to get others to do things that they would otherwise not do.*"[47] He then identifies three modalities of power which can be distributive (*over* others) or collective (*with* others), authoritative (*commanding*) or diffuse (unconscious acceptance), extensive (large numbers) or intensive (high commitment).[48] One of the main distinctions of Michael Mann's theory is between *despotic* and *infrastructural* political power.[49] To me, it is a distinction between the will to decide (volition), and the capacity to enforce the decision (execution). In political power the distinction is between how and what can be decided and what are the means of the state to enforce the decisions and make it real. Beyond the distinction itself, I want to stress its relevance to the four sources: "*We may distinguish between the despotic and the infrastructural powers of the state* (although the distinction could be applied to any power organization)."[50] And I would say this distinction is relevant to the other three sources.

Ideological power allows to distinguish between the ideas as they are conceived in minds, and the ways by which they are transmitted to other minds. Bunge puts it that way: "*As long as such a 'product' [of brain activity] remains inside the skull of its creator, it is only a brain process: it has got to be communicable to others in order to rank as a cultural object.*"[51] Michael Mann describes the ways by which early Christians managed to spread through the

---

[44] J. A. HALL et R. SCHROEDER (eds.), *An Anatomy of Power: The Social Theory of Michael Mann*, Cambridge; New York, Cambridge University Press, 2006; R. SCHROEDER (éd.), *Global Powers: Michael Mann's Anatomy of the Twentieth Century and Beyond*, Cambridge, Cambridge University Press, 2016.

[45] M. WEBER, *Economy and Society*, New York, Bedminster Press, 1968 (1922).

[46] D. SPERBER, *Explaining Culture: A Naturalistic Approach*, Oxford; Cambridge, Blackwell, 1996.

[47] M. MANN, *The Sources of Social Power*, vol. 3, *op. cit.*, p. 5.

[48] Ibid.

[49] Already in M. MANN, "The Autonomous Power of the State: Its Origins, Mechanisms and Results", cited note 42.

[50] M. MANN, *The Sources of Social Power*, vol. 3, *op. cit.*, p. 13.

[51] Mario BUNGE, *op. cit.*, p. 14.

whole Roman Empire and beyond, making Christendom a reality by using the same infrastructure of economic power (roads and maritime trade routes). Military power allows to distinguish between the making of war and battle plans, the giving of orders, and the *infrastructure* (men, weapons, equipment, and ammunitions) without which plans won't come together and orders won't be implemented, or objectives taken. Economic power allows to draw on the same line: a business plan won't make any profit without the *infrastructure*, that is the tangible and intangible things operated by the managers and workers of the firm.

That latter source of power allows to introduce a concept I forged during my PhD. To formalize what public legal entities *do* as owners, I transposed the equivalent concept of "business goodwill" ("fonds de commerce") in public law ("fonds administratif"). The idea is simple and built from the legal concept of "universality": a collection of things due to a common purpose or nature. Simple examples of universalities are a herd of livestock or a collection of books one can inherit or buy as a whole. It is possible to imagine a collection of all the things that are used to provide a public good, whether in the context of a public facility or a traditional government or administration. It is important to distinguish between such a universality, which remains a thing, and the person or persons involved in its management. So, the universality of things necessary to an activity is not an enterprise or entity, but only the complex system of things (and reified relations) that allow this activity to take place. Public action then becomes a formal model: public legal persons exercise their rights of authority and property, so they constitute and manage public universalities providing public goods and services.[52] But this reasoning is quite applicable to public or private activities and so to every of the four sources of social power.

Generalizing the argument, *social power is exercised by the coordination of authority and property rights, and the constitution of active universalities providing goods and services (whether in an economic market or not is of no relevance here)*. It means that social power of any kind always supposes to harness a part of reality, *i.e.*, of the material world so *things* are modified and/or exploited in a way that satisfy one's ambitions.

Ideological active universalities would be the books, journals, blogs, and so on. Economic active universalities obviously are any kind of business. Military active universalities are armies and warbands and every past or present *condottiere* knows how a very lucrative venture it can be. Political active universalities are political parties or other activist entities, sometimes mixed with ideological ones (sources, as ideal-types are never found in a pure form in real life), and Americans know more than French how property is involved in the electoral process.

---

[52] Goods and services refer to a common expression in European Union's law, goods and services being anything that is offered on a market, whether corporeal or not.

Infrastructural power on things so appears as the material basis of every power network, whatever the ideal-typical source is considered. Property is a fundamental mechanism of law because it is a fundamental mechanism of sociology. But it is so because the power on things is only a social representation of the material power, the material transformation of reality. That's how property, in the form of active universalities, can be linked to the concept of dissipative structures and to our anthropogenic ecological footprint, the latter to be discussed in the next section.

B. Property as the Basis of Our Anthropogenic Ecological Footprint

Property is the legal formalization of the infrastructural *social power* of individuals, groups, and institutions. But if we go deeper in understanding societies, we must acknowledge property as *material power* mobilized in the infrastructures of the four sources of power. Property is the fundamental mechanism of *social metabolism*, a concept I will present (1), before explaining our anthropogenic ecological footprint by the impacts of those infrastructural universalities functioning as *social dissipative structures* (2).

(1) Social Metabolism

A lot has been written about the material basis of civilization, their dependence on energy and the eventual collapse when it comes short. Joe Tainter[53] and Jared Diamond[54] must be mentioned, the latter with critics[55] who really should be ready to put his work in perspective. The main idea of these works is that societies are complex systems in interrelation with natural resources, made of fluxes of energy, materials, and information.

Forged by Abel Wolman,[56] widely used today as urban[57] or territorial,[58] *social metabolism* is a concept we can draw upon to make our point here.

---

[53] J. TAINTER, *The Collapse of Complex Societies*, Cambridge, Cambridge University Press, 1988.

[54] J. M. DIAMOND, *Guns, Germs, and Steel: The Fates of Human Societies*, New York, Norton, 2017 (1997); J. M. DIAMOND et C. MURNEY, *Collapse: How Societies Choose to Fail or Succeed*, New York, Penguin, 2005.

[55] P. A. MCANANY et N. YOFFEE (éds.), *Questioning Collapse: Human Resilience, Ecological Vulnerability, and the Aftermath of Empire*, Cambridge; New York, Cambridge University Press, 2010; on which I recommend reading the review by J. A. TAINTER, "Patricia A. McAnany and Norman Yoffee (eds.): Questioning Collapse: Human Resilience, Ecological Vulnerability, and the Aftermath of Empire", *Human Ecology*, 2010, vol. 38, No. 5, pp. 709–710.

[56] A. WOLMAN, "The Metabolism of Cities", *Scientific American*, September 1965, vol. 213, No. 3, pp. 178-190.

[57] LU, B. CHEN, "A Review on Urban Metabolism: Connotation and Methodology", *Acta Ecologica Sinica*, 2015, vol. 35.

[58] S. BARLES, «Écologie territoriale et métabolisme urbain: quelques enjeux de la transition socioécologique», *Revue d'Économie Régionale Urbaine*, 2017, No. 5, pp. 819-836.

It is the result of the combination of *sustainability science*[59] and the *theory of entropy of open or dissipative structures*[60] in the form of a new field of research: Social Ecology which "*offers a conceptual approach to society-nature coevolution pertaining to history, to current development processes and to a future sustainability transition.*"[61] The core idea is that "*the metabolism of social systems encompasses biophysical stocks, flows and – as we are gradually beginning to understand, certainly in analogy to organic metabolism – the mechanisms regulating these flows.*"[62] To put it in other words, "*all society is an assembly of phenomena pertaining to both dimensions: that of the* flows of mater and energy, *and that of the* flows of information *that organize, mould and give support to the latter in virtue of social conditionals such as institutions, legal rules and regimes, values, beliefs and knowledge.*"[63]

Social metabolism is the organized exchange of energy, materials, and information between nature and society that gives place "*to a new system (natural-social or social-natural) having a further higher complexity.*" Nature and society enter in relation with appropriation and excretion: "*Flow begins with appropriation, it then follows several courses within society, and it ends with excretion*"[64] (see their reproduced Fig. 1).

They identify five metabolic processes: appropriation, circulation, transformation, consumption, and excretion (see their reproduced Fig. 2). The "*primary mode of exchange between human societies and nature (...), the Apr process is always carried out by an appropriation unit, which can be a company (owned by the state or private), a cooperative, a family, a community, or a single individual (e.g., the capturer of solar energy).*"[65] Primary products are then transformed, transported, and consumed before ending as pollutions and wastes, if not recycled in a circular economic process.

The whole model of social metabolism aims at being "*isomorphic with the physical reality to which we belong*"[66] and I consider this statement in accordance with the *scientific materialism*, the central point being "*the thermodynamic consideration of social practice: all human acts require of appropriation and consumption of a given quantity of energy and materials from nature and*

---

[59] B. de VRIES, *Sustainability Science*, Cambridge, Cambridge University Press, 2013.

[60] G. NICOLIS et I. PRIGOGINE, *Self-organization in Nonequilibrium Systems: From Dissipative Structures to Order Through Fluctuations*, New York, Wiley, 1977.

[61] H. HABERL et al. (éds.), *Social Ecology*, Cham, Springer International Publishing, 2016.

[62] Ibid., p. 36.

[63] M. GONZÁLEZ DE MOLINA et V.M. TOLEDO, *The Social Metabolism: A Socio-Ecological Theory of Historical Change*, Environmental History, No. 3, Springer, 2014, p. 4.

[64] Ibid., p. 60.

[65] Ibid., p. 62.

[66] Ibid., p. 271.

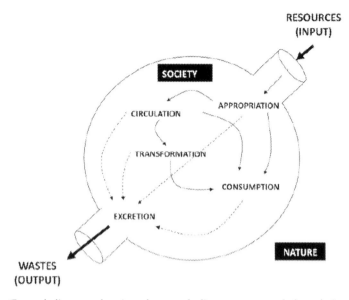

Fig. 1 General diagram showing the metabolic processes and the relation between society and nature

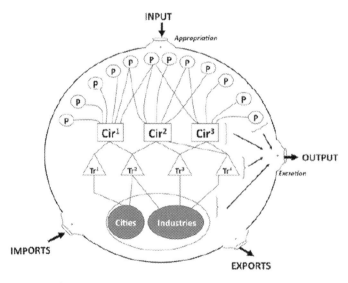

Fig. 2 Diagram showing the appropriation act in current societies, which is performed by a series of appropriation units (P). The appropriated materials and energy are later circulated (Cir) by different routes, in some cases transformed (Tr), and finally consumed by industry and cities. All these processes, in turn, generate a flow of wastes towards nature, or excretion (Exc). Along with domestic relations with nature (input and output), societies also import and export commodities to and from other societies

the expulsion of residues to nature."[67] In that sense society *is* nature, and culture is an evolutionary innovation due to the specifically human feature: "*exosomatic consumption of energy*"[68] which is consumption of energy by means of tools and machines outside the human bodies (endosomatic).

Human societies are part of the Universe and the product of the same processes that led to its appearance as a novel species. Human societies cannot escape their materiality and even if it would be grotesque to *explain* human history by the laws of physics, there is no historical fact that can violate them, and not the least the first and second laws of thermodynamics.

That's why property is so important to understand social metabolism.

(2) Property in Social Metabolism: Appropriation and Dissipative Structures

As we have seen in the reproduced figures, appropriation is the first mechanism by which human cultures start to build separate systems from the environment and nature. The specificity of mankind is not that relation, which is common to all living creatures and, more generally, to all structures that are far from the thermodynamic equilibrium (unlike diamonds). The specificity is in the scope of the social-natural systems. But the point is that *appropriation* is the first link between society and nature. It is the process by which producing units gather natural resources and services so they can be transformed and consumed, needing transportation (circulation) and ending by a variable amount of excretion.

To me, property is everywhere: in the five metabolic processes as it is in the four sources of social power. It is obvious in the case of appropriation, but that it is only the first step of human complexity which implies that secondary units of transformation and production provide the goods and services that will be consumed. The metabolic approach could be applied to every four sources of social power and that would lead to a thorough analysis of the flows of matter, energy and information involved in the functioning of the *infrastructure* of the considered power source. A legal approach would try to shed light on the institutions and rules that control the management of flows and stocks revealed by this analysis.

In a word, social metabolism is the description of the organized appropriation of nature. Property clearly is, in that perspective, power on things and through things, with consequences... social, economic, and ecological.

Besides appropriation, property appears in social metabolism in a form directly implied by the application of thermodynamics to societies: dissipative structures. They are autopoietic systems that use energy from the environment to organize processes that do work before ending as entropy, dissipated energy not able to do work. Galaxies and living beings are dissipative structures, and

[67] Ibid., p. 255.
[68] Ibid., p. 264.

"*from the biophysical perspective human societies can be considered as dissipative structures, or more precisely, can be considered as made up of dissipative structures exchanging energy, materials, and information with their environment;*"[69] "*dissipative structures of varied functionality that generate goods and services i.e., social order.*"[70]

I view those dissipative structures constituting societies as the universalities we have encounter. Universalities of production, whereas appropriation, transformation, or circulation, are active universalities providing goods and services to final consumers, one could conceive as passive universalities of consumption (individuals and their belongings, "patrimoine" as we would say in French) who also dissipate energy, but only produce wastes and pollutions.

Property is the power to harness things so they can serve one of the four sources of power, most often in the form of active universalities constituting the infrastructure enforcing the volitions associated with each form of "IEMP power." Those universalities are dissipative structures the sum of which constitutes the social metabolism of a given society. A metabolism that sustainability sciences allow to analyze: "*the magnitude of dissipative structures determines the amount of energy and materials consumed by a society, that is, its metabolic profile,*"[71] and its ecological impact.

Property is the core matrix of the Anthropocene because from individual possessions to the complexities of modern globalized economies, social dissipative structures made from the power exercised by persons on (and through) things, property, are the final causes of the ecological crises arising from our anthropogenic footprint. An individual manages a social dissipative structure in the form of his own domestic consumption: his universality is constituted with all the things he possesses and uses. An economic dissipative structure would be any enterprise and the active universality it manages: the coke, iron ore and furnace of a steel mill, the cars of rental ventures, the seeds, machines, and tools of a farmer. The point is that any economically passive (consumption) or active (production) social activity consists in the management of a dissipative structure that can be legally formalized as a universality. Hence, the idea that the fundamental mechanism at stake here is the power on things, property.

One might infer from this that to address the Anthropocene, something must be done regarding property. But the answer is not trivial. This first part intended to address the signification of property and its role as the fundamental mechanism of our ecological predicament. It plays this role because it is the legal and socio-political formalization of the physical and social power people exercise on reality, be it society or nature. The second part of this essay will engage in a more practical, though still theoretical discussion about how to deal with that fundamental power of property, if we want to limit the tragic consequences of the ecological impacts it generates

[69] Ibid., p. 268.
[70] Ibid, p. 272.
[71] Ibid.

I. Property, Central Issue of the Anthropocene

Property is power, the legally formalized social and physical power that, when exercised, dissipates energy, and impacts the environment. As the fundamental mechanism of the ecological crisis, property deserves some focus to solve the Anthropocene issues. Though, rushing against "private property" and "capitalism" would be an intellectual fault, to my humble opinion. First, we must acknowledge what challenge poses the Anthropocene to mankind: a deep and complex socio-ecological metamorphosis (A). Only then can we consider the "ecological appropriation" as the core matrix of a (more) sustainable metabolism (B).

A. Socio-Ecological Metamorphosis

If societies are complex systems where dissipative structures are the infrastructure of social power networks, then social change should imply a coevolution of both. A socio-ecological metamorphosis is an evolution of the social metabolism, a challenging socio-ecological transition (1) which can be conscious, when worldviews allow people to think about the process, as it is occurring (2).

(1) The Challenge of a Socio-Ecological Transition

In a 2011 article, Marina Fischer-Kowalski attempts a *"scientific treatment of sustainability transitions"* by labeling them as *"a shift between sociometabolic regimes."*[72] She stresses, though, how sensitive the perception of events is to the observer's choices: *"From a wider perspective something may appear as a continuous process, progressing steadily. But from a closer perspective the same process may appear as whimsical, sharply fluctuating."*[73] In a word, socio-ecological systems evolve in numerous ways where rupture and continuity are often a matter of interpretation, somehow like in Richard Wagner's colorful style of incremental generations of new themes from preceding ones.[74] That's the reason why Gonzalez de Molina and Toledo in their volume (see footnote 63) use the expression of "metamorphosis."

---

[72] M. Fischer-Kowalski, "Analyzing Sustainability Transitions as a Shift Between Socio-Metabolic Regimes", *Environmental Innovation and Societal Transitions*, 2011, vol. 1, No. 1, pp. 152–159.

[73] Ibid., p. 153.

[74] M. Bribitzer-Stull, *Understanding the Leitmotif: From Wagner to Hollywood Film Music*, Cambridge, Cambridge University Press, 2015. I suggest to listen to some excerpts of the *Ring der Nibelungen*, even only the very end when the Valhalla burns, to be rebuilt, or the masterpiece of this style which his *Parsifal*.

As Haberl and others put it in an article's highline: "*Society-Nature Interaction: Gradual* and *Revolutionary Change.*"[75] That ambiguous nature of social change and so-called transitions is problematic because of the contradiction between the deep nature of the transformations needed to shift toward a sustainable metabolic regime, and the urgency to achieve some of the most revolutionary parts of this transformation in a handful of decades, or less as some researchers maintain that 2030 is the time for changes to avert a catastrophe. Deep decarbonization is probably the most prominent part of all in that perspective, coupled with the energetic transition, of which more will be written. Stopping to accumulate atmospheric greenhouse gases implies to give up fossil fuels that amount to 80% of our net primary energy.

If it is true that "*a transition to a (more) sustainable state implies a major transformation, on a par with the great transformations in history such as the Neolithic or the Industrial Revolution,*"[76] then this contradiction appears in full light. The Neolithic Revolution is a process of millennial scale,[77] and the Industrial Revolution is probably one of the most ill-termed complex changes occurring in different forms in a wide range of countries and, even in Britain, through a lengthy intellectual, economic, and technical process during several centuries. The Net Zero Emissions objective is to be achieved in 60 years from now. Hence "*the presumption that a sustainability transition is both inevitable and improbable. It is inevitable, because the present socio-metabolic dynamics cannot continue for very long any more, and it is improbable because the changes need to depart from known historical dynamics rather than being a logical step from the past into a more mature future state.*"[78]

The challenge is huge. We can formulate it according to the preceding developments. In the coming decades, social power networks have to acknowledge the shift and operate it through the mobilization of, probably, all four sources of power: ideology especially if it includes worldviews and scientific knowledge, economics as the shift will clearly be the most visible in the active universalities acting as social dissipative structures when putting goods and services on the markets; military, sadly, because that shift is unlikely to avoid major geopolitical crisis; and political power, as many important decisions have to be made and enforced through all of the infrastructural power of governments at every scale of jurisdiction.

"*Impossible n'est pas Français,*" said Napoleon. Let us hope it is not human either. Nevertheless, once we accept the idea that a shift toward a sustainable

---

[75] H. HABERL et al., "A Socio-Metabolic Transition Towards Sustainability? Challenges for Another Great Transformation", *Sustainable Development*, 2011, vol. 19, No. 1, pp. 1–14.

[76] M. FISCHER-KOWALSKI, "Analyzing Sustainability Transitions as a Shift Between Socio-Metabolic Regimes", cited footnote 72, p. 153.

[77] J.-P. DEMOULE, *La révolution néolithique*, Paris, Éd. le Pommier, 2013.

[78] H. HABERL et al., "A Socio-Metabolic Transition Towards Sustainability?", op. cit.

decarbonized socio-ecological regime is a rapid and quite brutal restructuration of the current carbonized industrial regime, we still have to think about what it really means. What has to be changed? How deep and with what winners and losers? Facing which lock-ins and obstacles? The Grand Transformation can be thought with many different worldviews and that impact the manner we deal with property in the Anthropocene.

(2) Worldviews for the Anthropocene Crisis

Andrew Dobson, in his *Green Political Thought*, distinguishes between *environmentalism* and *ecologism*.[79] The first is not an ideology, only a slight amendment to the modernist and industrialist worldview in which science and technology will provide solutions to the ecological issues but "*without fundamental changes in present values or patterns of production and consumption.*"[80] Ecologism, "*holds that a sustainable and fulfilling existence presupposes radical changes in our relationship with the non-human natural world, and in our mode of social and political life.*"[81] This description of political thought is quite biased since it implies a political conversion to one radical ideology which is far beyond the thorough scientific assessment of the ecological situation of our current Society-Nature complex system of relationship: "*ecologism will suggest that climate change is not simply a result of inappropriate technologies for energy production, but rather that it is symptomatic of a misreading of the possibilities (or more properly here, constraints) inherent in membership of an interrelated biotic and abiotic community.*"[82] To me, climate change is a matter of greenhouse gas emissions, then a matter of energy, then a matter of technologies *as components of our metabolic regime*. But I would not adhere to the "new age" idea of an abiotic community… with rocks?

It is in line with a general but too simplistic answer, in my opinion, to the ecological predicaments. To some people, the ecological crisis is due to our industrial regime which is a consequence of anthropocentric modernity and "capitalism," so to them the solution is self-evident. If industrial ecological impacts are the symptoms, and if capitalism is the disease, as the economic counterpart of modernity, then the cure is an alternative to capitalism and a philosophical shift away from modernity. Let us be postmodern and get out of that capitalist Anthropocene: "*The ecological crisis, first and foremost, challenges the very foundational ideology of capitalism and modernity, the possibility of infinite growth on a finite planet.*"[83] But what would be a genuine alternative to capitalism? And what is capitalism in the first place?

---

[79] A. DOBSON, *Green Political Thought*, Abingdon, Routledge, 2012.
[80] Ibid., p. 2.
[81] Ibid., p. 3.
[82] Ibid., p. 4.
[83] U. MATTEI et A. QUARTA, *The Turning Point in Private Law, op. cit.*, p. viii.

Such "ecologists," wanting to avoid the anathema of being a mere "environmentalist," embrace an anti-modern (in line with Ivan Illich, Jacques Ellul and other radical critics of modernity, contemporary science and technology) and anti-capitalist ideology advocating biocentrism or anti-speciesism (all life is morally equal) and desirable degrowth. I disagree strongly with that very last idea because I see degrowth as an epochal structural recession with dramatic and violent consequences for people who will struggle and suffer from it.

An answer to that "mainstream" ecology can be found in people who might consider themselves as "radical environmentalists," the authors of the *Eco-Modernist Manifesto*.[84] Eco-modernists consider that *"Humans have long been cocreators of the environment they inhabit. Any proposal to fix environmental problems by turning away from technology risks worsening them by attempting to deny the ongoing coevolution of humans and nature."*[85] No doubt most of the radical movement would criticize such a statement as an illustration of the flaws of the modern worldview: *"Modernity conceals within itself a growing part of irrationality that it represses and denies under the garments of technical Rationality."*[86] And I can't help to reject a techno-optimism that does not see that, with or without science and technology, the future will be dire.

It reminds me, though, some bittersweet debates between communists and social-democrats labeled "social-traitors" by the former, not always without grounds. But a strategy that is doomed to fail (or turn badly) is another kind of treason to the cause. As I personally chose the "Bad-Godesberg option," that is to accept the market economy and refuse the dictatorship of the proletariat , it is the exact problem of political philosophy I deal with: how to conceive a reformist or social democrat plan of action for maybe the most rapid and brutal shift ever in social metabolism? In France, the *opportunists* who achieved to install the Republic as our form of government had a motto: *"we are moderate republicans, but not moderately republican."* It is quite the same to me regarding ecology: it is a matter of moderate *path* preferred to a *revolution*, but with the same goal of a not moderate ecological shift. But two questions are raised: Can a moderate shift bring radical consequences in only decades, if not years? Can an ecological revolution succeed? I do not have the answers.

If all humans behaved like monks (with no vow of chastity obviously), contenting themselves with vegetables from the garden and very few possessions, there would not be an Anthropocene. Would their metabolism be sustainable? Would this humanity be resilient enough to endure in that peaceful way for centuries, millennia? How would these humans and their simple life deal with the glaciation forthcoming in about 50,000 years? Or,

---

[84] https://www.ecomodernism.org/.

[85] M. SHELLENBERGER, T. NORDHAUS, *Love Your Monsters: Postenvironmentalism and the Anthropocene*, Oakland, Breakthrough Institute, 2011.

[86] A. FREMAUX et B. GUILLAUME, «L'horizon postdémocratique et la crise écologique», *Écologie politique*, 2014, vol. 49, No. 2, pp. 115–129.

confront their own obliteration in, at the latest, half a billion years when the Sun infatuates, burning almost all forms of life on Earth?

If only 7 million humans and not billions were living the American way of life, there would be no Anthropocene either. Resource depletion would be the major issue, with centuries to find a solution to a circular economy, not their ecological impact that would be localized and with no effect on the Earth System. Maybe their resilience would be far better, advanced medicine and technologies allowing to overcome glaciations. They might even have found a way—I consider the odds really against that probability—to escape at least a part of them from our cradle and grave planet. But most humans will live and die on Earth, until the very last ones disappear. Only philosophy, not technology, will be of any help to deal with these final moments of mankind.

There is an Anthropocene because humans have found an exosomatic metabolism that allows an important and growing part of a massive 8 billion-large population to harness reality, providing huge (unequally distributed) amounts of wealth, at the cost of a myriad of ecological impacts which, combined, disrupt the Earth System.[87] The only way to exit the Anthropocene with limited consequences from this disruption is to modify the human metabolism so its impacts level down. In practical terms, it means that the aggregate social dissipative structures (individual possessions and universalities) must impact the environment very much less.

A scientific materialist approach is a good one to assess what is at mankind's disposal for the coming decades. It should be able to discuss the exaggeration of techno-fanatics such as our "eco-modernists," without demanding everyone to convert himself to the dark green ideology. Assessing the true function of property, both inside the social metabolism and as a part of the socio-ecological metamorphosis, is an important element of the case for a moderate approach to this radical shift.

B. Ecological Appropriation

Social metabolism is the cultural appropriation of nature, made of interwoven dissipative structures, organized as universalities of tangible or intangible assets subjected to property rights under sociolegal arrangements (such as those encompassed in corporate law, contract law, labor law). In that perspective, appropriation is understandable in a twofold approach: first, as a matter of property rights and ecological ownership (1), and second, as a matter of dissipative structures and ecological political economy (2).

---

[87] M. FISCHER-KOWALSKI et C. AMANN, "Beyond IPAT and Kuznets Curves: Globalization as a Vital Factor in Analysing the Environmental Impact of Socio-Economic Metabolism", *Population and Environment*, 2001, vol. 23, No. 1, pp. 7–47.

(1) Property Rights and Ecological Ownership

"*Is a system of eco-private law compatible with private property?*", ask Ugo Mattei and Alessandra Quarta.[88] According to our approach, an ecological property is an ecological exercise of the power on things through property rights. It is, therefore, a matter of legal concepts, principles, and rules to favor such an ecological exercise. The question these authors ask is, therefore: Is it possible in the contemporary legal context or does it require a paradigm shift? In other words, does ecological ownership imply a set of reforms or a revolution?

The answer depends on the perception one has of the contemporary legal system of property law. If one considers that contemporary property is intimately linked to modernity and capitalism, then its core concepts are to be targeted as root causes of the Anthropocene. Gonzalez de Molina et Toledo, as Marxists, surely enter in that category, as Ugo Mattei and Alessandra Quarta, or French legal scholars who consider that "*the environmental issue is to radically transform the very concept of property.*"[89] If, as I do, one considers modern property as only a product of legal science and reasoning, like mathematics and geometry, core concepts are not necessary to be changed. I do think that what has been written in the very first section of this essay is as separated from political options as is the approach of social metabolism, which is, like Earth System Science, a product of modernity.

To justify their position of a radical shift from a sovereign extractive property founded on the right to exclude to a generative property founded on the principle of access and the philosophy of commons, Mattei and Quarta have to give credit to the idea of an absolute individual owner protected by an equally absolute right to exclude that permit every selfish behavior and the joint exploitation of humans and nature that produces the socio-ecological worst outcome: "*In the Seventeenth century, the theory of natural rights contributed to consecrating State sovereignty (imperium) and private property (dominium) as the foundational institutional structures of modernity (and of capitalist extraction).*"[90] Though, it is quite obvious that it is a hyperbolic exaggeration. The truth is that "*Blackstone's stirring talk (...) is ridiculed as a caricature of reality.*"[91]

As regarding the true meaning of the right to exclude, I can only enjoin to read Thomas Merrill articles[92]: "*What the exclusion thesis maintains is that if and when we recognize something as property, we will invariably find the right*

---

[88] U. MATTEI et A. QUARTA, *The Turning Point in Private Law*, op. cit., p. 31.

[89] S. VANUXEM et C. GUIBET LAFAYE, *Repenser la propriété, un essai de politique écologique*, Droits de l'environnement, Aix-en-Provence, Presses universitaires d'Aix-Marseille, 2015, p. 12.

[90] U. MATTEI et A. QUARTA, *The Turning Point in Private Law*, op. cit., p. 18.

[91] T. MERRILL, "Property and the Right to Exclude", op. cit., p. 753.

[92] T. MERRILL, "Property and the Right to Exclude", op. cit.; "Property and the Right to Exclude II", op.cit.

to exclude others."[93] My own school of thought also considers exclusion as the core criterion of ownership, only to say that without some kind—and nobody seriously asserts that the right to exclude is absolute—of right to exclude there is no ownership. With Thomas Merrill, we consider that the right to exclude is the first principle from which follows every other incident of property. It is not because we sacralize the mythical and fantasized "*sole and despotic dominion,*" but because if property is the legal recognition that one individual or legal subject is entitled to exercise (some) power on reality (according to law), then it is necessary a *dominion* that must be claimed and recognized as such. The right to exclude is nothing more than the power of the owner to legally claim his *dominion* and the title in virtue of which something as a part of material or social reality is in his power.

Consequently, the right to exclude has no positive consequence on what an owner is or is not entitled to do with his thing. That is a matter of legislation and case law: "*Because the exclusion thesis is analytical or interpretative, it is not a normative argument about the ends of property.*"[94] And there are countless cases where the law can enforce regulations conveying social values as the refusal of discriminations or the care for the environment: "*The world does not consist of islands of property, in the sense of large bundles of 'full ownership' rights, surrounded by a sea of unclaimed resources. Rather, it is a complex tapestry of property rights of different sorts (private, public, common) with different types and degrees of exclusion rights being exercised by different sorts of entities in different contexts.*"[95]

In that perspective, ecological ownership is not a matter of conceptual revolution but of rules that would implement an ecological exercise of *dominium*. There is no need of a revolution because modern law may convey the class interests of dominant people, call them capitalists lobbying a sometime captured state, but it is mostly a product of *reason*. And reason having understood that property is the recognition that things are in the power of a legitimate owner, reason can't help to conceive the right to exclude as its *sine qua non* principle.

"Shifting the default rules," if we had to consider this hypothesis, would have practical consequences that have been discussed, notably by Thomas Merrill who labeled it "forced sharing." Ugo Mattei and Alessandra Quarta stand against capitalist firms that have captured public authorities, and thus advocate a bottom-up revolution through the triumph of "commons." But it is "*forced sharing*" indeed "*to access the unmaintained garden of a neighbour for the purpose of landscaping, this being a generative action to increase the general beauty of the neighbourhood.*"[96] They acknowledge there would be

---

[93] Ibid., p. 3.
[94] T. MERRILL, "Property and the Right to Exclude II", *op. cit.*, p. 9.
[95] T. MERRILL, "Property and the Right to Exclude", *op. cit.*, p. 753.
[96] U. MATTEI et A. QUARTA, *The Turning Point in Private Law*, *op. cit.*, p. 152.

"*conflicts between access and exclusion,*" but do not elaborate, as Merrill, what implies such "*an ex post balancing of the interests of the owner and the parties seeking access,*"[97] whether made by the judges or by a "*Sharing Commission.*"

To Ugo Mattei and Alessandra Quarta, "*the position (they) take in this book is that the current state of affairs requires a vision that is radically transformative but compatible,* at least tactically, *with the current structure of the legal system.*" While it is a kind of Gramscian subversive tactics as "*(they) suggest a counter-hegemonic interpretation of existing property law,*"[98] we could take this position as social democrat approach of property confronting the Anthropocene. Maybe those two paths parallel the history of social rights as implemented in Soviet law after a revolution or, and surely under pressure, in Western law.[99] Without overthrowing the legal order, no one can deny that I live under social security and labor regulations my grandparents would never have dreamed of.

There are examples in French law that are in line with such an incremental evolution. Mathilde Hautereau-Boutonnet[100] examines two points of entry for the environment inside the Code civil (and not the Code of the environment): tort law as a statute of August 2016 (No 2016-1087) introduced the notion of ecological harm; and corporate law as a statute of May 2019 (No 2019-486) modified the article 1833 so that "*a firm is to be managed in its social interest, taking into consideration social and environmental issues of its activity.*"

The first statute modifies the Code so that article 1246 affirms "*every person liable for an ecological harm must remedy this damage.*" A new form of objective damage is a radical innovation considering the traditional "subjective damage" where someone holds liable someone else for the harm done. Here, non-human reality is the victim and article 1248 allows environmental associations to sue the offender for remedy. There is a lot to be said about the relative effectiveness of this legislation, but its innovative nature should not be understated. It does not necessarily mean biocentrism, but the mere protection of an ecological order that needs to be preserved, even if no human being is directly involved as a victim in its perturbation.

The second statute refers to harsh debates in case law regarding the legitimate interests to be considered in corporate law, whether the shareholders' only, or also the stakeholders'. Considering the social and environmental issues is a way to go beyond that second option and accept the idea that a commercial society must act according to some "social function" or, at least, conscious of its belonging to a greater society and then accountable for the consequences of its activities. A firm is not doomed to be cynically managed to maximize

---

[97] T. MERRILL, "Property and the Right to Exclude II", *op. cit.*, p. 23.

[98] U. MATTEI et A. QUARTA, *The Turning Point in Private Law, op. cit.*, p. 31.

[99] J. J. B. QUIGLEY, *Soviet Legal Innovation and the Law of the Western World*, Cambridge; New York, Cambridge University Press, 2007.

[100] M. HAUTEREAU-BOUTONNET, *Le Code civil, un code pour l'environnement, op. cit.*, pp. 22–23, 139–156 and 107–120.

profits even at the cost of brutal exploitation as was the very first, the East India Company.[101] Acknowledging this nature of a social dissipative structure, one becomes able to assess its real social value by considering what are its output and at what social and environmental cost they are produced.

A third point of entry is the ecological innovative interpretation of property law and its legislative evolution. In a word, besides some specific instruments like "*obligations environnementales*" which parallel, in a weaker form, "ecological easements" in common law, the main progress would be to rewrite article 544 Civil Code in the same way as the article 1833 after 2019, so that it would state: "*ownership is the right to enjoy and dispose of things according to law* and with due consideration of the ecological issues of its exercise" (our proposal).

Would it be enough though? Probably not. Would a legal rule inviting people to act altruistically be enough to get rid of selfishness? I doubt that. That's why beyond legal reforms, we surely need ideology to implement ecological ethics that would more efficiently make owners consider ecological issues in property. It is again a matter of worldviews and how to think and behave ecologically, as an individual or as a society.

(2) *Dissipative Structures and Ecological Political Economy*

Appropriation begins with the individual and collective exercise of property power by subjective rights, according to legal rules and, maybe most importantly, individuals' and corporations' ethics. It continues with the constitution of universalities that become dissipative structures which produce goods and services, or consume them, at the cost of wastes and pollutions. We thus require an ecological political economy that would consider the economy as the aggregate of universalities and social dissipative structures.

Energy is the core concept of dissipative structures. Energy that is synonym of *work*: 1 newton is the acceleration of an object 1 kilogram to 1 meter per second; 1 joule is the energy needed to move an object of 1 kilogram on 1 meter in 1 second; and 1 watt is the power needed to move an object of 1 kilogram at 1m/s. According to the law of thermodynamics, energy is never lost (1st law of conservation states there is always the same amount of energy in the system) but energy is never the same (2nd law of entropy states that dissipated energy will not be able to do work again, hence the "arrow of time" in physics).[102] So, it is *exergy* as useful work that is the most relevant concept here: the amount of energy effectively used to do work that was wanted, energy lost in the process being already as useless as dissipated energy after the process (entropy).

A dissipative structure is a complex system that is organized to maximize the useful work done by transforming energy into entropy and, in between, as

---

[101] N. ROBINS, *The Corporation that Changed the World. How the East India Company Shaped the Modern Multinational*. New Delhi, Orient Blackswan, 2006.

[102] A. BEN-NAIM, *Entropy Demystified: The Second Law Reduced to Plain Common Sense*, Hackensack, World Scientific, 2007.

much as possible exergy. The organized character of a dissipative structure is sometimes labeled negentropy because it delays the transformation of energy into entropy, allowing its transformation into exergy. That's why living creatures are typical dissipative structures: they are working systems between purely potential energy at low entropy and dissipated energy at high entropy. In other words, dissipative structures are systems that concentrate the dissipative process of energy to do some work and maintain the structure (*autopoiesis*).

Human civilizations are the combination of endosomatic dissipative structures (human bodies) and exosomatic dissipative structures (other living bodies or biophysical processes or machines). The main specific feature of human societies is that exosomatic structures are now without compare in exergy capacity than endosomatic ones. This process begun when hominids first mastered fire, then agriculture and herding, *i.e.*, concentrated solar energy through photosynthesis and alimentation, and eventually fossils fuels, hydraulic and nuclear power, and new renewable energies.

The Anthropocene originates in epochal shifts in social metabolism like the Neolithic Revolution, the Industrial Revolution and finally the Great Acceleration where almost all human societies were put on a trajectory toward the industrial metabolism. In terms of exergy, useful work, human metabolism transitioned from less than 1GJ/cap/year as hunter-gatherers, to an agrarian one of less than 5, and to an industrial one of 30–50.[103] The result is a demiurgic transformative power of the environment. The outcome has been socioeconomic "progress" in many forms from population size to life expectancy, and especially economic growth, but also major stresses on the Earth System. A concentration of atmospheric CO2 at levels unseen in more than 1 million years is the most evident indicator of this transformative capacity, putting the climate on a trajectory of +4°C, equivalent to the warming (+4/+6) that ended the last glaciation and begun the Holocene—though in 500 years instead of 10 000.

Neoclassical economics (NCE) is blind to the materialistic dimension of the economy and, besides the lack of pluralism in economics,[104] is flawed by a radical disconnectedness from biophysical world and the laws of physics. In the models of Samuelson, Solow, or Nordhaus, infinite growth is possible because of the substitutability of anything by something else, known or to be discovered by the infinite potential of the human mind: "*The economic process is driven not by the availability of physical resources, but rather by human ingenuity as depicted in the still widely used Cobb- Douglas function. The quantity of output produced (Q) is a function of only capital (K) and labor (L).*"[105]

---

[103] H. HABERL et al. (eds.), *Social Ecology*, op. cit., p. 61 Fig. 3.1.

[104] M. W. M. ROOS, F. HOFFART, *Climate Economics: A Call for more Pluralism and Responsibility*, Palgrave studies in sustainability, environment and macroeconomics, Cham, Palgrave Macmillan, 2021.

[105] C. A. S. HALL, K. KLITGAARD, *Energy and the Wealth of Nations: An Introduction to Biophysical Economics*, Cham, Springer International Publishing, 2018, p. 69.

Nicholas Georgescu-Roegen[106] was among the first to point out the anti-scientific nature of such a view. The main point he made was that economy is necessarily *inside* the global biophysical system and submitted to the laws of physics. NCE is, therefore, completely incompatible with the second law of thermodynamics (which precludes a perpetual motion machine such as the circular flow model), inconsistent with contemporary science and thus inacceptable to a scientific materialist.

The "pure technological change" or "total factor productivity" is the "miracle" constant that explains Solow's residue, unexplained growth that, among others, Robert U. Ayres has demonstrated to be explained by the ability to do work through *exergy*, and especially the use of fossil fuels.[107] Technology has mostly been a way to harness more energy and use it more efficiently. In other words, growth was for the most part a matter of *power* in the physical sense: work in energy units.

An approach of the economic process can be found in the works of Charles Hall and Kent Klitgaard and their idea of biophysical economics: "*economic production can be viewed as the process of upgrading matter into highly ordered (thermodynamically improbable) structures, both physical structures and information.*"[108] Biophysical economics appears as a complement to the social sphere perspective, "adding value" being the counterpart of "adding order" (and create disorder as well in the process). As "*it views most human economic activity as a means to increase (directly or indirectly) the exploitation of nature to generate more wealth,*"[109] it is to my opinion fully coherent with our idea of the socio-ecological metabolic appropriation of reality by human dissipative structures.

One could figure society as a house of cards: the first stage is the metabolic capacity of work (exergy) given the appropriation of energy sources and other raw materials. The upper stages would be the other metabolic processes and all the social structures of society. Without falling into reductionism and determinism, this first stage sets what is *physically possible* before each society evolves in what direction it considers *culturally desirable*. That's why energy and materials are not only factors of production but limiting factors of economy and society as well. This first stage sets the limits to the dimensions of the upper stages.[110]

---

[106] GEORGESCU-ROEGEN, *The Entropy Law and the Economic Process*, Cambridge, Harvard University Press, 1971.

[107] R. U. AYRES, "Exergy, Power and Work in the US Economy, 1900–1998", *Energy*, 2003, vol. 28, No. 3, pp. 219–273; R. U. AYRES, B. WARR, *The Economic Growth Engine: How Energy and Work Drive Material Prosperity*, Cheltenham; Northampton, Edward Elgar, 2009.

[108] C. A. S. HALL et K. KLITGAARD, *Energy and the Wealth of Nations*, op. cit., p. 77.

[109] Ibid., p. 83.

[110] V. SMIL, *Energy and Civilization: A History*, Cambridge, The MIT Press, 2017.

Our societies are built on a house of cards where the lowest stage is made of an energy mix with 80% or so of fossil fuels. With some carbon capture and other sequestration processes maybe some of these energy sources would be still allowed in a carbon neutral metabolism achieved between 2060 and 2080 at the global scale. But a conservative estimate would be sound to consider that most of this 80% must be abandoned—and will be, as finite resources are necessarily to be exhausted one day. By no means the development of alternative energies can compensate totally for the loss of fossil fuels in the coming decades, and that's why scenarios of energetic transition all show a reduction of energy consumption.[111]

Quantitatively, Olivier Vidal, among others, thinks "*we now appear to be entering a pivotal period of long-term production cost increases as we approach the minimum practical energy and thermodynamic limits for many metals.*"[112] Deploying the new energetic infrastructure will be a challenge. Qualitatively, Energy Return on Energy Invested (EROI),[113] is another limit to be considered. The new infrastructure, should it be fully deployed, will inevitably harness energy with less efficiency than the golden age of fossil fuels, whose EROI is also dramatically declining.[114]

This exergy depression would probably mean an economic depression as well since the economic process cannot be absolutely decoupled from material processes. Besides the empirical analysis that does show only debatable relative decoupling,[115] absolute decoupling would only be temporary, since wealth cannot grow toward infinity while energy consumption diminishes toward zero. Eco-modernists rely, here, on some misplaced optimism: "*even most catastrophic United Nations scenarios predict rising growth.*"[116] Those scenarios are built with a neoclassical paradigm and so growth is conceived

---

[111] IEA (2021), *Net Zero by 2050*, IEA, Paris https://www.iea.org/reports/net-zero-by-2050 shows an optimistic reduction of 8% in energy demand. The French Low Carbon Strategy shows an objective of 50% reduction in energy consumption by 2050 compared to 2012.

[112] O. VIDAL, *Mineral Resources and Energy: Future Stakes in Energy Transition*, Oxford, ISTE Press Ltd; Elsevier, 2018; O. VIDAL et al., "Modelling the Demand and Access of Mineral Resources in a Changing World", 2021.

[113] C. A. S. HALL, *Energy Return on Investment*, Cham, Springer International Publishing, 2017.

[114] L. DELANNOY et al., "Peak oil and the low-carbon energy transition: A net-energy perspective", *Applied Energy*, dec. 2021, vol. 304, pp. 1–17; L. DELANNOY et al., "Assessing Global Long-Term EROI of Gas: A Net-Energy Perspective on the Energy Transition", *Energies*, aug. 2021, vol. 14, No. 16, p. 5112.

[115] T. PARRIQUE et al., *Decoupling Debunked—Evidence and Arguments Against Green Growth as a Sole Strategy for Sustainability*, EEB—The European Environmental Bureau, 2019; T. VADÉN et al., "Decoupling for Ecological Sustainability: A Categorisation and Review of Research Literature", *Environmental Science & Policy*, 2020, vol. 112, pp. 236–244.

[116] M. SHELLENBERGER, T. NORDHAUS, "Evolve", *in* M. Shellenberger, T. Nordhaus (ed.), *Love your monsters, op. cit.*, §3.

as a necessary behavior of the economy. Besides, there is a political demand to economists in the working III of the GIEC to model growth, even a weak one. Degrowth is a political taboo, even recession probably is a very likely outcome of the energetic transition.

The agenda of an ecological political economy is, therefore, to ecologically manage the dissipative structures toward a metabolic transition. It needs sound macroeconomics[117] to cope with the economic and social consequences of the double quantitative and qualitative downgrading of our energy mix. Consequences such as: the management of social security in a recession,[118] but also social violence and perhaps wars. Should we refuse to engage in this shift, we would still have to cope with the—more, to my opinion—dramatic consequences of climate change. And energy should become abundant again (say with nuclear fusion), then we would have to worry about environmental impact since more power is more impact and entropy because of the second law of thermodynamics.

Law has a role to play. But maybe not in the ways some are too quickly advocating, that might be too radical to be accepted by the public opinions. Surely, all the branches might be concerned, from competition law to corporate law, but also public law as economic regulation. Though, legal traditions can deal with it without conceptual or philosophical revolutions in jurisprudence. Legal persons, limited to human beings and organizations, property rights, firms and legal relations are not to be replaced. Modern law is up to the task, because of its modernity, which is also the root of the better understanding of our Earth System and the origin of the very concept of Anthropocene.

This is not saying that the work to be done is easy. There is a revolution coming with our metabolism shift. Of the same magnitude than the Neolithic and Industrial ones. And even more astonishing if we consider its time scale of merely a handful of decades. This profound shift in social metabolism will probably need state interventionism to an extent unseen since the New Deal and wartimes, raising legal issues and needing a lot of legal innovations. My point here is to say that in every aspect of this legal engineering of the metabolic shift toward a neutral carbon economy, it is always taming property, the power to individually and collectively appropriate nature and exploit the physical possibilities of the world by our social dissipative structures. The fact that it can be done with the given political philosophy and legal thinking of the modern times should be reason for hope.

---

[117] T. JACKSON, *Prosperity Without Growth: Foundations for the Economy of Tomorrow*, London, New York, Routledge, 2017.

[118] C. CORLET WALKER et T. JACKSON, "Tackling Growth Dependency—The Case of Adult Social Care", CUSP. Working Paper Series, No 28, 2021.

Conclusion

Property is not only a field of law, divided in the common law tradition between real and personal property. Property is the power exercised on every dimension of reality. It is obvious regarding corporeal realities, and widely accepted regarding incorporeal properties, but it is also true regarding social relations (debts). Property is the mechanism of all form of physical and social power. It harnesses reality by managing dissipative structures that are the basis of every complex activity or venture, and that are the infrastructure of every source of social power.

As the power to harness physical and social reality, property appears as the fundamental mechanism of social metabolism. Societies are organized physical processes, fluxes of energy and matter from which emerge the economy. The Anthropocene is the result of a social metabolism of exponential growth in energy and matter harnessed by humans at the cost of also exponential impacts on the environment. Addressing the ecological predicament is the complex issue of taming property and dissipative structures so our metabolism does not destroy our "safe operating space"[119] while still allows as many people as possible not to suffer from deprivation, hunger and violence. And it is a matter of energy and its social counterpart, property and not in the only form of property rights.

The authors of *The Limits to Growth*, in its 30-year update, write they "*are much more pessimistic about the global future than (they) were in 1972.*" Behind this agreement, they also acknowledge "the great differences among the hopes and expectations of the three authors": "*Dana was the unceasing optimist*"; "*Jorgen is the cynic (...) sad to think that society will voluntarily forsake the wonderful world that could have been*" and "*Dennis sits in between,*" believing "*actions will ultimately be taken*" but "*the results secured after long delay will be much less attractive than those that could have been attained through earlier action.*"[120] My heart goes straight with Dana, while my reason hesitates between the others.

References

ADLER, J., "Introduction: Property in Ecology", *Natural Resources Journal*, 2019, vol. 59, No 1, p. x.

AYRES, R.U., "Exergy, Power and Work in the US Economy, 1900–1998", *Energy*, 2003, vol. 28, No 3, pp. 219–273.

AYRES, R.U. et WARR, B., *The Economic Growth Engine: How Energy and Work Drive Material Prosperity*, Cheltenham; Northampton, Edward Elgar, 2009.

---

[119] J. ROCKSTRÖM et al., "A Safe Operating Space for Humanity", *Nature*, 2009, vol. 461, No. 7263, pp. 472–475.

[120] D. H. MEADOWS, J. RANDERS et D.L. MEADOWS, *The Limits to Growth: The 30-year Update*, London, Earthscan, 2009, p. xvi.

BARLES, S., « Écologie territoriale et métabolisme urbain :quelques enjeux de la transition socioécologique », *Revue d'Économie Régionale Urbaine*, 2017, No 5, pp. 819–836.

BARNES, R., "The Capacity of Property Rights to Accommodate Social-Ecological Resilience", *Ecology and Society*, 2013, vol. 18, No 1.

BEN-NAIM, A., *Entropy Demystified: The Second Law Reduced to Plain Common Sense*, Hackensack, World Scientific, 2007.

BLACKSTONE, W. et al., *Commentaries on the Laws of England*, 2, Oxford; New York, Oxford University Press, 2016.

BRIBITZER-STULL, M., *Understanding the Leitmotif: From Wagner to Hollywood Film Music*, Cambridge, Cambridge University Press, 2015.

BUNGE, M., *Scientific Materialism*, Dordrecht, Reidel Publishing Company, 1981.

CAPRA, F., MATTEI, U., *The Ecology of Law: Toward a Legal System in Tune with Nature and Community*, Oakland, Berrett-Koehler Publishers, 2015.

CHAZAL, J.-P., « Le propriétaire souverain : archéologie d'une idol doctrinale », *Revue trimestrielle de droit civil*, 2020, vol. 1, pp. 1–33.

CORLET WALKER, C., JACKSON, T., "Tackling Growth Dependency—The Case of Adult Social Care", CUSP Working Paper Series, No 28, 2021.

CRÉTOIS, P., *La part commune: critique de la propriété privée*, Paris, Éditions Amsterdam, 2020.

CRUTZEN, P.J., "Geology of Mankind", *Nature*, 2002, vol. 415, No 6867, pp. 23–23.

DELANNOY, L. ET AL., "Peak Oil and the Low-Carbon Energy Transition: A Net-Energy Perspective", *Applied Energy*, dec. 2021, vol. 304, pp. 1–17.

"Assessing Global Long-Term EROI of Gas: A Net-Energy Perspective on the Energy Transition", *Energies*, aug. 2021, vol. 14, No 16, 5112.

DEMOULE, J.-P., *La révolution néolithique*, Paris, Éd. le Pommier, 2013.

DIAMOND, J.M., *Guns, Germs, and Steel: The Fates of Human Societies*, New York, Norton, 2017.

DIAMOND, J.M., MURNEY, C., *Collapse: How Societies Choose to Fail or Succeed*, New York, Penguin, 2005.

DOBSON, A., *Green Political Thought*, Routledge, 2012.

DOREMUS, H., "Climate Change and the Evolution of Property Rights", *UC Irvine Law Review*, 2011, vol. 1, No. 4, p. 1091.

DOREMUS, H., "Takings and Transitions", *Florida State University Journal of Land Use and Environmental Law*, 2018, vol. 19, No 1.

DROSS, W., *Droit Civil, Les Choses*, Paris, LGDJ-Lextenso éd, 2012.

EMERICH, Y., *Conceptualising Property Law*, Cheltenham; Northampton, Edward Elgar Publishing, 2018.

FISCHER-KOWALSKI, M., "Analyzing Sustainability Transitions as a Shift Between Socio-Metabolic Regimes", *Environmental Innovation and Societal Transitions*, 2011, vol. 1, No 1, pp. 152–159.

FISCHER-KOWALSKI, M., AMANN, C., "Beyond IPAT and Kuznets Curves: Globalization as a Vital Factor in Analysing the Environmental Impact of Socio-Economic Metabolism", *Population and Environment*, 2001, vol. 23, No 1, pp. 7–47.

FREMAUX, A., GUILLAUME, B., « L'horizon postdémocratique et la crise écologique », *Écologie politique*, 2014, vol. 49, No. 2, pp. 115–129.

GEORGESCU-ROEGEN, N., *The Entropy Law and the Economic Process*, Cambridge, Harvard University Press, 1971.

GONZÁLEZ DE MOLINA, M., TOLEDO, V.M., "The Social Metabolism: A Socio-Ecological Theory of Historical Change", *Environmental History*, No 3, Cham, Springer International Publishing, 2014.

HABERL, H. et al., "A Socio-Metabolic Transition Towards Sustainability? Challenges for Another Great Transformation", *Sustainable Development*, 2011, vol. 19, No. 1, pp. 1–14.

HABERL, H. et al. (eds.), *Social Ecology*, Cham, Springer International Publishing, 2016.

HALL, C.A.S., *Energy Return on Investment*, Lecture Notes in Energy, Cham, Springer International Publishing, 2017.

HALL, C.A.S., KLITGAARD, K., *Energy and the Wealth of Nations: An Introduction to Biophysical Economics*, Cham, Springer International Publishing, 2018.

HALL, J.A. et SCHROEDER, R. (eds.), "An Anatomy of Power: The Social Theory of Michael Mann", Cambridge; New York, Cambridge University Press, 2006.

HARDIN, G., "The Tragedy of the Commons", *Science*, vol. 162, No 3859, American Association for the Advancement of Science, 1968, pp. 1243–48.

HAUTEREAU-BOUTONNET, M., *Le Code civil, un code pour l'environnement*, Les sens du droit, Paris, Dalloz, 2021.

HEINBERG, R., *Power: Limits and Prospects for Human Survival*, Gabriola, New Society Publishers, 2021.

HELLER, M., "The Tragedy of the Anticommons: A Concise Introduction and Lexicon", *The Modern Law Review*, 2013, vol. 76, No 1, pp. 6–25.

HELLER, M., "The Tragedy of the Anticommons: Property in the Transition from Marx to Markets", *Harvard Law Review*, 1998, vol. 111, No 3, p. 621.

HERMITTE, M.-A., «La nature, sujet de droit ?», *Annales. Histoire, Sciences Sociales*, 2011, No 1, pp. 173–212.

JACKSON, T., *Prosperity Without Growth: Foundations for the Economy of Tomorrow*, London; New York, Routledge, 2017.

LU, Y. et CHEN, B.,"A Review on Urban Metabolism: Connotation and Methodology", *Acta Ecologica Sinica*, 2015, vol. 35.

MALM, A., *Fossil Capital: The Rise of Steam-Power and the Roots of Global Warming*, London, Verso, 2016.

MANN, M., "The Autonomous Power of the State: Its Origins, Mechanisms and Results", *European Journal of Sociology/Archives Européennes de Sociologie*, 1984, vol. 25, No 2, pp. 185–213.

MATTEI, U., QUARTA, A., *The Turning Point in Private Law*, Cheltenham, Edward Elgar Publishing, 2018.

MCANANY, P.A., YOFFEE, N. (eds.), *Questioning Collapse: Human Resilience, Ecological Vulnerability, and the Aftermath of Empire*, Cambridge; New York, Cambridge University Press, 2010.

MEADOWS, D.H., RANDERS, J. et MEADOWS, D.L., *The Limits to Growth: The 30-year Update*, London, Earthscan, 2009.

MERRILL, T., "Property and the Right to Exclude", *Nebraska Law Review*, 1998, vol. 77, No 4.

MERRILL, T., "Property and the Right to Exclude II", *Brigham-Kanner Property Rights Conference Journal*, 2014, vol. 3, p. 1.

MERRILL, T.W. et SMITH, H.E., *The Architecture of Property*, Rochester, NY, Social Science Research Network, 2019.

NICOLIS, G. et PRIGOGINE, I., *Self-Organization in Nonequilibrium Systems: From Dissipative Structures to Order Through Fluctuations*, New York, Wiley, 1977.

OSTROM, E., *Governing the Commons: The Evolution of Institutions for Collective Action*, Canto classics, Cambridge, Cambridge University Press, (1990) 2015.

PARRIQUE, T. et al., *Decoupling Debunked—Evidence and Arguments Against Green Growth as a Sole Strategy for Sustainability*, EEB—The European Environmental Bureau, 2019.

PENNER, J., *The Idea of Property in Law*, Oxford, Oxford University Press, 2000.

QUIGLEY, J.B., *Soviet Legal Innovation and the Law of the Western World*, Cambridge; New York, Cambridge University Press, 2007.

ROBINS, N., EAST INDIA COMPANY, *The Corporation that Changed the World How the East India Company Shaped the Modern Multinational*, New Delhi, Orient Blackswan, 2006.

ROCHFELD, J. et al., *L'échelle de communalité—Propositions de réformes pour intégrer les biens communs en droit*, Paris, GIP Justice, 2021.

ROCKSTRÖM, J., et al., "A Safe Operating Space for Humanity", *Nature*, 2009, vol. 461, No 7263, pp. 472–475.

ROOS, M.W.M., HOFFART, F., *Climate Economics: A Call for More Pluralism and Responsibility*, Palgrave studies in sustainability, environment and macroeconomics, Cham, Palgrave Macmillan, 2021.

ROSE, C., "The Comedy of the Commons: Custom, Commerce, and Inherently Public Property", *The University of Chicago Law Review*, 1986, vol. 53, No 3, p. 711.

SABLOFF, J. A., SABLOFF, P. L. W. (ed.), *The Emergence of Premodern States: New Perspectives on the Development of Complex Societies*, SFI Press seminar series, No 2, Santa Fe, SFI Press, 2018.

SCHMALTZ, B., *Les personnes publiques propriétaires*, Nouvelle bibliothèque de thèses, Paris, Dalloz, 2016.

SCHROEDER, R. (ed.), *Global Powers: Michael Mann's Anatomy of the Twentieth Century and Beyond*, Cambridge, Cambridge University Press, 2016.

SHELLENBERGER, M., NORDHAUS, T., *Love Your Monsters: Postenvironmentalism and the Anthropocene*, Oakland, Breakthrough Institute, 2011.

SMIL, V., *Energy and Civilization: A History*, Cambridge, The MIT Press, 2017.

SPERBER, D., *Explaining Culture: A Naturalistic Approach*, Oxford; Cambridge, Blackwell, 1996.

SPRANKLING, J.G., "Property Law for the Anthropocene Era", *Arizona Law Review*, 2017, vol. 59, p. 36.

TAINTER, J., *The Collapse of Complex Societies*, Cambridge, Cambridge University Press, 1988.

TAINTER, J.A., "Patricia A. McAnany and Norman Yoffee (eds): Questioning Collapse: Human Resilience, Ecological Vulnerability, and the Aftermath of Empire", *Human Ecology*, 2010, vol. 38, No 5, pp. 709–710.

TILLY, C., *Coercion, Capital, and European States, AD 990–1990*, Studies in social discontinuity, Cambridge, Blackwell, 1990.

VADÉN, T. et al., "Decoupling for Ecological Sustainability: A Categorisation and Review of Research Literature", *Environmental Science & Policy*, 2020, vol. 112, pp. 236–244.

VANDEVELDE, K., "The New Property of the Nineteenth Century: The Development of the Modern Concept of Property", *Buffalo Law Review*, 1980, vol. 29, No. 2, p. 325.

VANUXEM, S., *Des choses de la nature et de leurs droits*, Sciences en questions, Versailles, Quae, 2020.

VANUXEM, S., GUIBET LAFAYE, C., *Repenser la propriété, un essai de politique écologique*, Droits de l'environnement, Aix-en-Provence, Presses universitaires d'Aix-Marseille, 2015.

VIDAL, O., *Mineral Resources and Energy: Future Stakes in Energy Transition*, Oxford, ISTE Press Ltd; Elsevier, 2018.

VIDAL, O. et al., *Modelling the Demand and Access of Mineral Resources in a Changing World*, 2021.

VRIES, B. de, *Sustainability Science*, New York, Cambridge University Press, 2013.

WEBER, *Economy and society*, New York, Bedminster Press, 1968 (1922).

WOLMAN, A., "The Metabolism of Cities", *Scientific American*, 1965, vol. 213, No 3, pp. 178–190.

ZALASIEWICZ, J.A. (ed.), *The Anthropocene as a Geological Time Unit: A Guide to the Scientific Evidence and Current Debate*, Cambridge, Cambridge University Press, 2019.

ZENATI, F., *Essai sur la nature juridique de la propriété : contribution à la théorie du droit subjectif*, th. Lyon 3, 1981.

Environmental Politics and Theory in the Policy-Making Process

Ecosystem Policy and Law: A Philosophical Argument for the Anticipatory Regulation of Environmental Risk

John Martin Gillroy

THE SIGNIFICANT PROBLEMS WE HAVE CREATED CANNOT BE SOLVED AT THE LEVEL OF THINKING AT WHICH WE CREATED THEM…

ALBERT EINSTEIN[1]

The first part of the twenty-first century has been marked by wildfires, flooding, drought, rising sea levels and melting glaciers, all capped off by the COVID-19 global pandemic.[2] If this series of cascading crises tells us

---

[1] 1948. '*A Message To Intellectuals*".

[2] COVID-19, although superficially a health crisis, is also, fundamentally, an environmental risk dilemma as it involves the interaction of human and ecological systems where the stealth nature of the risk can render catastrophic results without proper, anticipatory, regulation.

This chapter is the product of thirty years of thought, the patience of Joel Kassiola and the expert editorial and publication skills of Margaret Murray, who reconceptualized and invigorated my ideas and the logic of their presentation for this chapter.

J. M. Gillroy (✉)
Lehigh University, Bethelehem, PA, USA
e-mail: jmg304@lehigh.edu

anything, it is that an environmental risk policy focused primarily on an efficient economy and only the *ex post* regulation of harm to the environment is catastrophically inadequate. Current environmental policies fail to protect and empower essential human and natural values which should be the foundation of planning and policy implementation. Such a basis for policy would prevent the global crises we are currently experiencing. The "economic" status quo has failed for decades to properly regulate climate change; much has been written about its limitations but it continues to be used. But the immediate threat and the drastic *ex post* measures required by the environmental crises and the pandemic provide vivid evidence that the conventional assumptions of our current public policy model for environmental risk are not just limited, but deadly to our planet and our existence.

A vital reason that these environmental dilemmas have escaped solution is that we continue to examine them within the context of modern theory, specifically positivism, the assumptions and presuppositions of which fail to grasp what is really at stake in these issues. To put it simply, the current methods are inadequate to solve the problem because they do not and cannot address the fundamental issues at the heart of many policy issues and the laws they support. This is especially true when one considered the underlying moral premises of environmental issues in particular. Positivism relies on a limited scope to consider problems and solutions: the present as defined by the empirical. To broaden that scope, we need to understand the essence, or full complexity, of the human being as a basis for deliberation and choice in policy and law.

While the nature of specific problems of law and policy cannot be effectively removed from their time and context-sensitive "reality," we can expand the limited options and possibilities of the dominant contemporary theoretical approach, and eliminate its inherent bias, by utilizing whole, systematic philosophical arguments about practical reason and human agency deciphered and written before the onset of social-scientific positivism in the mid-nineteenth century. The key to change is to employ policy paradigms deciphered from enlightenment philosophical arguments. By utilizing enlightenment arguments, the contemporary analyst can have access to a range of more universal, comprehensive and distinctly human-agent based paradigms. These can then be the foundation for a reconsideration of the "science" of public affairs and the reassessment of the "truth" of contemporary law and policy.

After decades of positivist-driven policy-making, the idea of considering such an approach as I suggest is challenging, because the status quo is more familiar and firmly enshrined in policy-making for both practitioners and critics of the status quo. But contemporary theoretical approaches to policy and law, of all types, in terms of both critical arguments against predominant status-quo methods (like efficiency) and constructive arguments for change (like sustainability) inevitably carry an inherent bias toward the categories, assumptions and predispositions of positivism as "social-scientific" method (this pertains even to arguments alleged to be non- or even anti-positivist).

However, we have arrived at a point in human history where we need to accept the challenge to "think anew."

Specifically, in direct contradiction to enlightenment assumptions, contemporary theory assumes the predominance of the empirical, and it dismisses metaphysics as absolutist and all substructural normative justification as "ideology". It offers only a narrow scope for theory and emphasizes the microdifferences in classification of the components of culture, law and policy over the universal nature of humanity-in-the-person. It views ideas such as dialectic as antediluvian and classifies anything and all things as "norms," indistinguishable from one another within fixed, overly defined and artificial disciplinary boundaries. It even dismisses science's original purpose in applying critical reason to both humanity and nature in pursuit of the truth about both (truth being a word that is assumed, by positivist practitioners, to be, at best, suspect and at worst totalitarian).

The only way past these prejudices is to seek a philosophical cosmology that expands positivism by predating it. Surprisingly, I am arguing that the way forward is to look back. It is only when we change the essential philosophical-moral premises of environmental risk policy and the codified laws that support it, that we can create *anticipatory ecosystem policy and law* that will prevent further damage and help us rebuild. It is in the arguments of pre-positivist Enlightenment philosophers, like Kant, and their logical maps of human nature, practical reason and moral agency, that a philosophical method can be found to synthesize prerequisite logics of concepts upon which modern theory, and its components, can be reconsidered. In this chapter, I will, as an example of how this may be achieved, replace the assumptions of the dominant market paradigm for policy choice with a Kantian paradigm for environmental risk analysis.

Introduction: Science, Morality and the Uncertainty of Environmental Risk

Environmental risk is characterized as a *zero-infinity* problem; that is, it threatens an almost *zero* probability of an *infinitely* catastrophic event. Normally, one distinguishes a risk from an uncertainty by the fact that a reliable probability number can be arrived at for the former, but not for the latter. However, the "science" of environmental risk estimation (Quantitative Risk Assessment) has ignored this distinction by taking what amounts to a pervasive uncertainty and assigning a probability number to it. Stating that environmental risk is characterized by "ignorance of mechanism," Page (1978) means that the physical and chemical processes by which risk agents make their way through the environment and integrate into what we might call a "risk soup," with health effects for human beings and nature, are not traceable or predictable through current scientific methodology.

The "certainty" of science does not begin with, or even involve, humanity or its social, political or moral dimensions, but begins and ends within natural systems' function and evolution. Ignorance of biological, chemical and physical mechanisms, compounded by the use of non-human findings to make human health estimates, results in policy decisions which are "choices under pervasive uncertainty" (Page 1978). This means that the numbers produced by quantitative risk assessment can be used only as data and not as definitive evidence for setting standards or regulating environmental risk (Graham et al., 1988). However, additional philosophical considerations can fill the gap between science and public policy choice to properly regulate risk.

To change this predisposition, we would need to acknowledge three postulates: first, humanity sets the terms of discourse and value for the assessment of humanity, nature and their interaction. Second, humanity's intrinsic value is moral in nature. Third, the intrinsic value of the environment has its roots in its capacity as a persistent living system which cannot be properly evaluated by the transference of human moral or intellectual attributes (e.g., thought, choice, strategy, rights, interest) to nature.

Like science, philosophical analysis is also a result of human design and execution (Bobrow and Dryzek 1987; Bohman 2021). We should be capable of morally evaluating nature on its own terms, without either importing human characteristics onto non-human entities or making humanity the only creature of moral value in the universe. If all human analysis is anthropomorphic in that it begins and ends with human categories, schema and, vocabulary, then the distinction of importance in environmental ethics is not between anthropocentric and eco- or bio-centric theory, but between *anthropocentric* moral theory and merely *anthropomorphic* analysis. This distinction defines *anthropocentric* policy as that which places only human instrumental value on nature while also placing humanity and its concerns as the core matter of importance in policy calculation. *Anthropomorphic* policy then becomes that which admits that all moral valuation is human, but that this, in and of itself, does not promote humanity to be the top policy priority.

By this distinction, all philosophical analysis is anthropomorphic, even that which places natural systems or the biosphere in the central place of value. The important question is not whether humans have decided the terms of analysis, but whether incommensurabilities are acknowledged in how distinct entities are evaluated, and in what terms they are valued.

Within present environmental risk law and policy, the economic models are both anthropomorphic and anthropocentric because nature is analyzed only as it has instrumental economic use to humanity. Nature is defined in terms of its "resource" value to individual preference, and its characteristics are described to facilitate human consumption (e.g., not a tree, or a component of a natural system, but 1000 board-feet of lumber). Humanity not only sets the terms of discourse for nature, but its economic preferences are the sole standard of value in this analysis.

The approach of science, on the other hand, is anthropomorphic, but not anthropocentric, because science attempts to understand the internal structure and function of nature, where humanity is seldom a character, never the central agent or the singular value (Abrahamson and Neis 1997; Odum 1975). The pervasive uncertainty in policy risk estimates appears when science tries to translate environmental, non-human data into commensurable human risks. In effect, risk analysis fails to alleviate uncertainty because it assumes a commensurability between natural and human health effects that may not exist. Policy evaluation and environmental ethics may exacerbate uncertainty by assuming a similar commensurability in the terms of ethical discourse and valuation. Perhaps science should not apply environmental data to judge human capacities, but philosophy ought not attribute human ethical designations to nature, as if it were also a moral agent.

Public policy, like science, may have to plead ignorance of exact mechanism in its evaluation of nature, but it must begin to consider nature within its own context and philosophically integrate it as a distinct but equally valuable component of our moral deliberations about good and bad policy, right and wrong public decisions. As a starting point, we should recognize that all philosophical deliberation, even that which places central and greatest moral value on the natural systems themselves, is anthropomorphic.

However, while we acknowledge our necessary role in the proper valuation of nature, we must also understand that our responsibilities to nature, and the duties that flow from these responsibilities, must be based on nature's functional characteristics and not quasi-human rights, utilities or interests. In addition, we need to realize that debating in human terms does not make humans the only proper subjects of moral duty. When making public choices, we should recognize the *intrinsic functional value* of the environment as a foundation that defines our duties to nature as the *intrinsic moral value* of humanity defines our duties to ourselves and other persons (Gillroy 2000; Kant MJ: 434–435).

Philosophically, if natural systems are to have a distinctive value and therefore a pride of place in policy analysis, we must put them there. Our central concern should be to distinguish between those policy principles and paradigms that can accommodate only nature's instrumental value to man, and those where the functional or intrinsic value of natural systems plays a role. Only within this second group of theories can nature have a place of equality or prominence in our moral consideration of what environmental risks are acceptable and which ones are not.

Environmental risk is not a single issue, law or policy, but a classification for a cross-section of environmental concerns characterized by pervasive uncertainty and zero-infinity management problems (e.g., climate change; COVID 19[3]). Using the characteristics of environmental risk as defined by Toby Page (1978) allows us to assess the failure of market assumptions and articulate the ethical-moral and administrative requirements of any paradigm seeking to more adequately regulate risk issues.

---

[3] See footnote 2.

First, we must identify the philosophical *substructure* of the uncertainty characteristic inherent in environmental risk. We do this by using the distinction between anthropocentric market assumptions and anthropomorphic philosophical theory and applying it to the essential dilemmas that involve debates over the distinctions between intrinsic and instrumental values and private and public goods. To do this, we will focus on two of Page's (1978) characteristics of environmental risk attributed to its uncertainty: *modest benefits* and their implications given both intrinsic and instrumental values involved in approaching uncertainty, and the *catastrophic implications* of environmental risk given a distinction between the public and private nature of the goods-decisions involved.

Second, we will consider the management *superstructure* of practical administrative dichotomies faced by the nature of environmental risk. Here, again, we will focus on two of Page's (1978) characteristics: the *stealth* quality of environmental risk, with its imperative for other than responsive institutions, and the distinction of its *internal* market benefits vs. its *external* environmental costs which imply the need for autonomy-based rather than efficiency-based policy.

Lastly, we will integrate philosophical substructure and policy superstructure by expanding our working definition of "ecosystem" to focus on the interface between human systems and natural systems as these interact and affect one another. This will provide a new definition of Ecosystem that will redefine the policy space and offer a substitute Ecosystem Policy and Law in place of economically-motivated environmental decision-making. To illustrate the differences in origins and outcomes, I will, first, derive a Kantian alternative policy paradigm to contrast with the market model.

Environmental Risk: A Moral Substructure

Instrumental vs. Intrinsic Value and Modest Economic Benefits

A key characteristic of environmental risk related to its uncertainty is what Page (1978) calls its "relatively modest benefits." A moral dilemma exists in how we judge the "modest" nature of the benefits of a zero-infinity dilemma and justify a decision that denies them to those with a market preference for them. In other words, the policy choice requires that we justify why we will use the coercive power of the state to make everyone do X regardless of their preferences (Gillroy and Wade 1992: vii; Gillroy 2000). In justifying a collective choice, coercion needs to bring cooperation without significant or systematic hardship or widespread rights depravation. Avoiding tyranny (Fishkin 1979) is of primary importance for political theory, but the justification of a decision that benefits some at the expense of others is also a necessary result of

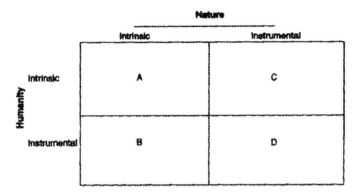

Fig. 1 Matrix of value: Humanity and nature

policy decision-making and the primary reason why economic policy design must move from the Pareto efficiency criterion to Kaldor efficiency.[4]

But the real inadequacy for the regulation of environmental risk policy is that economic policy design assumes the existence of only one type of value: instrumental. The concentrated focus on the standard of how much instrumental economic "benefit" (in terms of Kaldor efficiency, or wealth maximization) is accrued by choosing one alternative rather than another, as the major test of "good" policy, makes market decision-making ignore what is truly at stake in environmental risk issues (e.g., intrinsic value of both humanity and nature; need for anticipatory policy; etc.).

Alternatively, I suggest that we might consider that the uncertainty of risk issues is less critical to the decision process if we shift our analytic focus from the benefits themselves to the distinct types of values involved, and, more specifically, to tradeoffs that sort winners and losers. Instead of a world of trade between instrumental human values, where the metric of money and "willingness to pay" defines the difference between "modest" and "greater" benefits, let us expand the world to include both instrumental and intrinsic values to both humanity and the environment (Fig. 1).

Including the instrumental and intrinsic values for both humanity (moral) and the environment (functional) results in four combinations of risk trade-offs. They can produce potential winners and losers on two distinct levels of value and for two distinct "subjects" of concern. In cell A, the intrinsic value of humanity is weighed against the intrinsic value of nature. Here, we could be deciding between an old growth forest system and our need of lumber for

---

[4] Efficiency applied to public policy is not the standard Pareto Efficiency of microeconomics as it assumes that everyone is better or as well off in a new more efficient state-of-affairs. Since policy always produces winner and losers, a Kaldor-Hicks definition of efficiency was established for public policy that stipulates only that a new efficient state-of-affair is a potential Pareto improvement where the winners gain so much that they could "hypothetically" transfer wealth to make everyone as well or better off (see Gillroy and Wade, 1992: 6–3).

national security reasons. In cell B, the intrinsic value of nature is weighed against what has only instrumental value to humanity. Should we trade essential damage to the ozone layer for the convenience of propellants in spray cans? In cell C, what is of instrumental value to nature is weighed against what is of intrinsic value to humanity. For example, selective cutting of second-growth forests might be inconsequential to the persistence of the natural system, but essential to provide human shelter. Finally, in cell D, what is of instrumental value to both humanity and nature is traded. Shall we take an already used piece of urban land and build a mall?

In this model, the uncertainty of benefits and their implications for policy decision-making becomes more complex. But, simultaneously, uncertainty is also more definitive because we have anthropomorphically segregated what affects essential (intrinsic) value from what affects elective (instrumental) value, and are considering both humanity and nature in terms of their distinct characteristics and essential/elective requirements.

This creates a taxonomy of moral value replacing a single scale of instrumental welfare benefits. The uncertainty of symmetric outcome "benefits" is replaced with the certainty of intrinsic value and the responsibilities that occur with deciphering the essential from the elective and setting priorities between these. Consequently, the judgment and justification of "benefits" can be transcended to provide more options and arguments to the environmental decision-maker. In addition, the idea of "tyranny" now becomes intertwined with how one's public choices affect what is essential about humanity and/or the environment; concern for trading away intrinsic for instrumental value becomes a foundation of environmental risk decision-making. This segregates the most critical decisions, as in cell A where two intrinsic values, an old growth forest and human security, are the subject of policy from easier decisions, such as in cell B where the intrinsic value of the ozone layer is much more critical than propellants in spray cans.

This taxonomy of moral value also utilizes the attribution and evaluation of human-natural values as a moral means for the limiting of environmental risk uncertainty. In the same way that scientific "fact" can guide policy during a pandemic making decisions more certain, moral "value" arguments can create imperatives that promote the persistence and preservation of intrinsic value in the face of pervasive uncertainty. This makes environmental risk policy more flexible while giving it a non-contextual, non-time sensitive certainty that by providing a basis for anticipatory regulation, uses moral value to make decision-making more certain.

Private vs. Collective Goods and Catastrophic Results.

Another key dimension of uncertainty regarding environmental risk is its potential for catastrophic harm. In addition to a concern for the difference between intrinsic and instrumental value in public choice, consciousness of how a policy choice relates to degrees of private vs. public, or collective risk[5] and voluntary acceptance vs. involuntary imposition of risk, can further eliminate uncertainty by further defining "catastrophe" in environmental decision-making in terms of the context in which risk is being both generated and received. These should be factors in delineating different conditions of uncertainty if it is to be properly judged and regulated.

For market analysis, catastrophe has no meaning beyond that attributed to any considerable instrumental cost to humanity. Within a multi-dimensional tradeoff scheme, however, the catastrophe can be redefined by the distinction between private and collective goods. If we assume that a bad result is compounded if its effects are not only to intrinsic value, but also what economists call joint and nonexcludable "public" goods (Snidal 1979), then a catastrophe is more than just a greater than normal material cost to the individual. It compounds the potential harm to the essence of one's humanity and nature's fundamental functional integrity by the nature of the goods and choices involved.

In addition, if an environmental risk places everyone in harm's way, where one's subjection to it is joint with all, and where one suffers without consent or voluntary choice (which compromises their intrinsic value as a person), then one is discussing not merely benefit or cost, but harm that is collective, immoral and insidious, being unjustly and unknowingly inflicted on individuals without their knowledge. If one assumes, further, that harm to the intrinsic or essential value of humanity or nature is possible, and then contends that this harm may be collective and non-voluntarily transferred, this degree of harm can indeed be said to be potentially catastrophic in terms of harm to what is essential to being human. Uncertainty as to result may still prevail, but one can now place uncertainty about what is essential and collective in a separate class of considerations, *ex ante*, and both anticipate and justify their protection and empowerment in any environmental risk policy analysis.

By examining risk questions from within the dichotomy of collective and private goods transactions, we replace, first, the market's concern for the free-rider with a new consciousness of the non-voluntary risk imposed on the *imprisoned rider*[6] and, second, provide a priority for deciphering the various degrees of essential and collective harm involved in environmental risk. This new taxonomy may not lessen uncertainty, but it places it within a more complete and fitting context for the policy and legal matters at hand. For example, if we chart prospective risk policy decisions on a graph where the range from a fully collective choice to a fully private choice is on one axis, and a range from full voluntary acceptance to involuntary transfer of risk is

---

[5] While the distinction between public and private goods is well established in terms of jointness and nonexcludability, the idea of a collective good has more utility as it defines a good that is allocated or distributed by government, and justifies central regulation, regardless of its public/private characteristics.

[6] An imprisoned rider, as opposed to a free-rider is one who has harm (e.g., from environmental risk) imposed upon them without their ability to sense it and therefore to defend against it. (See Gillroy, 1972).

Fig. 2 Types of risk for policy choice

charted on the other axis, one can decide between more or less government regulation of risk by the degree the decision concerns collective involuntary environmental risk (like climate changes or a COVID pandemic.[7])

The southwest quadrant of this graph illustrates issues that involve voluntary private risks. Smoking cigarettes, for example, is a private choice that requires little government regulation as the risk is private to the individual smoking (in the closed room) and voluntarily accepted (by the adult smoker). The southeast quadrant of the graph shows risks that are still predominantly private, but which may be involuntarily transferred to the individuals involved. In these cases, like dealing with radon in a house, a bit more government involvement may be rational, at least to alert the individuals involved as to the private risks they are accepting by living in their home.

In the northwest quadrant where risk may be accepted voluntarily (sitting in a smoking section of a restaurant but not smoking oneself), risk is collective, affecting anyone in the vicinity who breathes the air. Because of this, we may require more government involvement to mitigate risk to otherwise innocent parties and protect collective interests.

Lastly, in the northeast quadrant of the graph, we find true zero-infinity dilemmas that involve collective involuntary risks, and collective non-voluntary exposure like the emission of radioactive air pollution or a pandemic. In these cases, catastrophe to the imprisoned rider is likely, and government

[7] See Note 2.

intervention to anticipate the risk, judge its necessity, and regulate it, is essential.

ENVIRONMENTAL RISK: A POLICY SUPERSTRUCTURE

Responsive vs. Anticipatory Institutions and Stealth

In addition to the pervasive uncertainty that characterizes environmental risk as a zero-infinity dilemma, it also provides management problems to the policy-maker who must justify the regulation of what amounts to an almost "zero" risk of an "infinitely" catastrophic exposure to harm. Among the most difficult policy issues regarding the administration of environmental risk, and the law that is codified from it, is the stealth quality of risk. Environmental risk affects individuals in intangible ways that make detection or individual defensive strategies nearly impossible. Dependence on the consciousness or preferences of the individual as a basis for policy, therefore, also becomes exceedingly problematic.

Preference is the currency of economic-market decision-making. Without the signals sent by preference, the market policy paradigm has no foundation for decision-making. Under the best of circumstances, it is difficult for an economist or policy-maker to distinguish between a true preference and a manifest preference (Sen 1982: Part II; Gillroy 1992a). Is the preference expressed by the person and made manifest to the policy-maker their real or true preference or does it involve strategic behavior or "gaming the system" to further self-interested ends? The surreptitious quality of environmental risk makes it even more difficult for a responsive government to decipher the true risk preferences of its' constituents (Page & MacLean, 1983), which is essential to policy formulation and implementation within the market paradigm.

The market paradigm assumes the sole function of government is to respond to the welfare preferences of consumers when markets cannot. However, even if an administrator could properly read individual preference, the stealth quality of environmental risk exacerbates the difficulties of using them as a basis for public choice. If one cannot sense environmental risk, then one cannot form a preference order with environmental risk as a factor. Even if informed of the existence of risk, it is difficult for one to measure cardinal or ordinal differences between risk preferences without an independent sense of each in terms of both presence and effect, which in any case is assumed to have a small possibility of producing catastrophic results.

Within the market paradigm, the stealth problem is further compounded by the connection between responsiveness and the burden of proof. In addition to the assumption that the only "ethical" and justifiable public decisions are based on the state's response to aggregate welfare preferences, market axioms also assume that the burden of proof is on the policy-maker to justify action that would intercede in the otherwise "normal" functioning of said market. The policy-maker should never intercede until specific market failure occurs. Even then, the state can only do what the market would have done; it cannot respond to unfulfilled preferences nor depose preference in favor of more secure non-instrumental values. To do so, within the market paradigm, would be an act of paternalism; it would create state action without authorization from consumers. For the market, responsive government, mimicking the market when it fails, is the only responsible government (Gillroy 1994).

The dilemma of waiting for preferences to form while environmental risk affects people collectively and irreversibly creates the collective, non-voluntary imposition of environmental risk characterized as the imprisoned-rider scenario (Gillroy 1992a, b). In order to overcome the stealth quality of environmental risk, and make policy that anticipates the imposition of environmental risk, we must move beyond the idea of "responsive" government as the only responsible government. The moral complexity of the concept of responsibility needs to be a recognized basis for policy choice so that the legal "burden of proof" in matters of regulation shifts to those who seek to impose environmental risk on society.

Specifically, in order to regulate *responsibly*, risk must be *anticipated* and *action justified ex ante*. The state, as the responsible party, must have a means to decipher what is of necessary value to its citizens and be enabled to protect or preserve these intrinsic environmental/human values from those who would place them at risk. The burden of proof in this approach to regulation changes from those regulating risk to those imposing it. Their responsibility is to justify the imposition of risk, to describe the potential effects on humanity and nature, and to gain political assent for each imposition of environmental risk.

To manage risk, a public decision-maker cannot rely solely on the expressed preferences of consumers, but must define and anticipate the moral values involved and the needs of citizens in terms of possible harm, in order to justify action *ex ante*. The state becomes more than a mere aggregator of market preferences. Anticipatory policy functions to define and provide the fundamental requirements of its citizens, as well as the natural world. Any actions that would infringe on or endanger these essential interests must be justified, not by cost–benefit analysis, but in terms of the intrinsic and instrumental values involved.

Efficiency-Based vs. Autonomy-Based Policy Instruments and Internal Benefits vs. External Costs

But how do we define the "essential" needs or interests of citizens? Confined to a responsive role, policy-makers have few tools at their disposal. In addition, the characteristics of environmental risk compound the question because most benefits are transmitted within the context of the market, while most costs are external and therefore not counted in market price computations (Page 1978).

Within the market paradigm, risk can only be addressed by internalizing the costs. But if these costs are unknown because of the stealth nature of risk, how can one internalize them? In addition, having the benefits of risk (e.g., spray cans, red apples) as immediate and tangible market factors prejudices the cost–benefit calculations because the superficial nature of economic benefits do not adequately take account of the stealth dimensions that characterize the essential costs or harms involved.

Efficiency-based policy instruments can only respond to expressed preferences and to the signals sent by price through the market. When we consider the factual characteristics of environmental risk policy, we face a serious dysfunction of these signals and a consequent exaggeration of economic benefits as opposed to the more essential harm to human and ecosystem values (e.g., the superficial need for cheap lumber against the keystone status of an old growth forest). We need a new set of signals that provide for an anticipatory state rather than the responsive market if the "external" costs of risk are too be appreciated in policy choice. These new signals ought to allow both the negative and positive effects of risk on essential citizen interests to be read properly and reported publicly *ex ante*.

Essential citizen's interests can be defined in terms of many philosophical concepts. For example, David Hume's philosophical argument defines the social utility of citizen's interests in terms of the evolution and persistence of social convention as a basis for a stable social order (Hume 1740); G.W.F. Hegel defines these interests in terms of individual freedom-through-social recognition and the creation of ethical life for a people within a state (Hegel 1820). Immanuel Kant creates the duty to the autonomy of the person as the core of their essential interests in terms of both their private ethical choices and the public or juridical choices of the state made on their behalf (Kant, MJ: 218–220). It is this final philosophical argument that I will import here as an example of how new moral-philosophical priorities could lead to an alternative paradigm of choice and anticipatory environmental risk policy.

I recognize that some might disagree with my characterization of philosophical argument and its utility as too broad and not applicable to contemporary policy issues.[8] However, I cannot stop to argue about the merits of finer textual analysis while the world is destroying itself. All I ask of the reader is that we agree on the need for a fundamental change to the moral standards used to define law and policy. We need to start somewhere, and the wealth of preexisting philosophical arguments can provide a basis for experimentation by the decision-maker. In the same way that "standards" provide a foundation for "jazz riffing," reconceptualizing the basic rhythm and melodies to see what else can be found or created, preexisting philosophical arguments can provide alternative conceptualizations of practical reason and moral agency as a basis

---

[8] They take a strict, contextual approach to philosophical literature. I consulted and studied with philosophers who have let me try out my ideas against their greater knowledge, pointed out my errors and validated my observations or insights where merited. I believe that philosophy has very concrete information to offer us on how to live and the fact that there is great work to be done to apply it to a contemporary context should not stop us.

for "policy riffing" that nonetheless pays respect to the essential "notes" of a dilemma like environmental risk.

To move from dependence on efficiency-based policy criteria and gain a foothold against the stealth quality of environmental risk and its external cost problems, we need an *autonomy-based* administrative system. Why autonomy-based? Because autonomy defines moral agency as capacity and the practical reason necessary to use it. It involves action, but can be defined to specifically address those actions that confirm and express what is essential to the person in terms of what empowers or protects their essential autonomy. Further, responsibility to autonomy requires the state to set an *ex ante* standard of regulation on the basis of what collectively provided goods are necessary to the material conditions of each citizen's moral autonomy, that is, what is in every human's essential interest as a being of moral-intrinsic value.

When applied to environmental risk policy, the concepts of essential interest and intrinsic value require the policy-maker to also consider the functional capacities of nature. The interfaces between moral agents and function-natural systems must become the focus of concern and the field where values must be assessed and tradeoffs contemplated. Public responsibility is no longer *responsiveness* to, but *ex ante responsible* for, protection and enhancement of human and natural capacities as these interact with one another.

As I have already argued, all human analysis is anthropomorphic. Yet we should construct a theory of human autonomy that includes what is unique and essential about nature without devaluing it. Anthropomorphically, nature can only have value as humans value it, or recognize duties toward it. But, because, as we will shortly see, of both its essential contribution to human freedom and its independent functional integrity, it ought to have equal functional-intrinsic value in policy deliberation and choice, even when no environmental crisis is looming on the horizon.

Autonomy-based criteria can account for environmental risk and all of its effects as part of a management decision. I suggest that to identify what is basic to all individuals, we focus on Kant's concept of moral integrity or "humanity-in-the-person" (Kant, MV: 388; 402–403; 448–450: C2; GW). If one asks what is most basic about one's humanity, it is one's internal capacity to act as an agent, or be autonomous. But not all choices or actions confirm one's humanity. A specific type of agency, that is, moral agency, needs to become the focus of our definition of essential interest and intrinsic human value in environmental risk policy. This necessitates that we take a closer look at Kant's philosophical argument.

KANT'S PARADIGM VS. MARKET PARADIGM IN A NUTSHELL

Here, I combine my knowledge of Kant's exegesis with my experience as a policy analyst. My goal is to integrate what I consider to be the necessary concepts and categories, or prerequisites, of proper policy choice with the components of Kant's philosophical argument for practical reason and human moral agency. I will compare and contrast these Kantian concepts with market assumptions about the same categories and concepts of policy-making. My purpose is not only to derive both the status-quo efficiency-based market paradigm for environmental risk policy but an autonomy-based Kantian alternative. With this alternative set of Kantian philosophical assumptions and values, we can create a philosophical substructure that is capable of transforming the management superstructure for environmental risk decisions toward what I will call "Ecosystem Policy & Law."

In order to decipher an alternative paradigm for policy choice from a preexisting whole philosophical argument, I've found that it's necessary to begin by identifying the three *fundamental assumptions* of any policy choice. First, the policy-maker needs to be aware of what they are assuming about the *individual* for whom they are making policy, that is, what definition the policy-maker uses to characterize the person's basis for choice and their motivations for agency. Second, how would policy respond to the *collective action* problems created by turning individual choices into public choices? Third, what role do the assumptions about the individual and collective action indicate about the proper *role of the state* in public policy decision-making.

From these fundamental assumptions, the next task is to derive a core *operating imperative*, or absolute presupposition of the paradigm structure. From the three assumptions and the operating imperative, the policy-maker moves on to identify the *material conditions* or empirical matter that will be required to implement the operating imperative. In addition, the core imperative provides a short-hand, policy and legal *priority* that can be employed to implement its recommendations.

Applying this approach to both the market and Kant will insure that the new values we have identified will be competitive with the status quo in justifying policy choice. This insures that the definitions of practical reason, as rendered through the operating imperative of *Kaldor efficiency* and its dictate to maximize wealth and *practical reason as autonomy* with the imperative to protect the intrinsic values of humanity and nature, will be sorted within a common framework that will be adequate to judge which is best applied to questions of environmental risk.

Before we proceed to detail this comparison, a fundamental distinction at the foundation of all policy and law must be posited. Specifically, approaching policy and law in this manner reveals an essential tension that permeates policy choice. Policy and law have a core dialectic structure[9] characterized by the tension between *process* ⇆ *principle*. This dialectic assumes that there are two fundamental categories of normative precepts in policy and law. The first and most basic is that which evolves from repeated human interactions that

---

[9] A philosophical-policy paradigm is assumed to be made up of dialectically interconnected ideas that overlap within a given philosophical system, while existing on a scale of forms, self-refining toward their essence over time through continued application and analysis. See, Collingwood, R. G. (1992/2005), 181–182; (1933/2005), 41–42; Gillroy (2000, 2009, 2013: Chapter 1, Gillroy, et al. 2008).

breed patterns of behavior and set expectations for the terms of cooperation and social stability: social conventions. This basis for moral value, as described by David Hume (1740: 486–488), creates the terms of social cooperation through practice and has three layers of progressive sanctions.

Sanctions assure cooperation and become more centralized as the size and complexity of the social group increases. The primary layer focuses on one's sense of honor and the need for social approbation (stage one); sanctions then become focused on a specific norm or convention of justice (stage two) that holds the system together as its moral-focal point and then, finally, sanctions produce formal law and institutionalized governance through contract-by-convention (stage-three). The point of the norms generated by this source of morality is to establish and maintain the stability of collective action based on a sense of the public good/utility. These are *process-norms* (Gillroy 2013: 12–13; 25–26), and they form that moral foundation for law that enables social coordination to stabilize cooperation over time. Social conventions informing/creating the positive law are easily entrenched as they become sensed as "legitimate" authority by those dependent on the legal system, and therefore as necessary to the fundamental stability or order of society. For our purposes, this helps explain why it is so difficult to change a set of essential values and their corresponding policy imperatives, especially when they have become conventional or traditional, as the market paradigm has in its application to environmental risk.

The second essential category of norms forming the moral foundation for the law is *critical principle*. These principles are transcendent of context and find their origins in argument from fixed metaphysical principle, generated and justified by human reason.[10] These *aetiological* norms or critical principles do not depend on their social context for legitimacy, but contain their own internal critical standard of validity that is inherently disruptive of social convention—primarily in the name of the status of the individual vis-à-vis the stability of social cooperation. Critical principles, while they inform both the substantive and procedural dimensions of the law, are primarily substantive because, instead of the stability of social process being the end-in-itself for these rules/ rights, it is the standing of humanity-in-the-person that provides their imperative.

Fundamental Assumptions: Sub-Dialectics Representing Process✕ ☞Principle

The three fundamental assumptions to policy analysis (individual, collective action, state), when examined through the lens of the *process✕☞principle* dialectic, further suggest three sub-dialectics. These represent the specific

---

[10] My project will rely on PPLD paradigms from Hegel and Kant for this normative category.

application of *process*✕☞*principle* to each fundamental assumption, so each can be deciphered more easily.

The first sub-dialectic, applied to defining the fundamental assumption of the individual, is between ***passion*✕☞*reason***. Human agency, our ability to act, is a balance, and sometimes a struggle, between these two characteristics derived from the proc*ess*✕☞*principle* dialectic. We assume that human agency contains both these components of character; the dialectic balance between them defines the person assumed to be the subject of policy and law.

The second sub-dialectic at play, related to the fundamental assumption defining collective action, is that between ***utility*✕☞*right***. Here, to establish and maintain public coordination, one asks how the transition from individual choice to collective outcomes balances the influence of the collective <u>utility</u> of social stability (*process*) against individual <u>right</u> (*critical principle*).

Lastly, the sub-dialectic related to the role of the state is the tension between ***active state*✕☞*passive state*.** Building on the assumptions about the individual and the collective action problem, an *active state* is charged not only with the protection of a private sphere, but the regulation of process in the name of critical principle; an active state supports not just the negative freedom of the person (i.e., freedom from interference), but the empowerment of their positive freedom (i.e., active moral agency in the world). Meanwhile, the *passive state* exists only to provide the legal background conditions of civil life, where only that negative freedom necessary to social cooperation is protected. Any further policy initiative is limited to mimicking the requirements of said process when they cannot be produced without public law.

Fundamental Assumptions: The Individual

A closer look at how these fundamental assumptions with their inherent dialectics play out in understanding and shaping policy is available through a comparative analysis of the market and my Kantian alternative. For the market paradigm and its foundations in classical economics, the individual is a "rational" consumer with welfare preferences where passion drives reason. The passion for wealth provides the core imperative for the individual. Kant's alternative provides the policy analyst with a more complex idea of the individual, one who is more than a collection of desires, or different "levels" of wants (Olson 1971; Sen 1974; Gillroy 2000).

Kant's individual is assumed to be a practical reasoner struggling to form an ethical character as an active moral agent on her own behalf. The internal character struggle is between the *predisposition* toward moral self-awareness in opposition to one's *propensity* to be a self-interested agent, which is the fundamental dialectic tension in Kantian Ethics (Allison 1990: 146–162; Kant RL: 22–3; 18–19; GW; C2; RL; PP; MJ; MV). This dual character describes a complex moral agent with the capacity and the predisposition to act in accordance with the maxim that she shall respect the moral autonomy or humanity in herself and others, but who is also a person in a material world of scarcity, uncertainty and fear which will condition her expression of autonomy.

The "reality" of being neither god nor animal but a combination of both defines the individual for Kant. The goals of the individual are not wants per se, but represent an overall imperative for self-empowerment through the expression of autonomy; Kant's person manifests their predisposition to act morally through the practical application of their moral agency (Kant, RL: BKI, 21–23; MV, 419–20 & Part I, Bk I, passim; C2; Allison 1990: 148–50). The focus for the policy-maker is now to create policy that supports the moral agency of the individual. By providing for this agency, the decision-maker protects and empowers the autonomy of all citizens.

Within Kant's definition of autonomy, or the moral integrity of the individual citizen, he also suggests that human existence within the "Realm of Freedom" is interdependent with the persistence of the "Realm of Nature" (Kant LT, 138–48: C3). Consequently, the policy-maker must also consider the functional integrity of Ecosystems made up of human and natural systems and their interactions (Gillroy 1996). Just as one's morally intrinsic value as an end-in-oneself is based upon practical reason and one's capacity, ability and purpose in becoming an autonomous person and a moral agent, the functional integrity or capacity, ability and purpose of nature as an end-in-itself must also be a policy priority.

This presents humanity with a primary obligation to use nature carefully and without waste (Kant, MJ: 443), because nature is critical to human autonomy and agency. This gives us what can be called Kantian conservationism (Gillroy 2000: 180). But Kant also suggests that the fundamental interdependence of humanity and nature requires respect for the intrinsic value or integrity of natural systems, because they are functional ends-in-themselves. This can be called Kantian preservationism (Kant RL: BK2,§1; LT: 143; OP: 21- 211; 22–549; MJ: 221–225; Gillroy 2000: 184). Kant's ideal is the "harmony" of the *Realm of Nature* with the human *Realm of Freedom* and because nature predates humanity as a functional and evolving end-in-itself, harmony can only be achieved if the unique intrinsic value of each Realm is recognized in policy and law.

Fundamental Assumptions: Collective Action

For the second assumption about identifying the nature of collective action, the market paradigm is skewed to the ***utility*** side of the utility⇆ ☞ right sub-dialectic. This concept of utility is no more than the aggregate sum of individual wants and preferences, while collective action is primarily concerned with establishing just enough cooperation to satisfy the material need of the most people possible. A Kantian view of the sub-dialectic examines the political community from the perspective of ***right*** over utility; that is, as a distinct collective entity where individuals (the moral building blocks of collective action) and the just state (which ratifies and enforces the collective terms of cooperation to insure the moral character of individuals) coordinate and reinforce one another. In Kant's alternative, the challenge with collective action is to encourage each individual's predisposition to act morally. The strategic situation is not a prisoner's dilemma, but an assurance game (Gillroy 1991, 2000: Chapter 6; Elster 1979) in which each citizen is assumed willing to cooperate in the production of morally-cooperative outcomes within a just state, which is more than the aggregation of welfare preferences (Kant, MJ: 255–56). Kant's state also has a duty to protect and treat each person as a moral agent. The political community, within a Kantian paradigm, is a necessary and vital actor.

Policy-makers using a Kantian approach to collective action should assume that when faced with the fear and uncertainty of exploitation in a community where no "public" regulation exists, the individual's underlying propensity toward "evil" or self-interested behavior will dominate one's predisposition to act morally (Allison 1990: Chapter 5). A policy-maker needs to be aware that the "externalities" of others' behavior and the subsequent fear of exploitation by these "external" market effects may cause the individual to ignore their predisposition toward cooperative action and move to protect their core of freedom in isolation, acting to exploit others before they are exploited themselves. This "mania for domination" (Kant, AT: 273) illustrates the political communities' failure to provide the conditions necessary to coordinate (an assurance game). But the Kantian decision-maker, anticipating this crisis with the priority to protect individual autonomy will regulate that economic behavior before it exploits essential moral capacity in some for the instrumental benefit of others. The collective action implications of the assurance game promote a distinct definition of "public interest" and are the reservoir of the moral precepts and principles that define the "right" as prior to the "good" and therefore the core of justice itself.

Fundamental Assumptions: The State

The place and function of the state, as the third fundamental assumption, are markedly different between a market paradigm and a Kantian paradigm. In a market paradigm, the *passive* state has only two functions: (1) to police and adjudicate property and contract matters, and (2) to provide a surrogate exchange system that can step-in when markets fail and allocations cannot be made without the involvement of a collective third party. For Kant's paradigm, the *active* state is a distinct entity, functioning independently and prior to economic markets, and existing to anticipate and regulate markets so they contribute to the "harmony of freedom" for all citizens (Kant, MJ: 230; TP: 297).

Within Kant's paradigm, each individual is assumed to have the capacity to recognize herself as a moral being. The resulting just state is for Kant a "Kingdom of Ends" (Kant, MJ:231; GW: 433–41) and is defined as that set of institutions and regulations that provide the assurance of those material conditions for the protection and empowerment of the moral capacities of its "active" citizens (Kant, MJ: §46). This responsibility requires the state to solve the assurance game and maintain cooperation over time. The "just" state empowers the individual's predisposition to act morally and defuses, or prevents, the "mania for domination" (Kant, AT: 273).

From this viewpoint, the state is more than the aggregate of individual preferences and is established to maintain an independent and duty-based "sense" of justice that insures its long-term existence and justifies its policy choices. The institutions in a just state are responsible to the individual; they support each citizen in seeking autonomy while protecting each against the immoral actions of others. The policy-maker can utilize Kant's assumptions about the individual, collective action and the just state to clarify their normative priorities and justify policy and law that anticipates environmental risk and regulates it with the dialectic interaction of the moral integrity of humanity and the functional integrity of nature at stake.

Policy-Law Operating Imperative

These foundational assumptions, taken together, render operating imperatives or distinctive absolute presuppositions for each paradigm that form the basis of policy choice for the decision-maker.

For the market paradigm, the protection and facilitation of each person's voluntary economic trade are of prime concern, and the principle of *Kaldor Efficiency* is used to support the maximization of aggregate social welfare. For Kant's paradigm, the principle of *autonomy* plays a similar role. Further, unlike efficiency, autonomy is divisible into three sub-principles, each corresponding to one of the levels of fundamental assumptions. The assumption about the individual relates to the principle of *freedom*, in both its negative and positive manifestations. Creating an assurance game to support collective action requires an adherence to the principle of *equality* before the moral law, that is, equality of all in terms of their being able to express rights or impose duties from the moral law on others. Finally, the attainment of a just and active state mandates recognition of the principle of civic *independence*. Freedom, equality and independence are what Kant calls the "juridical attributes" of the active citizen (Kant, TP: 290). They are the principles that the operating imperative of autonomy can offer to a policy-maker as standards for decision-making to assure autonomous citizens that they will not be exploited in their expression of moral agency and practical reason.

The three principles of freedom, equality and independence, like the categorical imperative (moral law) upon which they are based, all contribute to the same end: autonomy. Duty, individual rights and community interests are balanced by an active state attempting to protect and empower individuals as ends-in-themselves with intrinsic moral value. It is critical that a "thick" sense of moral autonomy (Gillroy 1992a) or "a higher order control over the moral quality of one's life" (Kuflik 1984) be supported by policy and law. It is also important that policy argument recognizes both the private and public dimensions of autonomy, in the three juridical attributes, as decision-makers strive to create a cooperative community of moral agents.

Material Instruments of the Policy-Maker

Operating imperatives require material conditions that can be used by the policy-maker to translate normative values and imperatives into practical goods and opportunities in the lives of citizens. In the market paradigm, the sole material condition that results from using the imperative of Kaldor efficiency is tangible property or wealth. For Kant, however, operating on the imperative of autonomy produces three instruments. Freedom of the individual is promoted by the *protection* of Ecosystem integrity (which is the consideration of both natural and human intrinsic value in policy choice). Equality within the political community is guaranteed through the *distribution and redistribution* of the physical property necessary to the widespread expression of autonomy through moral agency. Lastly, civic independence is empowered in the public provision of those *opportunities* necessary for anyone to apply their practical reason to both personal and political choice. The Kantian policy-maker is concerned with human and environmental integrity, collective and private property, and social opportunity, to provide for the "active" citizenship of their constituents.

Short-Hand Policy-Legal Priority

Operating with the imperative of Kaldor efficiency, the market paradigm requires the methodology of cost–benefit analysis so that the ends and means of public issues are submitted to the Kaldor efficiency test in order to render "efficient" policy. Using a Kantian approach, the decision-maker must protect integrity, distribute property and provide opportunity to empower active citizenship. This approach provides a *baseline* function as an alternative policy priority replacing dependence on Kaldor efficiency and cost–benefit.

This baseline function provides three alternative but interdependent variables for consideration in policy decision-making (Gillroy 2000: 276ff): Ecosystem integrity (in terms of collective goods (E^I); property (p_i); and opportunity (o_i). The baseline (E^I, p_i, o_i) represents Kant's argument that justice in collective policy choice requires government to consider, not an equal measure of each, but the basic protection of freedom, distribution of property and provision of opportunity to all. This baseline is necessary to solve the assurance situation and establish collective cooperation for the full expression of each person's practical reason and moral agency as active citizens. The charge of the policy-maker is to create those circumstances in which each citizen's predisposition toward acting morally is empowered, creating autonomy in each and a harmony of moral agency for all.

The baseline, required by a Kantian approach to law and policy, produces distinct policy choices and outcomes. For environmental policy, distributing the means to free expression, protecting equality and empowering independence move the burden of proof from the regulators to the creators of environmental risk. It also supports the use of standards to anticipate harm and protect basic agency and autonomous capacity in the face of collective assaults from market extraction, production and disposal.

In order to assure equal treatment of all citizens, the baseline requires the active state to make policy at the widest level of inclusion that is feasible (national or even international policy rather than local). It also, because of the interdependence of freedom and ecosystem integrity, requires the consideration of ecosystems as whole, integrated, interdependent systems with inherent functional integrity. This is an integrity that should be considered subject to harm whenever policy considers only one or some of its components for human use, non-holistically. The imperative of autonomy implies that regulation should be measured by the degree to which it provides for the flourishing of interdependent human ⇆ ☞ natural systems, rather than their minimal persistence as proscribed by the market maximizing wealth generation at the threshold of natural systems failure.

The baseline function illuminates the intrinsic functional value of natural systems and the intrinsic moral value of *humanity-in-the-person* and defines justice for the policy-maker in the balance of the intrinsic and instrumental values involved. Preference and/or the superficial freedom of consumer choice is consequently not as important as empowering those components of the baseline function that assure each and every person's freedom in terms of their essential capacity to choose. All policy choices now become subject to an ideal-regarding test: will this policy choice protect the intrinsic value of humanity⇆ ☞ nature?

For the policy analyst, utilizing a Kantian paradigm, the autonomy of the individual becomes the critical focus of policy. Human moral agency or autonomy is bound with environmental issues because the ability and capacity for human agency can be severely limited or eliminated altogether if nature's integrity is not also considered. The intrinsic value of both humanity and nature is essential in any environmental decision and must be protected and empowered by policy choice. Such an imperative produces duties and rights that take precedence over maximization of wealth and the processes that produce this wealth. In addition, the imperatives promote the citizen as practical reasoner over the consumer as rational maximizer. Considering the intrinsic value of the moral agent in policy choice requires that the community's assurance game be solved so that the cooperation of moral persons can persist across generations. Environmental degradation is a collective threat to the intrinsic value of individual autonomy and ecosystem integrity and should be considered a public phenomenon that can inhibit moral agency and trigger dysfunction in collective action. Under these conditions, the policy priority is to anticipate harm and regulate risk and other market externalities *ex ante* (e.g., harm from risk or contamination, commons problems).

The passive state promoted by the market paradigm can only approach risk after harm is detected, and only when the preferences for regulation become apparent to the policy-maker. This would not be true for policy made on the basis of Kant's imperatives. No longer driven by market assumptions and principles, the active state becomes independent in the provision and maintenance of collective action, and the protection and empowerment of individual moral agency. Risk becomes a "public" concern, not relegated solely to individual calculation. It is incumbent on the state to define what risks are collectively acceptable and which are not, based upon the requirements of the baseline function, to inform each citizen of the risks of any particular choice, and to provide for the regulation of those risks that would inhibit moral agency and therefore the individual struggle for autonomy.

Kant's paradigm helps us elevate the intrinsic value of humanity and nature over the instrumental value (i.e., price) of things and gives paramount importance to the creation of that public reality which empowers the expression of moral agency and the flourishing of natural systems integrity (Kant, MJ: 434–35). The maxims to protect freedom, distribute property and provide opportunity become our priorities when we consider the agency of the individual to seek their autonomy as central to policy and law. The environment of the political community, its capacity to persist and stop the exploitation of nature or of some of its citizens by others, may require that all risk-producing activity causing collective damage to environmental security be justified as supporting moral-baseline needs before it can begin or continue. This is a prescription for a *risk-conscious society*, where each collective risk is analyzed for its capacity to empower individual autonomy and support the functioning of natural systems before it is allowed into markets or the environment.

I have outlined the basic contrasting foundations of the status-quo market paradigm as it has been used to generate conventional environmental policy and offered an alternative paradigm, based on Kantian imperatives, as a more adequate response to the challenges of environmental risk. This fundamental change in the assumptions and values constituent of the philosophical substructure for policy design suggests a further change in the design approach to the superstructure of environmental management: *Ecosystem Policy and Law*.

Ecosystem Policy and Law

WE HAVE TRANSCENDED NATURE BUT BEAR DUTIES TO IT

The Ecosystem[11] design approach that I propose draws on a more inclusive definition of ecosystem than currently used. The sciences provide the predominant definition of ecosystem, attributed to Sir Arthur Tansley (1871–1955), an English botanist who "coined the term ecosystem for biotic and abiotic components considered as a whole" (Odum 1993: 38). Defined as a hierarchy within nature that includes both the organic (biotic) and inorganic (abiotic) components of the environment, ecosystems are assumed to be systematically interdependent (Odum 1975, 1993). For conservation biology, ecosystems are normally considered to be natural systems, that is, the environment within which animals, plants and other populations of organisms live. The distinction between human and natural systems provides the demarcation line for science between what they study and what they traditionally leave to the social sciences and humanities (Primack 1993).

To examine ecosystems as a tool of Kant's paradigm for autonomy-based policy design, one must transcend science because it does not give us enough information about the critical interface between humanity and nature. There are three ways in which one can understand the relationship between humanity and nature for the purposes of policy design. First, humanity is simply a part of nature, integrated into various ecosystems and considered as a component part. Taking the opposite view, ecosystems are the natural components of the biosphere, while humans and their artifacts are separate and distinct entities, outside the field of ecosystem studies. A third approach is to consider humanity as part of nature (i.e., humanity as interdependent with nature), but also transcendent, because our moral capacity and technological ability impose specific obligations upon us to preserve and protect the natural world from which we arose.

Placing humanity as merely another species within the natural world is the approach of many religious and tribal views of nature that assume a cosmology of man's awe and dependence upon nature as a point of departure (Grim 1983: Chapter 4). From this perspective, humanity is one among many species, all subject to extinction if they do not follow the "natural laws" of the ongoing ecological order, together with which they either survive or die. This picture of a dependent humanity as one within the myriad of species argues against technological evolution. This perspective rests on the truth that humanity originated in the natural world and remains a species of animal that, in the largest sense, depends upon the functioning of the biosphere for continued existence. However, this perspective fails to acknowledge that humanity, alone among the creatures of the earth, has moral capacity and extensive technological abilities which, together, impose specific moral and political responsibilities upon

---

[11] My idea for "Ecosystem Law" is not that created and applied without success by the US Forest Service in the 1990s under the Clinton administration. My argument is for fundamental change to the inherent principles at the core of environmental policy, not the reorientation of management based upon conventional, mostly market, assumptions. See Gillroy (2008: 202ff).

us as a particular species of animal with a unique capacity and ability to shape our environment.

A second common view of the relationship between humanity and nature is that humans are a "natural alien" (Evernden 1985). This view argues that, even though humanity has grown out of nature, we have established ourselves apart from the systemic interdependence of natural systems. Our independence is based upon the human ability to utilize technology to create our own environment "outside" the natural order.

> The consequences of technology are subtle but extensive, and one such consequence is that man cannot evolve with an ecosystem anywhere. With every technological change he instantly mutates into a new—and for the ecosystem an exotic—kind of creature. Like other exotics, we are a paradox, a problem for both our environment and ourselves. (Evernden 1985: 109)

The alienation of humanity from nature may be evaluated as either a good or a bad outcome. Most economists would view it favorably, since distinguishing humanity from nature and any interdependence, obligations or responsibilities, allows us to use nature, without moral strictures, toward the material enhancement of human life. For the market paradigm, evolving past nature to become "exotic" grants us our place at the top of the food chain, as that "creature" who has the power, and therefore the obligation, to make the best life possible utilizing current technology to exploit the environment as a resource.

Those who view the alienation of humanity from nature as a bad thing have a variety of reasons for this judgment. Some view this situation as a "truth" with moral ramifications, while others ignore the ethical dimensions, isolate humanity and focus on the ecosystems left behind by our exotic species as the primary subject of analysis. Among this latter group are most scientists who, when they speak of ecosystems, include humanity only as an afterthought or as a potential perturbation. Studies and experiments in natural science examine biological diversity of non-human plants and animals, and the interference of humanity in the otherwise ongoing evolution of natural systems, but do not fully integrate human and natural systems nor suggest that the former has specific obligations to the latter. The many sub-disciplines of the life and physical sciences studying the patterns and complexities of natural phenomena do so without specific concern for humanity's place (Abrahamson and Weis 1997), while fields like conservation biology examine humanity only in terms of our separate functioning as whole populations or societies (Primack 1993; Odum 1993; Chapter 6). We are never an integral part of a larger whole or unique individuals with moral capacities and value.

Acknowledging the moral ramifications of human alienation from nature provides a bridge to our third approach to the human-nature question: humanity has obligations to nature specifically based upon our capacity to evolve past its internal mechanisms.

> Man remains in nature even if the range of choice he enjoys seems incomparably greater than that of other species. Others are made to their world, while man must construct one with constant risk of error. (Evernden 1985: 118)

Our third approach provides a new definition of Ecosystem, based on Kant's argument that humanity is unlike any other product of nature on earth (Kant, C3: 429–439) because we are both moral and technological. Unique in the moral discourse of life, we alone hold ourselves to self-generated moral and legal strictures or standards of interrelations. We are also technologically singular, for even if other creatures can be recognized as using tools for construction (e.g., beavers, monkeys, ants), we have evolved the most sophisticated technology on earth, creating our own complex artificial environments.

This view of the humanity's ☞ nature dialectic also acknowledges that a complex two-way moral's ☞ functional interdependence characterizes our new sense of ecosystems. We are subject to natural constraints, but we place more complex and potentially devastating constraints on natural systems than any other species of animals. This dilemma has both empirical and normative implications for policy and law. We have transcended nature but bear duties to it, and our duties speak directly to how we utilize our knowledge, science and technology to create a life for ourselves on earth while respecting the pattern of natural systems that surrounds us and which produced us.

Understanding our duties can further assist policy-makers. But what provides the basis of our duties? Referring to the four-way table of values (Fig. 1), our obligations do not just concern the instrumental value of man to nature or nature to man, but must speak to what is essential, or intrinsically valuable, about ourselves and the natural world. Emphasizing the third perspective on the relation between humanity and nature, an Ecosystem from the Kantian approach needs to include the biotic and abiotic elements of both natural and human systems. This includes humanity's moral agency and its artifacts, machines, and social, political, legal, moral and economic constructions as well as plants, animals and their environment. The inter-systemic nature of this Ecosystem is located in the dialectic balance between human and natural systems, but the key to this "balance" lies in the moral responsibility specifically allocated to humanity to define its creativity and expressions of agency, conscious of both our own and nature's intrinsic value.

From a policy point of view, an Ecosystem is the dialectic intersection of human and natural systems which must be considered in public choice. Within a Kantian paradigm, our alienation forms an imperative to publically define our responsibilities as moral and technological creatures and to politically achieve that particular balance between human creativity and natural evolution that allows the essential intrinsic value of each to persist in harmony with one another.

Natural Systems: A Starting Point

Using Ecosystem Policy & Law, the policy-maker needs to consider the science of biological, chemical and physical systems and how they persist and evolve. But more importantly, they must be able to value them primarily for what they are, and not for what they can do for humanity.

To acknowledge the foundational imperative of natural systems is to acknowledge that all life begins with the evolution and progress of nature. The earth and its "life- support" systems (Drury 1998; Odum 1993) are a primary level of policy concern, not only because of the scientific or empirical functioning of these systems and processes, but for the purposes of making collective choices concerning them. Unlike the market paradigm with its foundation in the positivist distinction and devaluation of moral considerations in policy choice, both the *fact* and *value* of natural systems form the foundation of Ecosystem Policy. The market paradigm's fundamental devaluation of normative concerns for empirical analysis has its origin in David Hume's argument that what is (fact) cannot render what "ought" to be (value). This both distinguishes facts from values and promotes the former into prominence in policy analysis. However, Ecosystem Policy & Law arises from a diametrically distinct assumption. It posits that values can be derived from facts, that what "is" can directly determine what ought to be, that the normative imperatives and empirical facts of a policy are interrelated or dialectically interdependent. Therefore, in order for the Kantian approach to be operationalized, what has been called the "is-ought distinction" (Ryan 2021) must be overcome. This is a prerequisite to making risk regulation more anticipatory and inclusive of intrinsic values.

Hume contends that statements of fact cannot directly render value imperatives because one is distinct from the other (Hume 1740: 469). Applying Hume's contention to our subject-matter, the fact of an Ecosystem's persistence cannot lead to a moral imperative that it ought to be preserved. However, this perception of difference between is and ought is based on a very restrictive theory of morality which is not founded upon human reason, but on what Hume calls "the passions" (1740: Bk. II). Hume's premise, as I interpret it, assumes that morality is based in sentiment and that sentiment is alien to reason. Since reason is limited to the definition of empirical fact, while only the passions are associated with moral matters and agency, it follows logically that is and ought are constituents of distinct logical lines of thought process and argument and therefore one cannot be deduced from the other. However, because Kant promotes reason over passion and provides a non-sentiment-based foundation for the values involved, our Ecosystem design approach allows the policy-maker to move from situations of fact (e.g., a natural system is empirically unique) to moral imperatives (e.g., it ought to be preserved by public policy). To do so, we need to bypass the positivism inherent in Hume's assumptions and move toward the application of practical reason to public choice. Kant's paradigm allows for this move.

Hume assumes that reason applied to empirical reality is, by nature of the exclusive connection between moral value and sentiment, value-free. But is it? If practical reason not only delineates facts, but assigns value (based upon those facts), could a policy-maker not use a scientific assessment of a natural system to draw moral imperatives about that system? How might they do so? I suggest that the functioning and persistence of natural systems represent the moral value at issue. Nature is assumed to have no morality, which is an anthropomorphic attribute, but it can have value as a self-generated, self-perpetuating web of natural components. Within this interconnected systemic whole, each sub-system could be assumed to have value as it evolves through various states of homeostasis. Within Kant's paradigm, the fact of functionality then has essential or intrinsic value to the persistence of the whole. But here we are speaking of functional integrity, not moral integrity.

Nature is a functional entity that predates us, produced us and has the probability of continuing to exist long after us (Kant C3: §82–83). This functional independence is a fact, but a fact that compels a moral duty for humanity. Since we can disrupt this independent functioning, and also because we depend upon it, if we owe any duties to ourselves or others, then we also owe duties to nature, in terms of both the good of ourselves and the "good" of nature as an "other." Contrary to Hume's formulation, humanity's moral duties to nature can be deduced from the fact of natural systems' functionality (Gillroy 1996, 2000, 2013).

Pertinent from a policy point of view is the fact that natural systems function on their own. Morally, the primary value of these systems, internally or externally, requires that they be protected, empowered and allowed to persist over time. Since humanity is the only group of moral agents on the planet, it is our responsibility to assume this charge. The "facts" of nature define an intrinsic value that is not based on human contact or use of the environment, but is founded on the idea of nature as a functional end-in-itself. For the Kantian paradigm, the empirical persistence of natural systems and their internal functions have value to humanity, not because they are necessary to his economic prosperity, but because they are an essential prerequisite to his freedom. Practical reason assigns duty to action in support of the persistence of nature as an empirical world of cause and effect. Amoral within itself, nature is the subject of human moral responsibilities and obligations, due to our ability to cause disruptions in nature's empirical persistence over time.

If reason is a moral attribute of humanity, and it is possible that we can deduce an obligation from a set of facts, then both is and ought are subject to the same logical framework and can be part of the same policy argument. Hume's variant of the "naturalistic fallacy" (Moore 1903; Frankena 1967; Searle 1967) is, in reality, the fallacy that the sentiments are the only human capacity capable of moral motivation, action or evaluation.

Human Systems: Artifice and Obligation

Humanity deduces its obligations to nature from the fact of the empirical existence and functioning of natural systems. This contention assumes that humanity, as another level in the total Ecosystem policy argument, is capable of moral thought that combines reason and duty made imperative by critical moral principle. This suggests another dimension of the is-ought dilemma that is also addressed by Ecosystem Policy & Law.

All human creation is artifice, an addition to nature. For Kant's paradigm, the creation of the artificial, however, carries with it a moral capacity. First, we must set standards for our actions based upon our duties to ourselves and others (including nature) and, second, we must maintain these standards despite the exploitive possibilities suggested to us by our cumulative knowledge and our fabrication of technology. This ability to set standards suggested by duty and to maintain them reflects our capacity to decipher the difference between what *is* possible and what *ought* to be done. Specifically, our knowledge and ability to take a certain course of action do not automatically invoke it as a proper course to be chosen. The ability to construct nuclear weapons may be a matter of scientific and technical expertise, but the actual construction of these weapons and then their employment, in contrast to their scientific creation, are distinct *moral* choices.

To make this distinction between is and ought, a person must possess the moral capacity to judge the difference. But what is the nature of this moral capacity? From where does it arise? What obligations does it place on the moral agent? Using Kant's paradigm, I contend that this moral capacity is defined by the concept of autonomy, which is derived from and transcends our understanding of the persistent evolution of nature. Obligations arise from this moral principle which require the person to respect not only their moral autonomy and that of other humans, but the functional integrity of nature.

From nature, we learn that we are functional entities with an existence that has purpose as part of the ongoing evolution of the natural world. We begin to reflect about our needs and how we can continue to persist within the environment around us. We can also divide the normative world into its fundamental moral categories: social conventions or process-norms, and the critical principles based on non-contextual reasoning that amend these conventions (this is the aforementioned process ⇆ ☞ principle dialectic argued to be fundamental to policy and law).

We are first part and product of the causality of nature and this determinism dominates our thoughts, actions and beliefs. From this, we derive social conventions, as Hume pointed out. Our artifacts include justice as defined in terms of those conventions which allow us to anticipate the actions of others and cooperate toward shared goals (Hume 1740, 489–90). However, as Kant argues, in addition to the determinism of nature that is our ancestry, we also inherit the possibility of critical reflection that empowers us as agent-causes in the world. This is based upon the idea of freedom and the capacity to make decisions derived from one's own standards, not arising solely from convention, but from freedom which, for Kant, is a universal and necessary assumption for human reason (Kant, C1: B446-7; C2: 31–3). The human realization of the critical principle of freedom can be described as the true point at which the individual becomes the "natural alien" transcending the natural world. This is an ethical epiphany that results from the reflective process identified by Annas as originally Greek or Stoic.

> To live naturally, in fact, involves an inner change, without which outer changes are useless; we must start to reflect on what kind of being we are, what our needs are and so on. Clearly we cannot do this in a way which keeps ethics right out of it. The appeal to nature gives shape to a demand to come to terms with ourselves from an ethical point of view. . . ethical improvement is seen in terms of my increased rational ability to come to terms with and modify the given aspects of my life.
>
> (Annas 1993: 219–220)

In the inventory of needs and interests, we come to understand that, in addition to existing in the causal reality of the natural world, the person can also be a first cause himself in the satisfaction of needs and the creation of the world around him.

Kant's paradigm contends that, in order to be a human being, one must acknowledge freedom as the capacity to be an agent-cause in an otherwise deterministic world. This acknowledgment of one's "moral" capacity to have an effect on one's own and other lives evolves into one's capacity for autonomy. Autonomy then becomes the capacity to reason practically about one's existence and to act by intellectually approved principle in order to further one's ability to have a higher-order moral control over one's life (Kuflik 1984). Acknowledging freedom propels the person from the functional world of natural causality into the political and moral world of humanity and human communities. Because of the concept of freedom, one can assume obligations and duties, including those that utilize this freedom in ways that are not contrary to autonomy in the person or functionality in nature.

But how is nature part of human autonomy? As previously explained, all moral evaluation is human or anthropomorphic. I suggest that one can acknowledge the anthropomorphic fact of moral value without devaluing the non-human world. The idea of the evolution of morality and freedom evolving from preexisting nature gives us a conduit to establish a concept of autonomy that does just that: a theory that respects nature as part of human autonomy without devaluing it.

If nature is the point of origin for human ethical thought, it is because the existence of nature as an independent and self-generating entity gives it cause to continue, while it simultaneously gives us reason to protect and facilitate this evolution. If our realization of freedom comes from our reflection on this need, and this reflection itself begins with existence, then the first moral thought could be: because I exist, I ought to persist in that existence. This realization could not come without the experience of nature and our understanding of its systemic processes (e.g., life, death) from which we arose. Functional ecosystems are both necessary and sufficient for the continuation of natural wholeness or integrity. But while the same can be said of humanity physically, our ethical dimension makes functional existence necessary, but *not* sufficient, to the full moral agency or integrity of the person.

In Kant's paradigm, the advent of morality is the genesis of duty. Duty defines our obligations, but to what? If nature generates humanity, and we then transcend nature with our capacity to be a free agent-cause, then the subjects of duty fall nicely into two primary categories: our obligations to humanity and our responsibilities to nature. If, as Kant suggests, we are defined by our capacity to have and express freedom, and freedom is described in terms of autonomy as a higher-order control of one's life, then our duties to humanity would be to support autonomy in ourselves and in others. Our duty to nature could then be defined within either of these categories.

As a duty to ourselves, we might define nature as a necessary prerequisite to our freedom, as the context of our morality and the point of departure for both our physical and spiritual being. In these terms, nature is critical to who we are and therefore of vital importance to our essential freedom, or our capacity to be human. Our duty to nature is necessary to our autonomy; this instrumental value is not optional but essential. Nature is valued as much as autonomy; both are equal and interdependent in defining one's humanity.

However, our duties to nature are also independent of the instrumental value of nature to our autonomy, if we define obligation to nature as an external entity falling under our duties to others. The duty to other humans appears straight-forward as a duty to their autonomy. After all, if the standard of human morality in oneself is autonomy, it is reasonable that this same standard also marks others as moral ends in themselves and worthy of respect. Respect for nature however, since it lacks moral capacity or autonomy, must be defined by a distinct sense of intrinsic value, one that marks what is necessary and sufficient to the organic integrity, or whole intrinsic value, of nature. The functional integrity of nature offers such a definition.

I have argued that nature is a whole entity in terms of systemic function; it is a self-generating, persistent and evolving whole with many parts and subsystems. The wholeness of nature defines its intrinsic value and our moral obligation to respect the natural world independent of its instrumental use to us. To adequately evaluate and respect nature, the persistence and evolution of individual natural systems, apart from their human use, should become the gauge of our success.

Implementing Ecosystem Policy and Law

THE DOGMAS OF THE QUIET PAST ARE INADEQUATE TO THE STORMY PRESENT. THE OCCASION IS PILED HIGH WITH DIFFICULTY AND WE MUST RISE WITH THE OCCASION.

AS OUR CASE IS NEW, WE MUST THINK ANEW AND ACT ANEW.

WE MUST DISENTHRALL OURSELVES, AND THEN WE SHALL SAVE OUR COUNTRY.

ABRAHAM LINCOLN[12]

A Kantian alternative to the market approach to environmental risk is, alone, insufficient to establish an anticipatory, autonomy-based regulation of environmental risk. We have, perhaps, already imposed irreversible risks on ourselves and the biosphere. To reverse course on the way we approach the relationship between humanity and nature, we need to do two things: first, we need to formalize the thought process by which to derive, like the Kantian alternative, additional foundational paradigms for environmental regulation based on whole, systemic philosophical arguments. Second, with this method in hand, we need to drastically change our approach to environmental risk and environmental regulation in general; that is, we need to apply the appropriate paradigms to the appropriate dilemmas and "act anew" to save not just "our country" but the human race.

For 150 years, the social sciences have struggled to understand humanity from the vantage point of observing and experimenting with our superficial behavior and preferences. This phase of social science strove to understand the human status quo and worked from the obvious human character traits, basing its studies on such assumptions as self-interest, power and conflict. This mapping of the superficial human landscape was a critical prerequisite to an understanding of humanity, but, while necessary, it is not sufficient. It facilitates a basic knowledge of what one can observe about how we think and act, and the probable motivations and consequences of our actions. But it cannot adequately address more essential concerns about what constitutes human nature, how individual moral agency creates the social life we experience and what real constraints or incentives shape human life. Anyone who wishes to understand the human condition well enough to design policy and law that speaks to the intrinsic values involved and how our ideas become law through policy argument must not rely solely on science and positivism to limit uncertainty and justify legitimate policy. They must also understand the various pre-positivist philosophical arguments for definitions of practical reason and the concurrent imperative for agency that these philosophical systems demand, designing policy and law accordingly as the case requires.

---

[12] Address to Congress, December 1, 1862.

Unlike the positivist bias, that has made cost–benefit the primary point of departure for both critical and constructive environmental policy argument, this new approach to utilizing Enlightenment paradigms for policy choice creates a level of complexity necessary to understand both the status quo of what "is," and to make effective arguments about what "ought" to be. To facilitate change, we need this further depth of knowledge that allows us not only to decipher, but to effectively redesign the human condition, given the requirements of human agency that seem not just "real" but "required" to achieve a public end. This necessitates an understanding of the essence, or full complexity, of the human being and our natural context, as a basis for deliberation and choice in policy and law.

Specifically, social science, public policy and law have reached a level of maturity where they need to replace positivist methods and epistemology with a consideration of a completely distinct, prior, Philosophical Method.[13] This will more accurately illuminate not only what happens within the policy process, but what essential traits, principles and dialectics are involved in making policy and law and how these might be adjusted to initiate change and a more just or beneficial outcome. A new level of investigation is required, a philosophical investigation. A new philosophical method of analysis, complementary but prior to scientific method, will enable us to focus on the substructure of those physical circumstances and consequences that have been the preoccupation of social science to date. We need to lay out a fuller conceptual framework for empirical investigations. But uncovering the complexity of this philosophical substructure is a daunting task. Or is it?

Pre-positivist Enlightenment philosophers, like Kant, can provide logical maps of human nature, practical reason and moral agency. To empower effective change in human law and policy, we can, first, begin with these alternative arguments, or sets of premises, for human complexity. Second, we can harness them as alternative paradigms that allow us to translate essential understandings of the human experience into transformative policy and legal designs for the future. This is a call for a new era of *Philosophical-Policy & Legal Design* (PPLD).

With PPLD, we can implement Ecosystem Policy & Law through the use of many different paradigms with distinct definitions of practical reason generating different imperatives for agency, both in the service of evaluating and "designing" policy and legal change. Within this chapter, the characteristics of environmental risk have been examined and distinct Kantian goals have been set for an alternative design approach. Kant's PPLD suggests that a new design approach must be based upon the central role of anticipatory environmental regulation, autonomy-based administration and an expanded definition of Ecosystem in the policy decision-making process. Ecosystem has been redefined, not in terms of isolated natural systems and their interaction, but as the dialectic tension between natural and human systems. Having traced the

---

[13] See, Collingwood note 10.

origins of moral capacity in humans arising out of, and then transcending, nature, we subsequently suggested that there are two intrinsic values with which an Ecosystem approach must concern itself: human moral autonomy and the functional integrity of nature.

Environmental risk policy is riddled with uncertainty. That uncertainty is as much an ethical-moral, as a scientific, dilemma and understanding the full range of moral categories of value and the tradeoffs between those values is critical to making sound public choices. The administration of risk questions becomes a matter of sorting instrumental from intrinsic value and setting standards that protect the essential ethical and functional qualities of living things (e.g., keystone species, old growth forests, non-toxic air and water), while placing the burden of proof on those who would put them in jeopardy, especially for the promotion of elective market preferences.

The moral demands of uncertainty in risk decision-making require a theoretical paradigm that considers ethics as an anthropomorphic activity that regards nature in non-human terms without devaluing it. Further, uncertainty requires that policy design encompasses intrinsic as well as instrumental value and the tradeoffs between these. The policy choice should be characterized by its concern with collective goods and what action, at the baseline, will solve the coordination problems involved. Decision-makers should elevate both the facts of environmental dilemmas and the values involved, promoting the inherent values of acts over the consequences they produce (e.g., protecting bio-diversity through wise-use over classification of habitat by considerations of its economic benefit and cost).

Within the context of Kant's PPLD, we can define duties to both humanity and our intrinsic value, but also to the environment both as a necessary instrumental ingredient in human autonomy and as an independent functional end in-itself. No longer do we separate the environment into resources and sectors from the point of view of the economy. Instead, we design law and policy that divide the economy into extraction, manufacturing and disposal interfaces and examines all law that is relevant to a particular socio-economic function simultaneously (e.g., not distinct clean air and water acts, but a comprehensive pollution from manufacturing and use act).

The demands of Kantian environmental risk management are facilitated by this new design model. The requirements of risk anticipation and planning, within a context of autonomy-based policy standards, requires the economy to serve the environment and not the reverse. Policy is refocused onto the dialectic of the intrinsic values involved. The risk-conscious society requires that each potential risk be analyzed both scientifically and morally in order to replace pervasive risk with a degree of protection and certainty. This consciousness further requires that assessments be made to determine how human activity affects the interface between humanity & nature, *ex ante*, before a product or technology is allowed into the environment.

Considering the stealth, latency and irreversibility problems of environmental risk, our alternative paradigm offers a design model that defines the law in terms of a core focus on intrinsic value and the hierarchy of nature, state and then market. Kant's PPLD paradigm will provide an administrative model comprehensive enough to examine environmental characteristics and their economic requirements given the specific function at hand (e.g., extraction, manufacture or disposal). In this design model, the active state coordinates human collective action in terms of our individual and social duties to the balance of essential capacities between humanity & nature.

Only with a change in the foundational assumptions of our policy paradigm can legal design transcend the conventional design model where economic principle and assumptions determine the boundaries of politics and the uses of the environment on a media by media or resource by resource basis. In designing policy and law, Kant's PPLD provides a moral theory that can accommodate intrinsic as well as instrumental value, where the latter is not limited to economic definition. In order to answer the questions generated for the public manager (e.g., what can be designated a resource? What can be extracted? How will this be done?), Kant's PPLD sets distinct moral and non-economic standards for choice, making its philosophical capabilities more adequate than those of the market paradigm, for making policy in the complex world of environmental risk.

These questions make no sense in a simpler policy space where all natural material is subject to extraction by any efficient means given "willingness to pay." To the principle of Kaldor efficiency as the operating imperative of the market paradigm, everyone and everything is a resource and subject to extraction and use, limited only by the preferences and technology available. Such a paradigm cannot accommodate the preservation of a natural system from resource exploitation. It has no way of justifying a non-resource decision and it cannot argue that our use is trumped by our duty to the intrinsic value of a natural system. Comparing the market and Kantian alternatives through the lens of PPLD, one can see the differences based on sorting each model's essential values and assumptions into a common matrix (Fig. 3).

Kant's PPLD provides for a risk-conscious society rather than a no-risk or profit-risk evaluation standard. This new strategic model can be employed to provide a standard for identifying and defining acceptable resource use and how and if nature ought to be extracted, polluted or put at risk, given our dual responsibility to ourselves and to nature. The new design model provides a workable framework with which we can "think anew" on a foundation of autonomy-based ethics and Ecosystem Policy & Law and "act anew" in terms of the anticipatory-managerial demands of environmental risk. It only remains to ask: can we "disenthrall ourselves" from established practices and positivist conventions before it is too late?

PPLD: Market

The Individual *Passion ꟻ Reason*	Collective Action *Utility ꟻ Right*	Role Of The State *Passive ꟻ Active*
Self-Interested Consumer Where Passion Is Driving Reason	PRISONER'S DILEMMA: SOLVED BY INVISIBLE HAND: Collective Action In the Name of Social Welfare/Utility With Merely A 'Thin' Theory Of Right As Wealth Preference	MINIMAL STATE: Passive To Empower Markets & Mimic Them When Market's Fail

Policy/Law Operating Imperative = KALDOR-HICKS EFFICIENCY: Hypothetical Transfer As A Potential Pareto Improvement

Material Instruments of The Policy-Maker = PRIVATE PROPERTY-Goods Priced By Money in Terms Of Supply-Demand. Allocated To Maximize Or Optimize Wealth

Shorthand Policy-Legal Priority = COST-BENEFIT As Equation For Present Value

$$PV = \sum_{t=?} B_t - C_t / (1+r)^t$$

PPLD: Kant

The Individual *Passion ꟻ Reason*	Collective Action *Utility ꟻ Right*	Role Of The State *Passive ꟻ Active*
Moral Agent: Practical Reasoning Potentially Moral Agent With a Predisposition To Act Ethically & Predilection To Self-Preservation	Assurance Game: Duty To Facilitate Right Over Utility In The Empowering of Moral Agency	Active State: Anticipating Requirements of Justice From Autonomy In Providing Assurance For Active Citizenship

Policy/Law Operating Imperative = {Justice-From-Autonomy} Ethical ꟻ Juridical ⇒ Principle of Right

Material Instruments of The Policy-Maker = {Innate Right To Achieve Active Citizenship}
Freedom = Protecting Ecosystem Integrity
Equality = (Re) Distribution of Property
Independence = Providing Opportunity

Shorthand Policy-Legal Priority = The Baseline ⇒ (E^1, p_1, o_1)

Fig. 3 Comparative paradigms: Market vs. Kant's PPLD

References

Abrahamson, Warren G. and Arthur E. Weis. 1997. *Evolutionary Ecology Across Three Trophic Levels*. Princeton: Princeton University Press.

Aharoni, Yair. 1981. *The No Risk Society*. Chatham: Chatham House Publishers.

Allison, Henry E. 1990. *Kant's Theory of Freedom*. New York: Cambridge University Press.

Anderson, Charles W. 1979. "The Place of Principles in Policy Analysis." *American Political Science Review* 73: 711–723. Reprinted in Gillroy and Wade 1992, Pp. 387–410.

Annas, Julia. 1993. *The Morality of Happiness*. New York: Oxford University Press.

Aristotle (PL). 1905. *Politics* (Trans. by Benjamin Jowett) Oxford: Clarendon Press.

Andrews, Richard N.L. 1999. *Managing The Environment, Managing Ourselves: A History of American Environmental Policy*. New Haven: Yale University Press.

Aune, Bruce. 1979. *Kant's Theory of Morals*. Princeton: Princeton University Press.

Axelrod, Robert. 1984. *The Evolution of Cooperation*. New York: Basic Books.

Bohman, Brita. 2021. *Legal Design for Social-Ecological Resilience*. Cambridge University Press.

Bobrow, Davis B. and John S. Dryzek. 1987. *Policy Analysis By Design*. Pittsburgh: University of Pittsburgh Press.

Buchanan, Allan and Gordon Tullock. 1963. *The Calculus of Consent*. Ann Arbor: University of Michigan Press.

Buck, Susan J. 1991/1998. *The Global Commons: An Introduction*. Washington D.C.: Island Press.

Campbell-Mohn, Celia, Barry Breen and J. William Futrell. 1993. *Environmental Law from Resources to Recovery*. St. Paul, MN: West Publishing.

Collingwood, R. G. 1933/2005. *An Essay on Philosophical Method*. Oxford Clarendon Press, 1933.

———. 1992/2005. *The New Leviathan or Man, Society, Civilization & Barbarism*. Revised ed. Oxford: Oxford University Press, 2005.

Conservation Foundation. 1985. *Risk Assessment and Risk Control*. Washington, D.C.: Conservation Foundation Press.

DiIulio, John H. Jr. 1994. (ed.) *Deregulating the Public Service*. Washington, D.C.: Brookings Press.

Downs, Anthony. 1957. *An Economic Theory of Democracy*. New York: Harper and Row.

Drury, William Holland Jr. 1998. *Chance and Change: Ecology For Conservationists*. Berkeley, CA: University of California Press.

Elster, Jon. 1979. *Ulysses and The Sirens: Studies in Rationality and Irrationality*. Cambridge: Cambridge University Press.

Evernden, Neil. 1985. *The Natural Alien: Humankind and Environment*. Toronto: University of Toronto Press.

Fishkin, James S. 1979. *Tyranny and Legitimacy: A Critique of Political Theories*. Baltimore: Johns Hopkins.

Frankena, William K. 1967. "The Naturalistic Fallacy" in Foot. *Theories of Ethics*. Pp. 50–63.

Gillroy, John Martin. 1991. "Moral Considerations and Public Choices: Individual Autonomy and the NIMBY Problem." *Public Affairs Quarterly* 5: 319–332.

Gillroy, John Martin. 1992a. "The Ethical Poverty of Cost-Benefit Methods: Autonomy, Efficiency and Public Policy Choice." *Policy Sciences* 25: 83–102. Reprinted in Gillroy and Wade 1992. Pp. 195–216.

Gillroy, John Martin. 1992b. "Public Policy and Environmental Risk: Political Theory, Human Agency, and the Imprisoned Rider." *Environmental Ethics* 14: 217–37.

Gillroy, John Martin. 1992c. "A Kantian Argument Supporting Public Policy Choice" in John Martin Gillroy and Maurice Wade, eds. *The Moral Dimensions of Public Policy Choice: Beyond the Market Paradigm.* Pp. 491–515.

Gillroy, John Martin. 1993. (ed.) *Environmental Risk, Environmental Values and Political Choices: Beyond Efficiency Tradeoffs in Policy Analysis.* Boulder, CO: Westview Press. [Reissued by Routledge Publishing 2018].

Gillroy, John Martin. 1994. "When Responsive Public Policy Does not Equal Responsible Government" in Robert Paul Churchill, ed. *The Ethics of Liberal Democracy: Morality and Democracy in Theory and Practice.* Oxford: Berg Publishing.

Gillroy, John Martin. 1996. "Kantian Ethics & Environmental Policy Argument: Autonomy, Ecosystem Integrity and Our Duties To Nature." *Ethics & The Environment* 3: 131–58 (1998).

Gillroy, John Martin. 2000. *Justice & Nature: Kantian Philosophy, Environmental Policy, and the Law.* Washington D.C.: Georgetown University Press.

Gillroy, John Martin. 2009. "A Proposal for 'Philosophical Method' in Comparative and International Law." *Pace International Law Review.*

Gillroy, John Martin. 2013. *An Evolutionary Paradigm For International Law: Philosophical Method, David Hume & The Essence Of Sovereignty.* New York: Palgrave-Macmillan.

Gillroy, John Martin and Maurice Wade. 1992. (eds.) *The Moral Dimensions of Public Policy Choice: Beyond the Market Paradigm.* Pittsburgh: University of Pittsburgh Press.

Gillroy, John Martin and Breena Holland with Celia Campbell-Mohn. 2008. *A Primer For Law & Policy Design: Understanding The Use Of Principle & Argument In Environmental & Natural Resource Law.* West: American Casebook Series.

Gordon, George J. 1992. (4th ed.) *Public Administration in America.* New York: St. Martin's.

Graham, John, Laura Green, and Marc Roberts. 1988. *In Search of Safety: Chemicals and Cancer Risk.* Cambridge, MA: Harvard University Press.

Green, Donald P. and Ian Shapiro. 1994. *Pathologies of Rational Choice Theory.* New Haven: Yale University Press.

Hardin, Garrett. 1968. "The Tragedy of the Commons." *Science* 162: 1243–45.

Hardin, Garrett. 1993. *Living Within Limits.* New York: Oxford University Press.

Hardin, Russell. 1971. 1982a. *Collective Action.* Baltimore: Johns Hopkins University Press.

Hardin, Russell. 1982b. "Difficulties in the Notion of Economic Rationality." *Social Science Information* 23: 436–67. Reprinted in Gillroy and Wade 1992. Pp. 313–24.

Hargrove, Eugene. 1992. "Environmental Ethics and Non-Human Rights" in Eugene Hargrove, ed. *The Animal Rights—Environmental Ethics Debate.* SUNY Press.

Hegel, G.W.F. [1820] 2008. *Outline of the Philosophy of Right* (ed. T.M. Knox; Trans. S. Houlgate). Oxford: World Classics.

Henderson, David E. 1993. "Science, Environmental Values, and Policy Prescriptions" in John Martin Gillroy, ed. *Environmental Risk, Environmental Values, and Political Choices.* Pp. 94–112.

Hume, David. [1740] 1975. *A Treatise of Human Nature* (ed. by L.A. Selby Bigge) Oxford: The Clarendon Press.
Kant, Immanuel. Gesammelte Schriften. Berlin: Prussian Academy of Sciences. {All references are by Academy Page, Book [BK] Section [§] and/or Volume [V]}. The following initials, dates and volume numbers indicate particular texts.
Kant, Immanuel. (GW) 1786. Groundwork For A Metaphysics of Morals. [V4].
Kant, Immanuel. (C1) 1787. First Critique: Critique of Pure Reason. [V3(A)/V4(B)] (C2) 1788. Second Critique: Critique of Practical Reason. [V5] (C3) 1790. Third Critique: Critique of Judgment. [V5].
Kant, Immanuel. (TP) 1792. Theory and Practice. [V8].
Kant, Immanuel. (RL) 1793. Religion Within The Limits of Reason. [V6] (PP) 1795. Perpetual Peace. [V8].
Kant, Immanuel. (MJ) 1797. Metaphysics of Morals: Principles of Justice. [V6] (MV) 1797. Metaphysics of Morals: Principles of Virtue. [V6] (AT) 1800. The Anthropology. [V7].
Kant, Immanuel. (OP) 1803. Opus Postumum. [V21; V22].
Kuflik, Arthur. 1984. "The Inalienability of Autonomy." *Philosophy and Public Affairs* 13: 271–298. Reprinted in Gillroy and Wade 1992. Pp. 465–90.
Laver, Michael. 1981. *The Politics of Private Desires*. New York: Penguin Books.
Laver, Michael. 1986. *Social Choice and Public Policy*. Oxford: Blackwell.
Lehman, Scott. 1995. *Privatizing Public Lands*. New York: Oxford University Press.
March, James and Herbert Simon. 1958. *Organizations*. New York: John Wiley & Sons.
Moore, G.E. 1903. *Principia Ethica*. Cambridge: Cambridge University Press.
National Research Council. 1983. *Risk Assessment in the Federal Government: Managing the Process*. Washington, D.C.: NRC Press.
Odum, Eugene P. 1975. *Ecology: The Link Between the Natural and Social Sciences*. New York: Holt, Rinehart & Winston.
Odum, Eugene P. 1993. (2nd. ed.) *Ecology and Our Endangered Life-Support Systems*. Sunderland, MA: Sinauer Assoc.
Olson, Mancur. 1971. *The Logic of Collective Action*. Cambridge MA: Harvard University Press.
Ostrom, Elinor. 1990. *Governing the Commons: The Evaluation of Institutions for Collective Action*. Cambridge: Cambridge University Press.
Page, Talbot. 1973. *Economics of Involuntary Transfer*. Berlin: Springer-Verlag.
Page, Talbot. 1978. "A Generic View of Toxic Chemicals and Similar Risks." *Ecology Law Quarterly* 7: 207–244.
Page, Talbot and Douglas MacLean. 1983. *Risk Conservatism and the Circumstances of Utility Theory*. Pasadena, CA: California Institute of Technology. Mimeo.
Primack, Richard B. 1993. *Essentials of Conservation Biology*. Sunderland, MA: Sinauer Associates.
Rodricks, Joseph V. 1992. *Calculated Risks: The Toxicity and Human Health Risks of Chemicals in Our Environment*. Cambridge: Cambridge University Press.
Ryan, Philip. 2021. *Facts, Values, Policy*. Polity Press.
Schattschneider, E. E. 1960. *The Semi-Sovereign People*. Hinsdale, IL: The Dryden Press.
Searle, John R. 1967. How to Derive "Ought" from "Is" in Phillipa Foot, *Theories of Ethics*. Pp. 101–14.

Sen, Amartya K. 1974. "Choice, Orderings and Morality" in Stephen Korner, ed. *Practical Reason*. New Haven: Yale University Press, Pp. 54–67.

Sen, Amartya K.1982. *Choice Welfare and Measurement*. Cambridge, MA: Harvard University Press.

Shrader-Frechette, K.S. 1993. *Burying Uncertainty: Risk and the Case Against Geological Disposal of Nuclear Waste*. Berkeley, CA: University of California Press.

Simon, Herbert A. 1982. *Models of Bounded Rationality* Vols. 1 & 2. Cambridge, MA: MIT.

Smart, J.J.C. and Bernard Williams. 1973. *Utilitarianism: For and Against*. Cambridge: Cambridge University Press.

Snidal, Duncan. 1979. "Public Goods, Property Rights, and Political Organizations." *International Studies Quarterly* 23: 532–566. Reprinted In Gillroy and Wade 1992.

Taylor, Bob Pepperman. 1992. *Our Limits Transgressed: Environmental Political Thought In America*. Lawrence, KS: University Press of Kansas.

Viscusi, W. Kip. 1983. *Risk by Choice*. Cambridge, MA: Harvard University Press.

Viscusi, W. Kip. 1992. *Fatal Tradeoffs: Public and Private Responsibilities for Risk*. New York: Oxford University Press.

Willig, John T. 1995. *Environmental TQM*. New York: Wiley.

Tracing Instrumented Expert Knowledge: Toward a New Research Agenda for Environmental Policy Analysis

Magalie Bourblanc

> If sociology is to develop a deep and enduring interest in the relations between humans and our environment, one that is not heavily dependent upon the current level of societal attention to environmental problems, we need to overcome our deep-seated assumption that our species is separate from the rest of nature and exempt from ecological constraints […C]urrent efforts to analyse the constructions, and ignore the realities, of global environmental change make our field especially susceptible to […] trendiness [or whatever problem happens to be garnering national attention] (Dunlap and Catton, 1994, p. 24)

Nowadays, more and more scholars have started taking on board the epistemological, theoretical, and conceptual challenges that global environmental problems may pose to social sciences (Blanc, Demeulenaere, Feuerhahn, 2017). This has prompted some authors to reflect on the best way to somehow bring the material back in social sciences (Catton and Dunlap, 1978). At the same time, *Science and Technology Studies* (STS) have taught us that objects of nature are seldom accessible, unmediated, and are only perceptible or

---

M. Bourblanc (✉)
CIRAD (French Agricultural Research Center for International Development), UMR G-EAU, Université de Montpellier, Montpellier, France
e-mail: magalie.bourblanc@cirad.fr

CEEPA (Centre for Environmental Economics and Policy in Africa), University of Pretoria, Pretoria, South Africa

© The Author(s), under exclusive license to Springer Nature Switzerland AG 2023
J. J. Kassiola and T. W. Luke (eds.), *The Palgrave Handbook of Environmental Politics and Theory*, Environmental Politics and Theory,
https://doi.org/10.1007/978-3-031-14346-5_13

understandable through scientific knowledge. Hence, for STS, it is difficult to determine what is an environmental problem, since the sciences' practices do not limit themselves to represent nature in a neutral and objective way. Having emphasized how much knowledge production is a highly politicized process, even more so for expert knowledge, STS scholars working on the environment consider environmental issues primarily as social constructs.

However, because of their insistence on environmental scientific expertise as social constructs mainly, they essentially follow actors to determine the potential influence of such expertise and its technical objects within public policies. To move away from the self-referentiality of the social though, I posit that we need to take seriously the effects of such expert objects on environmental policies, sometimes independently of actors' preferences or interference.

Hence, I elaborated the concept of instrumented expert knowledge which allows me to promote a less anthropocentric approach of environmental policy analysis while at the same time recognizing the hybrid nature of instrumented expert knowledge characterized by both social and material influences.[1] This concept cross-fertilizes the insights of a critical sociology of knowledge on the one hand and the insights of public policy instrumentation on the other. I argue that it is a good entry point to study traces of the material dimension within environmental public policies.

In the subsequent sections, I, first, evoke other experience of "bringing the material back in" social sciences. Then, I focus on the different approaches that take seriously the production of knowledge and study the role of expert knowledge in particular within environmental public policies. Finally, I discuss the specific advantages of the concept of instrumented expert knowledge. On the one hand, alongside STS, I emphasize the way specific knowledge shape our understanding of environmental issues and ways to tackle them; on the other hand, alongside the policy instrumentation approach, I underline potential lock-in effects linked to expert reporting's equipment, ensuring that specific expert knowledge does not end up being sidelined. Thus, this cross-fertilization enables me to invest in the sociology of technical objects, a less developed tradition than the sociology of science within STS. It allows me to contribute to the recent "infrastructural turn" in STS (Star, 1999; Graham, 2009; Larkin, 2013).

THE "MATERIALIST TURN" IN ENVIRONMENTAL SOCIOLOGY AND POLITICAL ECOLOGY

In the 1970s, American Environmental Sociology held the theoretical ambition to transform mainstream sociology by rediscovering *"the material embeddedness of social life"* (Buttel, 1996). Indeed, contrary to Durkheim's demand to explain social phenomena only in terms of other social facts, they were

---

[1] As socio-technical devices or equipped knowledge, they can materialize in indicators, information systems, expert knowledge translated in some kind of metric systems, etc.

willing to grant biophysical variables a role in the explanation of social phenomena. Seeking to break free from what they called the "human exemptionalist paradigm," American environmental sociologists such as Dunlap and Catton (1978) advocated the need to rethink the interdependent relationship between human societies and their ecosystems. For years, heated debates have opposed proponents of this realist tradition to advocates of a more constructivist epistemology in the European environmental sociology. One recurring source of dispute revolved around the nature of environmental problems and to what extent they constitute specific social objects.

Indeed, in social problems theory, authors emphasize how much social problems depend on social mobilization to define some conditions as a problem (Spector and Kitsuse, 1977; Cobb and Elder, 1983). No objective conditions can be invoked to explain the reasons why a problem is set on the public agenda. For numerous environmental sociologists (Yearley, 1991), this also applies to environmental problems as well. In the most extreme cases, some scholars adopted a post-modern conception, even contesting the very existence of an environment outside of human experience. Nowadays, however, hardly any scholars within *European environmental sociology [would] doubt the existence of a biophysical reality, but what this reality is and what it means are always and only actualised through social practices and interpretative processes* (Lidskog et al., 2015, p. 346).

Opposition between these two schools has mainly receded in the 1990s. For Carolan (2008), such opposition could essentially be considered as a misunderstanding as far too often scholars *were speaking past each other. That is, the [social constructivists] were concerned primarily with making epistemological statements - e.g., 'Our knowledge of environmental problems is socially mediated!'–while [the realists] tended to emphasise ontological claims.–e.g., 'environmental problems are real!'*[2] (Carolan, 2008, p. 71). Being now dominated by European environmental sociologists, the opposition between the two traditions in the discipline has mainly phased out. So has, too, the theoretical ambition to include the *material* into environmental sociology's analytical framework. This is not surprising. As Mol bluntly put it, *whereas U.S. environmental sociologists were more worried about getting environment into sociology, European environmental sociologists were preoccupied with getting sociology into studies on the environment* (Mol, 2006, p. 11).

Partly, this can be explained by the fact that European environmental sociologists convincingly demonstrated the limits of a too narrowly positivist conception of environmental problems and their complexities. Drawing on the sociology of science and STS in particular, they have shown that the way we can apprehend environmental realities is often mediated by science, hence potentially subjected to bias. For them, the natural sciences and their

---

[2] Epistemology relates to the study of knowledge and ways of knowing about the world; ontology studies the nature of reality and of our existence.

representations should not be taken as neutral inputs into environmental sociology.

Critical Political Ecology has among other inspirations also built on the lessons learnt from STS and taken on board its awareness about the partly constructed nature of environmental sciences. Since its emergence in the 1970s, the primary objective of *Political Ecology* is precisely to (re)-politicize environmental studies and offer a political explanation to scientific narratives about environmental degradation. Its project has also embraced a natural turn. Following a kind of critical realism, it defines the environment: "*as having an ontological basis and a dynamic role as an agent in its own right, combined with our understanding of nature's agency as socially mediated [...]*" (Zimmerer and Bassett, 2003, p. 3). Hence, its ambition is to develop "*an integrated understanding of how environmental and political forces interact to mediate social and environmental change*" (Bryant, 1992, p. 12). For some authors though, *political ecology* merely consists nowadays of *a social science/humanities study of environmental politics*" (Walker, 2005, p. 73), i.e. a discipline that insists more on the social construction of nature or on the politics of ecology and less on the ecology in politics. Indeed, struggling to fulfill its agenda, it has been severely criticized for downplaying the natural variables in its explanatory framework.

Against such backdrop, what about the part of political science dealing with environmental issues? Just like mainstream sociology, political science had a delayed interest in environmental issues. When it eventually turned its attention to environmental affairs, it started by focusing on green parties and post-modernist values; adaptive and negotiated policy instruments; renewed policy-making modalities with participation; etc. As for environmental policy, political science has long acknowledged that environmental issues are a privileged terrain for the use of science within public policies wherein it takes center stage (Fischer, 1990), at the very least in agenda-setting. The next section will take up this question about the role of science within environmental public policies further.

The Central Role of Science in Environmental Public Policies

It is widely acknowledged that science and scientific expertise is holding a bigger role within decision-making to the point that nowadays scientific expertise is perceived to represent a new repertoire not only of social movements but of decision-making. As such, Robert (2008, p. 328) argues that the role of scientific expertise should not be considered as just another sub-field of research in political science but should rather be measured against the literature discussing the constitution of various forms of political authority and the modalities of power exercise. To the extent that nowadays, one may speculate that we might witness in the near future the redefinition of relevant political competence in public affairs, the so-called scientific objectivity competing with the principle of representation and/or the majoritarian rule (Deloye, Ihl, Joignant, 2013).

As for public policy analysis more specifically, its relationship to scientific knowledge is a complex one, partly depending on the epistemologic approach it adopts. Indeed, it is worth noting that, right before World War II in the United States, the sub-field of public policy analysis was born out of the positivist-inspired policy sciences (Lerner and Lasswell, 1951). With an assumed normative and prescriptive component, this field of study aimed at providing public decision-makers with knowledge about the public policy cycle in a bid to develop efficient and rational public policies in a typical speaking truth to power paradigm. Despite this willingness to utilize social sciences' expertise to improve public policy design, a better scrutiny of the policy-making process has revealed early on how limited such rationality of the public decision-making actually is (Simon, 1945). Breaking away from policy sciences, a sociology of public action, mostly characterized by a constructivist epistemology, has insisted instead on the different rationalities and interests of policy actors, focusing, in addition, on an extended group of actors and organizations intervening in the policy-making process. Often relying on a socio-historical perspective, this tradition shows how the policy-making process is influenced by policy actors' beliefs, values, and institutional routines. Therefore, it is not surprising that, for a majority of political scientists, policy actors tend to dominate exchanges involving experts and their recommendations, and sometimes even overlook expert reporting they had however ordered themselves. The attitude of policy actors toward scientific advice is far from being dismissive in most cases but remains nonetheless largely instrumental. In contrast, for STS scholars, expert knowledge holds a more foundational role within public policies. We examine these different views successively in the following sub-sections.

From Knowledge Utilization to Evidence-Based Policy: An Instrumental View

Anglo-Saxon and, particularly American, political science expresses an interest in the role of expert reporting early on in the 1970s and 1980s, realizing how much expertise had become an essential component of contemporary governance. The very dynamic school of knowledge utilization distinguishes between different *uses* of expert knowledge by decision-makers. Depending on the authors, the terminology may vary: Amara et al. (2004) evoke instrumental, conceptual and symbolic uses; Weible (2008) speaks about political, instrumental and learning uses. Peters and Barker (1993, p. 18) stress the importance of distinguishing three types of expertise according to its object: the one addressing the agenda-setting, the one targeting policy solutions and the one advising politicians (*politics expertise*). The influence of expertise is the strongest in the case of policy solution's advice.

For her part, S. Blum (2018) analyzes the various impacts on political decisions depending on the type of knowledge's use:

In the first type of instrumental or operational use, knowledge is assumed to give policy-makers information on policy interventions and alternatives, and be applied by them in rather direct ways. Informational or conceptual use is assumed to be less direct and rather directed at general enlightenment, often in the longer run ("learning"). Not least, knowledge can also be used symbolically or political–strategically, that is not for a content matter, but for political struggle and policy prospects. Often, the symbolic dimension of political–strategic use is highlighted, namely to "legitimate and sustain predetermined positions" [...], which may go as far as to purposefully misrepresent scientific evidence. But political–strategic functions can also include the use of research for mobilizing stakeholders, delaying the policy process through installing expert commissions, or serving as a scapegoat for unpopular reforms. (2018, p. 98)

Moreover, depending on the time span, the influence would be more or less pronounced. In the short term, expert knowledge influence on the policy design is perceived to be rather marginal compared to political considerations (Boswell, 2009). On the longer run though, learning effects can be observed (Weiss, 1979). In addition, Campbell Keller concludes that the role of scientists in policy-making changes with the stage of the policy-making process (2009, p. 9).

Hence, a fair amount of political scientists have started evaluating with an increasing critical eye the use of such expert knowledge within public policies (Hoppe, 2005). This explains the skepticism they manifest toward the supposed role of science in decision-making despite the ever-growing tendency to resort to scientific justification within public action over the past few decades. A lot of research within the realm of evidence-based policy-making has thus concentrated on the conditions under which expert discourses start to matter for politicians and decision-makers. Most work stress that policy uptake is seldom granted when expert reporting is directed to policy-makers without proper forms of interactions. Studying the science-policy interface, Sarkki et al. (2013) evoke three criteria conditioning the strengthening of links between expert knowledge and political decision: the perceived scientific legitimacy of the knowledge produced; the adequacy between the expertise produced and the socio-political demand and questioning; the degree of inclusivity of such knowledge, i.e., its ability to take into account the plurality of values, beliefs, and interests. Clark and Majone (1985) evoke the criteria of adequacy, value, legitimacy, and effectiveness as a way to appropriately format expert knowledge for its maximum uptake. Such criteria are often appreciated from an outside viewpoint though: actually, it is essentially experts whose credibility, relevance, and legitimacy are being evaluated. Expert and scientific *knowledge* itself largely remains a kind of black box in political science.

There are exceptions to such a statement. Political scientists inspired by STS and working on "regulatory science" (Jasanoff, 1990) in particular do not refrain from unraveling the way scientific facts are being "produced." The concept itself of "regulatory science" indicates a different position compared to the literature on knowledge utilization. If knowledge utilization and evidence-based policy preoccupied themselves with questioning the conditions and circumstances under which experts and their knowledge were likely to be used, "regulatory science" concentrates on the modalities through which expert *knowledge* takes over the regulatory role of politics. They speak about a co-production of science and society as we will see in the next sub-section.

FROM SCIENCE-POLICY INTERFACE TO SCIENCE-SOCIETY CO-CONSTITUTION: A FOUNDATIONAL VIEW

For Campbell Keller, it is clear that political science and STS disciplines hold sometimes opposite views over the effects of science in public policies: *Political scientists tend to discount the importance of scientists and science in policy-making. [...] This general disciplinary tendency is countered by scholars in the field of STS who argue for the importance of science and technology in shaping contemporary social interaction* (Jasanoff, 2004) (Campbell Keller, 2009, p. 3). This contrasting starting point in the assessment over the role of science can be explained by the fact that for long relatively few debates or dialogues have existed between the two disciplines. Moreover, as we have already seen, public policy analysis had to recover from a pure positivist epistemology. Hence, STS have tended to focus on the—largely politicized—production of knowledge while political scientists would mainly start their investigations when science has already been produced, thus concentrating on how knowledge is disseminated and used within political arenas.

This task's split between political science and STS, and the selective attention that results from it leads to major shortcomings. On the one hand, Wynne and Schackey's work on climate change has been criticized for its relative naive statement on the foundation of a scientific and political agreement, limiting it to the study of one national administration and forgetting the importance of broader inter-departmental and even inter-governmental relations (Boehmer-Christiansen, 1995). In other words, separating the scientific expertise agreement from broader political determinants.[3] On the other hand, political scientists tend to consider science as a black box and start their analysis when the game has already been settled scientifically, i.e., when the discussion has already been structured and biased toward a particular agenda and a particular topic reifying specific power positions. For instance, Callon (2006) criticizes Nelkin and Touraine for studying the controversy around nuclear power when almost nothing is left to be negotiated, when techniques have

---

[3] I have discussed elsewhere the importance of taking into account the influence and constraints of actors and organizations on the receiver end of knowledge production, particularly for the settlement of expert knowledge (Bourblanc, 2019). Indeed, I emphasized the fact that expert knowledge very often anticipates its circulation within public administrations, ensuring its adaptability and operability within public bureaucracies.

already solidified, when the supply-chain and its power have already settled in, and in other words, when we can no longer know what made the nucleocrats so strong and the protest movements so weak.

However, there are exceptions to such statement regarding a task's split, some political scientists deciding to broaden their focus when studying the role of expertise within public policies. Indeed, over the past few years, more and more political scientists have developed an interest in expertise, considering expert knowledge as a full-fledged social and political object. Some of them address the power of expert knowledge from a discursive and argumentative perspective. For instance, for Fischer (1990, p. 171), it is obvious that knowledge and expertise exert an increasing mediating power in political decisions. Political science specializing on environmental policies has very often pointed to the role of environmental scientists as policy brokers too. For Fischer, although experts might not take part in the final decision, they serve as intermediaries between political elites and policy target groups. This mediating power is expressed and enacted through scientific arguments. As for Haas' (1992) works on epistemic communities, he insists on the role of scientific knowledge production in the formation of environmental policy regimes and international cooperation around global challenges. STS scholars such as Jasanoff and Wynne (1998, p. 56) believe, though, that the existence of an epistemic community ought to be the dependent variable (what needs to be explained) rather than the independent variable (or what does the explaining) since the scientific consensus on which such epistemic community relies is very often difficult to obtain and hardly provided by natural necessity.[4]

Political scientists working within the realm of Jasanoff's regulatory science (2004) are another noticeable exception to the tendency of political scientists to leave aside the study of knowledge production. They have very convincingly sought to bridge the gap between political science and STS over the role of expertise. STS have shown that far from being only preoccupied with finding technical solutions to already given problems, scientists are engaged in struggles to determine what is problematic and what is not (Callon, 1986). In Callon's sociology of translation, a problem does not pre-exist its problematization process. In that respect, the sciences' practices constitute, reconstruct, and reconfigure environmental problematic conditions.

Likewise, for regulatory science too, environmental objects and problems are not self-evident; instead, scholars emphasize that such objects and problems are coproduced in relation to the (scientific) practices that seek to study, measure, and manage them. Basically, what Jasanoff means is that ways of knowing the world are inseparable from the ways of in which society seeks to organize and handle itself: *Scientific knowledge both embeds and is embedded in social identities, institutions, representations and discourses* (2004). In that

---

[4] I will go back to that point in the next sub-section, emphasizing how for STS *Nature does not alone underwrite scientific claims* (Pestre, 2004, p. 354).

respect, regulatory science scholars do not perceive science and society as two separate entities, at least analytically.

Accordingly, Jasanoff reinterrogates the category of regulatory science. Past research in the social sciences used the term to designate a type of science that is negotiated at the interface between policy-making and research. Yet, the very notion of interface leaves intact such categories of science and society when Jasanoff believes that these categories are both transformed in the process of co-production. Indeed, Jasanoff (1990) has demonstrated that categories of science, on the one hand, and policy, on the other, are blurred in reality. Hence, regulatory science cannot be considered as a specialized field taking place in-between domains of scientific practices and policy-relevant expertise. Regulatory science straddles both domains, with political considerations penetrating the production of expert knowledge and expert knowledge invading the space reserved for political regulation.

Nevertheless, because it takes its inspiration from STS' second wave, this version of regulatory science tends to verge on hyper-constructivism. In the next sub-section, I present a brief history of STS development as a discipline and propose a reinterpretation of STS' main contribution toward a material turn not always explicitly endorsed by its proponents.

THE NEED FOR A NEO-MATERIALIST RE-INTERPRETATION OF STS' HYPER-CONSTRUCTIVIST STANCE

For long, the sociology of science inspired by Merton's (1942) works was mainly interested in the communities of knowledge producers. Indeed, under Merton's (1942) tradition, works had focused on scientists and social factors surrounding scientific work, dissociating such social aspects from the cognitive conditions of knowledge production. It left aside the production of scientific knowledge itself whose analysis was undertaken by the epistemology and philosophy of science. In the 1970s, the *Sociology of Scientific Knowledge* (SSK) initially represented by the school of Edinburgh (Barnes and Mac Kenzie, 1979; Bloor, 1976) launched a second wave in STS, also called the social turn during which the analysis was recentered toward the explanation of scientific facts themselves. In doing so, SSK explicitly rejected the realist perspective on science.

In particular, Bloor's (1976) strong program represents a turning point in STS history. Bloor formulated several methodological principles (causality; impartiality; symmetry; reflexivity). The principle of symmetry in the explanation of scientific facts states successes and failures in establishing scientific claims should be studied and explained using the same methodology. When historians of science traditionally used to mobilize logical explanations for successful attempts at establishing scientific claims and resort to sociological factors to explain failures, Bloor argues that both success and failure have rational, irrational, sociological, etc., explanations. Bloor's strong program as much as SSK's field of study, however, has been heavily criticized for giving the impression that: *[...] reality has nothing to do with what is socially constructed and negotiated to count as natural knowledge* (Barnes, 1974, p. VII). In that respect, Collins declared in 1981: *the natural world has a small or non-existent role in the construction of scientific knowledge* (p. 3). British SSK promoters unrelently insisted on the idea that: *there was no 'natural necessity' behind scientific consensus (Nature does not alone underwrite scientific claims)* (Pestre, 2004, p. 354). According to Gingras, it might have mainly been a matter of stylistic effect, as sociologists of science were directing their arguments toward philosophers of science in a bid to defeat the positivist model of science the latter never contested, denying the importance of external (sociological) factors in scientific practice (Gingras, 1995, p. 12). In a sum, SSK defended that the production of the scientific truth is a social process and a matter of argumentation from which emerge winners and losers based mainly on a rhetorical performance that in a specific context, is able to get receivers' adherence in a precise situation of communication. In other words, the idea was to focus on other necessary properties of knowledge assertion, i.e., more social skills and translation skills, for instance.

Facing much criticism, Bloor never ceased to emphasize that he had been misunderstood and he had just formulated methodological principles (e.g., causality; impartiality; symmetry; reflexivity) that would allow space for a sociological explanation of the production of scientific facts. Nonetheless, Callon (1986) extended the symmetry principle into a generalized symmetry that would not dissociate between nature and society. Latour (2005) himself added the symmetry between subjects and objects (animated and non-animated objects). For some, this generalized symmetry proved to be even more problematic although it has strongly influenced the Anglo-Saxon field of STS. For Gingras, Callon and Law have been the most adamant in the defense of the original non-distinction between categories. Yet, for Gingras, *what is unclear is whether or not the impossibility to separate such categories comes from the fact that in practical terms (at the ontological level) they are all linked or intertwined or rather that analytical distinctions are not useful to understand such reality (at the epistemological level). Often, we can distinguish factors analytically that we cannot always separate in real life* (my translation, Gingras, 1995, pp. 3–4).

For Latour (2005), no factor, whether political, social, economic, scientific, prevails over the other in terms of explanatory power in the production of science, not the least because all these elements are undifferentiated and inextricably bound together in his analysis. Often prompting uneasiness, he was criticized for allegedly presenting scientific propositions as nothing more than localized narratives and as mere unverifiable social constructions. If Bloor was defending a kind of methodological skepticism, Callon and Latour seemed to be verging on cognitive relativism: to explain the production of science, we are left with the ability to convince and the performative effect of rhetoric (Caillé, 2001), the researcher essentially being a translator (Akrich, Callon, and Latour, 2006). Latour's ANT framework can also be criticized for its tendency to suggest that composite networks of human beings, non-humans, technological and material entities, etc., are orchestrated by (scientific) actors, hence defining rather unilateral enrollment mechanisms dominated by social actors (Granjou and Mauz, 2009).

Some authors have tried to reinterpret Latour's (2005) contribution though. Indeed, they argue that the radical phrasing of his proposal sometimes provoked interpretations opposite the position that he wanted to defend. They posit that with the Actor-Network Theory (ANT) promoted by Latour and colleagues, a kind of material turn has also emerged starting in the 1990s, although in a more ambiguous way as Latour's deconstructivist (or hyper-constructivist) approach would refuse to endorse such a qualification (Gingras, 2000).

What do these authors salvage of the heritage of ANT when reinterpreting it? If ANT undeniably seeks to include the role of objects in the making of social life, it is, however, through the use of the semiologic concept of actants, i.e., a more inclusive category of actors extending to non-humans. Indeed, not believing in a priori distinctions between social and natural actors, or between objects and subjects, Actor-Network Theory in particular would rather talk about actants where all these dimensions are meshed together. As Grossetti (2007, p. 4) reinterpreting Latour's contribution puts it: *"For Latour, the attempt to explain scientific facts with social factors failed and what the lesson he draws from it is not that there is a hard core of facts that would resist any social influence, but rather that we need to redefine the social to include actants without which scientific facts lose most of their significance. Facts are not socially constructed, they are constructed through the association between humans and non-humans [...] By reintroducing actants, Bruno Latour has proposed an interesting take to the issue of the construction of scientific facts: 'actants' are the ones making sure that researchers cannot do whatever please them. They are also the ones which, during controversies, can counterbalance effects of social positionings"* [my translation].

In the next section, building partly on this reinterpretation of STS' main contribution, I elaborate a new conceptual proposition granting center stage to material objects in a bid to promote a less anthropocentric approach of environmental public policy.

A Less Anthropocentric Environmental Policy Analysis: Toward a Renewed Research Agenda

New materialism emerged in the late twentieth century as a response to the post-modernist version of social constructionism. Inspired by Latour's and Descola's work, new materialist scholars also criticize nature and society dualism and would rather depict these two realms as always entangled and constitutive of one another. Similarly to ANT too, it bets on a relational ontology. Advocates of new materialism believe in a more-than-human acting capacity. Yet, taking stock of much criticisms addressed to ANT's so-called agency of actants, authors, such as John Wagner (2015), call to distinguish intentionality from agency. Researchers defending a vitalist perspective of matter, such as J. Bennett (2010), evoke the unpredictable behavior that somehow does not fail to take by surprise our social world often underestimating the constant transforming capacity of artifacts, technical devices, or living organisms. Talking about matter as living organisms, Australian philosopher Elizabeth Grosz strives to show natural objects' capacity to escape our human control and their ability to surprise us.

Considering how much the idea of an autonomous acting capacity can sound problematic for a number of scholars who perceive nature partly as a social construct, I would rather talk about the *effects* that more-than-human entities can produce. In that respect, the part of new materialism's tradition that emphasizes unpredictable effects is of utmost interest here. Yet, new materialism theories insist much more on contingent and relational processes; it emphasizes more change than stability. On the contrary, I believe the capacity of such entities to produce effects is linked to their durability and the difficulty to challenge them over time. Hence, the kind of unpredictability I envisioned here is different from the vitalist theory. What produces effects are not necessarily due to non-humans. Far from any animist belief, I would like to develop an inspiration borrowed from neo-institutionalist theories in the next section.

In this section, I examine the strengths and shortcomings of two schools of thought granting center stage to technical and material objects, i.e., a reinterpreted and rather critical sociology of knowledge and the public policy instrumentation framework. Indeed, for a long time, political sociology and public policy analysis have granted a primary explanatory role to variables such as interests, values, resources, and representations, often overlooking the role of technical objects and their political performativity. The literature around policy instruments sheds a new light on a kind of tacit influence of technical objects over policies.

Drawing on the complementarity of these two traditions, I elaborated the concept of instrumented expert knowledge to capture the specificity of environmental policies. I emphasize how such instrumented expert knowledge has its own force of action, not so much linked to the innovative use it can give birth to but, rather, to lock-in effects it provokes because of its ability to narrow down the space of what is visible and of restricting the field of the possible and/or the compatible options with its equipment.

Building on a Critical Version of STS: Expert Knowledge Production as a Medium Shaping Our Understanding and Perception of Realities

For STS' works, framing power represents the most compelling effects linked to natural knowledge production. This includes the ability to frame the issue on the agenda and the way such issue is being addressed. STS works also emphasize how much any scientific discipline inevitably shapes the way we see the world, enabling us to perceive parts of the reality out there and at the same time leaving other elements out of sight, rendering them somehow invisible. Indeed, all representation of nature always simplifies and reduces the complexity of the 'world out there'. Acting as a filter between nature and society, scientific and expert knowledge greatly influences the way the problems at stake are being tackled by decision-makers. Even works on the production of ignorance have shown that it is also the product of institutional and normative cultures that order everyday scientific practices. Kleinman and Suryanarayanan's (2013) study on colony collapse affecting honey bee populations in the United States argues that *concepts, methods, measures, and interpretations that shape the ways in which actors produce knowledge and ignorance in their professional/intellectual field of practice* explains the regulatory inaction so far toward a new class of pesticides. In that particular case, the dominant epistemic culture characterizing honeybee toxicology ended up favoring the interests of the agrochemical industry.

Within STS too Latour's work has been criticized for neglecting power issues within social relations, being influenced by a rather flat perception of society derived from its central concept of an actor-network, the idea of a network carrying a non-vertical, non-hierarchical, polycentric understanding of power. In that regard, it is not well equipped to study resource asymmetries and structural inequalities that network theories in general do not seem to be able to perceive. Moreover, according to the generalized symmetry principle, pre-existing *interests and other 'social factors' [...] cannot be used as causal items because they are consequences of negotiation and the effects of the settlement of disputes* (Shapin, 1988, p. 543).

Opposite Latour's radical relativism though, I want to demonstrate that the enrollment of collectives of humans and non-humans in the production of natural scientific knowledge is not undetermined: some given economic, social, political, etc., interests lie behind claims to scientific facts. I wanted to go even further than that and reintroduce various forms of constraints—rather than contingency—being exerted not only on knowledge production but on its dissemination, and its uptake in non-scientific arenas where such knowledge has to adapt and be rendered compatible with other regulatory principles of social, economic, and political relations. Indeed, another criticism of Latour has been linked to the fact that his study of scientific practices remains a rather micro-analysis centered around laboratory life and a micro-study of actors' agency (Pestre, 2004, p. 365). From this standpoint, it is difficult to determine the reasons why some attempts to build hybrid socio-technico-natural collectives succeeded and others failed. Gingras (1995, p. 12) stresses that if you do not consider past training and professional trajectory, you will not understand why in some countries engineers have better access than scientists to ministries, just because they share the same educational background with the civil servants populating the ministry of agriculture, of environmental protection, transportation, etc. In a word, scientists do not freely move from the lab to the plant or to the ministries, and some resources are better suited to get properly introduced and nurtured connections in these different governmental venues.

It is precisely the project of the 'new political sociology of science' to remedy this tendency to ignore power asymmetries and revisit STS' contribution toward more structural power (Frickel and Moore, 2005). In contrast, the new political sociology of science advocates for a revived attention paid to external institutions such as states' administrations, markets, social movements (Hess, 2005) and to unequal distribution of power and resources. Its objective is to launch *a collective call to bring social structure (and structurally attuned concepts such as interests, regimes, and organizations) 'back in'* to STS (Frickel and Moore, 2005, p. 9). Indeed, the new political sociology of science wants to re-position typical sociological variables and questionings at the center of the analysis, investigating, for instance, how social characteristics (age, gender, class, ethnic group, etc.) might be constituted into resources mobilized in the production and dissemination of scientific knowledge. For this school of thought, science represents a social field like any other. On the contrary, I deem important to withstand the focus on material objects while still paying attention to asymmetries of resources.

In the next sub-section, I, thus, turn to the scholarly tradition of Public Policy Instrumentation which also recognizes a specific role to technical instruments within public policies. Whereas ANT showed a reluctance to recognize pre-existing structures, the Public Policy Instrumentation has the merit to display a propensity to take into account more structural influences over the selection of technical instruments but also on the effects of such instruments outside the role that was first assigned to them.

Building on Public Policy Instrumentation: Instrument-As-Institution and Its Propensity for Emancipating from its Conceptors

Different perspectives exist around the policy instruments' approach. A lot of works have insisted on the articulation between actors and instruments. For instance, Bressers and O'Toole (1998) working on water policies show how the structures of policy networks' instruments determine the choice of policy instruments. Other works insist on a more reciprocal influence, emphasizing not only the capacity of actors to format policy instruments but also how such instruments tend to influence the selection of legitimate actors within the decision-making process. Recent literature on instrument constituencies demonstrates how a community of actors exist by and through specific instruments (Béland and Howlett, 2016).

Reflecting on two decades of academic production in the field of policy instrumentation, Hood (2007, p. 133) also emphasizes its wide variety:

Some of this work aimed to move beyond the largely classificatory style of studies in the Dahl and Lindblom (1953) style, and sought to use "tools" as part of an explanatory scheme in political science, using a physics-type language of causal analysis. Some of it, on the contrary, sought to move away from what was said to have been an excessively positivist style in the policy analysis of the previous generation, and focused instead on "frames" and argumentation (Schön and Rein 1994), often rejecting what were thought to be the oversimple means–ends distinctions of an earlier era. Some of it was concerned to put technological or institutional factors into the center of the stage. […].

For its part, the political sociology approach to public policy instruments promoted by Lascoumes and Le Galès (2007) emphasizes that public policies are defined as much by their technical instruments than by their political orientations and official goals. For these authors, what public policies effectively turn to be once implemented only partially correspond to the decision-makers' initial intention. This is the reason why it is important to track these socio-technical instruments in particular to account for public policy change effectively taking place. In contrast, most of the public administration literature devoted to instrumentation has been marked by a functionalist orientation, an orientation for which instruments are just means to achieve a pre-identified goal.[5] The public choice literature, for instance, is preoccupied by rationalizing and optimizing of policy instruments' choice. On the contrary, the public policy instrumentation school shows that instruments are not neutral devices: *they are bearers of values, fueled by one interpretation of the social and by precise notions of the mode of regulation envisaged* (2007, p. 4). It invites us to consider together the association of social, cognitive, and material dimensions around policy instruments. In that respect, Lascoumes and Le Galès were inspired by Linder and Peters (1990) approach on instrumentation underlining the subjective process leading to the selection of policy instruments. Subsequently, they develop a kind of *institutions-as-tools* approach around policy instruments. In that respect, they treat instruments as institutions in a broad sociological sense. Instruments partly determine the way in which actors within public policies are going to behave: *instruments allow forms of collective action to stabilize, and make the actors' behavior more predictable […]* (Lascoumes and Le Galès, 2007, p. 9). For these authors, a policy instrument can be defined as *both technical and social*. It is a device *that organizes specific social relations between the state and those it is addressed to, according to the representations and meanings it carries* (p. 4). Lascoumes (1996) also insists on the fact that the selection of policy instruments is done according to values promoted by the organization to which instruments' promoters belong, hence emphasizing the historical dimension of instruments and recycling phenomena within public policies. This stresses the importance of taking into account the already-there

---

[5] In that functionalist approach, public policy is conceived as pragmatic, aiming at solving problems via instruments.

of public policies, i.e., the fact that instruments respond to past routines and not only to the problem at stake.

The public policy instrumentation can also espouse a kind of politics-of-instrumentality approach, inspired by pioneering work from A. Desrosières (1998) on the link between statistical surveys and state-building or work from James Scott (1998) on metrics and rationalization techniques of states' action during the medieval era. These authors primarily focus on macro-social transformations, i.e., transformations affecting the polity. The Public Policy Instrumentation framework hence concentrates on evolving regimes of governmentality in the Foucauldian sense,[6] and on the politics/society relationship showing how *every instrument constitutes a condensed form of knowledge about social control and ways of exercising it*. For his part, Didier (2007) has insisted that rationalizing techniques of government aiming at disciplining the population might not be the only objective of policy instruments. Beyond the relationship between the governing and the governed, policy instruments might seek to reinforce the capacity of controlling bureaucracies and the state apparatus, in general. Hence, it might be even more important to not only study the impacts of such instruments on designated targets of public policies but also on the pilots of such public programs and on the competition between departments and services that characterize them. In other words, instruments would more likely help governing the government, disciplining its various administrations, competing administrative staff and fragmented public organizations. This could particularly apply to cross-cutting inter-sectorial environmental subjects involving several ministries at multiple scales. This suggests to adopt a more modest perspective on policy instruments studying its impact on public policies and policy sectors.

It is also worth mentioning that the Public Policy Instrumentation framework was not the first to pay attention to the role of technical instruments within public policies. Before it emerged, works in the sociology of quantification had established the power of statistics and numerical figures for instance. For such works, numbers offer the possibility of a common language that insures their circulation in different realms:

> the radical simplification of commensuration produces decontextualized, depersonalized numbers that are highly portable and easily made public. Numbers circulate more easily and are more easily remembered than more complicated forms of information. [...] *The capacity of numbers to circulate widely is enhanced by their abstractiveness, which appears to eliminate the need for particularistic knowledge in interpreting them*. We assume that the meaning of numbers is universal and stable. The erasure of context and people who make numbers is crucial for their authority. We trust that the methods that accomplished this erasure were *rigorous enough to constrain the discretion, politics, and biases of those who created them*.

---

[6] Combining the terms "Goverment" and "Rationality," "governmentatlity" designates activities that affect the conduct of people (*government*). Before things and people can be controlled and governed though, they ought first to be defined (*rationality*).

Because we think that numbers transcend local knowledge, we suppose they do not require the same sorts of interpretation and translation as others modes of communicating. [...] Numbers are easy to dislodge from local contexts and reinsert in more remote contexts. [...] as numbers circulate, we recreate their meaning by building them into new contexts and reinterpreting their meaning. We can use numbers to do new things without having to acknowledge that they now mean something different. (Espeland and Sauder, 2007, p. 18)

Unlike qualitative evaluation that can always give birth to different interpretations leading to a debate, numerical figures, on the contrary, tend to close discussion once adopted. In addition, the policy instrumentation approach can be distinguished from such works in the sociology of quantification in the sense that for Lascoumes and Le Galès, policy instruments may acquire a life of their own. Indeed, when Espeland and Stevens (1998)' works on what they call *new regimes of measurement* examine the social and political implications of commensuration exercises or when works of sociologist N. Rose (1991) on numbers show how they contribute to legitimating political authority, they also emphasize, like the political scientist, D. Stone (1997), that numbers far from speak for themselves. For them, there is no intrinsic value in figures. Defending an interpretative political sociology, D. Stone grants an important role to narratives: "*numerical analysis only makes sense when embedded in a larger and frequently normative narrative*" (Stone 2016). Likewise, Espeland and Stevens' work shows that the weight of measurement tools can only be revealed within context, since their objective and meaning are determined through their use.

In the field of the sociology of management tools, authors also contend that the exercise of power by indicators is neither univocal nor mechanical. To determine whether or not such tools or indicators can gain such power requires to reconstitute the configuration of power relations within which indicators make sense (Bezes, Chiapello and Desmarez, 2016, pp. 355, 356).

Against that background, Lascoumes and Le Galès go a bit further and claim that policy instruments specifically can change power equilibriums, policy instruments partly determining what resources can be used and by whom:

they create uncertainties about the effects of the balance of power; they will eventually privilege certain actors and interests and exclude others; they constrain the actors while offering them possibilities [...] The social and political actors therefore have capacities for action that differ widely according to the instruments chosen. (Lascoumes and Le Galès, 2007, p. 9).

More fundamentally, for these authors, instrumentation can also produce its own effects, independently of the stated objective pursued, i.e., the aims ascribed to policy instruments in the first place. The policy instruments' framework shows that instruments are not inert objects readily available to any kind of mobilization. Moreover, sometimes they might even escape the policy actors who promoted them in the first place. Lascoumes and Le Galès

(2007) mention some autonomization tendency of instruments, producing some original effects and unforeseen consequences. Such unexpected effects seem, however, largely to be linked to some innovative uses while circulating:

> *Once in place, these instruments open new perspectives for use or interpretation by political entrepreneurs, which have not been provided for and are difficult to control [...]* (2009, p. 9)

On the contrary, I believe such unexpected effects are not solely linked to the new uses they are submitted to. In the next sub-section, I develop further this aspect discussing lock-in effects, a less developed aspect of Lascoumes and Le Galès' work.

Lock-In Effects and the Specificity of Environmental Problems

Neo-institutionalist theories evoke the existence of reinforcing mechanisms linked to institutions. Against that backdrop, works on technologies of government were among the first to mention the importance of data series to sustain specific indicators.[7] Building on rational neo-institutionalism, they demonstrate how over time inventories accumulating information further entrench existing indicators, creating a path-dependence phenomenon and making more and more costly to change and develop new indicators.

Moreover, neo-institutionalist theories have demonstrated that a merit of institutions is to be able to reduce uncertainty. In previous works around environmental indicators (Bourblanc, 2011), I have also shown that policy instruments help contain complexity and operate as lock-in devices. It is because of such complexity that I stated a dependence to instruments more than to the path traced by institutions. Such potential is linked to their technical dimension and their link to materiality which make them more difficult to be manipulated or transformed. The argument is twofold. Indeed, as a transversal issue cutting across several policy sectors, environmental policies have to deal with a great number of actors, organizations, professional routines, etc., which can prove difficult to coordinate. In that respect, environmental policies have to manage such fragmentation and organizational complexity.

Another key characteristic of environmental policies is that they have to deal with unpredictability since they engage with living organisms. Managing such unpredictability, therefore, constitutes one of the primary objectives of environmental public policies. Being able to stabilize hybrid socio-technico-natural collectives encompassing both human and non-humans alike is the first challenge of environmental public policies. Highlighting such difficulty makes us better understand why, once adopted, instrumented expert knowledge is

---

[7] See Desrosières' works (1998) on statistics in particular.

not prone to easy change since they derive from hardly-negotiated compromises. In these negotiations, the non-humans never appear as an absolute constraint on public policies, but neither do they represent a tamed resource that collective action can mobilize almost effortlessly.

Considering the management of such complexity and unpredictability, I take the stance of not considering environmental problems and policies just like any other public policies. Moreover, I posit that the specificity of environmental problems is the most accurate when their link to territory is the tightest and their attachment to materiality the most tangible. In other words, when the mediation between natural objects and knowledge representing them is the slimmest, the verdict of nature and materiality is rendered more visible and hence more difficult to overlook by socio-political actors. In such a context, the stabilization of these hybrid entities into instrumented expert knowledge is the most difficult to obtain, hence its undoing is even more difficult—and risky—to pursue.

Against that backdrop, thinking in terms of instrumented expert knowledge offers the advantage of highlighting how through its equipment and technical instruments, instrumented expert knowledge has the ability of anchoring the cognitive categories produced by natural science into a more tangible reality. In previous works, I have demonstrated that new cognitive categories devoid of a metric apparatus or technical equipment of some sort can be more easily bypassed or sidelined by competitors. They also prove to be less trustworthy since they are less able to prevent reinterpretation along the way (Bourblanc, Bouleau, Deuffic, 2021).

Conclusion: The Concept of Instrumented Expert Knowledge to "Bring the Material Back In" in Environmental Public Policy Analysis

In this chapter, my ambition has been to explore new intellectual pathways to undertake the challenge of bringing the material back in the study of environmental public policies. This does not entail espousing a positivist epistemology as one has to recognize that pure nature does not exist: we are only ever dealing with socio-natures. In addition, we know that objects and subjects of nature are seldom accessible, unmediated and, often, are only perceptible or understandable through scientific knowledge. Studying environmental policies, thus, implies to reflect on the specific role held by scientific expertise within public policies.

First, I showed that most political scientists interested in expertise focus on experts rather than on expert reporting and its technical knowledge. When they do concentrate on the content of expert knowledge, they tend to not taking expert knowledge seriously enough, suggesting that actors are often able to manipulate what they conceive of as scientific information more than as a compelling argument. If they rightly underline that expert reporting has little to do with 'speaking truth to power', they, however, often consider that effects of expert knowledge over the decision-making process are easily controlled.

Adopting a more foundational view, STS works around regulatory science evoke the co-production of science and society. For STS, expert knowledge operates as a medium but unlike to the notion of science-policy interface, it does not presuppose two distinct realms. Expert knowledge shapes our understanding and perception of the environment around us, acting as a filter, making some aspects more visible, and making other become invisible. Yet, STS have been criticized for their hyper-constructivist—sometimes even relativist—stance. For the same reason, because of their insistence on environmental scientific expertise as social constructs, they would rather follow actors than hybrid collectives to determine the potential influence of such expertise and its technical objects within public policy. They do not neglect the long-lasting effects that environmental expert knowledge might exert on public policies, especially regarding its framing power but they are more reluctant to consider such effects independently of actors' strategies. However, to move away from the autoreferentiality of the social, I posit that we need to take seriously the effects of such expert knowledge and objects over environmental policies, beyond actors' play.

This entails better following traces of their material dimension. Hence, I intended to promote a less anthropocentric approach of environmental policy analysis, i.e., an approach that would cease to view humans as separated from and even above the rest of nature. I, thus, elaborated the concept of instrumented expert knowledge which cross-fertilizes the insights of a critical sociology of knowledge, on the one hand, and the insights of public policy instrumentation, on the other, in a bid to complement our understanding of the effects of expert knowledge within environmental policies with an attention to knowledge's technical equipment.

Thinking in terms of instrumented expert knowledge allows me to undertake a more relevant environmental policy analysis. When other authors evoke «discrete arenas» with closed and discretionary modes of decision-making (Henry, 2017), I would rather point to the discrete power of technical and expert instruments for their ability to select legitimated actors and participants to the discussion. Hence, if one wants to open up dialogue and broaden the topic of discussion, one has to dedicate specific attention to the inclusiveness of such instrumented expert knowledge.

Against such backdrop, most social scientists working on technical objects have insisted on the non-deterministic influence of objects of nature within public policies emphasizing that since they are social constructs, they do not have intrinsic properties ensuring such influence. They stress that such alleged influence relies heavily on contexts to determine their respective weight within public policies. Yet, I emphasize that they are not only social constructions but hybrid associations between the natural and the social, cross-cutting different influences. As hybrid objects between nature and society, I tried to demonstrate in the discussion above that such hybrid objects represent neither a fully tamed resource nor a definite constraint over human action. Capitalizing on the strengths of two schools, i.e., the critical sociology of knowledge on the one hand, and the insights of public policy instrumentation on the other, the concept of instrumented expert knowledge enables me to evoke the influence of such hybrid arrangements outside of their circle of production. A striking illustration of such power can be found in the COVID-19 recent crisis when epidemiological modeling tools predicting a saturation of public health services managed to obtain from decision-makers the unthinkable, i.e., for the first time to put part of the economy into an halt and society into lockdown. I believe this does not just set apart crisis periods but applies to more mundane times.

References

Amara N., Ouimet M., Landry R., (2004), "New Evidence on Instrumental, Conceptual, and Symbolic Utilization of University Research in Government Agencies". *Science Communication*, 26(1), pp. 75–106.

Barnes, B. (1974), *Scientific Knowledge and Sociological Theory*. London: Routledge.

Barry, A. (2021), "What is an Environmental Problem?". *Theory, Culture & Society*, 38(2), pp. 93–117.

Béland, D. and Howlett, M. (2016), "The Role and Impact of the Multiple-Streams Approach in Comparative Policy Analysis". *Journal of Comparative Policy Analysis: Research and Practice*, 18(3), pp. 221–227.

Bennett, J. (2010), *Vibrant Matter. A Political Ecology of Things*. Durham: Duke University Press, 176 pages.

Bezes, P., Chiapello, E., Desmarez, P. (2016), "Les tensions savoirs-pouvoirs à l'épreuve du gouvernement par les indicateurs de performance". *Sociologie du travail*, 58(4), pp. 347–369.

Blanc, G., Demeulenaere, E., Feuerhahn,W. (eds), *Les humanités environnementales, enquêtes et contre-enquêtes*. Paris: Presses de la Sorbonne.

Bloor, D. (1976), *Knowledge and Social Imagery*. Chicago: The University of Chicago Press.

Blum, S. (2018), "The Multiple-Streams Framework and Knowledge Utilization". *European Policy Analysis*, 4(1), pp. 94–117.

Boehmer-Christiansen, S.A. (1995), Comments on S. Shackley and B. Wynne paper "Global Climate Change: The Mutual Construction of an Emergent Science-Policy Domain". *Science and Public Policy*, 22/6, pp. 411–16.

Boswell, C. (2009), *The Political Uses of Expert Knowledge*. Cambridge University Press.

Bourblanc, M. (2011), "Emancipated Instruments. Dependence on Instruments in the Management of Agricultural Water Pollution in the Côtes-d'Armor Department (1990-2007)". *Revue française de science politique (English)*, 61: pp. 25–45.

Bourblanc, M. (2019), "Expert Assessment as a Framing Exercise. The Controversy Over Green Macroalgal Blooms' Proliferation in France". *Science & Public Policy*, 46(2), pp. 264–274.

Bourblanc, M., Bouleau, G., Deuffic, P. (2021), "The Role of Expert Reporting in Binding Together Policy Problem and Solution Definition Processes". In Zittoun,

P., Fischer, F., Zahariadis, N. (2021), *The Political Formulation of Policy Solutions*, Bristol, Bristol Policy Press, pp. 73–91.

Bressers, H.T.A and O'Toole, L.J. (1998), "The Selection of Policy Instruments: A Network Based Perspective". *Journal of Public Policy*, 18(3), pp. 213–239.

Bruno, I., Didier, E. (2013), *Benchmarking. L'Etat sous pression statistique*. Paris: La Découverte, 211 pages.

Bryant, R.L. (1992). "Political Ecology: An Emerging Research Agenda in Third World Studies". *Political Geography Quarterly*, 11, pp. 2–36.

Buttel, F. (1996), "Environmental and Resource Sociology: Theoretical Issues and Opportunities for Synthesis". *Rural sociology*, 61 (1), pp. 56–75.

Caillé, A. (2001), "Une politique de la nature sans politique". *Revue du Mauss*, 17(1), pp. 94–116.

Callon M. (1986). "Éléments pour une sociologie de la traduction. La domestication des coquilles Saint-Jacques dans la Baie de Saint-Brieuc". *L'Année sociologique*, 36

Callon, M. (2006), "Pour une sociologie des controverses technologiques". In Akrich, M., Callon, M., Latour, B., *Sociologie de la traduction*, Paris, Presses des Mines, pp. 135–157.

Campbell Keller, A. (2009) *Science in Environmental Policy. The Politics of Objective Advice*. MIT press.

Carolan, M. (2008), "The Multidimentionality of Environmental Problems. The GMO Controversy and the Limits of Scientific Materialism". *Environmental Values*, 17(1), pp. 67–82.

Catton, W. R., Dunlap, R.E. (1978), "Environmental Sociology: A New Paradigm". *The American Sociologist*, 13, pp. 41–49.

Clark, W. C. Majone, G. (1985), "The Critical Appraisal of Scientific Inquiries with Policy Implications". *Science, Technology and Human Values* 10(3), pp. 6–19.

Cobb, R., and Elder, C. D. (1983). *Participation in American Politics. The Dynamics of Agenda Building*. Baltimore, MA: John Hopkins University Press.

Deloye, Y., Ihl, O., Joignant, A. (2013), *Gouverner par la science. Perspectives comparées*. Grenoble: Presses universitaires de Grenoble.

Desrosières, Alain (1998). *The Politics of Large Numbers: A History of Statistical Reasoning*. Cambridge, MA: Harvard University Press.

Didier, E. (2007), "Quelles cartes pour le New Deal ? De la différence entre gouverner et discipliner". *Genèses*, 68, pp. 48–74.

Dunlap, R.E, Catton, W.R. (1994), "Struggling with Human Exemptionalism". *The American Sociologist*, 25(1), pp. 5–30.

Espeland W. N. and Stevens M. L. (1998), "Commensuration as a Social Process". *Annual Review of Sociology*, 24, pp. 313–343.

Espeland W. N. and Sauder M. (2007), "Rankings and Reactivity: How Public Measures Recreate Social Worlds". *American Journal of Sociology*, 113(1), pp. 1–40.

Fischer, F. (1990), *Technocracy and the Politics of Expertise*. Londres: Sage.

Frickel, S., and Moore K. (eds), (2005), *The New Political Sociology of Science. Institutions, Networks and Power*. Madison: University of Wisconsin Press.

Gingras, Y., (1995), "Un air de radicalisme. Sur quelques tendances récentes en sociologie de la science et de la technologie". *Actes de la Recherche en Sciences Sociales*, 108, Juin 1995, pp. 3–17.

Gingras, Y., (2000), «Pourquoi le'programme fort' est-il incompris?», *Cahiers internationaux de sociologie*, CIX, pp. 235–255.

Graham, S. (ed.) (2009), *Disrupted Cities. When Infrastructure Fails*. New York, Routledge.

Granjou, C., and Mauz, I. (2009), "Quand l'identité de l'objet-frontière se construit chemin faisant". *Revue d'Anthropologie des Connaissances*, 3(1), pp. 29–49.

Grossetti, M. (2007), "Les limites de la symétrie". *Sociologies*, http://journals.openedition.org/sociologies/712.

Haas, P.M. (1992), "Introduction: Epistemic Communities and International Policy Coordination". *International Organization*, 46(1), pp. 1–35.

Henry, E. (2017), *Ignorance scientifique et inaction. Les politiques de santé au travail*. Paris, Presses de Sciences Po.

Hess, D.J. (2005), "Historical and Institutional Perspectives in the Sociology of Science". In Frickel S., Moore K. (eds.), *The New Political Sociology of Science. Institutions, Networks and Power*, Madison, University of Wisconsin Press, pp. 122–147.

Hood, C. (2007), "Intellectual Obsolescence and Intellectual Makeovers: Reflections on the Tools of Government After Two Decades". *Governance: An International Journal of Policy, Administration, and Institutions*, 20(1): pp. 127–144.

Hoppe, R. (2005), "Rethinking the Science-Policy Nexus: From Knowledge Utilization and Science Technology Studies to Types of Boundary Arrangements". *Poièsis and Praxis. International Journal of Ethics of Science & Technology Assessment*, 3(3): pp. 119–215.

Jasanoff, S. (1990), *The Fifth Branch: Science Advisers as Policymakers*. Cambridge: Harvard University Press.

Jasanoff, S. and Wynne, B. (1998), "Science and Decisionmaking". In S. Rayner and E.L Malone (eds.), *Human Choice and Climate Change, 1, The Societal Framework*, Columbus: Battelle Press, pp. 1–87.

Jasanoff, S. (ed.) (2004), *States of Knowledge: The Co-Production of Science and Social Order*. London: Routledge.

Kleinman, D. L., and Suryanarayanan, S. (2013), "Dying Bees and the Social Production of Ignorance". *Science, Technology and Human Values*, 38(4), pp. 492–517.

Larkin, B. (2013), "The Politics and Poetics of Infrastructure". *Annual Review of Anthropology*, 42, pp. 327–343.

Lascoumes, P. (1996), "Rendre Gouvernable". In Chevallier, J. (ed.), *La gouvernementalité*, Paris, PUF.

Lascoumes P. and Le Galès, P. (2007), "Understanding Public Policy through Its Instruments—From the Nature of Instruments to the Sociology of Public Policy Instrumentation". *Governance: An International Journal of Policy, Administration, and Institutions*, 20 (1), pp. 1–21.

Latour, B. (1988), *Science in Action. How to Follow Scientists and Engineers Through Society*. Harvard University Press.

Latour, B. (2005), *Re-Assembling the Social. An Introduction to Actor-Network Theory*. Oxford: Oxford University Press.

Lerner, D. and Lasswell H.D. (1951), *The Policy Sciences: Recent Developments in Scope and Method*. Stanford University Press.

Lidskog, R., Mol, A. and Oosterveer, P. (2015). "Towards a Global Environmental Sociology? Legacies, Trend and Future Directions". *Current Sociology* 63(3): pp. 339–368.

Linder, S. and Peters, B.G. (1990), "The Design of Instruments for Public Policy". In *Policy Theory and Policy Evaluation*, ed. S. Nagel. Westport, CT: Greenwood Press.

Merton, RK. (1942), "Science and Technology in a Democratic Order". *Journal of Legal and Political Sociology*, 1, pp. 115–126.
Mol, A.P.J. (2006), "From Environmental Sociologies to Environmental Sociology?". *Organization & Environment*, 19(1), pp. 5–27.
Pestre D. (2004), "Thirty Years of Science Studies: Knowledge, Society and the Political". *History and Technology*, 20(4), pp. 351–369.
Peters, G. and Barker, A. (1993), *Advising Western European Governments. Inquiries, Expertise and Public Policy*. Edinburgh: Edinburgh University Press.
Porter, T.M. (1995), *Trust in Numbers. The Pursuit of Objectivity in Science and Public Life*. Princeton.
Robert C. 2008, "Expertise et action publique". In Borraz, O. and Guiraudon, V., *Politiques Publiques*, Paris, Presses de SciencesPo, pp. 309–335.
Rose, Nikolas. (1991). "Governing by Numbers: Figuring Out Democracy". *Accounting, Organizations and Society*, 16(7), pp. 673–692.
Sarkki, S., Niemelä J., Rob Tinch, Sybille van den Hove, Allan Watt and Juliette Young (2013), "Balancing Credibility, Relevance and Legitimacy: A Critical Assessment of Trade-Offs in Science-Policy Interfaces". *Science and Public Policy*, 41(2). pp. 194–206.
Schön, D.A. and Rein, M. (1994), *Frame Reflection: Toward the Resolution of Intractable Policy Controversies*. New York: Basic Books.
Scott, James C. (1998), *Seeing Like a State: How Certain Schemes to Improve the Human Condition have Failed*. Yale University Press.
Shackley, S., and Wynne, B. (1995a), "Global Climate Change: The Mutual Construction of an Emergent Science-Policy Domain". *Science and Public Policy*, 22(4), pp. 218–230.
Shapin, S. (1988), "Following Scientists Around". *Social Studies of Science*, 18, pp. 533–550.
Simon H. A., (1945), *Administrative Behavior: A Study of Decision Making Processes in Administrative Organization*. New York: The Free Press.
Spector, M., Kitsuse, J. I., (1977), *Constructing Social Problems*. Menlo Park, (CA): Cummings.
Star, S.L. (1999), "The Ethnography of Infrastructure". *American Behavioral Scientist*, 43(3), pp. 377–391, https://doi.org/10.1177/00027649921955326.
Stone D., (1997), *Policy Paradox. The Art of Political Decision-Making*. New-York: W.W. Norton and Company.
Stone D. (2016), "Quantitative Analysis as Narrative". In Bevir and Rhodes, *Routledge Handbook of Interpretive Political Science*, pp. 157–170.
Wagner J. R. (Dir.) (2015), *The Social Life of Water*. New York, Oxford: Berghahn.
Walker, P.A. (2005). "Political Ecology: Where is the Ecology?" *Progress in Human Geography*, 29(1), pp. 73–82.
Weible, C. M. (2008). "Expert-Based Information and Policy Subsystems: A Review and Synthesis". *The Policy Studies Journal*, 36(4), pp. 615–35.
Weiss C. H., (1979), "The Many Meanings of Research Utilization". *Public Administration Review*, 39(5), pp. 426–431.
Yearley, S. (1991). *The Green Case. A Sociology of Environmental Issues, Arguments and Politics*. London: Harper Collins Publishers.
Zimmerer K. and Bassett T. (2003). *Political Ecology: An Integrative Approach to Geography and Environment-Development Studies*. Guilford Publications.

The Rise of Environmental Health as a Recognized Connection and Academic Field

Corinne Delmas

The "control of nature" is a phrase conceived in arrogance, born of the Neanderthal age of biology and philosophy, when it was supposed that nature exists for the convenience of man... It is our alarming misfortune that so primitive a science has armed itself with the most modern and terrible weapons, and that turning them against the insects it has also turned them against the earth.

Rachel Carson, Silent Spring

I hope this committee will give serious consideration to a much neglected problem—that of the right of the citizen to be secure in his own home against the intrusion of poisons applied by other persons. I speak not as a lawyer but as a biologist and as a human being, but I strongly feel that this is or should be one of the basic human rights. I am afraid, however, that it has little or no existence in practice.

Rachel Carson

The "environment" is now considered as one of the determinants of health. The health risks posed by genetically modified organisms (GMOs), microwave antennas, pesticides, pollution (atmospheric, aquatic, etc.) or chemical substances emanating from everyday objects (bisphenol A contained

---

C. Delmas (✉)
Laboratoire Techniques, Territoires et Sociétés (LATTS UMR CNRS 8134), Gustave Eiffel University, Marne la Vallée, France
e-mail: corinne.delmas@univ-eiffel.fr

in baby bottles, water bottles and many kitchen objects, etc.) are now regularly debated in various arenas and media. The COVID-19 pandemic has exacerbated this concern about multiple and invisible dangers. As a zoonotic disease, the coronavirus has contributed to the debate on the close links between health and the environment; it has revived certain scientific and environmental warnings about climate risk and its impact on human health. This pandemic has also reminded us of the importance of prevention and the need to take into account environmental exposures, in contrast to public health approaches that focus on curative measures and the individualization of risk behaviors. It is also the aftermath of COVID-19 that is at the heart of our concerns, as the pandemic has reinforced a feeling of precariousness in the face of the threat that disruptions to our ecosystem pose to our lives, both locally and globally.

Although the link between health and the environment is a long-standing one, contemporary awareness of human damage to the environment and its health consequences has contributed to the emergence of new explanatory models of health. However, not all risks benefit from the same recognition and the analysis of the construction of environmental health problems highlights the diversity of their trajectories. This construction varies in particular according to the populations subjected to the risk, knowing that situations of injustice and inequalities, socio-economic, territorial, but also gender have been documented by the human and social sciences.

The Link Between Health and the Environment

The notion of environmental health was officially defined by the European office of the World Health Organization (WHO) at the Helsinki conference in 1994 as "aspects of human health [...] determined by physical, chemical, biological, social, psychological and aesthetic factors in the environment." This explanatory model of health is in line with old intuitions, according to which environmental factors would explain the appearance of specific ailments. It differs, however, in that the approach is now more global and takes greater account of the interaction between human beings and their environment, focusing in particular on the close links between human health and anthropogenic changes in our ecosystem.

Ancient Intuitions

Environmental health is part of several traditions, including that of Hippocrates, who studied the environment as an external cause of disease, and that of hygienism, which was eventually supplanted by a biomedical model (Annandale, 1998; Gaudillière, 2002).

In the fourth century B.C., Hippocrates invited us to take into account the influence of the seasons and the temperature on the "humors" irrigating the body (blood, bile, phlegm) (Kalachanis and Michailidis, 2015). A form of climatic and geographical determinism of diseases emerges with this thought.

In the Middle Ages, the Hippocratic theory of airborne miasmas was used to explain the spread of the plague. The geographical determinism of the Hippocratic vision permeates many publications, including Montesquieu's theory of climates in the eighteenth century (Shackleton, 1955). If Hippocratic medicine highlights the possible actions of the environment on health, it is necessary to wait several centuries to consider this environment in relation to human activities.

The development of modern medicine in the nineteenth century revealed the importance of germs, microbes and viruses in pathogenesis, but they could not explain the appearance of certain epidemics on their own. For example, Max von Pettenkofer, while studying a cholera epidemic around Munich, introduced the notion of predisposing factors (infected matter, hydrological conditions…) which are found in social medicine (J.S.H., 1901). It was during this period that human beings became truly aware of the link between their own choices and the deterioration of their health; this awareness is the basis of hygienism, whose etiological models closely link the health of the individual to his or her environment rather than to internal dysfunctions of the organism. It developed in the nineteenth century, before falling into disrepute in the following century; the preventive approach, which took environmental factors into account, then declined in favor of curative approaches.

Indeed, after World War II, a competing "biomedical" model gained ground (Gaudillière, 2002), built around the study of etiologies endogenous to the human body and the promotion of curative action. Contrary to the knowledge mobilized by hygiene, the knowledge developed by the basic sciences does not concern the individual's environment, but the body's internal dysfunctions and biological and physiological mechanisms. They come from genomics, molecular biology, immunology, endocrinology, etc. The sick body is broken down and divided by medical practitioners into as many fragments as there are specialties and types of medical interventions. The penetration of basic sciences and biology into the medical universe contributes to the construction of the biomedical regime; it occurs under the impulse of a reconfiguration of the relationships between the State, physicians, researchers and industrialist corporations (Clarke et al., 2010). In France, for example, State investments have contributed to the "molecularization" (Gaudillière, 2002) of knowledge about living organisms. This restriction of the medical viewpoint to very localized areas of the human body opens the way to the preponderance of genetic explanations of diseases in medical research and concerns, as well as to a form of genetic reductionism which limits the explanation to genetic factors and does not take into account, in particular, social and environmental factors.

During the 1980s, the decline of hygienism gave way to new approaches to public health which were characterized by the importance they gave to statistical thinking and the construction of health risks (Bourdelais, 2001; Guilleux, 2015). However, public health is rarely highly valued; in France, for example, its representatives are discredited or even marginalized within the medical profession; less remunerative, attracting left-wing oriented doctors, marked by May 68, it is often presented as an applied science whose social and transversal dimension of the objects of study and practices gives it a relatively subversive image. Moreover, its administrations have been discredited by a succession of crises and health scandals such as scandal of the blood contaminated by the HIV which had shown the failures of the institutions in charge of the collection of blood and had as a consequence the questioning of the responsibility of the minister of health of the time, scandal of the benfluorex, marketed under the name of Mediator having caused the death of 1,500 to 2,100 people, and side effects in many others, and which will have, as a result, the examination of several persons in charge of institutions of health (Delmas, 2011).

The multiplication of alerts and scandals caused by environmental disasters linked to industrial activities has, however, contributed to the emergence of a new explanatory model of health characterized by a global approach to the relationship between humanity and its environment, or even the emergence of an health ecology that questions human health as a relationship to the world and promotes a vision of the future that links living beings and the Earth (Waltner-Toews, 2004).

The Emergence of a New Explanatory Model of Health

If the hygienists were mainly worried about dangers perceptible by the senses (smells, black smoke…), environmental health crystallizes the emergence of new social representations of health dangers, invisible and impalpable: nanoparticles, endocrine disruptors, electromagnetic waves, etc. (Guilleux, 2015). Many pathologies have been requalified as "environmental" while human attacks on our ecosystem would have many deleterious effects for the human species (such as decrease in male reproductive capacities or adverse impact on intellectual capacities[1]); during the 1990s, several medical researchers on male fertility have integrated the explanatory hypothesis of endocrine disruptors which are natural substances and substances of industrial origin that resemble hormones sufficiently to mimic and disrupt their physiological effects.

Environmental health has not been built on the basis of traditional public health knowledge, the expertise of which has been criticized by researchers from the ecological and experimental sciences who are contributing to the development of a new way of thinking about toxicity. The approach is global; human health and environment are being perceived as interdependent.

Several works contributed to the emergence of these new approaches, including those of the biologist Rachel Carson, who denounced the destruction of natural environments by industrialization (Carson, 1962). In 1962,

---

[1] One example is the integration of endocrine disruptors on male fertility, during the 1990s, by several medical researchers. On this point, see infra, part 2 (on the construction of environmental health problems).

her book *Silent Sprint* revealed to the general public the presence of environmental dangers. It was echoed in new fears fueled by several environmental disasters, including the dumping of mercury, a potent neurotoxin, infamously killed hundreds and sickened thousands in Minamata, Japan, in the 1950s and the sinking of the supertanker Torrey Canyon off Great Britain in 1967 (Zio, Aven, 2013).

It raises the question of the growth of the world chemical industry and the extremely rapid synthesis of new molecules. This book by Carson that some scholars consider to have jumpstarted the modern environmental movement also highlights the presence in the environment of invisible dangers that weigh on animal and human health. Carson's denunciation of the impact of pesticides, in particular DDT (dichlorodiphenyltricholoroethane), is not isolated. Criticism of industrialization is found, for example, in René Dubos, a microbiologist and co-author in 1972 with Barbara Ward of the report *Nous n'avons qu'une terre* (We Have Only One Earth) intended for the preparation of the Stockholm conference (Ward and Dubos, 1972; Ward, 1973). To explain the development of diseases, he puts forward factors linked to modernization such as pollution, overeating of adults, psychological disorders linked to competitive behaviors, urban congestion, automated work and emotional loneliness. In France, the sociologist Claudine Herzlich (1969) had shown that the respondents, in order to explain the illness, referred to an "unhealthy lifestyle" (too fast a pace of life, noise, stress, air pollution, chemical food, etc.) which is detrimental to health itself, which is conceived as a balance and a harmony with the social environment.

The designation of health factors linked to lifestyles in industrial societies is also fed by a medical discourse that developed in the 1980s about the diseases of civilization (Stupfel, 1976; Saint-Marc, 1989), i.e., diseases specific to industrial societies and generated by urbanization, air pollution, radiation, artificial food, alcoholism, mental illnesses and drug addiction.

The exposure of multiple environmental scandals as well as the succession of pandemics in recent years, particularly the globalized pandemic of COVID-19, has contributed to alerting us today to the interdependence between human health and that of our ecosystem, on a planetary scale.

The Interdependence Between Human Activities and Health

Sociologists of modernity such as Ulrich Beck (1992) or Anthony Giddens (1999) have highlighted the emergence of a "risk society," a risk that would no longer be external but generated by society itself. The industrial society induces a multiplication of risk factors. Humanity would have become a "geological force," a thesis defended in particular by the chemist and meteorologist Paul Crutzen (2002) suggesting a new geological subdivision of the quaternary era, which he calls the "Anthropocene," characterized by the consequences of human activities. However, the impact of the environment on health is still largely studied in the context of localized research. From this perspective, the relationship between environment and health is established when a multitude of local changes—for example of agriculture and food cultures (Philippon, 2018)—lead to global changes. Health, understood as a system in equilibrium, depends on a combination of elements that make up an ecosystem: air, housing, workplace, genetic heritage, culture, etc. Human health depends on the health of this ecosystem. The Anthropocene consecrates the human being as a telluric force capable of modifying planetary balances (Crutzen, 2002).

The modifying actions are relocating; the result of an action is not necessarily observed where it was committed. They are also perceived as cumulative: the results of these actions end up adding up. This is illustrated, for example, by the increase in the concentration of carbon dioxide in the Earth's atmosphere, a consequence of successive industrial revolutions that were originally not global; and the massive release of chlorofluorocarbons which are abundant in aerosol cans or refrigeration equipment and were massively released by developed countries during the second half of the twentieth century causing the birth of components responsible for the destruction of stratospheric ozone (Sherwood Rowland, 1990). The related health consequences can be direct; for example, there is a statistical correlation between UV exposure and the rise of certain skin neoplasms. They can also be indirect; for example, increased UV affects the photosynthesis of some crop species (Whitmee et al., 2015). Anthropogenic changes in the chemical composition of the atmosphere contribute to the acceleration of climate change and induce indirect health impacts by altering the geographic distribution of some disease vector species. The survival conditions of insect vectors such as mosquitoes (of the genera Aedes or Anopheles) are closely linked to temperature and precipitation conditions (Jian et al., 2016).

Contemporary climate change is affecting this species and their distribution, leading to the emergence or re-emergence of tropical diseases outside their traditional location (Whitmee et al., 2015). In addition, rapid changes in land use, urbanization and massive deforestation are creating new breeding grounds and thus generating greater exposure to risk. The creation of megacities, road extensions and the incessant movement of population challenge certain ecosystems and bring the human species closer to the wild fauna, which can favor the transmission of infectious agents as well as the creation of intensive breeding (Whitmee et al., 2015). The generalized interconnection and the massification of international travel contribute, for their part, to global epidemics, as shown by the recent SARS, H5N1 or COVID-19 epidemics.

It is true that anthropogenic environmental changes have complex effects on the health of living beings. The interactions between environmental and human dynamics cannot be read in a linear fashion. Increased productivity has facilitated access to food; technological developments are more efficient in exploiting ecosystem services for provisioning. Health risks are linked to a greater sensitivity and vulnerability of populations to new and sometimes more frequent hazards, resulting in a stronger perception of potential threats (Beck, 1992).

For example, the increase in temperature generates a health risk that is more or less high depending on the population. Several health effects of environmental changes have been noted: malnutrition, resulting from a drop in production linked to soil erosion; an increase in gastrointestinal diseases, linked to a drop in drinking water supplies; ailments linked to exposure to pathogenic substances such as atmospheric pollutants, carcinogens, mutagens or chemical reprotoxins (Whitmee et al., 2015).

The industrial society finally induces a multiplication of risk factors which are all the more complex to grasp as there is undoubtedly a certain lag time between environmental changes and the deterioration of overall health, as attested by certain indicators such as the stagnation of disability-free life expectancy in certain countries such as France (INSEE, 2020). In addition, there are phenomena of ignorance and uncertainty that contribute to limiting the recognition and resolution of certain environmental health problems.

The Construction of Environmental Health Problems

Mobilizing researchers from various disciplines, environmental health is often analyzed in a localized manner, based on pathologies, risks and substances perceived as potentially dangerous for human health. However, they do not all have the same visibility and the trajectories of environmental health problems are very differentiated; certain phenomena of resistance are well documented in the context of "agnotology" in particular which refers to a field of research on the production of ignorance (Proctor, Chiebinger, 2008).

Pluridisciplinarity, Transversality and Non-Linearity of Construction Processes

Cross-disciplinarity characterizes research on environmental health issues. For example, the research initiated on nanoparticles or endocrine disruptors was launched by a multidisciplinary network of scientists (Krimsky, 2000; Langston, 2008, Guilleux, 2015). These substances, categorized according to their mechanism of action, are present in all environments (water, air, food). Their pathological effects are diversified and they potentially concern several families of chemical substances. They are therefore studied by physicians, toxicologists and ecotoxicologists, chemists, biologists, etc. They also crystallize questions about knowledge on the effects of low doses, the social conditions of production of expertise or the way uncertainties are treated by public authorities.

The category of endocrine disruptors was first used at the Wingspread conference (Wisconsin, United States) in July 1991 (Langston, 2008). This event, initiated by Theodora Colborn, an ecotoxicologist working at the time for the World Wildflife Fund, brought together about twenty researchers linked to environmental and environmental health mobilizations: toxicologists, immunologists, endocrinologists, cell biologists, psychiatrists, ecologists, zoologists and anthropologists, including Pete Meyrs, Mary Wolff, Ana Soto, John McLahan, Frederick vom Saal and Lou Guilette (Krimsky, 2000). However, this category was not built on the basis of traditional public health knowledge, the expertise of which has been criticized; in fact, for these researchers warning of the danger of endocrine disruptors, the dose does not make the poison (Langston, 2008). It is therefore important to detect a pollutant in small quantities, even if they are below the regulatory standards. According to them, these standards do not allow us to conclude that there is no danger because there may be synergistic and additive effects with other pollutants present in the environment (Krimsky, 2000; Langston, 2008), especially since the mimetic power of hormones, specifically sexual hormones, is translated into action at low doses on human, animal and plant organisms. This mimetic power would result in a non-linearity of the relationship between the doses and the effects which can be more important at low doses than at higher doses. It would lead to a tenfold sensitivity of organisms during certain phases of life such as embryonic development which are called "windows of exposure" (Brouwer et al., 1999). A single substance would have multiple effects depending on the time, the dose and the nature of the mixtures, since "cocktail" effects are particularly complex and irreducible to simple additions of impact; they would concern not only the people exposed but several successive generations because of genetic involvement (Crews et al., 2000).

This notion was a way of thinking about toxicity that had been formalized between the two world wars, in the field of industrial hygiene, in relation to the management of occupational risks (Fressoz, 2012). It was a precious resource for industrialists who could argue that thresholds were respected in the face of lawsuits brought by their employees. According to several authors, toxicology contributed to legitimize the idea that the presence of chemical substances in the environment is a normal consequence of economic development, the issue being to determine acceptable levels. The regulation of chemical substances, such as pesticides, was not intended to protect the health of populations, but to develop industry and intensive agriculture (Fressoz, 2012; Jouzel, 2019; Jas, 2010). Endocrine disruptors contribute to re-actualize the controversies around the health effects of low doses. Public health expertise would, moreover, be too compartmentalized, whereas living beings are in reality exposed simultaneously to multiple pollutants that accumulate (in the air, food, etc.) that result in the alterations in sexual development induced by chemistry.

Beyond the Wingspread conference, the network set up and its influence, it is necessary, as Sara Wylie (2012) reminds us, to place the emergence of endocrine disruptors in a longer history of substances that are now reclassified as such. Far from being new, as activist organizations often claim, the best-known substances were, for the most part, put into circulation between 1940 and 1970, or even much earlier as in the case of heavy metals. Thus, the first to be included in the new category were included because they had been the subject of significant research and expertise during the 1970s and 1980s that had begun to describe some of the specific properties and effects subsequently included in the paradigm (Wylie, 2012). In addition, these properties were identified in connection with consequent accidents—as in the case of DPCB (1,2-dibromo-3-chloropropane), keltane, kepone/mirex/chlordecone, DES (diethylstilbestrol), dioxins or PBBs (polybrominated biphenyls)—or with background contamination of the environment by ubiquitous substances such as DDT (dichlorodiphenyltrichloroethane) or PCBs (polychlorinated biphenyls), which raises the role of the regulation spaces (administrative, professional or legal) of chemical substances in the second half of the twentieth century (Laurent, 2011, Munck af Rosenschöl et al., 2014).

Moreover, far from being limited to the United States, the analysis must take into account the dynamics of international circulation, carried out by a wide range of actors, including political and regulatory bodies, industrialist corporations and militant organizations such as the «Réseau Environnement Santé» in France (Guilleux, 2015). Thus, as early as 1996, this hypothesis became an object of concern and action within the OECD (Organization for Economic Cooperation and Development) and the European Union (Ansell and Balsiger, 2009), and gave rise to international scientific initiatives outside the United States, and even, in the case of South Korea, to regulations specific to endocrine disruptors (Lee, 2010).

Contrasting Framing and Publicizing of Environmental Health Problems

Environmental health problems have variable trajectories. For example, they can be mediatized following a change in framing, as in the case of asbestos in France, which became a public health problem rather than an occupational health problem. Emmanuel Henry shows that it was when the asbestos risk was redefined as an environmental risk threatening the entire population that interest in this problem increased (Henry, 2007). Problems can, conversely, remain confined; for example, indoor air pollution receives little media coverage in France, due in particular to the fragmented nature of its media treatment, the juxtaposition of regulations and the particularities of mobilizations mainly carried by heterogeneous actors from different sectors and close to institutional spheres; the response comes mainly from the State, around the production of standards (guide values), informational tools and mobilizing market mechanisms (public notice and information, labeling), which has had the effect of channeling emerging concerns and mobilizations in this field of health environment (Crespin et al., 2015).

Trajectories can also vary from one country to another, as illustrated by the aforementioned case of endocrine disruptors, for which research, expertise and mobilizations have interacted to produce framings of the problem that are significantly different in France and Europe from that prevailing in the United States (Gaudillière, Jas, 2016). In the fall of 2015, the publication of a book on the actions of European lobbyists in this area (Horel, 2015) received a lot of media attention. In it, the author, an investigative journalist, shows how the European Commission came not to produce a regulatory definition of endocrine disruptors, following the mobilization of Brussels lobbyists representing various chemical industries. Lobbyists from non-governmental organizations, experts and the European Union (EU) institutions working for the regulation of these substances did not have comparable means (Horel, 2015).

The controversy over the regulation of endocrine disruptors resembles many other cases and debates about the chemicals massively produced by industry and disseminated in our environments and the threats they introduce to the health of workers, residents or consumers (Krimsky, 2000). However, endocrine disruptors, mobilizing multiple actors, including European scientists working on the decline of aquatic animal populations or on the decrease in male fertility, as well as American activists participating in mobilizations for women's health linking breast cancer and pesticide exposure (Krimsky, 2000), have been a target of debate and expertise in the arena of chemical regulation since 1996, in the United States as well as in Europe. It was only in the second half of the 2000s that they became a well-established object of research, while the controversies over Bisphenol A gave new scope to mobilization and activism (Gee, 2006).

Not all environmental health problems are necessarily immediately perceived as such, notably because they often first concern the world of work, and there are many possibilities for cross-fertilization between "occupational health" and "environmental health" (Bécot et al., 2021; Davies, Mah, 2020). Among the actors contributing to the emergence and framing of problems are employees or local residents who produce their own knowledge and help influence public action (Ottinger, 2013), without having the same resources as the polluting industries and lobbyists who sometimes play the role of "merchants of doubt" (Oreskes, Conway, 2010).

The Factory of Doubt and Scientific Ignorance

The diversity of actors, approaches and terms used ("health environment," "ecohealth," "environmental health," "sustainable health," "health and environmental risks," "sustainable health," etc.), as well as the conflicts that animate them, still make the perimeter and institutionalization of environmental health uncertain. On the one hand, situations of doubt and uncertainty persist concerning most of the risks linked to food, the manipulation of living organisms, environmental pollution or the emergence of new viruses. On the other hand, there are discrepancies between the knowledge of the problems, their visibility and the way they are handled by the public.

The fabrication of ignorance is the subject of a dynamic field of study (Jouzel, 2019; Proctor, Chiebinger, 2008; Proctor, 2012), agnotology, which gradually emerged in the first half of the 2000s. It takes as its object the cultural production of ignorance conceived as resulting from the active and organized work of dominant actors seeking to preserve their interests. It has essentially highlighted how industries and states seek to control the production and circulation of scientific knowledge on the deleterious health and environmental effects of industrial activities and products.

The concept was formalized by the historian of science Robert Proctor, who studied the controversies surrounding the causes of cancer and showed how certain industries, agencies and environmental organizations in the United States managed to direct the content of the research and knowledge produced on this issue (Proctor, Chiebinger, 2008; Proctor, 2012). The study of five million documents put online and indexed following a court decision allowed Proctor and other authors to reconstruct the strategies implemented by these industries to build and deliberately maintain, for several decades, the denial of the existence of harmful effects of tobacco on health, notably by directing research and expert opinions as well as by artificially maintaining controversies on the proven negative effects of tobacco, what the author called a "Golden Holocaust" (Proctor, 2012). Other authors (Michaels, 2008; Oreskes, Conway 2010) have also highlighted the organization of secrecy, the retention, selection and suppression of knowledge and information by industrialists.

These strategies partly explain the gaps between the knowledge of the problems, their visibility and their public handling. However, there are many reasons why ignorance is maintained and they are not limited to the actions of industrialists; in particular, we should mention the work of reduction, the factors of inequality between groups of actors leading to an unequal distribution of knowledge and an "undone science" (Hess, 2016; Frickel et al., 2010), and the organization of risk assessment work. Some "environmental health" issues remain quiet and suffer from ignorance while others make a big splash or do not face the production of doubt. The struggles around scientific (mis)knowledge based on the interplay of actors and the role of power relations between different stakeholders have been the subject of rich studies on issues with more or less discrete trajectories (Gilbert et al., 2009). These logics have been the subject of detailed analyses, for example in the context of comparative research (Gaudillière, Jas, 2016); we might mention investigations into the different approaches in the United States and in France to certain products and substances that are potentially dangerous from a health point of view—and their framing as an environmental health problem or not—such as GMOs, pesticides or endocrine disruptors, but also contrasting approaches to inequalities in the face of these risks.

Environmental Justice and Inequality

The effects of environmental factors on health are unevenly distributed, with some populations proving more vulnerable. Epidemiologist Richard Wilkinson (2005) has studied the consequences of economic inequalities on individual well-being and social relations using data from the US states, and has shown that health is strongly linked to the way people experience their place in the social hierarchy. This link between social and environmental issues is the subject in the United States of the reference framework of "environmental justice" (Mohai, 2018), whereas in Europe, it is more likely to be addressed through the issue of "environmental inequalities" (Lejeune, 2015) and will be discussed in the next section.

The Anglo-Saxon Approach to Environmental Justice

Environmental justice is a well-established field of research in the United States. The aim is to highlight the environmental injustices suffered by racial minority populations in the United States beginning in the late 1970s. This field of reflection is carried out by individuals, local committees and associations initially unfamiliar with activist practices. Its origin is often associated with two cases: on the one hand, *Bean v. Southwestern Management* in 1979, the first case to invoke racial discrimination in environmental matters before American courts and tribunals, brought by Professor Robert Bullard (Bullard, 1987; Bullard, 1990; Bullard et al., 2011); on the other hand, the study carried out in Warren County (North Carolina) by Reverend Benjamin Chavis, which aimed to establish a correlation between the presence of populations from ethnic minorities and the proximity of polluting installations on a national scale (Chavis, 1983).

North American environmental justice initially focused on forms of "environmental racism" and then broadened to include poor people in general. It has been progressively appropriated by various grassroots groups in the United States as well as by researchers and academics who seek to show that it is the racial minority population, as well as the poor, who suffer the most from environmental damage and pollution and live in the most degraded environments.

This field of research links social and environmental issues in a unique frame of reference. Environmental justice is gradually becoming the main way to address social and environmental issues together, mainly in American cities (McGranahan, Satterthwaite, 2002). It has received official recognition in the United States. Under pressure from social movements and coalitions of researchers who made their demands public at the First National People of Color Summit, held in Washington DC in 1991, President Clinton put the issue of environmental justice on the agenda and recognized it explicitly in an Executive Order entitled Federal Action to Address Environmental Justice in Minority Populations and Low-Income Populations in 1994 (Russell, 1996). All US agencies have since been expected to consider environmental justice when developing and implementing their policies. The environmental debate has truly opened up to the question of "environmental social justice."

The notion of environmental justice was taken up by a civic movement initiated by civil society in the United States which denounced the collusion between social and environmental injustices; the poorest people, particularly visible minorities, are more exposed to the risks and ecological nuisances linked to the development of industries and pollution, are disadvantaged in terms of quality of life and life expectancy, and have little legal protection. There is thus a rapprochement between environmental justice and local social justice. However, at the local level, a spatial approach to urban and peri-urban areas is dominant, seeking to demonstrate the inequalities in the situation and how populations mobilize or do not mobilize against urban planning and environmental choices. It is characterized by an approach centered on those who make their voices heard, rather similar to NIMBY ("Not in My Backyard"), providing an interpretation of citizens' responses to decisions to locate polluting industries, or LULUs ("locally undesirable/unwanted land uses") covering any new use deemed potentially harmful by residents and local associations. These movements can be understood in terms of environmental injustice or inequality; indeed, these mobilizations are often fueled by the proven or perceived existence of such inequalities or injustices (Schlosberg, 2007; Taylor, 2000).

NIMBYs or LULUs allow for a broadening of the scope of thinking to include those affected ("those affected" or "all affected") in the public decision-making process, including those who do not contest or generally have no voice (Lejeune, 2015). Environmental justice opens up the environmental debate to the question of social justice (Martinez Alier, 2003; Lejeune, 2015). It also makes it possible to include the phenomena of "non-mobilization" and their various potential causes. Finally, an approach based on environmental inequality or injustice opens up the reflection on the "ordinary environmental spaces of everyday life" (Whitehead, 2009) and on ordinary inequalities, through an in-depth analysis of the living and working environments of the people concerned: housing, residential setting, leisure spaces, access to services, etc. (Lejeune, 2015).

The European Approach to Environmental Inequalities

Gradually exported from the United States, environmental justice is being reappropriated differently in different national contexts, mainly under the name of environmental inequalities in Europe. In many European countries, researchers are trying to understand these inequalities in their national context and to raise awareness of this field of thought that is still largely unknown (Agyeman et al., 2003; Emelianoff, 2006; Kohlhuber et al., 2006; Cornut et al., 2007; Steger et al., 2007; Anguelovski, 2013; Lejeune et al., 2012).

This concept is used in Europe mainly to try to study situations of cumulative socio-economic and environmental inequalities at different scales: international (Martinez Alier, 2003), national, regional or local (Deboudt, 2010). This is because it is separated from the coexistence, which characterizes American environmental justice, of scientific research and collective mobilizations.

Two major conceptions of this notion of inequality prevail in Europe. For the Anglo-Saxon approach, inequality is assimilated to a difference in situation resulting from discrimination suffered by groups of individuals. From this conception stems a conception of public action that legislates new rights to correct differences in situation considered unjust. Approaches in continental Europe do not limit the analysis of inequalities to discrimination processes; they highlight the social and territorial dimension of these inequalities, as well as the role played, in their emergence or reduction, by public policies and the dynamics of territorial development (cf. Laigle, 2009).

The notion of "ecological inequality," on the other hand, refers rather to the consideration not only of the environmental risks to which populations are exposed, but also of the impacts generated by households (Dozzi et al., 2008), often the most affluent ones, whether on a local, national or international scale; it also takes into account an ecological justice centered on nature, and not on the relations between human beings and nature (Dobson, 1998).

Other terminologies make it possible to reflect on the issues raised by environmental justice: "urban spatial ecology," "environmental democracy," etc. Environmental inequalities are considered in France in close connection with urban and territorial dynamics. Indeed, it is mainly geographers (Emelianoff, 2006; Ghorra-Gobin, 2005) and economists (Laurent, 2010) who are interested in the issue, unlike in the United States where it is mainly sociology that has taken hold of it. However, these studies have not yet led to a real political agenda for environmental inequalities in France, even if some initiatives exist at the local level. The treatment of environmental inequalities is closely linked to urban heritages, public action conceptions and the local authorities involved in the implementation of sustainable urban development. Moreover, French sociology has not taken up the links between the environment and society in the same terms as the environmental justice movement in the United States, whose focus is to transform society by intervening in "new" inequalities. France is said to be late in integrating the environment into the social sciences and particularly in the study of social and political issues (Kalaora and Vlassopoulos, 2013), with the relationship between the environment, injustice and health occupying a marginal place in both scientific debate and collective action, as does the difficult appropriation of the concept of environmental justice. In France, it is other terminologies and approaches that have finally contributed to the growth of research and reflection on issues close to those raised by environmental justice: urban spatial ecology (Guermond, 2005), the sustainable city and environmental democracy (Barbier and Larrue, 2011).

It is through the prism of multiple themes, such as environmental health (Lavaine, 2010; Akrich, Barthe and Rémy, 2010), popular epidemiology (Martinez Alier, 2003, Brown, 1987), the quality of housing, environmental risks and transformations in society (Callon, Lascoumes and Barthe, 2001) or the relationship between justice and energy (Laurent, 2010; Da Cunha and Guinand, 2014) that a sociology of the environment was defined in France which was later open to issues of justice and equality and to their multiple facets.

Ecofeminism and Gender Inequalities

Despite the lack of a clear and stabilized definition of the terms "environmental justice," "environmental inequalities" and "environmental equity" (Bowen, 2002), the idea on which research on environmental inequalities is based is that people and groups do not suffer the same environmental burden and do not have equal access to urban and environmental services (Emelianoff, 2006).

There are multiple inequality factors: residential, socio-economic, age and gender. In this respect, these inequalities between men and women have been particularly highlighted in the context of the pandemic; for example, a large proportion of frontline workers have been women, working in largely feminized care and health professions. Inequalities in terms of health between men and women also concern teleworkers, as women may be more confronted with degraded working conditions due to the difficulty of combining work and home time. In addition, the lockdown measures have exacerbated intra-family violence, which has increased sharply during this phase of the crisis (Krishnadas and Hayat Taha, 2020; Malik and Naeem, 2020; Miani et al., 2022).

These gender inequalities in the face of environmental risks are at the heart of the ecofeminist approach (D'Eaubonne, 1974; Warren, 1990) which apprehends the aggravation of environmental destruction as the result of a masculine and Western cultural anthropocentric and androcentric logic based on the infinite exploitation of nature and all that is associated with it, including women. Ecofeminism has thus put at the heart of its reflection the connections that exist between the domination of men over nature and that which they exercise over women. It was a question of making women's voices heard within an environmental ethic that had until then been preoccupied with the relationship between man and nature, without asking which man it was. It must allow us to think about the intersecting oppressions of patriarchy on women's bodies and natural resources. Born at the end of the 1970s, particularly after the nuclear accident at Three Mile Island in the United States in 1979, this movement is composed of numerous groups of women mobilized in different contexts, national (Canada, the United States in particular, but also today in Europe or in Asia), European and international.

Ecofeminism which has emerged at national and international levean area of scholarly research since the seminal book *The Death of Nature* (Merchant, 1980), and, more broadly, perspectives that take into account the phenomena of domination as well as the multiple sources of inequality, propose to pay particular attention to environmental issues related to health and vulnerability, and question the autonomy of the economy that obscures its dependence on domestic space and the earth's environment. For, at the crossroads of an environment that is also social and health, environmental health implies taking into account the multiple sources (economic, cultural, gender, spatial…) of vulnerability to risks that the current pandemic context has made particularly visible.

Conclusion

Recognition of the links between the environment and health is long-standing; it can be traced back at least to hygienist movement. However, contemporary awareness of human damage to the environment and its health consequences has contributed to the emergence of new explanatory models of health and has helped to make environmental health a field of research in its own right, emerging in certain countries such as France, and older in others, such as the United States, where an approach in terms of environmental justice dominates. Exposure to risks varies from one population to another, with several factors contributing to situations of inequality or even injustice: economic and social resources, housing, location, age, occupation, being of racial minority populations, living conditions, etc. Moreover, not all risks are recognized in the same way, and an analysis of the construction of environmental health problems highlights the diversity of their trajectories. For example, some are not necessarily perceived as such because they primarily concern the world of work. Finally, not only do not all individuals and groups suffer from the same environmental risk, but access to urban environmental services is far from being equal. In this respect, ecofeminism highlights gender inequalities while more fundamentally questioning the cultural causes of the serious risks that a predatory and masculine approach to our ecosystem poses to the environment and, in turn, to human and animal health.

References

Agyeman J., Bullard R. D., Evans B., 2003, *Just Sustainabilities: Development in an Unequal World*. Cambridge, MIT Press.

Akrich M., Barthe Y., Rémy C., 2010, *Sur la piste environnementale: menaces sanitaires et mobilisations profanes*. Paris, Presses des Mines.

Anguelovski I., 2013, "New Directions in Urban Environmental Justice: Rebuilding Research, Addressing Trauma and Remaking Place". *Journal of Planning Education and Research*, 33 (June), pp. 160–175.

Annandale E., 1998, *The Sociology of Health and Medicine. A Critical Introduction*. Polity Press.

Ansell C., Balsiger J., 2009, "The Circuits of Regulation: Transatlantic Perspectives on Persistent Organic Pollutants and Endocrine Disrupting Chemicals". In Swinnen J., Vogel D., Marx A., Riss H., Wouters J., eds, *Handling Gobal Challenges. Managing Biodiversity/Biosafety in a Global World*, Leuven, Leuven Centre for Global Governance Studies, pp. 288–306.

Barbier R., Larrue C., 2011, *Démocratie environnementale et territoires: Un bilan d'étape Participations*, 1, pp. 67–104.

Beck U., 1992, *Risk Society: Towards a New Modernity*. London, Sage.

Bécot R., Ghis Malfilatre M., Marchand A., 2021, *Pour un décloisonnement scientifique de la santé au travail et de la santé environnementale*, Sociétés contemporaines, 1, n° 212, pp. 5–27.
Blanchon D., Moreau S., Veyret Y., 2009, *Comprendre et construire la justice environnementale*, Annales de géographie, 1 (665–666), pp. 35–60.
Boudia S., 2013, "From Threshold to Risk: Exposure to Low Doses of Radiation and Its Effects on Toxicants Regulation". In: Boudia S., Jas N., eds, *Toxicants, Health and Regulation Since 1945*, London, Pickering & Chatto, pp. 71–88.
Bourdelais P. ed., 2001, *Les Hygiénistes: Enjeux, modèles, pratiques*, Paris, Belin.
Bowen W., 2002, "Forum/An Analytical Review of Environmental Justice Research: What Do We Really Know?". *Environmental Management*, 29, pp. 3–15.
Brouwer A., Longnecker M., Birnbaum L., Cogliano J., Kostyniak P., Moore J., Schantz S., Winneke G., 1999, "Characterization of Potential Endocrine-Related Health Effects at Low-Dose Levels of Exposure to PCBs". *Environmental Health Perspectives*, August, 107, Supplément 4, pp. 639–649.
Brown P., 1987, "Popular Epidemiology: Community Response to Toxic Waste Induced Disease in Woburn, Massachusetts and Other Sites". *Science, Technology, and Human Values*, 12(3–4), pp. 76–85.
Bullard R. D., 1987, "Blacks and the New South: Challenge of the Eighties". *Journal of Intergroup Relations*, 15(2), (Summer), pp. 25–39.
Bullard R. D., 1990, *Dumping in Dixie: Race, Class, and Environmental Quality*. Boulder, Colo. Westview Press.
Bullard, R. D., Johnson, G. S. and Angel O. Torres., 2011, *Environmental Health and Racial Equity in the United States: Building Environmentally Just, Sustainable, and Livable Communities*. Washington, DC: American Public Health Association Press.
Callon M., Lascoumes P., Barthe Y., 2001, *Agir dans un monde incertain: essai sur la démocratie technique*. Paris, Seuil.
Carson R., 1962, *Silent Spring*. Houghton Mifflin Company.
Chavis B., 1983, *Psalms from Prison*. New York. Pilgrim Press, 1983.
Clarke A., Mamo L., Fosket J., Fishman J. and Shim J. ed., 2010, *Biomedicalization. Technoscience, Health and Illness in the U.S.* Duke University Press, Durham and London.
Cornut P., Bauler T., Zaccai E., 2007, *Environnement et inégalités sociales*. Bruxelles, éditions de l'Université de Bruxelles.
Crespin R., Ferron B., Hourcade R., Jamay F., Le Bourhis J.-P., Ollitrault S., 2015, *Air intérieur: Actions publiques et jeux d'acteurs*. Rapport CNRS ADEME.
Crews D., Willingham E. and Skipper J., 2000, "Endocrine Disruptors: Present Issues, Future Directions". *The Quarterly Review of Biology*, 75(3) (September), pp. 243–260.
Crutzen Paul J., 2002, "Geology of Mankind". *Nature*, 415, p. 23.
D'Eaubonne F., 1974, *Le Feminisme ou la Mort*. Paris: Pierre Horay.
Da Cunha A., Guinand S., 2014, *Qualité urbaine, justice spatiale et projet : ménager la ville*. Lausanne, Presses polytechniques et universitaires romandes.
Davies T., Mah A. (dir.), 2020, *Toxic Truths. Environmental Justice and Citizen Science in a Post-Truth Age*. Manchester, Manchester University Press.
Deboudt P., 2010, *Inégalités écologiques, territoires littoraux et développement durable*. Villeneuve d'Ascq, Presses du Septentrion.
Delmas C., 2011, *Sociologie politique de l'expertise*. Découverte, Repères.

Dobson A., 1998, *Justice and the Environment: Conceptions of Environmental Sustainability and Dimensions of Social Justice*. Oxford: Oxford University Press.

Dozzi J., Lennert M., Wallenborn G., 2008, *Inégalités écologiques : analyse spatiale des impacts générés et subis par les ménages belges, Espace, Populations, Sociétés*, n°1, pp. 127–143.

Emelianoff C., 2006, *Connaître ou reconnaître les inégalités environnementales ?*, ESO, *Travaux et Documents*, n° 25.

Fressoz J.-B., 2012, *L'Apocalypse joyeuse. Une histoire du risque technologique*. Paris, Seuil.

Frickel S., Gibbon S., Howard J., Kempner J., Ottinger G. & Hess D. J., 2010, "Undone Science: Charting Social Movement and Civil Society Challenges to Research Agenda Setting". *Science, Technology & Human Values*, 35(4), pp. 444–473.

Gaudillière J.-P., Jas N., 2016, *La santé environnementale au-delà du risque ? Perturbateurs endocriniens, expertise et régulation en France et en Amérique du Nord*, *Sciences sociales et santé*, 34(3), pp. 5–18.

Gaudillière J.-P., 2002, *Inventer la biomédecine. La France, l'Amérique et la production des savoirs du vivant (1945–1965)*. Paris, La Découverte.

Gee D., 2006, "Late Lessons from Early Warnings: Toward Realism and Precaution with Endocrine-Disrupting Substances". *Environmental Health Perspectives*, 114(1), pp. 52–160.

Ghorra-Gobin C., 2005, *Justice environnementale et intérêt général aux Etats-Unis*, *Annales de la recherche urbaine*, 99(15), pp. 14–19.

Giddens A., 1999, "Risk and Responsibility". *The Modern Law Review*, January, 62(1), pp. 1–10.

Gilbert C., Henry E. (dir.), 2009, *Comment se construisent les problèmes de santé publique*. Paris, La Découverte.

Guermond Y., 2005, *Pour une écologie spatiale urbaine, La ville durable, du politique au scientifique*, Cemagref, CIRAD, Ifremer, INRA, pp. 195–205.

Guilleux C., 2015, *L'institutionnalisation de la santé environnementale en France*, Sociology Thesis, Aix-Marseille.

Gusfield J., 1981, *The Culture of Public Problems: Drinking-Driving and the Symbolic Order*. Chicago, University of Chicago Press, University of Illinois Press.

Henry E., 2007, *Amiante: un scandale improbable*. Rennes, Presses Universitaires de Rennes.

Herzlich C., 1969, *Santé et maladie, Analyse d'une représentation sociale*. Paris, La Haye, Mouton & Co.

Hess D., 2016, *Undone Science. Social Movements, Mobilized Publics, and Industrial Transitions*. The MIT Press.

Horel S., 2015, *Intoxication. Perturbateurs endocriniens, lobbyistes et eurocrates: une bataille d'influence contre la santé*. Pa ris, La Découverte.

INSEE, 2020, *France, Portrait Social* https://www.insee.fr/fr/statistiques/fichier/4928952/FPS2020.pdf

Jas N., 2010, *Pesticides et santé des travailleurs agricoles en France »*, *Le Courrier de l'environnement de l'INRA*, 59, pp. 47–59.

Jian Y., Silvestri S., Brown J., Hickman R. and Marani M., 2016, "The Predictability of Mosquito Abundance from Daily to Monthly Timescales". *Ecological Applications*, December, 26(8), pp. 2611–2622.

Jouzel J.-N., 2019, *Pesticides. Comment ignorer ce que l'on sait*. Paris, presses de SciencesPo.
J.S.H., 1901, "The Work of Max Von Pettenkofer". *The Journal of Hygiene*, 1(3), July, pp. 289–294.
Kalaora B., Vlassopoulos C., 2013, *Pour une sociologie de l'environnement : Environnement, société et politique*, Paris, éditions Champ Vallon.
Kalachanis K., Michailidis I., 2015, "The Hippocratic View on Humors and Human Temperament". *European Journal of Social Behaviour*, 2(15), pp. 1–5.
Kohlhuber M., Mielck A., Weiland S. K., Bolte G., 2006, "Social Inequality in Perceived Environmental Exposures in Relation to Housing Conditions in Germany". *Environmental Research*, 101, pp. 246–255.
Krishnadas J., Hayat Taha S., 2020, "Domestic Violence Through the Window of the COVID-19 Lockdown: A Public Crisis Embodied/Exposed in the Private/Domestic Sphere". *Journal of Global Faultlines*, 7(1) (June–August), pp. 46–58.
Krimsky S., 2000, *Hormonal Chaos: The Scientific and Social Origins of the Environmental Endocrine Hypothesis*. Baltimore MD, Johns Hopkins University Press.
Laigle L., 2009, *Conceptions des inégalités écologiques : Quelle place pour/dans les politiques du développement urbain durable*, Marne-la-Vallée, CSTB.
Langston N., 2008, "The Retreat from Precaution: Regulating Diethylstilbestrol (des), Endocrine Disruptors, and Environmental Health". *Environmental History*, January, 13(1), pp. 41–65.
Laurent E., 2010, "Environmental Justice and Environmental Inequalities: A European Perspective". *Document de travail OFCE*, n° 5.
Laurent B., 2011, "Political Spaces for Nanomaterials". *European Journal of Risk Regulation*, 2(4), pp. 577–582
Lavaine E., 2010, "Atmospheric Pollution, Environmental Justice and Mortality Rate: A Spatial Approach". *Document de travail du Centre d'économie de la Sorbonne*.
Lee J., 2010, "Atopic Dermatitis and the Making of an Environmental Disease in Contemporary South Korea". In Boudia S., Jas N., eds, *Carcinogens, Mutagens, Reproductive Toxicants: The Politics of Low Doses and Limit Values in the xxth and xxIst Centuries*, Book of Papers, International Conference, Strasbourg, pp. 29–31.
Lejeune Z., 2015, *La justice et les inégalités environnementales*, Revue française des affaires sociales, N° 1–2, pp. 53–78.
Lejeune Z., Chevau T. and Teller J., 2012, *La qualité du logement comme variable environnementale : l'exemple de la région urbaine de Liège (Wallonie)*, Flux. Cahiers scientifiques internationaux Réseaux et Territoires, 89–90, pp. 30–45.
Malik S., Naeem, 2020, *Impact of COVID 19 Pandemic on Women: Health, Livelihoods & Domestic Violence*. Sustainable Development Policy Institute.
Martinez Alier J., 2003, *The Environmental of the Poor: A Study of Ecological Conflicts and Valuation*. Cheltenham, Edward Egar Publishing Limited.
Mc Granahan G., Satterthwaite D., 2002, "The Environmental Dimensions of Sustainable Development for Cities". *Geography*, 87(3), pp. 213–226.
Merchant C., 1980, *The Death of Nature: Women, Ecology, and the Scientific Revolution*. New-York, Harper & Row.
Miani C., Wandschneider L., Batram-Zantovoort S. and Razum O., 2022, "Covid-19 Pandemic: A Gender Perspective on How Lockdown Measures have Affected Mother with Young Children". In Kupfer A. and Stutz C., *Covid, Crisis, Care and*

Change? International Gender Perspectives on Re/Production, State and Feminist Transitions, Verlag Barbara Budrich, pp. 75–93.

Michaels D., 2008, *Doubt is Their Product: How Industry's Assault on Science Threatens Your Health*. Oxford, Oxford University Press.

Mohai P., 2018, "Environmental Justice and the Flint Water Crisis". *Michigan Sociological Review*, 32, pp. 1–41.

Munck af Rosenschöld J., Honkela N. and Hukkinen J., 2014, "Addressing the Temporal Fit of Institutions: The Regulation of Endocrine-Disrupting Chemicals in Europe". *Ecology and Society*, December, 19(4), pp. 30.

Oreskes N., Conway, E., 2010, *Merchants of Doubt: How a Handful of Scientists Obscured the Truth on Issues from Tobacco Smoke to Global Warming*. Bloomsbury Press, New York.

Ottinger G., 2013, *Refining Expertise: How Responsible Engineers Subvert Environmental Justice Challenges*. New York University Press, New York.

Philippon D., 2018, "Changing Food Cultures, Changing Global Environments". *Global Environment*, 11(1), p. 4–11.

Proctor R., 2012, *Golden Holocaust. Origins of the Cigarette Catastrophe and the Case for Abolition*. University of California Press.

Proctor R., Chiebinger L. (eds.), 2008, *Agnotology: The Making and Unmaking of Ignorance*. Stanford, Stanford University Press.

Russell C., 1996, "Environmental Equity: Undoing Environmental Wrongs to Low Income and Minority Neighborhoods". *Journal of Affordable Housing & Community Development Law*, Winter, 5(2), pp. 147–164.

Saint-Marc P., 1989, *Les catastrophes écologiques au quotidien*, Revue des Deux Mondes, décembre, pp. 102–112.

Schlosberg D., 2007, *Defining Environmental Justice: Theories, Movements and Nature*. Oxford, Oxford University Press.

Shackleton R., 1955, "The Evolution of Montesquieu's Theory of Climate". *Revue Internationale de Philosophie*, 9(33/34), pp. 317–329.

Sherwood Rowland F., 1990, "Stratospheric Ozone Depletion by Chlorofluorocarbons". *Ambio*, October, 19(6/7), pp. 281–292.

Steger T., Antypas A., Atkins L., Borthwick F., Cahn C., 2007, "Making the Case for Environmental Justice in Central and Eastern Europe", Budapest, Hungary, Health and Environment Alliance (HEAL), the Central European University, Environmental Justice Program, and the Coalition for Environmental Justice.

Stupfel M., 1976, "Recent Advances in Investigations of Toxicity of Automotive Exhaust". *Environmental Health Perspectives*, 17 (October), pp. 253–285.

Taylor D. E., 2000, "The Rise of Environmental Justice Paradigm. Injustice Framing and the Social Construction of Environmental Discourses". *American Behavorial Scientist*, 43 (4), pp. 508–580.

Waltner-Toews D., 2004, *Ecosystem Sustainability and Health*. Cambridge University Press.

Ward B., Dubos R., 1972, *Only One Earth: The Care and Maintenance of a Small Planet*. Norton and Co.

Ward B., 1973, *Only One Earth*, The UNESCO Courier: A Window Open on the World, 26 (1), p. 8–10.

Warren K., 1990, "The Power and Promise of Ecological Feminism". *Environmental Ethics*, 12, pp. 125–146.

Whitehead M., 2009, "The Wood of the Trees: Ordinary Environmental Injustice and the Everyday Right to Urban Nature". *International Journal of Urban and Regional Research*, 33, pp. 662–681.

Whitmee S., Haines A., Beyrer C. ed., 2015, "Safeguarding Human Health in the Anthropocene Epoch: Report to the Rockefeller Foundation—Lancet Commission on Planetary Health". *The Lancet*, 386, pp. 1973–2028.

Wilkinson R., 2005, *The Impact of Inequality How to Make Sick Societies Healthier*. London, Routledge.

Wylie S., 2012, "Hormone Mimics and Their Promise of Significant Otherness". *Science as Culture*, 21(1), 49–76.

Zio E., Aven T., 2013, "Industrial Disasters: Extreme Events, Extremely Rare. Some Reflections on the Treatment of Uncertainties in the Assessment of the Associated Risks". *Process Safety and Environmental Protection*, 91, pp. 31–45.

Sustainable Housing: International Relations Between Housing and the Environment Revisited

Sophie Nemoz

INTRODUCTION

More than forty years ago, at the first United Nations Conference on Human Settlements, growing environmental problems were identified (Habitat I: Vancouver, UN, 1976). Twenty years later, "sustainable housing" was proclaimed the planetary solution by the Habitat Agenda (City and Town Summit: Istanbul, 1996). At Habitat III, the third United National Conference on Housing and Sustainable Urban Development (Quito, 2016), delegates reaffirmed a "global commitment," despite this still remaining an understudied area of international relations. Over the last decade, much research has been done on sustainable housing and cities, with authors frequently describing the institutional impact of the UN conference cycle. While they highlight the predominance of environmental anchoring, few researchers have analyzed the construction of an "international regime" as such, namely the "implicit or explicit principles, norms, rules and decision-making procedures around which actors' expectations converge in a given area" (Krasner, 1983: 2; Haas et al., 1993).

What environmental regime applies to human housing? How should we understand both the continuity and cyclicality of the processes involved and

S. Nemoz (✉)
Laboratory of Sociology and Anthropology (LaSA, MSHE, UBFC), University Bourgogne Franche-Comte, Besancon, France
e-mail: sophie.nemoz@univ-fcomte.fr

process change? How do past actions and configurations influence future organizational dynamics and decisions?

This chapter revisits these questions in light of environmental politics. In the context of a proliferation of bilateral and multilateral agreements at various levels, from a regional to a global scale, research from an international regime-based perspective focuses on how each agreement is crafted, isolating the results obtained in relation to original objectives (Finnemore and Sikkink, 1998). My approach differs somewhat in this respect: rather than an accounting assessment of the efficacy of a particular international regime on sustainable housing, I am instead seeking a better grasp of the impact of time on the international relations connecting the environment and housing. This question is one already analyzed using a sociological approach, exploring the innovation underway at various levels (Némoz, 2009). International investigation revealed that the environmental regimes implemented in relation to the housing sector are not a monolithic object. The regimes were analyzed within the framework of the dynamics of an asynchronous process, carrying with it a jumble of traditions and fossilized representations as well as incorporating new knowledge, discoveries and a certain creativity with respect to rules.

This approach is one that has been pursued ever since, attentive to both the principles and modes of present norm creation and decision-making as well as to the influence of pre-existing mindsets. It follows Green Political Theory in that it always seeks to assess the impact of time and weigh its effects, and to report on political processes situated in contexts and configurations of unequal duration and form. Following on from the above and from the founding works of this interdisciplinary field of human and social sciences, the aim here is to examine in greater depth the time aspect of the international relations connecting the environment and housing, by studying the relationships that a "human group establishes between two or more processes, one of which is standardised to serve as a frame of reference and a standard of measurement for the others" (Elias, 1992: 15).

While my approach is thus resolutely comparative, the chapter combines this analysis of temporalities with a parallel approach to spatialities. I will begin by looking at this closer comparison of the time and space of the politics in question. Based on a review of existing research, I highlight the value of a mobile perspective on international sustainable housing regimes across decades and territories. I will elucidate its positioning within a theoretical context and its empirical anchoring, together with its supporting methodology. This approach will then allow me to analyze the historically constructed character of the UN's pronouncement on sustainable housing, demonstrating an uncertain convergence of international relations between transformation and reproduction. The analysis focuses on what the actors actually do, not on what they are supposed to do according to their institutional mandate. Lastly, through the field of Environmental Political Theory, I will focus on re- examining path dependence and change within international environmental regimes. The articulation between relationship to time and political action forms the core concern of this chapter, differentiating between the plurality of temporalities at play in the name of sustainable housing and the strategic and tactical forces steering its political practice.

The "Detours" of a Comparative Historical Analysis: A Decentered Approach to International Environmental Regimes

Questioning, rather than evaluating, modes of governance

My previous work sought, specifically, to question the transformation of modes of governance over the connections between the environment and housing (Némoz, 2009). This is a relatively unexplored space from a macrosocial point of view, remains mostly unknown to sociohistorical research and, yet, has regularly formed a key focus of international relations in recent decades. However, capacity to achieve the objectives established by the various agreements has been the subject of many evaluative approaches, resulting in the creation of databases of varying scope: regional, national and transnational (Cole, 2006; Winston, 2007; Janjua, Sarker, Biswas, 2019). While, for the sake of completeness, it would be desirable to produce an exhaustive inventory, the fact is that this type of approach dominates recent work, to the point of becoming almost unavoidable, whether in official documents, gray literature (materials and research produced by organizations other than commercial or academic publishing and distribution channels), or scientific publications. For a number of years, buildings, architecture, cities and territories have been regularly subjected to these accounting analyses of the environmental impacts of public policies, despite criticism of reductionism (Bunz et al., 2006; Moore, 2007; Némoz, 2009; 2016). Much work in silos and/or that is primarily quantitative in nature has been done on the question of sustainable housing without adopting an approach that takes effects and challenges into account.

The overall effectiveness of differential studies of a wide range of international regimes is starting to be more heavily qualified in the environmental field, where the most experienced commentators note the arduousness of such a task with some dismay (Conca and Dabelko, 2010). This review of the literature, which compares research work in both English and French, invites the re-examination of these regimes, and not only as institutions and as instruments of public policy at the international level, such as taxation, to name but one example (Sunikka, 2003), i.e., as means for governments and other stakeholders to better manage, collectively, problems perceived to be common to other parties. My analysis, which always aims to assess the impact of time on the political perception of the challenges associated with the organization of a shared world, forms part of the vast program of "political ecology" aimed at constituting a theoretical basis for understanding how societies collectively organize their economic and ecological functioning.

As far as "sustainable human settlements" are concerned, there is little work that questions the law. In addition, French political science has not yet employed an international relations-based approach to these subjects. Since the 1990s, research in this field has mainly explored the environment in relation to public action, political parties, the sociology of environmental mobilizations and participatory democracy (Lascoumes, 2018; Fourniau, Blondiaux, Bourg, Cohendet, 2022). Nowadays, the policy issues are becoming more pronounced with respect to standardization in the name of sustainable development, particularly in the housing sector where, from an international point of view, an increase in restrictions on the use of inhabited space has been observed by expert systems. The accumulation of construction standards is thus being driven by the technocratic power of governments in various countries: Argentine, Australia, Brazil, Canada, South Africa and United States (Manseau and Seaden, 2001). The acuity of this observation in terms of innovation was observed in the context of a multi-sited ethnography, which raised and explored the question of historical influences on sustainable housing governance in a manner decentered from the approaches traditionally used to study international regimes. Their analyses come from international political economy, which is characterized by a utilitarian and functionalist notion of the principles, rules and decision-making procedures around which actors' expectations converge in a given area of inter-state relations, without taking into account the regulations at play within national economies and the comparative factors related to their definition (Krasner, 1983; Gilpin, 1987).

Approaching the Time and Space of International Relations in Comparative Terms

International relations is a field of political science that focuses on relations among and between states and other institutions belonging to the international system. The latter was a topic I broached via field observations, interviews and archives when starting my thesis work (Némoz, 2009) at the French Ministry of Ecology, Energy, Sustainable Development and Land Use Planning (MEEDDAT), the year before the French multi-party round table on the environment—*Grenelle de l'Environnement*—took place. As a winner of the doctoral competition run by the French Environment and Energy Management Agency (ADEME) and being co-funded by the Urban Development, Construction and Architecture Plan (PUCA), I had a privileged vantage point over the governance of connections between the environment and housing. While the administrative merger of the Ministry of the Environment with the Ministry of Public Works was a key event, quite unprecedented in French history, other changes also took place during these years. The traditional role of *conseiller du Prince*—adviser to the Prince—was relegated in favor of the more modest but no less important role of *arpenteur du social*—social surveyor. In his preface to Max Weber's book *The Professor and the Politician*, Raymond Aron notes that: "Resistance of the social sciences to the intrusion of politics has always been more problematic than in the natural sciences... Nothing is easier or more tempting in political economics than confusion between mental schemes and reality. We attribute to the latter merits that, strictly speaking, belong only to the former" (1959: 23). Such a warning in this chapter dedicated to the political ecology of housing has the merit of paying attention to its dual dimensions, both symbolic and concrete, in order better to discern their interactions and not to confuse them.

Rather than bypassing the difficulty of axiological neutrality suggested by the Aron quotation in terms of subjective values and the scientific requirement of objectivation, it seems essential to analyze it. To avoid remaining confined to the investigation of current states of social phenomena, I have adopted a transhistorical approach. This involves comparing the differences and similarities between the various temporal and cultural contexts of modes of government. The aim of this approach was to forge a new knowledge-based relationship with the political issue of sustainable housing. The questioning not only of changes, but also of continuities, prompted this investigation of the past. The socio-history of relations between housing and the environment was established using several methods. In addition to almost daily observations of the French government's interactions with various institutions of the international system over a period of three years, the field research consisted of a corpus of material from nearly a hundred institutional, private and non-profit actors from across Europe, from supranational as well as national and local organizations. Biographical interviews were used to reconstruct the historicity of the processes of public actions. As well as involving a variety of informants and working on contrasting practices and representations, this methodology was accompanied by the study of archives in different countries.

The international comparison relates not only to time but also to space. By taking a historical and geographical step back, our dynamic perspective on the implicit or explicit principles, norms, rules and decision-making procedures linking the environment and housing in France, Finland and Spain involves "a detour," in the sense used by anthropologist Georges Balandier to defend intercultural distancing as the direct route to knowledge of our societies (1985). He further states that "this change of perspective better discerns how politics and power deal with the new and the unprecedented" (Balandier, 1985: 27). It is this distancing effect that we are seeking through mobile knowledge of international relations, the history of which continues to have an impact and which we continue to analyze within Europe and beyond, through the assiduous study of supra-, multi- and national texts and communications. For more than a decade, extensive research has aimed to "distinguish between observational effects and reality effects in order to be able to deconstruct normative arrangements" (Némoz, 2016: 33). It has certainly set us on a long march toward pragmatic science. This philosophical approach began in the late nineteenth century, in the United States. One of its earliest proponents, Charles Sanders Peirce, coined the expression from the Greek "*pragma*," which refers to what has been done or is being done (Peirce,

1878: 286–302). It thus signifies an attentive consideration of the action, of its practical effects and of its concrete consequences. According to William James, another founder of pragmatism, the attention paid to the experience interfaces with reflection on the multiple contexts that shape the action (James, 1907). It is with this open perspective toward what is happening, collectively and in relation to changes, that this article is attempting to examine the interactions between social relations with the environment and urban development policies.

A Pragmatic Assumption About International Regimes for Sustainable Housing Development

The sociology of risk, led by Ulrich Beck, has strongly criticized the intellectual constraints imposed by "methodological nationalism" (Beck, 2004). Although it can inspire to reflect transnationally on the regimes for the development of sustainable housing (Némoz, 2009; 2016), our research does not share its positions that turn away from state-centered approaches and focus on processes of individual "empowerment," relegating to the background of the social structures and national collective memories that preside over much of the mobilization involved (Grisoni and Némoz, 2017). By not opposing the observation scales, both temporal and spatial, of international relations from the long term at a macro (global) level, the medium term at a meso (multinational) level, to a more micro focus, sensitive to what is happening within one country, and even for its individual figures, my comparative historical sociological approach has sought to move beyond a linear and developmental understanding of innovations and their political history (Némoz, 2009).

Attentive to the meaning that actors attribute to their actions and to the categories they use in the context of speeches, official texts or even in gray literature, the hypothesis that is formulated can be described as *pragmatic* insofar as it concerns the capacity of actors to adjust to different situations, across time and space. It involves re-examining the convergence of expectations in international environmental regimes that are interacting, rather than considering them in isolation. This hypothesis also borrows the notion of "tactics" proposed by Michel de Certeau (1990).

A tactic is a calculated action determined by the absence of a proper locus. [...] The space of a tactic is a space of the other. Thus it must play on and with a terrain imposed on it and organized by the law of a foreign power. It does not have the means to keep to itself, at a distance, in a position of withdrawal, foresight, and self-collection: it is a maneuver "within the enemy's field of vision". [...] It operates in isolated actions, blow by blow. It takes advantage of "opportunities" and depends on them, being without any base where it could stockpile its winnings, build up its own position and plan raids (de Certeau, 1990: 60–61).

Tactics do not anticipate, they are defined in action. This notion allows us to account for the activity of an actor who constructs the meaning of their action over time, taking advantage of the opportunities that present themselves with no overarching vision. Strategies, on the other hand, rely on the resistance afforded by a global vision against the erosion of time. We hypothesize that this spatio-temporal relationship between strategies and tactics exists beyond the everyday, over the long term of international relations linking the environment and housing.

THE HISTORICAL CONSTRUCTION OF SUSTAINABLE HOUSING: AN UNCERTAIN CONVERGENCE OF INTERNATIONAL RELATIONS BETWEEN CHANGE AND CONTINUITY

From the Blurring of Traditional Reference Points to an Ideological Mask

In contrast to the rush to theorize the change and the positive potential of sustainable urban development, historical comparison has enabled the identification of both continuities and changes in international relations over the last fifty years. At present, it is as important to analyze the environment and its problems as it is to understand how the environment has become such a problem, i.e., to analyze the foundations of social interventions in this field, whether public or private. With this in mind, this international research focused on the first United Nations Conference on the Environment. Held in Stockholm's diplomatic quarter from June 5 to 16, 1972, the conference saw non-governmental, scientific and humanitarian organizations denounce the irreversible damage being caused to the planet by "population growth and human activities" (UN, 1973). The concept of "development", invented by the United States to launch the Marshall Plan in battered post-war Europe and then taken up by the economic and social policies of the countries of the Global North "in favor" of those of the South, lay at the heart of negotiations between these two categories of countries. It was of concern to Western governments due to its consequences in terms of the depletion of natural resources, and to the rest of the world because of its lack of effectiveness in addressing poverty. However, the chair of the conference, Maurice Strong, reached a consensus on the creation of the United Nations Environment Programme (UNEP), charged with expanding on the meaning of the term "eco-development", proposed by its secretary Ignacy Sachs. Sachs placed it "equidistant from excessive economism that does not hesitate to destroy nature in the name of immediate economic profits and the no less outrageous ecologism that sets up the conservation of nature as an absolute principle to the point of sacrificing the interests of humanity and rejecting the validity of anthropocentrism" (Sachs, 1980: 32). Many allusions were made to the heightening conflictual tensions between positions being presented as extremes. The rhetoric lands more on the side of the interests of human activities, the questioning of which is of secondary importance to achieving social pacification and consolidation of the established order.

It was not until four years later that the links between the spatial settlement of humans and the degradation of nature were explicitly highlighted at the United Nations Conference on Human Settlements, Habitat I, held in Vancouver in 1976. This seemed to indicate the first global "framing" of connections between the environment and housing (McAdam, Tarrow, Tilly, 2001: 6). Political science uses this term in reference to Erving Goffman's concept of "frame", defined as "definitions of the situation [that] are built up in accordance with the principles of organization which govern events (...) and our subjective involvement in them" (1974: 19). It refers to a dynamic of collective negotiations that lead to the official recognition of a situation that is problematic to the general interest. The conclusions of the United Nations Environment Programme were presented to the public at its session in 1974 in Cocoyoc (Mexico). These findings made the connection between human occupancy and the natural environment, as set out below:

(ii) The task of statesmanship is to guide nations toward a new system more capable of meeting the "inner limits" of basic human needs for all the world's people, and of doing so without violating the "outer limits" of the planet's resources and environment;
(iii) Human beings have basic needs: food, shelter, clothing, health and education (UNEP-UNCTAD, 1974).

The level of global demand for buildings was manifestly at the heart of the addresses made to the Habitat I Conference in Canada in 1976, with the text adopted at its conclusion, the Vancouver Action Plan, containing 64 recommendations for national action relating to "human settlements' (UN, 1976). By encompassing housing, construction and planning, this concept denotes a new reflexivity in terms of housing strategy. This reflexivity comes from knowledge of the effects of components, of their implementation and of the built environment on the natural environment, and from recognition of these effects, prompting changes in actions and consideration of impact. Four of the accompanying specifications clearly addressed the environmental impact of building production. They required greater attention to: energy consumption (Recommendation C.5. Energy), the long-term ecological burden of the technical choices made in construction operations (Recommendation C.6. Long-term cost of shelter, infrastructure and services), efficient use of water resources (Recommendation Water supply and waste disposal) and the emission of pollutants (Recommendation Waste management and prevention of pollution). From recriminations to recommendations, UN thinking on housing was moving toward greater consideration of the environment. This certainly emerged from the Vancouver Conference, which brought a patchwork of environmental items to the construction sector. And it was more than a passing concern for the United Nations Environment Programme which, following this event, in 1978, created a new institution dedicated to it: the United Nations Centre for Human Settlements (UNCHS).[1]

However, in the early 1980s, reconciliation between the environment and economic growth did not seem to be a given, with the United States being strongly opposed to the UN proposal for "eco-development" (Sachs, 1980). The quest for compromise was made imperative by the UN's inclusion of the term "sustainable development" in the WCED's mission statement. The WCED's task was to develop a new environmental theory of development. Examining the Brundtland Report entitled "Our Common Future," produced in 1987 following the 42nd Session of the United Nations and after an international consultation, there is no clear path mapped out for achieving the global change it called for, which involved challenging the prevailing principle applied in previous generations: Western-style consumption and the deprivation of basic satisfactions for those who do not participate. The brighter future heralded by the internationally recognized phrase "sustainable development" did not completely contradict the dominant growth-based economic system, but did require correction of its excesses for the social and environmental well-being of the global population. Based on this expression, the new concept of development was an oxymoron, a rhetorical form seeking to reconcile opposites while concealing their internal division. The ambiguities of this concept certainly lend themselves to divergent interpretations, as with the so-called weak and strong sustainability approaches. The first conception assumes the full substitutability of natural capital whereas the second approach argues that this substitutability should be damaged due to the presence of critical elements that natural capital provides for human existence and well-being (Neumayer, 2003). While sustainable development economists therefore emphasize the extreme malleability of this concept, this is evident sociologically speaking from the process of integrative negotiation of which the Brundtland report is the product. Its content is indeed very fragmented. It oscillates between environmental, market, social, political, philosophical and humanist considerations, ultimately incorporating the full constellation of ideas on environmentalism, reproducing their systemic vision of nature, but failing to take their inherent problems into account. Environmentalism is not defined by a single, coherent notion but covers a plurality of concerns; the antagonisms become sharper during this period, particularly between eco-, bio- and anthropocentric perspectives.

This UN declaration in favor of sustainable development seems to provide an ideological mask covering the connections between the environment and housing, which the sociology of innovation identifies as the first step in legitimizing a new technique. This shift in technical mindset can be observed from the very first UN statement on sustainable development and, more clearly, from the areas of uncertainty that linger over its definition. In effect, the

---

[1] Source: https://unhabitat.org/, accessed August 2017.

phrase is more evocative of a new worldview than of any specific transformation of world order. At the point of an international announcement, the change is tenuous and the ambivalence is to be noted without speculating on subsequent phenomena, later analysis of which will enable the discussion to be elaborated on through successive clarifications. By examining how its geopolitical interpretations swing back and forth, we will see how this international dynamic confers ideological legitimation on sustainable urban development, i.e., places value on it to the point of advocating the abandonment of any other type of development.

From Ideological Legitimation to Mobilisation, Uncertainity and the Control of a Political Temporality

Following the publication of the Brundtland Report (WCED, 1987), the sustainable development agenda was made official at the United Nations Conference on Environment and Development (UNCED). Also known as the Earth Summit, this international event took place in Rio from June 3 to 14, 1992. The words adopted there are official in the sense that this public space extended to recognized authorities in the geopolitical field. As well as 2,400 representatives of non-governmental organizations, it brought together 108 heads of state. Of the 25,000 or so recommendations listed in Agenda 21, the action plan for the twenty-first century, Chapter 7, entitled "Promoting sustainable human settlement development" (UNCED, 1993 [1992]), is of particular interest. It calls for "adequate shelter for all," "sustainable land-use planning and management," "integrated provision of environmental infrastructure: water, sanitation, drainage and solid waste management," "sustainable energy and transport systems in human settlements," "human settlement planning and management in disaster-prone areas," "sustainable construction industry activities" and "human resource development and capacity-building for human settlement development" (UNCED, 1993 [1992]: Chapter 7). These recommendations anchor the issue of housing within social, economic and environmental considerations with a more general than sector-specific scope. They do not detail how they should be applied within the residential sector. They merely urge this sector to apply them. It emerges from continued reading of this programmatic document on sustainable development that a mode of construction in harmony with the environment is imperative:

> The activities of the construction sector are vital to the achievement of the national socioeconomic development goals of providing shelter, infrastructure and employment. However, they can be a major source of environmental damage through depletion of the natural resource base, degradation of fragile eco-zones, chemical pollution and the use of building materials harmful to human health. The objectives are, first, to adopt policies and technologies and to exchange information on them in order to enable the construction sector to meet human settlement development goals, while avoiding harmful

side-effects on human health and on the biosphere, and, second, to enhance the employment-generation capacity of the construction sector. Governments should work in close collaboration with the private sector in achieving these objectives. (UNCED, 1993 [1992], *Agenda 21*: Chapter 7, section 7G).

This reasoning, which heralded double benefits for the economy and the environment in the construction sector, opened up a new sequence of association between the environment and housing. It therefore corresponds more to an ideological legitimation. The international order now valued "green" buildings to the extent that it advocated the abandonment of all other types of housing. Building according to the environmental precautionary principle, the polluter pays principle, and the principles of responsibility and citizen participation were officially recognized as the urban development policy to be implemented.

Following the Earth Summit, the United Nations Centre for Human Settlements decided to organize the City and Town Summit in 1996. Also known as Habitat II, this meeting of states in Istanbul aimed to reduce uncertainty in this industry sector. Urban development was becoming less of an inevitability and more of a challenge for all those involved in the building industry. There was a shift in perspective on urbanization, which was no longer simply criticized but seen as an opportunity to build in a different way. Local authorities could adopt a "Local Agenda 21" to promote a more coherent structuring of urban development, limiting sprawl that consumed too much land and energy through dense spatial planning of residential, administrative and commercial buildings, industrial plants and even leisure areas, and by avoiding scattered development along the outskirts of towns and cities. This setting of the agenda and, more broadly, the introduction of a cycle of UN Habitat conferences with fixed meetings every twenty years demonstrated the ambition to establish a political temporality governing international relations with a view to controlling the impact of human activities and the finite nature of the earth's resources. An "Industrial Agenda 21" was developed by the International Organization for Standardization and its ISO standards that give design recommendations. It contains methods for the analysis of construction products. These were aimed at companies and certified their ecological performance, in order to provide better information for occupants. At the turn of the twenty-first century, we saw the third stage of the legitimation of a new human establishment. An ideological mobilization can be perceived in its function of mobilizing the producers of the technology as well as its users.

The Latencies and Resurgences of a Market Framework at the Un Show

Historical comparison enables us to understand the framing processes at play during the constitution of international environmental regimes. By tracing this dynamic of convergence of expectations around implicit or explicit principles, norms and decision-making processes relating to the development of sustainable housing, changes and continuities can be discerned with regard to international relations. In the context we have just described, this statement on the global consideration of the impacts of human settlements revealed, through the legitimation of a new technique, a diffuse uncertainty at the same time as it announced a long-term control mechanism. Conferences tended to be repeated over the decades. Twenty years after Habitat II, the City and Town Summit, and forty years after Habitat I took place in Vancouver in 1976, the third United Nations Conference on Housing and Sustainable Urban Development, Habitat III, was held in Quito from October 17 to 20, 2016, on the theme "Sustainable urban development: the future of urban planning".[2] In addition to the cyclical nature of the content of this event, its organization followed the same pattern as the summits that had taken place two and four decades previously. Once again, it involved the promotion of a new urban development program. Assessing the achievements of the agenda established in Istanbul in 1996, a committee of ten UN Member States was again tasked with coordinating different types of written contributions: national reports, issue papers, policy papers, etc. At the end of this preparatory process and the conference, a further non-binding resolution proclaimed the implementation of a "New Urban Agenda." It essentially constituted a reiterative presentation of the political time of international relations between the environment and housing in the form of a recursive UN communications event deployed at a regular interval of two decades. While this phenomenon is commonly described as a "sounding board" aimed at accelerating public policies, it is also reminiscent of a show, or a charity fair, with Joan Clos, the Secretary General of the Habitat III Conference, reminding the audience at the last event of its beneficent and generous ambitions: "The Conference is a unique opportunity for re-considering the Urban Agenda in which governments can respond by promoting a new model of urban development able to integrate all facets of sustainable development to promote equity, welfare, and shared prosperity" (Clos, 2018).

By analyzing the historically constructed nature of international statements in favor of the development of sustainable housing and urbanization, I have highlighted the symbolic register of the technical ideology legitimizing a market framework. This is one of the key steps in a social process of innovation, the meaning of which can also be understood within the hegemonic context of neoliberal capitalism where the market economy is promoted more than political interventionism. Once this formulation was adopted at UN level, the reproduction of pre-existing patterns and certain configurations of actors took precedence over the quest for any change to economic growth in international relations. After a brief reassessment, it formed, to a greater extent, the focus of mechanisms for controlling uncertainty about the limits of the earth's resources, through the integration of targets and means of environmental action. In this convergence, a defining political temporality was detected around a recurring cadence. The tempo established by the United Nations through its conferences specifically dedicated to habitat—sustainable human

[2] Source: https://habitat3.org/, accessed August 2017.

settlements and urban development—has dictated the slow pace applied to the international processes relating to housing, with interim deadlines every twenty years. Highlighting the phenomena of latency and resurgence, this time-focused approach allows us to question, more broadly, the innovation of the processes at work and their circulation not only over the long term, but also within space, in this case within Europe.

European Circulation: Tactical Adjustments Rather Than Path Dependence

Since the birth of the ECSC (European Coal and Steel Community) in 1952, Europe's shared mission has been to align the laws, regulations and administrative provisions of its member countries that affect the freedom of their trade. Article 249 of the EC (European Commission) Treaty states that its directives define mandatory objectives but leave States free to choose the methods to be used, within a given time frame. (Moussis, 1995: 6)

In his *Handbook of European Union: Institutions and Policies*, Moussis presents the cross-border framework that spans this part of the world (Moussis, 1995), describing the degree of political timing required to reach the common horizon of inter-state ambitions. This invites us to question how the environmental issues inherent in the housing sector are included within this European framework. What connection is there between the international relations established by the United Nations and the supranational power of Brussels? I propose to investigate how and how regularly this growing convergence influences the agenda of different governments, both in the EU arena and within the Member States. Comparative historical sociology can be a springboard for critical reflection on the notion of "path dependence', particularly through differentiating tactical activities from strategic approaches.

The on-going clarification of how this distinction applies will improve the understanding of international relations about sustainable housing.

Normative Bricolage in the Absence of an EU Strategy

Two models of EU integration are commonly distinguished: positive integration based on political consensus and negative integration through the market. The extent of the reforms and the nature of the rules are the main criteria for differentiating these forms of supranational integration. Besides, according to many experts on the alliance of European nations, there is a division within the EU between economic and political projects. This division separates the housing and environment sectors in the European public policy space. As such, my study of its historical sedimentation process begins by shining a light on this division, in order to better analyze how this transnational government then overcomes it, in the name of "sustainable housing."

Since its political birth, the European organization has had no official mandate to legislate in the residential field, as housing policy in Europe is the responsibility of the states. Indeed, housing is not mentioned in either the Treaty establishing the European Economic Community, which, in 1958, established sectoral policies for industry, research, energy, transport, agriculture and fisheries, or the Single European Act, which, in 1987, created new supranational powers concerning the environment, research and technology (EC, 1987). During this preliminary period of the Common Market, the negative integration of the building sector is quite evident from Directive 89/106/EEC, known as the Construction Products Directive (CEU, 1988). This was the only EU legislation adopted for this sector. It is based on the New Approach principle, adopted by the European Commission from 1985 onwards and founded on two key ideas: the obligation of Member States to respect identical essential safety requirements for a series of industrial products and to refer to the European norm, harmonizing their technical specifications before international release (Pelkmans, 1987). We can therefore observe a certain contrast between this isolated measure and the first environmental action program undertaken from 1973 to 1978. Conversely, this latter seems to proceed from positive integration within the European framework. The *Manual of European Environmental Law* lists more than a hundred European directives for the period 1974–1992, specifying that all these measures suffer from one serious flaw: their poor enforcement, which varies greatly from one Member State to another (Kiss, Shelton, 1993: 89).

With regard to sustainable housing, the first statements of the Commission of the European Communities (CEC) were made in 1990, in its *Green Paper on the Urban Environment*, and advocated "mixed use of urban space, favouring in particular housing in inner city areas," in such a way as not to impair its natural surroundings (EEC, 1990: 34). This is a normative representation of relations between the environment and housing that shows how sustainable development in the residential sector should be approached. In the same year, UN bodies founded the International Council for Local Environmental Initiatives (ICLEI), the stated ambition of which was to promote sustainable urban development through operational actions on the ground and to create a network of "sustainable cities".[3] In 1994, the Aalborg Charter was signed following the first European Conference on Sustainable Cities and Towns, committing eighty local authorities to developing an Agenda 21 in their territories, including for housing stock. This managerial approach to the sustainable development of.human settlements was bolstered by a series of regulations and treaties, along with financial, epistemic and even operational instruments, through social housing for low-income households and experimental programs for the cost-efficient development and built implementation of environmental technologies (Némoz, 2009, 2016). Technical requirements

---

[3] Source: https://iclei.org/, accessed October 2018.

were established in particular by the European Directive on the Energy Performance of Buildings (CEU, 2003, Directive 2002/91/EC), which was recast in Directive 2010/31/EU on May 19, 2010, by the European Parliament and the Council (CEU, 2010).

The Treaty of Amsterdam of October 1997 confirmed this affirmative form of the enunciation of sustainable development, which was thereafter enshrined as one of the objectives of the European Union in Article 2 of the Treaty. However, while the draft European Constitution was supposed to ratify this inclusion, the protocol was finally rejected after heated debates within the Convention, chaired by Valéry Giscard d'Estaing, and opposing reactions from Member States.[4] In terms of the built environment, the supranational organization of European countries perceived, in its early days, an economic project involving new residential markets. As we have seen, its legislation consisted of facilitating the free movement of construction products within the Community area, by establishing a common denominator between the various requirements stipulated for these products by Member States. Of the six components of their European harmonization in 1989, one is called "Hygiene, health and the environment" (CEU, 1988, Directive 89/106/EEC). However, in 1997, when the CE marking was first applied to these types of products and this official EU logo demonstrated manufacturers' commitment to compliance with the requirements established by EU regulations, the ecological quality of these products was not specified. It therefore appears to be secondary in European environmental law.

The 1975 Directive on Waste detailed the various stages of waste processing and disposal, with the exception of domestic sorting. With regard to housing, it simply recommended "ensuring that landfills are not located in the vicinity of residential areas in order to limit dangers to human health" (CEU, 1975, Directive 75/442/EEC: Art. 6). These dangers were also referred to in EU drinking water legislation, particularly in relation to household distribution networks (Directive 76/464/EEC). It was also for the health of workers that the European Community published a directive on the risks of exposure to asbestos in 1983 (CEU, 1983, Directive 83/477/EEC). Although the European Council adopted a "strategy for sustainable development" in 2001 (COM (2001) 264), historical comparison of environmental regimes on housing reveals "bricolage," in the sense defined by Claude Lévi-Strauss: an "incidental movement (…) making do with 'whatever is at hand' and (…) not definable by a project" (Lévi Strauss, 1962). Housing did not fall within the European Union's supranational policy remit. It was a sector subject to Member States' power of intervention and on which measures taken in Brussels were applied indirectly through directives targeting other sectors, such as waste. It was not a matter of transferring, to the field of international relations, a paradigm of structuralist theory, dividing and hierarchizing societies.

---

[4] Source: https://archives.eui.eu/en/fonds/444976?item=SP-B, accessed September 2019.

It was about bringing some of its conceptual contributions to the notion of bricolage, which, in this definition, is clearly differentiated from science and technology by not requiring a specific and specialized tool. In terms of European policies, sustainable housing was initially covered by this type of instrumentation, borrowed from others in related fields such as urban development. Following an informal meeting of European ministers in Bristol in 2005 and the adoption of the Leipzig Charter in 2007, a generic reference framework for sustainable European cities was established in June 2010 *(Reference Framework for Sustainable European Cities*, RFSC[5]). It was presented as a tool for assessing the sustainability of sustainable housing projects and policies in different European countries.

Asymetries of National Implementation and Alignment Tactics

The European comparison that we performed between the Helsinki, Paris and Madrid regions provided an insight into the extent to which each country updates pre-existing power asymmetries, while converging toward a relatively similar socio-cognitive and socio-political shaping of relations between the environment and housing (Némoz, 2009, 2016). Originating in neo-institutional economics (Denzau and North, 1994: 3–30) and taken up in particular by sociology (Mahoney, 2001: 507–548), the notion of "path dependence" refers to the fact that with the sedimentation of political systems over time, the implementation of any new public action takes place within the context of previous ones and therefore tends to follow a certain trajectory, a typical way of dealing with problems. Our comparative approach across time and space within three European countries sheds new and informed light on these phenomena, through the analysis of institutional and normative adjustments relating to sustainable housing development.

As a vector of ideology and medium of governance over connections between the environment and housing, the international pronouncement on sustainable development was perceived differently by French, Finnish and Spanish policy stakeholders. The interviews conducted with them and the observations made alongside them, combined with the study of national archives, revealed contrasting understandings of international regimes. Although they did echo tradition in Finland, the institutionalization of the principles, norms, rules and decision-making procedures for sustainable housing development in Finland can be viewed as both lacking urgency and also taking place extremely fast. It was in these apparently paradoxical terms that the key national actors interviewed described this institutional dynamic. It can be observed in practice in 1983, with the creation of the first Finnish Ministry of the Environment, combined with the housing ministry. The weight of the past had a strong influence on the implementation of new public actions, with modes of living in contact with nature underpinning a

---

[5] Source: http://rfsc.eu, accessed May 2020.

sense of collective belonging to the nation, a pattern of urbanization that gave rise to the notion of "forest towns" in the mid-twentieth century (Meurman, 1972 [1947], and a "dilution of architecture in nature" (Nikula, 1993: 30). In the early twenty-first century, the integration of residential sector environmental quality into this extensive ministerial reconfiguration was based on close coordination with market players, as reported by Finnish interviewees from public, private and non-profit sectors, starting with the industrialization of construction using wood, the primary resource for national development before being considered an environmentally-friendly material.

Spanish discourse and measures in this area, on the other hand, describe an exogenous rather than an endogenous dynamic, with ways of living with the environment have largely been driven by processes alien to Hispanic culture. It was in reaction to Franco's dictatorship that the Spanish constitution of 1978 recognized many freedoms, including one that stated that "Everyone has the right to enjoy an environment suitable for personal development, as well as the duty to preserve it" (Boletín Oficial del Estado Español, 1978: art. 48). Political responsibility was initially entrusted to the local governments of the autonomous communities, with the introduction of the first national measures and the creation of a Spanish Ministry of the Environment not taking place until 1986 and 1996, respectively. Political responsibility was initially entrusted to the local governments of the autonomous communities, with the introduction of the first national measures and the creation of a Spanish Ministry of the Environment not taking place until 1986 and 1996, respectively. These dates coincided with Spain's probationary period of EU membership. The Technical Building Code adopted by Spain in 2007 in accordance with European Directive 2002/91/EC marked a change in procedures, defining common energy rules for all buildings within the country, which would no longer vary according to local laws.

In France, our interlocutors concentrated on the impact of the events of the May 68 protests, following which the resolution of issues was scheduled in a dirigiste and centralized manner by senior technical and administrative bodies. In response to this conflict, which was particularly acute in relation to the government's intense planning of territorial urbanization, unparalleled in Europe, many of these bodies introduced the integration of new qualitative targets, particularly in terms of energy, into existing housing policies. These were small-scale changes, through the prism of relatively new market instruments such as the "High Environmental Quality" public utility standard, created in the late 1990s by the major technical and administrative corporations of the French public system. This mode of governance is characteristic of a state edifice that is perfectly willing to replace the market as a purveyor of well-being. It was also in this sense that Gøsta Esping-Andersen identified France as a "conservative-corporatist welfare state" (Esping-Andersen, 1990). The certifications managed by the national association of public and private stakeholders are now in line with other international indicators, particularly European ones.

The heterogeneity of these intra-European assimilation processes vis-à-vis sustainable housing can be explained by a combination of various factors: differing representations of relations between housing and nature, differing paces of urban development and differing notions of the political/social relationship, as well as inherited models of the welfare state (Némoz, 2009, 2016). However, these national historical particularities have not prevented a growing convergence toward the same international regimes. Analysis of the institutional trajectories of relations between the environment and housing shows not only continuities but changing movements. These changes vary greatly from one country to another, depending on pre-existing regulations, the morphology of power and the arrangements required to approach the European "format" of sustainable housing (Némoz, 2009: 217 ff; Némoz, 2016: 40). Here, the idea of format refers to a set of socially legitimate requirements formulated with varying degrees of detail with respect to a category of objects. This analysis makes it possible to show how, in different cultures, ecological modernization dominates the institutional representations of the residential world resulting from this sector-specific framework, as well as the operations to identify, categorize and generalize, enabling the equivalence in time and space of scattered elements. While the term "strategy" has often been used in the various political arenas we have explored empirically, a comparative approach to the temporality of these adjustments to international regimes describes movements that are more tactical in nature, in which the national actors who have historically dominated building and planning policies have been able to reconsider their interests on a piecemeal basis to the rhythm of the European Union's deadlines and the UN conference cycle. While tactics refer to specific measures for achieving short- and medium-term results, strategies involve a master plan for a longer-term vision. The differentiation of these two practices on the political level shows how these international procedures apply via the varied relevance they give to temporalities, to possible combinations of heterogeneous durations and rhythms.

Conclusion

By questioning time and its influence on modes of government, this comparative approach is used to trace back the construction of the international regimes linking the environment and housing in a dynamic and inter-scalar manner. Resituated in different spatial contexts and time configurations, regimes undertaken in the name of "sustainable housing" do not emerge as a monolithic object. The detours of the comparison of international relations in terms of temporalities and spaces highlight an uncertain convergence between reproduction and transformation. While movements of confluence began with the blurring of traditional reference points more than forty years ago, diachronic analysis of the international genesis of a technical ideology of innovations sheds light on the latency and resurgence of a market framework.

The processes at play within international relations to frame sustainable development have alternated, over the past fifty years, between uncertainty over and characterization of residential assets, without ever questioning their increasing production, in order to better stimulate their renewal at market and institutional levels. Our exploration thereby reveals the cyclical continuity of the processes and trajectory of international relations, aligned with the twenty-year intervals of the UN conferences which resemble recurrent shows, punctuating the European circulation of the sustainable housing framework.

At other governmental levels, this supranational political temporality which is invested with a strategy encompassing an overarching vision, resistant to the erosion of time, is, in fact, pivotal to the normative bricolage that takes place with respect to connections between the environment and housing. Analysis of the regular rhythm over many decades sheds new and informed light on these processes of institutional and normative accommodation with respect to sustainable housing. Vectors of ideology and tools of governance over the forms and modes of environmentally-friendly housing, these processes continue the decision-making tempo without questioning the morphology of the power over how we live with nature. The latter is not, strictly speaking, a political project of the European Union. It is more of an incidental movement, meeting other international agendas and serving as a frame of reference and measurement standard. It is in this temporal and spatial slipstream that tactical adjustments have been observed from France, Finland and Spain, including asymmetries of empowerment, which are perpetuated. Compared with strategies, which anticipate long-term outcomes, tactics are actions expressed within a precise and limited temporality: "tactics rely on a clever use of time, the opportunities it presents and also the play that it introduces into the foundations of power" (de Certeau, 1990: 63). Thus, rather than concluding the existence of limitless path dependences, this mobile, transhistorical perspective on international sustainable housing regimes highlights the benefit of always assessing the impact of time on what specifically defines politics, namely the overall perception of the problems relating to the organization of a common world.

References

BALANDIER, Georges, (1985), *Le détour*. Paris, Fayard, coll. L'espace politique.
BECK, Ulrich, (2004), *Der kosmopolitische Blick oder: Krieg ist Frieden*, Vorwort, Herausgeber.
Boletín Oficial del Estado Español (BOEE), (1978), *Constitución Española*. Madrid.
BUNZ, Kimberly, HENZE, Gregor, TILLER, Dale, (2006), "Survey of Sustainable Building Design Practices in North America, Europe, and Asia". *Journal of Architectural Engineering*, 1(12), 33–62.
CERTEAU (de), Michel, (1990 [1980]), *L'invention du quotidien. 1/ Arts de faire*. Paris Gallimard, 10/18.
CEU (Council of the European Union), (1975), *Council Directive 75/442/EEC on Waste*. Brussels.

CEU (Council of the European Union), (1976), *Council Directive 76/464/EEC on Pollution Caused by Certain Dangerous Substances Discharged into the Aquatic Environment of the Community*. Brussels.
CEU (Council of the European Union), (1983), *Council Directive 83/477/EEC on the Protection of Workers from the Risks Related to Exposure to Asbestos at Work*. Brussels.
CEU (Council of the European Union), (1988), *Construction Products Directive* (council directive 89/106/EEC). Brussels.
CEU (Council of the European Union), (2003), *The Energy Performance of Buildings Directive* (council directive 2002/91/EEC). Brussels.
CEU (Council of the European Union), (2010), *The Energy Performance of Buildings Directive 2010/31/EU*. Brussels.
COLE, Raymond, (2006), "Shared Markets: Coexisting Building Environmental Assessment Methods". *Building Research and Information*, 34(4), 357–371.
CONCA, Ken, DABELKO, Geoffrey, (2010), *Green Planet Blues: Four Decades of Global Environmental Politics*, 4th ed. Boulder, CO, Westview Press.
CLOS, Joan, (2018), "Introduction" in UN-HABITAT, *The Quito Papers and the New Urban Agenda*. New York: Routledge.
DENZAU, Arthur, NORTH, Douglass C., (1994), "Shared Mental Models: Ideologies and Institutions". *Kyklos*, 47, 3–30.
EC (European Communities), (1987), "Single European Act". *Official Journal of the European Communities*, L(169), 2.
EEC, (1990), *Green Paper on the Urban Environment, Communication from the Commission to the Council and Parliament*. Luxembourg, Office for Official Publications of the European Communities.
ELIAS, Norbert, (1992), *Time: An Essay*. Oxford: Basil Blackwell.
ESPING-ANDERSEN, Gøsta, (1990), *The Three Worlds of Welfare Capitalism*. Princeton, New Jersey: Princeton University Press.
FINNEMORE, Martha, SIKKINK, Kathryn, (1998), "International Norm Dynamics and Political Change". *International Organization*, 52(4), 887–917.
FOURNIAU, Jean-Michel, BLONDIAUX, Loïc, BOURG, Dominique, COHENDET, Marie-Anne, (2022), *La démocratie écologique. Une pensée indisciplinée*. Paris, Hermann.
GILPIN, Robert, (1987), *The Political Economy of International Relations*. Princeton University Press.
GOFFMAN, Erving, (1974), *Frame Analysis: An Essay on the Organization of Experience*. Cambridge: Harvard University Press.
GRISONI, Anahita, NÉMOZ, Sophie, (2017), "The Social and Environmental Movements: Between Self-Reform and Class Relationships, Between National Histories and European Circulations", in GRISONI, Anahita, NÉMOZ, Sophie, eds., *Socio-Logos*, special issue: "Decompartmentalizing the Environment", 12/2017.
HAAS, Peter M., KEOHANE, Robert O., LEVY, Marc A., eds. (1993), *Institutions for the Earth: Sources of Effective International Environmental Protection*. Cambridge, MA: MIT Press.
JANJUA, Shahana Y., SARKER, Prabir K., BISWAS, Wahidul K., (2019), "A Review of Residential Buildings' Sustainability Performancessing a Life Cycle Assessment Approach". *Journal of Sustainability Research*, 1, e190006.
JAMES, William, (1907), *Pragmatism: A New Name for Some Old Ways of Thinking*. Londres: Longmans, Green and Co.

KISS, Alexandre, SHELTON, Dinah, (1993), *Manual of European Environmental Law*. Cambridge: Grotius Publications Limited.
KRASNER, Stephen D., ed. (1983), *International Regimes*, Ithaca: Cornell.
LASCOUME, Pierre, (2018), *Action publique et environnement*. Paris: PUF.
LEVI STRAUSS, Claude, (1962), *La pensée sauvage*. Paris: Agora.
MAHONEY, James, (2001), "Path Dependence in Historical Sociology". *Theory and Society*, 29, 507–548.
MANSEAU, André, SEADEN George, eds. (2002), *Innovation in Construction. An International Review of Public Policies*. London: Routledge.
MCADAM, Doug, TARROW, Sidney, TILLY, Charles, (2001), *Dynamics of Contention*. Cambridge: Cambridge University Press.
MEURMAN, O.-I., (1972 [1947]), *The Theory of Town Planning*. Helsinki: Museum of Finnish Architecture.
MOORE, Steven, (2007), "Models, Lists, and the Evolution of Sustainable Architecture", in TANZER, Kim, LONGORIA, Rafael, eds., *The Green Braid: Towards an Architecture of Ecology, Economy and Equity*. London: Routledge, 60–76.
MOUSSIS, Nicholas, (1995), *Handbook of European Union: Institutions and Policies*. European Study Service.
NÉMOZ, Sophie, (2009), *Eco-Housing, Sustainable Innovation... Sociological Analysis of Residential Ecology in France and With a Detour Through Finland and Spain*, Doctoral thesis, Université Paris Descartes – Sorbonne.
NÉMOZ, Sophie, (2016), "Political Eco-Housing: A Critical European Comparison About Sustainable Housing". *Sciences de la société*, special issue: "Sustainable Housing: Critical Approaches", 98, 31–43.
NEUMAYER, Eric, (2003), *Weak Versus Strong Sustainability: Exploring the Limits of Two Opposing Paradigms*. Northampton: Edward Elgar.
NIKULA, Riitta, (1993), *Architecture and Landscape. The Building of Finland*. Helsinki: Otava.
PEIRCE, Charles Sanders, (1978), "How to Make Our Ideas Clear". *Popular Science Monthly*, 12.
PELKMANS, Jacques, (1987), "The New Approach to Technical Harmonization and Standardization". *Journal of Common Market Studies*, Vol. XXV, 3.
SACHS, Ignacy, (1980), *Stratégies de l'écodéveloppement*, Paris, Les éditions ouvrières, coll. "Economie et humanisme".
SUNIKKA, Minna, (2003), "Fiscal Instruments in Sustainable Housing Policies in the EU and the Accession Countries". *European Environment*, 13, 227–239.
UN (United Nations), (1973), *Report of the United Nations Conference on the Human Environment*, New York, 1973.
UN (United Nations), (1976), *Report of Habitat: United Nations Conference on Human Settlements*, Vancouver, 31 May–11 June 1976.
UNCED (United Nations Conference on Environment and Development), (1993 [1992]), *Agenda 21. Rio Declaration on Environment and Development*. New York, United Nations.
UNEP-UNCTAD (United Nations Environment Programme), (1974), *The Cocoyoc Declaration Adopted by the Participants in the UNEP/UNCTAD Symposium on "Patterns of Resource Use, Environment and Development Strategies"* held at Cocoyoc, Mexico, from 8 to 12 October 1974.
WEBER, Max, (1959 [1919]), *Le savant et le politique*, introduction by ARON R., Paris, Plon, coll. 10/18, no. 134.

WINSTON, Nessa, (2007), "From Boom to Bust: An Assessment of the Impact of Sustainable Development Policies on Housing in Ireland". *Local Environment*, 12, 57–71.

WCED (World Commission on Environment and Development), (1987), *Our Common Future*. Oxford: Oxford University Press.

The Mobilization of the Philanthropic Sector for the Climate: A New Engagement?

Anne Monier

> "One individual cannot possible make a difference, alone. It is individual efforts,
> collectively, that makes a noticeable difference – all the difference in the world!"
> Jane Goodall

Social scientific research on global climate change has been extensive but loosely connected, and exchanges with the other social sciences and natural sciences have been limited. Ironically, until now the social sciences have played only a minor role in climate change reports and discussions because scientific research has been deeply entrenched in the natural sciences (Islamy & Kieuy, 2021). Yet the social sciences offer an important contribution to the understanding of the social causes and consequences of climate change as well as the way social change can be achieved.

The social sciences, especially sociology, have long studied the mechanisms by which large-scale cultural and social change is produced (Caniglia, Brulle & Szaz, 2014). Sociological scholarship, in particular, underlines the role played by the institutions of civil society in creating social change through citizen mobilization (McAdam, 2017; Calhoun, 1993; Tarrow, 1998; Skocpol, 2011). The study of social movements in the environmental field can help grasp the specificities of environmental social movements.

A. Monier (✉)
ESSEC Business School, Philanthropy Chair, Paris, France
e-mail: anne.monier@ens-lsh.org

In the last decade, concern over climate change has led to organized efforts that advocate for action to address this issue (Beddoe et al., 2009; Fisher & Nasrin, 2020; Caniglia, Brulle & Szaz, 2014). Scholarship has shown the development of environmental social movements as this topic was becoming a major political issue around the globe (Doyle & MacGregor, 2013; Caniglia, Brulle & Szaz, 2014).

A social movement is "a loose, non-institutionalized network of informal interactions that may include, as well as individuals and groups who have no organizational affiliation, organizations of varying degrees of formality, that are engaged in collective action motivated by shared identity or concern" (Rootes & Brulle, 2015: 1). The climate change movement is "an amalgam of loosely networked individuals, groups and organizations springing out of the environmentalist, development, anti-capitalist, and indigenous movements combined with a new wave of activists and groups that had no previous ties to other social movements" (Nulman, 2015).

Research on climate mobilizations has examined the particularities of mobilizing for such an important, urgent and global issue as climate change (North, 2011; Nulman, 2015; Tindalll & Robinson, 2017; O'Brien et al., 2018), showing the difficulties, like finding narratives (Jerneck, 2014), addressing such a complex issue (Aykut & Dahan, 2014) or getting the politics on board (Dahan, 2014). While the first mobilizations were conducted by environmental activists, the problem seems to be tackled today by a growing number of actors in different sectors who try and mobilize to fight climate change. Among them is the foundations world that is quite understudied.

This chapter aims at filling this important gap by analyzing the mobilization of the European philanthropic sector for climate, through the analysis of the coalitions of foundations for climate that have been created in several countries since 2019. I will show how philanthropy can be studied through the lens of environment movements, asking the question of its engagement, influence and even political role. Thinking of the mobilization of the philanthropic sector is also changing the common perspective of analyzing foundations engaged in climate just as funders. I will first quickly mention the development of climate philanthropy; then, the chapter will analyze this new movement that is emerging—the coalitions of foundations for climate; last, it will examine foundations as actors of change beyond their funder's role: their influence, political role or advocacy practices.

Methods and Fieldwork

This discussion is based on a new research that was launched in October 2020 on the mobilization of the philanthropic sector for climate in Europe, focusing on the coalitions of foundations. This research project poses three main questions: 1. how do philanthropic actors perceive / feel the problem and the possible solutions? 2. How do they engage and undertake their transition? 3.

How do they mobilize others and what are the consequences of this mobilization? This work is at the crossroads between several disciplines: political science to understand how these philanthropic actors mobilize for climate and the forms of this mobilization; sociology to study the actors of this mobilization, their profiles and motivations; anthropology to grasp the role of symbolic aspects (images, ideas, representations, knowledge) of this mobilization with also a focus on cognitive and emotional aspects.

The fieldwork is conducted within the five entities: three national associations of foundations that have created a coalition for climate—CFF (*Centre français des fondations*—French center of foundations) in France, AEF (*Asociación española de fundaciones*—Spanish Association of foundations) in Spain, ACF (Association of Charitable Foundations) in the UK, Philea (Philanthropy Europe Association) at the European level and WINGS (International Network) at the international level.

Our research relies on an inductive approach. It is a qualitative survey based on four main methods: ethnography; interviews; document and press analysis; and archives. At this stage of the project (March 2022), ethnography and document analyses have been undertaken as well as some interviews.

Philanthropy for Climate: An Understudied Topic

Regarding climate change, the role of philanthropy is quite understudied. Climate philanthropy constitutes an important gap in the literature. If scholarship on climate movements has shown the role of different climate movements—climate leaders movements (Acuto, 2013), youth activism (O'Brien et al., 2018), NGOs' mobilizations (Lal Pandey, 2015), etc.—or different sectors (political sector, corporate sector, etc.), few studies have focused on the role of the philanthropic sector. Among those few, there are those who have emphasized how little foundations have done so far (Morena, 2021). Researchers have underlined that the lack of funding is blatant (Mathieu, 2015). Indeed, funding for environmental issues only represent 2% of total spending of the philanthropic sector (Morena, 2016). Moreover, research on this topic concentrates on US foundations, whereas climate philanthropy in Europe has almost never been examined.

The few existing studies do not use a social movements approach and are in line with traditional climate research that is being conducted in social sciences. One interesting approach asks the question of the role of foundations on the international climate scene, showing the discrepancy that exists between their funding (quite low) and the legitimacy they have at the international level (Morena, 2021). Other articles stress how philanthropic foundations remain largely under the radar of environmental governance scholars (Betsill et al., 2021) and advance three important themes: the role of foundations in environmental governance, the outcomes of environmental philanthropy and the sources of foundation legitimacy (Betsill et al., 2021).

Another stream of research analyzes foundations as sources of funding for climate movements or counter-movements. Some works underline the role foundations have played in funding environmental justice activism, making elites and activists collaborate (McCarthy, 2004), in line with the work on social movement philanthropy. Researchers have also analyzed how foundations have participated in "institutionalizing delay" (Brulle, 2014) and "obstructing action" by funding climate counter-movement organizations in the United States (Brulle et al., 2021).

Beyond the field itself, it is also the philanthropic actors that are not often studied in the research on climate mobilizations. When looking at the different actors in the climate movements field, the focus is placed on citizen mobilizations, activists, scientists, NGO leaders, policy-makers and decision-makers, and a little less so on communities or grassroots actors, but there is almost no study on philanthropic actors, except to question them as funders of climate-related issues.

This is a crucial shift, as scholarship on climate change with a social or political sciences perspective often focuses on two types of actors: on the one hand, elites (decision-makers, policy-makers, etc.); on the other hand, the public (individuals, grassroots actors, NGO actors, etc.). It corresponds to a division between, on the one hand, those who are requested to change things and have the power to take decisions to do so (or not), and on the other hand, those who want things to change. Philanthropic actors are among the few actors that belong to both groups (wanting and having the power to change).

PHILANTHROPY FOR CLIMATE: A LONG HISTORY?

Climate philanthropy is both a recent and fairly peripheral phenomenon: it was only in the mid-to-late 1980s that foundations begin to seriously engage in climate (Morena, 2016). As Edouard Morena shows, it is only a handful of foundations that are actively involved in the climate field. Here are some of the initial foundations who engaged in climate:

	Date of creation
Rockefeller Foundation	1913
Ford Foundation	1936
Rockefeller Brothers Fund	1940
Packard Foundation	1964
Hewlett Foundation	1966
Mac Arthur Foundation	1970
Oak Foundation	1983
Bloomberg Foundations	1989
Betty and Gordon Moore Foundation	2000
Children's Investment Fund Foundation	2002
Sea Change	2006

(continued)

(continued)

	Date of creation

Source: Foundations' Websites

Many of these foundations were not climate foundations: they worked on other causes but created a climate program.

Besides this handful of climate funders, several foundations were created as "pass-throughs" to pool fund, i.e., different foundations created an entity to fund together some projects. It is the case of the foundations in the table below which were created later:

Pool foundations	Date
Energy Foundation	1991
Energy Foundation China	1999
Climate Works Foundation	2008
European Climate Foundation	2008
India Climate Collaborative	2020
The African Climate Foundation	2020

Source: Foundations' Websites

For example, the Climate Works Foundations is funded by several foundations and trusts, including the Bezos Earth Fund, Bloomberg Philanthropies, the Chan Zuckerberg Initiative, CIFF (Children's Investment Fund Foundation), the Packard Foundation, the Ford Foundation, the Ikea Foundation, the Hewlett Foundation and many others (27 in total).[1] The European Climate Foundation is funded by Porticus, Laudes Foundation, CIFF, Bloomberg Philanthropies, Climate Works Foundations, Ikea Foundation, Hewlett Foundation, Oak Foundation, RBF (Rockefeller Brothers Fund), Stiftung Mercator and others (16 in total).[2]

Nevertheless, if these major foundations are committed to fight climate change, as we mentioned earlier, climate funding only represents 2% of total philanthropic funding, which raises the question of the discrepancy that exists between the financial distribution and the legitimacy these foundations have at the international level in the climate change discussions (Morena, 2021).

A New Movement for Climate

If big funders and pass-throughs are the most well-known actors in climate philanthropy, there has been, in the last couple years, a new movement in the

---

[1] Source: The Climate Works Foundation's website.
[2] Source: The European Climate Foundation's website.

philanthropic sector with the creation of national coalitions of foundations in Europe.[3]

On November 6, 2019, a group of foundations from the UK created the Funders Commitment on Climate Change, which was then taken up by the Association of Charitable Foundations (ACF), the national representative organization of foundations in the country. The French and the Spanish national associations created their own movement, in parallel. On November 17, 2020, the Spanish national association of foundations, AEF (*Asociación Española de Fundaciones*—in English : the Spanish Association of Foundations), launched their movement "Fundaciones por el clima" (in English: foundations for climate), and on November 18, 2020, the French national association CFF (*Centre Français des Fondations*—in English : the French Center of Foundations) launched their "*Coalition française des fondations pour le climat*" (in English: the French coalition of foundations for climate). Besides these movements, national associations of other countries have created a movement, as in Italy, where ASSIFERO (*Associazione Italian delle Fondazione ed Enti della filantropia istituzionale*– in English: the Italian Association of foundations and institutional philanthropy bodies) launched a coalition in September, 2021.

All these national movements work in close contact with the European umbrella of these national associations—Philea (Philanthropy European Association), who created a European Philanthropy Coalition for Climate (EPCC) in Spring 2021. And the movement is taking today an international turn with the involvement of WINGS, an international network of philanthropic actors, who set up an "International Philanthropy Commitment on Climate Change," which has been launched in June 2021 and was officially announced at the COP26 in Glasgow in November 2021.The goal is to make it possible for foundations in countries which do not have a national coalition to sign the International Commitment on climate change. A Canadian coalition of foundations for climate was also created in September 2021.

It is also crucial to mention that there were already some attempts made by these national associations of foundations to create a movement around climate: it is the case in France, for example, around the efforts made with the COP21 which was organized in Paris in 2015. There were some meetings and gatherings organized for the occasion, some discussions, but the movement did not really work. Five years later, the context is much more favorable for these movements to thrive because there is more momentum around climate since the COVID-19 pandemic.

---

[3] All the information provided on this new movement (thus the information below) comes from a new research I am conducting on this movement.

A Climate Philanthropy Ecosystem

These funders and this new movement are part of a bigger ecosystem of philanthropic actors mobilizing to help fight climate change. Among them, there are some advisory groups and experts, like Active Philanthropy of Climate Leadership Initiative, but also networks like the Environmental Funders Network or Edge Funders Alliance. Philea made a try at mapping this complex ecosystem of climate philanthropy, as we can see below:

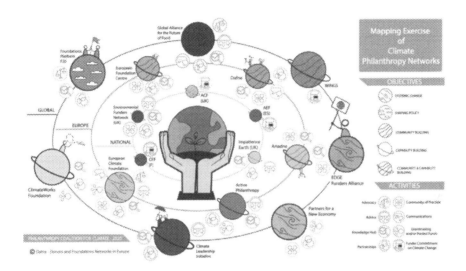

Source: Philea Website

There are different actors at different levels and with different objectives, interacting on the same topic. It is key to understand how all these actors interact and position themselves in the bigger picture. Beyond this climate philanthropy ecosystem, they are also linked to other climate actors beyond the philanthropic sector, like governments, activists, researchers, international organizations or structures—UNFCCC (United Nations Framework Convention on Climate Change) or IPCC (Intergovernmental Panel on Climate Change).

Getting Newcomers to Commit to Climate

The creation of these coalitions shows how foundations can contribute to social movements beyond their role as grant makers.

The role of these coalitions of foundations is to mobilize foundations beyond climate foundations or foundations already addressing climate issues. The idea is that foundations whose mission is art, poverty, education, health or other topics should also engage in fighting climate change as the crisis will have effects on all these fields and on the philanthropic sector itself. This is a change of paradigm for a sector that is deeply attached to working in silos within the causes that each foundation defends. To do so, they will focus on changing the way foundations work not only as funders, but also as investors (through their investments), as organizations (through their operations), etc.

To get engaged and join the coalition, foundations have to sign a commitment, that was written in each coalition (UK, France, Spain, but also at the international level). It is interesting to note that the manifest has different names, depending on the country, it is a "commitment" in the UK and at the international level, a "*manifeste*" in France and a "*pacto*" (pact) in Spain—as shown on their respective websites. The number of signatories varies depending on the country.

UK	88 signatories
France	128 signatories
Spain	144 signatories
Italy	72 signatories
Canada	41 signatories
International	34 signatories
TOTAL	**509 signatories**
Source: WINGS Website	

By signing the commitment, the foundations engage in the coalition which is a formal commitment. If a commitment can be considered as a "consistent behavior" (Becker, 1960), the nature of this act or state of commitment is not always specified.

In our case, one key element of the commitment of foundations is the fact that they have to sign a commitment to enter the coalition. It is interesting to note that one of the coalitions decided to organize a "signing ceremony" the day of the launch of the coalition, where people were signing in real time and others were watching it. As a zoom event, it had less of the decorum and ritualized aspect, but it was still very revealing of the importance they gave to this act.

A Community of Practice

When foundations sign the commitment, they join a community of foundations who previously signed the commitment and led by the national association (CFF in France, ACF in the UK, AEF in Spain). As for the community's name, it is interesting to note that this group is named differently depending on the country: it is a "coalition" in France, a "comunidad" in Spain and "framework" in the UK. One of the specificities is that these movements are for now quite undefined, as the only official goal is to make foundations get engaged for climate. The question remains unanswered: what kind of social movement are these groups of foundations? This question relates quite well to scholarship on climate movements which shows the loose networks and goals of some of these initiatives.

Umbrella organizations (ACF, CFF, AEF, Philea, WINGS) are infrastructures that help shape the collective action of these foundations. They work on the basis of memberships: they have members, who are foundations (for national associations) or associations/networks of foundations (for Philea and WINGS) and whom they represent. At each umbrella organization, they have a person (or two) in charge of this group around climate. This leading person is key to coordinate efforts, spread information and lead the movement toward a strategic direction. The role of these umbrella organizations is to make foundations sign the commitment and then help them do their transition, create "a community of practice" by fostering exchanges (events, peer group exchanges, resource exchanges, spread of the "best practices") and advocate for the coalition as well as getting them to report publicly on their progress, through an annual reporting process organized by national associations.

When foundations sign the commitment, they enter a community that is considered (and called) a "community of practice." This means that the national associations organize different events and meetings to foster exchange between foundations. This idea is also present in the words used by the national associations to present what they do ("cooperation," "collaboration," "together"). This community of practice is organized following a concentric circles model. A first circle is the leadership of the community, composed by the head or climate head of the national associations or the head of the network (for Philea or WINGS). Some coalitions also have a Steering committee and an Advisory Board who meet several times of year. A second circle is composed by leading foundations in the commitment for climate and who were often the first signatories. The third circle is all the other foundations who enter these coalitions.

The Birth of Sectorial Movements?

The mobilization of the philanthropic sector takes place in a context where several sectors are undertaking their transition. In the case of France, we have seen the rise of sectorial movements for climate, in particular in sectors that not usually very concerned about climate change, like the financial or the corporate worlds.

The *Convention des Entreprises pour le Climat* (CEC), which is the Convention of Companies (corporate world) for Climate, was created in 2020, inspired by the *Convention Citoyenne pour le Climat* and launched by Emmanuel Macron (France's president) in 2019.[4] The goal was to gather 150 company CEOs that have been selected (with a diversity of size, sector and regions) to work together during 11 months and make recommendations to advance the environmental transition of companies.

[4] Source: https://www.conventioncitoyennepourleclimat.fr/en/

Pour un Réveil écologique (For an Ecological Awakening) is the mobilization of the higher education sector. It emerged in 2018 after that the "Student Manifest for an Ecological Awakening" was signed by 30,000 students from 400 higher education institutions. Students and young graduates from different background are united to make things change in higher education but also within the companies they are now working at.[5]

Finance for Tomorrow was created after the adoption of the Paris agreement in 2015. The Green & Sustainable Finance was launched in 2016 by Paris Europlace, "which supports and develops the French financial industry and promotes Paris as an international financial centre." It was renamed "Finance for Tomorrow" in June 2017 and became a branch of Paris Europlace. In June 2020, it has more than 75 members. It aims at mobilizing the financial sector for a more sustainable finance.[6]

These sectorial movements take different forms but have common points. For many of them, in France, they started after the momentum brought by the COP21 and the Paris agreement, where many actors gathered to talk about the topic in their sector. And they have the same goal: to make things progress in their sector and open the way for the sector to transform itself. However, the way they work is diverse, based on different criteria: are they selective or not (do they welcome any member or do they have a selection process to enter the group?)? Do they offer recommendations / resources or not? What is their model of governance? A key aspect, as well, is the representativeness of the sector: how representative of the sector are these coalitions (e.g., how well does the "Finance for Tomorrow" group represent the financial sector?)?

If these coalitions reflect an important dynamic in different sectors, the lack of coordination at the national and international level can be a problem because they are created as parallel movements that do not have a lot of interaction with each other which makes it difficult to circulate information and makes them do the same things in different ways and not learn best practices from each other. This difficulty exists at the global level as one of the key challenges of climate action: coordination at the different levels and articulating scales, for example having carbon reduction goals at the sectorial level, but also at the organizational and individual levels, and understanding how these goals are linked.

Foundations as Influential Actors?

Is this movement a reflection of a growing role of the philanthropic sector on the climate international scene? Are foundations actors that can have an influence? As we have seen, Edouard Morena underlines the discrepancy that exists between the amounts of philanthropic spending that go to climate (2% percent) and the legitimacy that climate foundations have on the international

---

[5] Source: https://pour-un-reveil-ecologique.org/fr/qui-sommes-nous/

[6] Source: https://financefortomorrow.com/en/about-us/

climate scene because they are seen as intermediaries and catalyzers, as the example of the One Planet Summit that was organized in 2017 in Paris shows (Morena, 2016). For a long time, the power of foundations has relied on their capacity to be at the crossroads, influencing and pushing ideas, values and actions onto different fields. Scholarship has shown the influence foundations have been exerting in different contexts. In the edited book *L'argent de l'influence* (the money of influence) (2010), Ludovic Tournès and his colleagues showed how the Rockefeller Foundation, the Ford Foundation, the Open Society (Soros) and the Carnegie Endowment have been exerting their influence during the twentieth century, through their funding and actions, but also through the networks they created locally and internationally. It mostly regarded the intellectual space in Europe, as they were funding researchers, academics, teachers, scientists, academic institutions, and research projects in different countries, contributing to creating and disseminating ideas. Ludovic Tournès underlines how these foundations were participating in international diplomacy, a form of "philanthropic diplomacy," promoting values and ideas important in their countries (like democracy in the United States) through their philanthropic funding.

This role of foundations as actors exerting influence on the international scene has been tackled by other researchers and not only about American foundations. Dorota Dakowska (2014) analyzes the power of foundations and, in particular, their role as actors of the German foreign policy. She shows again the key role of networks, in particular between different elites, that make it possible to circulate information, ideas and knowledge that will help achieve the foundations' goals. The analysis of the institutional context brings up the question of gatekeepers and intermediaries, i.e., people who build bridges between fields. Social capital—how well connected people are—appears as a key element to understand the power of foundations and the way they function in the ecosystem of international action.

A Political Role?

Beyond the influence foundations can exert, these mobilizations of the philanthropic sector ask the question of the political role of foundations. Foundations have often been seen as apolitical (or not political), but recent research shows that this view can be nuanced or developed further.

Philanthropy and politics have always had a complicated relationship. As Olivier Zunz (2012) explains, the development of philanthropy in the United States came with the idea of a clear separation between philanthropic activities and political activities. He underlines that the State has opened possibilities for philanthropy, defining it and regulating it through a legal framework where the State aimed, in particular, at separating philanthropy and politics. The emergence of modern philanthropy at the end of the nineteenth century led to the definition of the acceptable sphere of philanthropy through the opening of its possibilities of intervention for a philanthropy with broader objectives and in tune with its time. It increased the political sphere that was within the reach of philanthropy: the distinction would no longer be between political and apolitical, but between "generally political" and "technically political," which was a major turning point in the relationship between philanthropy and politics—as Zunz explains (2012). He underlines that in the early twentieth century, in the US, the regulation of philanthropy changed hands from the courts to the federal government, through two major moments: the 1913 and 1934 Appropriations Acts. In 1913, the Finance Act was passed, which established the Income Tax (16th Amendment) and the State implemented a tax exemption policy, one of the objectives of which was to channel the political effects of philanthropy. A second turning point came with the 1934 Finance Act, which added a formal distinction between education and political activism. Olivier Zunz (2012) discusses this law in detail, showing us the importance of the terms.

But while defining the boundary between philanthropy and politics is still in process, some researchers show how some philanthropists do play a political role on the international scene. In his book (2021a), Peter Hägel analyzes what he calls "the hidden actors of world politics," aiming to show how the privatization of politics assumes a new dimension when billionaires wield power in international affairs. It was only in 2010 that the Supreme Court's *Citizens United* decision liberalized political finance in America which then allowed billionaires to spend hundreds of millions of dollars during election campaigns (Hägel, 2021a). In this sense, Theda Skocpol's invitation to study the role of philanthropy to understand politics is interesting because it shows that philanthropy can be of interest to political scientists for the mechanisms of power it entails: "Why political scientists should study organized philanthropy" (Skocpol, 2016). But Hägel shows that there is not much research that extends its horizons to world politics: within the study of International Relations, billionaires are more or less absent. Only in the specialized subfield of global health governance, the influence of the Rockefeller Foundation and the Bill & Melinda Gates Foundation is hard to ignore (Hägel, 2021b).

The enormous wealth of these billionaires is a highly fungible power resource to pursue their goals, especially through philanthropy. George Soros and Bill Gates have made attempts to transform the world's political and social configurations. Their wealth can give billionaires a privileged social standing that facilitates their access to political actors and the public sphere. Still, not all billionaires seem to be willing or able to turn their financial assets into political power, shifting from being entrepreneurs with economic power to becoming social and political entrepreneurs with social and political power. Hägel has also analyzed the role of Soros and Gates in politics, through their foundations (Hägel, 2021b). The Gates Foundation played an important part in international health policies, in particular, regarding vaccines. Soros participated, in its way, in the Rose Revolution in Georgia in 2003, by funding some movements and media that supported the candidate who won. Hägel underlines the role of networks and of specific individuals that are part of both fields (philanthropy / politics). At a moment when international relations are seeing the rise of non-governmental actors, foundations are becoming key actors of international politics despite being understudied by researchers and students of this academic field. But this political role still relies on the funding activities of foundations: what about their ability to mobilize as a sector?

THE DEVELOPMENT OF ADVOCACY

Beyond the influence and the political role of foundations, the emergence of these climate coalitions poses the question of the way foundations mobilize themselves for causes beyond their grant-making capacity. In this sense, there is an activity that foundations are developing in a quite open way and that reflects the willingness of the sector to be more engaged: advocacy.

If the role of philanthropy in influencing policy change has been recently well documented (Thomson, 2020; von Schnurbein et al., 2021; Lambin & Surender, 2021), the advocacy activity in itself is less tackled by researchers. A paper by Alexandra Williamson and Belinda Luke (2020) examines advocacy, agenda-setting and the public policy focus of private philanthropic foundations in Australia. Most research has indeed focused on advocacy and public policy influence in the United States (Williamson & Luke, 2020). The researchers advance the idea that in Australia, the sector may be characterized as "quiet philanthropy" rather than having a visible public presence. Foundations' advocacy focused on promoting philanthropy rather than altering or influencing public policy (Williamson & Luke, 2020).

Williamson and Luke mention how conceptions of advocacy in philanthropy are "dominated by the obvious, the outliers and the noisy." They remind us that foundations have power and influence beyond their grant-making which is often forgotten and that the influence of foundations, such as in the United States, has been criticized (Callahan, 2017; Reich, 2018). Prior to 2010, opposition to advocacy by philanthropic foundations came from regulatory bodies in Australia: foundations' timidity therefore arose from fear of sanctions, rather than from public censure or disapprobation (Williamson & Luke, 2020).

The two scholars insist: there is an "ongoing conservative political narrative against charity advocacy, especially by environmental charities including some foundations" (McGregor-Lowndes & Williamson, 2018). "The environmental foundations' path in a country with significant fossil fuel export production and environmental treasures such as the Great Barrier Reef may present an intriguing longitudinal case study of those that choose to protect the environment by direct remediation, and those that engage at the political policy level" (McGregor-Lowndes & Williamson, 2018, 1775). Williamson and Luke thus explain that foundations may fund charities to undertake policy advocacy through activities including research and dissemination, raising awareness, community organizing, etc.

Conclusion

The emergence of this new movement of foundations engaging for climate is quite new and complex to analyze, but it helps shift our perspective on the role of philanthropy on the climate scene. For a long time, foundations have mainly been seen as funders for climate projects or environmental projects, but there is today a shift in the way some foundations see themselves as actors fighting against climate change, as I have described in this chapter. In this sense, this discussion raises the question of the role of foundations regarding the climate change issue, in particular, and other environmental problems as well: what is their specificity of actions? what role can they play concerning the environmental crisis? It seems like this sector that has been considered for a long time as an apolitical sector is being more engaged on environmental problems, especially climate change. The development of advocacy in these cases is a reflection of the way foundations see their social role evolving, particularly in the face of urgent future environmental challenges. The future of this movement is yet to be determined and the need for future research on this topic will be important as we proceed through the various environmental problems in the twenty-first century. I hope this chapter has shown the need for social scientists and environmental activists to increase their understanding of foundations and philanthropy.

* * *

References

Acuto, M. (2013). "The New Climate Leaders?" *Review of International Studies* 39 (4), 835–857.

Aykut, S., and Dahan, A., (2014). "La Gouvernance du Changement Climatique. Anatomie d'un Schisme de Réalité," in D. Pestre (ed.) *Le Gouvernement des Technosciences*. Paris: La Découverte, 78–109.

Becker, H. S. (1960). "Notes on the Concept of Commitment." *American Journal of Sociology* 66 (1), 32–40.

Beddoe, R., Costanza, R., Farley, J., Garza, E., Kent, J., Martinez, L., Mccowen, T., Murphy, K., Myers, N., Ogden, Z., Stapleton, K., Woodward, J., Martinez, M., and Luz, A. (2009). "Overcoming System Roadblocks to Sustainability: The Evolutionary Redesign of Worldviews, Institutions, and Technologies." *Proceedings of the National Academy of Sciences of the United States of America*, 106.

Betsill, M., Enrici, A., Le Cornu, E., and Gruby, R. (2021). "Philanthropic Foundations as Agents of Environmental Governance: A Research Agenda." *Environmental Politics*, 1–22.

Bowman, B. (2019). "Imagining Future Worlds alongside Young Climate Activists: New Framework for Research." *Fennia—International Journal of Geography*, 197 (2), 295–305.

Brulle, R. J. (2014). "Institutionalizing Delay: Foundation Funding and the Creation of U.S. Climate Change Counter-movement Organizations." *Climatic Change*, 122, 681–694.
Brulle, R. J., Hall, G., Loy, L., and Schell-Smith, K. (2021). "Obstructing Action: Foundation Funding and US Climate Change Counter Movement Organizations." *Climatic Change, Springer*, 166 (1), 1–7.
Calhoun, C. (1993). "Civil Society and the Public Sphere." *Public Culture*, 5 (2).
Callahan, D. (2017). *The Givers: Wealth, Power, and Philanthropy in a New Gilded Age*. New York: Vintage, Knopf Doubleday Publishing Group.
Caniglia, B. S., Brulle, R. J., and Szaz, A. (2014). "Civil Society, Social Movements and Climate Change," in E. Dunlap Riley and J. Brulle Robert (eds.) *Climate Change and Society: Sociological Perspectives*. Oxford: Oxford University Press.
Dakowska, D. (2014). *Le pouvoir des fondations. Des acteurs de la politique étrangère allemande*. Rennes: Presses Universitaires de Rennes, coll. Res Publica.
Dahan, A. (2014). "L'impasse de la gouvernance climatique globale depuis vingt ans. Pour un autre ordre de gouvernementalité." *Critique Internationale*, 62, 21–38.
Doyle, T., and MacGregor, S. (2013). *Environmental Movements Around the World: Shades of Green in Politics and Culture*. Praeger.
Fisher, D., and Nasrin, S. (2020). "Climate Activism and Its Effect." *Wires Climate Change*.
Hägel, P. (2021a). *Billionaires in World Politics*. Oxford: Oxford University Press.
Hägel, P. (2021b). "Le pouvoir des milliardaires philanthropes dans la politique mondiale", in S. Lefèvre and A. Monier (eds.) *Philanthropes en démocratie*. PUF/VDI.
Islamy, M. D., and Kieuy, E. (2021). "Sociological Perspectives on Climate Change and Society: A Review." *Climate*, 9, 7.
Jerneck, A., (2014). "Searching for a Mobilizing Narrative on Climate Change." *The Journal of Environment and Development*, 23 (1), 15–40.
Pandey, C. L. (2015). "Managing Climate Change: Shifting Roles for NGOs in the Climate Negotiations." *Environmental Values*, 24 (6), 799–824.
Lambin, R., and Surender, R. (2021). "The Rise of Big Philanthropy in Global Social Policy: Implications for Policy Transfer and Analysis." *Journal of Social Policy*, 1–18.
McAdam, D. (2017). "Social Movement Theory and the Prospects for Climate Change Activism in the United States." *The Annual Review of Political Science*, 20, 198–208.
McCarthy, D. (2004). "Environmental Justice Grantmaking: Elites and Activists Collaborate to Transform Philanthropy." *Social Inquiry*, 74 (2), 250–270.
McGregor-Lowndes, M., and Williamson, A. (2018). "Foundations in Australia: Dimensions for International Comparison." *American Behavioral Scientist* 62 (13), 1759–1776.
Morena, E. (2016). *The Price of Climate Action: Philanthropic Foundations and the Global Climate Debate*. London: Palgrave.
Morena, E. (2021). "Les philanthropes aiment-ils la planète ? Capitalisme, Changement climatique et philanthropie", in S. Lefèvre and A. Monier (eds.) *Philanthropes en démocratie*. Presses Universitaires de France/la vie des idées.
North, P. (2011). "The Politics of Climate Activism in the UK: A Social Movement Analysis." *Environment and Planning. A, 2011–07*, 43 (7), 1581–1598.
Nulman, E. (2015). *Climate Change and Social Movements: Civil Society and the Development of National Climate Change Policy*. Birmingham: Palgrave Macmillan.

O'Brien, K., Selboe, E. and Hayward, B. M. (2018). "Exploring Youth Activism on Climate Change: Dutiful, Disruptive, and Dangerous Dissent." *Ecology and Society*, 23 (3).

Reich, R. (2018). *Just Giving: Why Philanthropy Is Failing Democracy and How It Can Do Better*. Princeton, NJ: Princeton University Press.

Rootes, C., and Brulle, R. J. (2015). "Environmental Movements." *International Encyclopedia of the Social and Behavioral Sciences*, 14 (2), 763–768.

Skocpol, T. (2011). "Civil Society in the United States", in M. Edwards (ed.), *The Oxford Handbook of Civil Society*.

Skocpol, T. (2016). "Why Political Scientists Should Study Organized Philanthropy." *Political Science and Politics*, 49, 433–436.

Tarrow, S. (1998). *Power in Movement. Social Movements and Contentious Politics*. Cambridge: Cambridge University Press.

Thomson, D. E. (2020). "Philanthropic Funding for Community and Economic Development: Exploring Potential for Influencing Policy and Governance." *Urban Affairs Review*, 1–41.

Tindall, D. B., and Robinson, J. L. (2017). "Collective Action to Save the Ancient Temperate Rainforest: Social Networks and Environmental Activism in Clayoquot Sound." *Ecology and Society*, 22 (1).

Tournès, L., L'Argent de l'influence. (2010). *Les Fondations américaines et leurs réseaux européens*. Paris: Autrement.

Von Schnurbein, G., Rey-Garcia, M., and Neumayr, M (2021). "Contemporary Philanthropy in the Spotlight: Pushing the Boundaries of Research on a Global and Contested Social Practice." *Voluntasi*, 32, 185–193.

Williamson, A., and Luke, B. (2020). "Agenda-Setting and Public Policy in Private Foundations." *Nonprofit Policy Forum*, 11 (1), 1–12.

Zunz, O. (2012). *La philanthropie en Amérique. Argent privé, affaires d'Etat*. Paris: Fayard.

Environmental Politics and Theory in the City

Cities and Nature: Conceptualizations, Normativity and Political Analysis

Nir Barak

There is no endeavor more noble than the attempt to achieve a collective dream. When a city accepts as a mandate its quality of life; when it respects the people who live in it; when it respects the environment; when it prepares for future generations, the people share the responsibility for that mandate, and this shared cause is the only way to achieve that collective dream.

Jaime Lerner (1937-2021), Brazilian politician, and former mayor of Curitiba

Introduction

Today, we are all environmentalists. However, we are all also urbanists: in high-income countries, 80% of the population are city dwellers, and the global 50% threshold was passed in 2007. People worldwide live in cities.[1] The amalgamation of these two aspects is usually addressed through the prism of "sustainable urbanism," which aims to transition cities into more environmentally benign patterns. Moreover, sustainable urbanism has become a political imperative for cities, states and global institutions, receiving significant attention in regional

[1] Percentage of urban populations in 2020: Africa – 43.5%; Asia – 51.1%; Europe – 74.9%; Latin America and the Caribbean – 81.2%; Northern America – 82.6%; Oceania – 68.82% (UN-Habitat 2020).

N. Barak (✉)
Department of Politics and Government, Ben Gurion University of the Negev, Be'er Sheva, Israel
e-mail: nirbarak@bgu.ac.il

and global agreements.[2] Given their key role in climate negotiations and their contribution to international agreements, such as the Paris Climate Accord, and city-based networks, such as the Global Covenant of Mayors, United Cities and Local Governments, and the Cities Climate Leadership Group (C40),[3] cities are clearly lead actors in the field of global environmental governance (Hale 2016; Bäckstrand et al. 2017; Bulkeley 2021). Moreover, this trend correlates with cities' current aspiration for greater political autonomy vis-à-vis the state: indeed, municipalities aspire to be increasingly autonomous and self-representing with regard to environmental governance. Environmentalism may in fact be one of the clearest fields in which relatively autonomous city-based action is realized (cf., Bouteligier 2012; Acuto 2013, 2016; Curtis 2014; Barber 2017; Barak 2020a; Fernández and Angel 2020).

However, the notion of "sustainable urbanism" is also open to interpretation (cf., Guy and Marvin 1999; Derickson 2018; Wachsmuth and Angelo 2018; Long and Rice 2019). It encompasses technological and economic factors as well as aesthetics; it provokes a multiplicity of interests among a range of actors, from industries and venture capital investment to city planners and landscape architects. Some core issues can nonetheless be said to characterize the contemporary practice of sustainable urbanism. The most common policies or policy recommendations focus on energy reduction, efficient management of resources, high-performance buildings and infrastructure, and the "smart" management of cities. Alongside these technological solutions, various models address ecological limits in cities, transportation, land use, open spaces, and energy and water use to foster safer, healthier, resilient, sustainable and more efficient cities (Worldwatch Institute 2016, Cohen 2017).

While these goals are attractive, their associated policies ignore some of the significant social and political implications for cities and their city-zens (i.e., citizens in cities). To be precise, the political aspect of urban sustainability usually focuses on the implementation and administration of reformative techno-managerial solutions in cities. While this is essential, it fails to fully realize the political aspects involved. A city's transition toward sustainability should also focus on how environmental issues are framed, the values that drive the policies and their implications for social and environmental justice.

While the political architecture of "sustainable urbanism" is constantly changing and evolving, some fundamental questions regarding the relationships between cities and the natural environment stand out: how are they

---

[2] For example, one of the primary objectives of the UN's 2030 Agenda for Sustainable Development (2015) is to "make cities and human settlements inclusive, safe, resilient and sustainable." The New Urban Agenda, also known as Habitat III (2017), elaborates on action-oriented commitments to sustainable urban development; the European-based "Pact of Amsterdam" (2016) responded with an enhanced Urban Agenda for the EU that includes a range of goals oriented to sustainable and inclusive growth.

[3] https://www.globalcovenantofmayors.org/; https://www.uclg.org/; https://www.c40.org/

conceptualized? To what extent are cities part of nature and what nevertheless differentiates them from nature? What are the normative and political implications of these relationships? Scholars offer varying responses, yet most share an opposition to city-nature dualism, which sees the relationship as oppositional and antithetical: the city is not part of nature and vice versa. While the dualistic standpoint entails inherent contradictions, its philosophical opposition, namely monism, also collapses. Monism and conceptions of the "natural city" regard cities as natural entities and nature as a higher moral order from which urban models ought to be derived. These two positions are analyzed and critiqued in the following section. The subsequent section focuses on two aspects of hybrid, socio-natural conceptions of city-nature relationships that highlight axiological and political challenges posed by urban sustainability. The chapter concludes by suggesting that *civic ecologism*, i.e., city-based politics of urban sustainability, offers fertile ground for addressing the normative and political implications raised in the analyses.

THE INADEQUACIES OF DUALISM AND MONISM

The dualistic standpoint opines that when in the city, we are no longer in nature and vice versa; it fully separates between the city's sociability and the *naturalness* of the natural environment. By contrast, the monistic standpoint views this relationship as somewhat synonymous; that is, cities and nature are inseparable. Both perspectives entail different presuppositions and implications; the following analyses highlight the inadequacies of both perspectives.

The Contradictions of City <> Nature Dualism

The dualistic notion of the city-nature relationship is epitomized in the opening statements of Wirth's *Urbanism as a Way of Life:* "Nowhere has mankind been farther removed from organic nature than under the conditions of life characteristic of great cities" (1938, pp. 1–2). Or, more generally regarding nature-culture dichotomies, Rolston III argues, "The architectures of nature and of culture are different, and when culture seeks to improve nature, the management intent spoils the wilderness.... The cultural processes by their very 'nature' interrupt the evolutionary process: there is no symbiosis, there is antithesis" (Rolston III 1998, p. 371). Anything *within* the confines of the city gates is depicted as a purely artificial entity; cities are portrayed as a radical split from nature. The latter, according to this conceptualization, encompasses all the nonhuman entities *outside* the city. Indeed, nature is depicted as the "ultimate other" in reference to cities and human society in general.

The notion of the city gate is revealing. Historically, many cities and city-states around the world had defensive walls and city gates that safeguarded local civilized citizens from outsiders—barbarians, savages and foreigners.

Outside the walls a "state of nature" exists; inside the city is a civilized polity. However, the "state of nature" here is double; it is not just a metaphorical thought experiment by philosophers of the social contract—the city was to be protected from nature per se. The city gates enabled the differentiation between citizens and non-citizens as well as the segregation and distinction between socially desired, domesticated and "good" nature *in* the city (i.e., urban parks, clean water and air) and the socially undesired, wild, untamed and "bad" nature to be left *outside* or removed (i.e., dangerous animals, sewage, pollution).

While clearly the "city gate" no longer physically separates the city from nature, Kaika (2005) argues that the *logic of dualism* nonetheless continues to apply, implying a double coding of both cities and nature.[4] For example, according to one version, the city is considered a hostile human environment and urban life is morally degraded and artificial in comparison with nature. This notion corresponds well with what Light (2001) defined as an urban "blind spot" in environmental ethics, which also correlates with a strong anti-urban bias in the history of environmental philosophy (cf., Jamieson 1984; de Shalit 1996, 2003; Gunn 1998; Fox 2000; King 2006; Booth 2008; Barak 2017; Epting 2018). In short, this bias views urban life as morally inferior to life lived closer to nature. For example, Arne Naess, one of the pioneers of Deep Ecology, argues that "humanity today suffers from a place-corrosive process. Urbanization, centralization, increased mobility…. Weaken or disrupt the steady belongingness to a place…There seems to be no place for PLACE anymore" (Naess 2008, p. 45).

It may be argued that this bias is not as dominant as it once was and no longer constitutes a "blind spot." And yet, the tendency to focus on "classical" environmental concerns, especially by proponents of non-anthropocentrism such as those involved in wilderness conservation, yields impressive metaphysical or ontological arguments regarding the value of nature or the essential relationship between humans and nature. However, it leaves us without an ethical, philosophical and political framework via which we can reflect on the interaction with the supposedly non-natural, artificial and urban world. Moreover, it relinquishes "brown" areas of the world—including their inhabitants.

In a second version of the logic of dualism, the city is considered humanity's greatest achievement and the hub of social and economic progress. Accordingly, nature is depicted as wild and dangerous: it must be tamed and controlled, thus giving rise to modernist conceptions of city planning. This notion, like the previous one, leads to normatively unwarranted conclusions in that it conceives of cities as "bounded social containers" isolated from the nonhuman world. Since this is a dualistic scheme, it necessarily implies that each of the two versions attributes moral superiority either to cities or nature, culture or nature. The first version necessitates the subordination of nature to

---

[4] See also Wapner (2010, chaps 3 & 4).

social needs—a kind of human hubris; the second version creates anti-urban romantic utopias. The former approach gives rise to a managerialist attitude that manifests in environmental reforms. The latter approach engenders anti-urbanism or what has been defined as "ruralism," voiding environmentalism of its social and political goals and correlating with nativist and racist attitudes (de Shalit 1996). Given the tensions, contradictions and drawbacks involved in dualistic conceptions of city-nature relationships, one intuition is that it may be possible to overcome dualism by pursuing monistic reasoning, as discussed in the next section.

The Fallacies of City=Nature Monism: Depoliticization

The Inadequacies of Dualism and Monism
In his critique of dualistic thinking, Callicott argues that:

> Nature as Other is over... Bluntly put, we are animals ourselves, large omnivorous primates, very precocious to be sure, but just big monkeys, nevertheless. We are therefore a part of nature, not set apart from it. Hence, human works are no less natural than those of termites or elephants. Chicago is no less a phenomenon of nature than is the Great Barrier Reef (a vast undersea coral polyp condominium) or limestone sediments formed by countless generations of calciferous marine organisms. (Callicott 1992, pp. 17–8)

The vision of the "natural city" (Stefanovic and Scharper 2012) follows this line of reasoning and conceives of cities as natural entities in which nature represents a higher moral order. The common thread associated with the "natural city" is that the city blends harmoniously with the natural world since it is part of the larger biotic community. Accordingly, the non-artificial (natural) features of the cityscape are recognized as foundational to the lives of the human community that inhabits the city. This understanding arises from a paradigm shift, resulting in a vision of a city that is "authentic or true to itself."

In the quest for a city of this type, Stefanovic asks: "Is the building of a 'natural city' simply a matter of integrating more parks into urban spaces? .... Moving forward will require more than assembling a compendium of such discrete initiatives. More important will be to consider a repositioning of fundamental values, paradigms, and world views that sustain these efforts in the long term" (2012, p. 18). The notion of authenticity referred to is the non-dualistic understanding that "calls us to find ways to live within the context given us by nature, destroying as little as possible. In this view, human life adjusts to its *natural context*. It seeks ways to improve its condition that also benefit its natural environment" (Cobb Jr. 2012, p. 191, my emphasis).

While some might find this vision attractive, its implicit suggestion that we derive our urban political models from nature backlashes. This line of reasoning overlooks a straightforward fact: while the naturalness of nature may be questioned, the social-political character of cities cannot be doubted. In direct response to Callicott's "Chicago" argument, Rolston writes: "It is only philosophical confusion to remark that both geese in flight, landing on Yellowstone Lake, and humans in flight, landing at O'Hare in Chicago, are equally natural, and let it go at that... Geese fly naturally; humans fly in artifacts" (2001, p. 268). Extending Rolston's argument, the qualitative difference between cities and nature is that nature, even when produced and manipulated, exhibits complexities that result from spontaneous processes, self-organization and autopoiesis; cities, in addition to their limited natural characteristics, also involve planning, deliberation, premeditation and politics, none of which are natural in the same way as the Great Barrier Reef is. This critique might seem to be a retreat into dualism. And yet, focusing on the difference, for example, between the "urban jungle" of New York City and the "natural" Amazon jungle does not imply a retreat into dualism. Rather, it calls for a more comprehensive political framework.

The most concerning inadequacy of "naturalizing" the city in this sense is that the city is valued almost exclusively in terms of its interrelatedness with nature and not for its human component. It seems to rest on one of several interpretations of non-anthropocentrism—as giving precedence to nature/the biotic community over the needs and rights of human beings if these two are considered in isolation. While this approach emphasizes "the integrity, beauty, and stability of the urban community, in communion with all the subjects [(i.e., human and nonhuman alike)] that dwell within and beyond city limits" (Scharper 2012, p. 100), it ascribes only minor significance to human affairs. Such a line of reasoning is unwarranted. Cities should be valued according to their degree of interrelatedness with nature but also, and primarily, for their cultural offerings, promotion of human rights, ability to provide educational opportunities, social heterogeneity, proximity to social and political institutions, or simply for being more liberal than rural settlements in their tolerance to difference and diversity, inclusive welfare policies and openness to immigrants and refugees (van Leeuwen 2010, 2018; Oomen et al. 2016; de Shalit 2018; Jeffries and Ridgley 2020). In the (direct) environmental context, cities should be valued for their density, for their ability to promote environmental policies better than other political entities (Owen 2010; Glaeser 2011; Barber 2013), and for their ability to foster a (human) community and political identity in an era of homogenizing global culture (Bell and de Shalit 2011).

Prescribing guidelines for the "natural city" instrumentalizes the concept of nature, promoting an ecocentric political agenda that is disguised as apolitical and acultural.[5] This attitude ignores the diversity of existing urban political cultures and identities. To elaborate, Cunningham (2012), for example,

---

[5] Conceptual instrumentalism is the use of one concept, nature in this case, to promote political ideas that are not necessarily related to it. It results from a hybrid of environmental philosophy and metaphysics (or alternative cosmology) with political theory (i.e., the natural city), using a single mode of reasoning. For an elaboration of conceptual instrumentalism, see (de Shalit 2000, chap. 2).

argues that the "natural city" is the antithesis of and antidote to contemporary "global cities," which are interpreted as cities designed ultimately and almost exclusively for the "global class." Accordingly, the "natural city" should alleviate the social and economic impoverishment resulting from global urbanization. However, just as Cunningham criticizes the global city for being exclusive to the global class, her vision/utopia of the "natural city" seems to be exclusive to the "naturalistic group"—a social group that can be characterized as having ecocentric attitudes.[6] It may be argued in turn that this group is inclusive, though this is true only to the extent of accepting the "repositioning of fundamental values, paradigms, and worldviews" (Stefanovic 2012, p. 18) suggested by the proponents of this model, i.e., ecocentrism. It proposes transcending contemporary (political) culture by following a form of critique that promotes an alternative worldview, questioning the values that are external to existing practices. While this form of utopianism might be attractive for some, it does not constitute a concrete political program. Moreover, this model is conceived by instrumentalizing the "natural context," thereby short-circuiting a less appealing and tougher political process whereby the values attributed to existing practices are constantly questioned, reaffirmed or altered.[7]

If this is not an attempt to short-circuit politics by instrumentalizing nature, then it assumes that urban political models can be derived from nature. This attitude was characterized by Meyer (2001, pp. 2–5, 47–50) as a derivative view of the relationship between nature and politics. Such an approach is methodologically wrong. Unlike natural processes, which are spontaneous, largely deterministic and involuntary (i.e., gravity, magnetism, thermodynamics, photosynthesis), city politics can take many forms. These political forms are evident in the diverse and mutating urban political cultures throughout the world. The methodological error results from the fact that the *nature* of nature is constantly questioned. The type of questioning I refer to is not the scientific study of nature, but rather the debatable and contested social and political implications of these scientific findings. Is nature harmonious or hostile? Is it hierarchical or egalitarian? The answers to these questions reflect the values that a person already holds. In other words, even if we agree that we ought to derive our political models and theories from nature (committing the naturalistic fallacy of deriving "ought" from "is"), we would still argue over what nature actually *is*, taking us back to politics.

A normative and pragmatic problem accompanies this methodological error. Even if we somehow derive political guidance from a "correct" conception of nature, there is no guarantee that it will lead to environmentally beneficial policies. Certain interpretations of nature can be disastrous for

---

[6] For elaboration on ecocentrism, "please see chapter on environmental ethics" in this volume.

[7] On the method of "critical political ecology" whereby values integral to current practices see Eckersley (2004). On the resonance dilemma, whereby contradictions between environmental motivations and quotidian practices emerge see, Meyer (2015).

humans (e.g., social Darwinism) or lead to ecologically destructive behavior (e.g., suburbanization) or a fortuitous mix of both (e.g., ruralism) (Meyer 2001; de Shalit 1996, 2000, pp. 99–103).

Finally, naturalizing politics in the implicit manner suggested by proponents of the "natural city" implies the *impossibility* of politics. Attributing naturalness to a certain political order suggests not only that this order is morally superior but also that it is permanent or subject exclusively to those "natural" pregiven "laws of politics." That is, if politics were natural, and assuming nature follows immutable principles, it follows that a person or a society has practically no ability to deliberately influence, reform or change any features of the political structures within which they live.

To conclude, the conceptual monism advanced by proponents of the "natural city" ignores the fact that concrete political institutions constitute the relationship between cities and nature and that a political process cannot be short-circuited or replaced by metaphysical conceptions. Therefore, in the subsequent section, we turn to analyzing warranted normative and political aspects of city-based politics of sustainability.

City-nature Nexuses: Embeddedness and Hybridity

While dualism yields inherent contradictions, monistic simplifications depoliticize urban sustainability. Therefore, the conceptual challenge may be framed as navigating between non-dualism on the one hand, without falling into the fallacies of monism on the other. The notions of embeddedness and city-nature hybrids adhere to this criterion and offer a more fruitful path for investigating the axiological (i.e., value related) and political challenges involved in transitioning cities into sustainable patterns. These aspects are analyzed in the following:

Axiological Considerations: Physical Interrelatedness of Cities and Nature

A primary path for overcoming both dualism and monism is to scrutinize cities' physical embeddedness in nature and nature as an important constituent of the city. Embeddedness involves recognizing that nature does not end at the "city gates" or municipal boundaries; it also implies that a city cannot be completely distinguished from nature. In this regard, embeddedness suggests that cities and nature are physically interrelated. Yet, interrelatedness does not imply a monistic simplification; rather it highlights the axiological challenges of environmental ethics in the built environment.

An example of the notions of city-nature interrelatedness is found in one of the principles of "green urbanism" articulated by Timothy Beatley, who writes that green cities are "designed for and function in ways analogous to nature" (2000, pp. 6–7). This analogy is limited to enhancing ecosystem services and the natural processes that constantly occur in cities, or rather the processes that were interrupted by the production of cities (like the flow of rivers or natural cleansing of water and air). The recognition that cities are embedded in nature and the importance of the natural processes within them gives rise to various planning approaches, among them ecological restoration, the implementation of nature-based solutions, replenishment and the nurturing of natural processes, and fostering biodiversity in cities (cf., Farr 2008, Lehmann 2010, Benton-Short and Short 2013, Beatley 2017, Kabisch et al. 2017, Xie and Bulkeley 2020).

The integration of natural elements in the city dents the dualism of the bounded social container. However, while these and other associated policies are integral to urban sustainability and yield important ecosystem benefits, they do not go uncontested. For example, urban forestry is a key practice in promoting sustainability and climate change adaptation in cities (Kabisch et al. 2017; Pearlmutter et al. 2017). Noted benefits of street-trees in urban spaces range from improving ecosystem services (e.g., reducing air pollution, moderation of extreme weather) to the improvement of mental health, especially among disadvantaged city dwellers (Baró et al. 2019; Marselle et al. 2020). And yet, urban forestry generates widely conflicting interests and opinions. On the one hand, some appreciate the ecosystem and social "services" it offers, such as shade, walkability and the reduction of urban heat-islands. On the other hand, some highlight the social disservice of this policy, for example blocked TV signals, damage to pavements, sap and bird droppings on cars (cf. Davies et al. 2017; Ordóñez et al. 2017, 2020). Tensions of this sort are accentuated, for example, with regard to wildlife in the city and urban interspecies relations (cf. Toger et al. 2018; Hunold and Lloro 2019; Hunold 2020; Shingne 2020; Collins et al. 2021).

The urban street-tree example highlights that the integration of natural elements in the city might be interpreted as inherent to the logic of dualism in that it narrows the natural elements of the city to designated spaces of "urban nature" (e.g., urban parks, green belt, etc.). To elaborate, we may distinguish between the "nature *of* cities," as implied by the notion of interrelatedness, and "nature in cities," which reaffirms the dualistic intuition. The former is prevalent throughout the cityscape (e.g., pavements, rooftops, etc.); the latter is manifest in designated "green spaces." The notion of "nature in cities" narrows the "nature of cities" to controlled locations of "urban nature" and thus reduces the interaction between city and nature to these designated areas. Reducing the nature of cities to "nature in cities" gives rise to the dialectic interplay of dualism, i.e., anything that escapes the designated "green space" may be considered a social burden. That is, urban nature is introduced in order to meet ecosystem services or recreational needs, but is immediately subordinated and controlled.

Therefore, the notion of embeddedness and interrelatedness of cities and nature implies that considerations of environmental ethics are neither limited to the landscape "exterior" of the city's built environment nor to "nature in cities." In addition to their instrumental benefits for ecosystem services, the policies of urban greening and environmental restoration call into question the city's socio-environmental norms and values. Problematizing a city's urban culture by challenging dualistic intuitions (that we are no longer in nature when in the city) is valuable for three main reasons. It causes us to reflect on the relationship between the built environment and the natural world; it allows us to evaluate nature from the experience of the built environment; and it may provide moral justifications for fostering environmentally friendly attitudes in cities that extend beyond "nature in cities" (cf. Barak 2017).

Fostering environmental awareness and literacy is usually considered a non-urban practice; when conducted in the city, this often relates to patches of "nature in cities." However, such practices should be extended to the "damaged and blemished" urban areas that are constitutive of the "nature of cities" no less than nature exterior to the city's boundaries. If, as Leopold (1986, p. 292) argues, "the weed in a city lot conveys the same lesson as the redwoods," then enhancing city-nature interrelatedness requires learning "the same lesson" as a civic community. This type of collective introspection is not limited to greenways, parks, or sidewalks and central plazas but embraces the city's civic culture and socio-environmental norms. An additional aspect of dualism involves demarcating borders between the social processes of the city and the natural processes "outside" of it. This aspect is challenged in the following section.

Inequality and Power Relations: The Socio-political Nexus of Cities and Nature

Nature does not only permeate the city, as discussed above, but is also found at a nexus of its socio-political and economic realities. For instance, a neighborhood or district that has undergone "green" retrofits (e.g., urban park, bike lane, access to public transport), or a city that subscribes to environmental sustainability, may lead to increased demand, higher costs and eventually to the displacement of disempowered and disadvantaged populations (i.e., green gentrification). While public policy can adequately address this matter by implementing rent control, public housing and more equitable development policies, these factors are often excluded from the planning process as a result of the greater political and economic power wielded by private investors, real-estate firms and economic elites in comparison with the local populations (Checker 2011; Wolch et al. 2014; Gould and Lewis 2016).

In this regard, social stratification, marginalization and unequal power relations are integral to the process of promoting urban sustainability. This notion underlies a variety of theoretical frameworks, such as the "green political economy," which critique orthodox ideologies of economic growth embedded in current sustainable development policies (e.g., Barry 2012); critiques of conceptions of a "nature" independent of socio-political realities (e.g., Vogel 2015); and environmental justice theories that analyze the inequitable distribution of environmental "goods" and "bads" in the city according to political power, class, race and gender (Agyeman 2005, 2013; Schlosberg 2007).

Additionally, the socio-economic and political interdependencies in the city-nature relationship are addressed by proponents of urban political ecology (hereinafter UPE). While political ecology typically focuses on the political economy of environmental degradation and its nexus with socio-economic inequalities "outside" the city, UPE integrates this analysis with critical urban geography to include the socio-ecological production of urban spaces. In his analysis of non-urban political ecology, Harvey claims:

> It is … inconsistent to hold that everything in the world relates to everything else, as ecologists tend to do, and then decide that the built environment and the urban structures that go with it are somehow outside of both theoretical and practical consideration. The effect has been to evade integrating understandings of the urbanizing process into environmental-ecological analysis. (Harvey 1996, p. 427)

Following up on this call, UPE emphasizes processes of urban metabolism and the unequal political and economic forces that drive them and are perpetuated by them (Heynen et al. 2006; Cook and Swyngedouw 2012; Angelo and Wachsmuth 2015).

In this framework, cities and artifacts (buildings, pipes, urban parks) are understood as nature transformed (and commodified) through human processes of production. Urbanization is not limited to the city's sociality (as understood by dualists)[8] but rather is a *hybrid* socio-natural metabolic process in which nature is transformed into new socio-spatial configurations (e.g., cities, dammed rivers, hydraulic conduits). Nature, therefore, is inconceivable as external to socio-economic dynamics and most explicitly without reference to processes of urbanization. Likewise, the socio-economic dynamic, and most unequivocally the inequalities enshrined in urban life, is inconceivable without reference to the production of these socio-natures. Perceiving the relationship between cities and nature in this manner breaks the dualistic schism and has led some authors to extend Haraway's (1991) concept of the cyborg: indeed, they regard the city as the outcome of *cyborg urbanization* in which nature, artifacts, technology, capital and socio-economic relations are assembled and transformed (Gandy 2005; Swyngedouw 2006; Luke 2014; White et al. 2016).

As a theoretical framework, UPE is not necessarily a *cityist* theory or even city-centered (Angelo and Wachsmuth 2015; Collard et al. 2018; Tzaninis et al. 2021). Rather, it focuses on urban and non-urban socio-natures, analyzing the metabolisms that enable and condition urban life. It concentrates on the political economy of produced environments such as dammed rivers and hydraulic conduits, air pollution, urban parks and wildlife ("nature in cities"), irrigated fields, etc., all of which involve transformative social processes as part of their production and yield uneven socio-environmental

---

[8] Cf. Wachsmuth's (2012) analysis of society-nature oppositions in urban sociology.

consequences based on uneven socio-economic power relations (Heynen 2014). And yet, UPE has clear implications for city-based politics of sustainability. The example noted above concerning the displacements that result from projects promoting environmental sustainability (i.e., green gentrification) indicates the nexus of issues of social equity and environmental policies. Social stratification and uneven distributions of political power are integral to the process of making the city's urban patterns more sustainable. Therefore, reducing questions of sustainable urbanism to techno-managerialism, or to an urban environmental ethics centered on "nature in cities," is at best inadequate and at worst intentionally misleading. More explicitly, disassociating social and political realities from the practice and general orientation of urban sustainability depoliticizes environmentalism, emptying it of its aspiration to equality, democracy and human rights.

Summary and Conclusions: City-based Ethics and Politics of Sustainability

At this junction, several conclusions may be drawn regarding the political and normative implications of different city-nature conceptualizations (see Table 1). Dualistic approaches, which see cities as artificial and radically split from nature, are not only empirically wrong but also entail contradictions that underlie normatively undesired conclusions. Dualism either justifies supposedly value-neutral managerialism or reaffirms an anti-urban bias and validates ruralism. Monistic approaches that seek to remedy the human chauvinism implied by dualism collapse due to the aspiration to derive urban political models from nature; in so doing, not only do they "naturalize" politics but also imply the impossibility of politics. Therefore, these two conceptual extremes, dualism and monism, fail to yield a comprehensive account of the relationship between cities and nature.

Navigating between these two conceptual extremes are two responses that offer fertile ground for city-based politics of sustainability. The notion of embeddedness and physical interrelatedness focuses on the "nature *of* cities," giving rise to moral justifications for fostering environmentally friendly attitudes that extend beyond managerialism or romanticism. In so doing, it problematizes a city's urban culture and calls into question its socio-environmental norms and values. The hybrid conception of cities as socio-natural entities highlights uneven power relations at the nexus of cities and nature. While these two approaches present different normative and political considerations, they may be seen as complementary.

Summarizing the analysis, it may be argued that cities and nature are physically interrelated and socio-politically interdependent, yet at the same time different. Building on this conclusion, I suggest an additional view that focuses on city-based politics of urban sustainability, which I term "civic ecologism" (Barak 2020a). This perspective concentrates on cities primarily as semi-autonomous polities that embody a political community of city-zens. Accordingly, a city-based politics of sustainability focuses on plural valuations of socio-natural relationships and on the normative and political considerations for cities and their city-zens outlined above.

On the one hand, environmental, social, economic and political aspects of urban sustainability are not necessarily limited to the city as a polity, and cities are not "self-enclosed political [territories] within a nested hierarchy of geographical arenas contained within each other like so many Russian dolls" (Brenner et al. 2003, p. 1). On the other hand, while cities are often characterized as "built environments" or as a sub-unit of the state that functions as a socio-economic node, cities are changing their role and becoming semi-independent political actors. As briefly discussed in the introduction, cities are gradually moving into policy fields traditionally associated with the state or the global political system and enhancing their autonomous capacities.[9] Examples of autonomous city initiatives range from awarding city-based marriage licenses to same-sex couples (Schragger, 2005) to policy tools designed to guarantee immigrants' rights within municipal boundaries (Bauböck 2006; Ridgley 2008; Varsanyi 2008; de Shalit 2018) or city-based policies during epidemiological crises. A recent example of the latter is the COVID-19 outbreak, where numerous cities throughout the world are making their own public health policies (OECD 2020a, 2020b) and exhibiting particular rates of contagion based on their unique socio-political attributes—one of them being trust in the central government (Barak et al. 2021). Thus, cities are increasing their political autonomy by enhancing their political and economic capacities and by filling lacunae that are neglected or insufficiently addressed by the state—including environmental and climate policies. In the environmental context, the common thread running through domestic and global initiatives driven by city-based networks is that they commit cities to goals of sustainable urban development almost independently of the state's action. Therefore, while the politics of urban sustainability are not limited to the city itself, this does not undermine the substantial role that cities, and city-zens, can and do play in transitioning to more sustainable urban patterns and in fostering ecological city-zenships (Barak 2020b).

The following considerations further justify the premise of "civic ecologism," i.e., city-based politics of sustainability. Contemporary urban political theories highlight cities' social and political attributes, such as their *distinctiveness* from the state and from one another (Barbehön et al. 2016), and the role cities play in cultivating civic virtues (Dagger 1997; Cunningham 2011). For example, Bell and de Shalit (2011) argue that some cities encapsulate *civicism*, a morally-thick collective political identity and ethos distinct from nationalism, which varies between cities. This idea ties in with Löw's *intrinsic logic* of cities,

---

[9] Cf. the analysis of the "new municipalist" movement by Russell (2019) and Thompson (2020) and a recent international multidisciplinary debate initiated by Rainer Bauböck and Liav Orgad (2020) on the question of whether urban citizenship should be divorced from nationality.

which argues that cities, besides being a completely distinct form of association (compared to the state), also "behave" differently because of their often implicit or hidden political, social, spatial, temporal and affective structures (Löw 2012) as well as their capacity to produce shared meaning for their city-zens (Löw 2013). Different cities encounter similar problems, but how these problems are interpreted and the methods and legislation used to address them varies according to the intrinsic logic of each city.

Moreover, Molotch et al. (2000) describe how a city's *character* and *tradition* manifest by analyzing how cities with similar socio-demographic backgrounds respond differently to the same exogenous factors, and how this persists over time. These aspects are reflected in what Dagger (1981, p. 729) terms *civic memory*: "The recollection of the events, characters, and developments which make up the history of one's city or town," and which are "a necessary condition for (responsible) citizenship." Together, a shared ethos, meaning, tradition and memory render the city a meaningful polity and its city-zens a political community. In the environmental context, these aspects are constituted spatially in what Cannavò (2007, pp. 26, 44) defines as a "sense of place" based on "strong familiarity with a particular place or set of places, including their complexity, small details, aesthetic qualities, and history," which "helps guard against [environmentally and socially] destructive behaviour."

Lastly, Bookchin's (1992) *Urbanization without Cities*, one of the earliest analyses of cities and citizenship in the context of environmental theory, illustrates this claim by distinguishing between *urbanization* and *citification*.[10] The city is a concrete polity encompassing lived experiences within an urban fabric, while urban and urbanism are the processes facilitating the socio-economic transformation of "nature" into a built environment. One need not share Bookchin's anarchism (i.e., municipal libertarianism/ municipal confederalism), nor agree with his lament concerning the decline of citizenship to see that his emphasis on cities' potential to nurture their inhabitants as city-zens, i.e., as active agents in a political community, is an adequate way to address the socio-political and environmental challenges discussed above.

Despite the richness and potentialities of civic ecologism, it is important to note that promulgating this form of city-based politics of sustainability does not signify a retreat to a "new localism" or a special privileging of the city. Nor does it suggest that addressing these issues only at the city level, without acknowledging the multi-scalar character of urban politics, especially in an age of "planetary urbanization" in a global political economy, is warranted or even possible (Wachsmuth et al. 2016; Angelo and Wachsmuth 2020). Seeing the city as the ultimate solution, instead of the given arena of operation, might be highly misleading, absolving the state and other polities of their responsibilities and overlooking the multi-scalar character of urban politics and sustainability. And yet, while cityism may not offer the ultimate solution, cities and city-based politics of sustainability can provide fertile ground for activism and institutional innovation that will help transition cities into a more desirable

---

[10] For elaboration see, White (2008, chaps 6 & 7).

Relationships between cities and nature

Political and philosophical views	Cities:	Nature:	Relationship between Cities & Nature	Political & Ethical Aspects
Dualism:	Not natural; artificial; radical split from nature	The ultimate "Other"	Antithesis	
	Elevate humans from the dangerous "state of nature";	Resource; Systemic services/ Untamed and uncivilized	Subordination and domination of nature	Managerialism
Dialectics of dualism	Cause of environmental and **social** "bads"/ Not *authentic*	Alleviates pressures of environmental and **social** bads	Alienation	Anti-urban bias, ruralism, romanticism
Monism	Natural entities	Higher moral order	Conceptual Monism	Politics is naturalized and thus neutralized
Physical inter-relatedness	Embedded in nature	Ecosystem services	Physically and geographically interrelated	Socio-environmental norms and values
Urban Political Ecology	Socio-natural entities; commodified nature	Mediated by (human) labor and incomprehensible without it	Social, economic, and political interdependence	Inequality; marginalization; environmental justice
Civic Ecologism	Semi-autonomous polities embodying a political community	A complex system of ecosystems to which we ascribe a complex system of values	Interrelated, interdependent, yet different	Plural valuations of socio-natural relationships; ecological city-zenship

future and address the inseparability of environmental, political, economic, social and cultural issues of urban sustainability highlighted when analyzing the relationships between cities and nature. Moreover, despite these drawbacks, cities' changing political role and aspiration for increased political autonomy underscore the need to foster a more profound city-based politics of urban sustainability and its inherent potential.

References

Acuto, M. 2013. *Global Cities, Governance and Diplomacy: The Urban Link.* London, New York: Routledge.

Acuto, M. 2016. "Give Cities a Seat at the Top Table." *Nature News*, 537 (7622), 611.
Agyeman, J. 2005. *Sustainable Communities and the Challenge of Environmental Justice*. NYU Press.
Agyeman, J. 2013. *Introducing Just Sustainabilities: Policy, Planning, and Practice*. London, UK: Zed Books.
Angelo, H. and Wachsmuth, D. 2015. "Urbanizing Urban Political Ecology: A Critique of Methodological Cityism: Urbanizing Urban Political Ecology." *International Journal of Urban and Regional Research*, 39 (1), 16–27.
Angelo, H. and Wachsmuth, D. 2020. "Why Does Everyone Think Cities Can Save the Planet?" *Urban Studies*, 57 (11), 2201–2221.
Bäckstrand, K., Kuyper, J.W., Linnér, B.-O. and Lövbrand, E. 2017. "Non-state Actors in Global Climate Governance: From Copenhagen to Paris and Beyond." *Environmental Politics*, 26 (4), 561–579.
Barak, N. 2017. "Hundertwasser—Inspiration for Environmental Ethics: Reformulating the Ecological Self." *Environmental Values*, 26 (3), 317–342.
Barak, N. 2020a. "Civic Ecologism: Environmental Politics in Cities." *Ethics, Policy & Environment*, 23 (1), 53–69.
Barak, N. 2020b. "Ecological City-zenship." *Environmental Politics*, 29 (3), 479–499.
Barak, N., Sommer, U. and Mualam, N. 2021. "Urban Attributes and the Spread of COVID-19: The Effects of Density, Compliance and Socio-political Factors in Israel". *Science of the Total Environment*, 793, 148626.
Barbehön, M., Münch, S., Gehring, P., Grossmann, A., Haus, M. and Heinelt, H. 2016. "Urban Problem Discourses: Understanding the Distinctiveness of Cities." *Journal of Urban Affairs*, 38 (2), 236–251.
Barber, B.R. 2013. *If Mayors Ruled the World: Dysfunctional Nations, Rising Cities*. Cambridge: Yale University Press.
Barber, B.R. 2017. *Cool Cities: Urban Sovereignty and the Fix for Global Warming*. Yale University Press.
Baró, F., Calderón-Argelich, A., Langemeyer, J. and Connolly, J.J.T. 2019. "Under One Canopy? Assessing the Distributional Environmental Justice Implications of Street Tree Benefits in Barcelona." *Environmental Science & Policy*, 102, 54–64.
Barry, J. 2012. *The Politics of Actually Existing Unsustainability: Human Flourishing in a Climate-Changed, Carbon Constrained World*. Oxford: Oxford University Press.
Bauböck, R., ed. 2006. *Migration and Citizenship: Legal Status, Rights and Political Participation*. Amsterdam: Amsterdam University Press.
Bauböck, R. and Orgad, L., eds. 2020. *Cities vs States: Should Urban Citizenship Be Emancipated from Nationality?* Retrieved from Cadmus, European University Institute Research Repository, at: EUI RSCAS, Global Governance Programme-386, [Global Citizenship], GLOBALCIT, [Global Citizenship Governance].
Beatley, T. 2000. *Green Urbanism: Learning from European Cities*. Washington, DC: Island Press.
Beatley, T. 2017. *Handbook of Biophilic City Planning & Design*. None edition. Washington, DC: Island Press.
Bell, D.A. and de Shalit, A. 2011. *The Spirit of Cities: Why the Identity of a City Matters in a Global Age*. Princeton, NJ: Princeton University Press.
Benton-Short, L. and Short, J.R. 2013. *Cities and Nature*. 2nd ed. New York: Routledge.

Bookchin, M. 1992. *Urbanization Without Cities: The Rise and Decline of Citizenship*. Montreal: Black Rose Books.

Booth, K.I. 2008. "Holism with a Hole? Exploring Deep Ecology Within the Built Environment." *The Trumpeter: Journal of Ecosophy*, 24 (1), 68–86.

Bouteligier, S. 2012. *Cities, Networks, and Global Environmental Governance: Spaces of Innovation, Places of Leadership*. London, New York: Routledge.

Brenner, N., Jessop, B., Jones, M., and Macleod, G., eds. 2003. *State/Space: A Reader*. Malden, MA: Blackwell.

Bulkeley, H. 2021. "Climate Changed Urban Futures: Environmental Politics in the Anthropocene City." *Environmental Politics*, 1–19.

Callicott, J.B. 1992. "La Nature est morte, vive la nature!" *Hastings Center Report*, 22 (5), 16–23.

Cannavò, P.F. 2007. *The Working Landscape: Founding, Preservation, and the Politics of Place*. Cambridge, MA: MIT Press.

Checker, M. 2011. "Wiped Out by the 'Greenwave': Environmental Gentrification and the Paradoxical Politics of Urban Sustainability." *City & Society*, 23 (2), 210–229.

Cobb Jr., J.B. 2012. "Sustainable Urbanization." In: I.L. Stefanovic and S.B. Scharper, eds. *The Natural City: Re-envisioning the Built Environment*. Toronto: University of Toronto Press, 191–202.

Cohen, S. 2017. *The Sustainable City*. New York: Columbia University Press.

Collard, R.-C., Harris, L.M., Heynen, N. and Mehta, L. 2018. "The Antinomies of Nature and Space." *Environment and Planning E: Nature and Space*, 1 (1–2), 3–24.

Collins, M.K., Magle, S.B. and Gallo, T. 2021. "Global Trends in Urban Wildlife Ecology and Conservation." *Biological Conservation*, 261, 109236.

Cook, I.R. and Swyngedouw, E. 2012. "Cities, Social Cohesion and the Environment: Towards a Future Research Agenda." *Urban Studies*, 49 (9), 1959–1979.

Cunningham, F. 2011. "The Virtues of Urban Citizenship". *City, Culture and Society*, 2 (1), 35–44.

Cunningham, H. 2012. "Gated Ecologies and 'Possible Uran Worlds'." In: I.L. Stefanovic and S.B. Scharper, eds. *The Natural City: Re-envisioning the Built Environment*. Toronto: University of Toronto Press, 149–160.

Curtis, S. 2014. *The Power of Cities in International Relations*. London, New York: Routledge.

Dagger, R. 1981. "Metropolis, Memory, and Citizenship." *American Journal of Political Science*, 25 (4), 715–737.

Dagger, R. 1997. *Civic Virtues: Rights, Citizenship, and Republican Liberalism*. Oxford: Oxford University Press.

Davies, H.J., Doick, K.J., Hudson, M.D. and Schreckenberg, K. 2017. "Challenges for Tree Officers to Enhance the Provision of Regulating Ecosystem Services from Urban Forests." *Environmental Research*, 156, 97–107.

Derickson, K.D. 2018. "Urban Geography III: Anthropocene Urbanism." *Progress in Human Geography*, 42 (3), 425–435.

Eckersley, R. 2004. *The Green State: Rethinking Democracy and Sovereignty*. Cambridge, MA: MIT Press.

Epting, S. 2018. "Philosophy of the City and Environmental Ethics." *Environmental Ethics*, 40 (2), 99–100.

European Commission. 2016. *Urban Agenda for the EU: Pact of Amsterdam*.

Farr, D. 2008. *Sustainable Urbanism: Urban Design with Nature*. Hoboken, NJ: Wiley.

Fernández, J.E. and Angel, M. 2020. "Ecological City-States in an Era of Environmental Disaster: Security, Climate Change and Biodiversity." *Sustainability*, 12 (14), 5532.

Fox, W., ed. 2000. *Ethics and the Built Environment*. London, New York: Routledge.

Gandy, M. 2005. "Cyborg Urbanization: Complexity and Monstrosity in the Contemporary City." *International Journal of Urban and Regional Research*, 29 (1), 26–49.

Glaeser, E. 2011. *Triumph of the City: How Our Greatest Invention Makes Us Richer, Smarter, Greener, Healthier and Happier*. Pan Macmillan.

Gould, K.A. and Lewis, T.L. 2016. *Green Gentrification: Urban Sustainability and the Struggle for Environmental Justice*. New York: Routledge.

Gunn, A.S. 1998. "Rethinking Communities: Environmental Ethics in an Urbanized World." *Environmental Ethics*, 20 (4), 341–360.

Guy, S. and Marvin, S. 1999. "Understanding Sustainable Cities: Competing Urban Futures." *European Urban and Regional Studies*, 6 (3), 268–275.

Hale, T. 2016. "'All Hands on Deck': The Paris Agreement and Nonstate Climate Action." *Global Environmental Politics*, 16 (3), 12–22.

Haraway, D. 1991. *Simians, Cyborgs, and Women: The Reinvention of Nature*. New York: Routledge.

Harvey, D. 1996. *Justice, Nature and the Geography of Difference*. 1 edition. Cambridge, MA: Blackwell.

Heynen, N. 2014. "Urban Political Ecology I: The Urban Century." *Progress in Human Geography*, 38 (4), 598–604.

Heynen, N., Kaika, M. and Swyngedouw, E. 2006. *In the Nature of Cities: Urban Political Ecology and the Politics of Urban Metabolism*. London, New York: Routledge.

Hunold, C. 2020. "Urban Greening and Human-Wildlife Relations in Philadelphia: From Animal Control to Multispecies Coexistence?" *Environmental Values*, 29 (1), 67–87.

Hunold, C. and Lloro, T. 2019. "There Goes the Neighborhood: Urban Coyotes and the Politics of Wildlife." *Journal of Urban Affairs*, 1–18.

Jamieson, D., 1984. "The City Around Us." In: T. Regan, ed. *Earthbound: New Introductory Essays in Environmental Ethics*. Philadelphia, PA: Temple University Press, 38–73.

Jeffries, F. and Ridgley, J. 2020. "Building the Sanctuary City from the Ground up: Abolitionist Solidarity and Transformative reform." *Citizenship Studies*, 24 (4), 548–567.

Kabisch, N., Korn, H., Stadler, J. and Bonn, A. 2017. "Nature-Based Solutions to Climate Change Adaptation in Urban Areas—Linkages Between Science, Policy and Practice." In: N. Kabisch, H. Korn, J. Stadler, and A. Bonn, eds. *Nature-Based Solutions to Climate Change Adaptation in Urban Areas: Linkages between Science, Policy and Practice*. Cham: Springer International Publishing, 1–11.

Kaika, M. 2005. *City of Flows: Modernity, Nature, and the City*. 1 edition. New York: Routledge.

King, R.J.H. 2006. "Playing with Boundaries: Critical Reflections on Strategies for an Environmental Culture and the Promise of Civic Environmentalism." *Ethics, Place & Environment*, 9 (2), 173–186.

van Leeuwen, B. 2010. "Dealing with Urban Diversity: Promises and Challenges of City Life for Intercultural Citizenship." *Political Theory*, 38 (5), 631–657.
van Leeuwen, B. 2018. "To the Edge of the Urban Landscape: Homelessness and the Politics of Care." *Political Theory*, 46 (4), 586–610.
Lehmann, S. 2010. *The Principles of Green Urbanism: Transforming the City for Sustainability*. London, Washington, DC: Earthscan.
Leopold, A. 1986. *A Sand County Almanac: With Essays on Conservation from Round River*. Reprint edition. New York: Ballantine Books.
Light, A. 2001. "The Urban Blind Spot in Environmental Ethics." *Environmental Politics*, 10 (1), 7–35.
Long, J. and Rice, J.L. 2019. "From Sustainable Urbanism to Climate Urbanism." *Urban Studies*, 56 (5), 992–1008.
Löw, M. 2012. "The Intrinsic Logic of Cities: Towards a New Theory on Urbanism." *Urban Research & Practice*, 5 (3), 303–315.
Löw, M. 2013. "The City as Experiential Space: The Production of Shared Meaning." *International Journal of Urban and Regional Research*, 37 (3), 894–908.
Luke, T.W. 2014. "Urbanism as Cyorganicity." In: D. Ibañez and N. Katsikis, eds. *New Geographies 06: Grounding Metabolism*. Cambridge: Harvard University School of Design, 39–53.
Marselle, M.R., Bowler, D.E., Watzema, J., Eichenberg, D., Kirsten, T. and Bonn, A. 2020. "Urban Street Tree Biodiversity and Antidepressant Prescriptions." *Scientific Reports*, 10 (1), 22445.
Meyer, J.M. 2001. *Political Nature: Environmentalism and the Interpretation of Western Thought*. Cambridge, MA: MIT Press.
Meyer, J.M. 2015. *Engaging the Everyday: Environmental Social Criticism and the Resonance Dilemma*. Cambridge, MA: The MIT Press.
Molotch, H., Freudenburg, W. and Paulsen, K.E. 2000. "History Repeats Itself, But How? City Character, Urban Tradition, and the Accomplishment of Place." *American Sociological Review*, 65 (6), 791.
Naess, A. 2008. "An Example of a Place: Tvergastein." In: *The Ecology of Wisdom: Writings by Arne Naess*. Berkeley: Counterpoint, 45–64.
OECD. 2020a. *The Territorial Impact of COVID-19: Managing the Crisis Across Levels of Government*.
OECD. 2020b. *Tackling Coronavirus (COVID-19): Cities Policy Responses*.
Oomen, B., Davis, M.F. and Grigolo, M. 2016. *Global Urban Justice*. Cambridge University Press.
Ordóñez, C., Beckley, T., Duinker, P.N. and Sinclair, A.J. 2017. "Public Values Associated with Urban Forests: Synthesis of Findings and Lessons Learned from Emerging Methods and Cross-Cultural Case Studies." *Urban Forestry & Urban Greening*, 25, 74–84.
Ordóñez, C., Threlfall, C.G., Livesley, S.J., Kendal, D., Fuller, R.A., Davern, M., van der Ree, R. and Hochuli, D.F. 2020. "Decision-making of Municipal Urban Forest Managers Through the Lens of Governance." *Environmental Science & Policy*, 104, 136–147.
Owen, D. 2010. *Green Metropolis: Why Living Smaller, Living Closer, and Driving Less Are the Keys to Sustainability*. New York: Riverhead Books.
Pearlmutter, D., Calfapietra, C., Samson, R., O'Brien, L., Ostoić, S.K., Sanesi, G. and Amo, R.A. del, eds. 2017. *The Urban Forest: Cultivating Green Infrastructure for People and the Environment*. Springer International Publishing.

Ridgley, J. 2008. "Cities of Refuge: Immigration Enforcement, Police, and the Insurgent Genealogies of Citizenship in U.S. Sanctuary Cities." *Urban Geography*, 29 (1), 53–77.

Rolston III, H. 1998. "The Wilderness Idea Reaffirmed." In: J.B. Callicott and M.P. Nelson, eds. *The Great New Wilderness Debate*. University of Georgia Press, 367–386.

Rolston III, H. 2001. "Natural and Unnatural; Wild and Cultural." *Western North American Naturalist*, 61 (3), 267–276.

Russell, B. 2019. "Beyond the Local Trap: New Municipalism and the Rise of the Fearless Cities." *Antipode*, 51 (3), 989–1010.

Scharper, S.B., 2012. "From Community to Communion: The Natural City in Biotic and Cosmological Perspective." In: I.L. Stefanovic and S.B. Scharper, eds. *The Natural City: Re-envisioning the Built Environment*. Toronto: University of Toronto Press, 89–103.

Schlosberg, D. 2007. *Defining Environmental Justice: Theories, Movements, and Nature*. Oxford: Oxford University Press.

Schragger, R.C. 2005. "Cities as Constitutional Actors: The Case of Same-Sex Marriage Democracy in Action: The Law & Politics of Local Governance." *Journal of Law & Politics*, 21, 147–186.

de Shalit, A. 1996. "Ruralism or Environmentalism?" *Environmental Values*, 5 (1), 47–58.

de Shalit, A. 2000. *The Environment: Between Theory and Practice*. Oxford University Press.

de Shalit, A. 2003. "Philosophy Gone Urban: Reflections on Urban Restoration." *Journal of social philosophy*, 34 (1), 6–27.

de Shalit, A. 2018. *Cities and Immigration: Political and Moral Dilemmas in the New Era of Migration*. Oxford, New York: Oxford University Press.

Shingne, M.C. 2020. "The More-than-human Right to the City: A Multispecies Reevaluation." *Journal of Urban Affairs*, 0 (0), 1–19.

Stefanovic, I.L. 2012. In Search of the Natural City. In: I.L. Stefanovic and S.B. Scharper, eds. *The Natural City: Re-envisioning the Built Environment*. Toronto: University of Toronto Press, 11–35.

Stefanovic, I.L. and Scharper, S.B., eds. 2012. *The Natural City: Re-envisioning the Built Environment*. Toronto: University of Toronto Press.

Swyngedouw, E. 2006. "Circulations and Metabolisms: (Hybrid) Natures and (Cyborg) Cities." *Science as Culture*, 15 (2), 105–121.

Thompson, M. 2020. "What's So New About New Municipalism?" *Progress in Human Geography*, 0309132520909480.

Toger, M., Benenson, I., Wang, Y., Czamanski, D., and Malkinson, D. 2018. "Pigs in Space: An Agent-based Model of Wild Boar (Sus scrofa) Movement into Cities." *Landscape and Urban Planning*, 173, 70–80.

Tzaninis, Y., Mandler, T., Kaika, M., and Keil, R. 2021. "Moving Urban Political Ecology Beyond the 'Urbanization of Nature'." *Progress in Human Geography*, 45 (2), 229–252.

UN-Habitat, ed. 2020. *World Cities Report 2020: The Value of Sustainable Urbanization*. Nairobi, Kenya: UN-Habitat.

United Nations General Assembly, 2015. *A/RES/70/1. Transforming our World: The 2030 Agenda for Sustainable Development*.

United Nations General Assembly, 2017. *A/RES/71/256: New Urban Agenda*.

Varsanyi, M.W. 2008. "Immigration Policing Through the Backdoor: City Ordinances, the 'Right to the City,' and the Exclusion of Undocumented Day Laborers." *Urban Geography*, 29 (1), 29–52.

Vogel, S. 2015. *Thinking Like a Mall: Environmental Philosophy After the End of Nature*. Cambridge, MA; London: MIT Press.

Wachsmuth, D. 2012. "Three Ecologies: Urban Metabolism and the Society-Nature Opposition." *The Sociological Quarterly*, 53 (4), 506–523.

Wachsmuth, D. and Angelo, H. 2018. "Green and Gray: New Ideologies of Nature in Urban Sustainability Policy." *Annals of the American Association of Geographers*, 0 (0), 1–19.

Wachsmuth, D., Cohen, D.A., and Angelo, H. 2016. Expand the Frontiers of Urban Sustainability. *Nature News*, 536 (7617), 391.

Wapner, P. 2010. *Living Through the End of Nature: The Future of American Environmentalism*. Cambridge: MIT Press.

White, D., Rudy, A., and Gareau, B., 2016. *Environments, Natures and Social Theory: Towards a Critical Hybridity*. Palgrave.

White, D.F. 2008. *Bookchin: A Critical Appraisal*. London: Pluto Press.

Wirth, L. 1938. "Urbanism as a Way of Life." *American Journal of Sociology*, 44 (1), 1–24.

Wolch, J.R., Byrne, J., and Newell, J.P. 2014. "Urban Green Space, Public Health, and Environmental Justice: The Challenge of Making Cities 'Just Green Enough'." *Landscape and Urban Planning*, 125, 234–244.

Worldwatch Institute, ed. 2016. *Can a City Be Sustainable?* Washington: Island Press.

Xie, L. and Bulkeley, H. 2020. "Nature-based Solutions for Urban Biodiversity Governance." *Environmental Science & Policy*, 110, 77–87.

Henri Lefebvre's "Right to the City:" Key Elements and Objections

Joshua Mousie

INTRODUCTION

In the past twenty years, environmental theorists and activists have repeatedly returned to Henri Lefebvre's concept "the right to the city" for a source of inspiration as well as an intellectual guide for conceptualizing the role of the built environment in our political life.

Specifically, the lack of power that most residents have over their built surroundings led Lefebvre (1996 [1967], 158) to describe the "right" as "a cry and a demand" for "a transformed and renewed *right to urban life*" (emphasis in original). Over fifty years old, Lefebvre's foundational work on the concept (*Le Droit a la ville*) was a timely publication. First published in 1967, a year prior to the May 68 student movement in Paris, the book had its finger on the pulse of the political moment in France. Yet his ideas—including the right to the city—only became more widely influential outside of France after his works began being translated into English in the 1990s (Aalbers and

---

Anne Le-Huu Pinneault, Teaching and Learning Librarian at Oxford College of Emory University, provided significant help with the research for this essay, for which I am grateful.

J. Mousie (✉)
Emory University, Atlanta, GA, USA
e-mail: joshua.mousie@emory.edu

Gibb 2014, 209). In the decades since, though, Lefebvre's concept has not only influenced academic writing on the politics of the built environment, but also the concept has been a galvanizing idea in global legislative bodies and conferences (Ibid.; Purcell 2014, 141–44; Bodnar 2013, 78–82).

The right to the city has ultimately become "a slogan with a life of its own" (Bodnar 2013, 73), as work on the concept has branched in several directions. Numerous studies apply the concept to local and national politics and case studies. These studies range from China's new urbanism (Qian and He 2012) to protests over public space in Kuwait (Al-Nakib 2014) to street vendors in Mexico (Muñoz 2018) and the politics of energy consumption in Germany and England (Becker et al. 2020). Arguably, most well-known are writings about Brazil's popularization of the term in their 2001 City Statute, which "formally incorporat[ed] the right to the city into national law" (Friendly 2020, 311; cf. Fernandes 2007). Additionally, there are erudite histories of the concept and its intellectual lineage (within Lefebvre's life and corpus as well as Marxist-Leninist philosophy) (Grindon 2013, Mendieta 2008, Purcell 2014) as well as those who discuss how the concept challenges traditional property rights (Dugard and Ngwenya 2019 and Friendly 2020).

Although Lefebvre's discussion of the concept is notorious for its opacity, thorough theoretical elucidation and elaboration of the concept are rare in the literature influenced by his work, and responses to critiques and possible objections are nearly nonexistent. More attention is paid to policing the term: purists defending Lefebvre's original intent, critiquing those authors and institutions that implement it in liberal democratic settings and conversations. Fuzzy, policed concepts tend to become dogmatic, unused concepts. The right to the city is a living concept that, if it is to have a future, needs a more careful assemblage of the several ideas that comprise its core meaning, which includes both how Lefebvre describes the concept and how others have developed it beyond his original intent (section one). As a living concept, those who employ it should avoid a dogmatic understanding of the concept and address outstanding, serious possible objections (section two). The term should not be abandoned because of these objections. Rather, theorists and activists should think of it as one possible tool that challenges the idea that a right to political life is separate from access to, and power over, built environments.

Section One: Four Key Concepts

A Right to Dealienation

First, consider the phrase's two main terms: "right" and "city." These two terms represent the theoretical, formal foundation of the phrase, while the final two terms discussed in this section provide further elucidation and contextualization. Regarding "right," Lefebvre is sparse on what exactly should be intended by the idea. The often-quoted view mentioned above is his claim that 1) it is a "cry and demand" for urban life. Putting aside, for the moment, what is meant by an "urban" life, equating right with an outcry, a protestation, signals that the term is active and agent-focused, for Lefebvre. That is, he did not intend, in contrast, a legal or natural view of rights. The right to the city is not egalitarian or universal, for the world's elite and wealthy already have power over, and access to, the city. Instead, here the right is something owed and demanded by those alienated and marginalized from hierarchical socioeconomic and sociopolitical life. It is a right exclusive to the excluded, a bottom-up effort to dealienate one's political life by means of the material condition of politics: the built environment. This meaning is implied in Lefebvre's work and is developed by other authors who utilize the phrase.

Lefebvre (1996 [1967], 179) implies an exclusionary right—a right of the oppressed and alienated—as he explains that "this right has a particular bearing and significance" for some and not others. Implementing a traditional Marxian distinction between proletariat and capitalist, Lefebvre identifies the working class as the city-alienated group "rejected from the center towards the peripheries, dispossessed of the city, expropriated thus from the best outcomes of its activity," and finding themselves with little control of their environs (Ibid.). Indeed, the working class find themselves in the position of an "inhabitant submitted to a daily life organized (in and by a bureaucratized society of organized consumption)" (Ibid., 178). The working class, for Lefebvre, inspired by Marx's view of labor, are subservient to others in their work, denied the freedom to choose the what, how, and when of the objects they produce, and this condition extends to their built environments insofar as it is capital and power that construct and develop the city.

Although a developed view of rights is missing from Lefebvre's account, authors elucidating and developing his concept in the past couple of decades have attempted to specify more carefully the nature of the right. David Harvey, for one, primarily echoes Lefebvre, but does describe the right to the city in terms of a human right. That is, Harvey also analyzes the right to the city through a class-based analysis: capitalists and their need to produce surplus value form the primary motor that alienates those excluded from the city. Yet, his vision goes further than Lefebvre by arguing that this right to dealienation amounts to a human right to the city. It speaks to "the freedom to make and remake our cities and ourselves," a form of collective decision-making that is "one of the most precious yet most neglected of our human rights" (Harvey 2008, 23–24; cf., 2013, 3–5). Adding "human" does not necessarily help to narrow in the concept, though, especially since Harvey does not address his understanding of human rights (a contested idea) and why it is significant to use this qualifier for the right to the city. His approach is at least more helpful than the add-all-rights-and-stir approach: "[The right to the city] is a new composite right that escapes the usual classifications; it is a mixture of claim rights and liberty rights, both negative and positive; it implies elements of political, economic, social, and cultural rights, but is none of them" (Bodnar 2013, 74). This rights-Frankenstein is as intractable as it is mythical.

As I have already suggested, the right seems best understood as an active, agent-focused right to dealienation. This approach has merit because a constant feature of the right to the city conversation is the suggestion that 1) it is needed by those who experience exclusion and marginalization and 2) it is enacted as a struggle against forms of power (capitalist markets, the privileges of inheritance) that are the sociopolitical and economic practices that distinguish those who have and those who don't have control over their built environments. Aalbers and Gibb (2014, 208), for example, home in on the active and continuous nature of this process, warning against the idea of thinking about the concept as "some unattainable utopia," and instead thinking of it as representing "the constant and continuous struggle to create urban space that is less alienated from the people who inhabit it."

Others have described the process of dealienation as a righteous struggle of the excluded, and this struggle is either thought of as an unavoidably institutional affair or one that asserts the right to basic, human deserts. The first approach is best seen in the work of Attoh (2011) as well as Mitchell (2003). Attoh analyzes the views of three authors in rights theory (Hohfeld, Waldron, and Dworkin) to pinpoint more specific notions of rights that should be used to clarify what authors mean by the phrase "the right to the city," and he suggests that these models can provide a way forward. Assuming that rights must be contextualized within an institutional context, Attoh (2011, 5) urges theorists using the phrase to "acknowledge that in a 'world characterized by scarcity and conflict' the institutionalization of any right poses trade-offs," meaning that in practice, for instance, claims for collective power over the built environment do not easily reconcile with the ability to challenge oppressive decisions (e.g., discriminatory laws) about the built environment that are formed by democratic majorities. That is, once these theories hit the ground, their effectiveness will necessarily have to negotiate conflicting demands and institutional processes (Ibid., 9).

Similarly, Mitchell (2003, 25) emphasizes that any rights talk involves "an institutional framework, no matter how incomplete, within which the goals of social struggle can not only be organized but also attained." In short, even though Lefebvre envisioned the right to the city as enacted against and outside of state institutions, as will become clear below, more recently authors have denied the plausibility of action abstracted from institutions in world that is laden (and policed) with them (cf. Qian and He 2012, 2802–04). Moving away from what might be thought of as Lefebvre's political idealism, these authors suggest that all social movements must consider how the state is often the only ground of protection for the marginalized, even when it solidifies domination, too (e.g., the institutionalization of rights; Mitchell 2003, 25–26). It seems that the state will remain a major form of power that must be dealt with and abandoning the conversation of institutionalism means allowing the state-capital relation to run free and continue to solidify its hegemonic relation.

The second approach is defended by Mark Purcell and also appears in Attoh's analysis. Purcell (2014, 149) suggests that "[i]n claiming a right to the city, inhabitants take urban space as their own, they appropriate what is properly theirs." The cry and demand that Purcell envisions is a taking back of the city, an assertion of just deserts and the unjust displacement of those lacking capital (social and monetary) by the wealthy and elite.

A similar notion of just deserts is found in Attoh's use of Dworkin's theory of rights as trumps, a model that affirms civil disobedience. He (2011, 5) argues that this model of rights can help us better contextualize the right to the city because it suggest that "[w]hen laws infringe upon our dignity or our equality," then we are within political right to challenge and openly disobey such laws. Dworkin's model, Attoh (Ibid.) argues, conceptualizes enacting rights as reclaiming our position in a body politic by direct action (i.e., asserting our trump card), even when majoritarian and democratic processes result in "unjust and discriminatory" laws. Both Purcell and Attoh's extension of the right to the city draw particular attention to the dealienation process as extending beyond the rules and legal procedure of state institutions, tapping into, while moving forward, Lefebvre's vision of political right and the built environment.

The Ontology of the City

Although it might seem more straightforward and obvious than what is intended by right, Lefebvre and others also employ a novel, sociopolitical ontology of the city. This ontology bolsters further the idea that exclusion from power over one's built environment is experienced as alienation from political life. There are two important ideas that define the city for these authors: "work" and "urban."

First, Lefebvre calls the city a "work," or *oeuvre*. Here, Lefebvre has in mind the idea that the city in its ideal and dealienated form is like a work of art: "And thus the city is an *oeuvre*, closer to a work of art than to a simple material product. If there is a production of the city and social relations in the city, it is a production and reproduction of human beings by human beings, rather than a production of objects" (Lefebvre 1996 [1967], 101). The distinction between human and objects is crucial to understanding what is meant by *oeuvre*.

The city that has alienated people is the contemporary capitalistic city of material production and exchange, which is created by, and serves, the wealthy. It is an "object" in the sense that it is reified—an abstract concept treated as a concrete entity—and used as an instrument (a means) to bring about the production of profit (an end). Hence, what gets built where, by whom and for whom, most often follows the logic of economic markets and profit. In this sense, the built spaces we are surrounded by are not the result of the free activity, ideas and cooperation of the inhabitants who live in these spaces. If so, then the city would be "close to a work of art," an *oeuvre*; that is, its production and development would not generally be driven by a single intention, but would always be the creative result of the historical and social moment; it would take shape according to the use-value determined by inhabitants instead of the exchange value of objects (e.g., housing) on a market (Ibid., 66). For Lefebvre, it is the proletariat who have been alienated from the *oeuvre* because of suburbanized capitalistic development projects (Ibid., 76–77), and by following the guidance of philosophy and art (messengers of "work" in an alienated world). These alienated groups can reconceptualize practices that might reestablish the city as an *oeuvre* (Ibid., 89;180).

Building on implied theories of the public and aesthetics in Lefebvre's concept, Mitchell (2003) and Grindon (2013) are two examples of how the city as *oeuvre* has been extended. For Mitchell, on the one hand, what drives the Marxian notion of the city as work is the idea, contrasted with "the *idiocy of rural life*" for Marx, that "cities were necessarily *public*—and therefore places of social interaction and exchange with people who were necessarily different" (2003, 18). The notion of work, that is, implies plurality and heterogeneity, mirroring the concept of a public in political theory. A public, Mitchell argues, always involves encountering difference and sustaining this difference between people and ideas in tension as struggle, rather than trying to overcome heterogeneity. It is "out of this struggle" of public life as heterogeneous that "the city as a work – as an *oeuvre*, as a collective if not singular project—emerges, and new modes of living, new modes of inhabiting, are invented" (Ibid.; cf. Fenster 2006).

On the other hand, Grindon homes in on how envisioning the city as an artwork implies that the right to the city is a right to a creative, aesthetic life with others. He sees in Lefebvre's concept a "broadening of the notion of creativity" that envisions collective city life as "a broad conception of immanent social creativity" (2013, 210–211). For Grindon (Ibid.), this creative work should be related to the Greek concept of *poiesis*, what he calls a "total human activity," which should be contrasted with the idea of work as labor. Instead an *oeuvre* is seen as a total form of living with others, the work of our lives with others instead of the work that we complete for pay and to care for life's necessities. Here, Grindon sees overlap with Lefebvre's view of the city as an *oeuvre* and his critique of everyday life. Grindon (Ibid., 211) draws this parallel, quoting an elucidating passage from Lefebvre's *Critique of Everyday Life*:

> In the future the art of living will become a genuine art. . . The art of living presupposes that the human being sees his own life—the development and intensification of his life—not as a means towards 'another' end, but as an end in itself. . . The art of living implies the end of alienation—and will contribute towards it.

Here, we see echoes of both the rejection of life activity guided by an ends-means relation and the thought that living artfully is equated with dealienation. Far from reducing this right to the city to a right to life in a metropolis, Lefebvre and others have been careful to stress that it is a right to *urban* life that is truly meant by "city." Lefebvre himself distinguishes between "urban" and the mere physical structure of a metropolis that we often refer to as a city (1996 [1967], 103). Additionally, Aalbers and Gibb (2014) explain that Lefebvre's theory is importantly classified as a right to urban life because this includes, for Lefebvre, the country as well as towns, since the aim of Lefebvre's critique is the built world shaped by capitalism, commodification and alienation. As the authors (Ibid., 209) discuss, "most of the inhabited world is urban as it is commodified and structured according to capitalist-bureaucratic logics. That is, the urban is generalized—the world is urban," which means that what Lefebvre is actually discussing is "a right to belong to a *place*" (emphasis added).

Moreover, Purcell (2014, 149) suggests thinking of a right to urban life as again a right to the artful heterogeneous life that encapsulates many of the different views of work in this section: the urban "is a space for encounter, connection, play, learning, difference, surprise, and novelty. The urban involves inhabitants engaging each other in meaningful interactions, interactions through which they overcome their separation, to come to learn about each other, and deliberate together about the meaning and future of the city." In sum, the right to the city has been considered the right of those who are dealienated to reclaim power over the built environment, not in order to profit or gain authority, but in order to see their life with others as a material and environmental endeavor. It supports the idea that many people are excluded from political life because they are positioned as a means for other people's endeavors due to their alienation from the built environment and powerlessness.

Participation, Use, Appropriation

The final two key concepts help to qualify further and contextualize what the right to the city might mean in practice, and they also build on points mentioned above. Accordingly, they do not require as lengthy an explanation as the first two concepts. First, the right to the city, in practice, is a right to participate in the use-value of urban life—that is, those alienated should, in practice, assert their position as one creator among other creators in the process of organizing and managing the *oeuvre*. Lefebvre (1996 [1967], 174) explains that "implied in the right to the city" is a "right to the *oeuvre*, to participation, and *appropriation* (clearly distinct from the right to property)," and authors influenced by his work tend to focus on "participation," "use" and "appropriation" as synonymous terms. These terms are, in sum, what it means to actively inhabit a built space in their model, and they carry both normative and participatory connotations.

Normatively, a broad right of appropriation for alienated people explicitly critiques a model of state-based, legal citizenship. That is, to appropriate as an inhabitant, for some authors, does not require official documentation or paperwork. Lefebvre (1991b; cf. Purcell 2002, 102–104) captures this tension by coining the word "citadin." This concept "fuses the notion of citizen with that of denizen/inhabitant," and the right to appropriation for the citadin is the right "to physically access, occupy, and use urban space." (Purcell 2002, 102–103). Thus, this right empowers everyone within their built spaces to consider themselves, in their actions with others, as rightful and deserving inhabitants, regardless of their national citizenship, a status that needs state-backing and recognition. Relatedly, some authors argue that the normative force of the citadin has important implications for women's empowerment. Considering gendered dimensions of access and appropriation, Levy (2013, 58) argues that the right to appropriate the city defended in the right to the city encourages "women's appropriation of the city through mobility in public space" in demanding better access to public transportation, which is a hopeful vision to "strengthen women's agency and autonomy," which she thinks has potential to have ramifications for interpersonal and familial relations. That is, since appropriating transportation in urban settings is a public act, it has the potential to bring political visibility and recognition to diverse groups of women, Levy argues, by highlighting the everyday choices women currently make in these settings and which choices they wish were accessible to them.

Building on the normative vision, the right to the city is an active, participatory political action that calls on all inhabitants, any citadin. For instance, Arcidiacono and Duggan (2020, 41) argue for a vision of participation in the right to the city as "the notion that cities should be spaces for all citizens to engage equally and collectively in practices pertaining to freedom of speech, movement, and access to a decent life in the city." Taken together, the normative vision of participation is radically inclusive and defines the inhabitant as literally any person who inhabits a space, and procedurally, this means that every person needs ready access to the decision-making and the physical production of these spaces. This localized participation is "an urban politics of the inhabitant" that is "exciting because it offers a radical alternative" to current theories of capitalistic, liberal citizenship, and yet "it is disconcerting because we cannot know what kind of a city these new urban politics will produce" (Purcell 2002, 100).

Autogestion/Self-managed

One may wonder, at this point, through what means citadins are expected to participate and claim their right to the city. Arguably, the least clear idea in the right to the city, Lefebvre suggests that the city as an *oeuvre* is only possible if an episodic uprising by concerned inhabitants morphs into a stable form of organization including all inhabitants. That is, *autogestion*—self-management—is the answer to dealienation for many thinkers in this tradition. In the brief mention of this idea in Lefebvre's 1967 text, he bemoans the episodic nature of most political participation, which does not result in inhabitant's becoming integrated into the power structures and the city's organization. Instead, "the ideology of participation" in liberal societies results in momentary inclusion only by the most concerned inhabitants, and "[a]fter a more or less elaborate pretense at information and social activity, [these inhabitants] return to their tranquil passivity and retirement" (Lefebvre 1996 [1967], 145). Contrasted with this state of affairs, "real and active participation" has only one name, according to Lefebvre, which is lasting self-management by and for inhabitants themselves (Ibid.). Elsewhere, Lefebvre (2009, 134–36) builds on these comments and claims that the anti-statist sentiment at the heart of autogestion is another word for "class struggle" and opposed to any sociopolitical passivity, it "must be continually enacted" (135). In more detail: "Each time a social group (generally the productive workers) refuses to accept passively its conditions of existence, of life, or of survival, each time such a group forces itself not only to understand, but to master its own conditions of existence, *autogestion* is occurring" (Ibid.).

Butler (2012, 144–45) explicitly connects the right to appropriation to a view of self-management that includes decision-making: the right to the city not only includes the right to appropriate the production of physical space but also "the development of capacities for self- management in two respects: first, the appropriation of space and, second, the development of forms of participation that permit the full engagement of inhabitants in decisions relevant to spatial production." Here "self-management" is best understood as a collective practice, not a call to individualism and isolationism. It resembles the idea of self-rule that is often referred to in political anarchism as well as the radical democratic tradition. A key idea at play in political self-management is that people are capable of managing power relations with others in mutualistic and cooperative ventures, and a healthy society is one that allows for self-management and finds opportunities for people to exercise it as well as develop their capacities for this activity. As Purcell (2014, 147) puts it, the right to city challenges the necessity of top-down decision-making responsibilities that impact so many dimensions of people's lives (bosses, state institutions), and especially where the state is involved, *autogestion* "insists on grassroots decision making and the decentralization of control to autonomous local units. And because it refuses to turn over responsibility to a managerial class, *autogestion* requires a great awakening on the part of regular people." In other work, Purcell also suggests that self-management of urban life by inhabitants can include participation in corporate board meetings and discussions about city policy and planning (Purcell 2002, 102).

SECTION TWO: FOUR POSSIBLE OBJECTIONS

The right to the city, as we have seen, defends the right of disenfranchised and marginalized people, seemingly any person who is confronted with a collective life in built spaces that are forced upon them, which is another way to say, using traditional language in political theory, that these spaces have "power over" them (domination) without the ready means, save protest and struggle, of the "power to" change (empowerment) how their environments are organized (cf. Hearn 2012, 6–7; Lukes 2005, 69–74). The experience is one of human alienation because the environments that have arguably the most immediate impact on the types of lives we are able to live and build with others fail to reflect the personal and collective blueprints and exercise of freedom of inhabitants. Similar to Marx's (1988 [1944], 69–84) objects of production, our built spaces are an accumulation of built objects that rarely reflect the will and mind of inhabitants and divide us into competitors (e.g., housing, transportation, amenities) more than they bring us together, collectively.

The idea is simple enough. Life being a material process and not reducible to speech acts, there is also good reason to support and see the value of this sentiment, especially in a culture that condones housing insecurity and dispossession. Similar to how Iris Marion Young (2011, chapter two) argues that liberal societies remain structurally unjust because they have come to expect (and therefore implicitly condone) violence against women and people of color, there is a fatalistic lack of surprise when another neighborhood is destroyed to be replaced with over-priced food, beverages and consumption-loving bean-to-everything goods. Nonetheless, a right to the city, when it hits the pavement, often lacks serious answers to serious problems, or one is underwhelmed with the resources it provides for moving forward in a specific direction. There are at least four concerns that theorists using the concept must be able to address for it to be relevant to actually existing environmental political life.

Theory of Alienation

First, the most basic issue that the right to the city faces in practice is legitimizing right claims. Let's remember the argument's logic: people who are alienated from the city are those who do not have power over their built environments. Primarily, they experience urban life as a commodity and object that is produced by others and which they must buy, consume and exchange in a capitalist market. Urban life is not a collective decision-making process where inhabitants—anyone residing in a place—craft their life together, forming a physical *oeuvre* that is participatory instead of an item for profit-making. On the one hand, there is good reason to follow Lefebvre and others, agreeing that there are serious forms of built environmental injustice that demand a social justice model that can respond to systemic forms of political harm that many people face. The right to the city is a provocative and inspirational approach. On the other hand, this model lacks a precise understanding of alienation that would help people make actionable steps toward socio-political justice. Although paradoxical, the concept tends to be both too narrow and too broad.

It is too narrow because it is a model that suggests only a traditional Marxian analysis of an empirical phenomenon that resists being reducible to class and economics. Certainly, control over the built environment and our ability to realize the built world as a collective *oeuvre* is partially a matter of who has capital and who does not. If nothing else, having capital means that exit is always an option: if your built environment is too damaging or isolating, you can relocate. Thus, having more capital gives you more control and options when choosing a place to live and inhabit in the first place. Nonetheless, not all people with capital can simply belong to a built environment that is their collective *oeuvre* with other people of a similar means. Not all people with capital are people who control urban life. The right to the city suggests an *experience* of alienation (facing a commodified urban world, not an *oeuvre*) that seems to fit the daily lives of the majority of people today, rich and poor. In other words, a major limitation of the theory is that it currently tends to suggest an economic cause of alienation yet it describes the experience and solution in terms of collective creation and decision-making. Again, it is correct that the rich and elite certainly have more freedom and opportunity, but the crux of the model is the ability to participate in a collective *oeuvre*, which is not the same as the right to exit or relocate.

Here, the right to the city must broaden its theoretical methodology beyond class and profit in order to home in on the more precise and specific features of alienation from the built environment in our contemporary world. For example, those using the model must also wrestle with the idea that the history of racism and settler colonialism is also a prominent feature of the distribution, and control over, built spaces, forms of power that include, but are not reducible to exchange.

The methodological narrowness of the right to the city therefore keeps it from a more precise model of alienation, especially since the experience of alienation it defines does not always map onto class distinctions, empirically. It therefore results in a notion of alienation that seems far too broad, one that encapsulates far too many people and doesn't track with its causal explanation. In part, this is because the determinates that shape political life include social, customary powers beyond class. As classic arguments in feminism, race and indigenous political theory argue (cf. Young 2011; Frye 1983; Bruyneel 2007), alienation from the world includes living with the threat of violence and types of racial and cultural imperialism that are somewhat indifferent to capital and class status. A more specific methodological lens for identifying alienation—one that questions a person's position within social, structural forms of power, for instance—would actually help to develop a narrower picture of those who are legitimately alienated from their environments. It would link better a causal and experiential model of alienation. Otherwise, in practice, the right to city includes contradictory cries and demands that make it nearly impossible to know how to move forward de-alienating a world of so many diverse, legitimate rights claims. A clearer and more developed theory of alienation will help distinguish legitimate from delegitimate rights to the city.

Normative Assumptions

Normative arguments about the right to the city are often question-begging; it is not clear why the city as *oeuvre* should be our normative vision for our built environment. The assertion that it should be our goal needs a defense, and more important, the assertion that alienated people should seek it seems to assume an already unified, normative vision shared by people alienated by their built environments. It is reasonable to assume that people who experience alienation do want political justice, but it is seriously questionable if these people all share (or should share) a vision of dealienation that looks anything like creating the city as an *oeuvre*. In short, the right to the city needs a more robust explanation of the diverse experiences and norms of those alienated by their built environments, it needs to take seriously the idea that not everyone seeks self-management and it needs further defense of the normative assumption that people wish to see themselves reflected in the organization of the city. All of these relate to a basic problem: those spoken of as alienated from the city are treated as homogenous (the "workers" or "proletariat"), when in reality they comprise a heterogeneous body with conflicting goals and norms for their political lives.

As discussed, Lefebvre was committed to a heterogeneous theory of the city. That is, the *oeuvre*, as an artwork of all inhabitants, is expected to be inclusive and diverse. As Mitchell (2003, 130) puts it, the theory of the public sphere assumed in the right to the city is not a rationalistic and orderly conception, a la a Kantian, enlightenment view. Instead, the view of public space being forwarded is that even "*how* and *when* people are to meet, under what conditions they are to do so, and what they are able to discuss are all themselves points of struggle." Yet, authors seem to assume that everyone alienated agrees, from the outset, that this construction of the city as an *oeuvre* is the preferred method that will generate a just city. There is good reason to object to such an assumption. Most important, there need to be thoughtful suggestions offered for how conflicts will be solved, assuming that those alienated from the city will not simply all be in agreement about how to move forward and create an alternative city Attoh (2011, 7) is worth quoting at length on this point:

> While we may embrace the right to the city's conceptual flexibility and its capacity to link disparate rights-claims under a unitary framework, we must still debate and grapple with the form that such rights may take. For example, our ability to successfully link a right to affordable transportation with a right to a clean and sustainable city will depend as much on political will as on the form that such rights take. If the right to affordable transportation is merely a call for expanding the private and individual ownership of cars, it is a right that may stand directly against a right to a clean and sustainable city. The right to the city framework offers us little help in navigating the way forward.

If those alienated from the contemporary city are not a homogenous political body, and if we think the city as *oeuvre* is clearly a superior model to the multiple alternatives, then those defending the concept need to provide a realistic model for political conflict resolution. Current accounts tend to assume that in the absence of states and state institutions, people will somehow, almost by magic, learn how to move from foundational and substantial conflicts to an actionable solution and non-violent collective life.

Furthermore, since we should think of those alienated from the city as a heterogeneous body, it is also reasonable to assume that not everyone will want autogestion. The work of self-management is onerous and burdensome, even if it is equally rewarding and fulfilling. Here again, those defending the right to the city tend to take too many liberties in assuming what people want and think of as beneficial, without sufficient explanation and defense of the merits of such political forms of organization. For instance, one can imagine that many people might prefer to continue to work to create more fair and more just institutional practices, building on the existing political conditions. They would be happy, we could imagine, if current institutions were improved rather than moving the burden of managing and organizing built spaces to them. After all, there are no political silver bullets and *autogestion* will also have limitations and serious complications (i.e., conflict as described by Attoh above). Therefore, it seems like if the two—more just and fair institutions that don't make decisions according to capital and autogestion—could theoretically produce similar ends (a dealienated built environment), and one is far easier to imagine growing out of already existing political conditions that dominate the globe (states and institutions), then it seems more an issue of political idealism and principle than practical political action guiding us if we insist on the latter.

These objections do not even broach the assumption that self-management will necessarily bring actionable results, an issue because it is described typically in negative terms, explaining what it doesn't involve (states) and doesn't look like (episodic protest and political organization). Nor do we have the space to deal with the larger assumption that people want to see themselves reflected in their built spaces and organizing, an aesthetic vision of life that needs defending. Yet, all of this makes clear that a convincing defense of the norms behind the right to the city does not currently have a strong theoretically footing or account that will persuade those not already sympathetic with its political vision.

Ontological Assumptions

A more abstract but nonetheless theoretically important objection is the persuasiveness of the right to the city's ontology of the city. For example, is it compelling to think of the ontology of our actually existing cities as reducible to exchange and profit? Surely, these norms (and the practices they encourage) often dominate how cities operate, but it seems both empirically inaccurate and politically limiting to think of the city in these simple terms.

Those defending the right to the city have to specify better which "city" they are actually talking about, theoretically. Urban theorist Doreen Massey (2015, 141), for instance, makes a compelling argument for cities having a "thrown-togetherness" that appears to defy reduction to singular forces or rules. Cities have no such coherence that would enable us to understand them through a single lens of sociopolitical organization: "There can be no assumption of pre-given coherence," but instead cities and places have a "uniqueness" that forces us to actually face and engage with the lack of coherence in the built world. For Massey, this is exactly the work of politics because in engaging built spaces and people without overly rigid assumptions about ontological causality, we take responsibility directly for the world and others we engage with therein, which is to deny us the simplicity of rules and ideological lens. The "thrownto-getherness" of cities and politics mirror each other because they include "not being able to reach for [eternal rules and principles]; a world which demands ethics and the responsibility of facing up to the event; where the situation is unprecedented and the future is open" (Ibid.).

If the right to the city is only meant as an equal right and opportunity to economic development, then it should specify and reclassify the right accordingly. That is, it might not be so much a right to *the* city that most authors are getting at (including Lefebvre) as it is a right not to be dominated by capitalistic markets that determine our shared interactions with the built world. It is true that most authors defending the view have something more visionary and provocative in mind. That is, they are not simply making arguments about the right to a certain form of physical production and economic benefit, but instead they envision and defend a collective, participatory physical construction of the world that tries to equal a radically democratic account of the discursive construction of our body politics. Even so, cities themselves do not seem to be the same thing as collective organization and decision-making, yet authors employing the right to the city tend to equivocate on this issue. Cities have often been heralded, similar to Massey, as intractable and home to the unexpected (cf. Young 2011, chapter eight; Jacob 1992, part two). With a far more pointed and specific "city" as its target—rather than assuming the city itself as its object—the right to the city would be more effective, politically.

The State and the City

Following my final point in the previous section, what is politically effective is often less important to authors employing the right to the city than is an unadulterated use and defense of the concept—that is, limiting the concept to Lefebvre's understanding, specifically regarding his view that the right to the city is completely separate from anything involving the state. Bodnar (2013, 79), for example, considers any approach that departs from Lefebvre's vision as truncated and in need of correction. And when the concept gained international recognition by being central to Brazil's City Statute, it was equally seen as a misstep by those faithful to Lefebvre.

Purcell (2014), for example, admits that the City Statute clearly helped to advocate habitation rights for people living in slums. Yet, this is followed by a Lefebvre-inspired critique of the City Statue: it "does not seek to move beyond property rights; it merely seeks to balance the interests of property owners with the social needs of urban inhabitants. The law asserts that property rights are no longer *more* important than other concerns. They are still firmly in place, but now they must coexist with social use value" (Ibid., 142–43). This result only seems to be a failure if every political action is to be judged according to pure, ideal standards. In this case, the movement fails ultimately because property is still intact and the state therefore involved. It is worth asking what the use of such a metric is in a decidedly non-ideal world, especially when the action in question is moving an alienated social group toward dealienation? More to the point, the right to the city, if it is to have a future, needs to explain why it is reasonable to think that a successful utilization of the concept involves an action that is completely abstracted from the state. This objection is not meant to suggest that we should not be anti-statists. The harms and injustices of states and their institutions, especially regarding their methods of political recognition, are widely discussed and documented, especially in indigenous political theory (cf. Jung 2008; Eisenberg et al. 2014). Yet, careful analysis and effective political mobilization require that an anti-statist politics deal directly with the state and challenge it because its ubiquity and serious, entrenched forms of power are not things that we can afford to ignore, as if these things will disappear and go away if we try to formulate action that is somehow outside the reach of the state (a dubious undertaking). Attoh, as quoted above, understands this point, as does Mitchell. The latter, in his book-length study of public space and the right to the city, argues that all social movements must address, face-on, that often the only ground for protection for marginalized people is the state, even when we also admit that the state also solidifies forms of domination. The state is a form of power that must be dealt with, and rights claims always assume some degree of an institutionalized setting; they can be a tool that checks and challenges the state-capital relations rather than letting it run free (2003, Chapter 1). Yet, such views are rare in the right to the city literature, and as mentioned above, views like Mitchell's are usually critiqued as an adulteration of the concept.

Nonetheless, a more careful analysis of state politics, especially in relation to anti-statist movements, gives us reason to support Mitchell's view and demand a more theoretically compelling case from those in the right to the city literature whose anti-statism, in the absence of such a case, tends toward political idealism theory that does not justify its principles in light of (or largely ignores the serious constraints of) actually existing political conditions that people live with and experience day-to-day. For instance, in her analysis of the Zapatista movement, Courtney Jung makes a compelling case for how the state both creates and determines most markers of political identity, yet this hegemony is exactly what is needed for forms of counter-hegemony to ground themselves in a political identity or movement that is effective. In fact, she argues that the hegemony of the state is the political antithesis that indigenous politics needs to concretize a salient identity within the heterogeneous groups that belong to these struggles:

> Indigenous people are partly constituted as a potential group because they occupy a common location of structural exclusion from the modern state, not because they possess a common language or culture—which they do not. It is this structural location that makes indigenous a group-in-itself. But such a structural location did not produce an indigenous rights movement, a group-for-itself, until the concept of indigenous rights developed sufficient traction to orient, and to open the political space for, indigenous politics. Markers develop political resonance when activists, like the Zapatistas or the Chartists, exploit an ideological opening to produce, mobilize, and circulate a discourse that orients a group of people to a recognizable set of political claims and alliances. (2008, 69)

In other words, Jung draws our attention to the fact that the markers and tools that the state has brought into existence, although hegemonic, are exactly what help create a space and political position for salient, effective and production resistance to the state, specifically because these markers are so ubiquitous and dominant. Without them, resistance efforts have their structural location in power relations, but have historically found it difficult to leverage this alone into politics that has purchase in the wider world (Ibid., chapters 1–4). In short, political markers like "[r]ights provide no guarantee, but by anchoring political identities they offer a framework for participation and voice" (Ibid., 73).

Echoing and widening Jung's view, Avigail Eisenberg argues that there is little reason to think that anti-statist and self-determination movements could plausibly escape the politics of the state and recognition. For example, groups of people that are successful in their effort to self-management and determine can find themselves with little to no resources since they have sought to create a space outside of state relations. "Political movements for self-determination may free people from oppressive relations with dominant states and empires only then to drive them deeper into poverty and increased dysfunction" (Eisenberg 2014, 298).

More important, however, is Eisenberg's understanding of the interconnectedness of politics in our contemporary world. First off, socio-political movements and states do not exist in a bubble, but rather they exist in an international and global context where they "coexist and are mutually reinforcing," and in many cases, especially in indigenous cases of self-determination, those anti-state struggles "rely on the recognition of others in order to build their capacity for self-determination" (Ibid., 299). That is, it is usually of "pragmatic necessity" that anti-statist groups still involve themselves with states, even if this means states external to the states within which they are located (Ibid., 300). It seems reasonable to assume that the *autogestion* favored by the right to the city shares important similarities with a general politics of self-determination, mainly because both seek self-managed sovereignty that denies the legitimacy of the state. Yet, since the types of political bodies that are in question are heterogeneous and have discursive and ontological identities that are contested internally and externally—whether indigenous or the alienated in the right to the city—it is crucial to follow this logic to its conclusion: "what is means to be self-determining is often controversial and this forces people to engage in a politics of recognition within their own communities" as well as within an international context with other states (Ibid.).

The right to the city needs a more compelling theory of the state. It is compelling to build a movement and theory that seeks political life not alienated, dominated and oppressed by state institutions and practices that rarely tolerate challenges to private property and capital. In our lived experiences and in the real, physical context of our political actions, however, we face multiple states and multiple forms of state power that we have to negotiate. Failure to have substantive answers for how we ought to address these real-life conflicts with domestic and international states results in a politics of the built environment that lacks persuasiveness (at best) or is strictly an academic conversation (at worst).

Conclusion: A Right to Build Political Life

Lefebvre's concept of the right to city was born out of political struggle and provided the foundation for many theorists and activists to reconceptualize how our built environments shape our collective lives. Although the core concepts and related terms that ground the theory (right, city, participation, autogestion) face several serious objections that remain unanswered, there remains good reason for further development of the terms, beyond the fact that so many people find the concept and theory useful.

The right to the city, therefore, puts a needed pressure on political theorists and activists to consider how we must think of the right to politics as a right to physically build collective life together. Primarily, we need frameworks that challenge us to think of politics as a literal process of building the material world around us. That is, if we support a right to politics and political participation for all people, a basic democratic sentiment found in a range of theories of governing, then this support needs to extend itself to models of power that includes access to, and control over, built spaces. People's sociopolitical lives cannot be abstracted from the material world wherein we live and work and reduced to ballot boxes. Nor can we support equal opportunity to economic resources without also considering the physical infrastructures (transportation, housing, food and water) of cities that help to determine whether people can make good on such opportunities. How the built environment is constructed—and who constructs it—is crucial to how politics gets realized and actually exists. The right to the city helps us to alter our political imagination and see the importance of designating and identifying rights owed to those alienated from their built environments, those removed from power over their environments to the extent that they rarely have the opportunity to exit. The right to the city calls our attention to such pernicious forms of domination, and with a more thorough development of its normative framework, its political ontology, and how to demand this right in a world dominated by state institutions, it might start to answer more effectively the cry and demand for built environmental justice.

REFERENCES

Aalbers, Manuel B. and Kenneth Gibb. 2014. "Housing and the Right to the City: Introduction to the Special Issue." *International Journal of Housing Policy* 14: 207–213.

Al-Nakib, Farah. 2014. "Public Space and Public Protest in Kuwait, 1928–2012." *City* 18: 723–734.

Archidiacono, Davide and Mike Duggan. 2020. *Sharing Mobilities: Questioning Our Right to the City in the Collaborative Economy*. London: Routledge.

Attoh, Kafui A. 2011. "What *kind* of Right is a Right to the City?" *Progress in Human Geography* 35: 1–17.

Becker, Sören, James Angel, and Matthias Naumann. 2020. "Energy Democracy as the Right to the City: Urban Energy Struggles in Berlin and London." *Environment and Planning A: Economy and Space* 52: 1093–1111.

Bodnar, Judit. 2013. *The Long 1968*. Bloomington: Indiana University Press.

Bruyneel, Kevin. 2007. *The Third Space of Sovereignty: The Postcolonial Politics of U.S.–Indigenous Relations*. Minneapolis: University of Minnesota Press.

Butler, Chris. 2012. *Henri Lefebvre: Spatial Politics, everyday life and the right to the city*. New York: Routledge.

Dikec, Mustafa and Gilbert Liette. 2002. "Right to the City: Homage or a New Societal Ethics?" *Capitalism Nature Socialism* 13: 58–74.

Dugard, Jackie and. Makale Ngwenya. 2019. "Property in a Time of Transition: An Examination of Perceptions, Navigations and Constructions of Property Relations Among Unlawful Occupiers in Johannesburg's Inner City." *Urban Studies* 56: 11165–1181.

Eisenberg, Avigail, Jeremy Webber, Glen Coulthard, and Andrée Boisselle, eds. 2014. *Recognition Versus Self-determination: Dilemmas of Emancipatory Politics*. Vancouver: UBC Press.

Eisenberg, Avigail. 2014. "Self-determination versus Recognition: Lessons and Conclusions." In *Recognition versus Self-Determination: Dilemmas of Emancipatory Politics*, Avigail Eisenberg, Jeremy Webber, Glen Coulthard, and Andrée Boisselle, eds. Vancouver: UBC Press: 293–306.

Fenster, Tovi. 2006. "The Right to the Gendered City: Different Formations of Belonging in Everyday Life." *Journal of Gender Studies* 14: 217–231.

Fernandes, Edéiso. 2007. "Constructing the 'Right to the City' in Brazil." *Social & Legal Studies* 16: 201–219.

Ford, Derek R. "Toward a Theory of the Educational Encounter: Gert Biesta's Educational Theory and the Right to the City." *Critical Studies in Education* 54: 299–310.

Friendly, Abigail. 2020. "The Place of Social Citizenship and Property Rights in Brazil's 'Right to the City' Debate." *Social Policy & Society* 19: 307–318.
Frye, Marilyn. 1983. *Politics of Reality: Essays in Feminist Theory*. Berkeley: Crossing Press.
Frye, Marilyn. 2008. "The Right to the City." *New Left Review* 53: 23–40.
Grindon, Gavin. 2013. "Revolutionary Romanticism." *Third Text* 27: 208–220.
Harvey, David. 2013. *Rebel Cities: From the Right to the City to the Urban Revolution*. London: Verso.
Hearn, Jonathan. 2012. *Theorizing Power*. New York: Palgrave Macmillan.
Jacobs, Jane. 1992. *The Life and Death of Great America Cities*. New York: Vintage Books.
Jung, Courtney. 2008. *The Moral Force of Indigenous Politics: Critical Liberalism and the Zapatistas*. Cambridge: Cambridge University Press.
Lefebvre, Henri and Kristin Ross. 1997. "Lefebvre on the Situationists: An Interview." *October* 79: 69–83.
Lefebvre, Henri. 1991a. *The Production of Space*. Translated by Donald Nicholson-Smith. Oxford: Blackwell Publishing.
Lefebvre, Henri. 1991b. "Les Illusions de la modernité." *Manières de voir* 13. Le Monde Diplomatique: 14–17.
Lefebvre, Henri. 1996. *Writing on Cities*. Translated by Eleonore Kofman and Elizabeth Lebas. Oxford: Blackwell Publishing.
Lefebvre, Henri. 2009. *State, Space, World*. Translated by Gerald Moore, Neil Brenner, and Stuart Elden. Minneapolis: University of Minnesota Press.
Levy, Caren. 2013. "Travel Choice refrAmed: 'Deep Distribution' and Gender in Urban Transport." *Environment & Urbanization* 25: 47–63.
Lopes de Souza, Marcelo. 2010. "Which Right to the City?: In Defense of Political Strategic Clarity." *Interface* 2: 315–333.
Lukes, Steven. 2005. *Power: A Radical View*. New York: Palgrave Macmillan.
Marcuse, Peter. 2009. "From Critical Urban Theory to the Right to the City." *City* 13: 185–197.
Marx, Karl. 1988 [1944]. *Economic and Philosophic Manuscripts of 1844*. Translated by Martin Milligan. New York: Prometheus Books.
Massey, Doreen. 2015. *For Space*. Los Angeles: Sage.
Masuda, Jeffery R. and Sonia Bookman. 2016. "Neighborhood Branding and the Right to the City." *Progress in Human Geography* 42: 165–182.
Mendieta, Eduardo. 2008. "The Production of Urban Space in the Age of Transnational Mega- Urbes." *City* 12: 148–153.
Mitchell, Don. 2003. *The Right to the City: Social Justice and the Fight for Public Space*. New York: The Guilford Press.
Muñoz, Lorena. 2018. "Tianguis as a Possibility of Autogestion: Street Vendors Claim Right to the City in Cancún, Mexico." *Space and Culture* 21: 306–21.
Purcell, Mark. 2002. "Excavating Lefebvre: The right to the city and its urban politics of the inhabitant." *GeoJournal* 58: 99–108.
Purcell, Mark. 2014. "Possible Worlds: Henri Lefebvre and the Right to the City." *Journal of Urban Affairs* 36: 141–154.
Qian, Junzi and Shenjing He. 2012. "Rethinking Social Power and the Right to the City Amidst China's Emerging Urbanism." *Environment and Planning A* 44: 2801–2816.

Young, Iris Marion. 2011. *Justice and the Politics of Difference*. Princeton: Princeton University Press.

Going Beyond the Great Divide Between Nature and Culture: The Concept of "Relocalized Society" to Account for the Local Agri-Food Networks

Clémence Nasr

> We grant that the land and the community must be reintegrated physically, that the community must exist in an agricultural matrix which renders man's dependence upon nature explicit […]. We can shift the center of economic power from national to local scale […]. This shift would be a revolutionary change of vast proportions, for it would create powerful economic foundations for the sovereignty and autonomy of the local community.[1]
>
> – Murray Bookchin, *Post-Scarcity Anarchism*

Introduction: From the Ecological Crisis to the Abandonment of the Concept of Society?

The ecological crisis we are going through poses a major challenge to political theory: that of taking a truly reflexive and critical look at the framings and concepts having a resolutely modern origin. Many of these concepts enter into the elaboration of what can be considered a central reflection of political theory: what makes a political society one and how to think and implement

---

[1] Murray Bookchin, *"Post-Scarcity Anarchism"* (Montreal-Buffalo: Black Rose Books, 1986): 134–137.

C. Nasr (✉)
The Paris Institute of Political Studies (Sciences Po), Paris, France
e-mail: clemence.nasr@sciencespo.fr

its "ordering".[2] The reflection on society is indeed properly modern insofar as this period of time, beginning at the end of the Middle Ages in Western Europe, is characterized by the unprecedented possibility of reflecting on the "self-institution"[3] and organization of the social body—a possibility that long-standing trends such as the "reconfiguration of temporal power" and the "questioning of traditional hierarchies" reflect.[4]

However, the fact is that this reflection is anchored in what has been labeled in particular by the anthropologist Philippe Descola and the sociologist of science Bruno Latour as a "Great Divide", that is a nature-culture dualism.[5] Western modernity was indeed built on the idea that "progress" for mankind consisted in "winning a decisive victory over nature".[6] In "the order of knowledge", this idea was translated into a "discourse of separation" between nature and culture.[7] The mistake that moderns made, according to Latour and Descola, was to end up believing that this separation in discourse corresponded to a "dissociation of things in themselves".[8] Actually, this dualism must imperatively be seen for what it really is: a mode of monitoring, in Western modernity, the "mediations between humans and non-humans",[9] in other words, a mode of taking charge of the "solidarity between human communities and the environment that affects them in their very constitution".[10]

---

[2] Wendy Brown, "At the Edge", *Political Theory*, 30, no. 4 (August. 2002): 571.

[3] Francesco Callegaro, "Le sens de la nation. Marcel Mauss et le projet inachevé des modernes", *Revue du Mauss*, no. 43 (January. 2014): 354. My translation. In speaking of the reconfiguration of temporal power, Callegaro refers to the particular trajectory of the instituting power, through which modern societies have emerged. Initially alienated in the 'state-empire' and manifesting itself as an 'extrinsic power' that enforces order through 'the pure discipline of bodies', power gradually becomes 'intrinsic': it allows society to constitute its unity, by 'appropriating the state' and placing it at the service of a normative organisat ion over which society has control.

[4] *Ibid.*, 340–342. My translation.

[5] See in particular Bruno Latour, *We Have Never Been Modern* (Cambridge, MA: Harvard University Press, 1993[1991]) and Philippe Descola, *Beyond Nature and Culture* (Chicago: The University of Chicago Press, 2013[2005]).

[6] Pierre Charbonnier, *Affluence and Freedom. An Environmental History of Political Ideas* (Cambridge: Polity Press, 2021[2020]): 13.

[7] P. Charbonnier, "La nature est-elle un fait social comme les autres? Les rapports collectifs à la nature à la lumière de l" anthropologie", *Cahiers philosophiques*, no. 132 (January. 2013): 85. My translation.

[8] *Ibid.* My translation.

[9] *Ibid.*, 76. My translation.

[10] *Ibid.*, 85. My translation. Charbonnier reminds us, however, that the nature-culture dualism and the dichotomy between non-human and human do not necessarily overlap – ethnographic anthropology has shown that, in Amazonian myths for instance, this overlap is absent. But even if we focus on the Western space, the non-overlap remains. Of course, if nature and culture are understood as 'concepts' designating a 'range of ideally locatable entities in the world', we will easily associate non-human entities with nature – made of animals, trees or climatic phenomena. But when we consider nature and culture as

If this dualism is back in the limelight today, it is because we realize that it is directly linked to the ecological crisis. As the philosopher Pierre Charbonnier points out, it is indexed to the ideal of freedom from nature and a "will to affluence;"[11] it thus encompasses a set of human activities that no longer guarantees the "reproduction of the living world".[12] Ironically, the dualism compromises "the conservation of a global climate regime close to the one that made this ideal possible".[13] In short, the articulation with nature that the "Great Divide" encompasses is now coming up against its limits and "seems more fallible than it ever was".[14] The difficulty lies for political theory in the fact that it has a specifically modern identity: it reinforces dualism by thinking of society as "ideally self-enclosed:"[15] the social is a "self-enclosed domain because the relations that unfold within it are [thought] as relations internal to that given domain of reality".[16] Such a conception of society denies the fact that natural reality, while "not truly human or social", is part of "the social dynamism itself", as mentioned above.[17]

How, then, should political theory relate to dualism? How should it question the nature-culture dualism? A part of political ecology takes the side of making the "Great Divide" disappear by dissolving culture in nature. This ecology points to an "environmental ethic movement" that can be either "biocentric" or "eco-centric".[18] The problem underlined by Charbonnier of this "moral conception of the environmental challenge", of which the dissolution of dualism is one of the principles, is that it constructs a priori a "system of radically new prescriptions" in contradiction with the "historical and social constitution of our intuitive moral reference points".[19] Actually, it is likely

---

'processes'—which they actually are—these two notions automatically become 'ontologically neutral, since any kind of being is likely to participate in properly social as well as natural mechanisms'. It is thus important to remember that the dualism between nature and culture covers 'socio-environmental relations that are constituted as a descriptive and normative issue from the outset'.

[11] P. Charbonnier, *Affluence and Freedom*, 23.

[12] *Ibid.*, p. 262.

[13] My translation of a sentence missing in the English edition of Charbonnier's book – p. 42 in the French edition.

[14] P. Charbonnier, "La nature est-elle un fait social comme les autres?", 91. My translation.

[15] *Ibid.*, 91. My translation.

[16] *Ibid.*, 80. My translation.

[17] *Ibid.*, 77. My translation.

[18] Catherine Larrère, "What the Mountain Knows. The Roots of Environmental Philosophies", *Books & Ideas* (April. 2014). Larrère explains that when it is biocentric, the environmental ethic holds that 'each living entity, taken in isolation', is a carrier of intrinsic value; when it is biocentric, the intrinsic value is rather for the community, *i.e.*, the 'whole' which gives value to its 'members'.

[19] P. Charbonnier, "La nature est-elle un fait social comme les autres?", 93. My translation.

that the questioning of the dualism should lead not to a moral absorption of culture into nature, but to a relationship that must be rethought in its entirety. In other words, questioning the nature-culture dualism is never in itself an answer, but the beginning of a problem.[20] And it is not impossible that in this work where political theory is forced to make a reflexive return on its centers of exclusion of the nonhuman, concepts as obvious as that of society are weakened.

But before we can rule on this point, let us specify at once the methodological content of this questioning of the dualism: the reflexive return that political theory must engage implies to circumscribe the social realities and thus practices, which, at the level of the economic, legal, scientific plans, etc., testify to a "form of rupture with the instituted forms of the Great Divide, of putting to the test the dominant schemes of the relationship to the nature".[21]

In what follows, I will argue that precisely the concept of society, encapsulating the idea that human beings are made to associate, seems to me to survive the reworking of the relationship between nature and culture, even though this concept is based on the idea of an exclusively human social sphere—and we saw how problematic this is. But it can only be maintained on the condition that it is significantly amended, *i.e.*, integrated into a normative perspective that is itself subordinated to the imperative of compatibility between human and nonhuman. There are practices today that, anchored in the economic and agricultural reality, renew our collective relationship to nature and consequently participate in the revision of dualism—precisely by making possible another conception of society. I will thus show that the development of local agri-food cooperation initiatives—initiatives that bring together food producers and food consumers who are geographically close—provides a space for the emergence of a normative perspective, subordinated to the horizon of a compatibility between humans and nonhumans, *i.e.*, to the ecological transition. Most importantly, within this normative perspective, a new representation of society can begin to take shape: a *relocalized society*.

Local agri-food networks are directly linked, as we shall see, to the strengthening of ecological awareness by targeting the need to reduce so-called "food miles" between producers and consumers, but also the need to "belong to a land"[22] one lives on. But this is only the first bridge, so to speak, between local agri-food networks and the ecological transition. There is a second bridge which is deeper and it refers to a more normative and theoretical order: I want

---

[20] With Lauriane Guillout (ULB-EHESS, France), we came to this conclusion during the preparation of a scientific conference "Political Ecology: Beyond the Great Divide" that took place in Brussels on January 2020, 30th and 31st.

[21] *Ibid.*, 93. My translation.

[22] B. Latour, *"Down to Earth. Politics in the New Climatic Regime"* (Cambridge: Polity Press, 2018[2017]): 53.

to argue that a new sense of what society is can be worked out from the analysis of local agri-food cooperation.[23]

This new sense crystallizes on one of the fundamental features of any human society: the spatial extension of the social body. I will show that the concept of a relocalized society makes it possible to make the spatial extension of a society a normative issue and that, linked to the reality of food relocalization, this concept suggests the possibility of shaping the contours of the political body according to a bounded geographical area delimited by the subsistence capacities it supports. I would like to make it clear from the outset that, however, I do not equate, as we shall see, territory with the idea of autarkic or self-sufficient communities.

The two bridges we will use are linked. In other words, if we are to propose a new idea of society based on food relocalization, this new idea is twofold. It concerns, of course, the ecological dimension of society, notably characterized, as we shall see, by an agriculture compatible with the reproduction of the living environment, social justice for farmers and social cohesion. But this idea also refers to the fact that society will have changed in its material form, in its morphology. Society will occupy a local territory—that is, at least smaller than that over which most nation-states now extend.

My argument will proceed as follows. To begin with, I will explore the content related to the first bridge I suggested, explaining the extent to which local and cooperative agri-food networks are linked to the ecological transition as a direct and tangible response to the urgency of the ecological crisis we are in. It will become clear, however, that this is just a stepping stone to the second bridge. In a second part, the analysis of local cooperation on food and agriculture will lead us into questions of social belonging and society—issues that are inseparable from those concerning the territory, its division, in short, the spatial bounds of the political body.

Finally, a few words on the geographical anchorage of my analysis. I focus here essentially on food relocalization as it is taking place in France, even if I also use social science work that focuses on local agri-food networks in the United States or in other Western societies because the similarities are

---

[23] In saying this, I maintain that this thesis on the relocalized society must be understood as a normative hypothesis, which in no way prejudges what work in social science will bring, and which can just as easily confirm this thesis as invalidate it. My thesis could be confirmed if empirical work came to light which would show for instance that individuals involved in local agri-food cooperative initiatives do indeed share a representation of society that shows signs of an aspiration to relocalize. According to Charbonnier, 'it is probably not philosophy's task to affirm by speculative means what will be the name and the exact form of this collective capable of establishing itself as the subject of the ecological counter-movement' (p. 257). But he also holds that 'the transformation of our political ideas must be of magnitude at least equal to that of the geo-ecological transformation that climate change constitutes' (p. 246). If it is therefore also a question of imagining major changes in our ways of thinking, I propose, in political theory, an attempt at this—I believe, moreover, that it is not purely speculative as it is based on the development of a social trend.

numerous. This focus is justified in more than one way: the political functioning of the country is based on strong centralization, which makes the aspiration to relocalize, whatever its purpose, particularly singular.[24] Secondly, France has a relatively strong cultural relationship with agriculture and rurality as this relationship is said to be trapped in an"impossible mourning of France's agrarian past".[25] However, I do not think that this focus on France weakens the proposal I am making around the concept of relocalized society. There is, to mention just one, a lot of thinking about the fact that local-food dynamics in the United States are driven by "segments of the population" which tend to "segue from the national to occupy new localist identity positions".[26]

THE CONTRIBUTION OF FOOD RELOCALIZATION TO THE ECOLOGICAL TRANSITION

Over the last few decades, a singular trend has emerged in the agri-food landscape of Western societies: local agri-food networks. But the story began elsewhere: the first *Teikei* appeared in Japan in the 1960s. Families, suspecting that foodstuffs were polluted, decided to form an association and buy directly from producers working in their locality. Shortly afterward, similar initiatives were set up in Austria, Germany and Switzerland.[27] In the United States, the first Community Supported Agriculture initiatives (CSA) date back to the mid-1980s: these are organizations linking producers and consumers in the same locality, in which the latter make a contractual commitment to support the former in the long term, while producers are obliged to do their utmost to provide them with fresh, quality food. In France, the first Association for the Maintenance of Peasant Agriculture (AMAP) was set up in 2001.[28] Again, this is a "local partnership" whereby consumers commit to a farmer over a more or less long period of time, receiving baskets paid for in advance; the former wish to link up with the producer through "common concerns"—good harvesting,

---

[24] On this topic, see Xavier Greffe, *La décentralisation* (Paris: La Découverte, 2005).

[25] Bertrand Hervieu, François Purseigle, *"Sociologie des mondes agricoles"* (Paris: Armand Colin, 2013): 118. My translation.

[26] Donal M. Nonini, "The Local-Food Movement and the Anthropology of Global Systems", *American Ethnologist*, 40, no. 2 (May. 2013): 274.

[27] Fabrice Ripoll, "Le concept AMAP: promotion et mise en pratique(s) d'une nouvelle norme d'échange entre consommateurs et producteurs agricoles", in V. Banos (ed.), *Espaces et normes sociales* (Paris: L'Harmattan, 2009): 99.

[28] AMAP corresponds to the French "Association pour le Maintien de l'Agriculture Paysanne". On the transnational dimension of AMAPs, see Fabrice Ripoll, "Le concept AMAP: promotion et mise en pratique(s) d'une nouvelle norme d'échange entre consommateurs et producteurs agricoles". For some figures on CSAs in the US, there were 50 in 1990. By 2010, the number had risen to more than 2500; see Patricia Allen, "Realizing Justice in Local Food Systems", *Cambridge Journal of Regions, Economy and Society*, 2, no. 3 (July. 2010): 297.

weather risks, soil-friendly practices and fair payment.[29] In France, these short food-supply chains are not new insofar as, contrary to the United Kingdom in particular, open-air markets and farm sales had never disappeared.[30] However, over the last twenty years or so, we have seen a diversification of the forms that these chains can take. The AMAPs have been joined by collective sales outlets (PVC[31]) managed collectively by producers, basket systems sold via an internet platform without commitment, etc. Collective catering institutions—school canteens and hospitals—when supplied by producers are included in a local agri-food network.[32] Local agri-food networks are a special type of short food-supply chain, based on geographical proximity between producers and consumers and not only on the number of intermediaries—this criterion being the one which allows defining short food-supply chains in general.

Promoting Ecological, Cooperative and Local Practices

Food relocalization appears clearly linked to a set of practices that point toward engagement in a form of ecological transition. But it is necessary to clarify here what is meant by "engagement". I refer deliberately to practices here—agricultural techniques, sales, distribution and consumption—which can be identified with a motto that has come back into fashion in the 1960s in France thanks to an "associationist revival" and is popular again today: that of the "here and now"—acting here and now to effect change from the ground up, on a small scale and in daily life.[33] To explain why I think it is important to focus on local agri-food network practices, I first need to look at how involvement in them is accounted for. Sociologists—whether they belong to rural sociology, economic sociology or more broadly to food studies who study short food-supply chains have pointed out that many actors who invest in local agri-food networks, whether farmers, craft manufacturers or consumers, stress their deliberate distance from political life, refusing to see their engagement as

---

[29] Claire Lamine, "Settling Shared Uncertainties: Local Partnerships Between Producers and Consumers", *Sociologia Ruralis*, 45, no. 4 (Nov. 2005): 339.

[30] Yuna Chiffoleau, *"Les circuits cours alimentaires. Entre marché et innovation sociale"* (Érès: Toulouse, 2019): 22. My translation.

[31] PVC for the French "point de vente collectif".

[32] There is no global data on short food supply chains – let alone on their economic weight in the world. For some economic indicators on the situation in France, we can look at the data collected by the last agricultural census, in 2010. At that time, one in five farms in France sold at least part of their production in short circuits. That same year, it was measured that, excluding winegrowing, for 40% of farms using short circuits, this method of sale represented more than 75% of their turnover (see Yuna Chiffoleau, *Les circuits courts alimentaires*, 56). Today, short food supply chains represent between 15 and 20% of the total food purchases of the French people; see Yuna Chiffoleau, "Les circuits courts: une réponse face aux crises?", *Sciences Humaines*, no. 338 (July. 2021): 42.

[33] Jean-Louis Laville, *Politique de l'association* (Paris: Seuil, 2010): 101. My translation.

activism. Consumers, marked by food scandals,[34] are seeking, for example, to alleviate their "health concerns" by accessing traceable foodstuffs. They may also simply be looking for products grown with fewer pesticides and more taste. And they may continue to buy some of their food in large- and medium-size supermarkets.[35] Producers, on the other hand, may be motivated by the "economic attractiveness" of their inclusion in a local agri-food network and focus on the "economic determinants" of that inclusion.[36] Of course, local-food networks do attract some activist individuals and groups associated with the left and political ecology above all. Some consumers, combining alter-globalization[37] with the ecological question, join local agri-food networks in the name of "challenging the agri-business model", which is accused of being unfair, of depleting the soil and of generating significant pollution.[38] Among farmers, there has been some reappropriation of the legacy of the *Confédération Paysanne*[39]—an agricultural institution that quickly became close to the alter-globalization movement and was committed to "ecological issues"—and some revival of thinking around the notion of the autonomous farmer.[40] Still on the ideological level of militant commitment, we should not forget that there is also a form of conservative activism closely bound up with the identity question. Conservatism, when it relates with food relocalization, conveys

---

[34] The 'mad cow' crisis (bovine spongiform encephalopathy), a disease first detected in the United Kingdom in 1986, for example, left a lasting impression in Europe.

[35] 5 Christine Aubry, Yuna Chiffoleau, "Le développement des circuits courts et l'agriculture périurbaine: histoire, évolution en cours et questions actuelles", *Innovations agronomiques*, no. 5 (2009): 61–64.

[36] Valérie Olivier, Dominique Coquart, "Les AMAP, des réseaux agroalimentaires alternatifs émergents en soutien à la petite agriculture locale? Quelles opportunités pour l'amélioration des revenus des agriculteurs?", 2èmes journées de recherches en sciences sociales – INRA, SFER, CIRAD (December 12th 2008): 14. My translation.

[37] Alter-globalization should be understood as a social movement bringing together actors and groups who criticise the consequences of economic globalisation. At the same time, they defend the need for cooperation and networking between societies on a global scale, particularly in order to face the challenge of the ecological crisis.

[38] Y. Chiffoleau, *Les circuits courts alimentaires*, 26. My translation.

[39] The Peasant Confederation or Confederation of Small Farmers. The rapprochement and association of the confederation with the alter-globalization movement was crystallised in the figure of the activist José Bové. In August 1999, although he had not been a member of the confederation's national bodies since 1990, he took part in the action to dismantle McDonald's in Millau, in the south of France. As a spokesperson for the confederation, he took a stand against 'junk food', North–South economic exchanges and the functioning of the World Trade Organisation. In any case, this event built up a certain image of the Confederation as a trade union that is attentive to international issues and developing countries, far from the perspective of the 'traditional closure of the peasant world to itself' (on this topic, see I. Bruneau, "La Confédération paysanne et le mouvement altermondialisation. L'international comme enjeu syndical", Politix, 17, no. 68 (October—November 2004): 124.

[40] Jean-Philippe Martin, *Histoire de la nouvelle gauche paysanne. Des contestations des années 1960 à la Confédération paysanne* (Paris: La Découverte, 2005): 194. My translation..

an idea of local or land-based identity which its proponents conflate with cultural identity and cast as timeless, existing through the ages and supposedly now threatened by globalization. The valorization of the *terroir*[41] as a "heritage inherited from the past"[42] rubs shoulders with "assumptions about the homogeneity and common interests of local places" in short, with a "defensive localization".[43] I shall return later to the fact that the local rooting of food networks triggers a conservatism based around identity and agrarian folklore and thus rightly generates mistrust of the local dimension: it turns out that this mistrust can sometimes hinder full understanding of the meaning of localism.

Despite the presence of these activisms, it must be remembered that conscious ideological commitment concerns only a minority of the actors involved in local agri-food networks. Finally, most of the actors are rather focused on "the day-to-day operations of the business or technical aspects of their work".[44] This is why some sociologists are justified in pointing out the "normative idealization" of local agri-food networks and calling for vigilance.[45] By this, they mean that food relocalization should not be seen as a strongly political tendency since many individuals join local-food networks for reasons that are not primarily ideological and which cannot be directly linked to any of the varied strands of political ecology. In my view, this vigilance must itself be guarded against being excessive, as it may well prevent us from accounting for 'the social and political scope'[46] of local cooperation around agriculture and food. As Chiffoleau points out, local agri-food networks are involved in the "transition of the agricultural model, but also of the dominant social model"—in short, the ecological transition. But the sociologist makes a very important point: it is all cooperative initiatives, not just those led by activists, that contribute to this transition. Even within nonactivist cooperative initiatives, one finds individuals engaging in "practices that resist the

---

[41] In the English-speaking world, the French word 'terroir' is exclusively linked to wine production. It refers to the natural characteristics of an area—soil, climate, environment—which would explain those of a particular wine. The meaning is broader in France and refers to the existence of areas that bring together lands that share similar agricultural aptitudes.

[42] Elise Demeulenaere, Christophe Bonneuil, "Cultiver la biodiversité. Semences et identité paysanne", in B. Hervieu et al. (eds), *Les mondes agricoles en politique. De la fin des paysans au retour de la question agricole* (Paris: Presses de Sciences po, 2010): 89. My translation.

[43] Clare C. Hinrichs, "The Practice and Politics of Food System Localization", *Journal of Rural Studies*, 19, no. 1 (January 2003): 37.

[44] Patricia Allen, "Alternative Food Initiatives in California: Local Effort Address Systemic Issues", *Centre for Agroecology & Sustainable Food Systems*, Research Brief #3 (Fall 2003): 8.

[45] E. Mélanie DuPuis, David Goodman, "Should We Go 'Home' to Eat?: Towards a Reflexive Politics of Localism", *Journal of Rural Studies*, 21, no. 3 (July 2005): 363.

[46] Sophie Dubuisson-Quellier, "Circuits courts. Partager les responsabilités entre agriculteurs et consommateurs", *Demeter, Economie et Stratégie Agricoles*, 16 (2009): 111. My translation.

pervasiveness of the dominant system, although they [...] are not connoted in explicitly political terms".[47] Thus, even though individuals do not conceive of their involvement in a cooperative and local agri-food initiative as bringing their daily lives into line with the rejection of certain "abstract political and moral verities"[48] and the adherence to others, this involvement "undermines [their] concrete loyalties"[49] to the contemporary agri-food system.

In order to understand how local agri-food networks form part of ecological transition—by attracting individuals who do not necessarily identify as ecologists by persuasion—we must thus first of all observe practices. Let us start with the most obvious: reducing so-called "food miles". One of the main aims of local-food networks is to reduce the distance between producers and consumers, *i.e.*, to link these actors when they live in the same locality. But the reduction of food miles is actually more convincing on paper than in reality: it is also on this count that the strategies deployed within local agri-food networks can backfire.

Studies have shown that where products are transported by rail (and to a lesser extent by sea) and the quantity of products transported is carefully managed, transporting a large quantity of products—even over large distances —may be less polluting than all the forms of transport used by local networks combined: for producers, car journeys for small quantities and vehicles returning empty, the other way round for consumers. The important variables are therefore "loading of the vehicle and return journeys without produce".[50] Moreover, "the energy consumption of transport of a product" could rather be determined by "the organization and volumes managed by the supply chain[51] Precisely on this point, local agri-food networks favor the return of diversification, with producers taking on the task of transforming the raw materials they produce. But, as stressed by Barbara Redlingshöfer, "the lower the volume transformed, the more energy required per unit of production since fixed energy consumption is divided across a smaller volume of produce".[52] Therefore, should we agree with the analysis delivered in 2013, in France, by the General commission for sustainable development?[53] Its analysis was as follows: "the benefits of local networks for sale of agro-elementary

---

[47] Edigio Dansero, Giacomo Pettenati, Alessia Toldo, "Si proche et pourtant si loin. Etudier et construire la proximité alimentaire à Turin", in P. Mundler, J. Rouchier (eds), *Alimentation et proximités. Jeux d'acteurs et territoires* (Dijon: Educagri, 2016): 310. My translation.

[48] Murray Bookchin, "On Spontaneity and Organization", *Anarchos*, 1972: §2.

[49] *Ibid*.

[50] Barbara Redlingshöfer, "L'impact des circuits courts sur l'environnement", in G. Maréchal (ed.), *Les circuits courts alimentaires. Bien manger dans les territoires* (Dijon: Educagri, 2008): 179. My translation.

[51] *Ibid.*, 178. My translation.

[52] *Ibid.*, 179. My translation.

[53] In French, *Commissariat général du développement durable*.

products are more socio-economic than environmental".[54] In order to rule on this issue, let us follow the path indicated—that of socio-economic benefits. We will see that these will ultimately lead us back to the ecological transition.

So, what are these socio-economic advantages? One might be tempted to mention, first of all, increased social cohesion, as an effect of the development of these initiatives in which producer–consumer geographical proximity facilitate face-to-face interaction between consumers themselves as well as between consumers and producers. However, apart from the fact that not all local agri-food networks operate on face-to-face interaction, several sociologists have insisted on the fact that tightening of social ties was not necessarily sought after by the actors and was even sources of pressure: some consumers "do not want to meet the producers any more than that"[55] and, above all, producers sometimes resent a "solidarity" that they feel is exclusively "top-down, from consumers to producers".[56] Thus, social cohesion is not the socio-economic benefit I would like to emphasize here. I would refer more to the fact that local agri- food networks allow some farmers to withstand exclusion from the food system—in short, the obligation to abandon their activity because they can no longer make ends meet—or remain solvent. Cooperative and local agri-food initiatives can indeed represent, for farmers, a solution for "economic and social survival"[57] in a very deteriorated context since the 1980s, during which the model based on productivism and global markets began to show its limits insofar as even those producers who had decided to embrace industrial and economic modernization saw their situation deteriorate at that time. Even if they do not generate increased profits, these networks at least allow producers to consolidate "a regular and stable income".[58]

But there is something else: a farmer who withstands the pressures of socio-economic exclusion thanks to food relocalization also often preserves of an area of land which is thereby saved from urban or suburban absorption and maintained as agricultural land. And it just so happens that, in France, the number of farms using direct sales for at least three quarters of their vegetable production is decreasing less than the number of all farms.[59]

---

[54] Quoted by Y. Chiffoleau, *Les circuits courts alimentaires*, 67. My translation.

[55] [55] Sophie Dubuisson-Quellier, Ronan Le Velly, "Les circuits courts, entre alternative et hybridation", in G. Maréchal, *Les circuits courts alimentaires. Bien manger dans les territoires* (Dijon: Educagri, 2008): 110. My translation.

[56] Patrick Mundler, "Les associations pour le maintien de l'agriculture paysanne (AMAP) en Rhône-Alpes. Entre marché et solidarité", *Ruralia*, no. 20 (2007): 16. My translation.

[57] Aurélie Dumain, Béatrice Maurines, "Composer les manières de gouverner", in A.-H. Prigent-Simonon (ed.), *Au plus près de l'assiette* (Versailles: éditions Quae, 2012): 217. My translation.

[58] Y. Chiffoleau, *Les circuits courts alimentaires*, 58. My translation.

[59] According to a survey realized in 2007 by the *Centre Technique Interprofessionnel des Fruits et Légumes* (CTIFL – Interprofessional Technical Centre for Fruit and Vegetables) reproduced by Y. Chiffoleau, "Les circuits courts de commercialisation en agriculture:

Here we find again an ecological perspective—which distinguishes between cultivated land and land covered by human buildings—rather than a purely economic or social one. However, it is important to realize that maintaining agricultural land requires joint action—that of all the components of a local agri-food network—and not only that of the actors who run a local agri-food initiative. Thus, we must take into account the work carried out by an association such as *Terre de*[60] via its satellite branch, *La*,[61] which acts in concert with certain local authorities that seek to contain the expansion of their urban area, *i.e.*, continuous building, whether for housing or industry. It is all actors, not just those involved in the operation of a local agri-food initiative, that clearly contribute to keeping farmers on the land and enabling organic and small-scale producers to settle on arable land. This is an important point to which I will return later.

This result—agricultural land maintenance—goes hand in-hand with another in the framework of food relocalization: "greening of farming practices"[62] and promoting biodiversity conservation. The French Agency for Ecological Transition (*ADEME*) has specified that the key factor in terms of environmental impact of food chains is production: "production methods and, in particular, the cultivation of seasonal produce for fruit and vegetables are much more decisive in terms of the environmental balance than the distribution method'".[63] We must therefore look at the production techniques associated with local agri-food networks and this certainly points toward ecological transition since associated agricultural practices are moving toward decreased use of artificial inputs and fertilizers whose production releases greenhouse gases, even, in some cases, toward seasonal growing and implementation of organic or agro-ecological techniques. For some types of produce (tomatoes, for instance), producers are also increasingly turning to heritage varieties—a noteworthy measure, since 75% of genetic biodiversity is estimated to have been lost in one century, according to the Food and Agriculture Organization (FAO).[64] Until now, producers in Europe have had to get around the law, since the European Commission ruled to allow the exchange of nonrecorded seed varieties produced by farmers only in 2020.[65]

---

diversité et enjeux pour le développement durable", in G. Maréchal (ed.), *Les circuits courts alimentaires. Bien manger dans les territoires* (Dijon: Educagri, 2008): 29.

[60] Land of Ties association.

[61] The Real estate branch of the Land of Ties association, which studies and seeks to have an impact on land use.

[62] Y. Chiffoleau, *Les circuits courts alimentaires*, 70. My translation.

[63] Les Avis de l'ADEME, "Les circuits courts alimentaires de proximité", April 2012. My translation. (https://www.ademe.fr/sites/default/files/assets/documents/avis_ademe_circuits_courts_alimentaires_proximite_avril2012.pdf).

[64] Y. Chiffoleau, *Les circuits courts alimentaires*, 78. My translation.

[65] *Ibid*. My translation. In France, there is a *Catalogue des espèces et variétés de plantes cultivées* ("Catalogue of Cultivated Plant Species and Varieties"). Only seeds listed in this catalogue can be sold. It was created in 1932, with the aim of clarifying the supply of

This brings me to the important point I wanted to make. These two elements—keeping producers on the land and moving toward ecological techniques—are in fact possible only because collectives, networks are established at the local level to enable and assist individuals pursuing these goals. I have already mentioned associations and groups that help to preserve farms and maintain agricultural lands. As Chiffoleau explains, in terms of greening practices and therefore on the side of the farmers, it is contact both between peers and with consumers that facilitates evolution (however fast or slow) toward production techniques that respect the environment. Indeed, consumers, particularly in initiatives where there is face-to-face interaction, do not hesitate to express to producers their expectations regarding production using fewer inputs and pesticides; similarly, contacts between producers, at open-air markets for example, allow those who might be hesitant to embark on a transition to take advice from those who have already started. Nothing would be possible without these small-scale local structures which are often essential in the challenging early years of a very young farm that hopes to sell its produce in local networks. Local solidarities are therefore crucial to maintaining small-scale agri-food enterprises on the land and to moving toward ecological techniques. In fact, if food relocalization can promote environmental advances, this is precisely because it also carries socio-economic advantages which we must primarily understand in terms of the local cooperative networks in which individuals take part. These two classes of benefits are far from opposed: actually, ecological benefits assume the social and economic benefits of cooperation without which the former would not be possible.

Cooperation in food relocalization means several things. The first is that this cooperation comprises the local agri-food initiatives themselves, because they are meeting places between peers and with consumers, but also the groups that gravitate around them—associations, local authorities, etc. The second is that, noteworthily this cooperation is rooted at the local level—which has everything to do with the geographical proximity between producers and consumers who commit to it. This is a connection somewhat overlooked in the study of local agri-food networks by Chiffoleau: while she stresses cooperation networks and their local dimension, she is apt to minimize the element of reduced geographical distance between producers and consumers. Yet this second element makes the first possible. There is an indication of this inseparability between local cooperation and geographical proximity between producers and consumers in another facet of Chiffoleau's analysis. Referring to the fact that interest in agriculture intended to supply local networks also eventually has an effect on urban space within towns and cities through

---

varieties and protecting users by guaranteeing them traced and quality seeds. This must be the case in other countries in Europe, as the European Commission has also banned the marketing of unregistered seed. The development of short food supply chains throughout Europe has undoubtedly contributed to this change in regulations.

urban,[66] she stresses a difference between these two strands—local agri-food networks involving farmers and urban agriculture. In reality, the connection with farmers as constructed within local agri-food networks assumes the existence of an agricultural space outside the city. And this space holds meaning for the connection between producers and consumers. Chiffoleau suggests that in a way, urban agriculture in fact contributes to "distancing local networks from the idea of reconnecting with nature".[67] I would argue that in addition to their role in production and food supply, local agri-food networks imply the imperative of building a new relationship with nearby agricultural space, that of our immediate surroundings. In other words, these networks resurrect the horizon—which tends, in France, to be erased by peri-urban expansion—of a space encompassing the cities and the surrounding agricultural land.

Returning to the Soil, Leaving Society?

In manifesting the importance of local cooperation networks and of connecting with truly agricultural nonurban spaces, food relocalization seems a clear instance of these "initiatives for returning to the soil[68] which Bruno Latour, in a book entitled *Down to Earth*, argues have been on the rise in recent times. They arise from a deep-rooted feeling of anxiety that has risen with intensified globalization: while the phenomenon has pushed hundreds of thousands of people to leave their countries of origin, it has also generated an acute "problem of belonging"[69] for the populations of Western industrialized societies, experienced not only in moral terms but also practically and spatially. Increased market flows and distances travelled by goods and the global interdependences forged through these processes generate the fear that "there is no longer that spatio-temporal frame to impart order to events and concerns each of us carry within ourselves".[70] Latour explains that the emergence of this feeling goes hand-in-hand with the collapse of an illusion—of which one facet has clearly to do with the land and the way we have inhabited it or made a territory of it. This Western and modern illusion is the following: that progress and development are a unidirectional pull inciting us to abandon our

---

[66] Urban agriculture includes the practices of growing plants and raising animals within cities. It serves several purposes: contributing to household food security, providing fresh food, creating jobs, recycling urban waste, making use of empty and derelict spaces, etc. However, it also involves health and environmental risks, one of which is the potential use of contaminated land and water.

[67] *Ibid.*, 79. My translation.

[68] B. Latour, *Down to Earth. Politics in the New Climatic Regime* (Cambridge: Polity, 2018[2017]]): 91.

[69] B. Latour, "La mondialisation fait-elle un monde habitable?", Territoires 2040. Prospectives périurbaines et autres fabriques de territoire, *Revue d'étude et de prospective*, no. 2, (2nd semester, 2010): 9. My translation.

[70] *Ibid.* My translation.

local roots and look ever further abroad, to stretch economic and social influences with no limits. In Latour's words, "it is towards the Globe", the world in its connected totality, equated "with wealth, freedom, knowledge and access to a life of ease" that the "project of modernization" was supposed to lead humanity.[71] Yet the ecological crisis shows us that the planet cannot bear this: there is no real globe for this globalization: in short, "the soil of globalization's dreams is beginning to slip away".[72]

This is why we are witnessing a return of borders and nationalism in today's world: we are trying to rediscover reassuring boundaries. Whether it is Brexit in Europe, Jair Bolsonaro in Brazil or Donald Trump's term in office in the United States, it is, for Latour, again an illusion. These nationalist leaders, by orchestrating real waves of disinformation, are merging two elements that until now have been in opposition: "the advance towards globalization and the retreat toward the old national terrain".[73] This shows that actually, these individuals have decided to go for broke. To understand this point and more specifically the need for a return of borders and reassuring boundaries, let's take a closer look at the reasoning of Latour.

The Local and the Global[74] are part of the same history: one cannot exist without the other to the extent that the Global, this "headlong flight forward imposed by modernization",[75] was to be opposed to something that was to be understood as its opposite, the Local, *i.e.*, "our native province, […] our traditions, […] our habits"[76] that modernization urged to abandon. According to Latour, both are fictitious. He argues that neither the Local nor the Global has any "lasting material existence",[77] which means that neither is adapted to the crisis we are going through and to the need for refoundation that this implies. These territories are not "livable".[78] At the same time, it is clearly the insoluble ideological opposition of these two fictitious territories that is the root of the anxiety and that best reveals the collapse of the illusion that the expansive development imagined by Western modernity is in fact possible.

The national dimension plays an ambiguous role in this Local–Global dichotomy. Today, the nation-state is "nothing more than another name for the local"[79]—but it is no longer the local that characterized the early days of modernization. This Local also promises "tradition, protection, identity and

---

[71] B. Latour, *Down to Earth*, 26.

[72] *Ibid.*, 4.

[73] *Ibid.*, 35.

[74] Latour uses capital letters on purposes because the Local and the Global should be understood as ideological benchmarks.

[75] *Ibid.*, p. 53.

[76] *Ibid.*, p. 27.

[77] *Ibid.*, 39.

[78] *Ibid.*, 30.

[79] *Ibid.*, 101.

certainty" but it encloses them "within national or ethnic borders".[80] But again, there is an interdependence between this Local-national and the Global. If we revisit the history of the progressive internationalization of the world, of the intensification of economic relations between different parts of the world, Karl Polanyi's analysis of the role of states in orchestrating this phenomenon is very useful. In *The Great Transformation*, he describes the near-simultaneous constitution of internal (*i.e.*, national) markets and the process of internationalization of the world. He explains that, from the fifteenth and sixteenth century, the "setting up of sovereign power" in "external politics" implied the emergence of a "territorial state", that is an actor capable of monitoring the "nationalization of the market"[81] and of inserting the latter in the "self-regulating world market".[82] Still today, states accompany economic globalization and are not its opposite: they do "enter into agreements to create international institutions, some of which have compromised their Westphalian sovereignty by establishing external authority structures".[83] In fact, this is why state administrations tend to weigh increasingly on civil societies—a way of compensating for the loss of one aspect of their sovereignty. So, as we can see, the idea that globalization makes states disappear is simplistic. On the contrary, the current reinforcement of the idea of a state that must affirm its power is inseparable from another—that of a nation which affirms the intangibility of its spatial contours on the international stage.

And to return to Latour's reasoning, this makes the nation another form of fictitious territory. In other words, believing that we can live in the world by constituting a multiplicity of nations is just as much of an illusion as believing that the Local and the Global are opposed to one another. Rather, we must return to reality and this is the meaning of the philosopher's call to "return to the Earth".[84]

This call implies a prerequisite: understanding that there is a link between, on one hand, the ideas about the relationship with the land as it has been constituted by the moderns and, on the other hand, the ideas about politics and the organization of power. If we accept this, the ideas that fall into these two categories must, according to Latour, be deconstructed simultaneously and in parallel. If we consider one category, we must also consider the other. And, for Latour, if the political organization of social life will also have to be profoundly transformed for an ecological society to be possible, then our entire relationship with the spatial surface will also change. Let us see why.

---

[80] *Ibid.*, 30.

[81] Karl Polanyi, *The Great Transformation. The Political and Economic Origins of our Time* (Boston: Beacon Press, 2001[1944]): 68–69.

[82] *Ibid.*, 188.

[83] Stephen D. Krasner, *Sovereignty. Organized Hypocrisy* (Princeton: Princeton University Press, 1999): 13–14.

[84] B. Latour, *Down to Earth*, 53.

The ideas of the local and the global,[85] of national interior and international exterior, are merely subspecies of a larger genus: the division of territory into continuous geographical units which fit into one another, from the smallest to the largest scale. Latour speaks of these "geographical maps [indicating] a series of interlocking units"; he insists on this "sort of zoom effect, which was fundamentally rather reassuring" because "people moved through areas of predictable dimensions that fitted into one other, from the neighborhood to the village or the world through a series of stages or rungs that were spatially and administratively bounded".[86]

Although these limits have always felt reassuring, Latour believes they have nothing to do either with the reality of how we are living in our territories now or how we should do so going forward. In terms of how we inhabit territories, geographical circumscription is an illusion since it fails to take into account mobility, migration, as well as the "networks of influence—political and commercial" which have always blurred this "purely geometric definition of belonging".[87] The philosopher thus believes such geographical-administrative divisions exist only for the purposes of "organization and projection of powers"[88] and the necessity of "landing somewhere"[89] forces us to distance ourselves from both.

Now let us turn to the way in which we should inhabit territories. Latour's use of this normative register he defends the necessary "project of landing someplace"[90] – is merely a call to lay ourselves open, so to speak, to the uncertainty of what he calls "the Terrestrial".[91] The Terrestrial, of which humanity is part, serves to name this uncertainty which is consubstantial to the relationship between a living being and the rest of what surrounds it. This uncertainty has been denied more and more as we have affirmed our identity as moderns. Latour thus does not give any precise description of the reality of a good relationship with the land. He does, however, stress the idea that now that we have entered an era in which the Terrestrial becomes visible, meaning that "the earth [has] stopped absorbing blows and [is] striking back with increasing violence",[92] we find ourselves forced to stop denying that we depend on the Terrestrial and everything that makes it up, that we are part of its very fabric and not separate to it as has been assumed in Western modernity. But if we were to achieve this interdependence with the Terrestrial, it would in no case lead to a fixed and permanent cartography. Cartography, as we saw, speaks

---

[85] Here, not as ideological benchmarks but rather as scales.
[86] B. Latour, "La mondialisation fait-elle un monde habitable?", 9. My translation.
[87] *Ibid.*, 10. My translation.
[88] *Ibid.* My translation.
[89] B. Latour, *Down to Earth*, 7.
[90] *Ibid.*, 13.
[91] *Ibid.*, 40.
[92] *Ibid.*, 20.

of the fundamental modern error, that of seeing space as a simple accessory to human action. Indeed, the modern conception of space into geographical units also reflects the fact that the trajectory of moderns has been built on the denial of the demands that arise from its occupation by other living species. To become the "Terrestrials"[93] we are and always have been—while having made sure to forget it—is thus, of course, to question the premise of dualism between the human and the nonhuman, progressively consolidated and reinforced by the hegemony of "positive sciences".[94]

In Latour's work, this fundamental change of perspective implied by embracing the identity of Terrestrials is crystallized in the notion of "territory", understood as a "dwelling place".[95] A territory is defined as a tangled web of dependence, no longer as a geographical unit: the territory implodes the geographical order insofar as for every individual it corresponds to the webs of dependence in which an individual's existence is caught. My territory is made up of elements that help to satisfy my needs, both physical and psychological. I am on the territory of another being as soon as the perpetuation or preservation of that being means that my actions or lack of them affect it. For example, Latour explains that he is on the territory of bluefin tuna in Japan as he consumes them: he thus points out that bluefin tuna "will or will not have a future depending on whether Japanese and Western consumers will stop eating it or not".[96]

This conception of territory is also inseparable of alternative conception of "emancipation", linked to a "process of plowing, a way to dig in".[97] Last, this process is founded on the recognition and maintenance of our "subsistence" since the territory of a living being includes all the resources it needs and depends on to live.[98] As we see it, territories—the emergence of which will be anything but peaceful as there will be evolving, moving "Critical Zones"[99] in which we will find ourselves caught because our subsistence needs will enter into conflict with those of other humans or other living beings—no longer have anything to do with the continuous surface, the "spatial [...] and geographical entity".[100]

I mentioned above that deconstructing ideas about our relationship to the land was inseparable from deconstructing the way we conceive of human coexistence and social life, i.e., the political. It was recalled above that spatial delimitation was an intrinsic part of the organization of power and thus of

---

[93] *Ibid.*, 83.
[94] *Ibid.*, 69.
[95] *Ibid.*, 95.
[96] B. Latour, "La mondialisation fait-elle un monde habitable?", 15. My translation.
[97] B. Latour, *Down to Earth*, 81.
[98] *Ibid.*, 95.
[99] *Ibid.*, 99.
[100] *Ibid.*, 95.

political organization. The promotion of the territory as a place of subsistence and the dismissing of the division of space into nested geographical units are extended into a questioning of the political structuring inherited from modernity. For Latour, we can no longer conceive of the political through the nation-state and the ideas associated with it: sovereignty, borders and so on. As has been said, the nation-state, for the philosopher, is one of the fictitious territories which, today, constitute the Local: nationalism only stands for a "rump territory" which is "unreal".[101] But, in his view, we must go further because the idea and existence of nation-states are merely the symptom of a deeper problem: the Western and modern representation of the political reveals that it designates an exclusively social bond, therefore a bond that excludes the nonhuman.

In a earlier work entitled *Reassembling the Social*, Latour explains that our understanding of the social has become inseparable from that of society, which Latour considers to be a "panorama",[102] *i.e.*, a horizon, constructed by[103] in the nineteenth century and masking two things. Firstly, society "has usurped", in a way, the "body politic".[104] In other words, society, by letting us think that it is always "there whether we like it or not",[105] prevents us from seeing that the "political task" of "progressively composing the common world",[106] *i.e.*, constructing the right conditions for coexistence is never complete. Second, society "keeps hiding" the "collective" that could result from the political task still to be pursued.[107] Here, Latour refers to the "premature closure of the social sphere",[108] that is to our blinding ourselves to anything that is not human. If one subscribes to all this, it is the very constitution of a body as society which is in question. More precisely it is the very idea of society itself—since it implies denying that the ultimate political task, that of constituting collectives, cannot be an exclusively human province.

Latour's call to return to the soil and to "land someplace"[109] goes hand-in-hand with a process of dismantling some of our most deeply ingrained ideas—specifically those that govern understanding of belonging. Those ideas tell us

---

[101] *Ibid.*, 30.

[102] B. Latour, *Reassembling the Social. An Introduction to Actor-Network Theory* (Oxford: Oxford University Press, 2005): 192.

[103] The 'sociology of the social', as he calls it: 173. The consubstantial limitations of the social were glorified by the founders of sociology when they argued that in order to understand the political fact we must ask not about abstract moral principles but go deep into the fabric of society, since the primordial political facts were societies and their cohesion.

[104] *Ibid.*, 164.

[105] *Ibid.*, 164.

[106] *Ibid.*, 189.

[107] *Ibid.*, 164.

[108] *Ibid.*, 260.

[109] B. Latour, *Down to Earth*, 99.

that on one hand, land can be clearly demarcated between geographical units and that, on the other hand, we belong to these units in the same way that as we belong to social entities, to societies. "Today, we have to restudy what we are made of"[110] —for the philosopher, attention to our subsistence and the dependencies that go with it must lead to the implosion of these closures and spatial limits precisely insofar as it must generate the awareness that an infinite number of nonhuman elements intervene in the process of constituting the social that modernity has defined as exclusively human.

The question I ask, is whether food relocalization, as a returning-to-the-soil initiative, constitutes a validation of the Latourian theoretical projections—the end of territory as a geographical unit, the end of society. In other words, does the philosopher's thinking allow for an account of local agri-food networks or does it fail to capture them? Food relocalization is about subsistence because it is about food. It is also a question of "the attentive care that [the soil] requires"[111] since one of the perspectives defended is that of agriculture and, if possible, reasoned or organic agriculture. But within food relocalization, there is also a strong focus on local space—even if its definition, whether by researchers or actors, is complex, as we shall see. Yet Latour states with certainty that the soil he refers to "has nothing to do with the Local".[112]

Renewing the Idea of Society: Founding Belonging on Local-Food Autonomy

In what follows, I want to show that food relocalization, in fact, invites us to reorient Latour's call. Relocalization has rather different implications in my view and amidst the welter of new ideas that are emerging from it, that of society seems to have to endure. The idea of society persists—while being modified—precisely through the idea of the local, understood as a re-established proximity between the producers cultivating a space and the consumers living within this space. The second part is thus devoted to the idea that food relocalization opens up an understanding of territory far less abstract than Latour's notion of territory as our ground in which we 'dig in'.

This conception of territory through food relocalization may not map onto geographical and administrative units, but it does not, in fact, break with the idea of spatial continuity which is an element of the geographical and modern definition of territory. Relatedly, I believe that food relocalization offers a vehicle for a new model of society, in other words, a vision of belonging, but also imparts new meaning to the term: a shift in our idea of society, currently linked to territory as it corresponds to nation-states, toward a social entity founded on local networks of economic cooperation, especially in the agricultural sector. Within our ideas about what society means, then, what changes

---

[110] B. Latour, *Reassembling the Social*, 248.

[111] B. Latour, *Down to Earth*, 92.

[112] *Ibid.*

in this new vision is the spatial scope of the social body—albeit for reasons of integration of the social, as we shall see below. I am talking about a reorientation of Latour's call, because there is indeed a positive presence of the local in his thinking: even though he dissociates the Local from the idea of the soil we "dig in," he holds that they should maybe be a "fraternization" between "supporters of the Local and supporters of the Terrestrial".[113]

I think that this line of thought needs to be explored further, which means something else. I tend to think that the Local "that reassures, that calms,"[114] that allows to "recover [...] an assured identity"[115] is not unreal, contrary to what Latour says. I even think that local cooperation around agriculture and food can bring about a norm of the local that frees it from its old demons: "ethnic homogeneity, a focus on patrimony, [...] nostalgia, inauthentic authenticity".[116]

Society as a Landmark

I need to start by explaining why I think the concept of society should be retained, before explaining how in the face of food relocalization it can be reshaped. Let us recall Latour's explanations regarding the replacement of the concept of society' with that of the "collective". The latter designates the linking of the multiplicity of various associations that emerge and evolve. Indeed, by asking the question of the collective rather than that of the body politic, *i.e.*, society—it is the sociologists who have sought to impose the latter—humans spontaneously embark on the path of collective existence, without prejudging what belongs to it, without limiting "in advance the sort of beings populating the social world".[117] But I tend to think that there is a kind of dissonance between the notion of the collective and that of territory as developed in *Down to Earth*. This dissonance crystallizes in the following questions: to whom is the question of subsistence addressed and who should answer it? Latour tends to use the impersonal pronoun "one", but we could be forgiven for thinking the response might be purely individual—in which case territories and collectives would basically refer to individual needs, or possibly, their aggregation. So, at what point could a territory—corresponding to a collective—come into being? Is there a territory that can bring the collective dimension to life, or give it material existence in some way? My feeling would be that if territory is to correspond to Latour's conception of it—a space breaking with the perspectives of geographical unity and perhaps even spatial continuity—then the possibility of a collective, of a "we", disappears as well. What I want to defend in the following lines is the relevance of the

---

[113] *Ibid.*, 53.
[114] *Ibid.*, 92.
[115] *Ibid.*, 42.
[116] *Ibid.*, 53.
[117] B. Latour, *Reassembling the Social*, 16.

concept of society even if it was built on the premature closure of the social sphere—if we are to continue to be able to say "we", as Latour himself does.

Latour's understanding of who "we" are is linked to his take on the political. The political is inseparably linked to society: the latter is a "body politic"; this means that it aims at "fusing the many into one and making the one obeyed by the many".[118] But, in Latour's view, the political is one of many possible types of "assemblage".[119] And the characteristic of this assemblage is that it is based on a conception of the social as specifically human. This is why the philosopher calls for the dissipation of the "confusion [...] between assembling the body politic and assembling the collective".[120] The collective, that of the "we", if it is not a body politic, results from a task that Latour considers political.

This having been recalled, I would like to point out two things which argue in favor of keeping the concept of society. First, I am not sure that Latour agrees that society is entirely replaced by the collective. Indeed, if "politics [...] is only one way of composing the collective",[121] could we not say that society is a facet of the latter, that it is included in a more global whole—which is called "collective"—and not erased by it? This would mean that humans would continue to form societies but, aware of the problematic human limitation of these societies, would revise their functioning as well as their relationship to their environment. Secondly, Latour clearly distinguishes between two projects arising from the redesign of[122]: unearthing the "deployment of actors" and allowing for the "unification of the collective into a common world acceptable to those who will be unified".[123] In this second part—the unification—we find the political task mentioned above.

All this being said, it seems to me that the closure of the social sphere is eminently problematic for the first project and is what sociology has to debunk. But I doubt that the political task that follows can be anything other than human. Latour expresses this well in *Down to Earth* when he holds that "there is no politics other than that of humans".[124] It seems to me that the inescapable fact is that the task of building a common world can be imagined only by humans. Associations are everywhere and permeate the living domain, but the power of reflexivity on the associative fact—which allows us

---

[118] *Ibid.*, 162.

[119] *Ibid.*, 128.

[120] *Ibid.*, 161.

[121] *Ibid.*, 171.

[122] Latour defends an abandonment of the sociology of the social and proposes to pursue the development of his actor-network theory.

[123] *Ibid.*, 256.

[124] B. Latour, *Down to Earth*, 85.

to consider and argue about assemblages, their nature and their right constitution—seems an exclusively human province.[125] Thus, when Latour says that "so many other entities are now knocking on the door of our collectives",[126] we are obviously not to understand that there is some sort of conscious will to build a shared sphere among other parts of the living world. Only humans enjoy this reflexivity, and it is they who must adjust their representation of this common world by working toward understanding the plethora of infinite interdependencies it holds and especially toward knowledge of other collectives in which humans are either included in one way or another, or are outsiders.

The proximity of the political task to human reflexivity—in the Western world, to which I will return immediately afterward—justifies, in my view, the retention of the concept of society, even if this does not exclude it being supplemented by that of the collective. As I just said, there is a problem of confusion between what we might describe as, on the one hand, the social domain's "crushing" of everything that is not social and thus becomes invisible and, on the other hand, the fact that a political collective will be, first of all, a social one. In saying this, I am not asserting that this collective will only be formed by humans, but that it will be conceived by humans and that only they will be fully aware of its existence and will be able to think and modify it. Because the collective seems to me a political assemblage and, as such, it is a special kind of one, whose aim—unlike other types of human association with religious, economic, etc., goals—means precisely the awareness that association is necessary in the first place. In other words, the political refers to an association where reflexivity on the associative fact is carried to its furthest lengths. This is why I think that the notion of collective does not erase that of social collective, any more than that of society.

But there is another reason why I think the concept of society should be maintained. I also contest its rejection on the grounds that the ideas associated with it cannot only evolve, but also remain useful, in the Western context, both

---

[125] I should emphasize here that, as Pierre Charbonnier points out, human or 'social reflexivity' is not necessarily understood in the same way everywhere. Indeed, this reflexivity goes hand-in-hand, in the Western world, with 'considering that collective autonomy is based on the exteriorization of something called "nature"', and there is nothing universal about this configuration. What is 'universal' is the 'ambition of self-preservation' and, in the Western world, this ambition has been structured as follows: 'reflexivity fixates on the concept of society' and thus makes nature the other of society. But this is far from being true everywhere. The 'animist collectives' show that 'cultural groups as worthy and respectable as ours' can be formed 'without going through a form of social reflexivity'. In other words, 'solidarity' and 'integration', through which a 'we' can be constituted, may involve a 'collective identification with nonhuman beings' (see P. Charbonnier, *Affluence and Freedom*, 220–221)—very far, therefore, from the closure of the social and its limitation to human beings. But this precision does not, in my opinion, call into question the fact that the collective must be imagined by humans and that this requires a reflexive faculty that only they possess. What this precision reminds us of is that this reflexivity has not been apprehended in the same way everywhere, has not had the same destiny everywhere.

[126] B. Latour, *Reassembling the Social*, 262.

for actors who have a clear ideological commitment and those who strive to modify their daily lives in practical ways without ideological commitments, as in the case of local agri-food networks. Let us explain this point.

Charbonnier stresses that the Latourian sociology—that of the actor-network theory—leads, by calling for an account of hybrid associations bringing together humans and nonhumans, to paradoxically obliterate the fact that, if the dualism between human and nonhuman, between culture and nature is indeed artificial when viewed from the point of view of a discourse that has been constructed, it is also a "historical reality" that has "very real effects, which concretely orient mental and practical attitudes".[127] Finally, what Descola's work teaches us according to Charbonnier is that the distinction between the natural and the social has a "consistency that is not limited to that of a simple discourse"[128] and has a strong "structuring potential".[129] What the anthropologist suggests is that we are increasingly able, through comparison with other cultural contexts, to perceive the "contingency" of this "classificatory and moral strategy".[130] But perceiving its contingency should not, on the contrary, take us away from the task of accounting for the "guiding patterns" that drive the "arrangement of mediations between the various beings composing the same world".[131] For precisely these patterns "impose or at least favor certain associations to the detriment of others".[132] To perceive the contingency of the patterns attached to the classificatory strategy proper to dualism, does not "orchestrate a critique of modernity elevated to the rank of ontological conflict", but instead makes it possible to "detect" within our "historical and social context the seeds of its transformation".[133]

I think that the concept of society is one of those guiding patterns whose structuring aspect we can keep while appropriating them in a more reflective way. Food relocalization is a trend that I believe makes this possible, as we will see later. In any case, Latour does not seem to disagree so much with Charbonnier's point since he argues that one of the major errors we can chalk up to modernity is the blanket discrediting of all notions of heritage and preservation and their knee-jerk association with conservatism. He says: "the perversity of the modernization front was that, by ridiculing the notion of tradition as archaic, it precluded any form of transmission, inheritance, or revival, and thus of transformation".[134]

---

[127] P. Charbonnier, "La nature est-elle un fait social comme les autres?", 87. My translation.

[128] *Ibid.*, 88. My translation.

[129] *Ibid.*, 89. My translation.

[130] *Ibid.* My translation.

[131] *Ibid.*, 90. My translation.

[132] *Ibid.* My translation.

[133] P. Charbonnier, *Affluence and Freedom*, 222.

[134] B. Latour, *Down to Earth*, 88.

I believe that side-lining the modern project of society perpetuates in the same error, even while this term can provide us with a landmark or beacon for the great transformation of modes of coexistence to which we are called. And some landmarks would certainly not go amiss, given that for the most part, we will have to "go forward by trial and error".[135]

'Society' for Naming the Edges of Local Cooperation

In what follows, I detail how, in my view, food relocalization can challenge the political theorist to consider the concept of society afresh. I will show that the society project supported by food relocalization goes beyond the economic, ecological and social improvements the latter brings about and also concerns society as a material form, especially its spatial extension. In order to embark on this path, two pitfalls must first be avoided: on the one hand, sidestepping the idea of society, as discussed above, and on the other, avoiding the idea of small-scale spatial bounding—the local in other words.[136] It should be clear that this double avoidance is not merely the sum of two separate parts. To question, on one hand, society and its representation, thus denying the possibility that some of its features may be able to evolve, to reject, on the other, the local dimension and local territory by the same token as the global, is in my view to rob ourselves of the potential hidden in marrying these two elements: society and the local. If we take into account, as we shall see, what the concept of society says about the social – namely that it can generate a level of integration that enables individuals to experience their group belonging as impacted by this integration process – and above all, in my view, what the local says about the social – that at this level the integration process encounters far fewer obstacles – then we can better grasp the potential of combining the two dimensions.

Let us first start by the local scale. In political philosophy, the idea of the local tends to be met with mistrust and rightly so. It has always provided a vocabulary for those who seek to promote ethnic and cultural purity among communities and contributed to making traditions and the status quo untouchable. The philosopher, Serge Audier, reminds us that the 'veneration of the *terroir*' is part of the reactionary or conservative score. In this score, one finds the idea of an interference of the local 'community' in a 'so- called

---

[135] Pierre Macherey, "Compte-rendu de B. Latour – Où atterrir ? Comments' orienter en politique, La philosophie au sens large", groupe d'études animé par Pierre Macherey (https://philolarge.hypotheses.org/2261), January 10th 2018. My translation.

[136] I cannot enter into the geographical debate about the definition of the local. I will retain here the minimal consensus that the local refers to a sub-national space (see, on this consensus, Béatrice Giblin-Delvallet, "La géographie et l'analyse du 'local': le retour vers le politique", in A. Mabilleau (ed.), *À la recherche du « local»* (Paris: L'Harmattan, 1993): 79.

natural matrix'; of an identity charged with a 'biological and vitalist meaning'.[137] The mistrust of the local can be traced back to the first reflections on modern society that accompanied the emergence of sociology as a discipline in nineteenth century Europe. In these reflections and in particular in those led by the French sociologist, Emile Durkheim, one can see clearly that the local is described as disappearing with the advent of modern society.

In his major work, *The Division of Labour in Society*, Durkheim holds that in premodern societies, solidarity is 'mechanical'[138] ,which means that individuals are locked into social functions that they do not choose and that the meaning of their existence is determined by their membership in the community; this society can be said 'segmentary' '[139] to the extent that it is made of 'group of segments linked by particular affinities'[140] In the earliest premodern societies, these segments are brought together by 'blood relationships, whether real of fictitious' and in the more recent premodern societies, the population is rather 'divided up [...] according to land divisions', which means that 'the segments are no longer family aggregates but territorial constituencies'.[141] The modern process *par excellence* thus refers to the fact that the division of labor has 'developed,[142] creating more and more connections between places and breaks down the barriers that used to keep small communities isolated. Therefore, for Durkheim, the local belongs to the premodern era; it is referred to "instinctive forces" such as "the affinity of blood, attachment to the same soil, the cult of [the] ancestors, a commonality of habits, etc".[143]

In my view, our times and particularly phenomena such as food relocalization call for a renewed attitude toward the local. While vigilance toward it is more than justified, it should not lead to its elimination as an important space for the development of social life, as I will show. But it is worth noting that if one makes the link with food relocalization, one realizes that eliminating the local would be tantamount to saying that any socio-economic effects offered by the delimitation of space—*i.e.*, the emergence of a territory constructed through economic cooperation, for example—have no weight or value. If one makes the link with the ecological crisis, refusing to consider the relevance of the local scale is also to miss the problematic dimension of the Latourian conception of territory which defines it as a web of flows that go in a multiplicity of directions, mobilizing an infinite number of scales, from the local

---

[137] Serge Audier, *La société écologique et ses ennemis. Pour une histoire alternative de l'émancipation* (Paris: La Découverte, 2017): 38. My translation.

[138] Emile Durkheim, *The Division of Labour in Society* (New York: Macmillan Press, 1984[1893]): 84.

[139] *Ibid.*, 128.

[140] *Ibid.*, 132.

[141] *Ibid.*, 135.

[142] *Ibid.*, 131.

[143] *Ibid.*, 219.

to the global, thus doing away with the idea of a space whose spatial continuity would be configured on a single scale. And it is important to see where this combination of scales poses a problem, in our context of ecological crisis. When the philosopher gives an example of the list of things we can understand as affecting our subsistence—which I should recall does not boil down to survival or purely nutritional subsistence in his view—and therefore as part of our territory, he gives the example, which concerns him, of an Australian wine. He says that it is when "that unknown label on a bottle of unfamiliar shape strikes us on a supermarket shelf that we can suddenly see that Australian vineyards now form part of the range of wines we want to sample"[144] and that these vineyards—if I extend Latour's thinking—are part of our territory.

However, it seems to me that the philosopher is sidestepping the imperative in the current ecological crisis of discriminating between things or activities that are legitimately attributed to a territory and others that belong to no territory other than an abstractly understood one, such as this territory bringing together the person of Latour and the Australian vineyards. If I take again his example of bluefin tuna, in which he explains that he is also part of the territory of bluefin tuna and vice versa, insofar as the survival of this animal depends, among other things, on the level of its consumption by the Japanese, but also by the Europeans. So, Latour, taking into account the alarm raised by environmentalists, says: "bluefin tuna should no longer be part of my territory because I am part of his".[145]

But I would say something quite different: in reality, bluefin tuna should no longer be part of our territorybecause it should never have been part of it; in other words, it should never have become a perfectly common foodstuff in Europe. This is why I think that agriculture and particularly local agriculture is important. Agriculture as an activity directly linked to the soil seems to be a good indicator of what can be part of our territory and what cannot. And agri-food networks based on local cooperation, by building on the proximity of producers and consumers, tend to bring back territories that, in a way, had disappeared. I will now specify what we can expect from this indicator, since it is not a question here of advocating food autarky and closure.

What I am going to do, therefore, in what follows, is to show the content of the combination between the concept of society and that of the local. First, I will start with what the local does to the concept of society and then, in a second step, I will explain what the concept of society does to the local. In fact, to combine the representation of society with the idea of a local territory is to affirm that there are benefits that can only arise from locally rooted economic cooperation—which does not mean that the local necessarily generates such cooperation—and that those must be recognized and made possible in the long term. These benefits are as follows: increased integration of the social domain and thus a clearer feeling of individual belonging. Here,

---

[144] B. Latour, "La mondialisation fait-elle un monde habitable?", 14. My translation.
[145] Ibid., 15. My translation.

I take up the Durkheimian perspective according to which the integration of the social in modern societies results from the increased interdependence of social functions as the division of labor increases: the "ties forged" by this phenomenon have everything to do with "relationships of mutual interdependence that unite functions".[146] In short, modern society is characterized, for Durkheim, by a high degree of integration that can be achieved or—which is rather the case—exists in potential. In other words, modern society faces the risk of "anomie",[147] a state in which, at the time Durkheim wrote, economic "forces [...] grow beyond all bounds, each clashing with the other",[148] thus compromising the "cohesion"[149] of society.

The emergence of social trends such as food relocalization invites us to take up Durkheim's perspective in a critical way—in fact, to delve into points that he himself raises. We can, for this, draw on the work of a more contemporary philosopher, Franck Fischbach. He explains that interdependence really does generate cohesion where it is "deliberate and conscious"[150] whereas the division of labor, if left in its "natural"[151] state, does bring about a "mutual dependence" but it is "suffered" by individuals.[152] Fischbach calls this voluntary and conscious dependence "cooperation".[153]

Anne W. Rawls gives us a very clear idea of what the concept of cooperation entails. She holds that if it is true that the division of labor makes solidarity possible, it is because it emerges from the practices deployed within increasingly specialized activities. Individuals, when invested in the same function, are obliged to shape and respect "rules"[154] without which they would not be able to achieve the results in the name of which the activity in question is undertaken: work is therefore steeped in "constitutive practices",[155] *i.e.*, regulated practices on which everyone agrees. In other words, individuals, within groups that correspond to their activity, can collectively decide on usages, objectives and rules. This is precisely what is happening in many different experiments with food relocalization. But this contemporary phenomenon highlights an aspect of thinking about cooperation that, it seems to me, is not highlighted enough: the local scale. It is when individuals or group of individuals are in

---

[146] E. Durkheim, *The Division of Labour in Society*, 101.

[147] *Ibid.*, XXXI.

[148] *Ibid.*, XXXII.

[149] *Ibid.*, XXXV.

[150] Franck Fischbach, *Le sens du social. Les puissances de la coopération* (Montreal: Lux éditeur, 2015): 190. My translation.

[151] *Ibid.* My translation.

[152] *Ibid.*, 223. My translation.

[153] *Ibid.* My translation.

[154] Anne W. Rawls, "Durkheim's Theory of Modernity: Self-Regulating Practices as Constitutive Orders of Social and Moral Facts", *Journal of Classical Sociology*, 12, no. 3–4 (September 2012): 481.

[155] *Ibid.*

contact with one another, or located at a reasonable distance from each other at any rate, that this elaboration of rules and thus the growing awareness of interdependence can emerge with the fewest obstacles. It is, therefore, at the local level that cooperation can be easiest and social integration, therefore, can be achieved to a relatively high degree.[156]

In my view, Rawls is not saying anything different when she refers to the "local and situated character"[157] of constitutive orders composed of constitutive practices that are themselves "locally organized".[158] In fact, if constitutive practices are to be understood as "mutual cooperation",[159] it is because they are "self-regulating"[160] in the sense that the rules that structure them cannot be imposed on them "at a distance". And finally, Durkheim also made this argument—but without drawing the consequences—when he recognized that a "state of anomie is impossible wherever organs solidly linked to one another are in sufficient contact and in sufficiently lengthy contact".[161] Food relocalization, because it is based on the renewed proximity between producers and consumers within a geographical space that allows for their cooperation, suggests that social integration—and with it, society—has something to do with the local scale. This is how the local does something to the concept of society: we can deconstruct the evidence attached to the national scales of societies and start thinking about the fact that the integration of the social is perhaps best achieved locally. The local in question is not, as will be seen later, similar to that which characterized the segments constituting premodern societies.

Society or the Institutionalization of the Territory of Cooperation

Let us now move on to the second step, which is to understand what the concept of society does to the local. Indeed, one might think that food relocalization only makes it possible to highlight the cooperative and therefore integrative potential of the local scale. However, we began by saying that if it is true that local economic cooperation brings about both economic and social benefits, then we must also explore how these can be maintained and sustained over the long term. It seems to me that the concept of society is one of the objects of this exploration. To speak of a relocalized society, indeed, means

---

[156] This does not mean, I repeat, that economic life at the local level necessarily adopts a cooperative form. It is only a possibility, but one that has the particularity of characterizing this scale and not another. There is a type of cooperation that is only possible at the local level.

[157] *Ibid.*, 497.

[158] *Ibid.*, 498.

[159] *Ibid.*, 486.

[160] *Ibid.*, 487.

[161] E. Durkheim, *The Division of Labour in Society*, 304.

suggesting that social cohesion, while locally generated through cooperative agri-food networks, should also give rise to a form of political consolidation. Put another way, the working rules of various local cooperative activities should converge toward a properly political—not purely social—institutional sphere which contains the ensemble of local cooperative economic networks, thus expressing their unification. For the political level, while not external to the social, nevertheless refers to institutions which have a particularly high consolidation force and which, in fact, have the power to perpetuate practices and networks of practices.

Why do I think the concept of society makes it possible to specify this requirement? To understand this, it is important to dissolve, as the philosopher Francesco Callegaro does, the evidence surrounding the concept of society in order to grasp the centrality of the political sphere and its specific contribution to social life. He explains that what characterizes a society is the presence of a government—not a state, which is only a form of government. A social body is a society to the extent that it has a government, in the sense that the integration processes running through the social are fully realized when there is a level at which the peculiarities of these processes—related to the different social groups, to the multiplicity of cooperative networks and dynamics of cooperation that make up these processes—are at once contained and encompassed. The government thus brings into existence "the plane of that higher-order totality of belonging"[162] to which the secondary groups refer—and it is in this that a political society exists. This diversity is included and encompassed in a totality because the government level, which Callegaro understands above all as a deliberative space, is one where rules applicable to all are formulated.

There is a second aspect of the concept of society to which I want to return and which is also linked to the presence of a political power, *i.e.*, a government. The latter validates the existence of a society and allows it to endure for a reasonable length time, notably by institutionalizing its material being—through the official existence of spatial limits, boundaries within which its jurisdictional power is applied. Thus, the existence of a government inevitably evokes the "sensitive contours"[163] of the social body—including its spatial extension—to which the actions of the government in question apply. What makes society, then, is above all its integration and internal cohesion. But what also makes a society is the governmental or political level which imparts institutional existence to the material contours of the integrated entity that constitutes a society. From this perspective, the relocalized society refers to a normative horizon in which there is a stronger connection between the

---

[162] F. Callegaro, "L'Etat en pensée. Emile Durkheim et le gouvernement de la sociologie", in B. Karsenti, D. Linhardt (eds), *Etat et société politique. Approches sociologiques et philosophiques* (Paris: Editions de l'EHESS, 2018): 200. My translation.

[163] Bruno Karsenti, "Une autre approche de la nation: Marcel Mauss", *Revue du MAUSS*, 36, no. 2 (February 2010): 290. My translation.
B. Karsenti, "Ethicité et anomie. De la philosophie sociale à la sociologie de l'Etat", in B. Karsenti, D. Linhardt (eds), *Etat et société politique*, 154. My translation.

political level of government—"this collective that de-particulates the other collectives"—and the social interaction of these collectives that make up the integrated social whole. And the strength of this connection would precisely be due to the fact that the spatial extension of the society would be more reduced compared to that of most nation-states. Furthermore, the horizon of a relocalized society reconnects with a specific understanding of territory—precisely that which Latour rejects: a social body is attached to a territory through a "territorial concentration",[164] which means that the cohesion of society is also based on the fact that it extends over a continuous spatial area. As a result, geographical delimitations and spatial divisions regain their meaning, but this has little to do with the illusion described by Latour, according to which territory merely conceals a vehicle for influence and power domination. If there are boundaries and territories, it is precisely because the soil is something which facilitates the advent of a society through cooperative networks which take root thanks to what it yields for human subsistence. We may imagine the possibility that power or government might be there merely to validate the near-spontaneous existence of a social entity born of local economic cooperation—above all in the agricultural sector, as we shall see.

The mention of soil, dear to Latour, allows me to focus, finally, on the link between the horizon of a relocalized society and food relocalization. This horizon is generated, as I explained above, by the combination of the concept of local and the concept of society and the way in which they work with each other. This concept is generated, as detailed earlier, by the combination of the concept of local and the concept of society and the way in which they work with each other. But it seems to me that, in terms of the link to be drawn between relocalized society and food relocalization, it is what the concept of society with its two aspects—brings to the local that is most significant since it touches on the way in which social innovations, such as local agri-food networks, can be sustained over time. Indeed, the horizon of the relocalized society encapsulates, above all, a questioning: how can the social life generated by local cooperative networks be maintained over the long term? For me, this presupposes that the contours of the geographical space that these economic networks have brought into existence can be institutionalized and that the economic life within it can be preserved, in some way, from dissolving into a larger space. Thus, the two aspects of the concept of society, which I recalled above, find an echo in food relocalization: the idea that there must be a totality of belonging where the imperative of collective decision is housed; and the idea that the political level is also what makes visible the material form that a society is just as much. Regarding the first idea, it can be said that food relocalization should also include a decision-making component. That is, it cannot be reduced to local production and consumption. Some

---

[164] Margaret Moore, *A Political Theory of Territory* (Oxford: Oxford University Press, 2015): 118.

of the decisions concerning agricultural and food issues, where local cooperative agri-food networks have developed, should be taken by the people they concern—and therefore locally. In other words, there should be a form of government on agricultural and food issues that can rise above the local agri-food networks and to which they would converge. The second idea can be transposed almost as it is to food relocalization: this local government of agricultural and food issues would make it possible for an agri-food territory to exist, with its limits, which, institutionalized by the existence of this government, would differentiate it from other territories.

As can be seen, the transposition of the concept of relocalized society to food relocalization—which, in my view, gives rise to it in the first place—does not lead to a conception as precise as that contained in the concept of the nation-state which itself shapes the representation of modern society. There are several reasons for this. The first is that we must avoid the trap of thinking that all we have to do is think smaller, relocalize in order to start the ecological transition. As Charbonnier points out: "one cannot have the same thing but smaller, a downsized industrial modernity, miniaturized to meet ecological demands".[165] The second is that the horizon of the relocalized society is not about the return of small, isolated, autarkic communities. As we will see in the conclusion, the agri-food territory is larger than that of the small local communities that modernity has erased. What I wanted to contribute, with this horizon of the relocalized society, is to reflect on what will be local in the ecological transition. Because, despite the vigilance we must have with regard to the risk inherent in the local—namely, the withdrawal of identity, there will be something local in the ecological transition and we must try to understand what it is. Thus, the transposition of the concept of relocalized society to food relocation at least allows us to highlight some key points. And the major avenue that seems to have opened up is that of the political translation—far from the concept of sovereignty—of a local material autonomy, not autarky, which would start with food but could extend to other aspects of productive life.

Conclusion: Situating the Local Using Agricultural Zoning

I would like to conclude by tackling a question that must have arisen on reading the above lines and which refers to the spatial delimitation that would allow us to determine this "local" which draws the contours of a relocalized society. In concrete terms, where is the local? How is space to be divided?

Before answering these questions by way of a detour through rural and agricultural geography, let us first return briefly to the role of agriculture in this projection of what a relocalized society might look like. As I said above, this is not about imagining society as a completely self-sufficient community.

---

[165] P. Charbonnier, *Affluence and Freedom*, 245.

It is fairly obvious that we do not live on food alone. But the aim is not even food autarky. *I am arguing that food subsistence could once again become a kind of indicator of the extension of the territory over which a social body could be deployed that provides its members with a sense of belonging based on cooperative inclusion and collective autonomy.*

This is what I believe current food relocalization could be heading toward: individuals, producers and consumers cooperating to make the most of the food resources in the space where they live because they are aiming for a form of material autonomy which, by being achieved through cooperative means, increases social integration. It seems to me Chiffoleau has an intuition for this possible fate of food relocalization when she speaks of those territories "where vegetables used to be grown;"[166] the cooperative dynamics around local networks aim to restore that same nutritional vocation by re-establishing crops that had disappeared from the land although it could still support them.

The idea is thus to re-establish the meaningful dimension, in a spatial sense, of food subsistence. It has already had this vocation: Latour refers to that era before the industrial revolution, but above all before agricultural modernization, when "one could scan in a single glance the territory that ensured one's subsistence".[167] While taking account of technical advances in agriculture, perhaps we should reconnect with the opportunity—necessarily linked to the human practice of transformation—that lies at the very heart of ecosystems to connect occupation of the land with cooperative networks that lead to strong integration. The normative horizon I propose—which is first thinkable before it is possible—makes agriculture the plane at which the cooperation attached to it could, more than anything else, draw the outlines of the social entity to which we belong. I would point out that this delimitation in no way calls into question the need for external exchange, either economic or social, or the fact that mobility and migrations would continue.

This importance of agriculture—as an indicator of spatial extension—in my thinking explains why I have repeatedly insisted that we should not dissociate the local nature of cooperative structures, in terms of local agri-food networks, from the reduction of producer–consumer geographical distance. If we separate these two facets, we risk seeing in this phenomenon only the beginnings of an evolution toward simple 'local-food,[168] to use Chiffoleau's expression. Governance is not so far from what I am suggesting but it is still much less ambitious than what local agri-food networks could help to develop. Chiffoleau herself seems to recognize this fact when she points out that when this governance is organized through projects and norms that issue from the central state, it comes up against obstacles and tends to be rejected at the

---

[166] Chiffoleau, *Les circuits courts alimentaires*, 98. My translation.
[167] Latour, *Down to Earth*, 97.
[168] Chiffoleau, *Les circuits courts alimentaires*, 125. My translation.

local level because these projects and norms are perceived as excessively "top-down approaches".[169] Because food relocalization is, first and foremost, about projects and norms that seek to establish greater local autonomy in terms of food, we can, therefore, see that there is more to local agri-food networks than an aspiration for broader governance with local ramifications: there is, through the production and exchange of agricultural products, a search for greater local autonomy.

Finally, I return to the question of delimiting space on the basis of agriculture and to the difficult question of where the local is located. For Chiffoleau, it is almost impossible to decide what the local is, considering the extreme diversity of the French agricultural space. According to her, this diversity makes it very hard to define what exactly is local and thus to divide space into local territories. But agronomy and geography have long been concerned in France with delimiting agricultural regions aimed at development and optimization of production. These regions also did not necessarily correspond to administrative limits or to geographical-administrative boundaries.

The history of this boundary-making is quite interesting, although I can only sketch it here. Though agricultural regions today are extremely specialized spaces—a model that is clearly out of alignment with the way I am suggesting we reconnect to the idea of food subsistence—this has not always been the case. Throughout the nineteenth century, the notion of the agricultural region more or less mapped onto what we call in France the *pays*, which would correspond to the current *terroirs*. Just after the war, in 1946, this notion was revisited and a new territorial zoning was envisaged, notably by an agronomist, André Cholley; he coined the notion of "agrarian system" as designating a recognizable and relatively defined space characterized by the "complex of elements borrowed from different domains, yet very closely interconnected" through agricultural activity: "the bio-climatic conditions of given space" as well as the type of agriculture, its techniques and other economic and social activity outside the agricultural domain.[170]

The characteristic of the agrarian system and the territorial checkerboard that it can found is that it endorses the role of the natural environment. So, there is a form of the natural given that is taken into account here. However, it does not evacuate the transformation, *i.e.*, the agricultural activity and it does even more, it encompasses this mode of activity in a much broader reality since it includes social, economic and political aspects—in short, the social body. This explains why the division of space into agrarian systems seems to have to be mobilized—and no doubt updated—by the perspective of food relocalization and relocalized society, as I have discussed in this chapter.

---

[169] *Ibid.*, 125. My translation.

[170] André Cholley, "Problèmes de structure agraire et d'économie rurale", *Annales de géographie*, no. 298, 1946, quoted by Hubert Cochet, "Origine et actualité du système agraire: retour sur un concept", *Revue Tiers-Monde*, 207, no. 3 (March. 2011): 98.

REFERENCES

ALLEN Patricia, 2003. "Alternative Food Initiatives in California: Local Effort Address Systemic Issues." *Centre for Agroecology & Sustainable Food Systems*, Research Brief #3.

ALLEN Patricia, 2010. "Realizing Justice in Local Food Systems." *Cambridge Journal of Regions, Economy and Society*, 3, no. 3: 295–308.

AUBRY Christine, CHIFFOLEAU Yuna, 2009. "Le développement des circuits courts et l'agriculture périurbaine: histoire, évolution en cours et questions actuelles." *Innovations agronomiques*, no. 5: 53–67.

AUDIER Serge, 2017. *La société écologique et ses ennemis. Pour une histoire alternative de l'émancipation*. Paris: La Découverte.

BOOKCHIN Murray, 1972. "On Spontaneity and Organization." *Anarchos*.

BOOKCHIN Murray, 1986. *Post-Scarcity Anarchism*. Montreal-Buffalo: Black Rose Books.

BROWN Wendy, 2002. "At the Edge." *Political Theory*, 30, no. 4: 556–576.

CALLEGARO Francesco, 2014. "Le sens de la nation. Marcel Mauss et le projet inachevé des modernes." *Revue du Mauss*, no. 43: 337–356.

CALLEGARO Francesco, 2018. "L'Etat en pensée. Emile Durkheim et le gouvernement de la sociologie." In B. Karsenti, D. Linhardt (eds.), *Etat et société politique. Approches sociologiques et philosophiques* (Paris: Editions de l'EHESS): 191–225.

CHARBONNIER Pierre, 2013. "La nature est-elle un fait social comme les autres? Les rapports collectifs à la nature à la lumière de l'anthropologie." *Cahiers philosophiques*, no. 132: 75–95.

CHARBONNIER Pierre, 2021. *Affluence and Freedom. An Environmental History of Political Ideas*. Oxford: Polity.

CHIFFOLEAU Yuna, 2008. "Les circuits courts de commercialisation en agriculture: Diversité et enjeux pour le développement durable." In G. Maréchal (ed.), *Les circuits courts alimentaires. Bien manger dans les territoires* (Dijon: Educagri): 19–30.

CHIFFOLEAU Yuna, 2019. *Les circuits courts alimentaires. Entre marché et innovation sociale*. Toulouse: Eres.

CHIFFOLEAU Yuna, 2021. "Les circuits courts: une réponse face aux crises?." *Sciences Humaines*, no. 338: 42.

COCHET Hubert, 2011. "Origine et actualité du système agraire: retour sur un concept." *Revue Tiers-Monde*, 207, no. 3: 97–114.

DANSERO Edigio, PETTENATI Giacomo, TOLDO Alessia, 2016. "Si proche et pourtant si loin. Etudier et construire la proximité alimentaire à Turin." In P. Mundler, J. Rouchier (eds.), *Alimentation et proximités. Jeux d'acteurs et territoires* (Dijon: Educagri): 307–321.

DEMEULENAERE Elise, BONNEUIL Christophe, 2010. "Cultiver la biodiversité. Semences et identité paysanne." In B. Hervieu, N. Mayer, P. Muller, F. Purseigle, J. Rémy (eds.), *Les mondes agricoles en politique. De la fin des paysans au retour de la question agricole* (Paris: Presses de Sciences po): 73–92.

DESCOLA Philippe, 2013. *Beyond Nature and Culture*. Chicago: The University of Chicago Press.

DUBUISSON-QUELLIER Sophie, LE VELLY Ronan, 2008. "Les circuits courts entre alternative et hybridation." In Gilles Maréchal (ed.), *Les circuits courts alimentaires. Bien manger dans les territoires* (Dijon: Educagri): 105–112.

DUBUISSON-QUELLIER Sophie, LE VELLY Ronan, 2009. "Circuits courts. Partager les responsabilités entre agriculteurs et consommateurs." *Demeter, Economie et Stratégie Agricoles*, 87–112
DUMAIN Aurélie, MAURINES Béatrice, 2012. "Composer les manières de gouverner." In A.-H. Prigent-Simonon (ed.), *Au plus près de l'assiette* (Versailles: éditions Quae): 215–231.
DUPUIS Mélanie E., GOODMAN David, 2005. "Should We Go 'Home' to Eat?: Towards a Reflexive Politics of Localism." *Journal of Rural Studies*, 21, no. 3: 359–371.
DURKHEIM Émile, 1984 [1893]. *The Division of Labour in Society*. New York: Macmillan Press.
FISCHBACH Franck, 2015. *Le Sens du social. Les puissances de la coopération*. Montréal: Lux éditeur.
GIBLIN-DELVALLET Béatrice, 1993. "La géographie et l'analyse du 'local': le retour vers le politique." In A. Mabilleau (ed.), *À la recherche du local* (Paris: L'Harmattan).
HERVIEUX Bertrand, PURSEIGLE François, 2013. *Sociologie des mondes agricoles*. Paris: Armand Colin.
HINRICHS C. Clare, 2003. "The Practice and Politics of Food System Localization." *Journal of Rural studies*, 19, no. 1: 33–45.
KARSENTI Bruno, 2010. "Une autre approche de la nation: Marcel Mauss." *Revue du MAUSS*, 36, no 2: 283–294.
KRASNER Stephen D., 1999. *Sovereignty. Organized hypocrisy*. Princeton: Princeton University Press.
LAMINE Claire, 2005. "Settling Shared Uncertainties: Local Partnerships between Producers and Consumers." *Sociologia Ruralis*, 45, no. 4: 324-345.
LARRERE Catherine, 2014. "What the Mountain Knows. The Roots of Environmental Philosophies." *Books & Ideas*.
LATOUR Bruno, 1993. *We have Never been Modern*. Harvard: Harvard University Press.
LATOUR Bruno, 2005. *Reassembling the Social. An Introduction to Actor-Network Theory*. Oxford: Oxford University Press.
LATOUR Bruno, 2010. "La mondialisation fait-elle un monde habitable?." Territoires 2040. Prospectives périurbaines et autres fabriques de territoire, *Revue d'étude et de prospective*, no. 2.
LATOUR Bruno, 2018. *Down to Earth. Politics in the New Climatic Regime*. Cambridge: Polity Press.
LAVILLE Jean-Louis, 2010. *Politique de l'association*. Paris: Seuil.
MACHEREY Pierre, 2018. "Compte-rendu de B. Latour – Où atterrir ? Comment s'orienter en politique, La philosophie au sens large." groupe d'études animé par Pierre Macherey.
MARTIN Jean-Philippe, 2005. *Histoire de la nouvelle gauche paysanne. Des contestations des années 1960 à la Confédération paysanne*. Paris: La Découverte.
MOORE Margaret, 2015. *A Political Theory of Territory*. Oxford: Oxford University Press.
MUNDLER Patrick, 2007. "Les associations pour le maintien de l'agriculture paysanne (AMAP) en Rhône-Alpes. Entre marché et solidarité." *Ruralia*, no. 20 [Online]

NONINI Donald M., 2013. "The Local-Food Movement and the Anthropology of Global Systems." *American Ethnologist*, 40, no. 2: 267–275.

OLIVIER Valérie, COQUART Dominique, 2008. "Les AMAP, des réseaux agroalimentaires alternatifs émergents en soutien à la petite agriculture locale? Quelles opportunités pour l'amélioration des revenus des agriculteurs?." 2èmes journées de recherches en sciences sociales – INRA, SFER, CIRAD.

POLANYI Karl, 2001 [1944]. *The Great Transformation. The Political and Economic Origins of our Time*. Boston: Beacon Press.

RAWLS Anne Warfield, 2012. "Durkheim's Theory of Modernity: Self-Regulating Practices as Constitutive Orders of Social and Moral Facts." *Journal of Classical Sociology*, 12, no 3–4: 479–512.

REDLINGSHÖFER Barbara, 2008. "L'impact des circuits courts sur l'environnement." In G. Maréchal (ed.), *Les circuits courts alimentaires. Bien manger dans les territoires* (Dijon: Educagri): 175–185.

RIPOLL Fabrice, 2009. "Le concept AMAP: promotion et mise en pratique(s) d'une nouvelle norme d'échange entre consommateurs et producteurs agricoles." In V. Banos (ed.), *Espaces et normes sociales* (Paris: L'Harmattan): 99–116

Environmental Politics and Theory in Specific International Regions

Environmental Justice and the Global Rights of Nature Movement

Chris Crews

> Indigenous people are the land. It provides us with language, ceremony, food – the land gives us a sense of belonging, tells us who we are.[1]
>
> Living with the earth is an intimate belonging, like the connectedness of a well-rooted tree. There is a solidness to it. There is also growth. And that growth is bound up with soil, water and sun in such a way that there is no separation. The tree is part of the earth and the sky. It protects and nourishes other life. It lives in harmony with everything around it.[2]

The Earth is in trouble. Everywhere you look things are falling apart. Since 2020 we have faced a global coronavirus pandemic, an expanding global climate crisis, growing authoritarian and racial violence, widening social inequity and collapsing social safety nets driven by disaster capitalism. And yet, all around the world social movements are fighting back against these bleak trends by offering a radically different vision for the future. This vision is rooted in a belief that all life on the planet is deeply and irrevocably interconnected. Advocates of this view point to an emerging paradigm shift in how we think about and relate to the Earth, a shift rooted in the belief that the

---

[1] Talaga (2020).
[2] Brown (1984, 14).

C. Crews (✉)
International Studies, Denison University, Granville, OH, USA
e-mail: chris@chriscrews.com; chris.crews@denison.edu

natural world has an inherent right to exist and flourish. In recent years, this belief became a rallying cry of the global Rights of Nature (RON) movement, a social movement that aims to expand the conceptual and legal boundaries of rights and justice. My central argument is that this movement marks the start of a new phase of global environmental justice politics, one that is finally moving away from the shackles of anthropocentrism and liberalism that shaped and defined earlier generations of environmental justice movements and associated politics.

Looking back at the early movements against environmental racism in the 1980s allows us to see just how far movements for environmental justice (EJ) have come, and why this new Rights of Nature movement might represent the next stage of movement evolution. Writing in 1994, environmental justice scholar Robert Bullard argued that "A new form of environmental activism has emerged in communities of color. Activists have not limited their attacks to well-publicized toxic contamination issues, but have begun to seek remedial action on neighborhood disinvestment, urban mass transportation, pollution, and other environmental problems that threaten public safety."[3] Discussing this "new" Environmental Justice Movement a few years later, Luke Cole and Sheila Foster noted what was significant about this movement was how "it transforms the possibilities for fundamental social and environmental change through redefinition, reinvention, and construction of more innovative political and cultural discourses and practices…and the forging of new forms of grassroots political organization."[4]

In the decades since these descriptions were written, the environmental justice movement has grown in important ways. Most visibly, it has expanding beyond the initial focus on toxins in urban communities of color in the United States, although this remains an important aspect of EJ movements today. Environmental justice movements today are far more diverse than those of the 1980s and they are focused on a wide range of global issues.[5] With this EJ movement landscape, the youth-led climate justice wing arguably represents the most radical and cutting-edge spaces where political innovation and activism are taking place, in part, thanks to the growing leadership role of Indigenous youth and women of color in these movements.[6]

The global Rights of Nature movement is at the forefront of the process of "redefinition, reinvention, and construction of more innovative political and cultural discourses and practices" described by Cole and Foster. In what follows, I explore how the rethinking of the boundaries of environmental justice challenges traditional understandings of, and organizing on, environmental justice issues. By challenging the long-standing liberal and anthropocentric conventions which treat nature as a static object to be acted

[3] Bullard (1994, 6).
[4] Cole and Foster (2001, 14).
[5] Tokar (2014).
[6] Whyte (2014), Harris (2016).

upon by humans, rather than as a rights-bearing subject, activists for the Rights for Nature are testing the definitional limits of environmental justice and political rights. Given the rapid growth of this movement, I argue scholars interested in environmental politics and environmental political theory should pay close attention to how this movement is reframing the discursive and legal landscape of environmental justice politics.

FROM ENVIRONMENTAL RACISM TO ENVIRONMENTAL JUSTICE

To better understand the impacts of this upstart movement, we need to look to the past. As our visions of a just future have evolved, so, too, have our ideas about what constitutes environmental justice. In the 1980s, environmental justice was chiefly concerned with discrimination and inequity in toxic industries, with community activists trying to show how environmental benefits and burdens were unequally distributed and deeply racialized.[7] The 1979 Whispering Pines landfill fight in Houston, Texas, and the 1982 fight over a PCB landfill in Warren County, North Carolina, are often pointed to as the origins of the environmental justice movement.[8] Similarly, the "Toxic Waste and Race in the United States" report released by the United Church of Christ Commission for Racial Justice in 1987 became a key document that made explicit the links between racial justice and environmental politics.[9] This frame continued to expand over the 1990s to include a broader critique of federal environmental policies and mainstream environmental groups, as well as a recognition that environmental justice was about more than just race and pollution.

Groups such as the Indigenous Environmental Network (IEN) were formed in the 1990s in response to many of the dynamics outlined in a 1987 UCC report. "At that time, a significant number of our tribal communities and villages were targeted for large toxic municipal and hazardous waste dumps and nuclear waste storage facilities and with industrial and mineral development in Indian country literally leaking and oozing out of the ground with toxic poisons. Organizing around environmental issues was relatively new to many of the tribal grassroots members and their tribal governments in the early 90's."[10] The 1991 First National People of Color Environmental Leadership Summit in Washington DC highlighted the expanding focus of environmental justice advocates and spoke to findings from the 1987 report. The Principles of Environmental Justice that came out of that meeting called on allies to build "a national and international movement of all peoples of color to fight

---

[7] Bullard (1990, 1998), Taylor (1997).
[8] Bullard (2001).
[9] United Church of Christ (1987).
[10] Indigenous Environmental Network (ND).

the destruction and taking of our lands and communities" and highlighted the need to "secure our political, economic and cultural liberation that has been denied for over 500 years of colonization and oppression, resulting in the poisoning of our communities and land and the genocide of our peoples."[11] This vision was reaffirmed in a 1996 meeting organized by the Southwest Network for Environmental and Economic Justice where participants drafted the Jemez Principles for Democratic Organizing, a short statement of principles to guide environmental justice movement building. "Groups working on similar issues with compatible visions should consciously act in solidarity, mutuality and support each other's work. In the long run, a more significant step is to incorporate the goals and values of other groups with your own work, in order to build strong relationships."[12]

These developments took place against the backdrop of an expanding anti-globalization movement in the late 1990s, a movement which came to a head at the 1999 WTO meeting in Seattle. These changes were part of the evolution of social movements during this period, with groups shifting from an anti-globalization to an alter-globalization framework that looked beyond critiques of neoliberalism to a vision summed up in the World Social Forum, namely that Another World Is Possible. The rapid expansion of neoliberalism under the guise of globalization in the 2000s made it clear that political resistance had to be global and intersectional. As environmental theorist, David Schlosberg, argues, "The central critique of the institutions of this new world economy is that they promote an inherently inequitable distribution of economic goods and related social and environmental bads. Social justice, environmental justice and ecological justice are tied together in these critiques, as the poor suffer both social and environmental inequity and nature is drained of resources for economic gain."[13] These political dynamics slowly gave birth to a new environmental movement that was more focused on the issue of climate change (or global warming in the language of the day).

It was not long before the term "climate justice" began to appear with more frequency. The first Climate Justice Summit was organized in November 2000 alongside the United Nation's COP6 climate meeting in the Netherlands and this event helped to push the idea of climate justice into public discourse. "While it may be hard to pin down the exact origins of the idea of climate justice, it is clear that the summit in 2000 was a turning point for its prominence in the climate change movement."[14] Rising Tide, a loose coalition of global anti-authoritarian and direct-action activists that had formed during this time, offered some insights into how individuals began to link their earlier work to this new idea of climate justice.

---

[11] United Church of Christ (1991).
[12] Jemez Principles (1996).
[13] Schlosberg (2009, 82).
[14] Whitehead (2014).

The [anti-road and radical environmental] activists argued: "A stable climate is much more important than saving some private interests, isn't it? It is an interest of all living entities on earth, uncomparable with business interests. Now that these environmental organisations are sitting around the table together with business and government they do tend to be considered the same as business in that they represent some private interest. As a consequence they have become an usage [sic] for those in power to legitimise practices by which ecology and people are subordinated for the sake of profit".

The anti-road protestors concluded that a radical answer to this top-level negotiations was highly necessary. And the only way to counter these practices is by organising ourselves, too. But not in a hierarchical and bureaucratic way as them. Rather on a grassroots level. In this way people and groups can be autonomous and choose their own priorities, while at the same time, on an international level, we represent a coalition for climate justice.[15]

As activists with Rising Tide noted, many environmental activists were not only concerned with the failure of global climate meetings to date (this was on the heels of the 1997 Kyoto Protocol, which had not yet gone into effect), but also because many mainstream green groups had become part of the entrenched system of neoliberal globalization these activists were fighting against.

Framing their movements through the lens of climate justice, then, was one way to signal that international climate frameworks like the UNFCCC were part of the same neoliberal system. As activist, Yin Shao Loong, pointed out in the Dissenting Voices zine, "The UN climate change conference at The Hague witnessed a complete integration of big business into the Kyoto Protocol process and confirmed the fears of many activists that the real goals of the Protocol are being undermined in the relentless pursuit of profits."[16] A similar concern was voiced at Klimaforum09, the 2009 people's climate summit organized in parallel with the COP 15 meetings in Copenhagen. In response, climate activists issued an official declaration titled "System Change - Not Climate Change" which raised similar issues to those from 2000.

> In Asia, Africa, the Middle East, Oceania and South and Central America, as well as the periphery of North America and Europe, popular movements are rising to confront the exploitation of their land by foreign interests, and to regain control over their own resources. A new type of activism has revitalized the environmental movements, leading to a wide variety of protests and actions against mining, dams, deforestation, coal-fired power plants, air travel, and the building of new roads, among others. There is a growing awareness about the need to change the present economic paradigm in a very fundamental way...The so-called strategy of "green growth" or "sustainable growth" has turned out to

---

[15] Dissenting Voices (2001, 5).
[16] Dissenting Voices (2001, 7).

be an excuse for pursuing the same basic model of economic development that is one of the root causes of environmental destruction and the climate crisis.[17]

This debate over whether market-based policy solutions can ever solve market-created problems continues to be a salient dividing line between mainstream and radical environmentalists today. Nearly a decade after the Klimaforum09 declaration, I saw many banners at the 2017 People's Climate March in New York City with the phrase "System Change, Not Climate Change."

For those pushing climate reforms rooted in market-based policies like carbon trading, REDD+ and sustainable development, the underlying logics of extractive global capitalism and endless growth are simply challenges to be managed by better technology and risk analysis. For those demanding climate justice, reliance on neoliberal reforms is seen as a kind of sickness.

Climate justice advocates have called for a new political vision not rooted in exploitation and domination, one that can bring about what many on the Left refer to as a Just Transition, a "vision-led, unifying and place-based set of principles, processes, and practices that build economic and political power to shift from an extractive economy to a regenerative economy."[18] Just Transition language now appears frequently in messaging from groups like 350.org, the Sunrise Movement, Extinction Rebellion and the Climate Justice Alliance.

This idea has also been expanded and deepened from its original labor union roots by Indigenous activists and scholars into what the Indigenous Environmental Network (IEN) calls its "Indigenous Principles of Just Transition." The IEN's conceptualization involves three core ideas: responsibility and relationship, sovereignty and transformation for action.[19] This shift beyond labor issues, in particular the emphasis on sovereignty and relationality, speaks to one new direction within environmental justice movements today, and these Indigenous perspectives on nonhuman relationality are one of the main intellectual drivers behind the global Rights of Nature movement. As Rights of Nature advocates have noted, one powerful way to undermine the extractive logics of capitalism is to give formerly excluded "natural resources" legal rights to challenge their exploitation. Central to this latest political battle is how we define "justice."

Boundaries of Justice

This rethinking of the boundaries of justice is a story about both political and philosophic changes. As our scientific understanding of the natural world has deepened, so too has our understanding of our interconnection with

---

[17] Klimaforum09 (2009, 7).

[18] Climate Justice Alliance (2019, 1).

[19] Indigenous Environmental Network (2019).

the Earth. While we have not erased the legacy of the European Enlightenment and its Cartesian dualism, its grip has weakened. That period famously signaled what Thomas Kuhn referred to as a paradigm shift, and I argue we are witnessing a new paradigm shift today involving how we think about the Earth. In its most broad sense, I would suggest we are seeing a new Earth-centered politics emerging, a trend I have suggested is linked to the rise of a new form of cosmopolitics shaped by the dynamics of the Anthropocene.[20] Part of this shift is a rethinking of the boundaries of justice from the perspective of an Earth-centered ontology, which is helping to drive this rearticulation of environmental justice.

These beliefs are manifesting in two different political forms. In its thin version, this idea is expressed as responsible management or stewardship of nature by humans. This approach is embodied in global efforts such as the United Nations declaring April 22 as International Mother Earth Day, a move with important symbolic weight, but no legal force. In its thick version, this idea is expressed as bonds of relationality or kinship that transcend the human-nonhuman divide and encompass everything from mountains and rivers to nature spirits and ancestral beings.

David Abram once described these place-based forms of kinship as expressing an allegiance "not to the human community, but to the earthly web of relations in which that community is embedded."[21] This sense of a direct kinship with other beings is common to many Indigenous communities around the world who never entirely lost a belief in kinship with the Earth and have continued to foster kincentric relations through everyday cultural practices and beliefs.[22] Discussing his own Indigenous Rarámuri (Tarahumara) community in northwest Mexico, Enrique Salmón described these intertwined relations as kincentric ecology. "Kincentric ecology pertains to the manner in which indigenous people view themselves as part of an extended ecological family that shares ancestry and origins. It is an awareness that life in any environment is viable only when humans view the life surrounding them as kin. The kin, or relatives, include all the natural elements of an ecosystem."[23]

Describing commonalities within Native American land relations, Annie Booth notes "reciprocal relationships with nature permeated every aspect of life from spirituality to making a living and led to a different way of seeing the world" for Indigenous peoples.[24] While rejecting a problematic discourse of "ecological Indians" used by some environmentalists, and also noting the many important differences in how Indigenous communities relate to the land and other than human kin, we can, nonetheless, argue these diverse

[20] Crews (2019).

[21] Abram (1997, 8).

[22] My usage here of the term "Indigenous peoples" follows the broad self-definition found in the UNDRIP.

[23] Salmón (2000, 1332).

[24] Booth (2003, 329).

Indigenous perspectives do have important political implications for rethinking environmental justice in a more than human world. For example, the material implications of these different cosmopolitical perspectives can be seen in the fact that many of the most biologically diverse regions in the world overlap significantly with Indigenous communities that have practices a variety of traditional land stewardship practices.[25]

In recent years, scholars and policy-makers have finally begun to recognize the value of what has been labeled "Traditional Ecological Knowledge" (TEK), in addition to ideas about how biocultural heritage and Indigenous knowledge systems have long responded to a changing climate.[26] Indigenous modes of relating to the land and other than human persons differ from the dominant settler colonial practices and have important ecological and political ramifications.

They are also serving as a model and inspiration for rethinking kinship relations within non-Indigenous communities, and it is this trend which is driving global environmental justice movements in recent years to take Earth-centered political ideas.

But, as Katie O'Bryan notes, care must be taken in how such knowledge is treated. Stripping Indigenous ecological practices from their biocultural context can unintentionally reinforce settler colonialism and feed into new forms of cultural appropriation.[27] Despite such caveats, it is significant that Indigenous and non-Indigenous allies are increasingly recognizing kinship with the Earth and putting these ideas into action. As Indigenous scholar, Nick Estes, (Lower Brule Sioux Tribe), argues in his book on resistance at Standing Rock, "Beyond the Dakota Access Pipeline, a growing international movement is fighting the expanding network of pipelines across North America…The appearance of each new flashpoint of struggle indicates a growing anti-colonial resistance, led by Indigenous peoples against settler colonialism and extractive capitalism."[28] Naomi Klein called this resistance to disaster capitalism "Blockadia." "Blockadia is not a specific location on a map but rather a roving transnational conflict zone that is cropping up with increasing frequency and intensity wherever extractive projects are attempting to dig and drill, whether for open-pit mines, or gas fracking, or tar sands oil pipelines."[29] I would argue this social resistance has expanded beyond just extractive projects, as the current surge in protests over racial justice and police killings of people of color make clear—young people are fed up and are increasingly willing to put their bodies on the line.

Despite these encouraging trends, the idea of kinship with other than human beings remains a marginal political view. This is a long-standing trend

---

[25] Toledo (2013), Berkes (2017), Begotti and Peres (2020).
[26] Berkes et al. (2000), IIED (2016).
[27] O'Bryan (2004).
[28] Estes (2019, 253).
[29] Klein (2014, 294).

among countries rooted in the Abrahamic religion which Lynn White Jr. famously called attention to when he pointed the finger at the Biblical doctrine of dominion as the root of environmental destruction.[30] There have been strides in the years since White's essay to reinterpret Abrahamic theologies through a green lens, from the "greening of religion" debates to celebrations of Pope Francis' *Laudato Si'* papal encyclical.[31] Despite such efforts, the religious injunction for humans to subdue and rule over nature remains the dominant political theology today. In response, many young people today are either looking outside their native religious traditions for alternative approaches or else they are creating their own hybrid, syncretic religious practices that are grounded in Earth-centered beliefs more open to a pluralistic cosmopolitics of justice not limited solely to humans.[32]

Another example of this can be seen in the popular outdoor survival book series written by naturalist and tracker Tom Brown. "Very quickly, while gathering wild edibles or picking an animal from a trap, you realize that you are as intimately connected to the web of life as the snake and the spider...If you can open your heart to the wisdom of the universe, you will come to know that man, animals, trees, rocks, rivers and skies all speak a common tongue. Sensing this, you cannot help but care more for the earth and all its creatures."[33] This growing interest in kinship with nature finds expression in what Bron Taylor calls, Dark Green Religion, as well as through renewed interest in New Animism, Paganism and related forms of Earth spirituality.[34] Such views are becoming more common in both environmental and climate justice movements.

Efforts to expand the boundaries of kinship, and by extension, to rethink the boundaries of justice, are about more than just recognizing the value of Indigenous perspective. They are also about respecting Indigenous culture and sovereignty. The push for formal recognition of Indigenous rights through international frameworks such as the International Labour Organizations' Indigenous and Tribal Peoples Convention (No. 169) and the United Nations Declaration on the Rights of Indigenous Peoples (UNDRIP) are part of the larger movement entangled with an Earth-centered paradigm shift that I argue is currently underway.

For example, Article 24 of the UNDRIP states "Indigenous peoples have the right to their traditional medicines and to maintain their health practices, including the conservation of their vital medicinal plants, animals and minerals."[35] Similarly, Article 25 notes that "Indigenous peoples have the

---

[30] White Jr. (1967).
[31] Taylor (2016), Tucker and Grim (2001).
[32] Taylor (2009, 2013), Bauman (2014), Pike (2017).
[33] Brown (1983, 10).
[34] Taylor (2009, 2013), Harvey (2015), Adler (2006), Starhawk (2005).
[35] UNDRIP (2008).

right to maintain and strengthen their distinctive spiritual relationship with their traditionally owned or otherwise occupied and used lands, territories, waters and coastal seas and other resources and to uphold their responsibilities to future generations in this regard."[36] Article 31 also asserts "Indigenous peoples have the right to maintain, control, protect and develop their cultural heritage, traditional knowledge and traditional cultural expressions, as well as the manifestations of their sciences, technologies and cultures, including human and genetic resources, seeds, medicines, knowledge of the properties of fauna and flora."[37] While neither ILO 169 nor the UNDRIP explicitly use the term "kinship", both call attention to and recognize the spiritual and material ties between Indigenous culture and ecological knowledge.

Such practices are a manifestation of the kincentric philosophies I have been discussing and are important (albeit imperfect) tools in the fight to expand recognition of Indigenous cultural and ecological kinship practices and to argue for the rights for nature based on these Indigenous worldviews (as the examples from Ecuador and Bolivia discussed below highlight). These legal frameworks are also important tools used by some biocultural rights advocates to show how arguments about nature and culture are interconnected and must be thought together.[38] Underpinning such issues is the question of who can make legitimate justice claims, which remains a long-running philosophical debate.[39] In recent years, this debate has increasingly turned to questions of multispecies or interspecies justice.[40] David Schlosberg and Danielle Celermajer, for example, have played active roles in advancing such discussions at the Sydney Environmental Institute and through the Multispecies Justice project. The SEI website described this research as challenging scholars to "reconceptualise justice in a way that is sufficiently capacious and fluid to accommodate the vast breadth of our multispecies world. This requires our imagining and including modes of representation and other political practices equipped to appreciate and accommodate the justice claims of all ecological beings – individuals, systems, and their relations."[41]

Like these earlier efforts, the Rights of Nature movement today is trying to rethink the boundary of the political, and by extension, questions of rights and justice. In doing so these debates build on earlier work by scholars and activists.[42] Some environmental scholars also point to early ecological thinkers

[36] UNDRIP (2008).

[37] UNDRIP (2008).

[38] Chen and Gilmore (2015).

[39] Singer (1975), Regan (1983), Nussbaum and Sunstein (2004).

[40] Livingston and Puar (2011), Pouliot and Ryan (2013), Weissenberg and Schlosberg (2014), Cochrane (2018), Meijer (2019), Blattner et al. (2020), Weaver (2020), Youatt (2020).

[41] Sydney Environmental Institute (2018).

[42] Stone (1972), Nash (1989), Burdon (2011, 2014), Boyd (2017), Celermajer et al. (2020).

like George Perkins Marsh, Aldo Leopold, Rachel Carson and John Muir as providing some of the intellectual underpinnings for an expanded view of ecological rights, with many genealogies citing Christopher Stone's 1972 reflections on legal rights for trees as an important early examples. While there continued to be calls for greater environmental rights in the intervening years following Stone's article, the specific idea of the Rights of Nature didn't catch on until the early 2000s. The growth of interest in this area of research is reflected in the growing body of allied concepts that have emerged since 2000, which now includes concepts such as Wild Law, Earth Rights and Earth Jurisprudence.

Global Rights of Nature Movement

Activists point to two early events as pivotal to the burgeoning Rights of Nature movement, both of which were grounded in community rights issues and movements for local democracy. The first was a fight over toxic waste (sludge and fly ash) in the rural Pennsylvania town of Tamaqua in Schuylkill County. After continued refusal by local and state regulatory agencies to address community health concerns about toxic waste, local residents, who had earlier formed "The Army For a Clean Environment" (ACE), took matters into their own hands and began organizing to stop more sludge dumping near their town.

In 2006, with the help of lawyers from the Community Environmental Legal Defense Fund (CELDF), Tamaqua passed the first-ever local ordinance (No. 612) granting legal protections to natural ecosystems and declaring local community rights superseded corporate profits and the supposed right to pollute.[43] The ordinance sought to "protect the health safety and general welfare of the citizens and environment of Tamaqua Borough by banning corporations from engaging in the land application of sewage sludge, by banning persons from using corporations to engage in land application of sewage sludge, by providing for the testing of sewage sludge prior to land application in the borough, by removing constitutional powers of corporations within the borough, by recognizing and enforcing the rights of residents to defend natural communities ecosystems."[44]

As one reporter noted, "At first glance it seems counterintuitive that such a seemingly radical approach to ecological conservation has sprung up in small-town PA - initiated and defended by voters who famously 'cling to guns or religion' - but perhaps it shouldn't. Religious groups in the United States and around the world have steadily adopted pro-environment positions, declaring a moral imperative to protect the environment and urging swift popular and legislative action."[45] The trend that Beale calls attention to here is part of a

[43] CELDF (2015).
[44] Borough of Tamaqua (2006).
[45] Beale (2009).

broader "greening of religion" movement and highlights the ways in which Community Rights is interwoven with the Rights of Nature in some areas. This early win in Pennsylvania emboldened community activists to attempt similar efforts elsewhere.

A second important origin point for the Rights of Nature movement took place in Ecuador a few years later, in 2008, when the country incorporated the rights of nature into their revised constitution. Mobilizations for Indigenous political rights and democratic inclusions, as well as popular support for the Andean Earth being known locally as Pachamama, were part of a resurgent Andean political landscape starting in the 1990s.[46] Here again CELDF was an important political player involved in legal efforts in Ecuador, where they offered insights from the successful legal efforts in Tamaqua, Pennsylvania in 2006. As Paulo Tavares noted about the Ecuadorian case, "Informed by the philosophy, politics, and ethics of the indigenous movement, the formulation of the Rights of Nature was also the product of debates that surged from within the field of modern environmental sciences, activism, and advocacy…Where James Lovelock's Gaia meets Pachamama, at the confluence of indigenous knowledge, modern environmental activism, and ecological/climate sciences, the politics of the Rights of Nature were gradually forged in Ecuador."[47] The case of Ecuador is more complicated than I can go into here, but the inclusion of the Rights of Nature into a national constitution made immediate shockwaves around the world.

Since then, interest in the Rights of Nature has become significant enough to even work its way into the pages of *Science*, usually a bastion of mainstream scientific thought. Now even scientists have taken up debates about Indigenous Earth goddesses and ecological kinship. As one article in the magazine noted, "Rights-of-nature thinking frequently blends Western rights concepts with nonWestern spirituality, sometimes as a means to remedy a previous usurpation of nature from another people's use. For example, New Zealand's recognition of the Whanganui River and surrounding area as the legal person, Te Awa Tupua, arose out of a treaty settlement with a Maori tribe and that tribe's spiritual connection to the river. Similarly, the Ecuadorian constitution recognizes the rights of Pacha Mama, an indigenous earth goddess."[48] As the movement grew it continued to build on these initial legal and constitutional development, which have now expanded to include Ecuador, Bolivia, Colombia, Mexico, New Zealand, Australia, India, Bangladesh, Uganda and the United States.[49] Movements in these countries have taken steps to grant legal rights to nature, although with varying degrees of success and at different political levels. Social movements in other countries are working to raise these

---

[46] Thomson (2002), Becker (2011).
[47] Tavares (2015).
[48] Chapron et al. (2019, 1392).
[49] Boyd (2017), Lim (2019).

issues, including Nepal and Ireland, and we are likely to see more examples in future as public support grows.

As the work of advocacy groups, like CELDF, makes clear, one entry point into the Rights of Nature movement is through the Community Rights movement and local battles that pit community decision-making and local democracy against outside corporate powers and state regulators. This example is especially clear when one looks at the interplay between, for example, the Daniel Pennock Democracy School organized by CELDF. At least one community member from Tamaqua attended the Democracy School training prior to the local ordinance being drafted in Pennsylvania, and there continues to be strong links between the work of CELDF on the Rights of Nature and a growing Community Rights movement.[50]

In the United States, resistance to the Rights of Nature movement has been strong among free market evangelicals (both figuratively and literally) who see any form of environmental rights as a threat to the sacrality of market capitalism. Some critics have even suggested such ideas are a threat to the very foundations of modern civilization.[51] But as the authors of the *Science* article quoted earlier noted, "The granting of legal rights to nonhumans is not in itself revolutionary or even unusual...Corporations, trade unions, and states are all nonhuman entities that have rights and duties under the law...The legal system has no difficulty adjudicating nonhuman rights."[52] As is often the case, the real issue boils down to who holds power.

At its core, I would argue the global Rights of Nature movement is a struggle to give birth to new ways of relating to the natural world, relations which are understood as in direct opposition to an instrumentalist view of nature. Craig Kauffman and Pamela Martin highlighted these connections in the development of Rights of Nature cases in the United States, Ecuador and New Zealand. In the case of the Whanganui River in New Zealand, efforts to secure legal rights for the river were about more than just protecting the river ecology. They also included taking seriously the idea of responsibility for the Whanganui, which the Māori *iwi* understand as being grounded in obligations of care and kinship.

> In both cases, the legal personhood language is expressly intended to reflect the Māori *iwi's* view that their respective ecosystems are living entities with intrinsic value that are incapable of being "owned" in an absolute sense and to enable them to have legal standing in their own right...Guardians charged with protecting not just legal rights but also spiritual and cultural rights is a unique feature of New Zealand's institutionalization of RoN. This relative focus on responsibility rather than rights reflects the inseparability of Māori people

---

[50] CELDF (2015).
[51] Vignelli (2019).
[52] Chapron et al. (2019, 1392).

and their respective ecosystems and the consequent responsibilities for taking care of ecosystems as kin, expressed in the concept of *rangatiratanga*.[53]

Expressions of kinship with water are evident in many cases. Consider the community response to destruction caused by the 2015 collapse of the Samarco iron ore tailings dam in Minas Gerais, Brazil. In discussing Brazilian responses to the event, restorative justice scholar Dominic Barter notes "There was such outrage at what these companies had allowed to happen. Local footage put shocking images of the destruction on everyone's screens. This disaster was not described simply by the number of deaths, or by the number of houses that were destroyed, it was defined as being 'the death of a river', of a being that was experienced as kin."[54]

Or consider the argument made by the Uttarakhand High Court in India in its 2017 ruling granting legal personhood status to the Ganga and Yamuna Rivers in which the court noted the sacred ties which bind Hindus to the Ganga River.[55] The Uttarakhand case drew on these explicit links between ecology and the sacred in India when the judges ruled that "the Rivers Ganga and Yamuna, all their tributaries, streams, every natural water flowing with flow continuously or intermittently of these rivers, are declared as juristic/legal persons/living entities having the status of a legal person with all corresponding rights, duties and liabilities of a living person in order to preserve and conserve river Ganga and Yamuna."[56] The Supreme Court later overturned the lower court ruling by arguing the Uttarakhand High Court had exceeded its authority to legislate across state borders, but not because of an argument in favor of the rights of the rivers.[57] Anthropologist, Georgina Drew, in her research on dams, ecological politics and the sacred in India, argues that "activists wanted to defend the integrity of a landscape that holds cultural and religious significance for its prominent role in Hindu understandings of India's sacred geography. Many of those involved in the [*Ganga Ahvaan*] campaigns to stop the dams were motivated by a notion that it was their religious duty (sometimes referred to as their *dharma*) to work on behalf of the Ganga and the Garhwal Himalaya."[58]

These examples help us see how the Rights of Nature movement has grown and continues to grow. Another recent example worth noting is the effort being led by the Earth Law Center and International Rivers to develop a Universal Declaration of the Rights of Rivers.

Several sections highlight how these expanded notions of kinship inform the Declaration.

[53] Kauffman and Martin (2018, 58).
[54] Biffi and Pali (2019, 55).
[55] Express News Service (2017).
[56] Mohd. Salim v. State of Uttarakhand & Others (2017).
[57] Times of India (2017).
[58] Drew (2017, 13).

Aware that all people, including indigenous communities and other local communities of all spiritual faiths, have long held through their traditions, religions, customs, and laws that nature (often called "Mother Earth") is a rights-bearing entity, and that rivers in particular are sacred entities possessing their own fundamental rights...

Convinced that recognizing the rights of nature, and in particular recognizing those river rights contained in this Declaration, will foster the creation of a new legal and social paradigm based on living in harmony with nature and respecting both the rights of nature and human rights, particularly with reference to the urgent needs of indigenous communities and the ecosystems they have long protected.[59]

The emphasis on water as sacred and the future vision of a "new legal and social paradigm" are examples of this Earth-centered paradigm shift mentioned earlier. Securing support for the legal rights to nature remains an uphill battle, but the fact that these examples can all be pointed to is illustrative of just how far we have come in the decades since scholars first raised the issue of granting legal rights to natural entities. While recognizing the limits of these movements, it is also important to celebrate the progress made.

Harmony with Nature at the UN

Another example of how ecological kinship and relationality have begun to shape environmental justice movements over the past decade can be seen in the idea of "harmony with nature" being promoted at the United Nations. Official UN support for this idea began in 2009 and was led by the Plurinational State of Bolivia under then-president Evo Morales. UN Resolution 63/278, which established April 22 as International Mother Earth Day, was adopted by the UN General Assembly that same year. The Resolution states that "Mother Earth is a common expression for the planet earth in a number of countries and regions, which reflects the interdependence that exists among human beings, other living species and the planet we all inhabit."[60] Morales offered the following remarks in his address:

> But 60 years ago [human rights] were recognized, and now we are convinced that, with today's declaration of International Mother Earth Day, planet Earth, Mother Earth, has also won her rights. And this new century, the twenty-first, should be the century of Mother Earth's rights. The decision we have taken today, under the leadership of the President of the General Assembly, is important. It is a singular, historic and unprecedented event for humankind, and those who have decided to support this great initiative will go down in history, the new history of awakening humankind.

---

[59] Rights of Rivers (2020).
[60] United Nations (2009).

That is why I reiterate once again that the twenty-first century is the century of the rights of Mother Earth and of all living beings. If we are to live in harmony with nature, we need to recognize that not only we human beings have rights, but that the planet does as well. Animals, plants and all living beings have rights that we must respect. What is happening to us now with climate change is the direct result of not respecting the rights of Mother Earth. The United Nations must ensure that the rights of Mother Earth and of all living beings are respected...

But I would also like to recall that in Bolivia, in particular among the local indigenous peoples that I am familiar with, Mother Earth is sacred to life. That is why we undertake sacred rituals of tribute to our rivers, forests, lakes and animals and have many tunes to express our respect for Mother Earth. I am convinced that Mother Earth is more important than humankind, and therefore the rights of Mother Earth are as important as those of any human being.[61]

Near the end of Morales' speech, he pointed out that for too long "human beings have been prisoners of development capitalism, which states that man is the sole owner of the planet," echoing Lynn White's critique, but with the finger pointing to capitalism rather than religion (although as White and others noted, it is impossible to disentangle the rise of Western industrial capitalism from its Protestant theological roots).[62] These Bolivian-led efforts continued the following year with the drafting of a "Universal Declaration of Rights of Mother Earth" during the World People's Conference on Climate Change and the Rights of Mother Earth in Cochabamba, Bolivia. This Declaration went even further in articulating an Earth-centered politics by arguing that "we are all part of Mother Earth, an indivisible, living community of interrelated and interdependent beings with a common destiny."[63] Efforts have been ongoing since 2010 to build support for its adoption. Organizers hope that one day UN member states will ratify the Declaration, giving it the same kind of customary legal weight that the Universal Declaration of Human Rights enjoys today.

As the UN's Harmony with Nature website notes, the "Interactive Dialogues of the General Assembly on Harmony with Nature to commemorate International Mother Earth Day have brought to the forefront the need to move away from a human-centered worldview - or 'anthropocentrism' - and establish a non-anthropocentric, or Earth-centered, relationship with the planet. Under this new paradigm, we recognize Nature as an equal partner with humankind."[64] A 2016 UN experts' report on Earth Jurisprudence, commissioned in 2015 and building on previous interactive and virtual dialogs under the UN framework of Harmony with Nature, highlighted this paradigm shift as key to the global movement to promote Earth-centered politics and the

---

[61] General Assembly (2009).

[62] General Assembly (2009).

[63] World People's Conference on Climate Change and the Rights of Mother Earth (2010).

[64] Harmony with Nature (ND).

Rights of Nature. "In the Earth-centred worldview, the planet is not considered to be an inanimate object to be exploited, but as our common home, alive and subject to a plethora of dangers to its health: this process requires a serious reconsideration of our interaction with nature as well as support for Earth jurisprudence in laws, ethics, institutions, policies and practices, including a fundamental respect and reverence for the Earth and its natural cycles."[65]

Expressions of ecological kinship were also evident in the closing remarks by the Moderator of the Interactive Dialogue Alessandro Pelizzon. A scholar of comparative law and legal anthropology, Pelizzon helped co-founded the Global Alliance for the Rights of Nature, one of the networks working to grow this movement. His remarks highlight the expansive views of kinship in Australian Aboriginal culture and the role of ancestors in kin-making practices.

> I would like to conclude this Ninth interactive Dialogue on the Harmony with Nature with a practice with which we, in Australia, begin all public gathering. That is, I would like to conclude by acknowledging the Country we inhabit today. In so doing, I want to pay respect to the Australian Aboriginal concept of Country, which transcends any mere geopolitical entity. As Professor Deborah Rose once suggested, Country is the nexus of identity and interconnectedness. As my mentors, Uncle Charles Moran and Aunty Barbara Nicholson, taught me, Country is the total sum of all the beings and phenomena with which we share this space today. Country is the total sum of all relationships with and among such beings and phenomena. And Country is the total sum of the stories within which these relationships are described and contained. With this in mind, I would like to pay my respects to the owners and custodians of such stories, the Elders who are, the ones who were, and the ones who are yet to be. And since Country is not just a place in space, but also a moment in time, I would like to acknowledge all of you for having shared this day.[66]

The Tenth Interactive Dialogue was originally scheduled for 22 April 2020 but was canceled due to coronavirus pandemic. A concept note circulated in advance of the meeting highlighted the importance of promoting Earth Jurisprudence, which was the thematic focus for the 2020 meeting. "The last decade has seen the growing use of formal and informal education curricula and the emergence of innovative legal developments to bring the underlying principles of Earth Jurisprudence to diverse public audiences. Earth Jurisprudence is the philosophical anchor for practical and multi-disciplinary approaches to creating change for living in harmony with Nature. From economics, to natural sciences, law and the arts, Earth Jurisprudence invites deep transformation from a human-centered to an Earth-centered paradigm by connecting our rational concern for biodiversity loss with responsibility for protecting our living planet."[67] In April of 2022, the Eleventh Interactive

---

[65] Harmony with Nature (2016).
[66] Pelizzon (2019, 1).
[67] Harmony with Nature (2020).

Dialogue was finally held, marking a return to the UN stage after the initial delay from COVID-19. The 2022 concept note called special attention to, "The values advanced by ecological economics and Earth-centered law, such as fairness, equity, justice, cooperation, dialog, inclusion, comprehension, agreement, respect, and mutual inspiration complement each other in the journey to move beyond the Anthropocene epoch.

Earth-centered approaches calls on humanity to listen to, take inspiration from, and care for the Planet that sustains us, and recognize that Nature is, and should be, a source of law, ethics and how we govern ourselves."[68]

As these examples highlight, the global Rights of Nature movement is an important part of the emerging environmental justice landscape, driven by growing support for the idea of ecological kinship with the Earth and continuing worries about the future of the planet. The fact that we've officially passed the ten-year mark on the Interactive Dialogues on Harmony with Nature speaks to their important role as a political space for further engagements to emerge. And while these are encouraging trends, I also want to be careful to not overemphasize the political power of these Interactive Dialogues and other UN efforts. From a geopolitical view, the impact of these Harmony with Nature efforts on state politics remains minimal—there are no real political risks to UN member states who designate April 22 International Mother Earth Day.

There is important symbolic power in such gestures, to be sure, and I don't want to downplay the importance of the symbolic realm for politics, but business as usual politics remains unchanged.

But what does this movement look like in a grounded context where the real political battles are being fought? For that story I turn to the case of the Lake Erie Bill of Rights. As I will show, the story of local citizen efforts in Toledo, Ohio, to protect their ecosystem through a Rights of Nature legal effort is illustrative of the kinds of political struggles taking place in diverse communities around the world and offers important insights into this global movement.

Lake Erie Bill of Rights (LEBOR)

Like many parts of the United States, the state of Ohio is no stranger to environmental justice issues. In the case of Toledo, the story begins with algal bloom on Lake Erie, one of the five Great Lakes that forms the northern border of Ohio. Harmful algal blooms, which are caused by excessive nutrients in the water (mainly nitrogen and phosphorous), have been a consistent problem for Ohioans living along Lake Erie and many other waterways. Starting in the 1990s, public health warnings about dangerous algal blooms on Lake Erie became a regular occurrence.

---

[68] Harmony with Nature (2022).

For residents of the city of Toledo, located in the state's northwestern farm region and borders Lake Erie, things came to a head in August of 2014 when the city's water supply, which was drawn from an intake station just off the Lake Erie shore, was shut down for several days due to the danger of these toxic algal blooms. A report noted "high microcystin concentrations were detected in drinking water from the lake. As a result, the water supply to 400,000 people in Toledo, Ohio, was shut down."[69] A three-day drinking water ban followed. The water was so toxic from the algal blooms that not even bathing was safe. The sudden shutdown sparked panic across the city as people rushed to find potable water. Local news coverage featured pictures of long lines of people outside stores looking for bottled water, juxtaposed with images of dead fish (and some unlucky pets) that had died from contact with the toxic water. In response, Republican governor John Kasich declared a state of emergency. After water was finally restored and the immediate panic subsided, the questions began. How had this happened? Who was responsible? What could residents do to prevent this from happening again?

This was not a new issue for environmental justice advocates in Ohio. Concentrated animal feeding operations, (CAFO), commonly referred to as "Big Ag" in organizing circles, are a leading source of water pollution in the United States, and a leading cause of algal blooms. To make matters worse, Ohio experienced industrial animal operations boom a decade prior, and many of these were unregulated operations in the areas around Toledo. As one Environmental Working Group (EWG) report noted, between 2005 and 2018 factory farms increased from 545 to 775 in Ohio (a 42% increase), with the number of animals in the watershed more than doubling (from 9 to 20.4 million), leading to an increase in manure use on farm fields from 3.9 to 5.5 million tons per year.[70] The Maumee River, where much of the northwest agricultural land drains into, flows past Toledo and directly into southwestern Lake Erie.

I spoke with one of the local organizers who became central to the Toledo LEBOR story, Markie Miller, about how she got involved with the local movement. According to Miller, residents and students in Bowling Green, a liberal arts college town 30 minutes south of Toledo, had been involved in a fight to stop construction of the Nexus gas pipeline slated to run across city land and near the city's water treatment facility. In response, Bowling Green students had organized a public event in the spring of 2015 with CELDF and other community rights advocates from Ohio who were involved in similar fights against gas pipelines and fracking, and Miller had attended. As she would recall about the origins of the Toledo Rights of Nature efforts, "LEBOR was born on a cocktail napkin" following the Bowling Green event.[71]

---

[69] National Science Foundation (2019).
[70] Environmental Working Group (2019).
[71] Miller (2020).

Community efforts to protect local rivers and Lake Erie had proved futile, ending in a mix of state inaction and voluntary self-monitoring by farmers. But the algal bloom scares in 2014 changed things. A coalition of local environmental groups began to organize around the idea of drafting an amendment to the city charter to gain greater control over industrial and agricultural pollution impacting the city and Lake Erie. As support for a bill of rights to protect Lake Erie gained traction, Miller and local activists began working with Advocates for a Clean Lake Erie (ACLE) and Toledoans for Safe Water (TSW), two groups already focused on agricultural pollution and water issues in Toledo.

By 2018 local organizing had created a citywide campaign to place an amendment to the Municipal Charter of the City of Toledo on the ballot that would grant legal rights to nature and allow city residents to sue polluters on behalf of the Lake Erie watershed. The proposed law read: "*We the people of the City of Toledo* find that laws ostensibly enacted to protect us, and to foster our health, prosperity, and fundamental rights do neither; and that the very air, land, and water – on which our lives and happiness depend – are threatened. Thus it has become necessary that we reclaim, reaffirm, and assert our inherent and inalienable rights, and to extend legal rights to our natural environment in order to ensure that the natural world, along with our values, our interest, and our rights, are no longer subordinated to the accumulation of surplus wealth and unaccountable political power."[72] As support for LEBOR grew, so did its organized opposition.

OPPOSITION TO LEBOR

Organized industry opposition to LEBOR soon went into high gear. Two shadowy political action committees (PACs) quickly emerged to try and undermine the ballot drive, Toledo Jobs and Growth Coalition and Real Solutions for Lake Erie. Campaign finance records would later reveal these two PACs were behind a well-financed PR campaign spearheaded by Strategic Public Partners, an Ohio-based lobbying firm with deep ties to the Ohio Republican Party. The bulk of funding for the anti-LEBOR campaign, which included radio and TV ads, targeted mailings and robocalls, came from a $302,000 wire transfer from BP Corp. North America, a US subsidiary of the global oil and gas giant British Petroleum.[73]

One political ad from Real Solutions for Lake Erie featured a mother with two small children, one of whom had a mix of fear and contempt on her face, alongside the following text: "THE LAKE ERIE BILL OF RIGHTS IS PUTTING US AT RISK!" It also claimed that LEBOR was "drafted by a group of out-of-state extremists who are: anti-development, anti- government, [and] trying to shut down family farms and family businesses."[74] Responding

---

[72] Lake Erie Bill of Rights (2019).

[73] Henry (2019).

[74] Real Solutions for Lake Erie (2018).

to revelations of outside political interference, LEBOR organizer Julian Mack told reporters "It felt like we had a noble cause and were fighting for what is right for our community...I'm glad the voters of Toledo recognized a lie when they saw it and that they weren't swayed by money and politics this time."[75] Some of these anti-environmental lobbying networks have since been dragged into the political spotlight following the July 2020 arrest of Ohio House Speaker and longtime Republican anti-environmental advocate Larry Householder. Householder was indicted as part of a massive political corruption scandals involving $60 million in bribes to ensure the passage of House Bill 6, a 2018 bill that provided a $1 billion taxpayer bailout for two failing nuclear power plants owned by FirstEnergy. At a press conference following the arrest, US Attorney David DeVillers called the arrest of Householder and co-conspirators "the largest bribery and money laundering scheme ever perpetrated against the people of the state of Ohio."[76]

In addition to the LEBOR ballot disinformation campaign, legal efforts were also being put in motion behind the scenes to challenge LEBOR should it succeed. The first legal challenge came in 2018 and attempted to stop LEBOR from even being placed on the ballot in the 2019 election, and the case went all the way up to the Ohio Supreme Court. The original petitioner, Josh Abernathy, had ties to the Allied Construction Trade union which represents the building industry in Ohio. He was joined by the State of Ohio as co-plaintiff, and *Amicus* briefs were filed in support by the Ohio Farm Bureau Federation, Ohio Corn & Wheat Growers Association, Ohio Pork Council, Ohio Soybean Association and the Ohio Dairy Producers Association (Supreme Court of Ohio 2018). The petitioners argued in their *Amicus* brief to the court that:

> The LEBOR could have a deleterious effect on the farming way of life, not only within the city of Toledo but also outside its limits. Provisions within the LEBOR would permit any Toledo resident to litigate frivolous claims against farmers in and around the Western Lake Erie Basin, include [sic] many members of amici. The "rights" granted by the LEBOR would open farmers up to significant legal expenses and de facto regulation-by- litigation as each farmer sought to avoid expensive-but-frivolous lawsuits. The LEBOR's provisions would extensively limit agricultural production, perhaps rendering agriculture practically impossible, because each farmer would have to reserve funds for challenges to secure their rights under licenses and permits issued by the state and federal governments.[77]

Two claims are noteworthy here.

First is the assumption that any legal challenge to business as usual agricultural practices would be frivolous. This claim is a bit odd since these same

---

[75] Henry (2019).

[76] Rosenberg and Evans (2020).

[77] Supreme Court of Ohio (2018, 6).

groups went to great length in their briefs to point out how concerned they were with protecting Ohio's natural resources. As the Ohio Corn & Wheat Growers Association stated, its "members take care in being stewards of arable land used for crop production. Many members are from families who have been raising crops in Ohio for multiple generations."[78] The Ohio Soybean Association similarly noted "Ohio's soybean farmers are committed to sustaining life while respecting the environment.

Water quality is a high priority for our organization and all Ohio farmers, who work hard every day to protect the soil and water, while also growing safe, nutritious food for our families and communities."[79] So both parties claimed to be concerned about protecting local livelihoods and Lake Erie, they just differed on how to best do that.

Advocates of LEBOR saw the Rights of Nature as the best strategy to achieve that goal. Opponents of LEBOR felt voluntary efforts by local farmers and industry was the best solution. That point was driven home by industry advocates in their brief when they highlighted how their members are actively engaged in supporting "common sense" community governance policies. "Farm Bureau members pride themselves on being educated voters and involved citizens who work together to find solutions to their communities' problems...Ohio Pork Council members pride themselves not only as active community members, but educated voters willing to work with interested parties in order to find common sense solutions to complex problems."[80]

While it may not have been intended as such, a second claim in the *Amicus* brief exposed a more plausible economic motivation for opposing LEBOR. "LEBOR's provisions would extensively limit agricultural production, perhaps rendering agriculture practically impossible."[81] Yet there is no reason that agricultural practices must harm the environment. As Nicole Rasul noted in her coverage of community responses to LEBOR, many farmers were aware that poor land management was a major source of water pollution in Ohio. "Joe Logan, a local farmer and president of the Ohio Farmers Union, is quick to acknowledge that Lake Erie's pollution problem is fueled by agricultural practices in the region. But he doesn't interpret the new measure to mean that farmers can't fertilize their fields. He tells producers who feel threatened by the Bill of Rights that their livelihoods are not in jeopardy if they aren't over-fertilizing their fields or applying manure haphazardly."[82]

Despite the concerted media campaign against LEBOR, and multiple legal challenges to prevent its appearance on the special election ballot, the Lake Erie Bill of Rights initiative was finally placed before voters in a special election on 26 February 2019. To the delight of supporters (and dismay of opponents),

---

[78] Supreme Court of Ohio (2018, 6).
[79] Supreme Court of Ohio (2018, 6).
[80] Supreme Court of Ohio (2018, 6).
[81] Supreme Court of Ohio (2018, 6).
[82] Rasul (2019).

LEBOR was approved by two-thirds of voters. As one commentator noted, LEBOR finally passed after "surmounting multiple challenges by the county board of elections and a cabal of corporate lobbying groups to keep it off the ballot and then funding an unsuccessful $320,000 effort to defeat it. Lobby groups included the Affiliated Construction Trades unions, the Ohio Chamber of Commerce, American Petroleum Institute, Ohio Oil and Gas Assoc., the Farm Bureau as well as hog, poultry and dairy factory lobbyists.

British Petroleum, N.A. Inc., contributed the lion's share of the 'Vote No' campaign."[83] The next day, Drewes Farms Partnership, a corporate farm near Toledo that identifies itself as a "fifth generation family farm," filed an appeal and a request for immediate injunctive relief. A local court granted the motion, once more sending LEBOR back to the courts.

It is worth noting that plaintiff Mark Drewes, who owns Drewes Farms, had previously opposed efforts by the Wood County Commissioners (which includes the southern tip of the Toledo metro area) to declare Lake Erie "impaired" by excessive nitrogen and phosphorus runoff from agricultural farms and CAFOs in the Lake Erie watershed. In a meeting in August 2016, Drewes argued that listing Lake Erie as impaired would trigger additional scrutiny of the sources of lake pollution, and "simple farmers" like him would be burdened with unnecessary paperwork and economic costs. It would be better, he argued, if agencies like the EPA stayed out of their business and left farmers to solve problems on their own without imposing more "Draconian" measures. "Impaired status will push agriculture to its breaking point," Drewes told the Commissioners at an August 2016 meeting "We are regulated beyond belief…Self-imposed controls are going to be more effective [and the] EPA scares the bejeebers out of me."[84] Given his close ties to Big Ag in the area, including his active involvement with the Ohio Corn Growers Board, Drewes was the perfect LEBOR protagonist: a hard-working, all-American Ohio family farmer who was trying to protect his farm livelihood from government overreach and environmental extremists. The Director of communications for the Ohio Farm Bureau, Joe Cornely, echoed a sentiment common among LEBOR opponents. "It's one of those things if it wasn't so serious it would be laughable."[85]

Kenneth Kilbert, a law professor at the University of Toledo and Director of the Legal Institute of the Great Lakes, noted in an opinion piece soon after LEBOR was passed that the law was unlikely to withstand court challenge, but the vote was nonetheless important because it "signals that the people of Toledo-in the immortal words of the Howard Beale character in the classic film Network-are mad as hell and they're not going to take it anymore…Last month's vote for LEBOR signals that the people of Toledo are tired of waiting for their state government to take action, so they are trying to take matters

---

[83] Ferner (2019).
[84] McLaughlin (2016).
[85] Begemann (2019).

into their own hands."[86] As both Miller and Kilbert noted at different times, the Lake Erie Bill of Rights was not just about protecting Lake Erie and its residents. It was also about a local community trying to assert its democratic right to self-governance, rather than having it dictated by outside corporate interests.

There was one final legal insult supporters of LEBOR would have to face. A May 2019 motion by Toledoans for Safe Water to intervene as co-defendants with the City of Toledo in the case was denied, but the same petition by the State of Ohio to join the Drewes Farm Partnership as co-plaintiffs was granted. Suddenly, the very group that had put years of work into drafting and advocating for the new law found themselves shut out of the upcoming legal battle to decide its fate. Hearings in the U.S. District Court began 28 January 2020, and members of Toledoans for Safe Water were forced to sit in the back of the courtroom as lawyers bantered back and forth about the group's motives in drafting the bill. "To be literally shut out is really frustrating," said Miller in response to the judge's decision. Local advocates were basically being told to "wait and be patient while the adults have a discussion."[87]

One moment from the legal cross-examination is worth highlighting as it relates to an earlier point about the challenges of implementing these Rights of Nature frameworks in our current legal environment. The legal ambiguity of the term "nature" was raised by the lawyers representing Drewes Farms when discussing the problems of implementing section 1(a) of LEBOR, which outlined the rights of the Lake Erie ecosystem. Section 1(a) of LEBOR read: "Lake Erie, and the Lake Erie watershed, possess the right to exist, flourish, and naturally evolve. The Lake Erie Ecosystem shall include all natural water features, communities of organisms, soil as well as terrestrial and aquatic sub ecosystems that are part of Lake Erie and its watershed."[88] As the lawyer for Drewes Farm argued, this ambiguity was a fatal flaw:

> Then you get, Your Honor, to the actual — to the rights that LEBOR in Sect. 1A purports to create, the rights to exist, flourish, and naturally evolve. None of those phrases, terms, are defined anywhere in LEBOR. What does the right to exist mean for soil? What does it mean for communities of organisms? Does it mean a ban on commercial fishing for example? Does it impose strict veganism in Lake Erie watershed? It's undefined what it means to the right to exist. What does it mean for mosquitoes or other pests, a community of mosquitoes what does it mean? What does the right to flourish mean for soil? For plants, does it mean no tilling, no manipulating the soil? Is it a ban on fertilizers period, or is it the opposite because phosphorus can improve the quality of soil. What does it mean for worms? Can you plant, or are you supposed to just let wild seed go wherever? Can you harvest the plants? Can you mow? Can you have a golf course? What does any of that mean as far as impacting

---

[86] Kilbert (2019).
[87] Henry (2020).
[88] Lake Erie Bill of Rights (2019).

any of the communities of organisms or soil's right to flourish within the Lake Erie watershed? What does it mean to naturally evolve? I don't know what that means...And it leaves—these words are malleable, they are twisted easily based on each individual's subjective moment-by- moment determination of what they think may violate and seek criminal liability against who they think is violating them.[89]

While Mr. Fusonie, the lawyer for Drewes Farm who made this statement, appears to be mocking LEBOR with his laundry list of hypotheticals, his argument highlights some important challenges that advocates for the Rights of Nature must wrestle with. How are advocates for the Rights of Nature anticipating and responding to such definitional questions? Can they carefully construct legislative language that resolves as many of these legal uncertainties as possible? "A fundamental question to effectively operationalize rights of nature is how to define the rights bearer. This ties in with the larger question of how to define nature. Examples of entities whose rights have been recognized include Mother Earth, Pacha Mama, rivers, ecosystems, natural communities, glaciers, species, and the animal kingdom. Each comes with its own definitional challenges."[90] There have been lively debates over whether strategies to expand legal rights within existing liberal, capitalist frameworks can ever succeed given the deep tensions between industrial capitalism, state sovereignty and care for the Earth.[91] Solon describes such debates within the movement as being "like a river made up of different streams that are flowing toward the ocean, but have not yet reached its shores."[92] The risk with a plurality of legal definitions is that when pressed in court, as those in Toledo witnessed firsthand, this ambiguity can undermine legal arguments for the Rights of Nature.

On 27 February 2020, U.S. District Judge Jack Zouhary issued his ruling and invalidated the new Toledo charter amendment, arguing "LEBOR is unconstitutionally vague and exceeds the power of municipal government in Ohio. It is therefore invalid in its entirety."[93] Initially the City of Toledo filed an appeal of the ruling in March of 2020, but it later withdrew its petition without comment. Exactly one year to the day had passed since LEBOR was approved by voters, ending what some locals now refer to as the Third Battle for Lake Erie.

While the LEBOR saga ends here, but there is one other aspect of this case I want to touch on briefly since it has implications for the larger Rights of Nature movement. We need to make a brief digression back to the summer of 2019 after LEBOR had been voted into law, but before the 2020 ruling.

[89] *Drewes Farm Partnership*, et al. *v. City of Toledo, OH* (2020, 44).
[90] Chapron et al. (2019, 1393).
[91] Solón (2014).
[92] Solón (2014, 1).
[93] Zouhary (2020, 8).

In the summer of 2019 Republican leadership, working in coordination with the Ohio Chamber of Commerce inserted a clause into House Bill 166, a bill authorizing the annual operating budget for the state. This specific amendment created a new section in the Ohio Revised Code (2305.11) that specifically prohibited future efforts to promote any Rights of Nature legal frameworks in Ohio.

> (1) No person, on behalf of or representing nature or an ecosystem, shall bring an action in any court of common pleas. (2) No person shall bring an action in any court of common pleas against a person who is acting on behalf of or representing nature or an ecosystem. (3) No person, on behalf of or representing nature or an ecosystem, shall intervene in any manner, such as by filing a counterclaim, cross-claim, or third-party complaint, in any action brought in any court of common pleas.[94]

We now know that the Ohio Chamber of Commerce played a leadership role in directing a secretive campaign against LEBOR as it was in the courts. The goal of the Chamber of Commerce was to undermine the underlying claims that nature has both intrinsic value and an inherent right to exist and flourish. As Don Kulak noted about the back and forth battle between activists and the state of Ohio, "Over a dozen rights-based measures that included recognizing rights for ecosystems have been systematically stymied from being voted on by illegitimate actions taken by the Ohio supreme court, secretary of state, appointed board of elections officials, and the state legislature."[95] Tish O'Dell, an Ohio organizer with CELDF, added that "It's not surprising that the Ohio legislature has the shameful distinction of being the first in the country to specifically name ecosystem rights – trying to quash them rather than taking the lead in recognizing them."[96]

Following an open records request by the Ohio Community Rights Network, it became clear just how active the Chamber was in attacking ecosystem rights. A series of e-mail exchanges on the impending 2019 budget omnibus amendment between Zachary Frymier, Director of Energy and Environmental Policy for the Chamber of Commerce at the time, and a legislative aid for Republican Representative James Hoops discussed draft language for the proposed new section 2305.11 in the Ohio Revised Code. Frymier wrote back to the aide after reading the draft language for the new section that it looked good, but needed one key addition:

> Generally the amendment looks good but I do have a concern with one of Aida's comments below. The statement that nature and ecosystems do not have standing currently contradicts what the Lake Erie Bill of Rights, which is current law in Toledo, states. Language in this amendment stating that they do not have

[94] Ohio Legislature (2019).
[95] Kulak (2019, 27).
[96] quoted in Kulak (2019, 3).

standing is essential to what we're trying to accomplish. If we could get that added I would be very grateful.[97]

Here is a clear example of how Ohio law was changed at the request of the Chamber to include a critical new sentence specifically targeting the Rights of Nature movement. And it was explicitly framed as "essential" to what the Chamber of Commerce wanted to accomplish. The new line from the Chamber of Commerce that made its way into HB 166 reads: "Nature or any ecosystem does not have standing to participate in or bring an action in any court of common pleas."[98]

Why was excluding legal standing for nature in Ohio's legal code "essential" to the Chamber's policy agenda? The emails provided through the open records request don't provide an answer, but we do know it was in response to the newly passed Lake Erie Bill of Rights in Toledo and ongoing efforts by the Chamber to forestall other Rights of Nature laws from emerging in Ohio. Importantly, several other Ohio cities had similar efforts underway to pass local Rights of Nature laws in response to fracking and extractive projects. But Toledo was by far the largest city engaged in such efforts, and a successful Rights of Nature law in a major metropolitan city could inspire others and set a new legal precedent, so legislation was needed to stop such a rights revolution from spreading to other Ohio cities. As a local Toledo organizer declared in a press statement responding to these legislative efforts, "The Ohio Chamber of Commerce is rewriting Ohio law, handing it off to supposed 'representatives of the people,' in order to try to ban a people's movement. But Rights of Nature is a movement whose time has come. The genie is out of the bottle, and not the state, not the courts, and certainly not the Ohio Chamber of Commerce can stop it."[99]

One final example highlights how the global Rights of Nature movement is intertwined with these local community struggles, and that is the inclusion of LEBOR at the UN's 2019 Ninth Interactive Dialogue on Harmony with Nature meeting, which took place 22 April 2019. One of the featured presenters at the event was Toledoans for Safe Water organizer Markie Miller, whose efforts in Toledo had led to the successful adoption of LEBOR into law a few months earlier. In her remarks at the assembly, Miller argued that "Climate justice will only be attainable when we make the conscious and humane decision to stand up and fight for it— winning legal protections for the environment—just as the people of Toledo voted to do…We realize that the system may rule against Lake Erie and our law, but the world must act on behalf of the rights of nature anyway and understand what Toledoans have learned: We don't lose until we quit—and we will never quit on Lake Erie

---

[97] Ohio Community Rights Network (2019).
[98] Ohio Legislature (2019, 492).
[99] Ohio Community Rights Network (2019).

or Mother Earth."[100] As Miller went on to note in her remarks, citizens in Toledo had grown tired of waiting on the state to fix an environmental injustice it had allowed to happen. For Miller and others, a primary concern was the inability of Toledo citizens to assert their rights to clean water and a healthy environment over and above the rights of corporate polluters to harm the Lake Erie ecosystem.

Conclusion

As the LEBOR case illustrates, local efforts to secure the Rights of Nature like those which took place in Toledo are part of a global movement, with local and global forces flowing back and forth constantly, a point driven home by Markie Miller's participation at the UN forum. It also highlights how some local struggles for the Rights of Nature are linked to the growing Community Rights movement. It's no coincidence that many Rights of Nature efforts in the US have emerged around struggles over fracking and fossil fuel expansion projects.[101] I didn't have time to explore these links in detail, but this is likely a promising avenue for further research.

These movements are also significantly influenced by Indigenous cosmopolitics, as evidenced by the social movement networks that have emerged around North American struggles such as Idle No More, Standing Rock, Wet'suwet'en and Mauna Kea, as well as the importance of Andean Indigenous groups in the early stages of the movement. Even in the traditionally conservative Midwest where Toledo is located, support for the Rights of Nature is growing, as is public resistance to extractive energy projects. As I have tried to show, this broad sense of ecological kinship is a growing part of global environmental justice movements. As the interconnection between the Toledo LEBOR case and the UN Interactive Dialogue on Harmony with Nature suggests, these global networks are an important space for dialog about ecological kinship, community rights (both human and nonhuman) and questions of ecological rights and justice. The birth of new organizations like Movement Rights, which has become an important space for facilitating Indigenous and non-Indigenous connections around the Rights of Nature movement and related environmental justice struggles, in addition to older networks like the Indigenous Environmental Network, is indicative of this growing cross-movement pollination that blurs the line between social and environmental justice movements and Indigenous rights struggles and speaks to ongoing EJ movement innovations highlighted by Cole and Foster.

As the 2014 Stillheart Declaration on the Rights of Nature stated, "All rights, including humans', depend on the health and vitality of Earth's living systems. All other rights are derivative of these rights. This requires an essential paradigm shift from a jurisprudence and legal system designed to secure and

---

[100] Miller (2019, 2).
[101] CELDF (2019).

consolidate the power of a ruling oligarchy and a ruling species, and to substitute a jurisprudence and legal system designed to serve all of the living Earth community."[102] It is through this dynamic interplays between the global and the local that the international movement for the Rights of Nature is growing.

As I have argued, the global Rights of Nature movement and efforts to expand the scope of environmental justice concerns is a result of the dynamic interplay of this growing movement of movements.[103] By uniting Indigenous and non-Indigenous allies, from Ecuador and New Zealand to Standing Rock and Toledo, the Rights of Nature movement is opening new frontiers of resistance and forging new alliances. As Nick Estes argues in reflecting on the lessons of Standing Rock and the fight for water embodied in the phrase *Mni Wiconi* (Water is Life), "It forces some to confront their own unbelonging to the land and the river. How can settler society, which possesses no fundamental ethical relationship to the land or its original people, imagine a future premised on justice? There is no simple answer. But whatever the answer may be, Indigenous people must lead the way."[104] Through their actions, these movements are pushing the boundaries of environmental justice and rights. By claiming that nature has an intrinsic right to exist and flourish, environmental justice advocates are positing a radical break with the liberal humanist norms that have historically underpinned mainstream social and environmental justice movements. The global movement for the Rights of Nature provides a space where movement activists can work together toward healing these historical traumas and ongoing injustices while striving to create a new share shared vision of a just transition to a better future, one built on the foundations of kinship and respect for all peoples and beings on the Earth.

WORKS CITED

Abram, David. 1997. *The Spell of the Sensuous: Perception and Language in a More-Than-Human World*. New York: Random House.

Adler, Margot. 2006. *Drawing Down the Moon: Witches, Druids, Goddess-Worshippers, and Other Pagans in America*. New York: Penguin.

Bauman, Whitney. 2014. *Religion and Ecology: Developing a Planetary Ethic*. New York: Columbia University Press.

Beale, Kate. 2009. "Rights of Nature: In PA's Coal Region, A Radical Approach to Conservation Takes Root." *Huffington Post*, February 2, 2009. https://www.huffpost.com/entry/rights-for-nature-in-pas_b_154842.

Becker, Mark. 2011. *¡Pachakutik!: Indigenous Movements and Electoral Politics in Ecuador*. Lanham: Rowman & Littlefield Publishers.

Begemann, Sonja. 2019. "Toledo OKs Lake Erie Bill of Rights." *Farm Journal*, March 2019, 142(4), p. 8. https://www.agweb.com/article/news-article/toledo-oks-lake-erie-bill-rights.

---

[102] Movement Rights (2017).

[103] Cox and Nilsen (2007).

[104] Estes (2019, 256).

Begotti, Rodrigo and Carlos Peres. 2020. "Rapidly Escalating Threats to the Biodiversity and Ethnocultural Capital of Brazilian Indigenous Lands." *Land Use Policy*, 96. https://doi.org/10.1016/j.landusepol.2020.104694.

Berkes, Fikret. 2017. *Sacred Ecology (4th Ed.)*. Routledge: New York.

Berkes, Fikret, Johan Colding, and Carl Folke. 2000. "Rediscovery of Traditional Ecological Knowledge as Adaptive Management." *Ecological Applications*, 10(5), (Oct. 2000), pp. 1251–1262. https://doi.org/10.2307/2641280.

Biffi, Emanuela and Brunilda Pali (Eds.). 2019. "Environmental Justice Restoring the Future: Towards a Restorative Environmental Justice Praxis." *European Forum for Restorative Justice*. https://www.euforumrj.org/sites/default/files/2020-01/digital_booklet_1.pdf.

Blattner, Charlotte E., Kendra Coulter, and Will Kymlicka (Eds). 2020. *Animal Labour: A New Frontier of Interspecies Justice?* Oxford: Oxford University Press.

Booth, Annie. 2003. "We are the Land: Native American Views of Nature." In *Nature Across Cultures: Views of Nature and the Environment in Non-Western Cultures*. Helaine Selin (Ed.), pp. 329–349. Dordrecht: Kluwer Academic Publishers.

Borough of Tamaqua. 2006. "Ordinance No. 612 of 2006." Tamaqua Borough, Schuylkill County, Pennsylvania. September 19, 2016. http://files.harmonywithnatureun.org/uploads/upload666.pdf.

Boyd, David. 2017. *The Rights of Nature: A Legal Revolution That Could Save the World*. Toronto: ECW Press.

Brown, Tom. 1983. *Field Guide to Wilderness Survival*. New York: Penguin Books.

Brown, Tom. 1984. *Field Guide to Living with the Earth*. New York: Penguin Books.

Bullard, Robert. 1990. *Dumping in Dixie: Race, Class, and Environmental Quality*. Boulder: Westview Press.

Bullard, Robert (Ed.). 1994. "Environmental Justice for All." In *Unequal Protection: Environmental Justice & Communities of Color*. San Francisco: Sierra Club Books.

Bullard, Robert. 1998. *Dumping in Dixie: Race, Class, and Environmental Quality*. New York: Taylor and Francis.

Bullard, Robert. 2001. "Environmental Justice." *International Encyclopedia of the Social & Behavioral Sciences*. Oxford: Elsevier, pp. 4627–4633. https://doi.org/10.1016/B0-08-043076-7/04177-2.

Burdon, Peter. (Ed.) 2011. *Exploring Wild Law: The Philosophy of Earth Jurisprudence*. Kent Town: Wakefield Press.

Burdon, Peter. 2014. *Earth Jurisprudence: Private Property and the Environment*. London: Routledge.

CELDF. 2015. "Tamaqua Borough, Pennsylvania." Community Environmental Legal Defense Fund. August 31, 2015. https://celdf.org/2015/08/tamaqua-borough/.

CELDF. 2019. "News Release: Ohio Community Members File Federal Civil Rights Lawsuit." February 4, 2019. https://celdf.org/2019/02/news-release-ohio-community-members-file-federal-civil-rights-lawsuit/.

Celermajer, Danielle, Sria Chatterjee, Alasdair Cochrane, Stefanie Fishel, Astrida Neimanis, Anne O'Brien, Susan Reid, Krithika Srinivasan, David Schlosberg, and Anik Waldow. 2020. "Justice Through a Multispecies Lens." *Contemporary Political Theory*, 15, pp. 475–512. https://doi.org/10.1057/s41296-020-00386-5.

Chapron, Guillaume, Yaffa Epstein, and José Vicente López-Bao. 2019. "A Rights Revolution for Nature." *Science*, March 14, 2019. 363(6434), pp. 1392–1393. https://doi.org/10.1126/science.aav5601.

Chen, Cher Weixia and Michael Gilmore. 2015. "Biocultural Rights: A New Paradigm for Protecting Natural and Cultural Resources of Indigenous Communities." *Intercontinental Cry*. September 3, 2015. https://intercontinentalcry.org/biocultural-rights-a-new-paradigm-for-protecting-natural-and-cultural-resources-of-indigenous-communities/.

Climate Justice Alliance. 2019. "Just Transition Principles." (Accessed on June 15, 2020). https://climatejusticealliance.org/wp-content/uploads/2019/11/CJA_JustTransition_highres.pdf.

Cochran, Alasdair. 2018. *Sentientist Politics: A Theory of Global Inter-Species Justice*. Oxford: Oxford University Press.

Cole, Luke and Sheila Foster. 2001. *From the Ground Up: Environmental Racism and the Rise of the Environmental Justice Movement*. New York: New York University Press.

Crews, Chris. 2019. "Earthbound Social Movements and the Anthropocene." *Journal for the Study of Religion, Nature, and Culture*, 13(3), pp. 333–372. https://doi.org/10.1558/jsrnc.39633.

Cox, Laurence and Alf Gunvald Nilsen. 2007. "Social Movements Research and the 'Movement of Movements': Studying Resistance to Neoliberal Globalisation." *Sociology Compass*, 1, pp. 424–442. https://doi.org/10.1111/j.1751-9020.2007.00051.x.

Dissenting Voices. 2001. "Climate Talks Den Haag November 2000." Platform London. (Accessed on Sept 4, 2020). https://issuu.com/platform-london/docs/dissenting_voices_cop_6_climate_tal.

Drew, Georgina. 2017. *River Dialogues: Hindu Faith and the Political Ecology of Dams on the Sacred Ganga*. Tucson: University of Arizona Press.

Drewes Farm Partnership, et al v. City of Toledo, OH. 2020. "Transcript of Motion Hearing Before the Honorable Jack Zouhary United States District Judge." US Court of Appeals, Sixth Circuit, Case: 3:19-cv-00434-JZ. https://www.courtlistener.com/docket/14573310/drewes-farm-partnership-v-city-of-toledo-ohio/#entry-61.

Environmental Working Group. 2019. "Explosion of Unregulated Factory Farms in Maumee Watershed Fuels Lake Erie's Toxic Blooms." (April 2019). https://www.ewg.org/interactive-maps/2019_maumee/.

Estes, Nick. 2019. *Our History Is the Future: Standing Rock Versus the Dakota Access Pipeline, and the Long Tradition of Indigenous Resistance*. New York: Verso.

Express News Service. 2017. "Uttarakhand HD Declares Ganga, Yamuna Living Entities." *The Indian Express*, March 22, 2017. https://indianexpress.com/article/india/uttarakhand-hc-declares-ganga-yamuna-living-entities-4579743/.

Ferner, Mike. 2019. "The Third Battle for Lake Erie." *Counterpunch*, May 15, 2019. https://www.counterpunch.org/2019/05/15/the-third-battle-for-lake-erie/.

General Assembly. 2009. "General Assembly, 63rd Session, 80th Plenary Meeting—A/63/PV.80." United Nations. https://documents-dds-ny.un.org/doc/UNDOC/GEN/N09/309/62/pdf/N0930962.pdf?OpenElement.

Harmony with Nature. N.D. "Harmony with Nature." United Nations. https://www.harmonywithnatureun.org.

Harmony with Nature. 2016. "Concept Note." United Nations. http://files.harmonywithnatureun.org/uploads/upload502.pdf.

Harmony with Nature. 2020. "Concept Note." United Nations. http://files.harmonywithnatureun.org/uploads/upload905.pdf.

Harmony with Nature. 2022. "Harmony with Nature and Biodiversity: Contributions of Ecological Economics and Earth-Centered Law—Concept Note." United Nations. http://files.harmonywithnatureun.org/uploads/upload1196.pdf.

Harris, Melanie. 2016. "Ecowomanism: An Introduction." *Worldviews*, 20(1), pp. 5–14. https://www.jstor.org/stable/26552243.

Harvey, Graham. 2015. *The Handbook of Contemporary Animism*. New York: Routledge.

Henry, Tom. 2019. "Campaign Finance Report: BP Backed Anti-Lake Erie Bill of Rights Effort." *Toledo Blade*, April 9, 2019. https://www.toledoblade.com/local/environment/2019/04/09/campaign-finance-reports-bp-financial-backer-anti-lake-erie-bill-of-rights/stories/20190409145.

Henry, Tom. 2020. "More at Stake Than Water with Lake Erie Bill of Rights Court Decision." *Toledo Blade*, January 25, 2020. https://www.toledoblade.com/local/environment/2020/01/25/more-at-stake-than-water-with-lake-erie-bill-of-rights-court-decision-toledo/stories/20200124136.

IIED. 2016. *Climate Change and Biocultural Adaptation in Mountain Communities*. Second International Learning Exchange of the International Network of Mountain Indigenous Peoples (INMIP). London: International Institute for Environment and Development. http://pubs.iied.org/14657IIED.

Indigenous Environmental Network. ND. "About." (Accessed on April 13, 2020). https://www.ienearth.org/about/.

Indigenous Environmental Network. 2019. "Indigenous Principles of Just Transition." (Accessed on September 15, 2020). https://www.ienearth.org/justtransition/.

Jemez Principles for Democratic Organizing. 1996. "Working Group Meeting on Globalization and Trade." *Southwest Network for Environmental and Economic Justice*. (Accessed on April 5, 2019). https://www.ejnet.org/ej/jemez.pdf.

Kaufman, Craig and Pamela Martin. 2018. "Constructing Rights of Nature Norms in the US, Ecuador, and New Zealand." *Global Environmental Politics*, 18(4), pp. 43–62. https://doi.org/10.1162/glep_a_00481.

Kilbert, Kenneth. 2019. "*Lake Erie Bill of Rights: Legally Flawed But Nonetheless Important*." *JURIST – Academic Commentary*. March 14, 2019. https://www.jurist.org/commentary/2019/03/kenneth-kilbert-lebor-important/.

Klein, Naomi. 2014. *This Changes Everything: Capitalism vs. the Climate*. New York: Simon & Schuster.

Klimaforum09. 2009. "People's Declaration – System Change – Not Climate Change." December 11, 2009. http://klimaforum.org/declaration_english.pdf.

Kulak, Don. 2019. "Lake Erie Bill of Rights Now Under Attack." 2019. *White Pine—The Sustainable Real Estate Journal*, (May–June 2019), pp. 27. https://solarwindspublishing.com/wp-content/uploads/2020/08/may-june-2019.pdf.

Lake Erie Bill of Rights (LEBOR). 2019. City of Toledo. https://www.utoledo.edu/law/academics/ligl/pdf/2019/Lake-Erie-Bill-of-Rights-GLWC-2019.pdf.

Lim, Michelle (Ed.). 2019. *Charting Environmental Law Futures in the Anthropocene*. Singapore: Springer Nature Singapore Pte Ltd.

Livingston, Julie, and Jasbir Puar. 2011. "Interspecies." *Social Text*, 29(1), pp. 3–14. https://doi.org/10.1215/01642472-1210237.

McLaughlin, Jan Larson. 2016. "Farmer Asks County to Not Declare Lake Erie 'Impaired'." *BG Independent News*, August 23, 2016. http://bgindependentmedia.org/farmer-asks-county-to-not-declare-lake-erie-impaired/.

Meijer, Eva. 2019. *When Animals Speak: Toward an Interspecies Democracy*. New York: NYU Press.

Miller, Markie. 2019. "Opening Remarks." Interactive Dialogue of the General Assembly on Harmony with Nature. United Nations. http://files.harmonywithnatureun.org/uploads/upload798.pdf.

Miller, Markie. 2020. Personal Communication, June 6, 2020.

Mohd. Salim v. State of Uttarakhand & Others. 2017. *High Court of Uttarakhand at Nainital*. https://www.ielrc.org/content/e1704.pdf.

Movement Rights. 2017. "Rights of Nature & Mother Earth: Rights-Based Law for Systemic Change." Movement Rights, IEN, WECAN. (Accessed June 3, 2020). https://www.movementrights.org/resources/RONME-RightsBasedLaw-final.pdf.

Nash, Roderick. 1989. *The Rights of Nature: A History of Environmental Ethics*. Madison: University of Wisconsin Press.

National Science Foundation. 2019. "Lake Erie's Toxic Algae Blooms: Why Is the Water Turning Green?" April 8, 2019. https://www.nsf.gov/discoveries/disc_summ.jsp?cntn_id=298181.

Nussbaum, Martha and Cass Sunstein (Eds.). 2004. *Animal Rights*. New York: Oxford University Press.

O'Bryan, Katie. 2004. "The Appropriation of Indigenous Ecological Knowledge: Recent Australian Developments." *Macquarie Journal of International and Comparative Environmental Law*, 1(1). http://www.austlii.edu.au/au/journals/MqJlICEnvLaw/2004/2.html.

Ohio Community Rights Network. 2019. "Anti-Rights of Nature Provision in Budget Bill Authored by Ohio Chamber of Commerce, Handed to State Representative." https://9b815260-3162-41cb-bb08-5944d5e83ed4.filesusr.com/ugd/df652d_0d7e6475af3a4aaf91a65c09ae053833.pdf.

Ohio Legislature. 2019. "Amended House Bill 166." 133 Session. (Accessed May 5, 2020). https://www.legislature.ohio.gov/legislation/legislation-documents?id=GA133-HB-166.

Pelizzon, Alessandro. 2019. "Concluding Remarks." Interactive Dialogue of the General Assembly on Harmony with Nature. United Nations. http://files.harmonywithnatureun.org/uploads/upload831.pdf.

Pike, Sarah. 2017. *For the Wild: Ritual and Commitment in Radical Eco-Activism*. Oakland: University of California Press.

Pouliot, Alison, and John Charles Ryan. 2013. "Fungi: An Entangled Exploration." *PAN: Philosophy Activism Nature*, 10, pp. 1–5.

Rasul, Nicole. 2019. "Factory Farms are Polluting Lake Erie. Will the Lake's New Legal Rights Help?" *Civil Eats*, April 9, 2019. https://civileats.com/2019/04/09/lake-erie-was-granted-legal-rights-could-it-change-the-farm-pollution-debate/.

Real Solutions for Lake Erie. 2018. "The Lake Erie Bill of Rights Is Putting Us At Risk!". (Accessed on April 3, 2019). http://realsolutionsforlakeerie.com/.

Regan, Tom. 1983. *The Case for Animal Rights*. Berkeley: University of California Press.

Rights of Rivers. 2020. "Universal Declaration of the Rights of Rivers." *Earth Law Center and International Rivers*. https://www.rightsofrivers.org/#declaration.

Rosenberg, Gabe and Nick Evans. 2020. "Ohio House Speaker Charged with Racketeering Conspiracy in Nuclear Bailout." *WOSU*, July 21, 2020. https://woub.org/2020/07/21/reports-federal-officials-arrest-ohio-house-speaker-in-60-million-bribery-case/.

Salmón, Enrique. 2000. "Kincentric Ecology: Indigenous Perceptions of the Human-Nature Relationship." *Ecological Applications*, 10(5), pp. 1327–1332. https://esajournals.onlinelibrary.wiley.com/doi/abs/10.1890/1051-0761%282000%29010%5B1327%3AKEIPOT%5D2.0.CO%3B2.

Schlosberg, David. 2009. *Defining Environmental Justice: Theories, Movements, and Nature*. Oxford: Oxford University Press.

Singer, Peter. 1975. *Animal Liberation: A New Ethics for Our Treatment of Animals*. New York: Harper Press.

Solón, Pablo. 2014. "The Rights of Mother Earth: Notes for the Debate." *Systemic Alternatives*, August 20, 2014. https://systemicalternatives.org/2014/08/20/notes-for-the-debate-the-rights-of-mother-earth/.

Starhawk. 2005. *The Earth Path: Grounding Your Spirit in the Rhythms of Nature*. New York: HarperOne.

Stone, Christopher. 1972. "Should Trees Have Standing?—Towards Legal Rights for Natural Objects." *Southern California Law Review*, 45, pp. 450–501.

Supreme Court of Ohio. 2018. "*State of Ohio ex rel. Josh Abernathy v. Lucas County Board of Elections*." Case No. 2018-1824. https://www.supremecourt.ohio.gov/Clerk/ecms/#/caseinfo/2018/1824.

Sydney Environmental Institute. 2018. "Concepts and Practices of Multispecies Justice." (Accessed April 18, 2019). http://sydney.edu.au/environment-institute/research/environmental-justice/developing-field-multispecies-justice/.

Talaga, Tanya. 2020. "Reconciliation Isn't Dead. It Never Truly Existed." *The Globe and Mail*, February 29, 2020. www.theglobeandmail.com/opinion/article-reconciliation-isnt-dead-it-never-truly-existed.

Tavares, Paulo. 2015. "Rights of Nature." *Social Text*, (March 8, 2015). https://socialtextjournal.org/periscope_article/rights-of-nature/.

Taylor, Bron. 2009. *Dark Green Religion: Nature Spirituality and the Planetary Future*. Berkeley: University of California Press.

Taylor, Bron. 2013. *Avatar and Nature Spirituality*. Waterloo: Wilfrid Laurier University Press.

Taylor, Bron. 2016. "The Greening of Religion Hypothesis (Part One)." *Journal for the Study of Religion, Nature, and Culture*, 10(3), pp. 268–305. https://doi.org/10.1558/jsrnc.v10i3.29010.

Taylor, Dorceta. 1997. "American Environmentalism: The Role of Race, Class and Gender 1820–1995." *Race, Gender and Class*, 5(1), pp. 16–62.

Thomson, Sinclair. 2002. *We Alone Will Rule: Native Andean Politics in the Age of Insurgency*. Madison: University of Wisconsin Press.

Times of India. 2017. "Supreme Court Stays Uttarakhand High Court's Order Declaring Ganga and Yamuna 'Living Entities'." July 7, 2017. https://timesofindia.indiatimes.com/india/supreme-court-stays-uttarakhand-high-courts-order-declaring-ganga-and-yamuna-living-entities/articleshow/59489783.cms.

Tokar, Brian. 2014. *Towards Climate Justice: Perspectives on the Climate Crisis and Social Change*. Porsgrunn: New Compass Press.

Toledo, Victor. 2013. "Indigenous Peoples and Biodiversity." In *Encyclopedia of Biodiversity*. Simon Levin (Ed.). pp. 269–278. Academic Press. https://doi.org/10.1016/B978-0-12-384719-5.00299-9.

Tucker, Mary Evelyn and John Grim. 2001. "Introduction: The Emerging Alliance of World Religions and Ecology." *Daedalus*, 130(4), pp. 1–22. n.

UNDRIP. 2008. "United Nations Declaration on the Rights of Indigenous Peoples." United Nations. https://www.un.org/esa/socdev/unpfii/documents/DRIPS_en.pdf.

United Church of Christ. 1987. *Toxic Waste and Race in the United States*. New York: United Church of Christ Commission for Racial Justice.

United Church of Christ. 1991. *Proceedings – First National People of Color Environmental Leadership Summit*. New York: United Church of Christ Commission for Racial Justice.

United Nations. 2009. "Resolution 63/278. International Mother Earth Day." United Nations. https://digitallibrary.un.org/record/652438?ln=en.

Vignelli, Guido. 2019. "Why Integral Ecology Will Destroy Civilization." The American Society for the Defense of Tradition, Family and Property. May 27, 2019. https://www.tfp.org/why-integral-ecology-will-destroy-civilization/.

Weaver, Harlan. 2020. *Bad Dog: Pit Bull Politics and Multispecies Justice*. Seattle: University of Washington Press.

Weissenberg, Marcel and David Schlosberg (Eds). 2014. *Political Animals and Animal Politics*. New York: Palgrave Macmillan.

White Jr., Lynn. 1967. "The Historical Roots of Our Ecological Crisis." *Science*, 155(3767), pp. 1203–1207. https://doi.org/10.1126/science.155.3767.1203.

Whitehead, Frederika. 2014. "The First Climate Justice Summit: A Pie in the Face for the Global North." *The Guardian*, April 16, 2014. https://www.theguardian.com/global-development-professionals-network/2014/apr/16/climate-change-justice-summit.

Whyte, Kyle Powys. 2014. "Indigenous Women, Climate Change Impacts, and Collective Action." *Hypatia*, 29(3), pp. 599–616. https://www.jstor.org/stable/24542019.

World People's Conference on Climate Change and the Rights of Mother Earth. 2010. "Proposal Universal Declaration of the Rights of Mother Earth." https://pwccc.wordpress.com/2010/04/24/proposal-universal-declaration-of-the-rights-of-mother-earth/.

Youatt, Rafi. 2020. *Interspecies Politics: Nature, Borders, States*. Ann Arbor: University of Michigan Press.

Zouhary, Jack. 2020. "Order Invalidating Lake Erie Bill of Rights." United States District Court for the Northern District of Ohio, Western Division. https://9b815260-3162-41cb-bb08-5944d5e83ed4.filesusr.com/ugd/df652d_bce7e7aa101c4fb98b172da404afe8e6.pdf.

Theorizing with the Earth Spirits: African Eco-Humanism in a World of Becoming

Anatoli Ignatov

Openness and attentiveness give us sensitivity to the world as 'alive, astir with responsive presences that vastly exceed the human'; they allow us to be receptive to unanticipated possibilities and aspects of the non-human other, reconceiving and re-encountering them as potentially communicative and agentic beings with whom we ourselves must negotiate and adjust…as active presences and ecological collaborators in our lives.

Val Plumwood, Environmental Culture

Introduction

This chapter seeks to address the conspicuous absence of African political thought from the field of environmental political theory.[1] It highlights African elders, soothsayers and earth priests as contemporary eco-theorists whose plural practices of theorizing make up enduring intellectual traditions and

---

[1] A notable exception is J. Baird Callicott's book *Earth's Insights*, in which he briefly explores Yoruba and San traditions of thought and characterizes Africa as a "big blank spot on the world map of indigenous environmental ethics" (J. Baird Callicott, *Earth's Insights: A Multicultural Survey of Ecological Ethics from the Mediterranean Basin to the Australian Outback*, [Berkeley: University of California Press, 1994], 158). Callicott sees very little potential for African traditions to offer any contributions to the

---

A. Ignatov (✉)
Appalachian State University, Boone, NC, USA
e-mail: anatoli@appstate.edu

alternative ethical ways of ecological life, which have been neglected by Euro-American political theory in favor of abstract theorizations that privilege anthropocentric and individualist perspectives.

The chapter challenges depoliticized readings of these African traditions of ecological thought. Drawing on the writings of African post-colonial leaders, philosophers and elders, as well as on my ethnographic research in Ghana, I argue that *African eco-humanism* presents us with an alternative frame of eco-centric political thought that plays out at multiple registers of governance outside the state in various more-than-human communities and alliances.[2] These traditions sensitize us to a nonhuman world swarming with "communicative and agentic beings with whom we ourselves must negotiate and adjust" in ways that resonate with Val Plumwood's philosophical animism.[3] They enlarge our understanding of what it means to be human in terms of the relational becoming of human beings in interdependence with land, plants, animals, spirits and ancestors. They also extend moral considerability to the future generations and the environment, and not just to the present human generation. These materialist traditions of African eco-humanism foster a dialogical relationship with the earth's vital agencies that repositions nature as the primary source of human systems of ethics, politics and governance.

field of environmental ethics due to these traditions' seemingly anthropocentric focus from which one can "distill no more than a weak and indirect environmental ethic, similar to the type of ecologically enlightened utilitarianism, focused on long-range human welfare" (Callicott, *Earth's Insights*, 158). For a critique of Callicott's representation of African ethics as anthropocentric, see Kevin Behrens, "The Imperative of Developing African Eco-Philosophy," in *Themes, Issues and Problems in African Philosophy*, ed. Isaac E. Ukpokolo (Cham, Switzerland: Palgrave, 2017).

[2] I borrow the term "eco-humanism" from the work of Michael Onyebuchi Eze, in which he highlights a widespread African notion of humanity articulated in terms of positive relationships with other human beings, animals, spirits, ancestors, biological life, nonbiological life and other vital forces that make up the environment: "I use the term *humanitatis eco* (eco-humanism) to qualify this Africanist view of the environment. What this means is that in the relationship between the human person and nature, neither the human person nor the environment is prior or superior in moral status and recognition. This relationship is holistic. Eco-humanism suggests a dialogic relationship with nature in its particular differences and uniqueness" (Michael Onyebuchi Eze, "Humanitatis-Eco (Eco-Humanism): An African Environmental Theory," in *The Palgrave Handbook of African Philosophy*, ed. Adeshina Afolayan and Toyin Falola [New York: Palgrave, 2017], 629). According to Eze, the Xhosa/Zulu aphorism "Umuntu Ngumuntu Ngabantu" (i.e. "a person is a person through other persons"), which I discuss in the section on *ubuntu* and *ukama*, best defines what constitutes African *eco-humanity* [my emphasis], i.e. a humanity expressed not only in relationship with other people, but also with living and nonliving beings, *and the environment* [my emphasis] (Eze, "Humanitatis-Eco," 629). Here Eze accentuates a key distinction between these African traditions and supremacist or exceptionalist versions of Euro-American "humanism" that are normally taken to mean the excessive valuing of humans to the exclusion of the environment. African traditions of humanist thinking include more than living humans. They are "eco-humanist" because they extend moral considerability and agency to the past and future generations and to the material environment.

[3] Val Plumwood, *Environmental Culture: The Ecological Crisis of Reason* (New York: Routledge, 2002), 195.

The Governance of the Spirits: The Ghanaian Chief of Medicine as an Eco-Political Thinker

To illustrate this tradition of eco-political theory, let's visit the chief of medicine *(BagenabaSɔ)* in Northern Ghana who upholds an ancestrally prescribed ethic for the entire polity through his jurisdiction on human-earth relations within a dynamic world of life forces.[4] *BagenabaSɔ* is the leader of a network of soothsayers and healers who has become renowned for his moral and political authority to counsel chiefs, *tindaanas* (earth priests) and other traditional rulers on environmental issues. Upon arrival at the chief's compound, we park our motors right next to the chief's colossal *tampugerɛ*, the rubbish heap in front of Gurensi earthen compounds whose size indicates the wealth of a household. A couple of vultures hop around the *tampugerɛ* and scrutinize our group closely. *BagenabaSɔ* is a wealthy and powerful man, an elder explains. People flock to his compound from all over Ghana and West Africa, soliciting his treatment for mental illness, epilepsy, blindness, stroke and other types of illnesses.[5]

BagenabaSɔ tells us that he inherited the spirits from the ancestors. The spirits selected him among his father's many children to take over the custody of the family gods. His knowledge of herbs is also a "god-given gift," whose healing powers are interdependent with the unceasing abundance and generosity of the giving savanna environment. The herbs can be obtained only from a distant forest and cannot be cultivated in the village: "The herbs are always available all the time in the forest…if you plant them here, they will not thrive, they will not grow but they are there all the time."[6] Since the herbs cannot be cultivated, the soothsayer promotes an ethic that instills biodiversity preservation principles in order to ensure that the earth continues to provide him with healing resources. There are "some sicknesses that herbs do not accept" and the chief refers such patients to the local hospital. For instance, sometimes he would send a patient to the hospital to get a blood transfusion, after which the person would come back for the rest of the treatment. Unlike many herbalists in Ghana, whose expertise is dismissed by medical doctors, *BagenabaSɔ* has a good working relationship with the hospital. Orthodox drugs and herbalists' drugs often heal in the same way, he explains, but when it comes to getting to know the depth of a sickness modern medications and the lab sometimes "cannot tell" because what they are dealing with are actually spiritual sicknesses. Within indigenous African cosmovisions, these spiritual sicknesses are not seen as individual organisms' states of affairs, but as disturbances of the social interrelationships between humanity, material nature and

---

[4] This section builds on my analysis of the chief of medicine's practices of theorizing in Anatoli Ignatov, "Ecologies of the Good Life: Forces, Bodies, and Cross-Cultural Encounters" (PhD Dissertation, Johns Hopkins University, Baltimore, United States, 2014).

[5] Interview with BagenabaSɔ by author and Jacqueline Ignatova, December 16, 2012.

[6] Interview with BagenabaSɔ by author and Jacqueline Ignatova, December 16, 2012.

spirituality.[7] The causes of such unhealthy conditions are usually attributed to nefarious influences exerted by agencies of the spirit world that require healers such as BagenabaSɔ to approach diagnoses with some form of divination or soothsaying.[8]

The chief of medicine is a medium, a bridge between the sick and natural agencies: "The spirits are like wind or water and speak to me in my mind, directing me what to do. I give herbs but they carry out the treatment… they are natural forces, I don't see them but yet I hear when they speak."[9] The spirits are air, wind and water, and only the chief can interpret what they are saying. In a follow-up conversation, he reiterates that it is not him who is doing the healing: "All this is done through the direction of the river god. Without their instructions, no healing can be attained."[10] It is nature who treats—river gods and earth agencies that he often personifies as the "mothers"—and all treatments take place at a "central healing shrine," where the sick go first for consultation. The shrine consists of a slender tree wrapped in numerous, blood-stained ropes and an earthen mound covered with the bones, shells, bangles and feathers of sacrificed animals.

According to the chief, the spirits instructed him to build the shrine and the tree. Once the spirits have diagnosed a patient, BagenabaSɔ is usually instructed to kill a fowl or perform some kind of sacrifice. During the sacrifice, he puts a tiny rope around the neck of the animal, which the slaughter stains with blood. The rope is removed and hung on the tree. The spirits then "pick the blood" and proceed with a treatment. The tree is the medical record of whoever comes in: "when someone is healed I go to write the name on that tree."[11]

The physical appearance of the tree illustrates a treacherous economy of replacement and exchange of vital forces: "the spirits take blood because, for example, when a witch gets someone they take the person's blood -- just that we human beings do not see it with our eyes but the spirits do -- this is why the spirits, too, take blood of the animals in exchange to save the person's soul."[12]

In such precarious environments, the violence of malevolent agencies remains ever-present and has to be repeatedly displaced by the conjoint efforts

---

[7] David Millar, "Ancestorcentrism: A Basis for African Sciences and Learning Epistemologies," in *African Knowledges and Sciences: Understanding and Supporting the Ways of Knowing in Sub-Saharan Africa*, ed. David Millar et al. (Barneveld, The Netherlands: COMPAS/UDS, 2006), 59.

[8] See Fara Jim Awindor, "A Quest for Sustainable Development: The Interface of Traditional African Medicine and Western Science-based Medicine" (PhD dissertation, University for Development Studies, Tamale, Ghana, 2018); P.A. Twumasi, *Medical Systems in Ghana: A Study in Medical Sociology* (Tema: Ghana Publishing Corporation, 2005).

[9] Interview with BagenabaSɔ by author and Jacqueline Ignatova, December 16, 2012.

[10] Interview with BagenabaSɔ, by author and Fara Jim Awindor, July 8, 2018.

[11] Interview with BagenabaSɔ by author and Jacqueline Ignatova, December 16, 2012.

[12] Interview with BagenabaSɔ by author and Jacqueline Ignatova, December 16, 2012.

of nature and the chief to keep witches at bay. Vultures feed in and fly over the chief's compound, guarding and watching over whatever treatments he may be carrying out. Another guardian bird—a white egret—remained perched on the top of the tree at the central shrine for more than two hours during our first visit, surveying attentively the inner courtyard and the shrine's surroundings.

BagenabaSɔ's dialogical relationship with the river, wind, vultures, egret and other vital earthen agencies highlights not only his status as a traditional healer, but also his role as a holder of eco-political and ethical authority in an interspecies community that extends across generations. He often speaks in parables and proverbs that require collective interpretation by knowledge holders. The objects that he enlists in the healing process (e.g., bones, bird feathers, herbs, amulets, dog tails, etc.) do not have pre-assigned meaning, but create occasions for interpretation and interaction with the earth spirits; they give visual and actual form to unseen forces and threats (i.e., threats that are now rendered visible and tangible). His practices of theorizing can be situated within a well-established tradition of spirit mediums, soothsayers and rainmakers in Africa whose authority status and "understanding of communal responsibility over the environment" expose "a trans-anthropocentric dimension of African thought."[13] In his analysis of rituals such as rainmaking ceremonies or *BagenabaSɔ*'s sacrifices, Madavo suggests that "African environmental ethics…goes beyond an anthropocentric concern, towards a broader and complex concern for the environment in itself."[14]

In contrast to Euro-American anthropocentric worldviews that treat nature as passive "objects" or "resources" of instrumental value for human beings, rainmakers and spirit mediums express "broader" concerns for the intrinsic value of the nonhuman world conceived as a dynamic collectivity of communicative beings that deserve to be granted moral worth, agency and respect. The wellbeing and interests of the nonhuman members of this "community of nature" are recognized as interdependent with those of humans.[15] This African notion of interdependence "holds out much promise for environmental ethics" because it "does not pit anthropocentric concerns and eco-centric concerns against each other;" it creates obligations to pursue peaceful modes of coexistence, reciprocity and cooperation with the nonhuman world.[16] It is expressed in one of the chief of medicine's main responsibilities to sustain and promote a certain moral and political ethic of care and reverence for the earth on behalf of entire clans and communities:

---

[13] Garikai Madavo, "African Environmental Ethics: Lessons from the Rain-Maker's Moral and Cosmological Perspectives," in *African Environmental Ethics: A Critical Reader*, ed. Munamato Chemhuru (Cham, Switzerland: Springer, 2019), 143, 150.

[14] Madavo, "African Environmental Ethics," 142.

[15] Behrens, "The Imperative of Developing African Eco-Philosophy," 198.

[16] Behrens, "The Imperative of Developing African Eco-Philosophy," 199.

> Our feet are the gods. The earth is the gods. Trees are gods, rivers are gods, and stones are gods. We owe our very lives to these mentioned elements of nature. Do you think you will live if you do not believe and take care of them?...If there are great ailments, catastrophes, as a result of our disobedience, I know where to pass to ensure that I appease the earth spirits and bring them home-together, at peace. If they (spirits) tell me to come home, I have to give this and that, the people are informed and the sacrifices are done. That is what it is.[17]

The destruction of ecological resources and the contemporary intensification of floods, droughts and illnesses are seen as indicators of moral and spiritual human-made transgressions that mobilize a series of propitiation rituals and measures meant to appease the earth gods. The environmental crisis, *BagenabaSɔ* insists, is a product of a collective disturbance of the shifting balances of human-nonhuman relations. It is a crisis that is at once ecological, ethical, spiritual and political. *BagenabaSɔ*'s integrated view of the crisis can serve as a counterweight to truncated Western anthropocentric worldviews that separate the ecological from the political or sideline culture, ethics and spirituality in the service of economic rationality. His ritual mediation of the relations between humans and the environment can be conceived as a form of political action that can inform collective responses to the crisis that reject narrow technocratic and managerial conservation approaches and center cultural and spiritual diversity.

Drawing on his authority derived from the invisible governance of the earth spirits, *BagenabaSɔ* has jurisdiction on human relations with sacred lands, trees and rivers. He emphasizes his complementary role to that of chiefs and earth priests in the service of peace:

> There are sacerdotal roles and rituals to be observed with regards to how shrines and river gods are to be upheld. For some of the rivers you do not fish in them or eat the fish caught in them, farming is forbidden along the river banks and the shores around water bodies' surrounding the rivers. When this is observed there is peace. In this there is mutual existence amongst the trees. Because the life of the trees depends on water. But if we extinct the trees through burning and logging, living contrary to societal and traditional lore, a time will come the society will fail to sustain itself... In my work, I help the earth priest and the chief in the physical and spiritual governance of the land. I serve them. They do not pledge allegiance to me.[18]

Peace is premised on maintaining correct relationships with natural forces through people's adherence to ancestral prescriptions such as taboos on fishing

---

[17] Interview with BagenabaSɔ, by author and Fara Jim Awindor, July 8, 2018.

[18] Interview with BagenabaSɔ, by author and Fara Jim Awindor, July 8, 2018.

in rivers and making noise at night, which have been put under pressure by the spread of Christianity, Islam and modernization.[19]

For *BagenabaSɔ*, the continual adherence to such prescriptions is tied to the continual recognition of the legitimacy of the conjoint jurisdiction of soothsayers, chiefs and earth priests in Northern Ghanaian communities:

> I ensure the peace of the land and propose ways of appeasing the gods when they are offended to avert any calamity. I communicate to the chief and he, in turn to the people. The chief does not tell the people I had prescribed the mode of appeasement. He makes the message his and delivers it so. Our relationship is like a husband and wife whose love is boundless. If the wife realizes her husband face imminent shame because he had to give a parting gift for a deserving guest, she privately calls her husband inside and gives him a gift...In this case, the husband gives out the gift as his but will not necessarily tell the guest how his wife had salvaged him in the act of generosity. The guest goes away with the gift thinking it came from the husband. Our work is confidential as it is complementary.[20]

The husband and wife metaphor suggests that even though power is distributed among these complementary registers of indigenous political authority, they are not equal. The chief may be the public face of political power in society, but it is the soothsayer who leads behind the scenes and mediates between human systems of governance and the dynamic earth-based order of ancestral jurisdiction, where the true locus of power lies.

---

[19] The chief of medicine attributes the ecological degradation of nature to the growing influence of Christianity and Islam in Ghana: "What has desecrated the land and the gods is the *pusiŋɔ* (prayers)" (Interview with BagenabaSɔ, by author and Fara Jim Awindor, July 8, 2018). The proliferation of excessively loud church services at night, for example, is commonly described as an insidious form of noise pollution that drives the earth spirits away and encourages Ghanaians to violate ancestral taboos: "...observing silence is key. Around here, pastors from different churches continue to make all forms of noise at any time of the day. The shrine (*tingane*) also goes out to scout its land at the time all are supposed to be quite and indoors, if the shrine encounters any being, all will not be well with him or her. Night belong to the spirits, day for humans. The shrine commune with God at night and such interferences from humans bring conflicts" (Interview with BagenabaSɔ, by author and Fara Jim Awindor, July 8, 2018). Ecological degradation is also attributed to a parallel disregard for the taboo on making noise from August to October, the time when the guinea corn pollination takes place: "The ancestors associated excessive noise during harvest time with heavy winds, and more trees with more rain. They knew little about science but understood the importance of not driving away pollinators. Today the noise of the 'ghetto blasters' scares the insects away" (Discussion with Members of Bolgatanga's *Tindanas*' Association, December 16, 2012). Likewise, earth priests and elders complain that taboos against cutting the trees in the *tingane* have become relaxed by Christian converts who have been told in church that nothing will happen if they encroach upon the earth gods. They also point out that Christianity promotes a "cheaper" form of worship than traditional sacrifice ("when you are praying direct to God you don't pay anything"), relocating the source of spiritual power away from the natural world and advancing anthropocentric ways of thinking about human-nature relationships (Interview with Elders by author, Jacqueline Ignatova and Christopher Azaare, October 13, 2012).

[20] Interview with BagenabaSɔ, by author and Fara Jim Awindor, July 8, 2018.

According to dominant understandings of "government" as the location of politics within Western political theory that tend to identify authority with a centralized "state's capacity to organize social life," the eco-political authority of the chief of medicine remains largely invisible.[21] Such indigenous forms of authority usually become relegated to the domain of ritual or "African traditional religion," which is treated as a separate sphere of social life from the spheres of politics and ecology.

This chapter seeks to trouble such depoliticized representations of African traditions that obscure registers of eco-political practice outside "apparent centres of political action" such as the ritual spaces of the chief of medicine's shrines.[22] The next two sections explore key concepts of human self and collectivity such as *consciencism*, *négritude*, *ukama* and *ubuntu*, which reveal political community as an "ecological" process of interdependent becoming involving the conjoint action of the living, the ancestors, the unborn and the earth's vital forces. I then highlight undervalued ecological aspects of the African indigenous concept of *tiŋa* as an ancestral trust. This concept of multigenerational trusteeship proceeds from the premise that land belongs to the ancestors and that the living are only temporary possessors, whose use of land is conditional upon compliance with time-honored ancestral prescriptions.[23] It generates moral obligations to the future generations and the environment that might enrich the contemporary debate about intergenerational justice and responsibility, especially with regard to climate change. The final section focuses on the ways in which these concepts of humaness and land are anchored in forms of sovereign oversight and eco-political authority that have important features in common across much of Africa. In particular, I examine the changing role of traditional authorities—e.g., chiefs, earth priests, soothsayers and clan heads—to act as trustees of land and promote peace through joint jurisdiction on human relationships with the environment.

My frequent references to these concepts and practices of theorizing as "African" should not be interpreted as saying that there is one single, monolithic African tradition of environmental political thought, or that all African eco-theorists commit to such conceptual orientations. I seek to highlight recurring strands of intellectual thought in sub-Saharan Africa that appear to share certain conceptual aspects in common. For instance, the concept of *ubuntu* is found among the Bantu-speaking people in Southern Africa, including the Nguni languages of Zulu, Xhosa and Ndebele spoken in South Africa, the Shona languages spoken in Zimbabwe and numerous other Bantu

---

[21] Amy Niang, *The Postcolonial African State in Transition: Stateness and Modes of Sovereignty* (London: Rowman and Littlefield International, 2018), 19–25.

[22] Niang, *The Postcolonial African State in Transition*, 25.

[23] S.K.B. Asante, *Property Law and Social Goals in Ghana 1844–1966* (Accra: Ghana Universities Press, 1975), 22.

languages.[24] The concept of *tiŋa* is widespread among the Voltaic- or Gur-speaking societies in the savannah region of West Africa such as the Gurensi, Boosi, Tallensi, Dagara, Kusasi, Kasena, Mossi, Dagomba and Mamprusi of Northern Ghana and Southern Burkina Faso. Both of these geographically diverse traditions of ecological thought share a notion of ethical community as a "triad comprised of the living, the living-dead (ancestors), and the yet-to-be-born."[25] Despite such commonalities, these traditions are fluid, internally diverse and contested. These earth-based orders of political thought and sovereign oversight coexist and contend with multiple registers of authority linked to colonial and state-making practices that have historically sought to diminish or negate the influence of the former on the constitution of moral order in African societies.[26]

Consciencism and *Négritude*: Human Becoming in a Universe of Vital Forces

A number of influential visions of "African humanism" were advanced by Africa's leading political thinkers in the context of decolonization in 1950s and 1960s. Rejecting the binary choice between communism and capitalism, these visions sought to formulate a "third way" for newly independent nations "within a liberated Africa based on the idea of strong moral "African traditions" that could show the way, and an "African personality" that could shoulder the change."[27] This concept of African personality, a term coined by Liberian statesman Edward Blyden at a speech in Freetown in 1893 and promoted by Ghana's independence leader Kwame Nkrumah, had been formulated in reference to a prevailing notion of what it means to be human that emphasizes the spiritual and communal elements in the formation of African personhood.

According to the latter, a human being becomes a human or a person a person, through others.[28] In this section, I turn to the philosophies of *consciencism* and *négritude* in order to highlight underappreciated ecological dimensions of these African political visions of interdependent humanity.

---

[24] Lesley Le Grange, "Ubuntu," in *Pluriverse: A Post-Development Dictionary*, ed. Ashish Kothari et al. (New Delhi, India: Tulika Books, 2019), 324; Munyaradzi Felix Murove, "An African Commitment to Ecological Conservation: The Shona Concepts of Ukama and Ubuntu," *Mankind Quarterly* 45, no. 2 (Winter 2004): 195–215.

[25] Mogobe B. Ramose, "Ubuntu," in *Degrowth: A Vocabulary for a New Era*, ed. Giacomo D'Alisa, Federico Demaria and Giorgos Kallis (New York: Routledge, 2015), 212.

[26] Niang, *The Postcolonial African State in Transition*, 21.

[27] Kai Kresse, "African Humanism" and a Case Study from the Swahili Coast," in *Humanistic Ethics in the Age of Globality*, ed. Claus Dierksmeier et al. (London: Palgrave Macmillan, 2011), 249.

[28] Kresse, "African Humanism," 250–251.

In *Consciencism*, Nkrumah develops the concept of "consciencism," a "new emergent ideology" that takes account of "the combined presence of traditional Africa, Islamic Africa and Euro-Christian Africa" and seeks to be "in tune with the original humanist principles underlying African society."[29] Consciencism is a new kind of socialism that views the "the cluster of humanist principles which underlie the traditional African society" as the basis for modern Ghana's cultural pluralism.[30] One such "cardinal ethical principle" is "to treat each man as an end in himself and not merely as a means," which is traceable to the egalitarian ethos and classless makeup of the traditional clan system.[31] For Nkrumah, the philosophical foundation of this principle of intrinsic equality of all mankind is the African monistic view of matter: "On the philosophical level, too, it is materialism, not idealism, that in one form or another will give the firmest conceptual basis to the restitution of Africa's egalitarian and humanist principles…It is materialism, with its monistic and naturalistic account of nature, which will balk arbitrariness, inequality and injustice."[32]

In contrast to Western conceptions of matter as inanimate and inert, Nkrumah grounds consciencism in the "traditional African idea of the absolute and independent existence of matter," which regards material nature as "alive with forces in tension" and as "a plenum of forces which are in antithesis to one another."[33] Matter is primary, independent and constantly transforming. This materialist eco-humanism does not involve abstract and dualistic conceptualizations of a transcendent, spiritual realm that is conceived to be separate from and superior to the natural, material world. In Nkrumah's view, many African societies ward off such transcendentalist frames of thought by "making the visible world continuous with the invisible world."[34] African personality unfolds in a world that is both immanent and material *and* spiritual and immaterial. The visible and invisible registers of existence interact in complex processes mediated by their common capacity to be shaped and reshaped in terms of particular relationships of forces in tension. Consciencism is thus grounded in the multiplicity, diversity, vitality and dynamic flux that is intrinsic to all of material nature.[35]

Nkrumah's vital materialist ontology in *Consciencism* resonates with the philosophical foundations of *négritude*, a vision of African eco-humanism

---

[29] Kwame Nkrumah, *Consciencism: Philosophy and Ideology for De-Colonization* (New York: Monthly Review Press, 1964), 70.

[30] Nkrumah, *Consciencism*, 79.

[31] Nkrumah, *Consciencism*, 69, 95.

[32] Nkrumah, *Consciencism*, 76.

[33] Nkrumah, *Consciencism*, 79, 97.

[34] Nkrumah, *Consciencism*, 12.

[35] Louise Du Toit, "'When Everything Starts to Flow:' Nkrumah and Irigaray in Search of Emancipatory Ontologies," *Phronimon* 16, no. 2 (2015): 11.

expounded by Léopold Sédar Senghor, Senegal's first postcolonial president. Senghor defines *négritude* as "relations with others, an opening out to the world, contact and participation with others" whereby the world is conceived "beyond the diversity of its forms, as a fundamentally mobile, yet unique, reality that seeks synthesis."[36] Suspending Western distinctions between subjects and objects, matter and life, he concurs with Nkrumah that the traditional African ontology views material nature as essentially alive. Matter is in a perpetual state of becoming: "As far as African ontology is concerned, too, there is no such thing as dead matter: every being, every thing—be it only a grain of sand—radiates a life force, a sort of wave-particle; and sages, priests, kings, doctors, and artists all use it to help bring the universe to its fulfilment."[37]

Unlike Nkrumah's philosophical emphasis on the unity and independence of self-moving matter, *négritude* is centered around the concept of "life force." Senghor sees life force as "the single reality of the universe: being, which is spirit, which is life force;" life force constitutes existence and is distributed across the material world of plants, animals and minerals, each of which exhibits various intensities and forms of this force.[38] To exist means to exercise a specific vital force in interdependent relationship to a material and dynamic multiplicity of other vital forces. It is within this continuum of life forces that human becoming unfolds:

> Each of the identifiable life forces of the universe—from the grain of sand to the ancestor—is, itself and in its turn, a network of life forces—as modern physical chemistry confirms: a network of elements that are contradictory in appearance but really complementary. Thus, for the African, man is composed, of course, of matter and spirit, of body and soul; but at the same time he is also composed of a virile and a feminine element: indeed, of several "souls." Man is therefore a composition of mobile life forces which interlock: a world of solidarities that seek to knit themselves together. Because he exists, he is at once end and beginning: end of the three orders of the mineral, the vegetable, and the animal, but beginning of the human order.[39]

As a "composition of mobile life forces which interlock" and "a world of solidarities" among spirits, animals, plants, minerals, water and wind, the human person is a microcosm of an interactive and vitalist universe. A person becomes a person through his or her direct involvement with the vital materialities of nature, and through these interactions, with the ancestors and the

---

[36] Léopold Sédar Senghor, "Négritude: A Humanism of the Twentieth Century," in *I Am Because We Are: Readings in Africana Philosophy*, ed. Jonathan Scott Lee and Fred L. Hord (Amherst: University of Massachusetts Press, 2016), 56, 58.

[37] Senghor, "Négritude," 59.

[38] Senghor, "Négritude," 58–59.

[39] Senghor, "Négritude," 59–60.

primary source of vital force, God.[40] According to Senghor, in the face of this vitalist order of existence, the African adopts an active ethical stance. To live ethically means "living according to his nature, composed as it is of contradictory elements but complementary life forces. Thus he gives stuff to the stuff of the universe and tightens the threads of the tissue of life…he transcends the contradictions of the elements and works toward making the life forces complementary to one another: in himself first of all, as man, but also in the whole of human society."[41] To live ethically means to give "stuff to the stuff of the universe" and reciprocate actively with the diverse forms of vital force that constitute the material environment. This kind of ethical conduct is illustrated by the Gurensi chief of medicine's collaboration with the river god. The chief's offerings of "the blood of the animals to the earth spirits" enable the spirits to reciprocate by taking the soul of the afflicted "to the water to heal it through soaking."[42] The goal is to harmonize the plurality of forces in tension both within the human self and within society at large in accordance with the force trajectories of other natural beings; it is to ensure the continuous increase of vital force within an interdependent collectivity.

Both Nkrumah's vision of *consciencism* and Senghor's philosophy of *négritude*, with their attendant concepts of African personality, have come under intense scrutiny within debates on African anti-colonial thought. The most notable critics of *négritude* include Frantz Fanon, Wole Soyinka and Stanislas Adotevi.[43] Together they have drawn attention to the nostalgic orientation of Senghor's vision toward African pre-colonial cultural worldviews; to the essentialization and over-simplification of African personhood in ways that are too ahistorical and monolithic to represent the diverse forms of everyday life and internal antagonisms of contemporary African societies; and to *négritude*'s indebtedness to European philosophy and defensive posture toward claims of Western intellectual superiority that inadvertently reproduce some of the same dichotomist structures of colonial thought that African humanism aspires to confront. Yet, as Abiola Irele observes, a purely historical or "realistic" view of *négritude* misses its most significant contribution "that it was also a genuine

---

[40] Donna V. Jones observes that for Senghor, the human status within this hierarchical web of vital forces enables humans to draw other beings such as animals, trees and winds into communicative fellowship. She illustrates this observation with her own translation of Senghor's 1939 essay "Ce que l'homme noir apporte," in which he writes: "People speak of their [Negro- African] animism; I will say their anthropo-psychism…all of Nature is animated by a human presence…Not only animals and the phenomena of nature—rain, wind, thunder, mountain, river—but also the tree and the pebble become men" (Donna V. Jones, *The Racial Discourses of Life Philosophy: Négritude, Vitalism, and Modernity* [New York: Columbia University Press, 2010], 140–141).

[41] Senghor, "Négritude," 60.

[42] Interview with BagenabaSɔ, by author and Fara Jim Awindor, July 8, 2018.

[43] See Frantz Fanon, *Les Damnés de la terre* (Paris: Maspéro, 1961); Wole Soyinka, *Myth, Literature and the African World* (Cambridge: Cambridge University Press, 1976); Stanislas Adotevi, *Négritude et Négrologues* (Paris: Editions 10/18, 1972).

rediscovery of Africa, a rebirth of the African idea of the black self."[44] In Irele's view, "this opening up of the African mind to certain dimensions of its own world which western influence had obscured...the way in which the best of these poets came to root their vision in African modes of thought has given a new meaning to the traditional African world-view."[45]

Here Achille Mbembe concurs with Irele that one of the most significant contributions of Senghor's philosophy is its rejection of abstract visions of the universal: "To affirm that the world cannot be reduced to Europe is to rehabilitate singularity and difference... The only universal is the community of singularities and differences, a sharing that is at once the creation of something common and a form of separation."[46] Senghor's concern for African personality, Mbembe argues, makes sense because it "opens the way for a reimagining of the universal community" and as such it is "indispensable for contemporary conditions" in the age of globalization and "multiple returns of colonialism."[47]

Today the underappreciated ecological aspects of Nkrumah's and Senghor's visions of reimagined universal community might also equip us with new intellectual resources to address the collective conditions in the Anthropocene. These philosophies of eco-humanism, rooted in ideas of self and community as dynamic concatenations of contending material forces, do not merely trouble familiar Cartesian nature-culture and matter-life divides. They also attest to neglected *relational* models of ethics and politics that extend the notion of political collectivity into the nonhuman world. And while remaining cognizant of the dangers of essentialization exposed by Senghor's critics, these models can be described as "African" insofar as they are expressions of "the most central strand of sub-Saharan ethical thought, which places relationality at the core of morality."[48] They prompt us to consider that what is most valuable about us as human beings is not our autonomy or some individualized attribute of rationality. Rather, it is our interdependence with other human beings and the environment and our participation in a dynamic ecology of constantly shifting force relations.

According to Bénézet Bujo, this strand of African environmental thought "does not define the person as self-realization or as an ontological act" but as "a process of coming into existence in the reciprocal relatedness of individual and community, where the latter includes not only the diseased but

---

[44] Abiola Irele, "Négritude—Literature and Ideology," *The Journal of Modern African Studies* 3, no. 4 (December 1965): 510.

[45] Irele, "Négritude," 510.

[46] Achille Mbembe, *Critique of Black Reason*, trans. Laurent Dubois (Durham: Duke University Press, 2017), 158.

[47] Mbembe, *Critique of Black Reason*, 158.

[48] Thaddeus Metz, "An African Theory of Moral Status: A Relational Alternative to Individualism and Holism," in *African Environmental Ethics: A Critical Reader*, ed. Munamato Chemhuru (Cham, Switzerland: Springer, 2019), 10.

also God."[49] This means that a "person becomes a person only through active participation in the life of the community" and "by means of 'relations,'" and not by virtue of membership in such a community.[50] Human becoming unfolds within a web of interactions defined by the "exchange of vital force" and the "continuous flow of life between all community members, including the deceased" whereby each member "must be conscious that his actions contribute either to the growth of life of the entire community or to the loss or reduction of its life."[51] This concern with the growth of vital force in the community and with the need to acknowledge the immanence of becoming of the human person within a dynamic field of ecological interactions is expressed in the practices of the Gurensi chief of medicine described in the introduction of this chapter. Both physical illness and ecological disasters are understood to be the visible manifestations of spiritual conflicts among the living and the dead at the invisible register of community life.[52]

There are many versions of the body at play in the therapeutic interaction and the chief of medicine's treatments do not necessarily work on discrete bodies bounded by skin or even on bodies at all. Rather, they focus on exchanges of vital force and relationships between ancestors, kin, family, malevolent agencies and spirits.[53] As a mediator between the spirits and the living, the chief of medicine upholds the "fundamental principle of African ethics" that "one becomes a human being only in a fellowship of life with others" and performs his responsibility to raise "the quality of the vital force not only for himself, but rather for the entire community, indeed for the whole of humanity."[54]

UKAMA, UBUNTU AND ANAMNESTIC SOLIDARITY: THE INTERDEPENDENCE BETWEEN PAST, PRESENT AND FUTURE

This principle that one becomes human only in relationship to other human beings and the more-than-human world is expressed in the Southern African concepts of *ukama* and *ubuntu*:

> Whilst the Shona word *Ukama* means relatedness, *Ubuntu* implies that humanness is derived from our relatedness with others, not only to those currently living, but also through the generations, past and future. When these two concepts are compounded in their togetherness they provide an ethical outlook

---

[49] Bénézet Bujo, *Foundations of an African Ethic: Beyond the Universal Claims of Western Morality* (New York: The Crossroad Publishing Company, 2001), 87.

[50] Bujo, *Foundations of an African Ethic*, 87–88.

[51] Bujo, *Foundations of an African Ethic*, 88.

[52] Bujo, *Foundations of an African Ethic*, 97.

[53] See also Stacey Ann Langwick, *Bodies, Politics, and African Healing: The Matter of Maladies in Tanzania* (Bloomington: Indiana University Press, 2011), 22–25.

[54] Bujo, *Foundations of an African Ethic*, 5.

that suggests that human well-being is indispensable from our dependence on and interdependence with all that exists, and particularly with the immediate environment on which all humanity depends.[55]

Rejecting atomistic notions of human self and agency, *Ukama* is a concept that advances an understanding of existence in terms of interdependence and relatedness.[56] According to Murove, "humanness - *Ubuntu* - is a concrete form of *Ukama* in the sense that human interrelationship in society is a micro aspect of relationality within the pervasive macro of the universe."[57]

Ukama's recognition that the wellbeing of humanity is dependent on the wellbeing of the broader community of nature within which human existence unfolds is an expression of a transgenerational model of environmental ethics. This model entails ethical obligations to treat other human generations and the nonhuman world with respect, care and gratitude.[58] For Murove, the goal of such ethical practice is to foster *ukama* between the living and the ancestors as a "communion which is actualised by acts that involve giving beer, water, snuff or meat to the ancestors as an expression of fellowship and remembrance."[59] Bujo uses the term "anamnestic solidarity" to describe this complex of ritual practices that seek to renew the relations of reciprocal interdependence between the living and the dead:

> African ethics are articulated in the framework of anamnesis, which involves remembering one's ancestors. A narrative community, fellowship here on earth renews the existence of the community of the ancestors. This re-establishing (*poiesis*) in turn implies the praxis which efficiently continues the remembrance of the ancestors and gives new dynamism to the earthly fellowship. Consequently, ethical behaviour in the Black African context always involves re-establishing the presence of one's ancestors; for one who takes the anamnesis seriously is challenged to confront the ethical rules drawn by the ancestors, in order to actualise anew the "protological foundational act" which first called the clan fellowship into life.[60]

---

[55] Murove, "An African Commitment to Ecological Conservation," 196.

[56] "Grammatically, *Ukama* is an adjective. As an adjective, its grammatical construction is *U-kama*...Taken as the stem, *-kama* becomes a word which means 'to milk a cow or a goat.' The idea of milking in Shona thought suggests closeness and affection...In its adjectival form, *Ukama* means being related or belonging to the same family. However, in Shona culture, as in many other African cultures, there is a sense whereby *Ukama* is understood as not simply restricted to marital and blood ties, because there is a tendency in this culture to see all people as *kama* or relatives" (Murove, "An African Commitment to Ecological Conservation, 196–197).

[57] Murove, "An African Commitment to Ecological Conservation," 196.

[58] Behrens, "The Imperative of Developing African Eco-Philosophy," 199.

[59] Murove, "An African Commitment to Ecological Conservation," 198–199.

[60] Bujo, *Foundations of an African Ethic*, 34–35.

Anamnesis can be understood as an articulation of *ukama* that enables the invisible community of the dead to engage in a series of communicative exchanges with the visible community of the living: "Through remembrance, the living are able to effect dialogue centred on values that have sustained the life of the community throughout the ages in the light or the challenges that are faced by the present community."[61] This communicative interaction is not understood in metaphorical terms as people's attempt to speak on behalf of or "represent" the dead; rather, the dead are perceived to be actually co-present thanks to the various rituals and ancestral prescriptions that have been laid down over multiple generations.[62]

These practices of "reestablishing the presence of one's ancestors" expand our understanding of the broader traditions of African eco-humanism that they exemplify in at least three significant ways. First, by calling anamnestic solidarity a form of "*poiesis*" Bujo suggests that these practices should not be equated with the uncritical acceptance of tradition or with a "break on the future of the community."[63] Rather, they are an adaptive, creative and dynamic form of reconfiguration of the relations between the past, the present and the future that "gives new dynamism to the earthly fellowship." When the elders, soothsayers and community knowledge holders gather at the ancestral shrine to recall clan histories and past events or to perform sacrifices to the spirits of the dead, their main concern is to consult tradition in order to address contemporary questions and issues. "The past is significant only when it proves to be the bearer of life for the present and the future" and the main goal is to forge an intergenerational discussion about how "to act under the changed circumstances *in the spirit* of the ancestors."[64]

This dynamism and fluidity of tradition is expressed in *Sankofa*, a symbol of the Akan people of Ghana. *Sankofa* is visually represented by a bird that has its feet firmly planted forward while looking backward with an egg (symbolizing the future) in its mouth. In the Akan language, the term is translated as *se wo were fi na wosan kofa a yenki*, which means "it is not taboo to go back and fetch what you forgot."[65] Going back for what has been left behind, however, does not mean blind fidelity to the ancestors, but taking a critical look at the past in order to correct mistakes and move forward. What is more, this interactive and dynamic process of renewal of *ukama* does not leave the ancestors unchanged. There is no end to human becoming as the ancestors' identities are constantly being made and remade: "an ancestor is a human being who

---

[61] Murove, "An African Commitment to Ecological Conservation," 199.

[62] Bujo, *Foundations of an African Ethic*, 49.

[63] Bujo, *Foundations of an African Ethic*, 49.

[64] Bujo, *Foundations of an African Ethic*, 49–50.

[65] Bernard Yangmaadome Guri and Wilberforce A. Laate, "Community Organisational Development in South-Western Ghana," in *Endogenous Development in Africa: Towards a Systematisation of Experiences*, ed. David Millar, Agnes A. Apusigah, Claire Boonzaaijer (Barneveld, The Netherlands: COMPAS/UDS, 2008), 77.

is called continually, even after death, to become a person; the living relationships that he maintains with his living descendants continue to shape his identity and personality."[66]

Second, these anamnestic interactions not only mutually enrich and transform the living and the ancestors, but they are also future oriented. *Ukama* underlies a moral commitment to the future generations: "In this conceptualization of ancestorhood, morality is not only for the individual's well-being at present, rather, one finds that a life lived virtuously at present promotes *Ukama* in the existence of the future generations."[67] In other words, the duties to ancestors are simultaneously duties to posterity. This is a recognition that our present wellbeing is at least partly an effect of what the ancestors have done on our behalf just as our ethical actions today constitute a positive or harmful inheritance for the yet to be born.[68] Human beings thus exist as ethical beings in a community of temporally interdependent generations. This obligation to ensure the wellbeing of posterity may be also expressed in terms of gratitude: "We ought to respect our ancestors by being grateful for how they ensured that we would inherit an environment in 'good shape', and we can express that gratitude by ensuring that we do the same for our descendants."[69]

Third, this shared multigenerational community identity is not limited to relations among humans: "the human person is only one of the many beings that are in constant symbiotic dialogue and interaction ...The earth, seas, rivers, forests, trees, animals, humans, spirits, forces, animate and inanimate things, everything together constitutes this eco-community."[70] Within such a dialogical eco-community, to further *ukama* between human beings and the ancestors means to further *ukama* between human beings and the natural world. Across sub-Saharan Africa, the ancestors are conceived to be lodged in groves, trees, lagoons, rivers, rocks, grasslands and other earthen forms, effectively reconstituting these communicative interactions and relations of care as relations between human beings and various natural beings and forces. These abodes of the spirits are considered sacred and their access and uses are governed by various taboos and prescriptions that sanction harmful human behaviors. The African concept of ancestorship also includes *totemic* relations,

---

[66] Bujo, *Foundations of an African Ethic*, 89, 94.

[67] Murove, "An African Commitment to Ecological Conservation," 202. Murove also turns to a Zulu proverb to illustrate this link between the past, the present and the future: "*Musa ukuqeda ubudlelwano maje kusasa uzofuna ukubuyela* - ("Do not disrupt harmonious relationships at present for tomorrow you might want to come back")...A moral lesson behind these proverbs is that one should not upset *Ukama* in the present because harmonious relationships in the present are a pledge for the future...one should not despise the past in preference for the future, as the past has a contribution to make to the future" (Murove, "An African Commitment to Ecological Conservation," 203).

[68] Behrens, "The Imperative of Developing African Eco-Philosophy," 200–201.

[69] Behrens, "The Imperative of Developing African Eco-Philosophy," 201.

[70] Eze, "Humanitatis-Eco," 627.

which signify long-term relations of kinship and interdependence between certain species of plants, animals and objects with human descent groups such as clans and lineages. For Murove, totemic ancestorhood is both a reminder of common belonging and of one's indebtedness to the wellbeing of posterity: "generations after generations will always exist in *Ukama* by virtue of their belonging to the common totemic ancestor. The totemic ancestor provides the primary link that unites members of the present community in *Ukama* with each other as well as with their ancestors."[71]

Totemic ancestorhood is an aspect that Southern African traditions of thought share with West African counterparts. According to Gurensi elder Christopher Azaare, many totemic relations in northern Ghanaian societies can be traced to historical events, in which an animal is believed to have saved an ancestor by leading him to water in the bush. Azaare recalls the most common version of such stories of human-animal cooperation:

> the founding father of a certain clan went out to the bush to do hunting but failed to get water. Fortunately, some animal happened to appear in the area and led the hunter to some water in a pond. The hunter on his return home narrated his experience with the animal and forbade his sons and all descendants of his from molesting, killing and eating the meat of that animal. This was later to become a totem to the man's clan.[72]

Another common account, Azaare shows, involves the view of certain animals and plants as emanations of pioneering ancestors: "…a powerful hunter may have been possessed by the spirit of some wild animal which he might have killed during hunting. Or this powerful hunter may have died and re-appeared as an animal. In both cases the animal is impersonated as a god and is looked upon as the family guardian providing the necessary protection against diseases and misfortunes."[73] These totemic relations are often enshrined in practices of naming. For instance, Azaare discusses the Gurene name "*Abaa*," which is derived from either the domestic dog or the leopard. With succeeding generations, the child's name may come to acquire the status of a clan's or a lineage's name (i.e., "*Abaabissi*"). The families of *Abaa* (*Abaabissi*) will observe the dog or leopard as their totem.[74]

Like Azaare, Michael Jackson observes that among the Kuranko of Northern Sierra Leone most narratives of how an animal became a totem involve the rescue of a stranded, hungry or thirsty clan ancestor from death in

---

[71] Murove, "An African Commitment to Ecological Conservation," 202.

[72] Christopher Azaare, *The Traditional Gurensi Religious Beliefs and Practices: Totems*, vol. 1 (unpublished manuscript, 2018, 24).

[73] Azaare, *The Traditional Gurensi Religious Beliefs and Practices*, 24.

[74] Azaare, *The Traditional Gurensi Religious Beliefs and Practices*, 24. According to Bujo, this should not be understood as a rebirth of the totemic ancestor in question but as a "special anamnesis" that requires the child "to develop in continuity with, but distinct from its ancestor" (Bujo, *Foundations of an African Ethic*, 90).

a far-away wilderness.[75] For Jackson, this is how the Kuranko "give enduring expression to moral and political choices which effectively enlarge the domain of recognized humanity" and "emphasize values that bind people together: cooperation, trust, mutual respect, conviviality."[76] In other words, totemism allows human beings both to affirm and to extend their humanity. The themes of "reciprocal altruism" and ability to empathize with others that run across these narratives imply a "transformation of linear dominance hierarchies into systems of balanced complementarity characterized by mutual aid, interdependency and merging of interests."[77] The narratives' emphasis on a sequence of events in which a being of a superior status, such as an ancestor or traditional ruler, becomes dependent on a being of a so-alled inferior status, such as a totemic animal or plant, not only facilitates a crossing over between nature and culture, but also constitutes a "reversal of the normal social order."[78] And it is this reversal of power relations through multigenerational associations with the nonhuman world that enables human beings to gain mastery over contingent events within a world of becoming over which they have limited control.[79]

This expanded understanding of humanity that involves an affirmation that human wellbeing cannot be realized outside *ukama* with the environment is articulated in the concept of *ubuntu*.[80] Mogobe Ramose observes that this indigenous concept of *botho/hunhu/ubuntu* is not easily translatable into Western categories of "humanism."[81] For Ramose, "humanness" offers a more appropriate interpretation because the concept refers to "being or the universe as a complex wholeness involving the multi-layered and incessant interaction of all entities."[82] In other words, *ubuntu* highlights an intrinsic order of the universe as a state of permanent, multi-directional movement and flux of becoming, rather than a state of being:

> It consists of the prefix *ubu-* and the stem - *ntu*. *Ubu-* evokes the idea of be-ing in general. It is enfolded be-ing before it manifests itself in the concrete form or mode of existence of a particular entity. *Ubu-* as enfolded be-ing is always oriented towards unfoldment, that is, incessant continual concrete manifestation through particular forms and modes of being. In this sense *ubu-* is always oriented towards - *ntu*. At the ontological level, there is no strict and literal separation and division between *ubu-* and – *ntu*...It is the indivisible one-ness

---

[75] Michael Jackson, *Allegories of the Wilderness: Ethics and Ambiguity in Kuranko Narratives* (Bloomington: Indiana University Press, 1982), 68–70.

[76] Jackson, *Allegories of the Wilderness*, 69.

[77] Jackson, *Allegories of the Wilderness*, 69.

[78] Jackson, *Allegories of the Wilderness*, 70–71.

[79] Jackson, *Allegories of the Wilderness*, 71, 74.

[80] Murove, "An African Commitment to Ecological Conservation," 213.

[81] Mogobe B. Ramose, *African Philosophy through Ubuntu* (Harare: Mond Books, 1999).

[82] Ramose, *African Philosophy through Ubuntu*, 155.

and whole-ness of ontology and epistemology. *Ubu-* as the generalised understanding of be-ing may be said to be distinctly ontological, whereas *–ntu* as the nodal point at which be-ing assumes concrete form or a mode of being in the process of continual unfoldment may be said to be distinctly epistemological.[83]

This definition of *ubuntu* as the co-articulation of being and becoming offers insight into the related term *umuntu* (human being) whose constitutive word *umu-* shares an ontological feature with *ubu-*. *Umuntu* can be thus conceived as a concrete manifestation of this universal tendency toward unfoldment. A human being can be conceived as "the specific entity which continues to conduct inquiry into be-ing, experience, knowledge and truth…an activity rather than an act."[84] In a world defined by ceaseless motion and transformation, being and becoming are two planes of the same reality, rather than incompatible opposites. *Ubuntu* is thus not an intrinsic attribute of pre-formed human identity, but something that manifests in our interactions with other human beings and the environment in an ongoing process of intergenerational identity formation.

The concept is best expressed in the maxim *Umuntu ngumuntu ngabantu* (*Motho ke motho ka batho*), which is usually translated as "a person is a person through other persons" or "I am because we are" or "to be a human be-ing is to affirm one's humanity by recognizing the humanity of others and, on that basis, establish humane relations with them."[85] According to Manzini, in order to ensure that the complex meaning of the maxim is not lost in translation or misinterpreted to be an anthropocentric reference to "we" or "others" as limited to the human species alone, its explanations should be grounded in its vernacular languages.[86] Ramose also clarifies that "here, the concept of others also includes all other entities that are not human beings and, therefore, the concept relates directly to a care and a concern for the environment."[87] In other words, *ubuntu* (*botho*) should be understood as a nonanthropocentric philosophy that "is anchored on the ethical principle of the promotion of life through mutual concern, care and sharing between and among human beings as well as with the wider environment of which the human being is a part."[88]

This maxim that human subjectivity finds its expression in association with other human beings and material nature means that any ecological disturbance

---

[83] Ramose, *African Philosophy through Ubuntu*, 50.

[84] Ramose, *African Philosophy through Ubuntu*, 50–51.

[85] Metz, "An African Theory of Moral Status," 13; Ramose, *African Philosophy through Ubuntu*, 52; Ramose, "Ubuntu," 212.

[86] Nompumelelo Zinhle Manzini, "African Environmental Ethics as Southern Environmental Ethics," in *African Environmental Ethics: A Critical Reader*, ed. Munamato Chemhuru (Cham, Switzerland: Springer, 2019), 117.

[87] Ramose, "Ubuntu," 212.

[88] Ramose, "Ubuntu," 212.

is also a disturbance of *botho* and of the universal process of perpetual generation and sharing of life forces.[89] Since everything in the cosmos is interrelated, it suggests that "when I harm nature, I am harmed." As such, the concept can be understood as "the manifestation of the power within all beings that serves to enhance life, rather than thwart it."[90] *Ubuntu* is anchored in the ethical imperative to promote life and care for the environment through various fertility rituals such as sacrifices to the land spirits or the observance of taboos and obligations to totemic ancestors. Its conception of eco-community as a triad composed of the living (*umuntu*), the living-dead (*abaphansi*) and the yet-to-be-born also suggests that the future generations have the same right to life and ecological wellbeing as the living.[91]

Tiŋa as a Multigenerational Trust and Communicating Environment

These ethical obligations to promote the wellbeing of future generations and the environment associated with the Southern African concepts of *ubuntu* and *ukama* resonate with the ecological ethic expressed in the West African concept of *tiŋa* as a multigenerational trust. This idea of trusteeship proceeds from the premise that land belongs to the ancestors and that the living are only temporary possessors of an intergenerational asset, whose use is conditional upon compliance with time-honored ancestral prescriptions.[92] In his manuscript on Gurensi land law, Ghanaian elder Christopher Azaare elaborates on this concept:

> According to the traditional Gurensi belief, the ancestors and spirits live under the land and in some of the trees that stand on it (*tingana*). Invariably land belongs to the ancestors (*yaabaduuma*) or *tingana* but administered by the *tindaana*. It is for this reason that the ancestors continue to take vested interest in the preservation of the land which they have left behind for their living descendants. The *tindaana* who acts as the link between the ancestors and the living controls and propitiates the lands and gods. He gives land on his own authority but he is not the owner…He is merely holding it for the community and neither he nor anyone else can sell it. Land, therefore is communal and sacred…attempts to destroy the groves by cutting them down, and taking soil without permission and the appropriate rites are often fiercely resisted.[93]

---

[89] Ramose, *African Philosophy through Ubuntu*, 57–58, 155–157.
[90] Le Grange, "Ubuntu," 324–325.
[91] Ramose, "Ubuntu," 212–213.
[92] Asante, *Property Law*, 22.
[93] Christopher Azaare, *Tindaanaship and Tindaanas in Traditional Gurensi (Frafra) Communities: Land Use and Practices* (Africa Local Intellectuals Series 90, no. 4, 2020), 41.

According to Azaare, it is the dead—i.e., the ancestors (*yaabaduuma*) and the earth gods (*tingana*)—who are the true "owners" of the land. Breach of trust, including the destruction of natural resources and commercialization of land relations, is tabooed and will incur ancestral sanctions. *Tindama* (earth priests) and *Yizuukɛɛma* (clan elders), and in a few rare cases *Naduma* (chiefs), are land trustees and custodians of the *tingane*, the shrine containing the spirits of the ancestors. But it is the skin, stool or the earth shrine—a multigenerational entity that is distinct from the individuals who make it up and that includes not only the living but also the dead and the unborn—into which the locus of ownership is lodged.[94] This notion of ancestral ownership vested in a skin or a stool, which regards the living as only temporary custodians of a multigenerational trust, thus excludes an absolutist conception of private property.[95]

Drawing on Akan traditions of thought, Kwasi Wiredu concurs with Azaare that land "belongs not to individuals, but to whole clans and individuals only have rights of use that they are obligated to exercise considerately so as not to render nugatory the similar rights of future members of the clan..."[96] Both the dead and the not-yet-born sectors of these clans have a "reality which exercises very considerable influence on thought, feeling and behavior" and Wiredu characterizes this duty to preserve the environment for posterity as a "two-sided conception of stewardship:"

> Of all the duties owed to the ancestors none is more imperious than that of husbanding the resources of the land so as to leave it in good shape for posterity. In this moral scheme the rights of the unborn play such a cardinal role that any traditional African would be nonplussed by the debate in Western philosophy as to the existence of such rights. In the upshot there is a two-sided concept of stewardship in the management of the environment involving obligations to both ancestors and descendants which motivates environmental carefulness.[97]

---

[94] In the South of Ghana, customary land is legally categorized as *stool land* in reference to the carved wooden stool which serves as a political symbol of the authority of Akan chiefs and a shrine of the spirits of their ancestors. In the north of Ghana, customary land is defined as *skin land* in reference to the animal skins that chiefs sit on or the skins that *tindama* wear. See also Kwamena Bentsi-Enchill, *Ghana Land Law: An Exposition, Analysis and Critique* (Lagos: African Universities Press, 1964), 23–32.

[95] For Asante, such "absolutist" conceptions of private property refer to the ascendance of individualistic conceptions of private property in Africa—e.g., the "individual freehold"—associated with Western jurist traditions such as Bentham's utilitarianism or Locke's and Smith's liberalism. These conceptions advance a view of exclusive human rights to the use of or access to land, including rights of acquisition and alienation, unencumbered by the rights or social identities of any other competing land owners (See Asante, *Property Law*, 176–189).

[96] Kwasi Wiredu, "Philosophy, Humankind and the Environment," in *Philosophy, Humanity and Ecology: Philosophy of Nature and Environmental Ethics*, ed. H. Odera Oruka (Nairobi, Kenya: African Centre for Technology Studies, 1994), 46.

[97] Wiredu, "Philosophy, Humankind and the Environment," 46.

This two-sided onus of "husbanding the resources of the land" is the primary ethical obligation enjoining land trustees to exercise their authority in accordance with the principles of sustainable land management. This is a positive and active obligation to "uphold the honour of the ancestors, to promote the prosperity of the kin group, and to ensure the security of generations unborn, insuring them against poverty or destitution."[98] The trusteeship's emphasis on intergenerational sharing and environmental preservation entails a prohibition of an absolutist and individualist conception of ownership or of any unfettered right to use and dispose of land, safeguarding "an equitable distribution of the community's resources."[99]

This does not mean that the concept of individual or private property is foreign to such traditions of African thought; on the contrary, across African customary systems the trusteeship has historically coexisted and continues to contend with a plurality of overlapping group and individual bundles of land relations. In Ghana's plural customary system, the hierarchy of such relations includes the usufruct/customary freehold, leasehold, and various tenancies and licenses.[100] The local terms describing these land arrangements vary from one traditional area to another, but all of these coexisting forms of ownership are derived from and constrained by the trusteeship, effectively limiting individual and acquisitive tendencies toward exclusive use of land.[101] In the case of Akan chiefs, these constraints are codified in a rule that denies them the ability to "hold or acquire private property, on the theory that such holding would compromise his fiduciary position as an administrator of stool property."[102]

In Asante's view, the concept of trusteeship also serves as a check on "exaggeration of the role of individual investment in creating property" and as such cannot be easily translated into Western conceptual frameworks such as Locke's labor theory of land (i.e., the theory that invests an individual with ownership of land into which he has incorporated his labor).[103] The notion of land as ancestral trust pre-supposes a social order that invests the individual with a right to use community land before the application of his or her labor.[104] The subservience of property to collective interest here means that "whatever an individual added to property by his own exertions" is "not relieved from the basic trusteeship idea;" on the contrary, the latter confers "a positive duty to augment community resources and to assure to posterity

---

[98] Asante, *Property Law*, 23.

[99] Asante, *Property Law*, 23–24.

[100] See The Project Secretariat of the ACLP, "Literature Review Report on Customary Law on Land in Ghana" (National House of Chiefs/Law Reform Commission, 2009); G.R. Woodman, *Customary Land Law in Ghanaian Courts* (Accra: Ghana University Press, 1996).

[101] Asante, *Property Law*, 24.

[102] Asante, *Property Law*, 24.

[103] Asante, *Property Law*, 13–16.

[104] Asante, *Property Law*, 15.

an increased heritage."[105] This duty requires respect for the rights of future generations in terms that transcend the individualistic frames of liberal rights discourses: "Since our obligation is one owed by our generation as a whole to future generations, as wholes, exactly which future people come into existence is of little relevance."[106]

There are limits, however, to relying exclusively on categories such as "property," "resources," "rights" and "ownership" to comprehend the eco-political dimensions of the concept of *tiŋa*. Within a theoretical framework imported from the West and built upon Western concepts of land law, custom and human-nature interactions, the diversity of Africans' lived relationships with a life-generating and communicative *tiŋa* recede behind anthropocentric vocabularies of "land management" and "land rights." Western beliefs about the land as an inert, passive and inanimate context for human action obscure African indigenous understandings of this concept as a web of dynamic exchanges among vital forces that inform the practices of political theorizing of Ghanaian elders or Nkrumah's and Senghor's worldviews.

Tiŋa in Gurene means (1) land, ground, earth and the physical world; (2) country, town and settlement; and (3) bottom and depths (n. *postpos.* under, beneath, below).[107] The concept troubles nature-culture divides as it refers to a collectivity of human beings, trees, stones, spirits, etc.; it is the ground "in which we bury."[108] The term *tingane* (earth god or earth shrine) is derived from *tiŋa* and *gane* (skin), a group of elders explained to me once, in reference to the *tindaana*'s regalia that signifies his earth-based authority. Others interpreted the term as derivative from *tiŋa* and *gaŋɛ* (exceed, be more than something), highlighting the *tingane*'s high rank in an ontological hierarchy of forces: the shrine "sends the sacrifice to god because that is the one which is nearer to the supreme, to the creator."[109] A third school of elders views *gaŋɛ* as a reference to "the skin of the earth," the delicate womb of our mother within which the ancestors are buried and interwoven into ecological cycles of renewal of life: "…*tiŋa* is like a womb. A woman has the womb and the *tiŋa* gives birth. And if you die, it means that you are nourishing the earth. That's why the *tiŋa* is the earth."[110] As a spiritual and material source of life, *tiŋa* is "the essence of fertility itself, including all of the natural elements that comprise, insure or demonstrate instances of fertility, like rainfall, sunlight, and the biological life in the soul. In this deity is combined a holistic ecology

---

[105] Asante, *Property Law*, 15–16.

[106] Kevin Gary Behrens, "Moral Obligations Towards Future Generations in African Thought," *Journal of Global Ethics* 8, nos. 2–3 (August-November 2012): 188.

[107] Kropp Dakubu, Awinkene Arintono, and Avea Nsoh, *Gurenɛ—English Dictionary* (Legon: Linguistics Department, University of Ghana, 2007), 170.

[108] Interview with Christopher Azaare, August 15, 2012.

[109] Interview with Tindaana and Elders, December 14, 2012; Dakubu, Arintono and Nsoh, *Gurenɛ—English Dictionary*, 52.

[110] Interview with Tindaana and Elders, December 14, 2012.

of life."[111] Respect for this ecology of life is paid through the observance of ancestrally prescribed rituals and taboos. For instance, in many *tindaanas*' compounds or *tingana* areas there is a taboo on wearing shoes, limiting one's footprint on the earth's delicate skin. This posture of respect and humility toward *tiŋa* is also expressed in the Gurensi saying *giim (viim, jiim, diim) n-dei tingane* (translated as "a cluster of trees makes a *tingane*" or "a thick grove of different plant species makes a *tingane*"). Here *giim* (or *viim, jiim, diim*) are all adjectives used to describe elements of nature such as shade from trees, clouds, etc. The saying not only draws attention to the unity of human cycles of intergenerational reproduction and growth with ecological processes, but also serves as an injunction that the sacred has to be treated as such and preserved (i.e., when we see *giim* or a cluster of trees we should be aware that it is a *tingane* that has to be approached with reverence).[112]

The *tingane* is also the active voice of the ancestors that communicates to the living what the earth requires from them in order to sustain peace and ecological integrity: "If we have a problem, we consult the *bakolgo* [i.e. soothsayer]. There the *tingane* speaks to us instructing us on the different forms of sacrifices to be made for health, peace, and rain."[113] Another *tindaana* at the same gathering explains that during the performance of sacrifices to the earth spirits he has to observe the following taboos: "My foremost taboo is that my *tingane* does not eat goat, any who brings a goat for pacification must find a substitute. Also, the shrine has black fish which none must catch. If you catch the fish, that is a score between you and the *tingane* to settle. Also we do not harm a python for it is our spirit and kindred."[114] It is the *tindaana*'s and *bakolgo*'s joint responsibility to communicate with the *tingane* and inform the community what the "land abhors."[115]

This dialogical relationship between the earth spirits and traditional rulers reveals that the concept of *tiŋa* cannot be confined to anthropocentric interpretations as an intergenerational set of "resources" that need to be held in trust and governed by imperatives for sustainable environmental management. The trusteeship is grounded in a communicative ethic of pragmatic environmentalism that treats ecological threats as shared experiences with a reciprocating "environment" that is engaged in the practice of giving as long

---

[111] J.P. Kirby, "The Earth Cult and the Ecology of Peacebuilding in Northern Ghana," in *African Knowledges and Sciences: Understanding and Supporting the Ways of Knowing in Sub-Saharan Africa*, ed. David Millar et al. (Barneveld, The Netherlands: COMPAS/UDS, 2006), 135.

[112] I am indebted to Joseph Aketema for this interpretation of the saying.

[113] Tindaana at Meeting of the Upper East Regional Tindama Council, Bolgatanga, July 6, 2018.

[114] Tindaana at Meeting of the Upper East Regional Tindama Council, Bolgatanga, July 6, 2018.

[115] Tindaana at Meeting of the Upper East Regional Tindama Council, Bolgatanga, July 6, 2018.

as human wants are respectful of nature's needs and biophysical limits.[116] This ethic suggests that addressing the global ecological crisis may require us to complement Wiredu's two-sided model of stewardship with the cultivation of better attunement to the ongoing exchanges among humans, plants, animals and other vital forces. It might mean to follow the lead of the *tindaana* to pursue the appeasement of the earth spirits so that we can transform our current state of ecological disequilibrium into a more desirable one via agonistic negotiation with offending agencies. Such dialogs may even encourage us to register floods, fires and droughts as political votes cast by the earth's planetary forces against the destruction wrought by extractivism and anthropocentric capitalist land relations. They may also allow us to perceive *tiŋa* as a political field of contesting force relations and overlapping registers of sovereignty. These animist concepts of land, collectivity and agency underlie practices of political oversight that are not exclusively human and it is to these earth-based grounds of political authority that I turn next.

TIŊA AS THE FOUNDATION OF INDIGENOUS POLITICAL AUTHORITY

Across sub-Saharan Africa, land is the foundation of indigenous political authority. Political rulers can exercise their governing functions "only in union with the invisible world" and "political authority draws its force from the support of both the people over whom it rules and from the ancestors."[117] This "invisible" basis of authority becomes evident by the widespread resistance to chiefs who were installed by colonial administrations and as such continue to be perceived as lacking the ancestral source of legitimacy and power of the political office among their subjects.[118] In this context, Amy Niang observes that our understanding of *Tenga* (the equivalent term to *tiŋa* among the Mossi of Burkina Faso and other Voltaic societies) should not be reduced to its meanings as a geographical location, a social or ethnic grouping or a female deity of fertility. For Niang, *Tenga* should be also conceived as a dynamic field of political action and a sovereign order that the African postcolonial state has been internally built against. This order consists of "a framework of interdependent relationships, a certain ecological homogeneity and a particular disposition against rigid hierarchy."[119] As a counter-discourse of centralization processes in the Voltaic regions of Africa,

---

[116] See Anatoli Ignatov, "The Earth as a Gift-Giving Ancestor: Nietzsche's Perspectivism and African Animism," *Political Theory: An International Journal of Political Philosophy* 45, no. 1 (February 2017): 52–75.

[117] Bujo, *Foundations of an African Ethic*, 95.

[118] Bujo, *Foundations of an African Ethic*, 96.

[119] Niang, *The Postcolonial African State in Transition*, 27.

It thwarted state claims for the monopoly of sociopolitical legitimacy while its tenants adhered, gave the impression of adhering to, the dominant order. Agency, within the realm of *Tenga* was not just a matter of reaction to a threat of extinction but rather the means of preservation of preexisting forms and structures, therefore denoting a constant tension between invention and convention, cooperation and destabilization...the agents of *Tenga* invested meaning...in their practice of resistance through a series of rituals and cultural practices that had to be constantly reworked in order to respond to anxieties generated by state power.[120]

Tenga is resituated as one register of sovereignty among a plurality of sovereign orders within a ubiquitous model of balance of power and sociopolitical organization among West African societies. This model involves "a fundamental social bipartition that distinguishes on one hand the 'people of power' associated with the state/power/war and the 'people of the earth' associated with belief/rituals/agriculture."[121] It reveals the traditional offices of chiefs and earth priests as complementary, rather than opposites, within a system where "both are manifestations of a sovereignty that sublimates conflict."[122] This complementarity was articulated in *BagenabaSɔ*'s metaphor of the husband and wife used to describe the relationships of power sharing that define the model of shared governance by chiefs, *tindaanas* and soothsayers. This is a model of balance of power between two "peoples," whereby priests, soothsayers and elders that serve as privileged intermediaries with the earth spirits through rituals become indispensable for the legitimation of royal/state power.

Niang's analysis allows us to explore the recent proliferation of land conflicts between chiefs and earth priests in Northern Ghana as conflicts over the inauguration of political authority. It alerts us to an indigenous notion of political power as a composite made of components lodged in different offices and orders of sovereignty, some of which derive their legitimacy from the earth, rather than human governance designs. In such conflicts, chiefs tend to conceptualize land as a political territory or administrative jurisdiction over the living governed via taxation, litigation, and provision of infrastructure and development: "The chief is the sole custodian of the land. If the Naba owns you as a person, how can he not own the land?"[123] For earth priests, in turn, land embodies a reciprocal social contract with the ancestors and the earth's spiritual agencies that must be continually renewed through rituals, sacrifices and rites: "I put on a calabash as a cap, the chief has a red cap. Can

---

[120] Niang, *The Postcolonial African State in Transition*, 28.
[121] Niang, *The Postcolonial African State in Transition*, 179.
[122] Niang, *The Postcolonial African State in Transition*, 180.
[123] Interview with Chief and Traditional Council, July 4, 2018.

he command rain with his red cap? It is the *tindaana* whose sacrificial role ensures the increase of the human race, not the chief."[124]

With the introduction of Islam and Christianity, the development of chieftaincy and colonialism "nowadays chiefs claim rituals as well as political possession of land in order to strengthen their authority."[125] The following statement by a chief in the Bongo traditional area illustrates these changes:

> ...the chief is taking responsibilities of the land to protect the land [and] that is why he is a trustee of the land...Now, the role of *tindaana* vary from area to area. In [our] traditional area, a *tindaana* is not a trustee of the land. The *tindaana*'s main responsibility is to sacrifice to the shrine...a *tindaana* is supposed to be a stranger who has come from somewhere to settle...the chief entrusts the shrines to him to perform, to sacrifice so that he will be able to earn a living...to take care of his family.[126]

This chief not only claims the trusteeship of land by virtue of being the descendant of the first settler, but also attributes the origin of tindaanaship to an indigenous arrangement between chiefs and *tindaanas* to delegate a portion of royal powers to *tindaanas* in order to ensure a mechanism for food security. This account is rejected by *tindaanas*, who, in turn, emphasize their own role as descendants of the pioneer settlers. In the advent of colonial rule, it was their ancestors who delegated power to chiefs and appointed the latter as "valets" to the White man:

> *Tindama* are the ancestors who were first to settle in the land. And anyone who comes after, and desires to settle, they will have to call on the *tindama* who shows them where to build and settle...the *tindaana* has never been in a competition for power, moving around with the Whiteman, which is why, when the Whiteman was coming...the *tindaana* appointed somebody to see the man through because it is not traditional to gird a loin cloth, slung a skin and move about with the Whiteman. He is in charge of the land and does not move about from place to place with his Earth Spirits. So this chap, sort of valet to the Whiteman, later called "chief," was nominated to move with the Whiteman.[127]

The *tindaana*'s account of the origin of chieftaincy resonates with the reports of colonial anthropologists and officials such as Rattray who sounded early warnings that many of the colonial chiefs in the Northern Territories

---

[124] Tindaana at Meeting of the Upper East Regional Tindaama Council, Bolgatanga, July 6, 2018.

[125] Member of the Upper East Regional Tindaama Council, Bolgatanga, August 11, 2017.

[126] Interview with Chief and Elders, July 4, 2018.

[127] Tindaana at Meeting of the Upper East Regional Tindaama Council, Bolgatanga, July 6, 2018.

were a British invention, "petty unconstitutional European-made Chiefs."[128] Even those chiefs who held existing traditional offices had been vested by colonial administrations with vastly extended and modified powers that Mahmoud Mamdani has termed *decentralized despotism*: "colonial powers salvaged a widespread and time-honored practice, one of decentralized exercise of power, but freed that power of restraint, of peers or people."[129] Indirect rule diminished the role of *tindaanas* and elevated the status of chiefs. The responsibilities expected of the latter far exceeded their traditional responsibilities and limited enforcement powers. The recently formed Upper East Regional Tindaama Council—an organization whose membership now includes more than fifty *tindaanas* from the Upper East Region of Ghana—is intent on rectifying some of these histories of chiefs' usurpation of tindaanas' prerogatives: "The *tindaana* and the chief have come a long way together. Government enskinned chiefs. Same cannot be said of a *tindaana* whose institutionalization is by the ancestor…government is in bed with the chiefs with the sole agenda of filching our authority…this association is formed to resist these attempts."[130] *Tindaanas* acknowledge that chiefs are better educated, well-resourced and connected to the government. They view their own exclusion from the political process in Ghana as a modern-day reflection of the colonial past: "As it exists now, nothing really has changed since the government of the day really acknowledges only the chiefs but not us."[131] Several members of the Council see colonialism as an ongoing process that today is being reproduced through the commodification of traditional leadership. They point out that while *tindaanas*' appointments are made by the ancestors and the earth spirits via consultations with the soothsayer, the mode of selection of chiefs is determined by human consideration of material wealth and status: "Wherever the *bakolgo* moves to, the crowd follows them until the *tindaana* is nominated. That is how a new *tindaana* is gotten. This is not same with chieftaincy where when one is dead people spend money competing."[132]

These competing narratives of the meaning of land and the origin and scope of authority reflect contested oral traditions that present the traditional political system as a product of an encounter between an imported order of chieftaincy/state and an indigenously conceptualized order of *tiŋa*. In Niang's view, this encounter "created a dichotomic opposition, violent

---

[128] R.S. Rattray, *The Tribes of the Ashanti Hinterland*, vol. 1 (London: Oxford University Press, 1932), xvii.

[129] Mahmood Mamdani, *Citizen and Subject: Contemporary Africa and the Legacy of Late Colonialism* (Princeton: Princeton University Press, 1996), 48.

[130] Member of the Upper East Regional Tindaama Council, Bolgatanga, July 6, 2018.

[131] Member of the Upper East Regional Tindaama Council, Bolgatanga, July 6, 2018. I am grateful to Hannah Cullen for drawing my attention to this framing advanced by a number of *tindaanas* while discussing their own historical marginalization within Ghana.

[132] Tindaana at Meeting of the Upper East Regional Tindaama Council, Bolgatanga, July 6, 2018.

at places, but which was over time transformed into a 'dual unity' consolidated into an ideology wrought in a specific political thinking and system."[133] This is a "bricolage, a syncretic system that wields indigenous and imported rituals into a new transcendent order" that redistributes political power and sublimates social conflict.[134] Niang's analysis is supported by Wyatt MacGaffey's observations that many Northern Ghanaian societies share a "common political culture" and political structures that are "in constant reorganization," making the boundaries between state and statelessness permeable and fluid.[135] He argues that "the relationship between chiefs and tindaanas was always intimately complementary and that some of the chiefs were originally *tindaanas*."[136] This view of complementarity is elaborated upon by a traditional leader who performs both the roles of a chief and *tindaana*:

> ...amongst the settlers who might have multiplied over time, the *tindaana* selects from one of them a leader (chief) to perform any function but rituals pertaining to the land. The *tindaana* performs sacerdotal roles and the chief governs the people... there is a boundary stating where my area of jurisdiction and that of the *tindaana*. The works of chiefs include solving misunderstanding between people...If the issue is also about governance and the people, the *tindaana* also consults the chief. They both play complementary roles. This is how the two come to exist...One can perform both.[137]

Even though the two offices are complementary, they are not presented as equal. The origin of chieftaincy is attributed to *tindaanas*' decision to transfer a portion of their power and authority into a complementary register of governance in order to address the needs of a rapidly growing polity and avoid becoming "valets" to the Whiteman. Across these political systems, chiefs have to establish and renew the basis of their political legitimacy through various rites and rituals performed by *tindaanas*. In the Upper East of Ghana, for instance, these legitimation rites include a visit by a newly appointed chief to the *tindaana*'s house, immediately after the chief is enskinned by the paramount chief:

> They bring him straight...to the place here. Then there are two rocks, or two stones, the chief sits on one, if he's to go through the rituals, and then the *tindaana* also sits on one. Now, [the chief] is not allowed to touch the earth, or the ground, because that is not his rule. He has – he doesn't have anything connected with the earth...The moment he touches the buttocks, or the body

---

[133] Niang, *The Postcolonial African State in Transition*, 158.

[134] Niang, *The Postcolonial African State in Transition*, 158.

[135] Wyatt Macgaffey, *Chiefs, Priests, and Praise-Singers: History, Politics, and Land Ownership in Northern Ghana* (Charlottesville: University of Virginia Press, 2013), 11.

[136] Macgaffey, *Chiefs, Priests, and Praise-Singers*, 12.

[137] Tindaana at Meeting of the Upper East Regional Tindaama Council, Bolgatanga, July 6, 2018.

touches the earth, then that's the day he will die. So they make sure that he sits straight…when you touch it that way… that means the land doesn't accept you as a chief.[138]

The ritual institutes the spiritual grounds of the political authority of the new chief and ensures that he is "accepted" by the earth as a legitimate political ruler. Political power becomes authorized only once it has been subjected to the procedures of the earth. And it is in the very act of legitimation of chiefly power through the installation ritual that "the state appropriates the political authority, hence power, that has always been part of the functions of the guardians of *Tenga*."[139] Rituals can be thus understood as political acts that activate indigenous mechanisms of redistribution of political power among overlapping ecology-based and human-based orders of sovereignty. They also reveal the earth and the ancestral world as the primary sources of human systems of politics and governance.

This role of ritual as a window into the hitherto neglected political functions of *tindaanas* can be also illustrated by their peace-making prerogatives. These prerogatives include *tindaanas'* duties to prevent war by casting their animal skins between combatants or to pacify the earth through "burying the blood" ceremonies in the case of violent conflict. In such conflicts, the infractions are not perceived to be merely among the warring parties, but also between the earth and the whole community that relies on it for sustenance.[140]

In my discussion of these duties, a *tindaana* and his elders in the Upper East described these rituals as "collecting the blood."[141] The *tindaana* is usually informed about any such "place where blood is shed, especially a place of conflict or a place which is observed to have the penchant of causing accidents" and then he proceeds with "the spiritual cleansing" of the place:

…in the traditional setting we say "blood has fallen"…To clean the blood is another cultural ceremony, you have the person who caused the bloodshed who has to provide certain things for them to clean the blood … we will now invite him telling him "this is what you are supposed to do as a result of the mistakes you've committed"… if he has agreed, he will do that and it will clean everything. The peace will win.[142]

The spilling of blood has unleashed a state of "pollution" on the eco-system and its fertility and vitality can be only restored by a set of purification rituals performed by the *tindaana* and his elders. The very fruitfulness of the earth is affected by such contagion and its regeneration depends on ending the

---

[138] Interview with Tindaana and Elders, May 15, 2015.
[139] Niang, *The Postcolonial African State in Transition*, 157.
[140] Kirby, "The Earth cult," 139.
[141] Interview with Tindaana and Elders, July 26, 2018.
[142] Interview with Tindaana and Elders, July 12, 2017.

animosity and reconciling the differences. If the pollution is not removed, the elders suggested, people will continue to be at risk from a range of natural disasters and ancestral punishments. What is needed is the exercise of the *tindaana*'s "ecology-based authority to the role of 'landowner' for the Chiefs."[143]

The peacebuilding process required here to address the spilling of human blood does not involve merely a two-dimensional negotiation between the immediate parties to the dispute. Rather, it is a three-dimensional reconciliation between the chief, the *tindaana* and, even more importantly, for all those in the community who are not directly involved in the dispute but will suffer its effects, reconciliation with the earth. What is at stake is the regeneration of life itself. Any breach in our relationship with one another in the visible world affects relationships with agencies in the invisible world: "...disruptions in the visible world, brought about by bloody conflicts, not only destabilize elements of the visible world horizontally, they also disrupt the vertical connection rendering a given territory infertile, unproductive and unable to sustain life. Thus the aim of dispute-settlement is...the holistic restoration of a visible and invisible ecology where the vertical relationships are the conduit for life itself."[144]

Conclusion

The pivotal role of *tindaanas* and soothsayers in processes of inauguration of political power and conflict transformation, as well as their shared jurisdiction over the "restoration of a visible and invisible ecology" of human-earth relations, reveals such forms of African indigenous authority as inherently eco-political. The dialogical relationship between these traditional leaders and the earth spirits eclipses anthropocentric interpretations of *tiŋa* as a passive set of intergenerational "assets" that require active human trusteeship. Rather, *tiŋa* emerges as the earth's dynamic order of sovereign activity that shapes and legitimates coexisting orders of human governance and sovereignty. In contrast to dominant Western understandings of sovereignty as unitary or linked to the centralized authority of a state, *tiŋa* reveals the sources of African indigenous political authority as multiple and diverse. And it is this multiplicity and diversity, expressed in the model of complementarity of power among chiefs, *tindaanas* and elders, that sustain it as a time-tested, alternative form of eco-egalitarian authority that wards off hierarchy. By taking into account the agency and expressivity of material nature, these ecology-based political systems have developed built-in checks not only on the centralizing tendencies of state-making and colonial processes, but also on the possessive materialism and commodification of land relations advanced by market capitalism, which fuel the contemporary ecological crisis.

[143] Kirby, "The Earth Cult," 138–139.
[144] Kirby, "The Earth Cult," 135.

Whether it is elders theorizing with the earth spirits under the shade of ancestral trees or postcolonial leaders advancing Pan-African models of eco-humanism, together these traditions of thought prompt us to realign human systems of ethics, politics and governance with the ecological flux of becoming of the earth. The concepts of *ubuntu*, *ukama* and *tiŋa* encourage us to focus on relationships of interdependence and reciprocity and to pursue a practice of environmental ethics and political theory that is more open to attunements that come from a communicative and agentic world of vital material forces. These traditions emphasize the multi-directionality of human ethical conduct with regard to the interdependent wellbeing of past, present and future generations and the nonhuman world. They seek to cultivate responsiveness and adaptiveness to the ceaseless flux of change in a world of becoming that is not pre-designed for humans, making them an invaluable resource for the practice of political theory in the age of Anthropocene dominated by anthropocentrism. Such a practice is essential today to pluralize a Eurocentric habit of theorizing that privileges written texts and abstract concepts over oral traditions and situated knowledges or focuses on the human exclusively and as superior to the environment. By highlighting human embeddedness within a multigenerational more-than-human community, it is also my hope that a productive extension of ideas of what it means to be human arises from the engagement with undervalued ecological aspects of African eco-humanism. Perhaps, its concepts of earth-centered authority, agency and collectivity can be viewed as part of broader processes of world-making that connect African thought to divergent practices of politics, which challenge hegemonic Western ideas of what counts as sovereign order. In the language of Plumwood's philosophical animism, these traditions "allow us to be receptive" to the urgent and proliferating messages of heavy rains, floods, landslides, fires and droughts. The ancestors have been yelling at us. Are we listening and acting in accordance with their time-honored advice? Such attentiveness may open up "unanticipated possibilities" to negotiate with the earth a different, more ecologically just future, for which our generation can be remembered and respected by posterity.

References

Adotevi, Stanislas. *Négritude et Négrologues*. Paris: Editions 10/18, 1972.

Asante, S.K.B. *Property Law and Social Goals in Ghana 1844–1966*. Accra: Ghana Universities Press, 1975.

Awindor, Fara Jim. "A Quest for Sustainable Development: The Interface of Traditional African Medicine and Western Science-based Medicine." PhD dissertation, University for Development Studies, Tamale, Ghana, 2018.

Azaare, Christopher Anabila. *The Traditional Gurensi Religious Beliefs and Practices: Totems*, vol. 1. Unpublished Manuscript, 2018.

Azaare, Christopher Anabila. *Tindaanaship and Tindaanas in Traditional Gurensi (Frafra) Communities: Land Use and Practices*. Africa Local Intellectuals Series 90, no. 4, 2020.

Behrens, Kevin. "The Imperative of Developing African Eco-Philosophy." In *Themes, Issues and Problems in African Philosophy*, edited by Isaac E. Ukpokolo, 191–204. Cham, Switzerland: Palgrave, 2017.

Behrens, Kevin Gary. "Moral Obligations Towards Future Generations in African Thought." *Journal of Global Ethics* 8, nos. 2–3 (August–November 2012): 179–191.

Bentsi-Enchill, Kwamena. *Ghana Land Law: An Exposition, Analysis and Critique*. Lagos: African Universities Press, 1964.

Blyden, Edward. "Study and Race: A Lecture to the Young Men's Literary Association of Sierra Leone." In *Black Spokesman: Selected Published Writings of Edward Wilmot Blyden*, edited by Hollis R. Lynch, 195–204. London: Frank Cass, 1971.

Bujo, Bénézet. *Foundations of an African Ethic: Beyond the Universal Claims of Western Morality*. New York: The Crossroad Publishing Company, 2001.

Callicott, J. Baird. *Earth's Insights: A Multicultural Survey of Ecological Ethics from the Mediterranean Basin to the Australian Outback*. Berkeley: University of California Press, 1994.

Dakubu, Kropp, Awinkene Arintono, and Avea Nsoh. *Gurenɛ—English Dictionary*. Legon: Linguistics Department, University of Ghana, 2007.

Du Toit, Louise. "'When Everything Starts to Flow:' Nkrumah and Irigaray in Search of Emancipatory Ontologies." *Phronimon* 16, no. 2 (2015): 1–20.

Eze, Michael Onyebuchi. "*Humanitatis-Eco (Eco-Humanism)*: An African Environmental Theory." In *The Palgrave Handbook of African Philosophy*, edited by Adeshina Afolayan and Toyin Falola, 621–632. New York: Palgrave, 2017.

Fanon, Frantz. *Les Damnés de la terre*. Paris: Maspéro, 1961.

Guri, Bernard Yangmaadome, and Wilberforce A. Laate. "Community Organisational Development in South-Western Ghana." In *Endogenous Development in Africa: Towards a Systematisation of Experiences*, edited by David Millar, Agnes A. Apusigah, and Claire Boonzaaijer, 75–89. Barneveld, The Netherlands: COMPAS/UDS, 2008.

Ignatov, Anatoli. "The Earth as a Gift-Giving Ancestor: Nietzsche's Perspectivism and African Animism." *Political Theory: An International Journal of Political Philosophy* 45, no. 1 (February 2017): 52–75.

Ignatov, Anatoli. "Ecologies of the Good Life: Forces, Bodies, and Cross-Cultural Encounters." PhD Dissertation, Johns Hopkins University, Baltimore, United States, 2014.

Irele, Abiola. "Négritude—Literature and Ideology." *The Journal of Modern African Studies* 3, no. 4 (December 1965): 499–526.

Jackson, Michael. *Allegories of the Wilderness: Ethics and Ambiguity in Kuranko Narratives*. Bloomington: Indiana University Press, 1982.

Jones, Donna V. *The Racial Discourses of Life Philosophy: Négritude, Vitalism, and Modernity*. New York: Columbia University Press, 2010.

Kirby, J.P. "The Earth Cult and the Ecology of Peacebuilding in Northern Ghana." In *African Knowledges and Sciences: Understanding and Supporting the Ways of Knowing in Sub-Saharan Africa*, edited by David Millar et al., 129–148. Barneveld, The Netherlands: COMPAS/UDS, 2006.

Kresse. Kai. "'African Humanism' and a Case Study from the Swahili Coast." In *Humanistic Ethics in the Age of Globality*, edited by Claus Dierksmeier et al., 246–265. London: Palgrave Macmillan, 2011.

Langwick, Stacey Ann. *Bodies, Politics, and African Healing: The Matter of Maladies in Tanzania*. Bloomington: Indiana University Press, 2011.

Le Grange, Lesley. "Ubuntu." In *Pluriverse: A Post-Development Dictionary*, edited by Ashish Kothari et al., 323–326. New Delhi, India: Tulika Books, 2019.

Macgaffey, Wyatt. *Chiefs, Priests, and Praise-Singers: History, Politics, and Land Ownership in Northern Ghana*. Charlottesville: University of Virginia Press, 2013.

Madavo, Garikai. "African Environmental Ethics: Lessons from the Rain-Maker's Moral and Cosmological Perspectives." In *African Environmental Ethics: A Critical Reader*, edited by Munamato Chemhuru, pp. 141–152. Cham, Switzerland: Springer, 2019.

Mamdani, Mahmood. *Citizen and Subject: Contemporary Africa and the Legacy of Late Colonialism*. Princeton: Princeton University Press, 1996.

Manzini, Nompumelelo Zinhle. "African Environmental Ethics as Southern Environmental Ethics." In *African Environmental Ethics: A Critical Reader*, edited by Munamato Chemhuru, 111–123. Cham, Switzerland: Springer, 2019.

Mbembe, Achille. *Critique of Black Reason*. Translated by Laurent Dubois. Durham: Duke University Press, 2017.

Metz, Thaddeus. "An African Theory of Moral Status: A Relational Alternative to Individualism and Holism." In *African Environmental Ethics: A Critical Reader*, edited by Munamato Chemhuru, 9–27. Cham, Switzerland: Springer, 2019.

Millar, David. "Ancestorcentrism: A Basis for African Sciences and Learning Epistemologies." In *African Knowledges and Sciences: Understanding and Supporting the Ways of Knowing in Sub-Saharan Africa*, edited by David Millar et al., 53–63. Barneveld, The Netherlands: COMPAS/UDS, 2006.

Murove, Munyaradzi Felix. "An African Commitment to Ecological Conservation: The Shona Concepts of Ukama and Ubuntu." *Mankind Quarterly* 45, no. 2 (Winter 2004): 195–215.

Niang, Amy. *The Postcolonial African State in Transition: Stateness and Modes of Sovereignty*. London: Rowman and Littlefield International, 2018.

Nkrumah, Kwame. *Consciencism: Philosophy and Ideology for De-Colonization*. New York: Monthly Review Press, 1964.

Plumwood, Val. *Environmental Culture: The Ecological Crisis of Reason*. New York: Routledge, 2002.

Ramose, Mogobe B. *African Philosophy through Ubuntu*. Harare: Mond Books, 1999.

Ramose, Mogobe B. "Ubuntu." In *Degrowth: A Vocabulary for a New Era*, edited by Giacomo D'Alisa, Federico Demaria and Giorgos Kallis, 211–214. New York: Routledge, 2015.

Rattray, R.S. *The Tribes of the Ashanti Hinterland*, vol. 1. London: Oxford University Press, 1932.

Senghor, Léopold Sédar. "Négritude: A Humanism of the Twentieth Century." In *I Am Because We Are: Readings in Africana Philosophy*, edited by Jonathan Scott Lee and Fred L. Hord, 55–64. Amherst: University of Massachusetts Press, 2016.

Soyinka, Wole. *Myth, Literature and the African World*. Cambridge: Cambridge University Press, 1976.

The Project Secretariat of the ACLP. "Literature Review Report on Customary Law on Land in Ghana." *National House of Chiefs/Law Reform Commission*, 2009.

Twumasi, P. A. *Medical Systems in Ghana: A Study in Medical Sociology*. Tema: Ghana Publishing Corporation, 2005.

Wiredu, Kwasi. "Philosophy, Humankind and the Environment." In *Philosophy, Humanity and Ecology: Philosophy of Nature and Environmental Ethics*, edited by H. Odera Oruka, 30–48. Nairobi, Kenya: African Centre for Technology Studies, 1994.

Woodman, G.R. *Customary Land Law in Ghanaian Courts*. Accra: Ghana University Press, 1996.

The Social Construction of International Environmental Policies in the Caribbean: The Case of Sargassum

Andrea Parra-Leylavergne

INTRODUCTION

One of the main features of international environmental policy seems to be its characterization as a policy under continuous construction. High-level meetings and their results are added to local public and private initiatives for the preservation of environmental resources. This implies a constant evolution of political systems that are under permanent construction. And the ones who build these systems are actors and agents coming not only from international organizations, regional organizations, but also from the civil society and from local governments. In specific regions such as the Caribbean, international environmental policy follows this same precept: it is a social construction. Being social, international environmental policy in the Caribbean is necessarily adapted to the specific phenomena of this region. These phenomena may be traditional, such as marine pollution, hurricanes, earthquakes, or less traditional, such as invasions of sargassum. Being social indeed, this international policy is subject in its elaboration, in its evolution and in the instruments of its

---

This work was supported by the FEDER CESAR and the ANR-19-SARG-0005

A. Parra-Leylavergne (✉)
National Center of Scientific Research (CNRS), Schoelcher, Martinique
e-mail: andrealeylavergne@hotmail.fr

improvement to the perception of these actors, to their interests and to their regional development objectives.

The objective of this contribution is to examine the manner in which environmental policies concerning the management of marine resources and ecosystems can be considered as being a social construction in the Caribbean as a region of the Global South. The analytical methodology is international constructivism. If I address these policies from this perspective, it is because I consider it to be the most adapted method to their rigorous conceptualization. Likewise, these policies are, at the same time, the result of an intersubjective interaction between diverse actors, related at several layers, whose interests and identities are in constant evolution. In order to illustrate the complexity of this approach, I have chosen to apply its various notional dimensions to the environmental policies mentioned (under construction/evolution) in the Greater Caribbean to the international management in this region of a fairly recent phenomenon: the massive stranding on the beaches of the Caribbean nations of sargassum algae.

Hence, the present contribution is articulated in three main sections. In the first one, I propose a discussion of the axioms, paradigms and conceptual elements that allow us to identify post-positivist approaches, including constructivism, as the most appropriate for offering an informed reading of environmental policies in the Caribbean. This reading will permit us to show how metatheoretical discussion enables us to conceptualize environmental policy in the Caribbean as a structure of the international system. Social structures are systemic units composed of three elements: shared knowledge, material resources and practices. These elements are interdependent and exist specifically as a function of their role in the structure (Wendt, 1995).

Considering environmental policy in the Caribbean as a structure implies that we can analyze it from the two fundamental perspectives of post-positivism: the ontological perspective (interested in how material resources, practices and knowledge interact within the structure) and the epistemological one (interested in what all these elements are made of). Thus, in its second section, this contribution proposes, from an ontological angle, a discussion concerning the manner in which the actors and their *agency* are determinants of the process of construction—evolution of the management system of environmental challenges in the Caribbean. This analysis draws on sargassum management processes to illustrate this chapter's points. Finally, our third section consists of an epistemological reading of environmental policies in the Caribbean. This analysis is epistemological because it requires discussing the approach to how this policy is to be understood. It allows us to glimpse the way in which the very structure of international environmental policy (in the broad sense) and in the Caribbean (in the narrow sense). It leads to its characterization as a Kantian anarchy in the constructivist sense in which the latter was described by Alexander Wendt in his works.

International Environmental Policies Concerning Sea Resources in the Caribbean: The Social Construction Process in a Material Context

Conceptualizing International Environmental Policies in the Caribbean

International environmental politics is a macrostructure of the international relations system (Onuf, 2013). This system is, at the same time, at the origin of and constructed by substructures (or "microstructures" in terms of Alexander Wendt, 1987). In the Global South, these substructures have hybrid configurations that become particularly complex when referring to the Caribbean systems of governance of sea and of maritime resources (Leff, 2021).

In order to conceptualize and explain this system, it seems relevant to adopt as a starting point for our discussion a rather traditional, yet structuring approach of the sociology of international relations: that of its three major metatheoretical debates (Dunne, 1995). These three debates are the following: firstly, that of the analytical level from which it is relevant to approach knowledge; secondly, the debate about the connection between the actors and the structure (Wendt, 1987); and thirdly, that of the interdependence between the first two.

The debate on the level of analysis (Singer, 1960; Taylor et al., 1960) implies that we should consider the angle from which we approach knowledge on the management of marine resources and ecosystems in the Caribbean. This level is global. It concerns an international area that takes part of the dynamics of the major international forums (O'Riordan and McCormick, 1990) for the environment resulting from the first human environment conference in Stockholm in 1972 (Vogler, 2018) and its results. After a first report on the Environment in 1968, a macrostructure of international relations began to take shape in the Stockholm Declaration of 1972 (United Nations [UN], 1972), articulated on 26 principles of environmental protection and an Environmental Action Plan.

If the level of analysis that takes into consideration this macrostructure of international relations is global, it is because we are referring to an institution whose purpose goes beyond the simple accumulation of the predetermined wills of States gathered in an international forum. It is constructed by addressing, from its origin, humanity in its modern characterization of Anthropocene (Lövbrand et al., 2020). The fact that this macrostructure is being created in the direction of the Anthropocene requires IR analysts to step outside the traditional framework for studying the international behavior of states, realist (Duroselle, 1962) and transnationalist theoretical approaches (Rosenau, 2018). The environment as a global issue (Taylor et al., 1960) requires the social sciences to examine the international system going beyond the idea of supranationality (Archer, 2010) and transnationality. The Stockholm Conference is considered by some authors as the very beginning of the global debate between industrialized and developing countries on the need to strike a balance between the notion of economic growth as a condition for human welfare and the need to preserve the land and marine resources exploited within the hegemonic neoliberal capitalist worldview.

Social structures are not reduced to the effects of their agents. They are consolidated on the basis of three elements: material conditions, interests and ideas (Wendt, 1999). Material conditions are both contextual and structural. They are contextual because they arise from—and are embedded in—social reality. They are structural because they embody the ideas and interests that constitute them (Wendt, 1987). The United Nations Environment Programme (UNEP), which emerged in 1972 from the Stockholm Conference, makes materializes/illustrates these two types of material conditions. In fact, UNEP appears to be a contextual element of this social construction in the sense that it constitutes a system that is itself embedded in a global one: the international system of environmental policy.

According to the theories of social constructivism, interests and ideas consolidate elements for the structures of the international system. In the Caribbean, interests and ideas about maritime management are embodied and conveyed by international actors that are institutions, such as the United Nations Environment Programme (UNEP), The Convention for the Prevention of Marine Pollution or "MARPOL" (International Maritime Organization [IMO], 1973), the United Nations Regional Sea Programs (launched in 1974), and the United Nations Convention on the Law of the Sea (UNCLOS) (UN, 1982). These institutions and international actors coexist and interact. They compose an interconnected substructure with the vocation of transforming shared ideas into a mechanism that shape actors' interests and actions and to mesh with other structures. They contribute to the establishment and development of a common cognitive framework considering the Wider Caribbean Sea as being humanity's collective resource that has to be protected. This common cognitive framework is constitutive of the system. It allows for an exchange of intersubjectivities between actors and agents.

In the Caribbean region, the maritime governance system is a structural element. As such, it produces two types of effects that are specific to IR structures: causal and constitutive effects on the corresponding macrostructure (Wendt, 1999).

The "causal effects" of macrostructures refer to the outcomes that may result from the occurrence of certain structures and actions. The existence of a broader UN framework for marine governance in the Caribbean results in the discursive positioning of issues at the center of the concerns of different actors of the region. Structures do not exist independently of their discursive dimension (Wendt, 1987, 2009). The latter contributes to the establishment and positioning of environmental diplomacy initiatives on topics such as the fragility of ecosystems in the Caribbean region, their vulnerability and the requirement to safeguard them.

The structures of this system also have constitutive effects. The Caribbean Environment Program (CEP), which emerged in 1981 from the UNEP

Regional Marine Program, led to the establishment in 1983 of the *Convention Pour La Protection et La Mise En Valeur Du Milieu Marin Dans La Région Des Caraïbes* (Convention for the Protection and Development of the Marine Environment in the Wider Caribbean Region (WCR) or Cartagena Convention, UNEP-CEP, 2012). The Cartagena Convention is one of the seven Regional Seas Conventions and Action Plans of the UNEP (UNEP). The Convention for Cooperation in the Protection, Management and Development of the Marine and Coastal Environment of the Atlantic Coast of the West, Central and Southern African Region (Abidjan Convention), the Convention for the Protection of the Marine Environment and the Coastal Region of the Mediterranean (Barcelona Convention), the Convention for the Protection, Management and Development of the Marine and Coastal Environment of the Western Indian Ocean (Nairobi Convention), the Framework Convention for the Protection of the Marine Environment of the Caspian Sea (Tehran Convention) and the Coordinating Body on the Seas of East Asia (COBSEA) and the Northwest Pacific Action Plan (NOWPAP). These UN conventions compose a structure focused on sea resources' protection in which they are embedded.

However, the Cartagena Convention is not only part of a whole. It also constitutes a system on its own. Its internal structure can be broken down into three protocols. The first one is related to the prevention and control of pollution generated by hydrocarbons. It is called Oil Spill Protocol (OSP) (UNEP-CEP, 2012). The second one concerns wildlife conservation in the region. It is known under the title of Protocol on Specially Protected Areas and Wildlife (SPAW) (UNEP-CEP, 2012). The third protocol addresses pollution from Land-Based Sources and activities (LBS) (UNEP-CEP, 2012).

These three protocols were initially conceived as being of a technical nature. Nevertheless, their concretization and transformation into discussion forums and work spaces for regional cooperation in the Caribbean have a high political component. The advances of their technical work are shared every two years in meetings of Contracting Parties (COPs) gathering diplomats and scientists in a hybrid social system in which the separation between political and scientific discourse is less and less visible. The nature of the analyses that can be made of such a social construction is then holistic (Reus-Smit, 2009). Holistic analysis within the discipline of international relations imply a type of scientific approach that considers that "parts exist only in relation to wholes" (Fearon and Wendt, 2012).

A holistic assessment of the systems of international relations proposes a methodological approach of considering the degree to which "the parts of a whole behave in the way that the whole requires it" (Hollis and Smith, 1994). It also proposes an ontological analysis in the sense that those environmental policies and the conceptual tools that conceive and study them impose decompartmentalizing the disciplinary discussions (Broadhead, 2002). Talking about environmental policies in the Caribbean compels IR theorists to be conscious of the fact that this region is regularly the scene of a superposition or combining of crises (Bull-Kamanga et al., 2003) related to violent and repeated climatic events (Bouchard-Bastien, 2020). Some of these events consist of natural catastrophies like the earthquake and tsunamis in Haiti in 2010 or the Hurricane Irma in Saint-Martin and Saint-Barthélemy in 2017. Others are slowly developing events like the damage to marine ecosystems due to the proliferation of invasive species, both fish and algae, and rising water levels which modify the habitat of citizens in coastal areas and coastal erosion.

The effects of these events are exacerbated when they affect Caribbean territories subject to an accumulation of structural vulnerabilities such as fragile road network, or access to potable water and sanitation networks. The last are particularly exposed or poorly protected from violent and constant climatic phenomena (Birkmann and Wisner, 2006; Blaikie et al., 1996). Moreover, entire neighborhoods and even villages are built in areas exposed to coastal erosion and to climatic challenges like hurricanes.

A huge part of these vulnerabilities are inherited from the colonial period (Amin, 1990; Tortosa, 2011). The composition of the economic tissue of these territories is inherited from that time in that they are highly dependent on the center of power that were their former colonial metropolis. Indeed, due to the effects of the distribution of production systems developed during the colonial period, industries have been developed in the former colonial metropolis. There is an almost total absence of industrial infrastructure in these territories. When it exists, it is not very diversified and is concentrated on two main activities: fishing and tourism. The rest of the economic production of these islands generally consists in agricultural exploitation (Yvars, 2019).

The effects of the combination of all these factors feed endogenous crises that are social, economic and migratory. They generate resistance and impose adaptations on the part of social actors. These social actors can be institutional like the representatives of councils, of local governments and even of decentralized national government offices or ministries. They also are non-institutional, like the representatives of informal associations defending the rights of citizens to a decent and safe habitat, members of civil society, federations of fishermen, farmers or tourism professionals. Thus, the policies related to the environment in this geographical space of the Caribbean region impose the challenge of establishing themselves in the interface of three strands of sciences as described by Elinor Ostrom (2010): natural sciences, economic sciences and political science.

Concretizing Environmental Policies Through International Instruments and Institutions

A thematic analysis of the three UN protocols associated with the Cartagena Convention and of the composition of their "regional activity centers" can provide an illustration about the way how the natural sciences can contribute to the construction and consolidation of the discourses that shape the work of the structures that host them. In this sense, while natural sciences allow for the measurement of pollution levels in the sea, the Convention on International Trade in Endangered Species (CITES) (1973) and the Samoa Agreement on Small Island Developing States SIDS (United Nations, General Assembly, 1994) are international agreements that serve as a backdrop for the political actions of actors and agents (Mahon and Fanning, 2021).

Also known as the Washington Convention, CITES convention, signed on 3 March 1973 and entered into force in 1975, gathered the representatives of 80 countries around a common objective. This common objective was to establish international parameters to control the international trade of approximately 34,000 species of living beings and their derived products (Convention on International Trade in Endangered Species of Wild Fauna and Flora—CITES, 1973). In concrete terms, this intergovernmental agreement establishes the international legal framework for regulating their trade. It places the international system in a dynamic of objectification of the principles of biodiversity conservation and reasonable exploitation of natural resources. This legal framework has been materialized in a control system that today is based on a document called CITES. CITES documents function as elements that provide a guarantee that a species in trade has been controlled by the authorities of the country of exploitation or any other competent authority and that its commercialization has been authorized because it takes into account the international standards established for its conservation (Cooney et al., 2021).

The Samoa Agreement established, in 1994, an international frame of reference that elaborates concrete guidelines for the elaboration of sustainable development policies specifically applicable to SIDS. This agreement enshrined global political actions such as the recognition in 1992 of small developing states as a specific group before the United Nations General Assembly. What makes this group specific is the recognition of a set of characteristics such as their small size, remoteness, limited access to resources for their development and strong exposure to climatic phenomena. This limits and undermines the sustainability of their development. This agreement gave rise to the Barbados Action Program. It had, among others, the following objectives: first, an informed analysis of the characteristics and trends of economic development in the various small island developing states; second, to examine the nature and extent of the vulnerabilities of these states; and, finally, to proceed to define the specific actions and policies to be developed in order to constitute sustainable development mechanisms that are adaptable and adapted to the vulnerabilities and indigenous realities of these territories (United Nations, 1994).

In this sense, these two agreements function as a backdrop for any political development in the region. They demarcate the space for political action because they define the criteria and parameters on the basis of which decision-makers transcribe into local and international rules the analyses and suggestions coming from the natural sciences. They influence the synthesis between resources' preservation and development sustainability to which decision-makers in the region are subjected when they elaborate their norms.

Hopefully, the next section's discussion of an illustration taken from environmental policy-making in the Caribbean will further illuminate and support these claims. It focuses on the case of the social construction for a sargassum management system in the Caribbean.

Illustrating International Environmental Policies in the Caribbean: An Ontological Analysis of a Sargassum Management System in Construction

The articulation between actors and structure (Wendt, 1987) broadens the scope of analysis up to the issue of structural complexity. This structural complexity is set up by actors of different nature, such as, independent states, local governments with special competences of central states (Martinica and Guadeloupe), regional organizations of economic integration, among others. It challenges IR scholars' observations because it requires an holistic Caribbean ontological analysis. In general terms, an ontological analysis explores "*what the social world is made of*" (Klöck and Fink, 2019). It compels IR scholars to move out of the field of positivist and mainstream approaches in IR analysis emphasizing: material results (rules and laws) of nation-states as done by the realist theoretical paradigm; the regional cooperation as a protective action for the preservation of their own resources as done by the liberal analysis or strategic actions for regional development as it is studied by the neo-functionalist approaches. It forces them to observe the regional process from a second point of view: the post-positivist one. Post-positivist analysis abandons the idea that reality is *something already given* (*or finished*) (Wendt, 2005).

Post-positivist scientists do not focus on the results of material facts and actions. They focus on processes. This former principle appears to be particularly relevant to the study of the construction of environmental policy. In fact, it supposes to be concerned with the evolutions constituting the social construction of international environmental public action. It supposes to approach the objectives, interests and identities of the international actors (States, local governments and international organizations) as being susceptible to evolve in the confrontation with those of other actors.

The Caribbean is a multidimensional space (Daniel, 2015) in which different substructures composed by all the actors formerly evoked. They are in constant formation, transformation, evolution and refinement. They can be defined by their essential characteristics such as their political and economic regimes, but also cultural. From the point of view of political status, the Caribbean is composed of two main kinds of state-actors. They are independent states, or outermost regions of European states including Dutch crown territories and those of the United States.

The independent states are Trinidad and Tobago, Saint Lucia, the Bahamas, Saint Vincent and the Grenadines. They belong to SIDS (United Nations, 1994). Their economic activities are particularly dependent on commercial resources from the sea. Their exchanges around the management of maritime resources are both state-centric (an exclusive competence of each respective State) and autonomous (not depending on any supranational entity). The group of outermost regions and territories is made up of territories with hybrid (non-independent) status. Puerto Rico and the US Virgin Islands are politically dependent on the United States (PR GOV, 2022). The British Virgin Islands, the outermost European territories such as Martinique, Guadeloupe or St. Barthélemy and the French territory of St. Martin are officially part of the overseas territories of the European Union (European Commission [EC], 2022). The territories of Curaçao and Bonaire are legally dependent on the crown of the Netherlands (López-Contreras et al., 2021). All these territories with different political status share, by their geographical situation within the Caribbean sea, a particularly important vulnerability to the effects and marine environmental problems of maritime phenomena. In the management of these resources, these Caribbean territories share what the social theory of international relations calls a "collective identity structure" (Wiener, 2006).

This collective identity structure is built on a shared post-colonial history that has produced interdependencies and shared languages that facilitate joint cultural identifications. This structure of international relations has favored regional configurations of cooperation as the creation of economic integration and cooperation organizations like the Caribbean Community and Common Market (CARICOM), the Organization of Eastern Caribbean States (OECS) and the Association of Caribbean States (ACS) (Galy et al., 2011).

My claim here is that this double consolidation of a collective regional identity co-constructed in the relations between the different agents (actors constituted into agents in function of their role in the structure) that constitute the Caribbean territories is a determining factor in their approaches to the environmental governance of maritime resources. The knowledge of this system is constitutive or creates constituents and is itself part of the latter (Braspenning, 2002). It is *"ideaist"* (issued from ideas and concepts) (O'Neill, 2016) in the sense that it no longer merely describes the system but constructs it from linguistic and political formulation of the collective knowledge constructed commonly by the Caribbean actors of their own role in international environmental policies (Wiener, 2006).

Since the beginning of the second decade of the twenty-first century, all these territories share a major challenge: the massive stranding of sargassum algae on their beaches (Van Tussenbroek et al., 2017). Sargassum is a specific type of floating or pelagic "macro algae" belonging to at least two different species: sargassum fluitans and sargassum natans (Oyesiku and Egunyomi, 2015). In fact, since 2011, large swathes of these types of algae have been making regular, massive and unpredictable occurrences in the waters of the Caribbean Sea (Sissini et al., 2017; Wang et al., 2019). Equally, regular, massive and unpredictable are their respective sargassum strandings on the windward coasts of the Caribbean islands. These strandings are blamed, especially by the local populations and local political leaders for causing numerous repercussions impacting the well-being of the island's human populations (Fraga et al., 2019; Ménez, 2019). These sargassum intrusions on Caribbean beaches are also the focus of attention of some scientists interested in analyzing their effects on ecosystems (Rodríguez-Martínez et al., 2019). These ecological effects present social and political challenges (Thirot et al., 2020) to authorities facing a multidimensional phenomenon that requires multiscalar processes (Compagnon, 2015) that should lead to the construction of a multilevel governance system (McConney and Oxenford, 2021) to manage and address the various threats to the Caribbean region as a result of sargassum beach encroachments.

In order to understand the various levels of this highly value-based multidimensional governance structure required to meet the sargassum challenge in the international system, it is necessary to consider the actors within the mentioned Caribbean structure: the process of the shaping of their identities and their historical development.

Caribbean Actors and Agents of the International Environmental Political System and Their Co-Constructed Identities

After an exercise of anthropomorphization (Wendt, 1999) I will refer, in the following lines, to Caribbean states and international organizations of the region as being "actors of Sargassum" in the Caribbean. These Caribbean states and international organizations have the three characteristics of a social actor. First, their existence is the result of an explicit intention. Second, they are collective human entities which implies a form of life. Third, they possess a determined collective consciousness and thus embody an intersubjective experience (Wendt, 2004).

Agency and Intentionality in the International Environmental System

These actors become agents through intersubjective exchanges with a structure that shapes and circumscribes their role (Wendt, 1992) as well as through exchanges with other actors who acknowledge and validate that agency (Bruce and Blumer, 1988). In dealing with sargassum management, decision-makers in local governments, states and international organizations in the Caribbean are increasingly becoming agents of a system that is constantly evolving since its inception: it did not exist before 2011 (UNEP, 2021).

We mentioned in our first section the manner in which the international instruments of the UN system have been designed by agents and recognized by their scientific or political authorities to shape and frame the management of Caribbean marine resources. Agents within the resulting cognitive structure take on the role of agents of the sargassum system progressively and intentionally (Soltani et al., 2014).

Embracing this role implies cognitive positionings that are shaped through the performative function of speech acts. Speech acts are defined as "performative" from their root verb "to perform," which means to act (White et al., 1963). This implies that talking about events and institutions entails an action whose objective is to construct them. Thus, in the international system, performative speech acts have a contractual meaning when they announce events: "We have no choice but to reach a global agreement on combating climate change" (ParisClimate2015, 2015, p.1) or declarative: "This Agreement, by enhancing the implementation of the Convention including its objective, aims to strengthen the global response to the threat of climate change, in the context of the climate change, in the context of sustainable development and efforts to eradicate poverty" (Paris Agreement, 2015a, art. 2).

These positionings are at the origin of the definition of both the actors' identities and their mandates for action. According to the radical approach of international relations theories (Ashley and Walker, 1990), the discourses that are expressed about reality are not only a cognitive reaction to it, but an integral part of its construction (Battistella et al., 2019). As a result, the placing of sargassum algae on the international agenda stems from the performative function of language exchanges (Fierke, 2002) among the agents of this system under construction and operation. Thus, a problematization of the process of construction of sargassum as a political problem starts with a set of very simple questionings. By what process does sargassum algae go beyond the domain of natural science only and become a political subject for study and decision-making? For which actors has sargassum become political? How do local, national, and regional actors define and perceive the sargassum challenge? What kind of actions do they expect to be taken by policy-makers to deal with sargassum strandings? Finally, at what level (local, regional, international) do these actors expect the corresponding responses? It is through the possibility of grasping reality by answering these questions that the process of publicizing and recognizing the problems presented by sargassum that transforms the system under construction into a social construction. The discursive exchanges between the different agents of the system are constitutive and constituent of any social reality. As Onuf (2013) noted, such exchanges contribute to shaping the identities of the agents in construction within this structure meaning that these exchanges about sargassum create collective social reality while also becoming a part of such a reality.

International Agents: Initiatives and Social Personalities Under Construction

Intentionality has been a determining factor in the construction of another agent of this structure: the SPAW sargassum working group on sargassum.

This new sargassum working group was established under the auspices of the SPAW Protocol of the Cartagena Convention in 2020 (UNEP, 2021a). The establishment of this group followed a bottom-up dynamic similar to the one that led to the inclusion of this algae in the public management agenda. They emerged, first separately in each stranding territory, as being a political issue to be addressed by local and national decision-makers in the Caribbean (International Conference on Sargassum, 2019). Then, they became an environmental political issue of regional scope.

It is in this same direction that the working group on sargassum was created within the SPAW structure. The expression of the concerns and demands of some representatives of the eighteen stakeholders affected by the sargassum strandings transformed this subject into a political issue (OECS, 2019). These same stakeholders then requested its positioning as a key priority task through the creation of this group. Formally established in January 2020 (UNEP, 2021b), this working group has the double mandate to coordinate collaboration with relevant regional and global initiatives in order to encourage synergies and to seek regional solutions to the proliferation of sargassum (STAC 12). It suggests: "cooperating with relevant partners to assess and merge information and best practices on the management of the sargassum influx impacting Caribbean countries" (STAC 8) and having scientific members with expertise in both Specially Protected Areas and Wildlife (SPAW) and Land-Based Sources of pollution (LBS) protocol topics, and thus examine potential health risks related to heavy metals while liaising with other Sargassum agents in formation, such as, the Atomic Energy Commission, UNESCO-IOC (on marine scientific research), CARICOM, UNDP or the Association of Caribbean States (UNEP, 2021b). This cooperation mandate has established a principle of networking that takes place at two levels. The first is internal. It questions the place and role of the political and scientific dimensions in the shaping of the international personality of this agent. The second is external. It questions its place in the general system of environmental agents in the region.

Constructing Social Personalities of International Environmental and Sargassum Actors and Agents in the Caribbean

An agent of the international relations system possesses, according to the social-constructivist approach, three different, interrelated personality types (Kowert, 2012). A psychological personality that refers to psychological and cognitive attributes. This internal component is constructed and consolidated based on the actor/agent's intersubjective exchanges, as well as between the actor's organic components. A moral personality (external component) that can be designated as that which constrains the behavior of the actors forcing them to adhere to a certain framework or moral code. The legal personality (also external) characterizes the social beings as beneficiaries of certain rights bound by obligations and enjoying prerogatives and privileges. These last two personalities are considered to be social conventions (Wendt, 2004).

The psychological personality of the Specially Protected Areas and Wildlife (SPAW) sargassum working group is shaped by its membership in the UN system. In fact, the sargassum working group is the most recent of the four groups that contribute to the SPAW internal functioning (UNEP, 2021a). One of the main characteristics of these groups is that they have been conceived as being adhoc groups (UNEP, 2001). In this sense, the contours of their psychological personalities are constantly tailored on the basis of their experience as part of the UN system. The case of the Sargassum Working Group is particularly illustrative of this assertion in that its two main mandatory tasks consist on developing "clear objectives and responsibilities for the Working Group" and establishing "coordination and collaboration with relevant regional and global initiatives in order to promote maximum impact of synergies and solutions to the sargassum outbreak" (UNEP, 2021c, art. 4). We can even observe that this group is in constant self-definition and that its only framework for this self-definition is the UN's system. In other words, one can think that its psychological personality is tailored by its membership to the UN.

In this system under construction and in constant evolution, the experts' meetings which consist of the functioning of the sargassum working group contribute also to the shaping of the psychological personality of the system itself. It seems to become reinforcing a dialogue between member states affected by sargassum that has been materialized by meetings like de Galveston Sargassum symposium in the United States, Cancun Sargassum Session in Mexico, Sargassum sessions in Colombia, Barbados Sargassum symposium in Barbados (Fardin, et al., nd). So collective construction of the cognitive attributes of this kind of personality is also the result of an international exchange of knowledge within this system between the Sargassum Working Group and the UN member countries. The contours and the scope of this relation are constantly being re-evaluated, according to the advances in the environmental discussions taking place within the system in which they are embedded as well.

The moral and legal personalities of this agent of the sargassum system are dependent on the context. They are subject to a network dynamic that operates in an interdependent way as if they were part of a gear within a mechanical system. In fact, the moral framework that determines the personality of this agent also contributes to shape the personalities of other agents. The Organization of American States (OAS) has been providing, since 2015 (Organisation of American States [OAS], November 2015) spaces for diplomatic discussion, for an eventual characterization—not yet possible—of sargassum. One of the options requested by the representatives of the Caribbean states within the OAS diplomatic spaces for such a characterization consists in considering sargassum as a cause of natural disaster.

One of the first written traces of this positioning is found in the speech of Joy-Dee Davis Lake, Alternate Representative of Antigua and Barbuda to the OAS in 2016. He proposed to consider sargassum as a disaster risk factor so that it can be managed under the Sendai Framework for Disaster Risk Management 2015–2030 (OAS, October 25, 2016). The Sendai Framework for Disaster Risk Reduction 2015–2030 was adopted at the Third UN World Conference in Sendai, Japan, on March 18, 2015 (UN, 2015b). The Sendai Framework is the successor instrument to the Hyogo Framework for Action (HFA) 2005–2015 and serves as the framework for action in the Caribbean in response to all types of disaster risks (Association of Caribbean States [ACS], 2022). The fact of introducing a discursive relationship between sargassum and the risks of catastrophe, within OAS diplomatic spaces, contributes to a cognitive positioning of this algae as a challenge (in the sense of a threat) for the Caribbean territories. Hence, the OAS acts both as a space for the production and circulation of shared knowledge on sargassum and as an actor of the management system of this algae. The OAS member states, by proposing such a direct relation between sargassum and disaster risk, by adhering to this approach and by exchanging on it, act as components of a system which they themselves make work as they interact with each other.

Other elements contributing to the definition of its moral personality as an agent of the sargassum are subsidies under the Reef Fix project of the Department of Sustainable Development of the OAS, intended to help countries affected by the sargassum (Huber, 2018), while pointing to its eventually negative effects on tourism in meetings with the decision-makers of its member states (OAS, 2019) or international forums (Inversión Turística, 2019). Thus the dialogue between member states that are organic actors and between OAS and those member states' representatives contributes to the consolidation of the OAS moral personality, especially in terms of the construction of shared knowledge about sargassum's characterization or definition.

The Association of Caribbean States (ACS) is another agent involving sargassum. At its initiative, a Symposium entitled "Challenges, Dialogue, and Cooperation towards the Sustainability of the Caribbean Sea" was held in 2015 in Trinidad and Tobago (ACS, 2015). Several agents of the system agree that this initiative was crucial for the dialogue on sargassum in the region. Organically speaking, the ACS relies on a "Caribbean Sea Commission (CSC)" to structure projects for the preservation of ecosystems in the Caribbean, such as the "Caribbean Sea Initiative" (CSI)—a proposal submitted by ACS Member States to the United Nations General Assembly (UNGA) in 1999, requesting the designation of the Caribbean Sea as a special area in the context of sustainable development.

Within the framework of its proposals, the ACS positions its regional legal personality as having a particular focus on the governance of the Caribbean Large Marine Ecosystem (LME). It positions itself as an actor and as a forum for the convergence of various elements related to the preservation of Caribbean resources, as in paragraph 25 on the protection of the oceans of the Barbados Program of Action, BAP, the United Nations Convention on the Law of the Sea (UNCLOS) and in particular, in terms of cooperation between coastal States bordering semi-enclosed seas (Article 123), the UN General Assembly Resolutions 54/225 seeking to promote "an integrated approach to the management of the Caribbean Sea in the context of sustainable development" (United Nations General Assembly [UNGA], 2001) and 63/214 entitled "Towards the sustainable development of the Caribbean Sea for present and future generations" (UNGA, 2012). The specific positioning of the ACS as an agent of the sargassum structure is done through two of its organic components. These are the Disaster Risk Reduction Directorate and the Caribbean Sea Commission. This approach toward treating sargassum crosses two cognitive variables: risk of disasters and the preservation of maritime resources. In this way, it contributes to positioning sargassum as a disaster risk factor for the different Caribbean territories. In the process of evolution of this agent of the system, it seems interesting to wonder about the place in terms of priorities that sargassum will take in the future in the agenda of priorities of the organization.

The Organization of Eastern Caribbean States (OECS) also functions as a sargassum agent in this system. A Caribbean player since 1980 (Treaty of Basseterre), this organization assumes its role as an environmental agent through its Sustainable Development Division. Decisions are taken by the Council of Ministers of the Environmental Sustainability (COMES) of the Member States. After preparatory meetings of experts in 2016, in 2018 (OECS, 2018) and in 2019 (OECS, 2019), this Council has declared itself in favor of a regional approach to the management and prevention of sargassum-caused damage. A call to the system resulting from the Cartagena Convention (the SPAW system) to integrate sargassum as a priority of its action appears as an element enabling us to open a way for a systemic analysis of a mutual shaping between the SPAW offices and the OECE of the legal personality of these two actors.

This institutional system is complemented by the action of the Caribbean Regional Fisheries Mechanism (CRFM), an autonomous institution belonging to the Caribbean Community and Common Market (CARICOM) system. Its mandate is articulated on the objectives of contributing to the effective management of marine and other aquatic resources under the jurisdiction of Member States through the provision of advisory services and technical advice to the fisheries divisions of these same states as well as cooperation between them (CRFM, 2002). Its vocation being technical, the CRFM collaborates with research programs and other regional projects aimed at improving the understanding of sargassum (plant and food research and survey on the impacts of the influxes in the region).

Consolidating International Environmental and Sargassum Actors and Agents' Identities in the Caribbean

Taking into account that, from our angle of analysis, social structures exist only in their processes (Wendt, 1995), we can underline that such coexistence in process contributes to shape the identities of the agents. An identity is consolidated in the practice of shared behaviors materializing the cognitive structures that are at its origin. Agents then develop a series of practices that, becoming repetitive, constitute patterns of social behavior generating collective recognition of the roles of the mentioned agents (Onuf, 2013). In social theory, the word "practice" is a legacy of Marx who defines it as a "sense of human activity" (Marx, 1959). Strictly speaking, "sense of activity" implies "direction." To choose one direction rather than another implies, naturally, to be capable of discernment. In other words, to make these social practices converge toward the construction and consolidation of a shared identity is a rational option of the social actors (Katzenstein et al., 1998).

Reciprocal typification takes place within the process of concretizing the sargassum system at the regional level in the Caribbean. It entails the co-construction of socially produced meanings within the structure. With their reiteration of acts related to this reciprocal typification, the actors acquire an awareness of the meanings produced. This collective cognitive process gradually leads to a history and a common meaning, participating in the institutionalization of a reality (Gardien, 2020).

It is precisely the interactionist nature (Wendt, 1992) of identity formation that may be challenging to the agents' positioning in the sargassum system. Their actions evolve through meetings, forums and research collaborations. Identifying their actions and achievements by other stakeholders beyond the system remains quite difficult. Their existence is barely recognized by agents other than the Caribbean inter-institutional network of highly specialized organizations. For example, local policy-makers and whistleblowers' representatives report a lack awareness on this set of agents and on some of their more relevant actions related to sargassum (CESAR project, 2022). In this sense, we can conclude to a kind of invisibility of the sargassum agents' system for local policy-makers and whistleblowers. This absence of visibility seems to make its own positioning and even the shaping of its own collective identity difficult. According to research in progress, one of the reasons for this difficulty of the sargassum agents' system identity consolidation can be in the absence of a legal and political instrument to consolidate its juridical personality. The following part presents a discussion of this absence of legal elements to define and to manage sargassum in the Caribbean. It raises one of the major discussions of constructivist analysis as a post-positivist approach applied to the international system. According to it, the state of anarchy is considered as an intrinsic characteristic of the international system. But contrarily to what is proposed by positivist analyses (realist, neo-functionalist and liberal ones), this state can take place in cooperative relationships, such as the one that takes place in the Caribbean with respect to the environment.

Looking for Definitions and Regulations in the International Environmental Policy System in the Caribbean

Social agents have a tendency to crave belonging to regulated environments (Onuf, 2013). This social behavior can be explained by the necessity to make the boundaries and obligations of one's role within institutions legible to actors, like States, international organizations and stakeholders. An institution in constructivist terms is defined as "a relatively stable set or structure of interests or identities" (Wendt, 1992). It is gradually consolidated through the pooling of intersubjectivities that formalize the shared knowledge and cooperative commitments of agents. Thus, the Organization of American States (OAS) provides a space of confrontation of the interests or identities of the actors that are local decision-makers and stakeholders. It contributes to the construction of a certain collective knowledge about sargassum that becomes synonymous with natural risk when its presence is evoked in diplomatic fora like in OAS or the Association of Caribbean States (ACS). It can be also been designated as an object of cooperation in the dialogue established among different international organizations of the region like the ACS, the OECS, Caricom or CRFM.

In the case of the Caribbean system of environmental and marine resource conservation policies, sargassum poses a major challenge to this dynamic of norm construction and, therefore, to the consolidation of the legal component of the cognitive structure in question. Their multidimensionality has prevented, to date, any juridical definition of sargassum itself (David, 2021). This lack of institutional definition has been an obstacle to any consolidation not only of a set of regulations, but also of a complete system around this seaweed. Ongoing analyses support this statement.

At present, beyond the discussions conducted in the various international scenarios provided by the actors mentioned previously, two initiatives can be identified as contributing to a possible adoption of regulatory standards: the creation in 2014 of an institution called the Sargasso Sea Commission and the project of UNESCO to inscribe the Sargasso Sea as an intangible world heritage of humanity.

The Sargasso Sea Commission was established through the signature of the Hamilton Declaration (2014). Its mandate is to encourage and facilitate the implementation of conservation actions regarding the Sargasso Sea within a so-called collaboration zone located largely on the High Seas of the Caribbean. The High Seas is defined as "all the waters that are not included in the Exclusive Economic Zone (EEZ), in the territorial sea or in the internal waters of a

State, or in the archipelagic waters of an archipelagic State" (United Nations Convention on the Law of the Sea [UNCLOS], 1982, Article 86).

The marine territory included in the collaboration area extends from the Exclusive Economic Zone (EEZ) and the Territorial Sea around Bermuda to the Azores (Hamilton Declaration, 2014). The Sargasso Sea is so named because it hosts sargassum rafts which, according to the perspective of the actor, carries both an ecological and even economic value and a possible interest in terms of Intangible Heritage for Humanity. In fact, UNESCO has been working since 2016 on a project aiming to extend the benefits of intangible World Heritage protection to territories outside the jurisdiction of States. Concretely, when a site is classified part of the World Heritage, it is the State in whose territory this one is located that is responsible for its preservation. But there are certain sites and territories that are not geographically located under the jurisdiction of any State because, for example, they are situated on the High Seas. This is the case of the Sargasso Sea. To be classified as intangible heritage, the sites must meet the same criteria as the sites that already belong to a list under the jurisdiction of the States. They must have an exceptional character because of their beauty or their richness in terms of biodiversity, for instance. After an in-depth examination of the characteristics of the Sargasso Sea, UNESCO proposed then the inclusion of the latter in the new list of Intangible Cultural Heritage of Humanity (Freestone et al., 2016). This inscription is currently in process (UNESCO, 2020).

The positioning of the Sargasso Sea Commission is to become a conservation and protection guardian for the resources contained in the sargassum rafts. UNESCO's positioning includes contributing to the protection and management of the "exceptional value" of the Sargasso Sea, which is located in a marine area beyond the limits of national jurisdictions (Freestone et al., 2016).

Even if no less than 16 different instruments that concern the Caribbean have been signed within the United Nations and are applied today in region (United Nations Treaty collection [UNTC], 2022), only that of the Cartagena Convention seems to be concerned by sargassum management. It has already been discussed earlier in this chapter. Nevertheless, one cannot fail to notice that the novelty of the phenomenon implies that the international system of sargassum management currently subscribes to the traditional notion (from international relations theories) that considers the international system as a fundamental and structural anarchy. In fact, it can be characterized as structural anarchy in that it lacks of supranational regulation and collective coordination among actors in the region. This international system as a structural anarchy is defined as "a structure lacking of centralized authority" (Wendt, 1999) or unique legal framework.

But the concept of anarchy, from a constructivist perspective, is very far from implying chaos or disorder in the international system of environmental politics (O'Neill, 2016). On the contrary, the state of anarchy is one of the essential characteristics of international systems, according to the constructionist viewpoint. Anarchy is, for constructionists, one of the characteristics of the international system (Wendt, 1992). It is something similar to the state of nature for this system. In this sense, it is worth mentioning that this anarchy can be separated into three types of relationships among the actors. They are Hobbesian anarchy, Lockian anarchy and Kantian anarchy (Wendt, 1992).

Hobbesian anarchy is defined from Hobbes' postulates. For him, the state of nature of the international system was that of the imminent and latent confrontation between its actors which for the purposes of his reasoning were only States (Little, 2007). These States were constantly in danger of demise, therefore, the only perspectives for interaction with the other actors of the system consisted of confrontation and war, either in search of territorial expansion or in defense. The actors of the system consequently perceived each other as enemies. The international system described from this perspective is one in which hostility to others prevails and in which cooperation is not a viable possibility (Brock, 2007).

However, the international system of environmental politics is structured, since its beginnings, by the possible cooperation between its actors and agents. The Lockian and Kantian perspectives of the structural anarchy of systems seem to be more suited to the analysis of the international environmental system and description of its construction.

Lockian anarchy conceives the actors in the international system as being essentially rivals (Battistella, 2012). Since this approach, the actors and agents of the international system have established peaceful or at least non-belligerent relations because it is in their interest to do so. The decision to cooperate, from this perspective, is a function of the "enlightened interest" (Battistella, 2012) of the actors. Relations between environmental actors have historically been marked by a relationship of informed exchange of interests on the part of the actors. John Vogler, in the introductory contribution to the book entitled *The Environment And International Relations* evokes environmental negotiations around pollution putting "actors of the North and the South" in a position of rivalry (Vogler and Imbert, 2017). The actors of the South demanded their right to the autonomous development of their economies in the wake of industrialized countries of the North's exploitative colonialism and how their desire to export greenhouse gas emission and pollution to Global South nations. The dialogue on environmental protection was thus the basis of a dialogue embedded in a system of Lockian anarchy.

In the Wider Caribbean, this state of regional Lockean rivalry is much less obvious to identify and, therefore, to characterize as such. The topics that focus the shared attention of actors seem to be part of a system of *Kantian anarchy* through the pooling of knowledge on common environmental vulnerabilities and obstacles to development (Battistella, 2012). Based on a reflection on Kant's "project of perpetual peace," Wendt proposed in his approaches to the social construction of international systems, the existence of a culture of association where agents are called to cooperate (Wendt, 2003). From this perspective, this cooperation is possible because the different agents perceive the other Caribbean agents as counterparts capable of sharing a certain understanding of the territorial reality, and, therefore, legitimate enough to cooperate in the construction of strategies for the preservation of ecosystems. In the sense of the construction of regional environmental standards, as quite often is the case in the exercise of conceptualizing certain international regimes (Morin, 2015) around the environment, uncertainty remains—for the moment—a major element of this system. Political and natural scientists from international organizations in the Caribbean should be tasked with providing recommendations based on their expert knowledge of potential scenarios for sargassum management strategies. Caribbean States' decision-makers, for their part, will have to endorse the responsibility to ensure that the legislative system of this management evolves. However, the answer will only be found in the dialogue between the two groups; in the construction of a collective knowledge about the sargassum and on the invention of a regional system for its governance.

Conclusion

Throughout this contribution, I demonstrated that theories approaching the phenomena of the international system as social constructs seem to be the most likely to identify and describe the dynamics and components of environmental policy in the Caribbean. New challenges, such as that of sargassum, produce both the need and possibility of improving the constructing and consolidating of effective cognitive structures whose positioning is constantly evolving, and hopefully improving.

The debate about the interdependence of agents and the structure for the evolution of systems leads us to note that it is the intersubjective exchange of actors and agents that shapes their interests, identities and norms within the international system of environmental policies in the Caribbean. The latter are social constructions because of their double character: they are both material and social (Checkel, 1998). They are material because these social constructions are materialized by international instruments and institutions, such as: the instruments referred to in this chapter in the Caribbean environmental system: the United Nations Environment Program (UNEP); the system related to the Cartagena Convention; the Caribbean Sea Commission of the Association of the Caribbean States; or the Caribbean Regional Fisheries Mechanism (CRFFM) within the Caribbean Community and Common Market (CARICOM) system; or the Ocean Governance and Fisheries Unit (OGFU) and the Environmental Sustainability Division in the Organization of Eastern Caribbean States (OECS) who are agents involved in this international socially constructed system. These are social constructions because they depend on their own intersubjective exchanges (Adler, 2005) to ensure construction, evolution and effectiveness (Vogler, 2018; Wendt, 1999) of the system they compose and in which they evolve.

It was my aim in this chapter to show the benefits of the social constructionist theory to making beneficial international environmental policies, in general, and, in particular, regarding sargassum in the Caribbean Sea region.

My purposes in this paper were multiple. On the one hand, I sought to examine the way in which environmental policies related to the management of marine resources and ecosystems could be considered as social constructions in the Caribbean. To this regard, this document employs the main axioms of constructivist analysis to show how these policies can be seen to be the product of intersubjective interaction between various international actors such as representatives of international organizations interested in the environment, regional organizations present in the Caribbean and the governments of the different Caribbean States. On another hand, my aim also was to demonstrate that only post-positivist theoretical filters were capable of analyzing and explaining the multidimensionality of the environmental policy under construction. It was established that international constructivism is the most complete analytical toolbox (among all available metatheoretical discussions) to conceptualize environmental policy in the Caribbean. Thus, it has been allowed to determine that this policy constitutes a structure of the international system. This type of conceptualization enabled us to illustrate that social structures are systems composed of three specific elements: the knowledge shared among the agents of the system, the material resources and the practices that lead to the elaboration of rules and that maintain the system. As a further challenge, we aimed to demonstrate that environmental actors in the Caribbean and their agency determine the construction of natural resource management and preservation systems in this region.

On the basis of the case of the public action related to the progressive construction of a management system for sargassum seaweed, an ontological approach was used to delve into the underlying processes that are progressively leading to the establishment of such a system.

Finally, an epistemological analysis provided useful insights into the way in which another of the axioms of constructivism, that of anarchy as the state of nature of the international system, can be used to characterize the absence of a regional standard for the management of this algae in the Caribbean. Of course, the evolutionary character of this system imposes the challenge of following up on it, in order to determine if it follows the rules of every social constructions or if it would be adapted to a specific Caribbean type of working.

Works Cited

Adler, E. (2005). "Communitarian International Relations: The Epistemic Foundations of International Relations". *Communitarian International Relations: The Epistemic Foundations of International Relations*. Palgrave. London and New York. https://doi.org/10.4324/9780203022443

Amin, S. (1990). "Maldevelopment: Anatomy of a Global Failure". *Maldevelopment: Anatomy of a Global Failure*. Zed Boks. London and Tokyo. https://doi.org/10.2307/524876

Archer, M. S. (2010). "Morphogenesis Versus Structuration: On Combining Structure and Action". *British Journal of Sociology*, 61(SUPPL. 1). https://doi.org/10.1111/j.1468-4446.2009.01245.x

Ashley, R. K., and Walker, R. B. J. (1990). "Introduction: Speaking the Language of Exile: Dissident Thought in International Studies". *International Studies Quarterly*, 34(3). https://doi.org/10.2307/2600569

Association of Caribbean States. (November 23–24, 2015). *Challenges, Dialogue and Cooperation Towards the Sustainability of the Caribbean Sea*. Port of Spain. http://hdl.handle.net/1969.3/29090. Accessed 4 March 2022.

Association of Caribbean States. (2022). *The Implications of the Sendai Framework for Disaster Risk Reduction 2015–2030 for the Greater Caribbean Region. Introducing the PITCA and UN-GGIM Projects*. http://www.acs-aec.org/index.php?q=disaster-risk-reduction/the-implications-of-the-sendai-framework-for-disaster-risk-reduction-2015-20. Accessed 5 March 2022.

Battistella, D. (2012). "*Constructivism in International Relations. The Politics of Reality*". *Études Internationales*, Cambridge, Cambridge University Press, 289 p. 35(3). https://doi.org/10.7202/009910ar

Battistella, D., Cornut, J., and Baranets, É. (2019). *Théories des relations internationales*. Presses de Sciences Po. https://doi.org/10.3917/scpo.batti.2019.01.

Birkmann, J., and Wisner, B. (2006). "Measuring the Un-Measurable. The Challenge of Vulnerability". *Source*, 5.

Blaikie, P., Cannon, T., Davis, I., and Wisner, B. (1996). *Vulnerabilidad. El Entorno Social, Político y Económico de los Desastres*. First Edition: Julio de 1996.

Bouchard-Bastien, E. (2020). García-Acosta V. et Musset A. (2017). "Les catastrophes et l'interdisciplinarité. Dialogues, regards croisés, pratiques". Louvain-la-Neuve, Academia-L'Harmattan, coll. « Investigations d'anthropologie prospective », 228 p., bibliogr. *Anthropologie et Sociétés*, 44(1). https://doi.org/10.7202/1072785ar

Braspenning, T. (2002). Constructivisme BRASPENNING. "Constructivisme et Reflexivisme En Théorie Des Relations Internationales". *Annuaire Français de Relations Internationales*, 3, 314–329.

Broadhead, L.-A. (2002). *International Environmental Politics: The Limit of Green Diplomacy*. L. Rienner. Boulder.

Brock, S. (2007). *Realism and Anti-Realism*. Routledge. London. https://doi.org/10.1017/UPO9781844653645

Bruce, S., and Blumer, H. (1988). "Symbolic Interactionism: Perspective and Method". *The British Journal of Sociology*, 39(2). https://doi.org/10.2307/590791

Bull-Kamanga, L., Diagne, K., Lavell, A., Leon, E., Lerise, F., Macgregor, H., Maskrey, A., Meshack, M., Pelling, M., Reid, H., Satterthwaite, D., Songsore, J., Westgate, K., and Yitambe, A. (2003). "From Everyday Hazards to Disasters: The Accumulation of Risk in Urban Areas". In *Desenredando*. (Vol. 15, Issue 1). http://www.desenredando.org/

Caribbean Regional Fisheries Mechanism. (February 4, 2002). *Agreement Establishing the Caribbean Regional Fisheries Mechanism*. Volume 2242, I-39916. http://www2.ecolex.org/server2neu.php/libcat/docs/TRE/Full/En/TRE-001814.pdf

Checkel, J. T. (1998). "The Constructive Turn in International Relations Theory". *World Politics*, 50(2), 324–348. https://doi.org/10.1017/S0043887100008133

Coastal Environnement under Sargassum Presure. (2022). *Research Project in Progress*. https://anr.fr/Project-ANR-19-SARG-0005

Compagnon, D. (2015). "Réalité multiscalaire et articulations multiniveaux dans la gouvernance environnementale globale". In *L'Enjeu mondial*. https://doi.org/10.3917/scpo.gemen.2015.01.0127

Convention pour la protection et la mise en valeur du milieu marin dans la région des Caraïbes (Cartagena de India, 24 mars 1983). (1994). *Revue Juridique de l'Environnement*, *19*(1). https://doi.org/10.3406/rjenv.1994.3045

Convention on International Trade in Endangered Species of Wild Fauna and Flora - CITES. (March 3, 1973). https://cites.org/sites/default/files/eng/disc/CITES-Convention-EN.pdf. Accessed 5 March 2022.

Cooney, R., Challender, D. W. S., Broad, S., Roe, D., and Natusch, D. J. D. (2021). "Think Before You Act: Improving the Conservation Outcomes of CITES Listing Decisions". *Frontiers in Ecology and Evolution*, *9*. https://doi.org/10.3389/fevo.2021.631556

Daniel, J. (2015). "La mise en discours et en politique du développement durable dans l'espace caraïbe". *Natures Sciences Sociétés*, *23*(3), 280–288. https://doi.org/10.1051/nss/2015043

David, V. (2021). Interventions in the Framework of the Project CESAR, National Center for Scientific Research (CNRS) – Institut of Research for Development (IRD).

Dunne, T. (1995). "The Social Construction of International Society". *European Journal of International Relations*, *1*(3), 367–389. https://doi.org/10.1177/1354066195001003003

Duroselle, J.-B. (1962). "Paix et guerre entre les nations: la théorie des relations internationales selon Raymond Aron". *Revue Française de Science Politique*, *12*(4). https://doi.org/10.3406/rfsp.1962.403400

European Commission. (2022). *EU and Outermost Regions*. https://ec.europa.eu/regional_policy/en/policy/themes/outermost-regions/#8. Accessed 5 March 2022.

Fardin, F., Fontaine, A., Vanzella-Khouri, A., Mc Donald Gayle, K., (n.d.). *Regional Cooperation on the Sargassum Influx Around the Wider Caribbean*. Sargassum Regional Conference Session 3, Moskito Island, BVI. https://www.car-spaw-rac.org/IMG/pdf/sargassum-workshop_spaw-rac_presentationreduite.pdf. Accessed 2 March 2022.

Fearon, J., and Wendt, A. (2012). "Rationalism v. Constructivism: A Skeptical View". In *Handbook of International Relations*. SAGE Publications Ltd. London, New Delhi. https://doi.org/10.4135/9781848608290.n3

Fierke, K. M. (2002). "Links Across the Abyss: Language and Logic in International Relations". *International Studies Quarterly*, *46*, 331–354. https://doi.org/10.1111/1468-2478.00236

Fraga, J., Robledo, D., Hernández, L., and Ménez, F. (2019). "Arribazones de sargazo en el Caribe: Interacción Humano-Medioambiente". *Avance y Perspectiva*, 5.

Freestone, D., Laffoley, D., Douvere, F., and Badman, T. (2016). *World Heritage in the High Seas: An Idea Whose Time has Come*. UNESCO. Union Internationale pour la Nature. https://unesdoc.unesco.org/ark:/48223/pf0000245467. Accessed 2 March 2022.

Galy, K., Jos, Réno, F., Macdissi, Ch., and Gémieux, Fr. (2011). *Les nouvelles tendances de la coopération et de l'intégration régionales dans l'espace Amérique-Caraïbes*. Cujas, Actes and études. Paris.

Gardien, E. (2020). "Pairjectivité: des savoirs expérientiels ni objectifs, ni subjectifs". *Éducation et Socialisation*, 57. https://doi.org/10.4000/edso.12581

Hamilton Declaration. (March 11, 2014). *Hamilton Declaration on Collaboration for the Conservation of the Sargasso Sea*. Hamilton, Bermuda. http://www.sargassos eacommission.org/storage/Hamilton_Declaration_with_signatures_April_2018.pdf. Accessed 2 March 2022.

Hollis, M., and Smith, S. (1994). "Two Stories About Structure and Agency". *Review of International Studies*, 20 (3), 241-251. https://doi.org/10.1017/S0260210500118054

Huber, R. (2018). *ReefFix: An Integrated Coastal Zone Management (ICZM) Ecosystem Services Valuation and Capacity Building Project for the Caribbean*. https://www.oas.org/en/sedi/dsd/biodiversity/ReefFix/ReefFix%203rd%20Quarterly%20Report%20July-October2018.pdf. Accessed 2 March 2022.

International Conference on Sargassum.(2019). *Final Statement*. http://www.acs-aec.org/sites/default/files/declaration_of_the_international_conference_on_sargassum.pdf. Accessed 24 March 2022.

International Maritime Organization. (1973). *International Convention for the Prevention of pollutionfromShips*. https://www.imo.org/fr/about/Conventions/Pages/International-Convention-for-the-Prevention-of-Pollution-from-Ships-(MARPOL).aspx. Accessed 5 March 2022.

Inversión Turística. (January 23, 2019). *Sargazo en playas de la cuenca del Mar caribe, tema en Conferencia Iberoamericana en el marco de FITUR*. https://inversion-turistica.com/2019/01/23/sargazo-en-playas-de-la-cuencua-del-mar-caribe-tema-en-conferencia-iberoamericana-en-el-marco-de-fitur/. Accessed 5 March 2022.

Katzenstein, P. J., Keohane, R. O., and Krasner, S. D. (1998). "International Organization and the Study of World Politics". *International Organization*, 52 (4), 645–685. https://doi.org/10.1017/s002081830003558x

Klöck, C., and Fink, M. (2019). "Dealing with Climate Change on Small Islands: Towards Effective and Sustainable Adaptation?". In *Dealing with Climate Change on Small Islands: Towards Effective and Sustainable Adaptation*. 1–15. Göttingen University Press. Göttingen. https://doi.org/10.17875/gup2019-1209

Kowert, P. A. (2012). "Conclusion: Context and Contributions of the Ideational Alliance". In *Psychology and Constructivism in International Relations*.

Leff, E. (2021). *Political Ecology: Deconstructing Capital and Territorializing Life*. Palgrave Macmillan. Cham.

Little, R. (2007). "Kenneth N. Waltz's Theory of International Politics". In *The Balance of Power in International Relations: Metaphors, Myths and models* (pp. 167–212). Cambridge: Cambridge University Press. https://doi.org/10.1017/CBO9780511816635.006

López-Contreras, A. M., van der Geest, M., Deetman, B., van den Burg, S., and Brust, H. (2021). *Report: Opportunities for Valorisation of Pelagic Sargassum in the Dutch Caribbean*. Wageningen Food and Biobased Research. Wageningen. https://doi.org/10.18174/543797

Lövbrand, E., Mobjörk, M., and Söder, R. (2020). "The Anthropocene and the Geo-Political Imagination: Re-Writing Earth as Political Space". *Earth System Governance*, 4, 1–8. https://doi.org/10.1016/j.esg.2020.100051

Mahon, R., and Fanning, L. (2021). "Scoping Science-Policy Arenas for Regional Ocean Governance in the Wider Caribbean Region". *Frontiers in Marine Science, 8,* 1–17. https://doi.org/10.3389/fmars.2021.685122

Marx, K. (1959). "Economic and Philosophic Manuscripts of 1844". *Economica, 26* (104). https://doi.org/10.2307/2550890

McConney, P., and Oxenford, H. (2021). "Caribbean Sargassum Phenomenon: Complexities of Communicating". *The Journal of Caribbean Environmental Sciences and Renewable Energy, 3* (2). https://doi.org/10.33277/cesare/003.002/02

Ménez, F. (2019). " La télé est morte". *Techniques and Culture. 72,* 184–199. https://doi.org/10.4000/tc.12359

Morin, J.-F. (2015). "Les régimes internationaux de l'environnement". In *L'Enjeu mondial.* 113–123. Presses de Sciences Po. Paris. https://doi.org/10.3917/scpo.gemen.2015.01.0113

Organization of Eastern Caribbean States. (2019). https://pressroom.oecs.org/opening-of-the-1st-international-conference-on-sargasso. Accessed 5 March 2022.

Oneill, Christopher. (2016). "Taylorism, the European Science of Work, and the Quantified Self at Work". *Science, Technology and Human Values,* 42. https://doi.org/10.1177/0162243916677083.

Onuf, N. G. (2013). *Making Sense, Making Worlds: Constructivism in Social Theory and International Relations.* Routledge. London and New York. https://doi.org/10.4324/9780203096710

Orani, A., Vassileva, E., Azemard, S., and Alonso-Hernandez, C. (2020). "Trace Elements Contamination Assessment in Marine Sediments from Different Regions of the Caribbean Sea". *Journal of Hazardous Materials, 399* (122934), 1–15. https://doi.org/10.1016/j.jhazmat.2020.122934.

Organization of American States. (2015). *Ocean Conservation and Climate Change in the Americas.* http://www.oas.org/en/sedi/nl/1015/4_en.asp. Accessed 17 February 2022.

Organization of Eastern Caribbean States. (1980). *Treaty of Basseterre Establishing the Organization of Eastern Caribbean States Economic Union.* http://foreign.gov.vc/foreign/images/stories/Foreign_Affairs/Article_pdf/revised-treaty-of-basseterre.pdf#:~:text=%E2%80%9CTreaty%20of%20Basseterre%201981%E2%80%9D%20means%20the%20Treaty%20establishing,Caribbean%20States%20done%20at%20Basseterre%20on%2018thJune%201981. Accessed 5 March 2022.

Organization of Eastern Caribbean States. (July 18, 2018). *Statement by the OECS Council of Ministers on Environmental Sustainability on the Sargassum Challenges and Opportunities.* Fifth meeting *Council of Ministers on Environmental Sustainability* (COMES 5). Brades, Montserrat. Accessed 4 March 2022.

Organization of Eastern Caribbean States. (May, 2019). *Sixth Meeting of the Council of Ministers of Environmental Sustainability (COMES 6).* Fort de France, Martinique. Accessed 5 March 2022.

O'Riordan, T., and McCormick, J. (1990). "The Global Environmental Movement: Reclaiming Paradise". *Transactions of the Institute of British Geographers, 15* (3), 383–384. https://doi.org/10.2307/622684

Ostrom, E. (2010). "Polycentric Systems for Coping with Collective Action and Global Environmental Change". *Global Environmental Change, 20*(4), 550–557. https://doi.org/10.1016/j.gloenvcha.2010.07.004

Oyesiku, O., and Egunyomi, A. (2015). "Identification and Chemical Studies of Pelagic Masses of Sargassum Natans (Linnaeus) Gaillon and S. fluitans (Borgessen)

Borgesen (Brown Algae), Found Offshore in Ondo State, Nigeria". *African Journal of Biotechnology. 13*(10), 1188–1193.

ParisClimate2015. (June 8, 2015). *Paris Declaration*. OCEAN Objectives. Proposals for a blue economy. Paris_Declaration_June8_sign_201511251.pdf (gcft.fr)

PR GOV. (2022). *Constitución de Puerto Rico*. https://www2.pr.gov/SobrePuertoRico/Pages/Constituci%C3%B3ndelEstadoLibreAsociadodePuertoRico.aspx

Reus-Smit C. (2009). "Constructivism and the English school". In: Navari C. (eds) *Theorising International Society*. Palgrave Studies in International Relations Series. Palgrave Macmillan, London. https://doi.org/10.1057/9780230234475_4

Rodríguez-Martínez, R. E., Medina-Valmaseda, A. E., Blanchon, P., Monroy-Velázquez, L. v., Almazán-Becerril, A., Delgado-Pech, B., Vásquez-Yeomans, L., Francisco, V., and García-Rivas, M. C. (2019). "Faunal Mortality Associated with Massive Beaching and Decomposition of Pelagic Sargassum". *Marine Pollution Bulletin, 146*. https://doi.org/10.1016/j.marpolbul.2019.06.015

Rosenau, J. N. (2018). *Turbulence in World Politics a Theory of Change and Continuity*. Princeton University Press. Princeton. https://doi.org/10.2307/j.ctv301hg5

Singer, J. D. (1960). "International Conflict Three Levels of Analysis". *World Politics. 12*(3), 453-461. https://doi.org/10.2307/2009401

Sissini, M. N., de Barros Barreto, M. B. B., Szechy, M. T. M., de Lucena, M. B., Oliveira, M. C., Gower, J., Liu, G., de Oliveira Bastos, E., Milstein, D., Gusmão, F., Martinelli-Filho, J. E., Alves-Lima, C., Colepicolo, P., Ameka, G., de Graftjohnson, K., Gouvea, L., Torrano-Silva, B., Nauer, F., Marcos De Castronunes, J., … Horta, P. A. (2017). "The Floating Sargassum (Phaeophyceae) of the South Atlantic Ocean - Likely Scenarios". *Phycologia.,56*(3), 321–328. https://doi.org/10.2216/16-92.1

Soltani, F., Jawan, J. A., and Ahmad, Z. B. (2014). "Constructivism, Christian Reus-Smit and the Moral Purpose of the State". *Asian Social Science, 10*(10), 153–158. https://doi.org/10.5539/ass.v10n10p153

Taylor, R. W., Waltz, K. N., Weiss, P., and Polyani, M. (1960). "Man, the State and War: A Theoretical Analysis". *Midwest Journal of Political Science, 4*(2), 194–197. https://doi.org/10.2307/2108708

Thirot, M., Palany, P., Gros Désormeaux, J.-R., and Tupiassu, L. (2020). "La mise en place du Parc naturel marin en Martinique: un révélateur du rapport inégalitaire entre le local et le global". *VertigO, Volume 20 numéro 1*. https://doi.org/10.4000/vertigo.27812

Tortosa, J. M. (2011). "Maldesarrollo como Mal Vivir". *América Latina En Movimiento, 445.*

United Nations. (June, 1972). *Declaration of the United Nations Conference on the Human Environment*. Documents Gathering a body of global agreements. http://www.un-documents.net/unchedec.htm. Accessed 5 March 2022.

United Nations. (1982). *Convention on the Law of the Sea (UNCLOS)*. Article 86. https://www.un.org/depts/los/convention_agreements/texts/unclos/unclos_e.pdf.

United Nations. (1983). *United Nations Convention on the Law of the Sea*. https://www.un.org/Depts/los/convention_agreements/texts/unclos/unclos_e.pdf. Accessed 5 March 2022.

United Nations. (1994). *Global Conference on the Sustainable Development of Small Island Developing States—SIDS*, Bridgetown, Barbados. United Nations publication,

Sales No. E.94.I.18 and corrigenda. Chap. I, resolution 1, annex II. Accessed 5 March 2022.

United Nations. (2015a). *Paris Agreement*. https://unfccc.int/sites/default/files/english_paris_agreement.pdf. Accessed 5 March 2022.

United Nations. (2015b). *Sendai Framework for Disaster Risk Reduction 2015–2030*. Sendai. https://www.unisdr.org/files/43291_sendaiframeworkfordrren.pdf. Accessed 3 March 2022.

United Nations. (2019). "Report of the Secretary-General on SDG Progress 2019: Special Edition". In *United Nations Publications*. http://www.un-documents.net/unchedec.htm. SDG Progress

United Nations. (2022). *UN Treaty Collection*. https://treaties.un.org/pages/Treaties.aspx?id=27andsubid=Aandclang=_en. Accessed 14 January 2022.

United Nations Environment Programme. (September 25, 2001). *Report of the Meeting. First Meeting of the Contracting Parties (COP) to the Protocol Concerning Specially Protected Areas and Wildlife (SPAW) in the Wider Caribbean Region Havana*. Cuba. UNEP(DEC)/CARIG.20/7. Accessed 14 December 2021. https://wedocs.unep.org/bitstream/handle/20.500.11822/10045/UNEP%28DEC%29CAR%20IG.20-7-en.pdf?sequence=1andisAllowed=y. Accessed 3 March 2022.

United Nations Environment Program-Caribbean Environnement Program. (2012). *Convention Pour La Protection et La Mise En Valeur Du Milieu Marin Dans La Région Des Caraïbes*. Kingston. http://www.cep.unep.org/cartagena-convention/text-of-the-cartagena-convention. Accessed 24 November 2021.

United Nations Environment Programme—Caribbean Environment Programme. (2021a). *Sargassum White Paper—Turning the crisis into an Opportunity. Ninth Meeting of the Scientific and Technical Advisory Committee (STAC) to the Protocol Concerning Specially Protected Areas and Wildlife (SPAW) in the Wider Caribbean Region*. Kingston, Jamaica.

United Nations Environment Programme. (2021b). *Rapport du Groupe De Travail Du Stac SurLesSargasses*. https://www.car-spaw-rac.org/IMG/pdf/unep_depi_car_wg.42.7_rapport_gt_sargasse_-_fr.pdf. Accessed 20 October 2021.

United Nations Environment Programme. (March 9, 2021c). *Terms of Reference of the Spaw STAC Ad Hoc Working Group. Ninth Meeting of the Scientific and Technical Advisory Committee (STAC) to the Protocol Concerning Specially Protected Areas and Wildlife (SPAW) in the Wider Caribbean Region*. UNEP(DEPI)/CAR WG.42/INF.12. http://gefcrew.org/carrcu/SPAWSTAC9/Info-Docs/WG.42-INF.12-en.pdf. Accessed 20 October 2021.

United Nations General Assembly, (February 3, 2001). "Promoting an Integrated Management Approach to the Caribbean Sea in the Context of Sustainable". *A/RES/55/203*. https://digitallibrary.un.org/record/430047. Accessed 20 October 2021.

United Nations General Assembly, (August 15, 2012). "Towards the Sustainable Development of the Caribbean Sea for Present and Future Generations: Report of the Secretary-General". *A/67/31*. https://digitallibrary.un.org/record/734472?ln=fr. Accessed 20 October 2021.

Van Tussenbroek, B. I., Hernández Arana, H. A., Rodríguez-Martínez, R. E., Espinoza- Avalos, J., Canizales-Flores, H. M., González-Godoy, C. E., Barba-Santos, M. G., Vega-Zepeda, A., and Collado-Vides, L. (2017). "Severe Impacts of

Brown Tides Caused by Sargassum spp. on Near-Shore Caribbean Seagrass Communities". *Marine Pollution Bulletin*. 122(1–2), 272-281. https://doi.org/10.1016/j.marpolbul.2017.06.057

Vogler, J. and Imbert, M. (2017). *The Environment and International Relations*. Global Environment Change Program.

Vogler, J. (2018). *Global Environmental Politics*. Routledge, London. https://doi.org/10.4324/9781315179537-2

Wang, M., Hu, C., Barnes, B. B., Mitchum, G., Lapointe, B., and Montoya, J. P. (2019). "The Great Atlantic Sargassum Belt". *Science*, 364(6448), 83–87. https://doi.org/10.1126/science.aaw7912

Wendt, A. (1992). "Anarchy is What States Make of It: The Social Construction of Power Politics".*International Organization*, 46(2), 391–425. https://doi.org/10.1017/S0020818300027764

Wendt, A. (1995). "Constructing International Politics". *International Security*, 20(1), 78–81. https://doi.org/10.2307/2539217

Wendt, A. (1999). *Social Theory of International Politics*. Cambridge University Press, Cambridge. https://doi.org/10.1017/cbo978051161218

Wendt, A. (2003). "Why a World State is Inevitable: Teleology and the Logic of Anarchy". *European Journal of International Relations*, 9(4), 491–542. https://doi.org/10.1177/135406610394001

.Wendt, A. (2004). "The State as Person in International Theory". *Review of International Studies*, 30(2). https://doi.org/10.1017/S0260210504006084, 289–316.

Wendt, A. (1987). "The Agent-Structure Problem in International Relations Theory". *International Organization*, 41(3), 335–370. https://doi.org/10.1017/S002081830002751X

Wendt, A. (2009). "The Agent-Structure Problem in International Relations Theory". *Uluslararasi Iliskiler*, 6(23).

White, A. R., Austin, J. L., and Urmson, J. O. (1963). "How to Do Things with Words". *Analysis*, 23, 58–64. https://doi.org/10.2307/3326622

Wiener, A. (2006). "Constructivism and Sociological Institutionalism". In *Palgrave Advances in European Union Studies*. Palgrave Macmillan, London https://doi.org/10.1057/9780230522671_3

Yvars, B. (2019). "La Caraïbe face aux défis du xxie siècle". *Études caribéennes*, L'économie de la Caraïbe. Volume 43–44, August–December 2019. https://doi.org/10.4000/etudescaribeennes.16724

Contemporary Youth Environmental Activism: Lessons from France and Italy

Paolo Stuppia

"Why are we striking?

Response is simple: we strike, because we have no choice. […] Climate change is already here. We saw it this summer, with forest fires and heat waves. Scientists warn us since 50 years. […] Its effects will impact everyone, rich and poor, but they will be more devastating for most vulnerable people, i.e. poorest and youngest. Collective action is the only response to this crisis. Taking to the street, doing civil disobedience actions: all is useful to alert public opinion and to constrain power-holders to resolve it! We are the only generation that can stop this crisis!"[1]

Fridays For Future Italy, 2022

Introduction

Contemporary ecological protests led by young generations in the wake of the Fridays for Future (FFF) movement raise many questions. During the two years preceding the COVID-19 pandemic, weekly students' strikes

---

The original version of this chapter was revised: Author's name corrected from "Paola Stuppia" to "Paolo Stuppia". The correction to this chapter is available at https://doi.org/10.1007/978-3-031-14346-5_28

[1] https://fridaysforfutureitalia.it/perche-scioperare/. Accessed March 15, 2022.

P. Stuppia (✉)
European Center of Sociology and Political Science (CESSP), University of Paris, Paris, France
e-mail: paolo.stuppia@univ-paris1.fr

and demonstrations, posts on social networks under hashtags #climateaction, #savetheplanet or #thereisnoplanetb, and more radical performances such as traffic obstructions, urban camps or occupations of construction areas of projects reported as land consuming or polluting, known under the acronym ZAD (defense zones[2]), have multiplied. So have collectives and networks— Extinction Rebellion (XR), "La Bascule[3]..., legally registered or spontaneous, acting online and offline, focused specifically on climate change.

Global warming is a scientific, well-documented reality since at least the 1970s. The first United Nations (UN) conference dedicated to this question took place in Stockholm in 1972. The independent experts' pool Intergovernmental Panel on Climate Change (IPCC) was created in 1988, and has since published 45 reports, including 5 complete assessments, alerting the dangers of rising temperatures of more than 1.5–2 °C for 2100. After the third UN Rio Earth Summit of 1992, the Conference of Parties (COP) was set up in 1995. These international annual meetings aimed originally to reach legally binding treaties, following IPCC recommendations. However, since the 2000s, there have been successive reverses: Kyoto Protocol (COP3, 1997) has come into force not before 2005; 2009 Copenhagen COP15 negotiations for more strict rules aborted; 2015 Paris COP21 decided new measures without establishing any sanctions for reluctant countries; and 2021 Glasgow COP26 has just reaffirmed the Paris Agreement goal of "limiting global temperature rise to 1.5°", with no supplementary ambitions [Maréchal 2016; Mathieu 2022].

According to researchers, Stefan C. Aykut and Amy Dahan [2015], these failures are due to "slowness" and "schism of reality" of policy-makers unable to resolve a well-established problem both for economical and for strategic reasons. Thus, the problem persists, and even worsens, giving young generations the feeling that their future is sacrificed: Swedish student Greta Thunberg, aged 15 at the time, has precisely expressed this sensation when she started a solitary action in front of the Swedish Parliament in August, 2018 with a simple placard stating "school strike for future". In a record time of few months, FFF has become "the largest environmental global campaign in history" [Bessant et al. 2021: 2], surrounded by an immense media hype. Most journalists and commentators have insisted on the spreading of a new environmental conscience among "Millennials and generation Z,"[4] their "higher willingness to stand for the climate compared to other generations (X, boomers, and others),"[5] or even on the rising of a "new activist

---

[2] *"Zones à Défendre"* in French.

[3] A "citizen lobby" movement created in France in 2019 by an organic business owner and several students.

[4] https://www.diariovasco.com/antropia/jovenes-espanoles-buscan-20211220162137-ntrc.html. Accessed October 11, 2021.

[5] https://www.pewresearch.org/science/2021/05/26/gen-z-millennials-stand-out-for-climate-change-activism-social-media-engagement-with-issue/. Accessed October 11, 2021.

generation", challenging the "stereotype that they are apathetic."[6] Calls to "change the system, not the climate" seen on banners during COP and FFF marches would represent a perfect synthesis of this landscape: according to sociologists, Karen O'Brien, Elin Selboe and Bronwyn M. Hayward, indeed, we could currently be witnessing the rising of a new "age of dissent" led by young generations, a term describing "a conscious expression of disagreement with a prevailing view, policy, practice, decision, institution or assumption that is exacerbating climate change" when one knows that they "will have to live with and manage future risks and uncertainties associated with" global warming [2018].

Other scholars tend, however, to nuance this vision, revealing novelties as well as continuities with the past. First, a series of academic works have confirmed a "generational renewal" on relations to politics and democracy in general since the last 25 years [e.g., Norris 2011; Lardeux and Tiberj 2021]. Their research has shown that a growing portion of young people are rejecting traditional structures (associations, parties, trade unions, etc.) and questioning the vertical organization of the society, while elders are still marked, in their immense majority, by a "deference citizenship" to socio-political elites [Tiberj 2017]. Compared to the former generations, these young activists also tend to challenge more norms and authorities, and to defend larger causes—international solidarity, preservation of Earth, etc. [Inglehart 2018]. "Alter-globalization" protests at the turn of the millennium, joining a series of ecological, social and indigenous claims, appear particularly important to understand this shift. "Social Forums" and counter-summit marches, such as in Seattle in 1999 against the World Trade Organization (WTO) or in Genoa (Italy) in 2001, against the G8, reunited a series of traditionally divided components, from NGOs to radical groups and created a new activist youth (sub-)culture, seeking for more horizontal, participatory and inclusive social relations [Juris and Pleyers 2009].

Secondly, environmental struggles do not only refer to climate change, although this question appears nowadays central and, of course, interconnected with many others. They designate a wide range of local, national and international causes (defense of biodiversity, food production and consumption, recycling of waste, pollution, renewable energies, etc.) which have historically been dealt with by different individuals and groups with approaches varying from scientific expertise to court cases, from lobbying to demonstrations and from street actions to rural settlements since the emergence of a political ecology in the 1960s [Armiero and Sedrez 2014; Blatrix and Gervereau 2016]. Therefore, it is necessary to try to understand contemporary youth environmental activism in this long and diversified history by identifying its connections with former protestors' claims, resources [Agrikoliansky et al.

---

[6] https://www.ypulse.com/article/2020/07/14/this-is-how-gen-z-millennials-have-changed-activism/. Accessed October 11, 2021.

2010], repertoires of collective actions/performances [Tilly 1986, 2008] and cultures, as this chapter will seek to do.

Finally, the media *cliché* of a "climate generation" taking over environmentalism and renewing engagement in a supposed wasteland needs to be deconstructed. Conjunctures of intense youth public expression, nowadays both in real and in virtual spaces, often leads commentators, and in some cases, scholars to reactivate a recurrent myth: the uprising of an "involved generation" [Becquet and Stuppia 2021] sharing the same concerns, values, and experiencing a higher level of activism, even for a short period of time, compared to others. Depending on contexts, this vision fits well with the "Long Civic Generation" that came of age during the Depression and World War II in the U.S. [Putnam 2000] and with the "1968 generation", i.e., "baby boomers" who participated to 1960s and early 1970s protests globally. For both cases, scholars tend to emphasize elevated levels of civic and political participation, although they are far from being generalized: a survey made in Italy in 1969–1970 shows, for example, that only 22% of 15-to 24-year-old respondents claim to be "involved" in organizations and social movements [Cavalli et al. 1984: 82].

Regarding more particularly the "generation of 1968", mainstream representations also focus on a quick shift from radical activism to individualism and promotion of neoliberal values, following reconversions to capitalism of most popular ex-student leaders—Jerry Rubin (United States), Daniel Cohn-Bendit (France), Josckha Fischer (Germany), etc.—during the 1980s. "Hippies to yuppies" is a stereotype about this generation challenged by sociologist Julie Pagis: rather than considering careers of a few well-known individuals, she analyzes hundreds of life trajectories of "anonymous participants to May 1968 protests" in France. While they share similar age, their social and political backgrounds, motivations, *savoir-faire*, and action during the events, but also effects on their later professional and personal careers, vary greatly: she concludes to the existence of "micro-units of generation" instead of a unique "1968 opportunist generation" [2014: 16–17]. This point is crucial not only to consider differences within activist cohorts, but also to link two research fields "seldom interconnected – *mobilizations* and *socialization*" [Fillieule 2013].

Lastly, Pagis' work is important to show how powerful are nostalgia, blame and injunctions when one compares youth involvements in the past and in the present: on the one hand, young people are often accused of being "passive" in relation with their predecessors. On the other hand, when they act, they are constantly invited to be "reasonable" and warned about their unavoidable destiny on the other side of the barricade [Becquet 2014].

This chapter focuses on three examples of environmental networks—FFF, XR and ZAD—that are particularly invested by students in two European countries, France and Italy, for reasons that are mentioned below. It combines two perspectives—the first, coming from the sociology of social movements based on changing performances and cultures; the second, coming from the sociology of youth centered on age and (units of) generation—to try to determine who are the young environmental protestors today, how do they involve themselves, what are their relationships with former ecological struggles and what are the lessons from transformations or renewals they have contributed to introduce in the activist scene. Not all youth are engaged and "rebellious" when it comes to environmental questions, only a specific fraction of them, with particular backgrounds. Within these "units of generation", one can identify a common preference for spontaneous performances and horizontal, leaderless and issue-related "Do-It-Ourselves (DIO) politics" rather than conventional involvements. However, it is also possible to argue that these questions specifically matter for the most politicized protestors, often radicals, sharing the same culture, called "alter-activism", inherited by alter-globalization protests, and finding specific expressions online and offline.

Focus and Methodology

As it is impossible to study the whole spectrum of groups that are currently led by youth environmental activists, I have selected three different networks that have sparked a high interest among media, academics and officials worldwide, especially since 2018.

My first example is FFF, also known as—depending on countries—"Youth4Climate", "Climate Strikes" or "School Strikes for Climate." As remembered before, this global student coalition has been established following Thunberg's action in 2018. According to their website, FFF goal "is to put moral pressure on policy-makers, to make them listen to the scientists and then to take forceful action to limit global warming."[7] They also propose that their movement "is independent of political parties", even if their members are accepted: for example, German FFF figure Luisa-Marie Neubauer belongs to Alliance90/The Greens; Italian FFF spokeswoman Martina Comparelli has been candidate in 2019 European Parliament elections for *Europa Verde*. Indeed, anyone who agrees with a charter signed in Lausanne in 2019[8] can claim to be part of the movement and freely acts in its name [Chaillou and Monti-Lalaubie 2020]. Therefore, the coalition is decentralized and horizontal, but media leaders—such as those mentioned above, Uganda high-school activist Vanessa Nakate and naturally Greta Thunberg—have emerged. They have been invited to meet or speak in front of political and business representatives during major events such as COP conferences or the 2020 Davos World Economic Forum [Doherty and Saunders 2021].

---

[7] https://fridaysforfuture.org/. Accessed October 12, 2021.

[8] Containing three demands: (1) Keep the global temperature rise below 1.5 °C compared to pre-industrial levels; (2) Ensure climate justice and equity; (3) Listen to the best united science currently available. Charter also states that movement is nonviolent and must respect diversity and plurality. Document was signed by 400 activists from 38 countries in August 2019 [Chaillou and Monti-Lalaubie 2020].

The first international FFF, on 30 November 2018, gathered around 27,000 demonstrators in 9 Anglo-Saxon and Northern European countries. Three and a half months later, they were more than 2 million in 133 countries, including Global South and authoritarian States (China, Turkey, etc.). A record participation of 3 million has been registered on 20 September 2019: school strikes and marches have been reported in 4400 cities from 156 different countries. The spreading of COVID-19 pandemic in 2020 has naturally slowed down FFF, but thousands of youths have gathered again in Milan and Glasgow in the fall of 2021 to protest COP26.[9]

My second example is XR, another international network (now in more than 50 countries) close to FFF, but including other—units of—generations within it, and practicing more radical and spectacular civil disobedience actions. XR was founded in United Kingdom in 2018 by experienced activists in alter-globalization protests and Occupy movements that followed the 2008 financial crisis in 2011–2012. Roger Hallam, Gail Bradbrook and Simon Bramwell are the most well-known figures by the public [Scappaticci-Martin 2019]. Young people have quickly joined them, by participating either in the main network or, especially in the UK, in local youth groups. The XR logo, an hourglass in a circle, symbolizes the limited amount of time at disposal before a "6th mass extinction" happens. Subsequently, they invite governments to respond to three main demands: "tell the truth by declaring a climate and ecological emergency; act now to halt biodiversity loss and reduce greenhouse gas emissions to net-zero by 2025; and create, and be led by the decision of a citizen's assembly on climate and ecological justice."[10]

Theatrical and nonviolent XR actions are often privately prepared on secured Web platforms, then broadcasted after they start. By taking part in them, activists must accept to be arrested as a supplementary media coverage strategy [Wall and Bijeard 2020].

The movement has started with the occupation of Greenpeace offices in London 17 October 2018, to invite members of this NGO to join. A "declaration of rebellion" was proclaimed in Parliament Square on October 31 of the same year. During the following months, XR UK also organized several bridge occupations and traffic stops in the main cities of the country. A first "international rebel week" took place in April, 2019 (actions reported in 33 countries), followed by a second in October, 2019 (protests in 60 cities worldwide[11]). During the pandemic, civil disobedience acts diminished without coming to a complete stop [Biard et al. 2020].

---

[9] All quoted statistics are available on: https://fridaysforfuture.org/what-we-do/strike-statistics/. Accessed October 12, 2021.

[10] https://extinctionrebellion.uk/the-truth/demands/. Accessed November 8, 2021.

[11] https://www.lemonde.fr/climat/article/2019/10/07/lancement-d-une-rebellion-internationale-pour-le-climat-dans-soixante-villes_6014498_1652612.html. Accessed November 8, 2021.

My final example is what the French call the ZAD, particularly Notre-Dame-des-Landes (NDDL), near the town of Nantes on the Atlantic coast. These land occupations aim at transforming a rural or a natural site intended for building projects into an intentional community composed of cabins, food gardens, meeting spaces, etc., preventing and effectively halting construction works [Barbe 2016]. This practical target is compounded by strong anarchist, eco-socialist or eco-feminist influences: indeed, ZADs are hosting many radical environmentalists, dubbed *zadists*, with a remarkable presence of far left youth equally involved in other groups—*antifa*, pro-migrants rights "no-borders", etc.. If they appear less connected to global warming than FFF and XR activists, they have seized this question "as a lever to create a counter-power," according to sociologist, Sylvaine Bulle [2021].

The first "defense zone" was set up in NDDL in 2008 after the reactivation of a 1970s building plan for the new Nantes airport: around 60 people, supported by local dwellers, started to live there. Government sent riot police to evacuate them in 2012, but they failed: quite the opposite happened, and around 200 more *zadists* joined the site, squatted new lands and made new cabins [Verdier 2018]. This stalemate situation forced the government to withdraw law enforcement and the ZAD expanded even more during the following years. In 2016, according to scholar, Geneviève Pruvost, NDDL was composed of "around 60 living spaces and one hundred hectares of farm lands. Local people, farmers, squatters, punks, anarchists" continued to struggle together in what she defined as "the most extended anti-capitalist experimental place in Europe" at the time [2017: 37]. The airport project was abandoned in 2018 and several agreements were signed to legalize former squatted lands and buildings.

It was the second time that a ZAD had succeeded, but for the first time without being marked by a tragedy: in 2014, a dam plan in the Tarn County (Southern France) had already been abandoned after the killing of a young activist during clashes with the police nearby the "Sivens defense zone" [Subra 2016]. Around 15 ZADs have been established since the decade of 2010 in France, Belgium and Switzerland against numerous projects (shopping centers, burial sites for nuclear wastes, etc.). Some of them have been evacuated, others are still ongoing [Sommier 2021].

Reasons to focus on France and Italy are multiple. The two neighboring countries share many similarities, especially regarding student activism. Movements are generally self-organized, both locally and nationally: horizontal structures open to all (assemblies, coordinations, etc.) and are more popular than student unions since 1968 [Legois et al. 2021]. School strikes, marches and occupations are recurrent, even if participation and intensity vary. For example, in the 2000s, massive youth protests regularly challenged neoliberal agendas in the wake of the alter-globalization movement, reuniting up to 3 million demonstrators against a new work contract in France [Stuppia 2020]; during the following decade, however, they have slightly decreased, particularly in Italy [Lo Schiavo 2021]. 2019 marked a comeback: according to the FFF Website, French and Italian youth represented about half of total participants to the first "Global School Strike" (GSS) Friday 15 March 2019 (approximately 500,000 in Italy and 550,000 in France).

27 September 2019 constituted a "record day" for FFF in Italy, with 1.2 million students marching for the climate. Even during the pandemic, despite variations due to restrictions, some Fridays reached great levels of participation: 7 May 2021, 115,000 strikers were reported in France, and 100,000 on the other side of the Alps on September 24, 2021.[12]

XR operations have also been important in both contexts, especially during the "international rebellion week" of October, 2019, known in France under the acronym RIO ("Rebellion Internationale d'Octobre"): in Paris, activists first occupied a mall (Italie2) for a whole Saturday, then established a camp in Châtelet, the most central square and crossroad of the capital. They left this camp, symbolically renamed "Châtelet ZAD", five days later.[13] In Italy, several die-ins were organized in Rome and 10 XR activists started an hunger strike in front of Parliament, an action repeated in February, 2022, in a new campaign called "Ultima generazione" (the last generation). Since 2019, they regularly update the list of their operations on a Web Magazine and a podcast, reporting banner deployments, traffic stops, etc., in the main cities of the country.[14]

Finally, a similar movement to French ZADs is an ongoing Italian protest called no-TAV (no to high-speed trains), initiated mid-1990s by Susa Valley dwellers to protest against the construction of a high-speed rail link across the Alps, and symbolizing "a fight against consumerism, arbitrary expropriation, and waste of public resources" [Calabrese 2014: 197]. As in NDDL, locals were rapidly joined by young radicals from universities and urban squats where they usually organize political initiatives, called "centri sociali" [Pitti 2021]. All the attempts to stop the works and establish camps have been violently repressed by law enforcement, nevertheless, and today skirmishes with riot police protecting construction sites continue.[15]

My methodology is based on a variety of sources (oral interviews, observations and statistics) gathered in France and in Italy during different surveys I have conducted between 2008 and 2021 on student mobilizations, radical utopias—"back-to-the-land" intentional communities and eco-villages—and youth activism in formal and informal organizations. More specifically, since 2019, I started to collect observations during intergenerational climate

---

[12] Quoted statistics are available on: https://fridaysforfuture.org/what-we-do/strike-statistics/. Accessed November 12, 2021.

[13] https://www.leparisien.fr/societe/paris-extinction-rebellion-leve-le-camp-a-chatelet-d-autres-actions-imminentes-11-10-2019-8170938.php. Accessed March 16, 2022.

[14] https://extinctionrebellion.it/media/podcast and https://extinctionrebellion.it/XR-Magazine/. Accessed March 6, 2022.

[15] For example: https://www.torinoggi.it/2022/03/10/leggi-notizia/argomenti/cronaca-11/articolo/scontri-no-tav-pugno-duro-contro-leader-e-militanti-di-askatasuna-13-misure-cautelari.html. Accessed March 16, 2022.

marches, FFF demonstrations and XR actions (particularly their "international rebel week" in Paris in the fall of 2019). The spreading of COVID-19 pandemic has shifted part of this work online: for example, interviews with Italian FFF network members on the eve of Glasgow COP26 have been conducted by using different applications (Zoom, Whatsapp, etc.) in the fall of 2021.[16] A review of the literature and results of a workshop I have co-directed during the 9th Congress of the French Sociological Association (7 July 2021) complete the sources of this chapter.

A "Climate Generation" Divided

Although FFF participants, as well as younger XR and ZAD (or no-TAV) activists, are often lumped together, from less to more radical side, in a larger "Climate Generation" [Flandrin et al. 2019], similarities and differences do not only refer to their age and to their political positioning. Beyond the question of repertoires of actions and cultures, examined afterward, they also involve socio-economic backgrounds, gender and race relationships—i.e., men/women, sexual and ethnic minorities [Druez 2022].

Ages, Classes, Genders and Colors of Dissent

Regarding age, as many other European countries, French and Italian FFF have been launched by "extremely young people, often at their first activist experience" [Haeringer et al. 2020: 159]. Indeed, before the spring of 2019, only a few dozen high-school and university students, most of them undergraduates, refused to enter their classes on Fridays. A survey conducted in Paris, Nancy and Lille by a group of scholars, "Quantité Critique", during the first GSS (March 15, 2019) shows that high school students represented around 60% of demonstrators, and university students the remaining 40%. Since then, proportions have been reversed: undergraduates and graduated students constituted more than the half of the strikers during the second GSS in September 2019 [Gaborit 2021]. This phenomenon has been observed in other countries: in Belgium, for example, part of 14–19 years old, which represented the great majority of participants to the first GSS, reduced to approximately a quarter of total six months later [Lensing 2021].

Conversely, as seen before, XR and ZADs have been initiated by activists of different ages, often experienced, but youth have immediately played a key role in these networks: students already represented most participants for the first XR large-scale action in France, an occupation of the Sully Bridge in Paris on 28 June 2019 [Hardy 2020]. Images of riot police throwing tear gas on young people peacefully sitting down to stop traffic shocked opinion

---

[16] A radio broadcast about this topic is available: https://www.franceculture.fr/emissions/cultures-monde/de-berlin-a-milan-en-decoudre-pour-le-climat. Accessed March 7, 2022.

and helped to popularize the movement. Four months later, during the RIO, most of Italie2 and Châtelet occupiers were again teenagers and young adults. Among nonstudents, observations that I gathered in the mall and in the camp include various professions: WWOOFers,[17] computer engineers, "third sector" employees,[18] organic grocery store cashiers, school professors, etc.

Youth also marked their own age identity within these movements: in Italy, a "Komitato Giovani no-TAV" (Youth No-Tav Kollective [*sic.*]) was created in 2011 by high-school students to support Susa Valley struggle. Since, they have organized several demonstrations around the railroad construction area [Collective Askatasuna 2012]. Missing in both countries, XR Youth was founded in the UK in February, 2019, "out of a need to create a platform for the young person's voice", i.e., "those born after 1990 – a generation that has never experienced a stable climate and whose lives will be wholly affected by the climate and ecological crisis", according to their website.[19]

Focusing on XR and FFF, research conducted in France and in Great Britain in 2019 confirms the young age of protestors, generally ranging from 11 to 34 years old. Differences can obviously be established between those beginning secondary education and those departing young adulthood that year, but, following authors, all these activists share a common socialization during a period of overlapping crises (financial, environmental, etc.) shaping their political values, attitudes and behavior [Pickard et al. 2022]. These crises have also affected the quality of their lives: compared to previous generations, they are experiencing—or, for younger, they will experience—higher levels of precariousness at work, increasing their feeling of ordinary insecurity and, thus, their distrust and disappointment about conventional politics [Toscano 2007]. However, scholars also suggest a certain homogeneity regarding their socio-economic backgrounds: they observe, for example, that university students' "biographical availability" to protest is reduced today, either because many of them are obliged to work to earn money or because they are subject to strong educational pressures. Therefore, despite supporting demonstrations, lower profiles are often unable to engage [Pickard et al. 2022].

These conclusions are similar to the "Quantité Critique" survey on the French FFF: since the first GSS, they observe a correlation between class and participation to climate marches [Gaborit 2021]. Data gathered on 15 March 2019 shows that only a few demonstrators study at technical high schools, while prestigious colleges and universities are over-represented. Most strikers state that they get "good results" at exams and describe an "easy" progression in their educational pathway. Above all, the majority of them come from middle-upper class families: more than 50% of demonstrators

---

[17] WWOOF is an international movement allowing people who want to work in agriculture to learn skills by visiting and working in different organic farms. To get more information: https://wwoof.net. Accessed March 16, 2022.

[18] Nonprofit and nongovernmental sector, including charities, cooperatives, NGOs, etc.

[19] https://xryouth.uk/about/. Accessed March 8, 2022.

have a father occupying an executive position, while less than 10% are children of manual workers. The data do not reflect the general composition of public of the French educational system [Erlich 1998]; conversely, they fit in with responses gathered in questionnaires submitted to adults during intergenerational climate marches: environmental issues seem to be middle and upper class-related, lower socio-economic profiles aim at protesting less against global warming than higher profiles [Gaborit 2021].

ZAD participants represent exceptions in the landscape: permanent occupiers of NDDL include a variety of social categories, with a significant proportion of precarious situations such as seasonal workers, job seekers or beneficiaries of RSA allowance[20] [Verdier 2018]. But one can point out that this phenomenon is not new: the back-to-the-land movement in the 1970s already involved people coming from higher classes, who voluntarily moved to rural areas despite material needs and people coming from lower classes who associated their choice with material necessities, like finding a place where to live and produce food [Stuppia 2016]. Moreover, an (intentional) community-based organization and representations of political autonomy are primal in ZADs: the coexistence of diverse socio-economic background can be analyzed as directly related to their will to transform the society "here and now" (see below).

Besides class-related questions, young environmental activists seem to attach a central importance to gender today. Indeed, young women are at least as involved as men in FFF, XR and ZADs, with an increasing portion of people "that define themselves nonbinaries" [Haeringer et al. 2020].

Feminization is particularly clear regarding FFF figures, despite the formal rejection of leadership. In Italy, for example, FFF has virtually elected six national spokespersons in 2021, two men and four women, all aged from 14 to 27 years old, "following two guidelines: to respect gender equality and territorial equality."[21] The media have focused on a single figure, Martina Comparelli, since she has integrated the FFF delegation (composed by her, Greta Thunberg and Vanessa Nakate) which has been privately received by Italian Prime Minister during the 2021 pre-COP meeting in Milan, i.e., the final ministerial meeting ahead of the Glasgow COP26. Spokespersons consider that this emergence is "a problem, in a horizontal movement, but also an opportunity" to access "mainstream media and correctly inform the public about the ongoing climate crisis;"[22] according to Comparelli, young women, "who have to deal everyday with labels and stereotypes, have an extra oomph"[23] to play this role in the organization.

---

[20] A French social benefit.

[21] Collective interview conducted on October 13, 2021.

[22] *Idem*.

[23] https://milano.corriere.it/notizie/cronaca/21_settembre_27/conferenza-clima-milano-martina-comparelli-leader-fridays-for-future-donne-guida-marcia-piu-65f9bed8-1f5b-11ec-b908-b44816b61f2f.shtml. Accessed October 12, 2021.

Regarding other gender identities, activists are as attentive as possible to include all the LGBTQI+: in NDDL, *zadists* often use the gender-neutral first name "Camille" to preserve their anonymity when they are interviewed by journalists, but also to show that they outreach classic male/female divisions; announcements of FFF Italy often replace the final letter of gendered names and adjectives with neutral signs, creating new grammatical forms.[24] This consideration of gender and sexual minorities sometimes contrast with reality, encouraging the creation of nonmixed groups: in NDDL, for example, a feminist circle has been established in 2012 to blame "sexism, sexual assaults, appropriations of working tools by men and overrepresentation of women in the ZAD legal teams and medical teams. They also built a non-mixed cabin" [Pruvost 2017: 117] equally welcoming trans people who denounced several aggressions.

Lastly, contemporary environmental activism is confronted with questions of race and color blindness. Indeed, excluding Global South, FFF marches, XR actions and ZADs have often been described as "white led", combining color and economic privileges such as living in a wealthy neighborhoods [Shapiro 2006].

In the UK, for example, XR has been criticized in 2019 not only because they mostly involved middle-upper class activists, but also because their "tactic of encouraging mass arrests ignored the reality of police racism" and "made the protests the preserve of privileged white people."[25] In 2020, in the aftermath of the Black Lives Matter (BLM) movement, they have published a statement on their website recognizing that their previous strategy "has made it easier for people of privilege to participate" and that their "behaviors and attitudes fed into the system of white supremacy", suggesting several improvements (reviewing training workshops, setting up cooperations with other groups, etc.).[26]

Taking our two cases, France is quite exclusively concerned by these social issues today. Indeed, most of the youth with a migrant background arrived in Italy too recently to be socialized there or they just belong to the second generation, two factors explaining why they are overall more politically inactive than their native peers [Riniolo and Ortensi 2021]. Conversely, on the other side of the Alps, a longer colonial and migrational history allows us to consider their (non-)participation in environmental movements.[27] As in the rest of Global North countries, only a few French youth with a migrant

---

[24] *Tutti* (all, which is masculine in Italian) is for example transformed in the new form, *tutt**.

[25] https://www.theguardian.com/environment/2020/aug/04/evolution-of-extinction-rebellion-climate-emergency-protest-coronavirus-pandemic. Accessed March 8, 2022.

[26] https://extinctionrebellion.uk/2020/07/01/statement-on-extinction-rebellions-relationship-with-the-police/. Accessed March 9, 2022.

[27] Even if this subject is highly controversial in a country forbidding ethnic statistics.

background are involved in FFF, XR or ZADs.[28] However, their degree of withdrawal varies: they were more likely to demonstrate during FFF marches than to participate to XR actions observed in Paris in 2019, excluding the occupation of Italie2 mall. That day, XR invited other groups to join their protest, trying to connect social, anti-racist and environmental claims under a banner stating: "nature is not for sale: people's ecology". Some youth racialized activists from the "Collectif Adama", an organization struggling against police violence and racism in poor neighborhoods, effectively joined the action and spoke in front of the shopping center.[29] But this junction appears as quite an isolated act: white protestors have been over-represented again in the rest of actions of the RIO, especially those including a good chance of being arrested, such as during the attempt to occupy the Concorde Bridge in front of Parliament at the end of the rebel week. This means that the higher are the costs and the risks of collective action [McAdam 1986], the lower seems to be the probability for nonwhite youth participating in environmental protests, at least in European countries. Explanations in terms of systemic racism overlap here those in terms of socio-economic profile.

The description of the "micro-units of generation" acting for the climate outlined in the previous pages can be completed by focusing on their political preferences.

(DIO) Politics of Dissent

Available surveys show variations in terms of political positioning inside the three networks, but one point seems to be shared by all activists. When protesting in FFF, XR or ZAD, participants "do it themselves and do it together, in order to bring about change", a way to relate to politics defined by researcher Sarah Pickard as "Do-It-Ourselves (DIO) politics" [2019, 2022]. This concept refers to the fact that youth tend today to "take politics into their own hands metaphorically and literally through participatory actions outside institutional politics", particularly parties, trying at the same time "to influence politicians and other power-holders. This produces feelings of belonging, agency and self-efficacy". DIO politics also describe the inclination of "young generations to mobilize around issue politics – e.g., MeToo, March for Our Lives and BLM – rather than to adhere to traditional ideologies" [Pickard et al. 2022].

FFF constitutes the network presenting the most diversified positioning within it. According to scholar, Maxime Gaborit, since the first marches in 2019, the media have spread the idea that participants are "apolitical", i.e., "unable, or refusing to situate themselves on the political spectrum", due to

---

[28] According to scholar, Margot Verdier, permanent occupiers of NDDL share similar "race [backgrounds], but not the same socio-economic profiles" [2018: 172].

[29] https://reporterre.net/L-action-d-Extinction-Rebellion-a-eu-lieu-a-Paris-au-centre-commercial-Italie-2. Accessed March 16, 2022.

their young age and their involvement in a "civic movement, out of cleavages". Quite the opposite is revealed by questionnaires submitted to participants to the two GSS in March and September, 2019, that show that an immense majority of demonstrators declare a political preference, around 40% for the left and 30% for the far left, often in conformity with family choices [2021]. However, they do not always indicate a predilection for a party and the number of responses increase with age. In Italy, quantitative data are missing, but interviews gathered in 2021 seem to confirm these findings. Responses range from denials—e.g., "this is not a question anymore!"[30]—to precise, strong statements. All the protesters affirm that FFF is "totally independent from parties", as their "goal is to include as many people as possible;"[31] nevertheless, they also all reject the mainstream media frame, defining their movement "political, but a-partisan."[32] "What's more political than a fight to enable mankind to have a future on this planet,"[33] a FFF city organizer asks? If they recognize that lots of strikers are closer to the left and the far left due to their family backgrounds and/or their volunteering in different associations, "youngsters, who can't vote already, haven't often very, very precise ideas,"[34] and many elders feel "disillusioned, disinterested or not so represented by one party, or another,"[35] even when they express a firm preference on the political spectrum. They declare, therefore, to be more interested by a (doing-it-themselves) politics of "small and concrete actions,"[36] both individual and collective, than standard electoral and party politics.

In this landscape of variating positioning, the absence of the far right is particularly remarkable. Indeed, recent polls indicate that these ideas are progressing among young people in both countries,[37] while far-right organizations are "greening" their programs, associating re-localization of production and anti-immigration narratives. FFF students seem to be almost unanimous in rejecting them, especially in Italy. Spokespersons declare for example: "we

---

[30] FFF university student in Turin (man, 21 years old), interview conducted on 11 October 2021.

[31] Collective interview with national spokesperson conducted on 13 October 2021.

[32] FFF university student in Turin (woman, 20 years old), interview conducted on 13 October 2021.

[33] FFF organizer working in an association in Cuneo (man, 28 years old), interview conducted on 12 October 2021.

[34] FFF high-school student in Cuneo (woman, 16 years old), interview conducted on 12 October 2021.

[35] FFF organizer working in an association in Cuneo (man, 28 years old), interview conducted on 12 October 2021.

[36] FFF high-school student in Cuneo (woman, 17 years old), interview conducted on 14 October 2021.

[37] For example: https://www.lemonde.fr/politique/article/2021/04/05/le-rassemblement-national-premier-parti-des-25-34-ans_6075574_823448.html and https://www.affaritaliani.it/politica/sondaggi-analisi-del-voto-pd-al-top-tra-gli-over-55-i-giovani-scelgono-fdi-763217.html. Accessed 13 February 2022.

share an alternative vision of the future, and we are struggling for a fair ecological transition. This can't fit with the far-right. We have concrete propositions to decarbonize the economy, and among them we include the promotion of human rights, and of our values. Anti-fascism is one of them."[38] Another high-school student affirms: "school strikes for climate aren't originally born as anti-fascists, but they are inspired by people with a particular sensitivity for social themes. Fascism is undemocratic, anti-constitutional, and immoral. We the environmentalists refer to an ecology connecting the climate sphere, and the anthropological [*sic.*] sphere, i.e. a system of solidarity and mutual cooperation between persons, living in harmony with Nature."[39]

As already highlighted by a study on climate justice movements before 2018, the main division that seems to cross FFF is a cleavage between a reformist side, seeking "change within the existing political and economical arrangements, and a radical side that rejects the existing order", even if this polarization is "fluid and subject to change based on the steady worsening of the climate crisis and individuals' and groups' experiences within it" [Foran et al. 2017: 375].

"Quantité Critique" survey develops precisely this last perspective, combining collective and private levels, to identify three main groups of positioning among FFF protesters: the "radicals", the "reformists" and the "centrists."[40] The first group is composed of youth that generally describe themselves as "very left wing". They try to match their environmental convictions and practices, in activism and in everyday life. They also declare that they participate in XR actions, or at least to be ready to do it. The second group is composed by students mostly "left wing", who support civic disobedience, but less inclined to take part to XR operations; finally, the third group encompasses youth of all political sides, demonstrating during marches, but refusing XR methods. Even if they usually affirm that the ongoing climate crisis can be resolved by individual behaviors (green investment, reduce meat consumption, etc.), they rarely put their theories into practice [Gaborit 2021].

These analyses of student strikers' divisions open an interesting research field regarding XR and ZAD, by suggesting a greater homogeneity of political positioning. In both cases, indeed, radical ecology [Dufoing 2012] rallies anarchist and libertarian traditions. According to Sylvaine Bulle, XR "rebels" and *zadists* share nowadays a similar perspective of political autonomy from parties, state and capitalism, and common "anxieties" toward climate change [2021]. They express them either by setting up blitz operations aimed to alert opinion to the ongoing ecocide and an upcoming "collapse" [Servigne and

---

[38] Collective interview with national spokesperson conducted on 13 October 2021.

[39] FFF high-school student in Cuneo (woman, 17 years old), interview conducted on 14 October 2021.

[40] Based on a multivariate analysis of matches between declared values (e.g., "anti-capitalism", "freedom of commerce", etc.) and everyday life individual practices (e.g., "boycott a company", "become vegan", etc.) of more than 2000 signatories of a Student manifesto.

Stevens 2015], regarding XR, or by creating intentional communities experimenting alternative socio-environmental models called to spread into society, concerning ZAD.

Both networks also criticize concurrent mainstream environmental groups, especially those that, dialoging with authorities and economic stakeholders, are accused to support "greenwashing". Conversely, they often seek alliances with other radical networks struggling for different causes (migrants' rights, social justice, etc.). For example, during 2019 RIO, XR invited German ship captain Carola Rackete to speak at Châtelet. She formerly volunteered in Mediterranean Sea to rescue migrants with Sea-Watch organization. They also welcomed a "yellow vest" gazebo—a grassroots movement demanding social and fiscal justice—in the name of the "convergence of struggles".[41]

As noticed by Sarah Pickard, Benjamin Bowman and Dena Arya, taking the example of XR UK, the two networks are equally invested by youth (who are sometimes the same people who demonstrate in FFF, as mentioned before), defining themselves as "part of a radical community, breaking social norms" [2020: 253], and aspiring to a complete shift in politics as well as in their life. They frequently consider their involvement on a pre-figurative register [Polletta and Hoban 2016], i.e., as an immediate reflect of the future dreamed society, and as a "day-by-day experience" [Pleyers 2016], especially when establishing camps in urban areas or over construction sites (see below).

Compared to FFF, inner variations of positioning are consequently existing in XR and ZAD, but, on the one hand, they cover a smaller portion of political spectrum, and, on the other hand, they are continuously subjects of collective and individual reconfigurations by lived experiences in action.

Despite these similarities, two main differences appear between XR and ZAD "radicalness", and will be discussed in the next section.

What "Radicalness" Means: Differences Between XR and ZAD

The first distinction is linked to historical background. According to scholar, Isabelle Sommier, XR refers to a radical ecology that spread at the turning of the millennium in the alter-globalization protests, with attempts to peacefully invade "red zones" during international summits (WTO, G8, etc.) and disobedience actions such as the voluntary mowing of GMO crops. This repertoire has been inspired, in turn, by a series of twentieth-century figures (Gandhi, Martin Luther King, etc.) and movements (1960s US civic rights and French anti-Algeria war, draft-dodging, etc.) [2021: 144–146].

ZADs are rather connected to 1970s radical environmentalism, especially Bookchin's "communalism", and anti-nuclear power plants protests, including both peaceful and direct actions. In the 2000s, anarchist "black bloc"

---

[41] All the elements reported in this paragraph come from my observations.

[Dupuis-Déri 2005] renewed this repertoire, trying to create so-called TAZ—"Temporary Autonomous Zones"[42]—eluding State control during anti-global demonstrations, often by using force. ZADs can be analyzed, therefore, as attempts to form a "definitive autonomous zone" where experiencing another life "here and now" is made possible [Sommier 2021: 147–148]. Thus, these analyses slightly contrast with Bulle's reading, "autonomies" defended by XR and ZAD partly differing.

The second contrast can be described following sociologist Anne Muxel's separation between a "protest radicalness", encompassing expressions of disagreement with ongoing policies ranging from vote for "extremist" parties to collective nonviolent actions, and a "disruptive radicalness", including revolutionary ideas, and behavior which justifies violence [2018: 203].

"Quantité Critique" survey shows, for example, that French participants to the first GSS seem to be generally marked by a strong "protest radicalness", 4 out 5 of them declaring to support or to be ready to "block polluting infrastructures", but a very weak "disruptive radicalness", only 10% of them envisioning to take part to these actions, if they include material destructions [Gaborit 2021].

The same goes for XR young activists, but not for *zadists*. During XR operations, the only use of force approved is "the use of force by occupying protesters against their own bodies, in the form of restraints against removal" [Pickard et al. 2020: 264]. In a few cases, activists caused limited material damages: graffiti, glued posters and soiling of walls of Ministry of Ecological Transition in Rome 1 February 2022, electric scooters barricades to defend the "Châtelet ZAD" entrances in October, 2019, etc.[43] This is why an XR activists interviewed by Sarah Pickard, Benjamin Bowman and Dena Arya in the UK, affirmed: "we are radical in our kindness", which is also the title of their study [2020]. By contrast, in Susa Valley and in NDDL, pacific demonstrations coexisted from the beginning with other contentious performances based on direct action, such as the constitution of "black blocs" attacking law enforcement. While participants to XR operations are asked to accept a "rebel agreement" stating that they must "show respect to everyone", including "the government and police", and that they practice "nonviolence, physical or verbal,"[44] participants to ZAD are invited to accept another principle: the "diversity of tactics" [Piven 2006; Gelderloos 2018], asking everyone to respect different choices than their own, including the use of violence, without judgments or condemnations.

During the 2019 RIO, these two orientations have been regularly debated, both in Italie2 and Châtelet occupations: in the mall, which was transformed

---

[42] See also: Bey [1991].

[43] For example: https://extinctionrebellion.it/press/2022/02/03/ultima-generazione-smentisce-che-ci-siano-state-violenze-al/. Accessed March 16, 2022.

[44] https://extinctionrebellion.uk/wp-content/uploads/2019/10/XR-REBEL_AGREEMENT_A6_Flyer-FINAL-NO-CROPS.pdf. Accessed February 15, 2022.

in a hub for different radical protests, XR strategy was soon criticized by other groups—yellow vests, *antifa*, etc.—closer to *zadists* through a series of humorous graffiti inscribed on walls and store fronts: "Greta [Thunberg] wearing a K-way" (a direct reference to black bloc), "Harry Potter himself is violent", "please, break this store front", etc. A group of occupiers also actively resisted a riot police attempt to evacuate the shopping center in the evening; eventually, all the demonstrators left peacefully a few hours later, without being arrested. To avoid the repetition of this situation, XR dispatched "peacekeepers" wearing an orange vest since the first day of Châtelet occupation. They also distributed leaflets containing the "rebel agreement" and stating that peacekeepers "could invite any person who disrespects rules to leave the action". Some tensions followed, especially with invited yellow vests, but the action continued without major clashes until the next day, when XR activists observed a minute of silence in memory of three policemen killed in a terrorist attack, and deployed a banner "against all violence".[45] Radical collectives that previously occupied Italie2 published an open letter to XR activists,[46] denouncing their attitude to put on same level an Islamist murder and an ACAB[47] graffiti and accusing them of ignoring police violence against lower social classes, minorities, etc., renewing "classist" and "color blindness" accusations previously mentioned.

Changing Activism in Climate Change Era

The analysis of prevalent political positioning in XR, ZADs and, to a lesser extent, FFF reveal common threads between the sequence inaugurated in 2018–2019 and the one opened by alter-globalization protests twenty years earlier, developed since in anti-neoliberal and anti-austerity student movements—e.g., 2008 "Abnormal Wave" in Italy, 2012 "Penguins" in Chili—and social justice place occupations, such as Arab Springs, Occupy (US, UK...) and Spanish "Indignados" gatherings in 2011–2012. Analogous traditional and innovative performances (e.g., sit-in and FabLab,[48] boycott and hacktivism) coexist in all these mobilizations, even if the question of the emergence of a "third repertoire of collective actions" [Cohen and Rai 2000] outreaching national anchorages is more debated [Offerlé 2008]. In the same way, activists' individual practices know reciprocal influences and imitations worldwide. Sociologist Jeffrey S. Juris and Geoffrey Pleyers conclude to the rise of a new

---

[45] All the elements reported in this paragraph come from my observations.

[46] https://paris-luttes.info/lettre-ouverte-aux-militant-e-s-d-12726. Accessed February 15, 2022.

[47] "All Cops Are Bastards".

[48] Fabrication Laboratory, small self-organized workshops offering digital fabrications.

activist culture,[49] called "alter-activism", based on "lived experience and process; a commitment to horizontal, networked organization; creative direct action; the use of new Information and Communication Technologies (ICTs); and the organization of physical spaces and action camps as laboratories for developing alternative values and practices" [2009: 58].

Most of these features match with the notion of "DIO politics": both describe contemporary youth preferences for more individualized modes of involvement compared to the twentieth century, an "increasing support for leaderless, decentralized, horizontal, fluid, networks and movements that claim internal democracy" [Pickard 2022], and personal lifestyles placing "ethics at the core of activism: [...] it is not only about changing society, but about building oneself as a person who changes society, to change both life and the world. The relationship to oneself is central to this form of involvement, alongside a sense of personal responsibility and a quest for consistency between the practices and values defended" [Pleyers 2021]. Whether using the concepts of "alter-activism" or of "DIO politics", a majority of Italian and French FFF, ZAD and XR members seem to share these new relationships to collective and personal participation in social movements.

As their predecessors at the turn of the millennium, first, young *zadists*, school strikers and XR "rebels" are impacted by globalization process, but not only negatively: they are "shaped by precarious working conditions on the one hand, and constant exposure to and participation in global information streams on the other hand" [Pleyers and Glasius 2013: 550]. This fosters the emergence of an intense online activism, at three stages, benefiting from the latest ICTs developments.

At the first stage, the Internet is used to produce self-controlled information, in line with Indymedia[50] slogan "don't hate the media, be the media" popularized during alter-globalization counter-summits. Social networks, forums, apps and websites are nowadays employed both to relay alternative contents from mainstream media and to diffuse scientific assessments on climate change (FFF, XR) or ecological impacts of construction work (ZADs, no-TAV), in order to raise awareness and support. For example, a banner at the top of FFF Italy Homepage includes a timer from years to seconds left to limit global warming to 1.5° ("deadline") and another counter showing positive data such as the evolution of percentage of renewable energies in the world ("lifeline").[51] Hashtags and smartphones greatly help this "consensus mobilization" [Klandermans 1984] work: one of the most common processes to create alerts about global warming consists in tweeting a selfie picture while

---

[49] Defined as "a logic of action based on a coherent set of normative orientations, a conception of the world, of social change and of the nature and organization of the social actors who bring about this change" [Pleyers and Capitaine 2016: 107].

[50] An open publishing activist network created in 1999, particularly active during 2000s alter-globalization protests.

[51] https://fridaysforfutureitalia.it/. Accessed February 20, 2022.

holding an handwritten placard and some hashtags, taking inspiration from 2018 solitary Thunberg's action. Several FFF figures, especially women, have emerged through this technique worldwide. This specific use of the Web is also fit into a discursive register combining science and emotions mentioned at the end of this chapter.

At the second stage, digital technologies are used for "action mobilization" [Klandermans 1984] efforts, i.e., to transform approval and understanding previously gathered in concrete, "in-real- life" performances. Forms greatly vary, ranging from mailing lists to encrypted instant messages in order to invite supporters to join; moreover, during 2020 and 2021 COVID lockdowns, Italian and French activists have elaborated innovative hybrid operations to be heard despite restrictions. On 24 April 2020, for example, FFF Italy Facebook page gave four instructions to take part to the fifth GSS: "1. Hang a blanket, a towel, or whatever else green out of the window; 2. As usual, prepare a placard and take a selfie. Use the hashtags #24A and #FFF; 3. Be ready for the Tweet storm; 4. Plant a seed for a different future for the planet and its dwellers."[52]

At the third stage, new ITCs are employed as inner organization platforms. In France, XR actions are prepared on an Internet forum, called "La Base", with five distinct access levels, from "new user" to "organizer", rooted on peer-to-peer trust. Each level opens new functionalities: "new users" reach the status of "XR member" by completing three tasks; if they visit *La Base* during 3 days, receive a "like", introduce themselves and read at least 10 threads, they become "XR activists". Following their implication in "real" local or national work groups (logistics, artivism [*sic.*], etc.), they can eventually integrate the two higher levels, giving access to more information, and private spaces.[53] Currently, around 8,000 users have completed step 2, giving an approximate estimation of number of French XR supporters. On the other side of the Alps, school strikers use independent messenger service Telegram to coordinate actions nationwide, and several other channels locally. Therefore, FFF and XR can be considered as "two examples of new 21st Century form of movement organization in which the communication networks become the political organization" [Doherty and Saunders 2021: 253], at least in ordinary times.

Outside of the Web, action camps are the most significant illustrations of the spreading of an alter-activist culture among young environmentalists, particularly regarding ZAD and XR members. Indeed, NDDL and Châtelet can both be analyzed as "direct confrontation and experimentation spaces, marked by a concrete implementation of democratic and ecological practices and by the intensity of lived experiences" as they were "Nuit Debout"—a monthlong occupation of République Square in Paris—in 2016, Occupy London

[52] https://m.facebook.com/events/506195466669748/. Accessed February 22, 2022.

[53] https://base.extinctionrebellion.fr/. Accessed February 20, 2022. According to scholar, Hervé Brunon, this forum adopts "the principle of holacracy, a horizontal governance form based on a common distribution of responsibilities, allowing the dissemination of decision making" process [2021: online].

in 2011–2012, and alter-globalization campsites before them. In all these occasions, activists "combine private and public [spheres], friendship and political commitment, fun and resistance, happiness and fight for a better world. The contentious dimension often takes second place, compared to alternative practices and to sociabilities [established there]. Participants self-organize by distributing tasks, relaying on the involvement of everyone. Hence, they are confronted with very concrete problems: how to democratize the decision making process, how to deal with delegation, how to manage their political and cultural differences, etc." [Pleyers 2016: 38]. Beyond official targets, e.g., withdraw a construction project, campers strive to demonstrate "*in situ* that another world is possible – a direct reference to alter-global most popular slogan – by a complete reversal of norms" [Pruvost 2017: 36]: this prefigurative register guides the entire organization of the action, be it in space or in time, from the most common tasks such as preparing food to assemblies.

During 2019 RIO in Paris, for example, XR activists first established six blocking points all around Châtelet Square to stop traffic, laying down around DIY tools fashioned to lock their arms into the ground and delay law enforcement interventions. Within the "ZAD" perimeter, they quickly pitched tents, built cabins, including a "regeneration bubble" yurt to take a rest and meditate, and created other spots: a kitchen, a library, dry toilets, a "convergence of struggles" space hosting a yellow vest gazebo, a XR hourglass logo soil-straw sculpture, a "permaculture garden" composed by some planters and flower pots, etc. They eventually set up specific areas for general assemblies, and DJ parties in the evening.[54]

According to their skills, participants were asked to take specific responsibilities, but also to rotate from risky, unpleasant or thankless duties (e.g., sleeping on a blocking point during the night, collecting wastes, etc.) to more satisfying efforts such as animating a debate, with a particular attention to gender relationships—avoiding the stereotype of "men at war, women at stoves" [Mann 2019]—and horizontality. Each of six access points to the square self-organized its own daily gathering, both to argue about climate change and to action related issues; a spokesperson, different from one day to the next, was designated to report conclusions and ideas in front of the camp general assembly. During these meetings, pre-figuration reached its highest point: in line with alter- globalization Social Forums, audience was invited to listen and respect any intervention, signifying approbation or rejection by silent hand movements. Peacekeepers tried to resolve conflicts by communication techniques, e.g., individual dialogs with yellow vests booing a speech supporting the banner "against all violence" previously described. Gender, and above all ethnic minorities—despite their low participation mentioned before, were explicitly invited to express their position, in order to promote inclusivity.

[54] All the elements reported in this paragraph come from my observations.

Lastly, rather than votes, participants sought to elaborate consensus, taking more time and debates.[55]

All these elements fit with the experimentation of new social relationships in action camps, translating protesters' long-term ambition to "live in a world where hierarchies no longer exist,"[56] and to build a society based on "real democracy, dignity, and justice" [Pleyers and Glasius 2013: 560] prototyped in these temporary intentional communities. Contrasts with reality—sexism as in NDDL, privileges regarding XR campground, etc.—foster continuous self-critical and "self-educational" [Stuppia 2016] works, both at individual and at collective stages, to realize that errors have been made, to identify their causes and to explore solutions. "La Base" hosts, for example, a "debriefing" thread, where activists exchanged about the RIO week. On the one hand, all highlighted the positive welcoming initially received to their proposition of "convergence of struggles", both from "Collectif Adama" and from yellow vests; on the other hand, they also recognized mutual misunderstandings. A forum member writes, regarding Châtelet: "[XR and YV] shared some akwardness: when we [XR] saw them [yellow vests] making braziers, drinking alcohol"—as the "rebel agreement" forbids its use during the action time, as well as drugs—"and moving tree grates, [but also] when them saw us leaving the square on Friday and left them alone, when we erased their graffiti or when they had the feeling to be "lectured" on violence and alcohol. This was a real test for XR: we declare ourselves inclusive, but are we able to find permanent cooperation methods with other movements such as yellow vests?" A series of positive suggestions for future joint actions followed—to better identify common points and differences, to work more on dialog, etc.[57]

Finally, in laboratory camps as well as in their everyday life, all the young ZAD, XR and radical FFF protesters seem to pursue their quest for matching practices and values mentioned by Pleyers. Observations gathered in Châtelet and interviews with Italian FFF members concluded that these activists share similar wishes "to do concrete things, rather than to replicate politicians' "blah, blah, blah""[58]—referring to a sentence pronounced by Greta Thunberg at the end of September, 2021, similar consumption decisions (to reduce plane travels, to adhere to a community-supported agriculture association to get food, etc.), re-embedding politics in their lives "even by organizing very simple actions [such as] a night movie, a clothes exchange in order to promote

---

[55] *Idem*.

[56] https://extinctionrebellion.it/chi-siamo/extinction-rebellion/. Accessed February 24, 2022.

[57] https://base.extinctionrebellion.fr/t/debrief-7-octobre-occupation-pour-la-suite-du-monde/35141/7. Accessed February 23, 2022.

[58] FFF university student in Turin (man, 21 years-old), interview conducted on October 11, 2021.

a circular economy, a campaign to collect and recycle trash at the river."[59] In the same way, they seem to show a notable sensitivity to digital "ethics" and "sustainability". Both "La Base" and FFF Italy websites are, for example, built on open-source technologies, use renewable energies, and are hosted by certified net-zero emissions servers. Administrators justify their choices in specific areas, asking users to follow suit. Italians write: "we realize that [the most popular social networks] finance fossil industries, steal data, and even foster the election of dictators. Although we feel like David standing in front of Goliath, we have decided to do our part by promoting the use of alternative platforms, even if we cannot exclude" Facebook, Twitter, etc. They consequently separate "sustainable" (green label) and "nonsustainable" (black label) services, inviting school strikers to prefer the use of "ecological and ethical social media"[60] such as RSS, Mobilizon or Peertube.

Conclusion

In France and in Italy, as in the rest of the world, the sequence opened in 2018 by FFF powerful demonstrations, XR spectacular actions and the multiplication of movements like ZADs and no-TAV leave the feeling that youth are firmly committed to rise up when facing climate change that condemns them to a bleak future. Indeed, their intense activism, both online and offline, could lead us to think that a new and united "climate generation" has been born. The elements in this chapter show, nevertheless, that not all youth are involved and "rebellious" when it comes to environmental questions, but rather only a specific fraction of them, with particular backgrounds. Gender is a key point: women are, at least, or more, engaged than young men and ecological claims often go hand-in-hand with LGBTQI+ and egalitarian issues. Racism is another important aspect to be considered, as ethnic minorities, fearing more police interventions than white people, tend to shy away from most radical environmental protests such as XR operations, or ZADs.

Within these "micro-units" of the "climate generation", one can identify a common preference for spontaneous performances and DIO politics rather than conventional involvements. This confirms a global shift observed since the latest 25 years "from the traditional forms of political participation to a series of unconventional, de-structured, horizontal and unprecedented actions" [Riniolo and Ortensi 2021: 925]. Thus, all the young environmental activists seem to fit in with scholarly analyses that highlight a definitive takeover of "in-movement student citizenships" on "institutional student citizenships" [Legois et al. 2021] during the decades of 2000 and 2010 from both sides of the Alps. They also appear in line with "real democracy" social movements, e.g., Arab Springs, or Occupy, that followed alter-globalization protests, aimed

[59] FFF high-school student in Cuneo (woman, 16 years-old), interview conducted on October 12, 2021.

[60] https://fridaysforfutureitalia.it/sostenibilita-digitale/. Accessed February 24, 2022.

at "formulating a demand which the largest number of citizens could adhere to" while "respecting a triple requirement: neither leader"—we could specify, for our cases: except for the media, "nor program"—despite general claims, for example, those expressed by XR or FFF, "nor party affiliation" [Montoni-Rios 2015: 213].

However, it is also possible to argue that these questions specifically matter for the most politicized protesters, by introducing supplementary divisions within the "micro-units of generation" as previously identified. Indeed, one can distinguish several wings of climate mobilizations, especially in the most diversified collectives such as FFF: the "radicals" look for a coherence between practices and values, and experiment utopias with them in laboratories—ZADs, street camps, etc.—where they prefigure personal and collective change "here and now". They also seek a convergence with other social movements and share the same alter-activist culture when involving in contentious performances as well as in their everyday, real and digital, life. Conversely, they diverge on tactics to reach expected goals and, in particular, on the question of violence: while *zadists* and no-TAV generally accept the use of physical force, most of school strikers and XR members reject it, except against their own bodies to resist arrests. In this case, XR France has asked them to be "resilient" and promised a complete legal and psychological support; they also launched a campaign in 2021 pledging to plant a tree for each "rebel" placed in custody.[61]

Focusing on personal experiences such as those in the campgrounds or in police stations imply to consider activists' emotions and, first and foremost, the "emotional work" [Hochschild 2003: 32] which is usually shaped by social movement organizations to turn individual feelings into political involvement [Sommier 2010]. This perspective is almost entirely missing in this chapter's findings. It appears, however, very important to situate the evolution of contemporary youth environmental activism into four stages.

First, it could help to understand why, nowadays, young people seem to prefer to "personify involvement as whistleblowers" on climate change [Becquet 2019: 40], often explaining that their will to stand for the planet results from a "moral shock" inspired in turn by single figures (e.g., Greta Thunberg) or lived situations (e.g., a river pollution), rather than being organized in and by traditional structures such as NGOs or green parties.

Secondly, regarding more specifically the FFF, it could explain the increasing distrust of their members in "adult" politics, and, to a lesser extent, their distance with former ecological protests, rarely mentioned as references (excluding, in my interviews with Italians, alter-globalization[62]).

---

[61] https://extinctionrebellion.fr/actions/2021/02/26/une-arrestation-un-arbre.html. Accessed February 26, 2022.

[62] Interviews have been conducted in a particular context, the 20th anniversary of G8 protests in Genoa, and during the second year of COVID Pandemic: the media have long focused on Genoa Social Forum claims 20 years earlier, especially environmentalists

Indeed, school strikers frequently describe themselves to be part of a generation let down by "older people, including politicians" [Pickard et al. 2020: 273]. Referring to conservatives, progressives or even environmentalists, they are most of time unanimously designated as responsible for the ongoing global warming. FFF, indeed, "addresses adults, in adult arenas, to point out their own immaturity. Greta [Thunberg] reminds them that they have been talking [...] for 30 years, and nothing has changed. It's now young people who will take over. So, she reverses the roles. This posture works well in the media, and it also speaks to other young people" [Haeringer et al. 2020: 159].

Thirdly, FFF protesters, XR activists and *zadist*s tend to use a discursive register combining emotions and science, online and offline. On the one hand, they ask to "listen to the scientists" and to "open the eyes" on their alerts (IPCC reports, articles on *Nature* review, etc.). On the other hand, they point out power-holders' deafness and blindness to gather support. Requests of "telling the truth"—the first of XR's demands—are shared by all the three networks. One can argue that, locally, public risk assessments have been conducted on building projects such as TAV in Italy or ZADs in France, determining their environmental compatibility. If so, protesters mobilize independent scientific studies, either to emphasize the partiality of state expertises, or to "suggest valid technological alternatives" for a more sustainable future [Calabrese 2014: 203].

Lastly, considering negative feelings like, indignation or deception, could contribute to the study of "radicalization" in climate movements. It could also clarify why, after two decades of "mainstreaming the ecology movement", the successive failures of COP and of national policies to limit global warming have contributed, in the second half of the 2000s, to the renewal of an "apocalyptic environmentalism" [Gaborit 2021]. Indeed, since the first Stockholm UN conference in 1972, activists and researchers already adopted a "disaster perspective" which persisted until the 1990s, supported by anti-nuclear protests, both for military and for civil uses, Three Mile Island (1979) and Chernobyl (1986) catastrophes, as well as other mass pollutions, such as 1976 Seveso chemical accident in Italy, or 1978 oil slick shipwreck of Amoco Cadiz, near French coasts. However, with the achievement of Cold War, 1992 Rio UN summit opened a new, more optimistic era of around 20 years. This age seems today to be totally exhausted, and feelings about "the end of the world" are back again, especially in the most recent geopolitical context of overlapping COVID pandemic, Ukraine war, global warming and rise of oil prices revealing the contemporary dependency on fossil fuels. Youth express them with a particular intensity, combining contradictory emotions, already perfectly synthesized by placards seen on 2019 FFF demonstrations, multiplying pop references and science, hope and fear, and rage and love [Lensing 2021], wishing us a "Happy Apocalypse!".

and for a 100% free and public healthcare system. Thus, it is not surprising that the alterglobalization movement has been mentioned as a reference by most of strikers interviewed.

REFERENCES

Agrikoliansky, E., et al. (2010). *Penser les mouvements sociaux: conflits sociaux et contestation dans les sociétés contemporaines*. Paris, La Découverte.

Armiero, M. & Sedrez, L. (eds.). (2014). *A History of Environmentalism. Local Struggles, Global Histories*. London, Bloomsbury Academic.

Aykut, S. & Dahan, A. (2015). *Gouverner le climat? Vingt années de négociations internationales*. Paris, Presses de Sciences Po.

Barbe, F. (2016). "La "zone à défendre" de Notre-Dame-des-Landes ou l'habiter comme politique." *Norois*, 238–239, 109–130.

Becquet, V. (2014). *Jeunesses engagées*. Paris, Syllepse.

Becquet, V. (2019). "La génération climat monte au front." In: Flandrin, A., et al. (eds.). *Génération Climat*. Paris, Le Monde/Hors Série, 38–43.

Becquet, V. & Stuppia, P. (2021). *Géopolitique de la jeunesse. Engagement et (dé)mobilisations*. Paris, Le Cavalier Bleu.

Bessant, J., et al. (eds.). (2021). *When Students Protest. 3. Universities in Global North*. Lanham, Rowman & Littlefield.

Bey, H. (1991). *TAZ: Temporary Autonomous Zone*. New York, Autonomedia.

Biard, B., et al. (2020). "Penser l'après-corona. Les interventions de la société civile durant la période de confinement causée par la pandémie de Covid-19 (mars-mai 2020)." *Courrier hebdomadaire du CRISP*, 2457–2458, 5–130.

Blatrix, C. & Gervereau, L. (eds.). (2016). *Tout vert! Le grand tournant de l'écologie (1969–1975)*. Paris, Éd.Musée du Vivant/AgroParisTech.

Brunon, H. (2021). "Extinction Rebellion: pratiques et motivations de l'activisme écologiste radical." *Métropolitiques*. https://halshs.archives-ouvertes.fr/halshs-035 00907/.

Bulle, S. (2021). "Cartographie des collectifs engagés dans le tournant écologique." In: Stuppia, P. et al. *Engagement des jeunes sur des causes environnementales: formes de rupture et/ou de convergence générationnelle?* 9th AFS Congress, July 7, 2021. https://webtv.univ-lille.fr/video/11506/engagement-des-jeunes-militants-sur-des-causes-environnementales-rupture-etou-convergence-generationnelle-.

Calabrese, A. (2014). "Dissent, Counter-Knowledge, and Cosmopolitanism in NO TAV Movement." In: Caraus, T., Parvu, C. (eds.). *Cosmopolitanism and the Legacies of Dissent*. London, Routledge. 193–213.

Cavalli, A. et al. (1984). *Giovani oggi*. Bologna, Il Mulino.

Chaillou, A. & Monti-Lalaubie, M. (2020). "Jeunes pour le climat: en coulisses, ça continue!." *Revue Projet*, 375(2), 44–49.

Cohen, R. & Rai, S. (eds.). (2000). *Global Social Movements*. London, The Athlone Press.

Collective Askatasuna. (2012). *A sarà düra. Storie di vita e di militanza no tav*. Rome, Derive/Approdi.

Doherty, B. & Saunders, C. (2021). "Global Climate Strike Protesters and Media Coverage of the Protests in Turo and Manchester." In: Bessant, J., et al. (eds.). *When Students Protest. 3. Universities in Global North*. Lanham, Rowman & Littlefield, 251–268.

Druez, E. (2022). "Quel 'Nous' dénonce l'injustice? Politiser les discriminations par le biais d'identifications raciale ou urbaine à Paris et à Londres." *Critique internationale*, 94, 99–121.

Dufoing, S. (2012). *L'écologie radicale*. Paris, Folio.

Dupuis-Déri, F. (2005). "L'altermondialisme à l'ombre du drapeau noir." In: Agrikoliansky E., et al. (eds.). *L'Altermondialisme en France. La longue histoire d'une nouvelle cause*. Paris, Flammarion, 199–231.

Erlich, V. (1998). *Les nouveaux étudiants: un groupe social en mutation*. Paris, Armand Colin.

Fillieule, O. (2013). "Quelques réflexions sur les milieux étudiants dans les dynamiques de démobilisation." *European Journal of Turkish Studies*, 17. http://ejts.revues.org/4834.

Flandrin, A., et al. (eds.). (2019). *Génération Climat*. Paris, Le Monde/Hors Série.

Foran, J., et al. (2017). "'Not Yet the End of the World': Political Cultures of Opposition and Creation in the Global Youth Climate Justice Movement." *Interface*, 9, 353–379.

Gaborit, M. (2021). "Le mouvement pour le climat: une mobilisation générationnelle?." In: Stuppia, P., et al. *Engagement des jeunes sur des causes environnementales: formes de rupture et/ou de convergence générationnelle?* 9th AFS Congress, July 7, 2021. https://webtv.univ-lille.fr/video/11506/engagement-des-jeunes-militants-sur-des-causes-environnementales-rupture-etou-convergence-generationnelle-.

Gelderloos, P. (2018). *Comment la non-violence protège l'Etat: essai sur l'inefficacité des mouvements sociaux*. Paris, Éditions Libre.

Haeringer, N. et al., (2020). "Un mouvement mondial de la jeunesse: les grèves du climat." *Mouvements*, 103(3), 156–163.

Hardy, A. (2020). "Quel discours face à l'urgence écologique?." *Esprit*, 6/2020, 135–140.

Hochschild, A. R. (2003). "Travail émotionnel, règles de sentiments et structure sociale." *Travailler*, 1, 9, 19–49.

Inglehart, R. (2018). *Cultural Evolution, People's Motivations Are Changing, and Reshaping the World*. Cambridge, Cambridge University Press.

Juris, J. & Pleyers, G. (2009). "Alter-Activism: Emerging Cultures of Participation Among Young Global Justice Activists." *Journal of Youth Studies*, 12(1), 57–75.

Klandermans, B. (1984). "Mobilization and Participation: Social-Psychological Expansions of Ressource Mobilization Theory." *American Sociological Review*, 49(5), 583–600.

Lardeux, L. & Tiberj, V. (eds.). (2021). *Générations désenchantées? Jeunes et démocratie*. Paris, La Documentation française.

Legois, J. P., et al. (eds.). (2021). *Démocratie et citoyennetés étudiantes après 1968*. Paris, Syllepse.

Lensing, A. (2021). "De Berlin à Milan, en découdre pour le Climat." *France Culture*, October 14, 2021. https://www.franceculture.fr/emissions/cultures-monde/de-berlin-a-milan-en-decoudre-pour-le-climat.

Lo Schiavo, L. (2021). "*Student Protests against Neoliberal Education Policies in Italy: Three Student Organizations*." In: Bessant, J., et al. (eds.). *When Students Protest. 3. Universities in Global North*. Lanham, Rowman & Littlefield, 105–122.

McAdam, D. (1986). "Recruitment to High-Risk Activism: The Case of Freedom Summer." *American Journal of Sociology*, 92(1), 64–90.

Mann, C. (2019). "XXI. Hommes à la guerre, femmes aux fourneaux. Représentations et réalités." In Baecher J. (ed.). *Guerre et Histoire*. Paris, Hermann, 265–281.

Maréchal, J. (2016). "L'Accord de Paris: un tournant décisif dans la lutte contre le changement climatique?." *Géoéconomie*, 78, 113–128.

Mathieu, C. (2022). "Un bilan de la COP26." *Politique étrangère*, 1/2022, 13–16.

Montoni-Rios, A. (2015). *Radicalisation de l'action collective et jeunesse populaire: construction du politique et ré sistances au Chili*. EHESS, PhD dissertation in Sociology.

Muxel, A. (2018). "Radicalité politique: entre protestation et rupture." In: Galland, O. & Muxel, A. (eds.), *La tentation radicale: Enquête auprès des lycéens*. Paris, PUF, 203–265.

Norris, P. (2011). *Democratic Deficit. Critical Citizens Revisited*. Cambridge, Cambridge University Press.

O'Brien, K., et al. (2018). "Exploring Youth Activism on Climate Change: Dutiful, Disruptive, and Dangerous Dissent." *Ecology and Society*, 23(3), 42. https://doi.org/10.5751/ES-10287-230342.

Offerlé, M. (2008). "Retour critique sur les répertoires de l'action collective (XVIIIe–XXIe siècles)." *Politix*, 81, 181–202.

Pagis, J. (2014). *Mai 68, un pavé dans leur histoire. Evénements et socialisation politique*. Paris, Presses de Sciences Po.

Pickard, S. (2019). *Politics, Protests, and Young People*. London, Palgrave Macmillan.

Pickard, S. (2022). "Young Environmental Activists and Do-It-Ourselves (DIO) Politics: Collective Engagement, Generational Agency, Efficacy, Belonging and Hope." *Journal of Youth Studies*, special issue, to be published.

Pickard, S., et al. (2020). "'We Are Radical in Our Kindness': The Political Socialisation, Motivations, Demands and Protest Actions of Young Environmental Activists in Britain." *Youth and Globalization*, 2(2), 251–280.

Pickard, S., et al. (2022). "Youth and Environmental Activism." In: Giugni, M. Grasso, M., *Routledge Handbook of Environmental Movements*. New York, Routledge, 521–537.

Pitti, I. (2021). "Student Activism in Bologna: Old Fractures, Emerging Alliances, and Use of Depoliticization as a Repressive Strategy." In: Bessant, J., et al. (eds.). *When Students Protest. 3. Universities in Global North*. Lanham, Rowman & Littlefield, 233–250.

Piven, F. F. (2006). *Challenging Authority: How Ordinary People Change America*. Lanham, Rowman & Littlefield.

Pleyers, G. (2016). "De la subjectivation à l'action. Le cas des jeunes alteractivistes." In: Pleyers, G. et al., *Mouvements sociaux: quand le sujet devient acteur*. Paris, Éditions MSH, 27–48.

Pleyers, G. (2021). *Conclusion du séminaire alteractivisme et subjectivation dans les mouvements contemporains*. https://www.fmsh.fr/fr/projets-soutenus/conclusion-du-seminaire-alter-activisme-et-subjectivation-dans-les-mouvements.

Pleyers, G. & Capitaine, B. (2016). "Introduction: Alteractivisme: comprendre l'engagement des jeunes." *Agora débats/jeunesses*, 73, 49–59.

Pleyers, G. & Glasius, M. (2013). "The Global Movement of 2011: Democracy, Social Justice and Dignity." *Development and Change*, 44, 547–567.

Polletta, F. & Hoban, K. (2016). "Why consensus? Prefiguration in three Activist Eras." *Journal of Social and Political Psychology*, 4, 286–301.

Pruvost, G. (2017). "Critique en acte de la vie quotidienne à la ZAD de Notre-Dame-des-Landes (2013–2014)." *Politix*, 117, 35–62.

Putnam, R. (2000). *Bowling Alone: The Collapse and Revival of American Community*. New York, Simon & Schuster.

Riniolo, V. & Ortensi, L.E. (2021). "Young Generations' Activism in Italy: Comparing Political Engagement and Participation of Native Youths and Youths from a Migrant Background." *Soc Indic Res*, 153, 923–955.

Scappaticci-Martin, A. (2019). "Extinction rebellion appelle à la désobéissance civile de masse." *Alternatives Non-Violentes*, 192, 16–20.

Shapiro, T. M. (2006). "Race, Homeownership, and Wealth." *Washington University Journal of Law & Policy*, 20, 53–74.

Servigne, P. & Stevens, R. (2015). *Comment tout peut s'effondrer*. Paris, Seuil.

Sommier, I. (2010). "Les états affectifs ou la dimension affectuelle des mouvements sociaux." In: Agrikoliansky, E., et al. (2010). *Penser les mouvements sociaux: conflits sociaux et contestation dans les sociétés contemporaines*. Paris, La Découverte, 185–202.

Sommier, I. (2021). "Les "sociétaux": une violence en devenir?." In: Sommier, I. et al. (eds.), *Violences politiques en France*. Paris, Presses de Sciences Po, 135–156.

Stuppia, P. (2016). ""La révolution dans le jardin". Utopies communautaires et expériences néo-rurales françaises après Mai 68." *Education et Sociétés*, 37, 49–64.

Stuppia, P. (2020). *2006: une victoire étudiante? Le mouvement "anti-CPE" et ses tracts*. Paris, Syllepse.

Subra, P. (2016). *Géopolitique locale: Territoires, acteurs, conflits*. Paris, Armand Colin.

Tiberj, V. (2017). *Les citoyens qui viennent. Comment le renouvellement générationnel transforme la politique en France*. Paris, PUF.

Tilly, C. (1986). *La France conteste: de 1600 à nos jours*. Paris, Fayard.

Tilly, C. (2008). *Contentious Performances*. Cambridge, Cambridge University Press.

Toscano, M.A. (ed.). (2007). *Homo instabilis. Sociologia della precarietà*. Milan, Jaca Books.

Verdier, M. (2018). *La Perspective de l'autonomie: la critique radicale de la représentation et la formation du commun dans l'expérience de l'occupation de la ZAD de Notre-Dame-des-Landes*, University Paris Nanterre, PhD dissertation in Sociology.

Wall, D. & Bijeard, R. (2020). "Luttes écologistes et anticapitalistes au Royaume-Uni." *EcoRev'*, 48, 75–85.

Environmental Politics and Theory in the Anthropocene

The Anthropocene New Stage: The Era of Boundaries

Florian Vidal

INTRODUCTION–THE ANTHROPOCENE QUESTION UNDER PLANETARY BOUNDARIES

The acknowledgment that human activities alter Earth's climate has driven debate concerning "planetary boundaries" required to keep the anthropogenic forcing within "safe operating limits."[1] The introduction of this concept comes after the eve of the Anthropocene, coined as a new geologic epoch, where humanity's collective actions become the dominant driver for planetary changes.[2] In the last two centuries, the impacts of humans on the global environment have escalated. Because of these anthropogenic carbon emissions, global climate may depart significantly from natural behavior for many millennia to come. Global warming and climate change, driven by CO_2 emissions, are the most dramatic representation of the impact of civilization on the planet.

---

[1] See Barnosky (2012); Lenton (2008); Brook et al. (2013); Lenton, (2008); Rockström et al. (2009); Steffen et al. (2015).
[2] See Crutzen (2002); Steffen et al. (2011).

F. Vidal (✉)
LIED (CNRS) (Laboratory on Interdisciplinary Studies on Energy, French National Center for Scientific Research), Paris Cité University, Paris, France
e-mail: florian.vidal@live.no

As Frank rightly put into perspective, our societies must drop the question of whether or not the human species created climate change. Instead, our species must assume that we, a planetary civilization, transformed the planet's climate.[3] That said, the discussion is to qualify the nature of the Anthropocene: Is it an event or an era? For Gibbard et al., postulating the Anthropocene event would define the current process "in a similar way to globally significant transformations that have previously affected the Earth's biosphere" (e.g., the Palaeoproterozoic Great Oxidation Event, c. 2.4–2.0 billion year; the Great Ordovician Biodiversity Event, 485–455 million years).[4] Given that fact, the Anthropocene is a process rather than a condition; it is a dynamic that drives our planet into new physical and chemical conditions. From this dynamic, new boundaries are emerging that may contest human adventure on Earth on existential terms. In a broader context, the Anthropocene creates boundaries that alter the conditions of existence on the planet for humans but also living organisms. As defined by Lenton et al., tipping points operate as a critical delimitation in considering trajectories for the coevolution of coupled civilization-planet systems.[5] Frank et al. specify that "a tipping point is a critical threshold where small perturbations can alter large-scale state or evolution of a system."[6]

Frank and his colleagues set the terms of the debate in a singular and essential way for the future of humanity as a whole,

> Earth's entry into an anthropogenic era poses challenging questions for the long-term sustainability of global human civilization. It is, in fact, not clear if a planetary civilization as energy-intensive as ours can be sustained for centuries.[7]

From social sciences and humanities (SSH) perspective, the question of appropriating the debates led by astrobiologists is central to defining the beginning of new governance by taking into account new paradigms hitherto ignored in the organization and material structure of our human societies. While the Anthropocene concept has being widely referred to in natural sciences, it has stimulated new thinking frameworks in humanities and social sciences. The numerous debates on the origin of the Anthropocene, the responsibilities of this trajectory, as well as the political and institutional solutions to bring a long-term response to new planetary conditions, have

---

[3] See Frank (2018).
[4] Gibbard et al. (2022), 2.
[5] See Lenton et al. (2019).
[6] Frank et al. (2018), 515.
[7] Frank et al. (2017), 14.

been broadly discussed.[8] Even further, ongoing discussion casts Anthropocene scenarios and possible futures for human societies.[9] Henceforth, the debate should focus on human artifacts and their imbrication in the feedback processes of nature. The artificialization of the planet caused by humankind reflects its technical and economic expansion and its growing use of planetary resources. Even though the human footprint on Earth relentlessly amplified through millennia, the planetary entanglement reminds "our relative unimportance, and our complete dependence upon a biosphere that has always had a life entirely its own."[10] According to Frank et al., given its global scale, might the shift represented by the Anthropocene be a common feature of any planet evolving a species that intensively harvests resources regarding the technological civilization course?[11]

In that respect, the astrobiological perspective intends to stand as the first step "in our maturation and our ability to face the Anthropocene."[12] From this point of view, there is an urgency to address critical nodes that human civilization is about to face. First of all, the relationship to time between the human and geological scales needs to be rethinking. Indeed, Earth trajectory opens new and unknown perspectives in human history. It highlights unthinkable facts because the planet background is divergent from our own fate and opens the debate on habitability. If we think about human civilization over time, we must now begin the physical and biological planetary transformations that the consequences of the Anthropocene are bringing about. Lastly, once the diagnosis has being made in the face of the emergence of this new planet, our human civilization should address two issues: the use of resources and their sustainability; and the need to rethink our infrastructures in light of these new planetary conditions.

NEW EARTH'S TRAJECTORY: THE HABITABILITY VERTIGO

Anthropogenic activity is changing Earth's climate and ecosystems in ways that are potentially dangerous and disruptive to humans. In this setting, the Earth System is branded by nonlinearity and complexity, which means, it has low predictability.[13] Unprecedented in human history, our trajectory on Earth faces "a spatiotemporal discontinuity and uncertainty in a geological timescale."[14] In the absence of rapid and substantial reductions in greenhouse gas emissions, projections made past 2100 point to large areas of the Earth

---

[8] See Berkhout (2014); Biermann (2014); Fagan (2019); Knight (2015); Latour (2018); Lewis and Maslin (2015); Lövbrand et al. (2020); Malhi (2017).

[9] See Rothe (2020); Tyszczuk (2021).

[10] Sagan and Margulis (1997), 157.

[11] See Frank et al. (2018).

[12] Frank (2018), 224.

[13] See Knight (2015).

[14] Hamilton (2017), 593.

will change in ways that lessen their capacity to support the large-scale human occupation. The long-term effects of twenty-first-century warming will be felt for centuries to come, even if emissions are limited in the future.[15] Thus, it is necessary to initiate original conceptual frameworks by calling upon the natural sciences. To nurture and consolidate the approach of the humanities and social sciences to this issue, understanding from the perspective of the natural sciences remains indispensable because they allow the grasp of the ongoing transformations of the Earth. This should include an assessment of the extent of the challenges to the governance and organization of our societies. That is to say, human societies should oppose an ontological shift in which the Anthropocene entangled the nature-human dynamic into a new course. To reconsider our relationship with nature, we must overcome the Cartesian dualism between nature and human beings. This paradigm has resulted in the exclusion of human beings from object nature.[16] Therefore, a new ethical dimension may redesign human interactions with non-humans and their impact on the functioning of the Earth System.

In this respect, Lovelock's contribution, assisted by Margulis, was crucial to redefining our approach to the Earth. As Lenton et al. remind us, Lovelock's work allowed us to frame a new thought in which Life emerges as a living entity.[17] From this new perspective, Life as a comprehensive and complex living entity holds the ability to alter its global environment. As a result, this entanglement becomes so intense that attempts to disconnect life boundaries from its environment are difficult.[18] Behind this paradigm, a fundamental question arises: Can life on Earth become extinct? In other words, without interference from outside the planetary process (i.e., meteorite impact), what are the mechanisms that allow us to define the thresholds of habitability that create optimal conditions for life to emerge and prosper?

From the limits that exist for conditions favorable to the existence, Lineweaver and Chopra suggest that Earth and life have coevolved, but warn us that they cannot adapt infinitely.[19] The presence on Earth of places devoid of life demonstrates its fragility and exposes the notion of boundary, so vital to the continuation of the human adventure. The limits are in what composes it: the processes that regulate the supply of water and carbon to a planet control its habitability. And from four billion years of uninterrupting evolution, life has limits on Earth which rely on the water-based carbon chemistry. The state of balance of this feature is very thin that it can tip over at any moment and jeopardize all living things on this planet. The planetary shift in the Anthropocene event is altering atmospheric carbon dioxide concentrations at a rate that commonly exceeds those known in our planet archives. In other words,

[15] Lyon et al. (2021), 356.
[16] See Haila (2000).
[17] See Lenton et al. (2020).
[18] Ibid.
[19] See Lineweaver and Chopra (2012).

the biophysical boundary crossing may trigger a drastic reconfiguration of the biosphere parameters of our planet. In such an event, the whole Earth System could tip over into conditions where life is no longer sustainable and could disappear.

Therefore, according to the astrobiological context, the condition humanity finds itself at the entry into the Anthropocene is likely not unique. In this regard, the study of habitability (also known as habitology) is a new, immature cross-disciplinary synthesis of facts and theory from Earth and planetary sciences, biology, and astronomy. As rightly pointed out by Lineweaver and Chopra, the scientific research on habitability zone (HZ) and, extensively on "extraterrestrial life is a search for ourselves and our place in the universe."[20] Against the backdrop of the laws of thermodynamics hold for all planetary systems, Frank et al. assessed that feedback associated with the Anthropocene must have happened somewhere else.[21] Put another way, current processes on Earth are not unique and are a reminder of the unstable nature of any planet.

In light of the ongoing planetary shift, one critical feature that defines emerging climate conditions is the ever-growing greenhouse gas concentrations in the atmosphere. This alteration of the atmospheric composition ensures that such changes will be felt well beyond 2100 and plan to linger many centuries. According to Lyon et al., there is no other option to enforce deep and swift reductions in greenhouse gas emissions to control our lane departure, while climate change will go along with the human species for centuries.[22] Although the response of our thermo-industrial civilization will depend on social, cultural, and many other features, Frank et al. caution that any effective solution solely lies in the realm of planetary-scale thermodynamics.[23] Due to the complexity of the physico-chemical processes at work, the utter efforts to extend projections beyond 2100 are made difficult.

By separating Earth from outer space, the atmosphere is a vital boundary for life on the planet. As Olson et al. hint, the atmosphere is not static.[24] On the contrary, throughout Earth's 4.5. billion years (4.5 Gy) history, the atmosphere has changed and evolved in near aspect. Some of these fluctuations have been fundamental in maintaining our planet in the HZ. All along the geological course, these evolutions prompt us of the precarious state of the atmosphere both physically and compositionally. Therefore, the current status of the atmosphere is not representative of the extended Earth trajectory, nor its terminal state with 4 Gy remaining before the final extinction. To put it simply, our planet's atmosphere and biosphere "will continue to coevolve with

---

[20] Ibid., 616.
[21] See Frank et al. (2017).
[22] See Lyon et al. (2021).
[23] See Frank et al. (2017).
[24] See Olson et al. (2018).

its solid interior and the Sun."[25] Hence, the geological scale of the Earth's total life expectancy is based on a basic triptych that combines the Solar-mass, the Moon's mass, and Earth's mass.[26] These joint three-time scales represent nearly 10 Gy. This is the length of Earth's existence in the universe, but it does not correspond to the capacity to support life on it, let alone human experience. The window of opportunity for the human project is restricted in the geological time of our planet. The habitable lifetime for our intelligent species is limited in time, while it took 4.5 Gy to appear on Earth. Ultimately, the Sun will influence the fate of our planet at ~ 6 Gy once it becomes too luminous to allow life on Earth.[27] This order-of-magnitude re-emphasizes the gap between the human time scale and the geological time scale. If their trajectory is parallel to the human perspective, they do not have the same temporality. In this respect, the interrogation of Sagan on the condition of the human adventure in a few million years echoes the non-existence of this same species a just few million years ago.[28] While the Earth does exist since 4.5 Gy, multicellular life forms did appear around 600 million years ago, which tallies for more than 10% of our planet's lifetime until now.

Nevertheless, the Earth is not what astrobiologists call a Doomed or an Extreme planet.[29] Its evolutionary history demonstrates the absence of transformation on very massive scales in the biological and chemical composition of the biosphere. The five mass extinction events that occurred on our planet reflect the planetary resilience of its habitability. In this regard, Earth's geological record indicates that the planetary trajectory has undergone dizzying episodes in its continuity in the HZ. Indeed, scientists have highlighted evidence of extreme climate disturbances that may have come very close to extinguishing all life.[30] As Nicholson et al. argue, most used models concerning planetary habitability and boundaries focus on the physical processes that are taking place on the planets.[31] Based on these investigated and outlined processes, it is possible to determine the HZ. According to Dartnell, "the actual upper survival limit for temperature is defined by chemical constraints and the molecular stability of the components of life."[32] Once again, the definition of an HZ is not automatically correlated to the presence of life on a planet, only conditions favorable to its emergence and maintenance.[33] Based on this work, astrobiologists can consider the construction of

---

[25] Ibid., 2842.
[26] See Waltham (2019).
[27] See Waltham (2017, 2019).
[28] See Sagan (1980).
[29] See Nicholson et al. (2018).
[30] See Tyrrell (2020).
[31] See Nicholson et al. (2018).
[32] See Dartnell (2011), 1.28.
[33] See Preston and Dartnell (2014).

a model on the HZ and the factors necessary for the emergence of life. If astrobiological research focuses on life search on exoplanets, their work mirrors the critical conditions of life on Earth.[34] So far, the dual human and planetary trajectory prevented tipping into the moment when Earth became too hot or too cold and consequently lost life on it.[35] Moreover, these studies finally reveal that the fluke factor participates in preserving the HZ.

The habitability vertigo reframes the relations between human timescale and geological timescale, which will be even further entangled. As previously stated, Earth's geological history proved the thin line that separated the tipping point between a planet habitable for living organisms and one that sees all life extinguished. In the Gaia hypothesis, life can influence feedback mechanisms that directly occur in the Earth System, enabling it to achieve a self-regulating system.[36] As a result of this critical influence, life on Earth let its signature in the planetary chemical composition of Earth's atmosphere, oceans and soil. The paradox in the state of our scientific knowledge leads us to have a deeper knowledge of Mars or Venus than of our planet. In many ways, human understanding remains ultimately superficial.[37] This situation leaves our vulnerability to control the geological trajectory of our home in the immense universe exposed.

Hence, the human trajectory faces a new direction in light of the Anthropocene event. The essential issue of this new phase is whether or not to maintain our planet in the HZ as anthropogenic climate change opens up unique perspectives on the geological history of the planet. Given the rapidity of a massive transformation of planetary conditions, the human project is reprogramming the Earth System and the definition of the conditions of its maintenance in the HZ. While our planet is unlikely to head into extreme conditions able to extinguish all living things upon it, these emerging conditions will raise several challenges for humanity and its civilizations, inheritors of the Holocene era.

Redrawing the Boundaries of Our Planet

Heikkurinen et al. recall that "human agency was largely unbounded in its effects on the fabric of the Earth and human life upon it" in the early stages of the Anthropocene.[38] The introduction of the planetary boundaries framework, in which encompasses nine Earth System thresholds, gives operative tools for comprehending the standing conditions for life on Earth and the

---

[34] An exoplanet is a planet located outside the Solar System.
[35] See Tyrrell (2020).
[36] See Lovelock and Margulis (1974); Lovelock (2000); Lenton (1998).
[37] See Waltham (2019).
[38] Heikkurinen et al. (2019), 2.

consequences of crossing one of them.[39] Indeed, global human activities in the two-last centuries impacted the Earth System such as climate change, and related sub-global processes, including land and water use. That being said, human activities are so closely interlocked with the Earth System dynamics that they are bounded within planetary constraints and their potential nonlinear feedbacks.[40] The maturing Anthropocene epoch entails a complex, interconnected and unstable world characterized by globalized and manufactured risks.[41] Today, such hazards directly jeopardize the life-upholding systems upon which human settlements lie. However, the transgression of planetary boundaries is a sign of failure of the thermo-industrial civilization to recognize the Anthropocene and what it involves.

The Physical Earth's Metamorphosis

Our civilization project began about 11,000 years ago when the last ice age ended, then our planet's climate converted into warmer and wetter conditions. From this point of view, humans started to settle and create cities, while stopping their wandering.[42] Even if some human communities maintained a nomadic life in remote and hostile areas until now, the human settlement process spread and became the norm throughout the planet. Against the background of the relatively temperate Earth, the human civilization that expanded under the Holocene conditions succeeded making its way into flourishing and enlightening societies for several millennia. However, these favorable conditions are progressively sliding into the past in light of the changing planet as a result of global anthropogenic CO_2 emissions. In 2021, the global annual level of CO_2 concentration in the atmosphere was 416.45 parts per million (ppm). Therefore, this stands as an increase of 2.21 ppm compared to the year 2020 (i.e., 414.24 ppm).[43] If we keep this trend, the level of CO_2 concentration in the atmosphere should reach 500 ppm by 2060. The possibility of reaching such a level within a few decades is dizzying and the consequences of such a change for planetary conditions are set for several millennia. Brannen soberly restates that "the story of the planet Earth is not the story of Homo sapiens."[44]

As Lineweaver and Chopra argue, regional deserts do not match critical requirements for life, namely low rainfall, low temperatures or low nitrate levels.[45] These areas on the planet are outside the optimal range for life in

---

[39] See Rockström et al. (2009); Steffen et al. (2015); Steffen et al. (2021).
[40] See Malhi (2017).
[41] See Lövbrand et al. (2020).
[42] See Frank (2018).
[43] See NOAA Earth System Research Laboratory (2022).
[44] Brannen (2017), 15.
[45] See Lineweaver and Chopra (2012).

which total biomass is low. To put it another way: these places are hostile to any human settlement. Despite these constraints, some human communities succeeded to adapt in water deserts and polar regions for millennia. And with technological advancement thanks to the industrialization era, humans settled into new territories such as the Antarctic (i.e., permanent scientific bases). In other words, human communities must cope with geographic spaces on Earth that are inhospitable to maintaining a sustainable sedentary civilization. Remote terrestrial spaces constitute frontiers, in the same way as the marine world, in the arrangement of the living spaces of human settlements.

Now, the transformation of global conditions under the impulse of anthropogenic climate change is challenging the physical borders of the planet. The emergence of these new frontiers will completely reshape our relationship with territories and our capacity to maintain a presence there in the coming centuries. What borders are we discussing? On the one hand, it is a question of putting forward the climatic mechanisms that disrupt the physical conditions of the planet and its adjoining geographical spaces. On the other, the political consequences of this new reality in gestation will unfold in the coming decades and most likely last for centuries.

As Lövrand et al. suggest, "the new geographies of danger presented by melting glaciers, rising sea levels or more extreme weather feed into this re-territorialization of global affairs."[46] The advent of new physical boundaries induces significant geopolitical effects on human societies. This phenomenon poses the interdependence created between human and geological time scales. In other words, their fate is correlated while bringing low predictability due to the nonlinear nature of the Earth's trajectory. Therefore, planetary metamorphoses should increasingly influence human societies and their dynamics. Despite the fact the dangers of increased conflict over water use are considerably discussed, the geopolitical effects of sea-level rise on coastal areas worldwide are little anticipated.[47] Under the current path, cities like Bangkok (Thailand), Dhaka (Bangladesh) and Miami (United States) are under threat going down the drain in the long term. Similarly, the repeated effects of extreme weather events threaten the viability of maintaining certain human settlements in some areas (e.g., Caribbean islands are highly vulnerable to storms). Rising temperatures and high humidity put entire regions in an untenable position for any human being. These conditions endanger the physiology of the human body, which cannot withstand the combination of extreme temperatures and high humidity.[48] For instance, Pakistan and India are among countries extensively exposed to this risk in the coming decades. While the threshold of a wet-bulb temperature of 35 °C would regularly be crossed in the coming decades, Im et al. explain that human beings in South Asia will face disruption of their body's ability to regulate the temperature that could

---

[46] See Lövrand et al. (2020).
[47] See Dodds (2021).
[48] See Raymond et al. (2020).

immediately impair physical and cognitive functions.[49] More broadly speaking, these increasing planetary physical shifts need to be anticipated on political, economic, social and cultural levels as they may sustain further domestic and regional destabilization.

The Coming Boundary of the Living Things

If climate change is reshaping the physical appearance of the planet, its ecosystem composition is also changing. Whereas nature is now deteriorating worldwide at an unparalleled pace in the human timescale, the scientific community extensively discusses the hypothesis of a Sixth Massive Extinction event. In 2019, the Intergovernmental Science-Policy Platform on Biodiversity and Ecosystem Services (IPBES) released its report assessing that around one million fauna and flora species are under impending extinction.[50] On this basis, some argue that "the Sixth Mass Extinction may have not occurred yet, but heightened rates of extinction and huge range and population declines have already occurred."[51] To put it simply, this means that the mechanisms for mass extinction are in place. However, the nature of this extinction diverges from previous extinction phases. According to Turvey and Crees, the profound nature of the extinction crisis diverges from previous episodes in planetary history in terms of both patterns and drivers.[52]

The biodiversity loss and ecosystem degradation are working together to create a dry and depleted planet that will affect humans and other species. The overall prognosis for the survival of a large proportion of existing species appears gloomy. Although there is undeniably a biodiversity crisis, the question of the Sixth Mass Extinction is still being debated. Though a part of the scientific community appears pessimistic about the fate of the Earth's biodiversity, the tragedy regarding the ongoing dynamic is that "much of which is going to vanish without us ever knowing of its existence."[53]

In light of the new stage reached in the Anthropocene era, Cooke underlines that the most fundamental challenge remains "the imminent destruction of web of life, the life-generating and life-sustaining ecosystems that constitute the planet Earth."[54] The present process is characterized by extinction without replacement "with an operational timeframe of rapid ecological time instead of longer-term evolutionary time."[55] Several factors are enumerated to explain

---

[49] See Im et al. (2017).
[50] See IPBES (2019).
[51] Cowie et al. (2022), 654.
[52] See Turvey and Crees (2019).
[53] Cowie et al. (2022), 657.
[54] See Cooke (2020), 1.
[55] Turvey and Crees (2019), 985.

the historical process underway: overexploitation, habitat extinction or introduction of non-native species. All of these drivers come from anthropogenic activities on Earth. Global changes could thus precipitate this anthropogenic ecosystem loss. This perspective reinforces the trajectory of an atrophied Earth System—i.e., a substantive curtailment of life in the biosphere and, by extension, the loss of human habitat territory—leading to a complete refoundation of the journey of humankind.

The implementation of a biological frontier with the collapse of many ecosystems (e.g., marine species, invertebrates, plants) therefore weakens the conditions of existence of the human species on the planet. Although extinction rates can vary between categories, global biodiversity loss could reduce the food supply for the human population. Indeed, food chains are critically disrupted under pressure from human activities (e.g., land-use change, overexploitation, pollution) and now climate change. The biodiversity loss holds straight impact on food security, since connections between biodiversity and human well-being are apprehended in the concept of "ecosystem services."[56] According to Newbold et al., contributions that provide the natural environment and biodiversity to human well-being can be classified as follows: "provisioning (e.g., crop production, clean water, timber, fuelwood, non-timber forest products), regulating (e.g., carbon storage and sequestration, pollination, disease regulation), and cultural services (e.g., aesthetic, spiritual, or recreational value)."[57] Or, global human activities directly affect the ability of particular ecological communities to provide ecosystem services. This disruption has been specifically identified in pollination, in which pollinating species are in extensive decline due to land use and climate change.

In light of the COVID-19 crisis, that is part of past and future epidemics, the 2020 global pandemic recalls our thermos-industrial civilization that everything is interconnected, the health of animals, the health of nature, and the health of humans.[58] Specifically, the pursuit of human activities generating further habitat loss and degradation will open the door to other global health crises. In this regard, Lapola highlights that Amazonia stands as the perfect spot for the next global pandemic.[59] Under the direct effect of biodiversity loss and climate change, the disruption of ecosystem balances within the Amazonia rainforest increases the risk of a virus outbreak. From such a perspective, the global society appears at risk from pandemic threats and food insecurity. All in all, the close intertwining of the dynamics of living organisms with those of our species underlines that massive biodiversity loss will be at the expense of human civilizations.

In the end, the double phenomenon of atrophy (i.e., reduction of physical space and extinction of living organisms) that seems to be at work

[56] See Newbold et al. (2019).
[57] Ibid., 212.
[58] See Schmeller et al. (2020).
[59] See Lapola (2020).

highlights the urgency of redefining the political and institutional instruments to elaborate governance in front of this emerging Earth, unknown to humans. As Malhi highlights, "human social, economic, and political decisions have become entangled in a web of planetary feedback."[60] Henceforth, the redefinition of our actions must integrate the human–environment relationship, including the human-material and material-environment dynamics. The Anthropocene is nearing a new stage where resources and infrastructures are at the heart of ontological refoundation. Undeniably, this questions the founding principles that the human trajectory on Earth has known since its emergence and the acceleration that has accompanied it for two centuries.

Resources and Civilizations: The End of Boundless Human Structures?

The emerging planetary conditions will sharply redefine human existence conditions. Considering this transformation, the sustainability issue remains a fundamental factor if the human civilization aims to drive off collapse. From the astrobiological perspective, the quest for human civilization management relies on the necessity to shift toward sustainable development in order to avoid its demise. While the planetary system is transitioning to a new state unknown in human history, the relationship of human civilizations to resources becomes even more essential. The challenge that the importance of resources represents for our societies is multiple and complex. Against this backdrop, this confronts us to innovate to manage them sustainably and move away from an economic model that leads to overexploitation and depletion such as minerals.[61] Taken metaphorically, the idea that the Earth would be the spaceship for human civilization underlines the need to chart the boundaries for a safe planet.[62] In these circumstances, the scientific and popular interest over the past few decades in the fate of Rapa Nui mirrors our anxieties as a planetary civilization isolated in the universe.

Rapa Nui and the Magnifying Effect

Without a doubt, the American scientist Jared Diamond is the one who propelled Rapa Nui and popularized this interest in light of our condition by attempting to shed light on the mechanisms of the collapse of the Pacific civilization. For decades, Rapa Nui (Easter Island, Chile) remained at the heart of the scientific debate about the uniqueness of its geographical location in the middle of the Pacific Ocean, surrounded by more than 1,000 miles of ocean in all directions. As a matter of fact, the emergence of its civilization, the tempo

---

[60] Malhi (2017), 97.
[61] See Norther et al. (2018); Prior et al. (2012); Tilton (2003, 2018).
[62] See Sterner et al. (2019).

of investment in monumentality (i.e., megalithic platforms (*ahu*) and multi-ton statues (*moai*), and its final demise were broadly investigated. As described by Basener and Ross, the small island was settled between 400 and 700 AD by a small group of at most 150 humans to reach a population of approximately 10,000 sometime between 1200 and 1500 AD.[63] In the meantime, the Rapa Nui peoples demonstrated their ability to sustain a resilient societal organization for more than 1,000 years. For centuries, their inhabitants built a culture that was artistically and technologically sophisticated enough to construct and transport their iconic stone statues.

Despite all the geographical constraints (i.e., tiny volcanic island, remote location in the ocean system), the present archaeological record of Rapa Nui boasts hundreds of *ahu* and nearly 1000 *moai* illustrating evidence of a sophisticated social organization.[64] While the scientific debate remains tense about the details of its history, many studies stress that Rapa Nui's inhabitants depleted their resources.[65] Apart from the discussed collapse, the most intriguing in this case relies upon, as stated by Diamond, to be "the clearest example of a society that destroyed itself by overexploiting its own resources."[66] The wood formed the fundamental resource for the organization and functioning of this society, the destruction of the forest surface, and all of its tree species went extinct. The loss of the timber resources drove cascading events that critically disturbed its functioning, including an end to transport and the erection of statues.[67]

This example cannot overlap in our contemporary society. Nevertheless, more than the episode of collapse, the management of resources on this island alerts us to the need to consider physical boundaries. Such boundaries challenge the natural ecosystem and its use by human communities demonstrated throughout the history. The intensive use of wood, in this instance, was the cement in the socio-economic organization of the island. But this condition precipitated the depletion of the resource without the possibility of restoring it. Thus, it rightly emphasizes the dependence of our societies on resource availability, access, quantity and their capacity to regenerate or not. As rightly pointed out by Lima et al., the overpopulation and continuing climate change are severe difficulties facing modern societies when the historical course that Rapa Nui met provides vital insights regarding food security and ecological resilience in a similar way.[68]

---

[63] See Basener and Ross (2005).
[64] See DiNapoli et al. (2020).
[65] See Frank et al. (2018).
[66] Diamond (2021), 77–78.
[67] Ibid.
[68] See Lima et al. (2020).

Despite the fate of Rapa Nui, persistently dwells "a paragon of societal collapse," this case has been lauded as a stark lesson for global sustainability.[69] Accordingly, the Rapa Nui case offers a valuable example for understanding our trajectory as a globalized human society closely entangled with evolving planetary conditions. Therefore, a reflection on the sophisticated management of resources for human civilization is yet to be engaged, since it constitutes a determining factor for the sustainability of its organization and its adjoining functions in the long term. Such considerations imply the elaboration of models that could also be adapted to questions of the coupled evolution between human civilizations and their environments.

In order to address the imposition of this physical resource boundary, Rees highlights the need to perform an extension of energy and material flow assessment.[70] To this end, an eco-footprint analysis (EFA) would achieve a comprehensive inventory of the annualized energy and material flows generated by the subject population to feed, clothe, house, transport and otherwise maintain and grow itself. The prospect to create such a model could allow for a biocapacity assessment (or "natural capital"), which is required to meet that population's consumptive and assimilative demands.[71] In other words, the inception of an instrument to measure the population's ecological footprint would facilitate the piloting of a policy focused on the necessary resources management and their availability. In addition to this approach, Dockstader et al. suggest maturing sophisticated socio-ecological models enabling to envision human growth and resource consumption in light of the study of the role of metapopulation effects. In the meantime, we have to cope with a socio-economic regime which led our planet to the brink.[72]

The "Great Acceleration" triggered in the post-1945 era saw a globalization process of this disrupted trajectory. This process enables an increasing speed of the artificialization of the planet thanks to the human footprint. In that context, technology development has contributed to the globalization of natural resources management. This process has led industrial societies to cause environmental impacts that are no longer felt by them only, but rather by geographically distant populations, weakening the short-term coupling between humans and their environment.[73] This comprehensive process is apprehended by, for instance, the linkages between local deforestation and high pressure for international agricultural exports. As a result, the wide-ranging entanglement of human activities has created a rift between continued technological advancement and the grip on natural resources on the planet. In another way, our development has brought us to cross all the limits that planetary physics imposed on us. However, new resource exploitation in isolated

---

[69] DiNapoli et al. (2020), 8.
[70] See Rees (2012).
[71] Ibid.
[72] See Dockstader et al. (2019).
[73] Ibid.

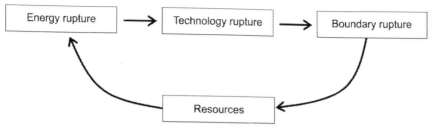

Fig. 1 Resources and human entanglement

geographical areas, which is critical for the upholding of our civilization, will require even more energy and resources. As Just et al. explain, technological uncertainty influences both the rate of depletion of the resources and the rapidity of implementation of the existing backstop technology.[74]

Technological development accompanies an exponential need for the use of resources. In this sense, the industrialization process, which started in the UK at the end of the eighteenth century, engaged our civilization in a massive complexification of our structures since then. This civilizational course is a rupture. In this equation, the intensive use of fossil fuels has propagated this trajectory of the thermo-industrial civilization. As a result, a triple disruption was formed, generating an ecosystem specific to the functioning of the planetary human society breaking down by energy, technology, and boundary (see Fig. 1). Namely, access to new energy resources is one of the prerequisites to lead to a technological break. Once the technological breakthrough has been initiated, it allows the crossing of physical boundaries inherent to planetary conditions. Therefore, these crossed boundaries enable access to new resources that may bring a further dynamic to the rupture loop (see Fig. 1).

The key question lies in the ability of our planetary civilization to succeed once again in "complexifying" in response to an unprecedented challenge and ensuring the survival of global human societies until the next challenge comes along. In light of dominating unsustainable patterns of human consumption, there is an urgency to redefine the Anthropocene to allow future generations to grow in a decent and habitable world. According to Stoll-Kleemann and O'Riordan, the sustainability flagship comes with structural change in our society that should deal with fewer natural resources and a different organization of human life.[75]

In light of the transformation of the energy system, the question of resources will become even more critical to the functioning of our society. As Vidal pointed out, the energy transition has an enormous cost in terms of resource requirements, which will lead to a phase of overconsumption. Although the development of technologies may seem viable at this stage, they may lead us to a dead-end without reflection on the energy model.[76] As

---

[74] See Just et al. (2005).
[75] See Stoll-Kleemann and O'Riordan (2018).
[76] See Vidal (2018).

yet, the dilemma around resource management poses the risk of an implosion of the thermo-industrial civilization, which is a real possibility.[77] The planetary limits are threatening the equation between resource use and the energy system supporting the technological development that our civilization has reached. In other words, there is an urgent need to think about the functioning and organization of the thermo-industrial civilization and the adjoining societies that must lead to a truly out-of-the-box predicament.[78] To this end, it seems imperative to conduct a complete reassessment of our technical–economic environment. Diogo and his colleagues highlight how "the relations between nature, society, and technology are at the very core of many thorny issues of the debate on the Anthropocene."[79] What is at stake is the requirement of a deeper engagement with materiality, which is unavoidable to achieve a paradigm shift. This will require comprehensive simulations to get a full treatment of a resource-intensive civilization and its feedback on the host planet.

All in all, the rise of the thermo-industrial civilization completed the belief in infinite growth, thus becoming hegemonic throughout the planet. Adeney Thomas et al. recall that this process structured "postwar economic and political institutions in capitalist and communist countries, in non-aligned nations and international organizations."[80] These conditions propelled the "Great Acceleration" in the post-1945 context. Considering the acceleration phase of the human adventure on Earth, this hegemonic model marginalized peripheral human communities (e.g., Arctic indigenous peoples, indigenous communities in Brazil, and Pacific indigenous communities). While the notion of endless growth sticks as an objective since the nineteenth century, calls to shift away from this economic rationale turn urgent, it goes through an exit from the current nature/human entanglement in the economic and social system. In this respect, Ehrlich summed up perfectly the magnitude of the issue: "a new ecological-economic paradigm must be constructed that unites nature's housekeeping and society's housekeeping, and make clear that the first priority must be given to keeping nature's house in order."[81] In a nutshell, the natural ecosystem must subdue human societies in the way of organizing and functioning on Earth. Since resources feed our energy system, our thermo-industrial civilization needs even more energy to extract resources in hostile parts of the planet. The latter are increasingly geographically distant from the core of our societies intending to respond to the depletion of resources. In order to cope with this phenomenon, a technological and economic race is engaged to ensure the viability of our energy model, even though it is hardly sustainable. In this complex equation, technology breakthrough remains speculative. From this perspective, the perspective of any sustainable resource use model cannot be shaped by an unforeseen factor. Ultimately, a physical hurdle

[77] See Dockstader et al. (2019).
[78] See Rees (2012).
[79] Diogo et al. (2017), 31.
[80] Adeney Thomas et al. (2020), 136.
[81] Ehrlich (1989), 14.

stands in the way of human civilization's path on Earth, questioning the essence of its trajectory since its inception. It is in this that the resources are determinant in the link to the materiality in which human beings are locked up. To answer the issue of resource availability, a structural transformation is necessary for the feasibility of a sustainable techno-economic model.

In brief, such a transformation embeds all characteristics of a functioning civilization (i.e., cultural, social, political, and economic), while a reflective alteration is required to achieve an ontological shift enabling to ensure a human civilization in tandem with nature.

The Design of an Infrastructure Governance

The management of human infrastructure is arguably the biggest governance challenge of the coming decades. First of all, we must stress that infrastructure, which reflects human and demographic structures, provides critical services for our society (e.g., heating, clean water, communication, etc.) that are seen as indispensable.[82] The density of our complex and integrated infrastructure on the planet today is unprecedented in human history. The increasing effects of climate change on these artificial facilities directly threaten the viability of our societies. Indeed, the extent of challenges to adapting current infrastructure is due to two factors: these systems were designed for conditions that no longer exist when used technology is deemed outdated.[83] Based on this observation, a reflection must be conducted on the implementation of tools that can support the governance of infrastructure in the future. On the one hand, it is a question of anticipating the consequences of climate change on all critical infrastructures, but also the major urban centers of the planet. In other words, we need to manage the existing infrastructure despite these changing conditions. The need to modernize aging infrastructures is, thus, unavoidable in the socio-economic stability of our societies and the management of these legacies of the thermo-industrial civilization. That is the case, for instance, with the extensive investment plan of the United States to the tune of $1.2 trillion.[84] On the other, the new infrastructures induce new practices that are indispensable to resist the evolving conditions of the planet. Clearly, though, Hallegatte emphasizes that decision-making on the construction of new infrastructure (e.g., urbanization plans, infrastructure development for water management, transportation, norms, etc.) is established on long-term commitment and has an influence on a long time scale. In other words, it is about changing our relationship to time and projecting our infrastructures on a long-term horizon (i.e., 2100–2200).[85]

---

[82] See Dawson et al. (2018); Chester et al. (2019).
[83] See Shortridge and Camp (2019).
[84] See The White House (2021).
[85] See Hallegatte (2009).

Climate change impacts on our infrastructures are diverse depending on their climatic and geographical situations. Nevertheless, the increase in the effects of climate change on infrastructure is unavoidable. This pressure will lead to more disruptions, a growing potential for disaster risks and increased requirements for resources needed to restore or retrofit structures.[86] Extreme heat, hurricanes, heavy rainfall and drought already have dramatic impacts on worldwide infrastructure systems, while they were not specifically designed in the past to withstand extreme conditions associated with climate.[87] The regularity and intensity of extreme weather events are now becoming part of everyday life, and all point toward a systemic shifting of human living conditions on Earth. In the Arctic, the effects of thawing permafrost are threatening the infrastructure of a large land portion of the coastal countries. In the meantime, the Arctic is warming three times faster than the rest of the planet.[88] In these circumstances, Russia is undoubtedly the country most exposed to the phenomenon. Although permafrost accounts for 65% of the Russian territory, it is thawing in some places and the presence of several human settlements threatens to be unsustainable in the long term.[89] The mining town of Norilsk, for example, is witnessing these changes with increased fragility of its buildings which implies the implementation of new construction standards.[90]

The consequences of climate change threaten to reconfigure global physical boundaries and directly put human settlements at risk. For instance, port cities (e.g., Shanghai, China; Mumbai, India; Lagos, Nigeria; New York, United States) are on the frontline of the impending massive scale change that has been never experienced before in human history. In light of the risk of coastal flooding, Moretti and Loprencipe underscore that adaptation strategies and actions should be undertaken to cope with Sea-Level Rise (SLR).[91] Coastal areas and small islands around the world are boundaries splitting ocean and land that may vanish and directly affect 70 countries. As Dodds recalls, "changing sea level will obliterate the lower-lying communities and territories of the world."[92] In those human communities, the SLR risk is existentially threatening to destroy all infrastructure making it impossible to maintain them in these areas.

To enable sustainable governance of our infrastructure, Auld highlights the need to "identify gaps in current adaptive capacity to existing climate variability and extremes."[93] For instance, the prevention of meteorological disasters is

---

[86] See Auld (2008).

[87] See Shortridge and Camp (2019).

[88] See AMAP (2021).

[89] See Suter et al. (2019); Hjort et al. (2022).

[90] See Poroshina (2018).

[91] See Moretti and Loprencipe (2018).

[92] Dodds (2021), 97.

[93] Auld (2008), 286.

an essential tool to anticipate and prevent any potential impacts. This involves the implementation of structural measures as well as non-structural measures to avoid loss of life and limit material damage. In other words, infrastructure governance should rely on climate change models, including their spatial and temporal resolutions, and support adaptation options able to be developed over time through "adaptation learning."[94] To this end, a systemic approach will require comprehending the multiple points of connection, feedback and feedforward paths that describe a region's complex infrastructure interdependencies. Arguably, without such a comprehensive overview, our infrastructure network will not be able to deal with amplifying life-threatening weather conditions. According to Clark et al., "the potential for an unmanaged feedback loop may create conditions that amplify a small disruption originating in one system to cause tragic collapse of critical services in others."[95]

Given that fact, most current infrastructure design remains based on past conditions, and thus despite increasing evidence to the contrary. Chester et al. pointed out the necessity for an engineer to handle the issue of robustness of infrastructure in light of ever more frequent and stronger extreme events.[96] There is an urgency to address the complexity of infrastructure design and management standards that needs to deal with unpredictable patterns and negative feedback loops regarding Earth's shifting conditions. In terms of institutional arrangements, addressing infrastructure resilience must be complementary and autonomous when implementing these public policies. In response to these challenges, Chester et al. underline that different policy levels (i.e., towns, cities, regions, states and countries) should codify altogether the level of risk that their infrastructure should be able to endure in the coming decades.[97] To support the implementation of such public policies, Dawson and his colleagues emphasize the need to generate a database of "the location, function, design, and condition of assets," but also "a record of any adaptation to these assets in order to provide a reliable assessment of current and future infrastructure performance."[98] Although climate change will increasingly alter and disrupt infrastructure networks, legal, political and technical answers should be embedded in a long-term and flexible design framework.

What We Can Learn from the Japanese Way?

Japan stands out for its leadership position when it comes to addressing infrastructure. The country has set the issue of standards as an utter index for

---

[94] Ibid.
[95] Clark et al. (2018), 2.
[96] See Chester et al. (2020).
[97] Ibid.
[98] Dawson et al. (2018), 16.

determining the quality of infrastructure in which it determines its investments.[99] From this requirement, the establishment of infrastructure governance must develop a new ethic that allows for the design of a decision-making architecture. In that respect, this designed structure intends to reverberate within the whole of society, namely its organization and the physical layout of its infrastructures. To allow for adequate governance in the face of future climate shocks, Japanese society must reconsider the principles of its relationship with nature. As Takahashi reminds us, "the relationship between humans and nature is interactive" based on coevolution throughout history.[100] These intrinsic links lay the foundation for a shift closely connecting respect for the integrity of the ecosystem and humans. In doing so, an understanding of the Earth's geographic and geological conditions is an indispensable vehicle for designing this planetary ethic. In the case of Japan, these conditions are the source of both beneficial and destructive effects, as evidenced by the natural disasters encountered throughout its history. Most recently, climatic conditions become an absolute parameter with the transformation of the biosphere. If the processes are global, on the other hand, the chains of actions and reactions are visible at the local level. From this general perspective, Japan must learn and manage global transformations and their direct impacts on its territory. The tsunami and the Fukushima nuclear disaster of March 2011 ensued a catastrophic combination highlighting all the intertwining and ramifications for the infrastructure laid out along the coastal area of the Tōhoku region. In the wake of climate change, such a scenario is once again possible, putting the territory's infrastructure network in disarray.

On the national level, Japanese authorities have fostered a culture of natural disasters among the population, for instance, by introducing the Disaster Prevention Day (*Bōsai no hi*) in 1960. More recently, the Japanese government presented a new comprehensive strategy, so-called Society 5.0, attempting to tackle the complex challenges in the coming decades, including climate changes.[101] In light of the shockwaves ahead, the country's public policy must integrate technological and innovative solutions in order to back disaster and climate change management efforts. Against the background of this technological tropism, disaster prevention and management are not incorporated under a single nationwide system. While setting up a Ministry of Disaster Prevention has been proposed for several years by the Japanese Governors' Association and several experts, the successive Japanese governments have not been convinced so far.[102] Undeniably, such a centralized structure would facilitate administrative and territorial organization throughout the country. Despite the lack of an efficient and centralizing institutional body, Japan has implemented an innovation policy to design infrastructures able to resist

[99] See Aizawa (2019).
[100] Takashi (2017), 227.
[101] See Mavrodieva and Shaw (2020).
[102] See Heimburger (2022).

the numerous hazards (e.g., tsunamis, earthquakes, typhoons). Since the Ise Bay typhoon (1959), the level of protection infrastructure has improved significantly with the construction, in particular, of breakwaters. However, Heimburger underscores that several issues came along with these infrastructures affecting the natural ecosystem, including the alteration of the coastal landscape, the vanishing of beaches and mudflats, and the reduction of the marine fauna and flora.[103]

In spite of the high rate of urbanization, meteorological disasters in urban areas are the most critical threat in the country. Difficulties lie in maintaining services and infrastructure networks in the affected territories. Japanese cities have introduced, in recent years, infrastructure solutions to prevent urban flooding by setting up anti-intrusion panels and watertight doors. Meanwhile, they have also built rainwater storage tanks to avoid saturation of the drainage system.[104] Introduction of green infrastructures (e.g., "rain gardens") as part of long-term and sustainable solutions are other beneficial measures to limit the impacts of heavy rainfall. But, these preventive measures are not sufficient to fully eradicate urban flooding in Japanese cities. Among them, Tokyo counts as one of the most populous metropolitan regions worldwide. As Krishnan et al. recap, the capital has a long tradition with multi-natural hazards as the whole national territory.[105] If the Great Kantō earthquake was the most catastrophic event that the Japanese capital faced, the frequency and intensities of extreme weather conditions have been rising in the archipelago. In fact, Tokyo authorities have framed a holistic and integrated response to meet hazard risks, such as flood control measures, emergency evacuation systems and a river management system. Concerning infrastructures, the city has built hydraulic structures (e.g., levees and gates) to tackle flood risk, but there are inadequate to deal with emerging climate conditions. Furthermore, the Tokyo metropolitan area relies on underground discharge channels and reservoirs. According to Nakamura and Oosawa, the use of underground space reveals significant advantages in terms of cost and land acquisition.[106] To consolidate the metropolitan flood defense infrastructure network, the construction of "super levees" (a high standard river embankment that is about 30 times as wide (around 300 m) as it is high) would better integrate risk management measures.[107] Nevertheless, such mega projects raise democratic and ethical issues as an integral part of governance matters, including exceptionally high costs, long construction timelines and displacement problems.

All in all, the case of Japan remains symbolic, but it can nevertheless guide many countries to integrate a systemic vision in the interaction between our infrastructures and their environment. In order to achieve infrastructure

[103] Ibid.
[104] Ibid.
[105] See Krishnan et al. (2019).
[106] See Nakamura and Oosawa (2021).
[107] See Krishnan et al. (2019).

governance in line with global developments, the integration of uncertainty is undoubtedly an utmost factor in all long-term decisions. In the meantime, the challenge of infrastructure governance will lie in the ability to keep the distinction between new infrastructure and existing infrastructure in the management and development strategy. This approach initiates another way of looking at infrastructure projects better fitted for the future. Against the backdrop of anticipation of the nonlinear and complex climate change effects, impacts will remain lower and more manageable. Hallegate suggested that "an anticipatory adaptation strategy can buy us the time we need to wait for (still-to-be-implemented) mitigation policies to become effective."[108] More than being a model to be copied, the interest of the Japanese case lies in the multifaceted approach that the country nurtures to the human-nature dynamic. Over the centuries, Japan has succeeded in feeding a risk culture in the midst of extreme disasters, which provides lessons for other human societies. Best practice sharing and efficient knowledge circulation between these societies shall contribute to delivering sustainable improvements in the disclosure of infrastructure governance. This governance shall be designed to withstand impacts that will be more intensive and more recurrent than ever before. From this capability to cope with these shockwaves, this governance will be then set up in a regenerative way. This paradigm shift should facilitate the advent of institutions directly plugged into Earth System dynamics and enable them to maintain governance functioning and continuity.

Conclusion: The Human Downturn

The singularity of the human trajectory is now opposed to the physical realities of planetary boundaries. If these planetary boundaries have been extensively explained, new ones will emerge and impose themselves on our societies due to the nature of the functioning of the thermo-industrial civilization. Far from simple management of possible shortages (e.g., food, water, fossil fuels, critical raw materials), it is a question of integrating new constraints. In this respect, current government decisions and pending global governance, specifically under the guidance of the UN Sustainable Development Goals, embrace transition policies to shift from one energy model to another, but without any paradigmatic shift in proportion with the ongoing planetary trajectory.

Frank et al. explain the scale of the looming transformation in light of crossed tipping points: this is not a shift, but instead of a bifurcation.[109] In other words, the planetary conditions are changing system parameters which propel Earth "into a transition to distant attractors."[110] Hence, the status quo for humankind can no longer be sustained over the long term, while the structural element of materiality now requires a paradigmatic evolution that

---

[108] Hallegate (2009), 246.
[109] See Frank et al. (2018).
[110] Ibid.

directly impacts the relationship between humans and non-humans. That is an ontological factor of the human condition, essential to the sustainability of the organization of post-thermo-industrial human civilization. Consequently, this condition will allow our relationship to the use of resources to be reshaped (i.e., from their extraction to their integration into the artificialized human universe). In this way, human civilization has to reconsider its long-term relationship to energy and technology. That in turn will evolve to embed boundaries that the Anthropocene consubstantially carries.

Frank et al. argue the expansion of technological civilization may continue that movement. However, it remains to be seen if human trajectory could ultimately rely on solar power through efficient conversion of solar photos to useful powers (i.e., today's photovoltaics, other technologies in the future).[111] On the one hand, this imaginable world based on solar energy is part of the alternatives that the Anthropocene is opening up. On the other, recycling economies provide new possibilities to embed our relationship with materiality in a circular life cycle and no longer in a linear one. After all, the idea is doing with what already exists in a manner that saves the Earth's resources. As Roth unveils, the Anthropocene pluriverse leaves space "to envision a different kind of planet politics" acknowledging "the plurality of worlds in and the heterogenous temporality of the Anthropocene."[112] In a nutshell, the planetary limits will define the room for maneuver in which our societies will have to set in the future.

Even though long-term trajectories remain fundamentally speculative, the human trajectory as civilization appears, more than ever, deeply entangled in the planetary course, which is on a much-extended timescale. Crutzen emphasizes that humans will continue to be a central and inescapable environmental force for the next several millennia.[113] In light of new boundaries, humans should set their limits to maintain the Earth in a safe habitability zone. Far from the hypothetical and potentially counterproductive projects that geoengineering promises, humankind could consider fallback governance to cope with the enormity of the forthcoming shock and safeguard its very existence on Earth. As Morton points out with his concept of Dark Ecology, it is thus a matter of reconsidering the "logistic" program that has been in use since the Neolithic.[114] To that extent, social metabolism should be considered a critical tool for implementing this ontological shift that human beings face.[115]

[111] Ibid.
[112] Roth (2020), 162.
[113] See Crutzen (2009).
[114] See Morton (2016).
[115] See González de Molina and Toledo (2014).

REFERENCES

Adeney Thomas, J., Williams, M., & Zalasiewicz, J. (2020). *The Anthropocene: A Multidisciplinary Approach.* Cambridge: Polity.

Aizawa, M. (2019). "Sustainable Development Through Quality Infrastructure: Emerging Focus on Quality over Quantity." *Journal of Mega Infrastructure & Sustainable Development, 1*(2), 171–187.

AMAP. (2021). *Arctic Climate Change Update 2021: Key Trends and Impacts. Summary for Policy-Makers.* Tromsø: Arctic Monitoring and Assessment Programme (AMAP).

Auld, H. (2008). "Adaptation by Design: The Impact of Changing Climate on Infrastructure." *Journal of Public Works & Infrastructure, 1*(3), 276–288.

Barnosky, A., Hadly, E., Bascompte, J., Berlow, E., Brown, J., Fortelius, M., Getz, W., Harte, J., Hastings, A., ... & Smith, A. (2012). "Approaching a State Shift in Earth's Biosphere." *Nature, 486*(7401), 52–58.

Basener, B., & Ross, D. (2005). "Booming and Crashing Populations and Easter Island." *SIAM Journal on Applied Mathematics, 65*(2), 684–701.

Berkhout, F. (2014). "Anthropocene Futures." *The Anthropocene Review, 1*(2), 154–159.

Biermann, F. (2014). "The Anthropocene: A Governance Perspective." *The Anthropocene Review, 1*(1), 57–61.

Brannen, P. (2017). *The Ends of the World: Volcanic Apocalypses, Lethal Oceans, and our Quest to Understand Earth's Past Mass Extinctions.* New York: HarperCollins.

Brook, B., Ellis, E., Perring, M., Mackay, A., & Blomqvist, L. (2013). "Does the Terrestrial Biosphere have Planetary Tipping Points?" *Trends in Ecology & Evolution, 28*(7), 396–401.

Chester, M., Markolf, S., & Allenby, B. (2019). "Infrastructure and the Environment in the Anthropocene." *Journal of Industrial Ecology*, 1–10.

Chester, M., Underwood, S., & Samaras, C. (2020). "Keeping Infrastructure Reliable Under Climate Uncertainty." *Nature Climate Change, 10,* 488–490.

Clark, S., Chester, M., Seager, T., & Eisenberg, D. (2018). "The Vulnerability of Interdependent Urban Infrastructure Systems to Climate Change: Could Phoenix Experience a Katrina of Extreme Heat?" *Sustainable and Resilient Infrastructure.* https://doi.org/10.1080/23789689.2018.1448668.

Cooke, M. (2020). "Ethics and Politics in the Anthropocene." *Philosophy and Social Criticism,* 1–15.

Cowie,R.H., Bouchet, P., & Fontaine, B. (2022). "The Sixth Mass Extinction: Fact, Fiction or Speculation?" *Biological Reviews.* https://doi.org/10.1111/brv.12816.

Crutzen, P.J. (2002). "Geology of Mankind." *Nature, 415,* 23.

Crutzen. P.J. (2009). "Can We Survive the 'Anthropocene' Period?" *Project Syndicate.* https://www.project-syndicate.org/commentary/can-we-survive-the--anthropocene--period. Accessed 25 April 2022.

Dartnell, L. (2011). "Biological Constraints on Habitability." *Astrobiology, 52,* 1.25–1.28.

Dawson, R., Thompson, D., Johns, D., Wood, R., Darch, G., Chapman, L., Hughes, P,..., & Hall, J. (2018). "A Systems Framework for National Assessment of Climate Risks to Infrastructure." *Philosophical Transactions of the Royal Society A, 376.* https://doi.org/10.1098/rsta.2017.0298.

Diamond, J. (2021). *The Last Tree on Easter Island.* London: Penguin Books.

Diogo, M., Louro, I., & Scarso, D. (2017). "Uncanny Nature: Why the Concept of Anthropocene is Relevant for Historians of Technology." *ICON: Journal of the International Committee for the History of Technology*, 23, 25–35. https://www.jstor.org/stable/26454974.

DiNapoli, R., Rieth, T., Lipo, C., & Hunt, T. (2020). "A Model-Based Approach to the Tempo of 'Collapse': The Case of Rapa Nui (Easter Island)." *Journal of Archaeological Science*, 116. https://doi.org/10.1016/j.jas.2020.105094.

Dockstader, Z., Bauch, C. T., & Anand, M. (2019). "Interconnections Accelerate Collapse in a Socio-Ecological Metapopulation." *Sustainability*, 11(7), 1852. https://doi.org/10.3390/su11071852.

Dodds, K. (2021). *Border Wars. The Conflicts that will Define our Future*. London: Ebury Press.

Ehrlich, P. (1989). "The Limits to Substitution: Meta-resource Depletion and a New Economic-Ecological Paradigm." *Ecological Economics*, 1, 9–16.

Fagan, M. (2019). "On the Dangers of an Anthropocene Epoch: Geological Time, Political Time and Post- Human Politics." *Political Geography*, 70, 55–63.

Frank,A. (2018). *Lights of the Stars: Alien Worlds and the Fate of the Earth*. New York: W.W. Norton & Company.

Frank, A., Kleidon, A., & Alberti, M. (2017). "Earth as a Hybrid Planet: The Anthropocene in an Evolutionary Astrobiological Context." *Anthropocene*, 19, 13–21.

Frank, A., Carroll-Nellenback, J., Alberti, M., & Kleidon, A. (2018). "The Anthropocene Generalized: Evolution of Exo-Civilization and their Planetary Feedback." *Astrobiology*, 18(5), 503–519.

Gibbard, P., Walker, M., Bauer, A., Edgeworth, M., Edwards, L., Ellys, E., Finney, S., Gill, J. L., Maslin, M., Merritts, D., & Ruddiman, W. (2022). "The Anthropocene as an Event, Not and Epoch." *Journal of Quaternary Science*, 1–5. https://doi.org/10.1002/jqs.3416.

González de Molina, M., & Toledo, V.M. (2014). *The Social Metabolism. A Socio-Ecological Theory of Historical Change*. Springer: Cham. https://doi.org/10.1007/978-3-319-06358-4_3.

Haila, Y. (2000). "Beyond the Nature-Culture Dualism." *Biology and Philosophy*, 15, 155–175.

Hallegatte, S. (2009). "Strategies to Adapt to an Uncertain Climate Change." *Global Environment Change*, 19, 240–247.

Hamilton, S. (2017). "Securing Ourselves from Ourselves? The Paradox of 'Entanglement' in the Anthropocene." *Crime, Law and Social Change*, 68, 579–595.

Heikkurinen, P., Clegg, S., Pinnington, A.H., Nicolopoulou, K., & Alcaraz, J.M. (2019). "Managing the Anthropocene: Relational Agency and Power to Respect Planetary Boundaries." *Organization & Environment*, 1–20.

Heimburger, J.-F. (2022). *Le système de prévention et de gestion des catastrophes météorologiques au Japon*. Paris: Foundation for Strategic Research (FRS).

Hjort, J., Streletskiy, D., Doré, G., Wu, Q., Bjella, K., & Luoto, M. (2022). "Impacts of Permafrost Degradation on Infrastructure." *Nature Reviews Earth & Environment*, 3, 24–38.

Im, E.-S., Pal, J., & Eltahir, E. (2017). "Deadly Heat Waves Projected in the Densely Populated Agricultural Regions of South Asia." *Science Advances*, 3(8). https://doi.org/10.1126/sciadv.1603322.

IPBES (2019). *Summary for Policymakers of the Global Assessment Report on Biodiversity and Ecosystem Services of the Intergovernmental Science-Policy Platform on Biodiversity and Ecosystem Services*. Bonn: IPBES secretariat.

Just, R., Netanyahu, S., & Olson, L. J. (2005). "Depletion of Natural Resources, Technological Uncertainty, and the Adoption of Technological Substitutes." *Resource and Energy Economics*, 27, 91–108.

Knight, J. (2015). "Anthropocene Futures: People, Resources and Sustainability." *The Anthropocene Review*, 1–7.

Krishnan, S., Lin, J., Simanjuntak, J., Hooimeijer, F., Bricker, J., Daniel, M, & Yoshida, Y. (2019). "Interdisciplinary Design of Vital Infrastructure to Reduce Flood Risk in Tokyo's Edogawa Ward." *Geosciences*, 357(9). https://doi.org/10.3390/geosciences9080357.

Lapola, D. (2020). "Futuras pandemias poderão começar no Brasil." *Folha de São Paulo*. https://www1.folha.uol.com.br/opiniao/2020/05/futuras-pandemias-poderao-comecar-no-brasil.shtml. Accessed 25 March 2022.

Latour, B. (2018). *Down to Earth. Politics in the New Climatic Regime*. Cambridge: Polity. Lenton, T. (1998). Gaia and natural selection. *Nature*, 394, 439–447.

Lenton, T. (1998). "Gaia and Natural Selection." *Nature*, 394, 439–447.

Lenton, T. (2008). "Tipping Elements in the Earth's Climate System." *PNAS*, 105(6), 1786–1793.

Lenton, T., Rockström, J., Gaffney, O., Rahmstorf, S., Richardson, K., Steffen, W., & Schellnhuber, H. J. (2019). "Climate Tipping points—Too Risky to Bet Against." *Nature*, 575, 592–596.

Lenton, T., Dutreuil, S., & Latour, B. (2020). "Life on Earth is Hard to Spot." *The Anthropocene Review*, 7(3), 248–272.

Lewis, S., & Maslin, M. (2015). "Defining the Anthropocene." *Nature*, 519, 171–180.

Lima,M., Gayo, E. M., Latorre, C., Santoro, C. M., Estay, S. A., Cañellas-Boltà, N., Margalef, O., ..., & Stenseth, N. (2020). "Ecology of the Collapse of Rapa Nui Society." *Proceedings of the Royal Society B*, 287. https://doi.org/10.1098/rspb.2020.0662.

Lineweaver, C.H., & Chopra, A. (2012). "The Habitability of our Earth and Other Earths: Astrophysical, Geochemical, Geophysical, and Biological Limits on Planet Habitability." *Annual Review of Earth and Planetary Sciences*, 40, 597–623.

Lovelock, J., & Margulis, L. (1974). "Atmospheric Homeostasis by and for the Biosphere: The Gaia Hypothesis." *Tellus*, 26(1–2), 2–10.

Lovelock, J. (2000). *Homage to Gaia: The Life of an Independent Scientist*. Oxford: Oxford University Press.

Lövbrand, E., Mobjörk, M., & Söder, R. (2020). "The Anthropocene and the Geo-Political Imagination: Re-writing Earth as Political Space." *Earth System Governance*. https://doi.org/10.1016/j.esg.2020.100051.

Lyon, C., Saupe, E., Smith, C., Hill, D., Beckerman, A., Stringer, L., Marchant, R., J. McKay, J., Burke, A., O'Higgins, P.,..., & Aze, T. (2021). "Climate Change Research and Action must Look Beyond 2100." *Climate Change Biology*, 28, 349–361.

Malhi, Y. (2017). "The Concept of the Anthropocene." *Annual Review of Environment and Resources*, 42, 77–104.

Mavrodieva, A., & Shaw, R. (2020). "Disaster and Climate Change Issues in Japan's Society 5.0—A Discussion." *Sustainability*, 12, 1893. https://doi.org/10.3390/su12051893.

Moretti, L., & Loprencipe, G. (2018). "Climate Change and Transport Infrastructures: State of the Art." *Sustainability*, 10. https://doi.org/10.3390/su10114098.

Morton, T. (2016). *Dark Ecology: For a Logic of Future Coexistence*. New York: Columbia University Press.

Nakamura, H., & Oosawa, M. (2021). "Effects of the Underground Discharge Channel/Reservoir for Small Urban Rivers in the Tokyo Area." *IOP Conf. Series: Earth and Environmental Science*, 703. https://doi.org/10.1088/1755-1315/703/1/012029.

Newbold, T., Adams, G., Robles, G. A., Boakes, E, Braga Ferreira, G., Chapman, A, ..., & Williams, J. (2019). "Climate and Land-Use Change Homogenise Terrestrial Biodiversity, with Consequences for Ecosystem Functioning and Human Well-Being." *Emerging Topics in Life Sciences*, 3(2), 207–219.

Nicholson, A., Wilkinson, D., Williams, H., & Lenton, T. (2018). "Gaian Bottleneck and Planetary Habitability maintained by Evolving Model Biospheres: The ExoGaia Model." *Monthly Notices of the Royal Astronomical Society*, 477, 727–740.

NOAA Earth System Research Laboratory. (2022). *Mauna Loa CO2 Annual mean Data [Text file]*. Retrieved from https://gml.noaa.gov/webdata/ccgg/trends/co2/co2_annmean_mlo.txt.

Northey, S., Mudd, G., & Werner, T. (2018). "Unresolved Complexity in Assessments of Mineral Resource Depletion and Availability." *Natural Resources Research*, 27, 241–255.

Olson, S., Schwieterman, E., Reinhard, C., & Lyons, T. (2018). "Earth: Atmospheric Evolution of a Habitable Planet." In *Handbook of Exoplanets*, ed. H. Deeg and J. Belmonte, 2817–2853. Cham: Springer.

Preston, L.J., &. Dartnell, L.R. (2014). "Planetary Habitability: Lessons Learned from Terrestrial Analogues." *International Journal of Astrobiology*, 13(1), 81–98.

Prior, T., Giurco, D., Mudd, G., Mason, L., & Behrisch, J. (2012). "Resource Depletion, Peak Minerals and the Implications for Sustainable Resource Management." *Global Environmental Change*, 22(3), 577–587.

Poroshina, S.S. (2018). "Rasteplenie vechnomerzlykh gruntov pod zdaniiami v Noril'ske [Permafrost Soils Thawing under Buildings in Norilsk]." *Gradostroitel'stvo I Arkhitektura [Urban Construction and Architecture]*, 8(2), 65–70.

Raymond, C., Matthews, T., & Horton, R. (2020). "The Emergence of Heat and Humidity too Severe for Human Tolerance." *Science Advances*, 6(19). https://doi.org/10.1126/sciadv.aaw1838.

Rees, W. E. (2012). "Cities as Dissipative Structures: Global Change and the Vulnerability of Urban Civilization." In *Sustainability Science: The Emerging Paradigm and the Urban Environment* ed. M. P. Weinstein and R. E. Turner, 247–273. New York: Springer.

Rothe, D. (2020). "Governing the End Times? Planet Politics and the Secular Eschatology of the Anthropocene." *Millennium: Journal of International Studies*, 48(2), 143–164.

Rockström, J., Steffen, W., Noone, K., Persson, Å., Chapin, F., Lambin, E., Lenton, T., et al. (2009). "Planetary Boundaries: Exploring the Safe Operating Space for Humanity." *Ecology and Society, 14*(2). http://www.ecologyandsociety.org/vol14/iss2/art32/.

Sagan, C. (1980). *Cosmos*. New York: Ballantine Books.

Sagan, D., & Margulis, L. (1997). "Gaia and Philosophy." In *Slanted Truths*, 145–157. New York: Springer.

Schmeller, D., Courchamp, F., & Killeen, G. (2020). "Biodiversity Loss, Emerging Pathogens and Human Health Risks." *Biodiversity and Conservation*, 29, 3095–3102.

Shortridge, J., & Camp, J. S. (2019). "Addressing Climate Change as an Emerging Risk to Infrastructure Systems." *Risk Analysis*, 39(5), 959–967.

Steffen, W., Grinewald, J., Crutzen, P., & McNeill, J. (2011). "The Anthropocene: Conceptual and Historical Perspectives." *Philosophical Transactions of the Royal Society A*, 369(1938). https://doi.org/10.1098/rsta.2010.0327.

Steffen, W., Richardson, K., Rockström, J., Cornell, S., Fetzer, I., Bennett, E., Biggs, R., Carpenter, S., De Vries, W., ..., & Sörlin, S. (2015). "Planetary Boundaries: Guiding Human Development on a Changing Planet." *Science*, 347(6223). https://doi.org/10.1126/science.1259855.

Steffen, W., & Morgan, J. (2021). "From the Paris Agreement to the Anthropocene and Planetary Boundaries Framework: An Interview with Will Steffen." *Globalizations*, 18(7), 1298–1310.

Sterner, T., Barbier, E., Bateman, I., van den Bijgaart, I., Crépin, A.-S., Edenhofer, O., Fischer, C., Habla, W., Hassler, J.,... & Robinson, A. (2019). "Policy Design for the Anthropocene." *Nature Sustainability*, 2, 14–21.

Stoll-Kleemann, S., & O'Riordan, T. (2018). "Biosphere Reserves in the Anthropocene." *Encyclopedia of the Anthropocene*, 347–353.

Suter, L., Streletskiy, D., & Shiklomanov, N. (2019). "Assessment of the Cost of Climate Change Impacts on Critical Infrastructure in the Circumpolar Arctic." *Polar Geography*, 42(4), 267–286.

Takahashi,T. (2017). "Disaster Prevention as an Issue in Environmental Ethics." In *Japanese Environmental Philosophy*, ed. B. Callicott and J. McRae, 227–241. New York: Oxford University Press.

Tilton, J.E. (2003). "Assessing the Threat of Mineral Depletion." *Minerals & Energy*, 18, 33–42.

Tilton, J.E. (2018). "The Hubbert Peak Model and Assessing the Threat of Mineral Depletion." *Resources, Conservation and Recycling*, 139, 280–286.

Turvey, S., & Crees, J. (2019). "Extinction in the Anthropocene." *Current Biology*, 29, 982–986.

Tyrell, T. (2020). "Chance Played a Role in Determining Whether Earth Stayed Habitable." *Communications Earth & Environment*, 1(61). https://doi.org/10.1038/s43247-020-00057-8.

Tyszczuk, R. (2021). "Collective Scenarios: Speculative improvisations for the Anthropocene." *Futures*, 134. https://doi.org/10.1016/j.futures.2021.102854.

Waltham, D. (2017). "Star Masses and Star-Planet Distances for Earth-Like Habitability." *Astrobiology*, 17, 61–77. https://doi.org/10.1089/ast.2016.1518.

Waltman, D. (2019). "Is Earth Special?" *Earth-Science Reviews*, 192, 445–470.

The White House. (2021). *Fact Sheet: The Bipartisan Infrastructure Deal*. Washington: The White House Briefing Room. https://www.whitehouse.gov/briefing-room/statements-releases/2021/11/06/fact-sheet-the-bipartisan-infrastructure-deal/. Accessed 25 March 2022.

Vidal, O. (2018). *Matières premières et énergie: les enjeux de demain*. London: ISTE Editions.

The Anthropocene and Global Environmental Politics

Philipp Pattberg and Michael Davies-Venn

Introduction

The Anthropocene hypothesis continues to generate global debates, both in academia and among practitioners. The basic claim of the concept is that human activities have transformed planet Earth and its atmosphere. Subsequently, humans have evolved into a geological force. The transformation implies planetary-scale changes, evidence of which is provided in the scale and scope of environmental challenges that have significantly broadened and deepened and continue to threaten the very processes—from a stable climate to biodiversity—on which life on earth in general and human development in particular depend. These developments provide pause for thought on the important nexus of human activities and their subsequent impacts on planet Earth; however, a number of important questions remain unanswered.

As Pattberg and Zelli (2016, 1) observe on the Anthropocene, "no agreement exists concerning a number of important issues, including the exact start date and appropriate stratigraphic markers, its normative implications and political consequences." Similarly, Pattberg and Davies-Venn (2020) argue that the global framework used to decide and define global units used for

P. Pattberg (✉) · M. Davies-Venn
VU University of Amsterdam, Amsterdam, The Netherlands
e-mail: philipp.pattberg@vu.nl

M. Davies-Venn
e-mail: m.daviesvenn@vu.nl

© The Author(s), under exclusive license to Springer Nature Switzerland AG 2023
J. J. Kassiola and T. W. Luke (eds.), *The Palgrave Handbook of Environmental Politics and Theory*, Environmental Politics and Theory,
https://doi.org/10.1007/978-3-031-14346-5_25

dating planet Earth—the International Union of Geological Sciences (IUGS), the International Commission on Stratigraphy (ICS) along with the Quaternary Stratigraphy Sub-commission and its Anthropocene Working Group (AWG)— is faced with serious challenges that must be addressed within the AWG and its associated bodies before the concept is formally adopted. Far from being a "neutral" scientific concept, the Anthropocene is loaded with political meaning; debates—and eventually decisions—about the appropriate start date, markers and related narratives carry political implications that require further scrutiny. In this chapter, we provide an overview of key concepts and definitions (see also annex A) related to the Anthropocene, engage in a brief summary of key arguments supporting the Anthropocene hypotheses, and introduce the scientific bodies tasked with deciding on and defining the units that date our planet. Finally, we provide reflections about implications for governance and politics before concluding with the main findings and outlining avenues for further research.

Box: Key Concepts
Anthropocene: Is a proposed hypothesis that, in part, posits that through human activities planet Earth has been permanently transformed

Anthropocene Working Group (AWG): is a group of scientists established to assess the concept for the purpose of using the Anthropocene as division in the Geological Time Scale

Chronostratigraphical Record: indicates the age of rocks

Epoch: is a unit of measurement used to express an age in the history of planet Earth

Geology: is a branch of science focused on understanding the history of planet Earth, for example by studying rock sediments

Global Standard Stratigraphic Age (GSSA): is a historical reference point that defines boundaries between dates in geology, such as Epochs or Periods

Golden Spike: is a metal driven into the Earth to mark a change in geological time

Geological Time Scale (GTS): is a chart showing chronological dates and their relationships in our planet's history. The time are indicated in Periods, Epochs and Age.

Geoscientific Record: will include sediments, magnetic information or fossils

Geological Time: is a record in the GTS, such as an Epoch

Holocene: Much as the hour hand on clock indicates a certain time of day, the Holocene Epoch indicates a time in the history of planet Earth

International Chronographic Chart: shows units, such as eons of which there are four, used to date planet Earth. For example, within the Phanerozoic Eon, there are three Eras, four Periods, Epochs and Age

International Union of Geological Sciences (IUGS): Represents geoscientists worldwide from its headquarters in Beijing, China

International Commission on Stratigraphy (ICS): is a group of geoscientists within the IUGS who define units to date planet Earth

> **Stratigraphy**: is a branch of geology whose specialists including, stratigraphers, study rock sediments, for example, to determine when they were deposited and thus infer an age in our planet's history
> **Start Date**: A disputed concept that should indicate the start of an epoch, for example, the Anthropocene **Quaternary Stratigraphy Sub-commission (SQS)**: is a group within the ICS constituting geoscientists focused on dating the planet from 2.6 million years ago

THE ANTHROPOCENE HYPOTHESIS: SUGGESTED START DATES AND POLITICAL IMPLICATIONS

Background

The Anthropocene hypothesis was first proposed in the International Geosphere–Biosphere Programme (IGBP) newsletter. Paul J. Crutzen, an atmospheric chemist, and Eugene F. Stoermer, a biologist, supported the proposed concept by noting the impacts of "the expansion of mankind" (2000, 17) on planet Earth and its atmosphere. Human footprints on the planet include, among others: a tenfold increase in population; urbanization; the relative short period, over few generations, that mankind has taken to exhaust fossil fuels that were generated over several hundred million years and subsequent pollution; up to 50% of land transformation; including "loss of 50% of the world's mangroves"; species extinction "by thousand to ten thousand fold in the tropical rain forests" including from fishing; and substantial "green-house" gases increases in the atmosphere. These planetary scale changes, induced by human activities, led Crutzen and Stoermer to conclude that "it seems to us more than appropriate to emphasize the central role of mankind in geology and ecology by proposing to use the term 'anthropocene' for the current geological epoch" (Crutzen and Stoermer, 2000, 17). Population growth has been a driver to changes within the "Earth System" (IGBP(a)., n.d.), which constitutes "human society, social and economic systems" (ibid.,(a)., n.d.) and the structural and biophysical parts, such as "land, oceans, atmosphere and poles, the planet's natural cycles—the carbon, water, nitrogen, phosphorus, sulphur and other cycles—and deep Earth processes"(ibid(a).,n.d). The Anthropocene is a way to recognize that "human-induced global change has pushed the Earth system into a no-analogue state" (IGBP(b).,n.d.).

Should the IUGS formally adopt the Anthropocene, it will be a most fitting description of a period in geological history that is characterized by far-reaching impacts of human activities on our planet. However, consensus on the exact start date of the Anthropocene is only slowly emerging within the geosciences, two decades after the concept was proposed. Crutzen and Stoermer suggest a "latter part of the eighteenth century" as the start date to the Anthropocene. In so doing, they left unresolved a key requirement necessary to adopt the concept, claiming that it is "somewhat arbitrary to assign a more specific date to the onset of the anthropocene" (2000, 17). This has induced debates within and outside the scientific community, because "deciding on an appropriate start date and related marker is more than a technical-administrative act" to be completed by the IUGS and its administrative framework (Pattberg and Davies-Venn, 2020, 130). This is because of the encompassing scope of the concept and political implications from making a determination on when humans started affecting planetary scale changes.

When Did the Anthropocene Begin?

No fewer than five start dates have been suggested and contended, including a suggestion to indefinitely defer formally dating the Anthropocene because "extant geological changes does not reach the thresholds necessary to define a new epoch" (Santana, 2018, 6). Another proposal is to combine the present Holocene Epoch with the proposed Anthropocene into a "single geologic time span," (Certini and Scalenghe, 2015, 246). It is argued that doing so "would provide the Anthropocene with a climatic justification" (ibid., 2015). Further, based on analysis of "anthropogenic soil" (Certini and Scalenghe, 2011), two suggested start dates are rejected—Industrial Revolution and Early Anthropocene—the former includes the period suggested by Crutzen and Stoermer (Crutzen, 2002; Crutzen and Stoermer, 2000). As a result, debates on a start date for the Anthropocene continues (Marlon et al., 2013; Steffen et al., 2011; Lewis and Maslin, 2015; Balter, 2013), also within and across disciplines (Abrams and Nowacki, 2015; Fischer-Kowalski et al., 2014; Vidas, 2010). Complicating this important aspect further, deciding on earlier geological history of planet Earth has primarily been an intra-disciplinary process within the geosciences. However, compared to the Holocene Epoch, the Anthropocene is different because human activities and impacts are crucial to defining the concept. This implies that the status of humans has changed because, "human influence has become a principal agent of change on the planet, shifting the world out of the relatively stable Holocene period into a new geological era" (IPPC, 2018, 53)—the Anthropocene. This should mean that considerations, including on the start date, which may lead to formal adoption of the Anthropocene, must be more inclusive, particularly within the AWG. Broadening the AWG's focus beyond collecting and assessing only chronostratigraphic records on the Anthropocene would allow deeper understanding of the many ways in which humans have evolved into "a geological force on a planetary scale" (Ludwig and Steffen, 2018, 56). Evidently, geoscientific concepts are much too focused and as such are not sufficient for capturing, explaining and making inferences on all human activities, regardless the onset. The concept of an Earth System advanced by the IGBP newsletter, which originally proposed the Anthropocene, clearly suggests that the *system* is far too complex for isolated conceptual tools to adequately explain, even without considering the complexities of "human society" and impacts within the Earth System. Decisions on the Anthropocene, particularly its formal adoption, must therefore be informed by contributions from outside the geosciences. Such should come from any discipline, or sub-discipline, focused on understanding humans and the natural environment. For example, societies have developed in tandem with technologies, which have played a major role in "shaping macrosocial dynamics" (Reynolds and Krivo, 1996, 95) to an extent, "technological innovation continues to be the most basic underlying force responsible for societal change and development" (ibid., 1996, 95). Thus, a history of this parallel evolution in societies could usefully contribute to understanding factors that geoscientists cannot explain, such as emergence of energy systems—how and why they evolved and multidimensional concepts on why societies continue to use them—for example, could provide useful perspectives for addressing environmental problems while also taking note on the geological record of our planet.

Among the proposed dates to mark the start of the Anthropocene are, Paleoanthropocene and Fire, Early Anthropocene, Orbis Spike, Industrial Revolution and Great Acceleration. This last date is understood to span from 1945 to 2000 + (Steffen et al., 2011) and is characterized by some of the more recent contributory indicators to support the concept suggested initially by Crutzen and Stoermer, including population increases, economic activities, globalization, scientific knowledge, technological innovation, motor vehicles, international travel, consumption, environmental pollution, particularly CO_2 emissions, economic growth, increases in electronic communication and urban growth. Often understood as barometers that measure human progress, these sometimes rapid changes have moved the Earth System "outside the envelope of Holocene variability" (Steffen et al., 2011, 850). The cyclical effects of these indicators is noted, for example, in growing demand for food to feed an ever growing global population during this period, led to agricultural practices that severely damaged "the nitrogen economy of the planet [that] will persist for decades, possibly centuries" (Canfield et al., 2010, 192).

The proposed Industrial Revolution start date to the Anthropocene is not entirely divorced from the foregoing debate. At the core of this hypothesis is a global "growing energy bottleneck" (Steffen et al., 2011, 848), and political and economic factors. For example, fossil fuel discovery and extraction allowed for a sharp increase in human energy use and this meant "industrial societies used four or five times as much energy as their agrarian predecessors, who in turn used three or four times as much as our hunting and gathering forebears" (ibid., 848). Increases in energy use is matched with population growth from about one billion to six billion between 1800 and 2000; surface land use increases from about 10 to about 25–30%; and rise in greenhouse gases (ibid., 848). Thus, "the advent of the Industrial Revolution around 1800 provides a logical start date for the new epoch" (Steffen et al., 2011, 842).

Whereas energy use increases would have contributed to greenhouse emissions, the reverse effect that resulted in different "spike" is observed in 1610 and marks another proposed start of the Anthropocene, 150 years earlier to the Industrial Revolution. This happened between 1570 and 1620, during which Lewis and Maslin (2015) note a decline in atmospheric CO_2 concentrations. The drop was fostered, in part, from impacts from what they described as the meeting of the "Old and New World human populations" and is preceded by a "suite of changes," including global trade, coal use and European expansion into the Americas, which resulted in "catastrophic decline in human numbers" (ibid., 175–176), to the tune of some 50 million indigenous people between 1492 and 1650 due to "exposure to diseases carried by Europeans, plus war, enslavement and famine" (ibid., 175). This widespread loss of human life contributed to re-growth of abandoned agricultural lands and induced two related outcomes with significant atmospheric impacts. First, near-cessation of farming and reduction in the use of fire; and second, regeneration of over 50 million hectares of forest (ibid., 175). These contributed to the "Orbis spike"—a noticeable decline, within a half century, in atmospheric CO_2 concentrations between 7–10 ppm (ibid., 175). These events contributed to a "swift, ongoing, radical reorganization of life on Earth without geological precedent" (ibid., 174).

The Early Anthropocene hypothesis marks yet another reverse in the timeline used in debates to justify the beginning of the Anthropocene. This later date reverses the emission decrease that characterized the previous proposed date and is marked by an "anomalous increase 8000 years ago" (Ruddiman 2003, 261) of CO_2, along with another greenhouse gas. During this period, between 8000 BP and 1800 AD, there was also increase of CH_4 5000 years ago (ibid., 2003, 261). Further support for the "Early Anthropocene" date is based on "paleoclimatic evidence" produced by "early agriculture in Eurasia….including the start of forest clearance by 8000 years ago and of rice irrigation by 5000 years ago" (ibid., 2003, 261). Technological and social innovations at the time enabled "deforestation by humans" that was helped by "innovations in agriculture" (ibid., 273). In addition, megafauna extinction during the Pleistocene and human agency therein is cited in support of an Early Anthropocene start date. Evidence is collected from 28 sites on the Australian Continent (Roberts, 2001, 1888); and in North America from "random hunting, and low maximum hunting" (Alroy, 2001, 1893). Comparable dates to the "Early Anthropocene" hypothesis include the "Anthropocene soil" by (Certini and Scalenghe, 2011), which suggests using widespread changes in the pedosphere is "the best indicator of the rise to dominance of human impacts on the total environment" (ibid., 1269).

The paleoanthropocene and Fire period is perhaps the oldest proposed start date for the Anthropocene. Starting from a "minimum age of > 1.8 million years (Ma) ago" (Glikson, 2013, 91), it encompasses the Early Anthropocene, the "Middle Anthropocene" when extensive grain farming developed [and] late Anthropocene "with the onset of combustion of fossil fuels" (Ibid., 2013, 91). The paleoanthropocene is "a transitional period, which is not easily fixed in time" (Foley et al., 2013, 84). It is characterized by "the mastery of fire" (Ibid., 2013, 89) which facilitated mass megafauna extinction and burning of forests. The former outcome resulted in extinction of mammoths, albeit through hunting, and is suggested as "the first human-induced global warming" (Doughty et al., 2010), leading to a conclusion that "the onset of the Anthropocene should be extended back many thousand years" (ibid., 2010, 5). The claim is furthered by the suggestion that "anthropogenic fire has been a factor in shaping plant communities through human prehistory" (Pinter et al., 2011, 269). Marlon et al., (2009, 2519) suggest human-induced abrupt climatic changes started during the last glacial-interglacial transition, between 15 and 12 thousand years before the present, and noting as evidence the "clear links between large climate changes and fire activity" (ibid.). Other evidence from humans' use of fire includes, "rapid decline of rainforest gymnosperms" fauna and so-called megaherbivores extinctions, along with "devegetation" (Pinter et al., 2011 269); and "lack of lightning ignitions for most of the eastern US" (Abrams and Nowacki, 2015, 44). As well, using global "charcoal records" (Marlon et al., 2013, 9) show that "novel anthropogenic sources of ignition," used by small groups such as hunter gatherers, transformed landscapes and ecosystems (Pinter et al., 2011, 270).

Political Implications of Various Start Dates for the Anthropocene

Climate change perhaps offers the most widely publicized example of how human activities have transformed planet Earth and its atmosphere in the Anthropocene. However, next to the Industrial Revolution, other possible start dates have been suggested, each with its own implications. We discuss three possible start dates in this section, mainly to illustrate their diverging political implications.

The Industrial Revolution narrative is essentially one of fossil fuel extraction. Consequently, the agents in this story are the early industrializing countries in the European heartland and the implications are largely synonymous with sources of emissions that have led to climate change. As the consequences of climate change are feared by societies around the globe, and are becoming increasingly visible, the Industrial Revolution start date arguably entails the strongest call for immediate action to halt unsustainable developments approaches which started in England during the Industrial Revolution. Given that impacts from this period cannot reasonably be divorced from climate change, they invariably introduce notions of climate justice and (in)equity, and thus contribute to existing challenging transnational environmental governance conditions. Governance in an Industrial Revolution Anthropocene would probably evolve around drastic, sustained and system-wide mitigation and adaptation measures that should have transformative impacts, and normatively around ideas of responsibility and compensation (with the 2015 Paris Agreement as a blueprint).

The "Orbis hypothesis" (Maslin, 2015) is promising from a social sciences perspective, as the observed atmospheric changes coincide with the emergence of the capitalist world system in general (Wallerstein, 1974) and the emergence of the plantation system more specifically (Haraway, 2015). The meeting of European and American cultures and the related decline in atmospheric CO_2 concentrations (with the start of CO_2 reduction in 1610 as a possible marker) illustrates a complex and unpredictable outcome of human-nature interactions. Agents in this story are European explorers and by extension, early modern European capitalist dynasties (see also Parenti and Moore, 2016; on Anthropocene or Capitalocene?). Direct outcomes included death and destruction, and a broader systemic impact was the emergence of an unequal global capitalist system that has mainly maintained its structure with defined "winners" and "losers" along the Global North and Global South divide. Subsequently, the "Orbis hypothesis" also introduces questions of global justice and inequity within environmental politics, and projects the Anthropocene consequently, first and foremost, as a social phenomenon. The "Orbis hypothesis" offers a promising opportunity to properly address the challenges of global environmental change. Governance emanating from this interpretation might address global power inequalities, principally between developing and developed regions, much more centrally than more techno-optimist narratives such as the "Great Acceleration." In addition, the differentiated nature of responsibility becomes more visible in this narrative, aligning with postcolonial studies, which also do not imply a collective "we" as agency in the Anthropocene, but emphasize a diverse and differentiated humankind (Dürbeck, 2019).

With the Early Anthropocene (Ruddiman, 2003) narrative, the start of the Anthropocene coincides with the emergence of social, cultural and technical transformations that we often equate with modern civilization: the beginnings of agriculture, sedentary lifestyles and early urban cultures. Surplus energy generated by farming resulted in greater societal division of labour, specializations, arts and culture and the scientific enterprise. Agents of the early Anthropocene are the early advanced civilizations who managed to maintain political and economic control over larger territories for longer time periods. Like the paleoanthropocene, but different from all other suggested dates, the "Early Anthropocene" seems the most positive in its implied balance between impacts and gains. Consequently, governance might be understood as unnecessary, as the Anthropocene is tangent to evolution of human civilization and progress toward higher forms of development, notably social, technological and cultural. This also makes it difficult to define a start date for the proposed Anthropocene.

The Scientific Bodies on the Anthropocene

Units used for dating planet Earth have historically been suggested and decided on by scholars in the geosciences. The IUGS is at the core of the international framework for this process and serves as its ultimate authority. The union has political power within the geosciences to reject or accept scientific findings, such as those assembled by the AWG. The IUGS aims to "unite the global geological community" through scientific studies, promote education, awareness and participation in geoscience; facilitate and encourage "from all parts of the world" interaction and participation among and between scientists, "regardless of race, citizenship, language, political stance or gender" (IUGS, 2016). Tasked with developing "global standards for the fundamental scale for expressing the history of the Earth" (ICS n.d.), the ICS is crucial to this international framework. As a sub-commission within the ICS, the SQS established the AWG with the primary task to critically "examine the status, hierarchical level and definition of the Anthropocene as a potential new formal division of the Geological Time Scale" (AWG, 2009, 1).

The scientific case by the AWG to formally study the Anthropocene for possible adoption by the IUGS and entry into the Geological Time Scale was primarily made in three British publications. The first, *The Anthropocene: A New Epoch of Geological Time?* (Zalasiewicz et al., 2011), was published by the Royal Society of Great Britain in 2011. Following that was *A Stratigraphical Basis for the Anthropocene* (C. N. Waters et al., 2014), published by the Royal Society of London. In 2019, *The Anthropocene as a Geological Time Unit, A Guide to the Scientific Evidence and Current Debate* (Zalasiewicz, Waters, Williams, et al., 2019), was published by Cambridge University Press. It summarized possible evidence for the Anthropocene (AWG, 2014, 4) and formed "the basis for the AWG's recommendations to the ICS" (AWG, 2015, 11).

The AWG took the unusual approach of voting twice on its recommendations. The first was at a group meeting in August 2016 during an International Geological Congress in Cape Town, South Africa. It was agreed that "the Anthropocene possesses geological reality"; that it is best considered at epoch/series level; that it is best defined beginning in the mid-twentieth century with the "Great Acceleration"; that it should be defined by a GSSP ("golden spike"); and for "a formal proposal to forward for consideration, initially, to the Subcommission of Quaternary Stratigraphy" (AWG, 2017, 3). It was recognized that "our recommendations are more divided over the primary signal that should be used to define the Anthropocene" (Ibid., 2017, 3). However, following guidance provided the AWG by its parent bodies—the SQS and ICS (AWG/SQS, n.d.)— the AWG voted again in 2019 "to affirm some of the key questions that were voted on and agreed at the IGC Cape Town meeting in 2016" (AWG/SQS, n.d.).

Group members were asked to cast votes on two questions: "should the Anthropocene be treated as a formal chrono-stratigraphic unit defined by a GSSP?" (AWG/SQS, n.d.) and "should the primary guide for the base of the Anthropocene be one of the stratigraphic signals around the mid-twentieth century of the Common Era?" (AWG/SQS, n.d.). The results of the binding vote on both questions were consistently the same; 29 members voted in favor and 4 against (AWG/SQS, n.d.). The ballot had a response rate of 97%, meaning the required 60% threshold was reached. Thus, the Anthropocene concept was approved for further study, paving the way for its formal adoption by the IUGS. However, lingering disagreements remain within and outside the AWG, including on key scientific questions and procedural issues. Further, there remains dispute, even within the group, concerning the start of the Anthropocene. For example, the "Great Acceleration" hypothesis was challenged during the Cape Town meeting by an AWG member who provided "evidence of a pre-Industrial Revolution metal smelting signal" (AWG, 2017, 13), suggesting an earlier date to the start of the Anthropocene. This date, "at around either 3000 BP or 2000 BP" (Wagreich and Draganits, 2018, 15) supports the "Early Anthropocene" hypothesis and is argued to conform "to standard stratigraphical procedures" (ibid., 2018). The AWG is also divided over whether to accept the globally distributed signature of the nuclear "bomb spike" of the 1950s and early 1960s as the primary marker of the Anthropocene (AWG, 2017, 3). In its 2018 report to the IUGS, the ICS mentioned a marine site in China and "meromictic varved deposits" in Crawford Lake, Southern Ontario, in Canada, as two possible GSSP sites to be used as markers for dating the Anthropocene (ICS, 2018). According to ICS rules, the AWG exceeded its initial term limit, having been operational for more than 8 years (ICS, 2017). Subsequently, the group must be "dissolved and then reconvened" (ICS, 2017).

Since its inception and until now, the AWG continues to acknowledge interests in the Anthropocene concept beyond the geoscientific community. However, the scientific work by the group does not reflect interdisciplinary research on the concept. Even after reconvening, the AWG maintained a geoscientific focus. During 2020, it sought only new members who have "specialisms not currently covered by the group [and] that are necessary for analyzing proxy markers or links to potential GSSP sites" (AWG, 2019). During 2020, the group was also focused on consolidating "work on potential GSSP site" and "analysing and articulating the utility of the Anthropocene as a formal part of the International Chronostratigraphic Chart" (AWG, 2019).

Who Speaks and Who Decides on the Anthropocene? A Closer Look at the AWG

Origin: The AWG's geographical reach in its membership has primarily been limited to the Global North. Starting at the group's inception in 2009, 50% of members were from the UK. Around 2015, more than 30% of AWG members were still from the UK. The United States contributes the second largest group. Comparatively, representation from the Global South between 2009 and 2020 was disproportionately low. Whereas the UK had eight members at the start, the continent of Africa had only two. There was a member each from Kenya, South Africa, China and Brazil. Despite overall increases, membership of the AWG did not improve to reflect the "worldwide representation" (AWG, 2009, 1) in its composition. In more than a decade, only one new member from the Global South was added to the group in 2020, when a scholar from China joined. By this time, only one of the two scholars from Africa was still with the AWG, as well earlier, a founding member from Brazil left in 2012. In 2020, a scholar from China and another from Africa are the only two from the Global South among the group's total of 42 members, most of whom are from the UK and US. The continent of Africa, where all "modern human species" are said to have evolved some 200,000 years ago into Eurasia, (Stewart and Stringer, 2012, 1317–1318; Roberts, 2014) currently has a single member from Kenya.

Leadership: This imbalance in geographic representation is extended to the AWG leadership. As early as 2009, members were informed that the role of the chair and secretary, constituting the group's leadership, "will rotate among the Working Group over time" (AWG, 2009, 3). But this rotation, since 2009 until 2020, has been done only among male group members from the UK. Also, there have been three AWG secretaries during this period, all from the UK.

The extent to which this low level of "interaction and participation" between scientists from all over the world has impacted the scope and depth the scientific evidence assembled by the AWG may never be fully known. However, a unique opportunity for advancing IUGS objectives, including to "unite the global geological community from all parts of the world" through the AWG is, thus far, lost and rather unfortunately on a topic that is acknowledged within the group to have appeal beyond specialized communities. Failure by the AWG to properly reflect a global scientific community does not result from a lack of geographic representation within the IUGS. In fact, the IUGS boasts worldwide "adhering organizations" which are "national committees for geology or a national academy of sciences, designated by the appropriate authority to represent in the Union the geological scientists of a country or of a geographically defined area" (IUGS, 2012, 10). There are 122 such committees, including for example, Vietnam, Lebanon and India (Ibid., 2012, 10).

Gender: In principle, the IUGS discourages gender inequity among members in its scientific groups. However, in practice, within the AWG, the ambition of the IUGS to foster interaction between scientists irrespective of gender was not advanced throughout the AWG's history, during which only nine female scientists joined the group. It was only in 2013, after the group was formed including just one female scholar from South Africa who left in 2015 that three others joined. Three more joined the next year, and one in 2020. Domination of male scholars in the AWG's membership continues up to 2020, reflected by 36 male compared to six female scholars. Whereas the IUGS professes gender neutrality, this objective is not reflected in the AWG membership. Between 2019 and 2020, no female scholar ever served in any of the leadership positions at the AWG. The AWG, contrary to stated aims of its parent body the IUGS, maintains and reflects "traditionally male-dominated" (Thornbush, 2016, 7) inequity in the geosciences.

Discipline: The fact that the factors that led to the emergence of the Anthropocene differ from those of the Holocene in substantive ways is not one that registers well in the AWG. This is even though the group recognized early that "the Anthropocene is still a young concept, with much to develop both in its 'narrow' stratigraphic analysis and as regards its wider relationships with other studies and other communities" (AWG, 2014, 3). Gender inequity within the AWG is matched only by a reduced diversity in the disciplines within the group. The late Crutzen, who won a noble prize in chemistry, and a law scholar, were the only two non-geoscientists at the start of the group's scientific deliberations. A journalist, who left in 2015 had joined after 2009, followed by a historian in 2011. And then one each from anthropology and archaeology, the former left within a year. Finally, another historian joined in 2014, and a law intern in 2018. Including the intern, there have only been 8 non-geoscientists within the AWG between 2009 and 2020. By 2017, when the AWG submitted its summary evidence, there were only five, two historians, two archeologists and a lawyer in the group, the rest were geoscientists. However, the AWG claimed in its summary evidence and recommendations on the Anthropocene that "from the beginning, the AWG represented a broader community than is typical of ICS working groups" (Zalasiewicz et al., 2017, 56). How broad of a community is perhaps debatable, but considering all scientific disciplines, that a number under 10 is represented in the AWG is noteworthy. To date, the AWG is constituted primarily by scholars from the geosciences, in particular geologists, stratigraphers and palaeontologists.

GLOBAL GOVERNANCE CHALLENGES IN THE ANTHROPOCENE

The Anthropocene hypothesis, if formally adopted by the IUGS, will provide further confirmation of global environmental challenges, notably anthropogenic climate change. Compared to the Holocene, ratification of the Anthropocene should indicate the urgency with which global politics, notably environmental politics, should approach addressing sources and causes linked to environmental problems that are growing in tandem with "the expansion of mankind" (Crutzen and Stoermer, 2000, 17).

A central challenge in global environmental governance in particular, and global politics in general, in the Anthropocene is to find solutions for development that are sustainable and that fully recognize interdependent characteristics of human development approaches and direct impacts on the Earth's natural systems. And that also properly account for differences and divisions between political regions, such as the Global South and Global North. Negotiations on climate change solutions, notably the Paris Climate Agreement, provide a preface to the degree of complex governance challenges certain to emerge in the Anthropocene. Virtually all anthropogenic contributions from human activities linked to the Anthropocene are associated with earlier modern human development practices and choices, primarily in the Global North. Overtime, these have produced, perhaps in ways not imagined, a myriad set of complex and linked problems, such as biodiversity loss, fresh water scarcity, nitrogen excess and chemical pollution. Linkages between biodiversity loss and the COVID-19 pandemic (Turney et al., 2020; Roe et al., 2020) and subsequent impacts on the global economy is the latest example of complex interactions between impacts from human activities and the environment. Furthermore, the Anthropocene calls for a careful geopolitical balancing act between states' right to development on the one hand, such as provisioned in articles 1, 2 and 3 of the Declaration on the Right to Development (OHCHR, 1986) and article 22 of the African Charter on Human Rights and Peoples' Right (ACHPR, 1981), and global impetus toward sustainable development on the other, envisioned in the UN's Sustainable Development Goals.

It has been argued that the Anthropocene "emphasizes that all of us are collectively responsible for the future of the world" (Leinfelder, 2014, 1). Such claim fails to consider complex conditions associated with the geologic change over time. Assumption of responsibility in the Anthropocene is a legal (Chiro, 2016) issue; a dilemma of ethics—encompassing "non-human" considerations (Cooke, 2020) and morality, as "values and principles of moral responsibilities toward the environment and its boundaries are crucial in the Anthropocene era" (Mannu, 2020, 101); and an issue that fosters "deeper notions of egalitarian politics" (Burdon, 2020, 92). To imagine effective global governance of complex environmental problems in the Anthropocene, a critical first step may be assumption of responsibility. However, based on current norms in global environmental governance, it may be challenging for agents to accept responsibility for anthropogenic contributions to the Anthropocene. This implies certain duties on certain agents as "the Anthropocene was not created equally; it was made by a specific subset of humans, namely, those on the frontlines of modernization: white, wealthy, rich males of European heritage" (Harrington, 2016, 6). Thus, the North–South geopolitical divide in relation to responsibility cannot be ignored in the Anthropocene, regardless of whether a clear date for the proposed epoch is determinable, or whether certain impacts on the Earth System such as biodiversity loss, can be attributed to one or another agent. Confirmation of the epoch thus "provides an opportunity to raise questions regarding the regional differences, social inequities, and uneven capacities and drivers of global social–environmental changes" (IPCC, 2018, 54).

Legal principles of fairness and equity (Pittel and Rübbelke, 2013) are central normative cornerstones of global politics, particularly in the relations between developed and developing regions and in global environmental governance. This has led to conceptualizations such as "climate change justice" (Page, 2013, 231). When these issues are raised during global environmental negotiations, it is usually to address questions on whether it is fair and equitable for developing regions to assume associated costs of climate change, such as for climate mitigation and climate adaptation (Hall and Persson, 2018). Or, whether it is fair that regions that contributed far less to climate change are likely to suffer more from its impacts (Toulmin, 2009). Or, on how to establish and enforce global emission budgets and the associated equitable allocation of emission rights among states (Page, 2013, 232).

As we noted above, certain agents have greater responsibility than others for the anthropogenic contributions that have impacted the Earth System. The development trajectory that produced those contributions cannot be repeated in the Anthropocene. This means that developing regions are to embark on sustainable development approaches. It will be necessary to properly address questions of fairness and equity in relation to constraints that sustainable development approaches and practices impose on developing regions. For effective governance in the Anthropocene, differences in historic contributions, generally between the Global North and the Global South, should be recognized and accounted. For example, the US greenhouse gas emissions between 1850–2010 amounted to 18.6% of the total, compared to 5.7% for the continent of Africa, excluding South Africa and Nigeria (den Elzen et al., 2013).

Ethical and legal norms employed within a "holocene mindset" do not fully reflect realities of the emerging epoch. As the Anthropocene is not reversible, concepts of fairness and equity should become permanent and fundamental principles in global environmental governance. Considering economic and development differentiation between states, equity and fairness are essential for states' collaboration and proper environmental policy coordination in the Anthropocene.

Conclusions

In this contribution, we have suggested to explore the meaning of the Anthropocene beyond its more narrow technical-geological implications, and focus on broader socio-political formations in the Anthropocene. Three important topics have been identified that call for further scrutiny by the academic community in the years to come.

First, current seemingly technical debates about the appropriate start date of the Anthropocene (related to the science-internal process of officially declaring the Anthropocene as a new geological epoch) carry deeper political implications via the underlying narratives that "justify" a particular start date. For example, whereas the "Orbis" hypothesis supports political interventions that might address global power inequalities, the "Early Anthropocene" might support political narratives in which governance might be understood as unnecessary, as the Anthropocene is tangent to evolution of human civilization and progress toward higher forms of development, notably social, technological and cultural. Future research could investigate in more detail how these foundational narratives on the Anthropocene become policy-relevant.

Second, the scientific bodies that are deciding on the Anthropocene are not neutral, but carry their own biases. We have in particular discusses bias in terms of geographic origin, gender and disciplinary background of members of the AWG, but deeper unconscious biases might exist that are not as easy to detect. Future research could study these biases in more detail, contributing to a more inclusive and participatory debate about what the Anthropocene is and how it might be recognized.

Third, the Anthropocene calls for new governance approaches that take into account three specific challenges: the need to balance development with planetary limitations; the question of responsibility; and the related normative considerations of fairness and equity. While there is no one-size-fits-all approach, governance that aims at being inclusive and participatory, taking into account the unique experiences of various stakeholders, seems most appropriate for navigating socio-ecological co-evolution within planetary boundaries. Future research could scrutinize the conditions for such participatory and inclusive governance arrangements to develop and thrive.

Annex A: Key Concepts and Definitions

Some key concepts the reader will come across in this chapter are listed and defined in this annex.

Global Boundary Stratotype Section and Points (GSSP): This is a physical location that must exist in a "specific geographical location" (Cowie et al., 1986, 5). A GSSP marks the lower boundary points of chronostratigraphical divisions of planet Earth. For example, between the present Holocene Epoch and the start of the proposed Anthropocene. Six requirements must be met for a GSSP to be accepted by the ICS and subsequently ratified by the IUGS. In the words of Cowie and colleagues (1986, 7): "One of the main aims of the boundary stratotype procedure of lCS is to attain a common language of stratigraphy that will serve geologists worldwide and avoid the dissipation of energy in petty argument and unproductive controversy."

Global Standard Stratigraphic Age (GSSA): Like the GSSP, this is a global chronological benchmark used for dating planet Earth. But unlike the GSSP, wherein physical locations are used, the GSSA is used to denote "boundaries defined by an agreed age" (SQS, n.d.). A consensus on the most suitable method has not been reached to date within the geoscientific community (ICS, n.d.; Walsh, Gradstein and Ogg, 2004).

Golden Spike: A "Golden Spike" is metal used to mark a "unique place where a specified point in time is indicated" (ICS, n.d.). The spike is used at GSSP locations to denote physical separation between two points, or boundaries, in the history of planet Earth. Once a GSSP is identified, "the golden spike, which represents a point in the rock section and an instant in geological time, is then driven into the section" (Harper, 2019, 25).

Source ProGeo

Geological Time Scale (GTS): This is a chart used for classifying and chronologically dating planet Earth. The scale shows the geological history of the planet since formation to the present Holocene Epoch. It indicates, for example, the duration of the Jurassic Period, noted between 199.6 to 145.5 million years (Ma). It also shows divisions, such as between the Triassic-Jurassic boundary, dated at about 201.3 Ma (Ogg, Gradstein and Hilgen, 2012, 182).

International Chronographic Chart: This document comprises Earth's history from the Precambrian to the current Phanerozoic, and shows geologic divisions and subdivisions of the planet, such as Eons, Era, Ages, Periods and Epochs. It provides "numerical ages" for each division, showing upper, middle and lower boundaries. For example, the present Holocene Epoch is within the Quaternary Period and within its upper limits is the Meghalayan Age (Cohen et al., 2013). The chart is produced by the International Commission on Stratigraphy (ICS).

Start Date: This date denotes the start of an Eon, Era, Age, Period or Epoch. It remains a contentious issue concerning the Anthropocene because of the centrality of human activities to the proposed Epoch which suggests "man as geological agent" (Waltham and Forster, 2001, 1). One main argument is that the start of the Anthropocene should coincide with that of humans on our planet. This is simply because the activities responsible for the geological changes that define the Anthropocene concept cannot be logically divorced from humans. Stratigraphic principles and practice, however, with emphasis on geoscientific records do not accommodate this reasoning. The AWG rejects the logic, arguing impacts by earlier human activities were "relatively small" (Zalasiewicz et al., 2017, 9).

International Union of Geological Sciences (IUGS): Founded in Paris, France in 1961, the IUGS is a global scientific organization representing geoscientists and is a member of the International Council of Scientific Unions. The IUGS is concerned with solving geological problems, "especially those of world-wide significance" (IUGS, n.d.). Other functions include, promoting geoscientific research and education and influencing public policy. The IUGS also sets international standards in the geosciences and provides information, such as the International Chronographic Chart. Further, the union fosters "dialogue and communication among the various specialists in earth sciences around the world" (ibid., n.d.). From its headquarters in Beijing, China, the IUGS also serves "as a vital link in solving problems requiring interdisciplinary input from other international scientific unions" (Ibid., n.d.). The International Commission on Stratigraphy (ICS) is one of seven scientific commissions of the IUGS.

International Commission on Stratigraphy (ICS): The ICS is constituted by "expert stratigraphers" for the purpose of "promoting and coordinating long-term international cooperation and establishing and maintaining standards in stratigraphy" (ICS, 2017). A primary task "is to precisely define global units," used to date planet Earth such as "systems, series, and stages, periods, epochs and age" (IUGS, n.d.; ICS, 2017). And are then used to develop the International Geological Time Scale and the International Chronostratigraphic Chart. Functions of the ICS are primarily carried out by 17 sub-commissions, "each responsible for a specific period of geological time" (ICS, 2017) and for which they standardize stratigraphic units, document and communicate major stratigraphic data and promote international stratigraphic cooperation. Sub-commissions are managed by a "chair, a secretary and one or two vice-chairs" (ICS, 2017). The Quaternary Stratigraphy Sub-commission (SQS) is a sub-commission of the ICS.

Quaternary Stratigraphy Sub-commission (SQS): The Quaternary represents a period, ~ 2.6 Ma, or 2.6 million years, in the history of planet Earth, and includes the Holocene and Pleistocene Epochs (Gibbard and Pillans, 2012, 1). Sub-commission members are focused on studying this period with a main objective of establishing "a standard, globally-applicable stratigraphical scale" (SQS, n.d.). This is done through promoting and coordinating co-operation and integration across the world and "by applying the statuary scientific goals of the ICS to the Quaternary" (SQS, n.d.). This sub-commission produces the Global chronostratigraphical correlation table and geologic time scale. Scientific functions of the SQS are carried out by Four Working Groups, one of which is the Anthropocene Working Group (AWG).

The Anthropocene Working Group (AWG): The group was established to scientifically assess the proposed Anthropocene hypothesis. The group presented and held its first of two ballots on its preliminary findings and recommendations during a meeting in Cape Town, South Africa. The stratigraphic evidence assembled on the Anthropocene by the group is the first formal step that may contribute toward the IUGS formally adopting the Anthropocene as a geological unit. This will replace the Holocene Epoch, said to be "the most recent interval of Earth history and includes the present day" (Gibbard and Pillans, 2012, 1000). That decision means the Anthropocene will be entered into the Geological Time Scale.

REFERENCES

Abrams, Marc D., and Gregory J. Nowacki. 2015. "Exploring the Early Anthropocene Burning Hypothesis and Climate-Fire Anomalies for the Eastern U.S." *Journal of Sustainable Forestry* 34 (1–2): 30–48. https://doi.org/10.1080/10549811.2014.973605.

"African Commission on Human and Peoples' Rights." 1981. Text. African Commission on Human and Peoples' Rights. 1981. https://www.achpr.org/legalinstruments/detail?id=49.

Alroy, J. 2001. "A Multispecies Overkill Simulation of the End-Pleistocene Megafaunal Mass Extinction." *Science* 292 (5523): 1893–96. https://doi.org/10.1126/science.1059342.

Anthropocene Working Group. 2009. "Newsletter of the Anthropocene Working Group Newsletter 1." Anthropocene Working Group of the Subcommission on Quaternary Stratigraphy (International Commission on Stratigraphy). http://quaternary.stratigraphy.org/wp-content/uploads/2018/08/Anthropocene-Working-Group-Newsletter-No1-2009.pdf.

———. 2014. "Newsletter of the Anthropocene Working Group Vol. 5: Report of Activities 2013–2014." Anthropocene Working Group of the Subcommission on Quaternary Stratigraphy (International Commission on Stratigraphy). http://quaternary.stratigraphy.org/wp-content/uploads/2018/08/Anthropocene-Working-Group-Newsletter-Vol-5.pdf.

———. 2015. "Newsletter of the Anthropocene Working Group Volume 6: Report of Activities 2014–2015." Anthropocene Working Group of the Subcommission on Quaternary Stratigraphy (International Commission on Stratigraphy). http://quaternary.stratigraphy.org/wp-content/uploads/2018/08/Anthropocene-Working-Group-Newsletter-Vol-6-release.pdf.

———. 2017. "Newsletter of the Anthropocene Working Group Volume 7: Report of Activities 2016–2017." Anthropocene Working Group of the Subcommission on Quaternary Stratigraphy (International Commission on Stratigraphy). http://quaternary.stratigraphy.org/wp-content/uploads/2018/08/Anthropocene-Working-Group-Newsletter-Vol-7-release.pdf.

———. 2019. "Volume 9: Report of activities 2019." Anthropocene Working Group of the Subcommission on Quaternary Stratigraphy (International Commission on Stratigraphy). http://quaternary.stratigraphy.org/wp-content/uploads/2020/09/Anthropocene-Working-Group-Newsletter-Vol-9-final.pdf.

Balter, M. 2013. "Archaeologists Say the "Anthropocene" Is Here--But It Began Long Ago." *Science* 340 (6130): 261–62. https://doi.org/10.1126/science.340.6130.261.

Burdon, Peter. 2020. "Rethinking Global Ethics in the Anthropocene." In *The Crisis in Global Ethics and the Future of Global Governance: Fulfilling the Promise of the Earth Charter*, edited by Peter Burdon, Klaus Bosselmann, and Kirsten Engel. Cheltenham, UK; Northampton: Edward Elgar Publishing.

Canfield, Donald E., Alexander N. Glazer, and Paul G. Falkowski. 2010. "The Evolution and Future of Earth's Nitrogen Cycle." *Science* 330 (6001): 192–96. https://doi.org/10.1126/science.1186120.

Certini, G., and R. Scalenghe. 2015. "Holocene as Anthropocene." *Science* 349 (6245): 246. https://doi.org/10.1126/science.349.6245.246-a.

Certini, Giacomo, and Riccardo Scalenghe. 2011. "Anthropogenic Soils Are the Golden Spikes for the Anthropocene." *The Holocene* 21 (8): 1269–74. https://doi.org/10.1177/0959683611408454.
Cohen, K.M., S.C. Finney, P.L. Gibbard, and J.-X. Fan. 2013. "The ICS International Chronostratigraphic Chart." *Episodes* 36 (3): 199–204. https://doi.org/10.18814/epiiugs/2013/v36i3/002.
Cowie, J.W., W. Ziegler, A.J. Boucot, M.G. Bassett, and J. Remane. 1986. "Guidelines and Statutes of the International Commission on Stratigraphy (ICS)."
Cooke, Maeve. 2020. "Ethics and Politics in the Anthropocene." *Philosophy & Social Criticism* 46 (10): 1167–81. https://doi.org/10.1177/0191453720903491.
Crutzen, Paul J. 2002. "Geology of Mankind." *Nature* 415 (6867): 23. https://doi.org/10.1038/415023a.
Crutzen, Paul J., and Eugene F. Stoermer. 2000. "Global Change Newsletters No. 41-59 – IGBP." Text. May 2000. http://www.igbp.net/publications/globalchangemagazine/globalchangemagazine/globalchangenewslettersno4159.5.5831d9ad13275d51c098000309.html.
Di Chiro, Giovanna. 2016. "Environmental Justice and the Anthropocene Meme." Edited by Teena Gabrielson, Cheryl Hall, John M. Meyer, and David Schlosberg. Vol. 1. Oxford University Press. https://doi.org/10.1093/oxfordhb/9780199685271.013.18.
Doughty, Christopher E., Adam Wolf, and Christopher B. Field. 2010. "Biophysical Feedbacks between the Pleistocene Megafauna Extinction and Climate: The First Human-Induced Global Warming?: Biophysical Feedbacks of Extinctions." *Geophysical Research Letters* 37 (15): n/a-n/a. https://doi.org/10.1029/2010GL043985.
Dürbeck, Gabriele. 2019. "Narratives of the Anthropocene From the Perspective of Postcolonial Ecocriticism and Environmental Humanities." In *Postcolonialism Cross-Examined: Multidirectional Perspectives on Imperial and Colonial Pasts and the Neocolonial Present*, edited by Monika Albrecht, 1st ed. First edition. | New York, NY: Routledge, 2019. https://doi.org/10.4324/9780367222543.
Elzen, Michel G. J. den, Jos G. J. Olivier, Niklas Höhne, and Greet Janssens-Maenhout. 2013. "Countries' Contributions to Climate Change: Effect of Accounting for All Greenhouse Gases, Recent Trends, Basic Needs and Technological Progress." *Climatic Change* 121 (2): 397–412. https://doi.org/10.1007/s10584-013-0865-6.
Fischer-Kowalski, Marina, Fridolin Krausmann, and Irene Pallua. 2014. "A Sociometabolic Reading of the Anthropocene: Modes of Subsistence, Population Size and Human Impact on Earth." *The Anthropocene Review* 1 (1): 8–33. https://doi.org/10.1177/2053019613518033.
Foley, Stephen F., DetlefGronenborn, Meinrat O. Andreae, Joachim W. Kadereit, Jan Esper, Denis Scholz, Ulrich Pöschl, et al. 2013. "The Paleoanthropocene – The Beginnings of Anthropogenic Environmental Change." *Anthropocene* 3 (November): 83–88. https://doi.org/10.1016/j.ancene.2013.11.002.
Glikson, Andrew. 2013. "Fire and Human Evolution: The Deep-Time Blueprints of the Anthropocene." *Anthropocene* 3 (November): 89–92. https://doi.org/10.1016/j.ancene.2014.02.002.
Gradstein, Felix M., James G. Ogg, and Frits J. Hilgen. 2012. "On The Geologic Time Scale." *Newsletters on Stratigraphy* 45 (2): 171–88. https://doi.org/10.1127/0078-0421/2012/0020.

Hall, Nina, and Åsa Persson. 2018. "Global Climate Adaptation Governance: Why Is It Not Legally Binding?" *European Journal of International Relations* 24 (3): 540–66. https://doi.org/10.1177/1354066117725157.

Harper, David A.T. 2019. "The Golden Spike Still Glitters: The (Re)Construction of a Global Chronostratigraphy." *Acta Geologica Sinica - English Edition* 93 (S3): 24–27. https://doi.org/10.1111/1755-6724.14234.

Harrington, Cameron. 2016. "The Ends of the World: International Relations and the Anthropocene." *Millennium: Journal of International Studies* 44 (3): 478–98. https://doi.org/10.1177/0305829816638745.

Haraway, Donna. 2015. "Anthropocene, Capitalocene, Plantationocene, Chthulucene: Making Kin." *Environmental Humanities* 6 (1): 159–65. https://doi.org/10.1215/22011919-3615934.

"ICS - International Commission on Stratigraphy - Statutes." 2017. ICS - Statutes of ICS. 2017. http://www.stratigraphy.org/index.php/ics-statutesofics.

"ICS - International Commission on Stratigraphy - Annual Report." 2018. ICS - Annual Report 2018. https://stratigraphy.org/files/ICS_AnnReport2018.pdf.

"IGBP (a) - International Geoshpere-Biosphere Programme." (n.d.). "Earth system definitions." http://www.igbp.net/globalchange/earthsystemdefinitions.4.d8b4c3c12bf3be638a80001040.html.

"IGBP (b) - International Geoshpere-Biosphere Programme." (n.d.). "Earth as a complex system." http://www.igbp.net/globalchange/earthacomplexsystem.4.1b8ae20512db692f2a680001681.html.

"IUGS - Statutes and Bylaws of the International Union of Geological Sciences." 2016. http://Iugs.Org/Index.Php?Page=statutes-Bylaws. 2016. http://iugs.org/.

"IUGS – International Union of Geological Sciences Book of Facts 1961 – 2011." 2012. https://www.researchgate.net/profile/Alberto-Riccardi/publication/280041640_IUGS_Book_of_Facts_1961_-_2011/links/55a5127a08aef604aa041dfa/IUGS-Book-of-Facts-1961-2011.pdf

"IPCC (Intergovernmental Panel on Climate Change)." 2018. *Global Warming of 1.50 C. An IPCC Special Report on the Impacts of Global Warming of 1.50 C Above Pre-Industrial Levels and Related Global Greenhouse Gas Emissions Pathways, in the Context of Strengthening the Global Response to the Threat of Climate Change, Sustainable Development, and Efforts to Eradicate Poverty,* SR15.

Leinfelder, Reinhold. 2014. "Assuming Responsibility for the Anthropocene: Challenges and Opportunities in Education." Application/pdf, February, 21 Pages, 423.04KB. https://doi.org/10.5282/RCC/6211.

Lewis, Simon L., and Mark A. Maslin. 2015. "Defining the Anthropocene." *Nature* 519 (7542): 171–80. https://doi.org/10.1038/nature14258.

Ludwig, C., and W. Steffen. 2018. "The 1950s as the Beginning of the Anthropocene." In *Encyclopedia of the Anthropocene*, 45–56. Elsevier. https://doi.org/10.1016/B978-0-12-809665-9.09940-7.

Marlon, Jennifer R., Patrick J. Bartlein, Anne-Laure Daniau, Sandy P. Harrison, Shira Y. Maezumi, Mitchell J. Power, Willy Tinner, and Boris Vannière. 2013. "Global Biomass Burning: A Synthesis and Review of Holocene Paleofire Records and Their Controls." *Quaternary Science Reviews* 65 (April): 5–25. https://doi.org/10.1016/j.quascirev.2012.11.029.

Marlon, J. R., P. J. Bartlein, M. K. Walsh, S. P. Harrison, K. J. Brown, M. E. Edwards, P. E. Higuera, et al. 2009. "Wildfire Responses to Abrupt Climate Change in North America." *Proceedings of the National Academy of Sciences* 106 (8): 2519–24. https://doi.org/10.1073/pnas.0808212106.

Mannu, Giuliana. 2020. "Ethic and Responsibility in the Anthropocene Era." In *Environment, Social Justice, and the Media in the Age of Anthropocene*, edited by Elizabeth G. Dobbins, Luigi Manca, and Maria Lucia Piga. Environment and Society. Lanham: Lexington Books.

"OHCHR" – Office of the High Commissioner for Human Rights "Declaration on the Right to Development". 1986. Text. Office of the United Nations High Commissioner for Human Rights. 1986. https://www.ohchr.org/EN/Professional Interest/Pages/RightToDevelopment.aspx.

Page, Edward. 2013. "Climate Change Justice." In *The Handbook of Global Climate and Environment Policy*, edited by Robert Falkner. Chichester: John Wiley & Sons.

Pattberg, Philipp, and Michael Davies-Venn. 2020. "Dating the Anthropocene: Why Deciding on a Start Date for the Most Recent Geological Epoch Matters." In *The Anthropocenic Turn: The Interplay between Disciplinary and Interdisciplinary Responses to a New Age*, edited by Gabriele Duerbeck and Philip Hupkes. Routledge Interdisciplinary Perspectives on Literature. New York: Routledge.

Pattberg, Philipp H., and Fariborz Zelli, eds. 2016. "Environmental Politics and Governance in the Anthropocene: Institutions and Legitimacy in a Complex World." *Routledge Research in Global Environmental Governance*. London; New York, NY: Routledge.

Parenti, Christian, and Jason W. Moore, eds. 2016. *Anthropocene or Capitalocene? Nature, History, and the Crisis of Capitalism*. Oakland, CA: PM Press.

Pillans, B., and P. Gibbard. 2012. "The Quaternary Period." In *The Geologic Time Scale*, 979–1010. Elsevier. https://doi.org/10.1016/B978-0-444-59425-9.00030-5.

Pinter, Nicholas, Stuart Fiedel, and Jon E. Keeley. 2011. "Fire and Vegetation Shifts in the Americas at the Vanguard of Paleoindian Migration." *Quaternary Science Reviews* 30 (3–4): 269–72. https://doi.org/10.1016/j.quascirev.2010.12.010.

Pittel, Karen, and Dirk Rübbelke. 2013. "International Climate Finance and Its Influence on Fairness and Policy: International Climate Finance." *The World Economy* 36 (4): 419–36. https://doi.org/10.1111/twec.12029.

Roberts, Neil. 2014. *The Holocene: An Environmental History*. Third Edition. Hoboken, NJ: Wiley-Blackwell.

Roberts, R. G. 2001. "New Ages for the Last Australian Megafauna: Continent-Wide Extinction About 46,000 Years Ago." *Science* 292 (5523): 1888–92. https://doi.org/10.1126/science.1060264.

Roe, Dilys, Amy Dickman, Richard Kock, E.J. Milner-Gulland, Elizabeth Rihoy, and Michael 't Sas-Rolfes. 2020. "Beyond Banning Wildlife Trade: COVID-19, Conservation and Development." *World Development* 136 (December): 105121. https://doi.org/10.1016/j.worlddev.2020.105121.

Ruddiman, William F. 2003. "The Anthropogenic Greenhouse Era Began Thousands of Years Ago." *Climatic Change* 61 (3): 261–93. https://doi.org/10.1023/B:CLIM.0000004577.17928.fa.

Reynolds, John R., and Lauren J. Krivo. 1996. "Change in Societal Technology and Development, 1950–1990." *Social Science Research* 25 (2): 95–124. https://doi.org/10.1006/ssre.1996.0005.

Santana, Carlos. 2018. "Waiting for the Anthropocene." *The British Journal for the Philosophy of Science*, March. https://doi.org/10.1093/bjps/axy022.

Steffen, Will, Jacques Grinevald, Paul Crutzen, and John McNeill. 2011. "The Anthropocene: Conceptual and Historical Perspectives." *Philosophical Transactions* of the *Royal Society A: Mathematical, Physical and Engineering Sciences* 369 (1938): 842–67. https://doi.org/10.1098/rsta.2010.0327.

Stewart, J. R., and C. B. Stringer. 2012. "Human Evolution Out of Africa: The Role of Refugia and Climate Change." *Science* 335 (6074): 1317–21. https://doi.org/10.1126/science.1215627.

"SQS - Subcommission on Quaternary Stratigraphy." n.d. Chronostratigraphy. http://quaternary.stratigraphy.org/stratigraphic-guide/chronostratigraphy/.

"SQS - Subcommission on Quaternary Stratigraphy." n.d. About SQS. http://quaternary.stratigraphy.org/.

Thornbush, Mary. 2016. "Introduction to the Special Issue on Gender and Geoethics in the Geosciences." *International Journal of Environmental Research and Public Health* 13 (4): 398. https://doi.org/10.3390/ijerph13040398.

Toulmin, Camilla. 2009. *Climate Change in Africa*. African Arguments. London ; New York: Zed Books in association with International African Institute, Royal African Society, Social Science Research Council: Distributed in the USA exclusively by Palgrave Macmillan.

Turney, Chris, Anne-Gaelle Ausseil, and Linda Broadhurst. 2020. "Urgent Need for an Integrated Policy Framework for Biodiversity Loss and Climate Change." *Nature Ecology & Evolution* 4 (8): 996. https://doi.org/10.1038/s41559-020-1242-2.

Vidas, Davor, ed. 2010. *Law, Technology and Science for Oceans in Globalisation*. Brill | Nijhoff. https://doi.org/10.1163/ej.9789004180406.i-610.

Walsh, Stephen, Felix Gradstein, and Jim Ogg. 2004. "History, Philosophy, and Application of the Global Stratotype Section and Point (GSSP)." *Lethaia* 37 (2): 201–18. https://doi.org/10.1080/00241160410006500.

Waltham, Tony, and Alan Forster. 2001. "Man as Geological Agent." *Geology Today* 15 (6): 217–20. https://doi.org/10.1046/j.1365-2451.1999.00005.x.

Waters, Colin N., Jan A. Zalasiewicz, Mark Williams, Michael A. Ellis, and Andrea M. Snelling. 2014. "A Stratigraphical Basis for the Anthropocene?" *Geological Society, London, Special Publications* 395 (1): 1–21. https://doi.org/10.1144/SP395.18.

Wallerstein, Immanuel. 1974. "Dependence in an Interdependent World: The Limited Possibilities of Transformation within the Capitalist World Economy." *African Studies Review* 17 (1): 1. https://doi.org/10.2307/523574.

Wagreich, Michael, and Erich Draganits. 2018. "Early Mining and Smelting Lead Anomalies in Geological Archives as Potential Stratigraphic Markers for the Base of an Early Anthropocene." *The Anthropocene Review* 5 (2): 177–201. https://doi.org/10.1177/2053019618756682.

Zalasiewicz, Jan, Colin N. Waters, Colin P. Summerhayes, Alexander P. Wolfe, Anthony D. Barnosky, Alejandro Cearreta, Paul Crutzen, et al. 2017. "The Working Group on the Anthropocene: Summary of Evidence and Interim Recommendations." *Anthropocene* 19 (September): 55–60. https://doi.org/10.1016/j.ancene.2017.09.001.

Zalasiewicz, Jan, Colin N. Waters, Mark Williams, and Colin Summerhayes, eds. 2019. *The Anthropocene as a Geological Time Unit: A Guide to the Scientific Evidence and Current Debate*. 1st ed. Cambridge University Press. https://doi.org/10.1017/9781108621359.

Zalasiewicz, Jan, Mark Williams, Alan Haywood, and Michael Ellis. 2011. "The Anthropocene: A New Epoch of Geological Time?" *Philosophical Transactions of the Royal Society A: Mathematical, Physical and Engineering Sciences* 369 (1938): 835–41. https://doi.org/10.1098/rsta.2010.0339.

Foucault's Biopolitics and the Anthropocene: Making Sense of Ecopower

Pierre-Yves Cadalen

Anthropocene and the Defeat of Biopolitics

"Man is Nature becoming conscious of itself," the French geographer Elisée Reclus once wrote (Reclus, 2015). Political perceptions are intimately linked to the configurations which unite history, collective autonomy and the objective conditions of their deployment. At the interface of personal desires and humankind future, politics is particularly central to what have been coined as the *Anthropocene*. Indeed, if human economic activities are responsible for the fast and brutal transformation of our *milieux* (environment), human actions are required to untie this destructive dynamic. Even though the lines of responsibilities follow those of capitalist production (Malm, 2017; Moore, 2017), this centrality of human action remains (Chakrabarty, 2017, 2009).

Politics is thus key to the Anthropocene problems and solutions. If, as it appears sometimes a fatality, contemporary structures of production do not evolve, humankind would be in a paradoxical situation by which collective awareness of the necessity to implement a massive shift on a global and multilevel scale does not meet political, economic and social dynamics at all. As stated by Joel Wainwright and Geoff Mann though, "this may seem paradoxical, but history is replete with illustrations of highly unequal and apparently contradictory social-political orders ruled by elites who remained

P.-Y. Cadalen (✉)
The Paris Institute of Political Studies (Sciences Po), Paris, France
e-mail: pierreyves.cadalen@sciencespo.fr

hegemonic for a considerable duration (typically with violent consequences), despite lacking answers to fundamental problems. As Gramsci, writing between the world wars, once put it: "The crisis consists precisely in the fact that the old is dying and the new cannot be born; in this interregnum a great variety of morbid symptoms appear"" (Wainwright and Mann, 2018, p. 24).

This morbidity, as well as its contrary, are precisely the objects of the following pages. Anthropocene encourages discussions about the fundamental question of our species survival, and beyond our species of the *milieux* (Berque, 2014) surrounding us. Indeed, the "morbid symptoms" at stake here are not only linked to the possible resurgence of fascism or the deepening of authoritarian liberalism (Schmitt and Heller, introduced by Chamayou 2020). They are fundamentally plugged to the issue of survival. Politicizing the issue of survival implicates to refuse the fatalism which would only play the Cassandra as if we were already dead and to look for the crucial reasons why the apparent paradox raised in this introduction works as a power mechanism, reproducing the destructive dynamics which make both climate change and biodiversity loss huge threats in an already quite initiated century (IPCC, 2022; Broswimmer, 2010; IPBES, 2019).

This chapter is a prolongation of Foucault's biopolitics through the lens of the Anthropocene. What have become of biopolitics at the age of the Anthropocene?

First, it is necessary to come back to the intimate link between life and the very concept of biopolitics. Then, we formulate the thesis that biopolitics is twisted by the Anthropocene. This is precisely what leads us to our core concept, ecopower as the new form of power, succeeding and completing Biopower, one point where life politics meet death politics (Thomas, 1978).

LIFE AS THE KEY TO BIOPOLITICS AS A GOVERNMENT: GOVERNING POPULATIONS

Our path must begin with a little *detour* by the prior links and interrelations between Anthropocene and Biopolitics. Biopolitics, as a major and widely discussed concept, is anterior to the Anthropocene. It has been coined and defined by Michel Foucault in his lecture *Sécurité, territoire, population*, (Security, Territory, Population), given at the Collège de France in 1977–1978 (Foucault, 2009).

The Relevance of Biopolitics to the Anthropocene

Why are we interested, in such a handbook, in biopolitics? The way Michel Foucault invents and defines this concept is of first interest for environmental theory. Biopolitics is the politics moved by Biopower as a radically new form of power, opposed to sovereign power, and born later as well. Biopower, to put it bluntly, is the power over life.

In the *Will to Knowledge* Michel Foucault already sketches such an opposition between sovereign power and what would become, in his posterior works, Biopower. "The right which was formulated as the 'power of life and death' was in reality the right to take life or let live. Its symbol, after all, was the sword" (Foucault, 1978, p. 136). The evolution is depicted by the philosopher in the direct aftermath of this definition of sovereign power: "One might say that the ancient right to take life or let live was replaced by a power to foster life or disallow it to the point of death. This is perhaps what explains the disqualification of death which marks the recent wane of the rituals that accompanied it. That death is so carefully evaded is linked less to a new anxiety which makes death unbearable for our societies than to the fact that the procedures of power have not ceased to turn away from death" (Foucault, 1978, p. 138).

Biopower, in its early formulation, is described as a power to *foster life*, which implies both a transversal and statistical management of population, and a rejection of death as a non-political procedure, which differs a lot from the sovereign use of power by the King for instance. The sociological mystery of suicide is, in this perspective, a scandal for "a society in which political power had assigned itself the task of administering life" (Foucault, 1978, p. 139). The object of this Biopower is life itself, which is the core argument present in Foucault's works: life, in all its dimensions, whether it serves public interest or manifests private intimacy, life in its great variety becomes the central object of power.

This power over life directly deals with non-human dimensions. Life is not only men and women, but rather what provides them with the conditions of their reproduction. The first lesson of *Sécurité, territoire, population* at the Collège de France is quite enlightening regarding this dimension:

> The sovereign is no longer someone who exercises his power over a territory on the basis of a geographical localization of his political sovereignty. The sovereign deals with a nature, or rather with the perpetual conjunction, the perpetual intrication of a geographical, climatic, and physical milieu with the human species insofar as it has a body and a soul, a physical and a moral existence; and the sovereign will be someone who will have to exercise power at the point of connection where nature, in the sense of physical elements, interferes with nature in the sense of the nature of the human species, at the point of articulation where the milieu becomes the determining factor of nature. This is where the sovereign will have to intervene, and if he wants to change the human species, Moheau says, it will be acting on the milieu. I think we have here one of the axes, one of the fundamental elements in this deployment of mechanisms of security, that is to say not yet the appearance of a notion of milieu, but the appearance of a project, a political technique that will be addressed to the milieu. (Foucault, 2009, p. 23)

The *milieu* is central in the conceptual construction of this new form of power coined as Biopower. Governing life is not only equivalent to the statistic management of populations, as it also gathers very conditions to transform the *milieux* where these populations are meant to live. Indeed, this paragraph might sound quite abstract to the student or auditor listening this lecture the 11th of January 1978. As for a contemporary specialist of environmental matters, it rather sounds as a definition of the Anthropocene. Indeed, this redefinition of sovereignty under a Biopower regime is precisely located at the interface between the changes physical *milieux* cause to societies, and societies to physical *milieux*. Sovereign is who attempts to master this infinite complexity, in "this point of interconnection." It is particularly difficult, as social sciences and even governments, are currently used to separating humans from their *milieux*, to draw all the conclusions implied by this new sovereign.

Let us consider that this definition of sovereignty makes quite obsolete the discussion about whether Anthropocene is a self-human-centered concept or, on the contrary, a concept which would imply a non-human point of departure. It is, necessarily, both at the same time. Such an interpretation of Foucault at the light of contemporary debates seems to fit the invitation formulated by Timothy Morton to adopt an "ecological thought" in which strong and definitive differentiations would no longer work to understand the dynamics our societies are engaged with (Morton, 2010). In this perspective, bodies, and their relation to their *milieux*, are central to the reproduction of power mechanisms.

Reading Foucault this way is an invitation to nuance the conceptual newness of the Anthropocene. Indeed, the position of humans regarding their *milieux*, the relation between economy and political power and the possibility for the people to change one or the other are not new. Paradigms can change, as some dominated paradigm can, for a long historical period, coexist with a dominating one before replacing it (Kuhn, 1996). The emergence of Anthropocene as a paradigm (Beau and Larrère, 2018; Bonneuil and Fressoz, 2013) can be the sign of self-consciousness about the object of sovereignty, this "point of articulation where the milieu becomes the determining factor of nature."

This nature is understood both as human condition and Nature in its physical dimensions: the Anthropocene implies, indeed, one and the other.

History and Duration of the Anthropocene

Biopolitics could then work as a conceptual anticipation, as far as political theory is concerned, of what the Anthropocene implies. This point is even more striking if we consider the historical dimensions of both concepts. As indicators of paradigm changes, they are not neutral about history, even though their relation to history can be neutralized. However, it is clear that dating the beginning of the Anthropocene is an object of intense debates, not only within the geologists' community, but also among the vast social scientists' community (Lewis and Maslin, 2018).

Foucault clearly explained the historical roots of Biopolitics, defined in *Dits et écrits* (Words and Writings), as "the way by which rationalization of problems, addressed to governmental practice by phenomenon characteristic of living beings constituted in population, has been attempted since the eighteenth century: health, hygiene, nationality, longevity, races" (Foucault, 2001, p. 818, my translation). Biopower is then definitely situated as a historical phenomenon. Is it even a concept which basically addresses new historical problems, such as the demographic expense and the unprecedented densification of populations over the nineteenth and twentieth centuries. Biopower offers this opportunity to be both a philosophical and a historical concept, which is aligned to the Deleuzian definition: a concept addresses specific problems raised by specific historical conditions, which is particularly the case of biopolitics according to Gilles Deleuze (Deleuze, 1990; Deleuze and Guattari, 1991). Anthropocene has also such characteristics, as it specifically points out the necessity to both understand and potentially act on the destructive processes which affect our *milieux*.

The recent work by Jean-Baptiste Fressoz and Fabien Locher give an important echo to such a thesis which would unite Foucauldian Biopolitics and the Anthropocene. In their latest book, *Les révoltés du ciel* (The Revolt of the Sky), they basically argue that the very idea of governing the climate—in order to improve conditions of life, harvests and the general welfare of the populations—emerged by the end of the eighteenth century, which is coherent with the possible beginnings of the Anthropocene in the seventeenth or eighteenth centuries (Fressoz and Locher, 2020). Governing climate is a historically rooted idea, and as such perfectly fits the dynamics of Biopolitics as defined by Michel Foucault. As stated by Fressoz and Locher, "men of the eighteenth century live in an Anthropocene: for them, Earth and social temporalities are one.

The human climate action is one modality of this intertwinement, a clue of this time concordance" (Fressoz and Locher, 2020, p. 45). Anthropocene and Biopolitics share the same historical origins, whether they are scientific, economic, political or even religious. Anthropocene politics is Biopolitics.

As for one or the other, some suggest earlier beginnings. Anthropocene could have started with the disappearance of large mammals, and large impacts humans have gained over their environment, which is basically one of the underlying thesis defended by James Scott in his famous *Against the Grain* (Scott, 2017). Similarly, Giorgio Agamben clearly advocates for a temporal extension of Biopolitics whose judicial and political roots can be analyzed from the Antique Roman practice of the *sacred* man who can be killed by anyone after being declared so (Agamben, 1997). However, this temporal extension of both concepts likely leads to their historical dilution. The simultaneity of capitalist expansion, government techniques widely applied to vast populations and the attempt to govern climate dynamics is a sufficient correlation to stick to the historical dimensions suggested by Michel Foucault, Jean-Baptiste Fressoz and Fabien Locher.

Forms of Power Matter

The historiographical debate is far from being the most important one regarding the links between Anthropocene and Biopolitics. What matters most is the theoretical value of Foucault's reading of power dynamics, which implies the creation of new forms of power as well as their possible concomitance. The invention of Biopolitics does not necessarily imply the abolition of sovereign power. There is a crucial importance not to neglect this complementarity between forms of power. Indeed, Biopolitics can work without suppressing the possibility of a direct exercise of sovereign power, as it is depicted in the *Will to Knowledge*. Marc Abélès makes this statement, as the historical dimension of Biopolitics produces the risk of a diachronic analysis of power. Forms of power function, on contrary, in synchrony (Abélès, 2007). He then crosses the path of Gilles Deleuze, who was considering the possibility of a disciplinary society, inherited from one dimension of Biopolitics, to coexist with a control society, both of which were and are fundamental obstacles for human kind to self-determine their collective action and history (Deleuze, 1990).

This last element allows us to conclude clearly on the links between Biopolitics and Anthropocene. Indeed, what matters is that sovereignty as the power to kill one individual, and let the others live, is still a possible horizon of State power, as wars or arbitrary uses of power might regularly recall us. Still, Biopolitics has taken a growing importance as the general management and administration of populations. As it was mentioned in the Introduction, Biopolitics has become, through the nineteenth and twentieth centuries, the dominant paradigm of power, as the power "to foster life or disallow it to the point of death."

"Fostering life" is a key element to the definition of Biopolitics. Biopolitics is not thought or defined as a mainly or uniquely repressive apparatus of power: it is positively producing life, and reproducing life. Let us come back to the first lecture of *Sécurité, territoire, population* at the Collège de France:

> Finally, the milieu appears as a field of intervention in which, instead of affecting individuals as a set of legal subjects capable of voluntary actions—which would be the case of sovereignty—and instead of affecting them as a multiplicity of organisms, of bodies capable of performances—as in discipline—one tries to affect, precisely, a population. I mean a multiplicity of individuals who are and fundamentally and essentially only exist biologically bound to the materiality within which they live. What one tries to reach through this milieu, is precisely the conjunction of a series of events produced by these individuals, populations, and groups, and quasi natural events which occur around them. (Foucault, 2009, p. 21)

The "quasi natural" event is neither fully human, nor fully natural. Overall, it makes us wonder what the conditions would be to reach a concrete possibility to govern this *milieu*. Indeed, even though Michel Foucault is a critical thinker, Biopolitics as a power technique to govern the *milieux* appears to be a concept forged to understand and explain a historical evolution which is basically, on a transversal basis, coined as the Anthropocene today.

Anthropocene, as it has now become an acceleration of climate and *milieux* changes, questions Biopolitics as a valuable technique of power for humankind. Indeed, it appears that the form of power that emerged in order to "foster life," leads us to a situation of destruction and disarray which in return raises a key interrogation on whether a self-determining sovereignty can be reactivated, and on which basis. But let us first examine the twist current and contemporary evolutions due to the Anthropocene imply.

The Horizon of Disappearance and the Twist of Biopolitics

Anthropocene puts at stake the survival of the human species itself. One of the reasons it does so is precisely the uncertainty underlying our belonging to this time of fast climatic and biological changes. We do not know if the future conditions of life will be compatible with human life. But what we do know is that the speed and wide scope of those changes strongly depend on human actions, and particularly our capacity to engage a fast and deep shift of our modes of production and consumption.

The Failure of Biopolitics

Biopolitics, according to the Italian philosopher Roberto Esposito, "has no other object than the maintenance and expansion of life" (quoted by Levinson, 2010, p. 240). Basically, the main problem of Biopolitics in this perspective is its political reductionism. Everything is reduced to the biological dimension of our lives, which is considered as a risk of moral and political wreck of humankind. This political implications of Biopolitics are a clear source of decisive discussions regarding the human *devenir* (future).

Still, Anthropocene being the age of a possible disappearance of humankind's basic conditions of life, life reductionism is not the sole problem anymore, at least not as Roberto Esposito meant. Life annihilation is a horizon of possibility directly connected to the ability or the incapacity of human structures—*i.e.*, human agency in the same movement[1]—to move toward the possibility of preserving humankind from this horizon. In a way, Biopolitics has become much more serious, as a concept pointing out a reality, since it is quite obviously failing.

---

[1] The classic divide between structures and agency is more than arguable indeed, and is before all a mental projection, representation whose main purpose is to facilitate the collective apprehension of the social and political dynamics. Pierre Bourdieu, for instance, was more than skeptical about the interest of such dichotomies. This interview is particularly enlightening regarding this perspective: https://www.franceculture.fr/sociologie/pierre-bourdieu-lhabitus-ce-nest-pas-le-destin-le-fatum.

Indeed, a political regime which "has no other object than the maintenance and expansion of life" and which at the same time is producing unprecedented destruction worldwide, of the very *milieux* that are conditions to human life, is a failure, especially regarding its particular and specific objective. If we consider Biopolitics as regime whose main obsession is life, the ongoing process of *milieux* destruction is particularly not fitting Michel Foucault's definition. This new power emerged out of the Anthropocene and no longer "fosters life and/or disallows it to the point of death."

Just as the pretention to ideally govern the climate failed, Biopolitics as a project to foster life also failed: the concomitance between Anthropocene and Biopolitics is to be found here again. Biopolitics has then an intimate and little discussed relation to Anthropocene: both engage with the possible end of humankind. The post-apocalyptical fictions enter with us in this new age. Even if their function might be, as stated by Jean-Paul Engélibert (Engélibert, 2019), to conjure this possible end, they still indicate a new and massive form of collective preoccupation for the future of humankind.

Biopower does not work as a reassuring power which would guarantee the survival of the species from an obscure bargaining process in which people would have lost their political freedom. The irruption of a new geological age and the simultaneous scientific and political consciousness about the Anthropocene is definitively undermining the conditions of possibilities an eventual Biopolitical society would lie upon. As for the shift from sovereign power to Biopower, the fact this paradigm is being contested does not mean its main manifestations would immediately vanish from our social and political reality—as it is implied by the coexistence of paradigms mentioned earlier. Instead of imagining such a situation, it seems far more interesting to draw the new arrangements of sovereign power and Biopower in the light of these radically new stakes. That is why I offer to discuss a new concept, that of "ecopower."

Ecopolitics and "Ecopower"

We define "Ecopower" as the power to perpetuate life conditions of human species and the vertebrate living, or to produce their destruction. In other words, Ecopower comes as a new form of power which makes the sovereignty-Biopower duo a trio. As Biopolitics seems to have considerably failed in the move to preserve life, a radically new political stake appears as the modifications of our *milieux* accelerate abruptly: our preservation as a species, as well as the preservation of species which rely on the same basic conditions of life as humans do.

Ecopower works dialectically. Indeed, sovereign power is "to take life or let live" individuals who are subjected to the sovereign, while Biopower works as the power "'to foster life or disallow it to the point of death." As contemporary politics are deeply transformed by ecological challenges, it appears that one new form of power has emerged, one which would reconcile both former ones, as Ecopower is the form of power that can "take life or let live" at the scale of the species. This extension of Foucault's works implicate a material dimension, given that this new form of power resides in the possibility to govern the physical dynamics which jeopardize the vertebrate living, those very same dynamics that have been launched from the human-capitalist realm (Angus, 2018; Charbonnier, 2013).

Would it be possible to reorientate those dynamics, such that political and social reality would remodel the logics of power as the sole possibility of preventing the end of human life. This would tend to be a factor of strong political polarization. As depicted by the French philosopher, Pierre-Henri Castel, the radicality of the stake implies new forms of moral determination. As he writes in *Le mal qui vient* (The Coming Evil), if some ruling elites basically bet on the end of humankind as a way that their power and wealth remain untouched, these elites will not arouse political moderation from the populations as a response to their morally and concretely destructive behavior (Castel, 2018).

Ecopower shares one central characteristic with Biopower: it is polycentric. As Michel Foucault assumes in the *Will to Knowledge*, "in political thought and analysis, we still have not cut off the head of the king" (Foucault, 1978, pp. 88–89). Even if Ecopower signals a return to sovereign power logics, basically because it implies the possibility to annihilate life conditions and makes death central again in power dynamics, this concept we propose does not mean we would put the head back on the King's shoulders as far as political theory is concerned. It only means that we cannot consider power to be exclusively transcendent—as criticized by Michel Foucault, as in the conception of an ideally "absolute" monarchy—nor totally immanent—power being produced without any forms of hierarchization or power asymmetry which could be identified. Power production has many centers and the ecological matters imply to catch these centers at the multiple crossroads between politics and economics, as once suggested about the State by the Greek philosopher, Nikos Poulantzas (Poulantzas, 2013).

The "sociological mystery" identified by Michel Foucault as concomitant to the emergence of Biopolitics, was suicide: "It is not surprising that suicide—once a crime, since it was a way to usurp the power of death which the sovereign alone, whether the one here below or the Lord above, had the right to exercise—became, in the course of the nineteenth century, one of the first conducts to enter into the sphere of sociological analysis; it testified to the individual and private right to die, at the border and in the interstices of power that was exercised over life" (Foucault, 1978, pp. 138–139).

This mystery is renewed at the light of Ecopolitics and Ecopower. Indeed, the various trajectories taken by the distinct societies over the globe are, in their vast majority, destructive of our *milieux*. Their economic unity, despite of their cultural and political diversity, is a key element to explain this phenomenon (Wallerstein, 2004). It leaves us, though, with a crucial interrogation: why are we, as a species, taking a direction that could be a close analogue to collective suicide? Of course, the polycentricity and structural dynamics at stake in the Anthropocene prevent the analogy to be strictly applied, from personal suicide to collective self-destruction. However, this power seems to escape both from the sovereign, as there was no explicit decision to jeopardize our species as such, and from Biopower, as it is aiming at the preservation of life. This fundamental interrogation, which can once again be an extension from Foucault's works, invites to the reflection about the links between Ecopower, the possibility of our *devenir (future)* and the Anthropocene. Collective self-destruction is not the only way out for humankind, which is one of the interests implied by the concept of "Ecopower," according to my point of view of this concept and phenomenon.

The Grey Zone of Anthropocene and Ecopower as a Determined Uncertainty

Biopolitics is twisted by the Anthropocene and we ought to analyze this shift in human history. Indeed, our species is coming to the edge of a new space–time agency which involves the short-term possibility of disappearance. The status of human activity, particularly the status of politics, is to become central in the gray zone of our contemporary history. Such a situation leads our thought to reconsider Michel Foucault's statement according to which "one can certainly wager that man would be erased, like a face drawn in the sand at the edge of the sea" (Foucault, 1994, p. 387). Paradoxically, the current ecological situation overthrows this intellectual dynamic analyzed by Foucault, as humanism is partly resuscitated by the very possibility of collapse. Here, the conclusion of Dipesh Chakrabarty's article on climate change is a useful *contrepoint* (counterpoint) to Michel Foucault's thesis:

> Yet climate change poses for us a question of a human collectivity, an us, pointing to a figure of the universal that escapes our capacity to experience the world. It is more like a universal that arises from a shared sense of a catastrophe. It calls for a global approach to politics without the myth of a global identity, for, unlike a Hegelian universal, it cannot subsume particularities. We may provisionally call it a "negative universal history." (Chakrabarty, 2009, 222).

Humanism, defined as the ideological production of human political unity is reborn out of the possible ashes of our civilization. Indeed, this creation of a "negative universal history" is based not on the intellectual possibility of man's vanishment, but rather on the very concrete possibility of humanity being self-erased. One partial conclusion is that humanism as a fully abstract figure is not only on the path to disappear, but that present times urge to build a concrete humanism which would integrate both material conditions of ecological dynamics' deployment, on the one hand, and human particularities on the other (Césaire, 1956). As man can effectively vanish as a fugitive trace on the sand, a concrete humanism seems to emerge at the light of Ecopower relations. Biopolitics reject human autonomy out of the social machinery, as it is not required for the management of populations.

On the contrary, Ecopower urges for human politics and human action to take back the control over the course of history, on the grounds of this new "negative universal history." This is precisely the point to which we are conducted by climate change and the ongoing sixth mass extinction. Our current historical situation strengthens Gunther Anders's thesis which puts a particular emphasis on human responsibility under an atomic age (Anders, 2007). Indeed, history and the relation between politics and time have already deeply changed since the introduction of nuclear weapons as a massive means of human self-destruction. One strong paradox pointed out by Anders in this apocalyptical philosophy is linked to the irreversibility of this new historical condition for humankind. Even if humanity would reach a global agreement on full nuclear disarmament, the technical conditions of possibility of its existence remain, so that this horizon definitely belongs to human history.

Under the Anthropocene, these conditions of uncertainty are even truer (Petit and Guillaume, 2016). What was right about nuclear armament has become true for our economic conditions of production and consumption. However, there is an additional difficulty, which resides in the intersection between political and economic changes and the mitigation of Anthropocene destructive phenomenon: will it be enough to perpetuate human conditions of life? Will it, all things remaining unchanged, effectively threaten human conditions of existence as a species? Ecopower, as the power to perpetuate or to eradicate the human conditions of existence, is a power exerted on a determined uncertainty of history.

The negative universal evoked by Dipesh Chakrabarty can actually be extended to other species that are jeopardized by human activities and by our current mode of production. Capitalism does not only produce strongly unequally distributed wealth, it produces an assured disaster (Angus, 2018). This extension to other species and the living, as undefined this latter category is, is clearly of a strong importance considering the consequences of a destructive Anthropocene which would pursue the path of the erosion of our conditions of life. Still, regarding the responsibilities, socioeconomic human structures are first in chain and first to be changed as far as causes are regarded. In this situation of indetermination, the conceptualization of Ecopower aims at usefully completing the Biopolitics. First, because Biopolitics is indeed overcome and twisted by the current situation. Secondly, because Ecopower points at a fundamental power and political intersection between human societies, our *milieux* and the physical dynamics of transformation affecting and produced by both. In a way, Ecopower as a concept born out of an imaginary dialogue with Foucault aims at shedding some light on the gray zone of indetermination that characterizes, at least for a few decades, the Anthropocene.

Ecopower and Ecopolitics: Perpetuating or Destroying Life

Anthropocene is, as I write, still indeterminate regarding the stakes risen by Ecopower. Much scientific evidence, from the IPCC successive reports to the recent Intergovernmental Science-Policy Platform on Biodiversity and Ecosystemic Services (IPBES) report, shows that the twenty-first century is particularly central regarding the possibilities to adapt or to be wiped out in the middle-run, historically speaking. Some studies receiving notice pay more attention to the positive feedback loops which could accelerate the dynamics of climate change and sixth mass extinction (Motesharrei et al., 2014; Steffen et al., 2018). It is not a scientific impossibility that we would have already ceased to be the actors of this history if tipping points are already reached (Federau, 2017). This statement is not proveable though, nor the most likely, and the possibility to mitigate changes and to adapt to them so that Anthropocene would be compatible with human life is still a thinkable possibility, and as such, belongs to the realm of politics. That is precisely where the gray zone becomes much more determined, as the action taking place in our contemporary history settles this new landscape to come.

Ecopower, Foucault and Polycentricity

It is worth noting Foucault's conception of power is rightly coined as immanent. There is no power transcendence or clear hierarchies which could explain social and political dynamics. Each domain of power has its own power dynamics which are basically working on a horizontal plan, far from any pyramidal conception of power. This conception is particularly clear in the following passage:

> It seems to me that power must be understood in the first instance as the multiplicity of force relations immanent in the sphere in which they operate and which constitute their own organization; as the process which, through ceaseless struggles and confrontations, transforms, strengthens, or reverses them; as the support which these force relations find in one another, thus forming a chain or a system, or on the contrary, the disjunctions and contradictions which isolate them from one another; and lastly, as the strategies in which they take effect, whose general design or institutional crystallization is embodied in the state apparatus, in the formulation of the law, in the various social hegemonies. (Foucault, 1978, pp. 92–93)

This conception of power was basically linked to the will Foucault had to excerpt his works out of the Marxist sphere, which he accused to put too much emphasis on the State as the key to all powers in society. As it is obvious reading Foucault, the State apparatuses only "embody" the "institutional crystallization" of power strategies specific to each sphere of power. In order to understand and complete what we want to introduce here with the concept of

Ecopower, it seems quite worthy debating this approach of power. The existence of multiple spheres of power is undeniable and that is a common ground to Sovereign power, Biopower and Ecopower.

However, there is no certainty about the fact that power logics are "immanent" in these spheres. Indeed, this idea raises two fundamental interrogations. The immanence cancels the analysis by the asymmetry and hierarchy between social forces, on the one hand, and it traps the political analysis as well as political actions in the deceitful perspective of collective action paralysis, on the other hand. In other words, even if it is clear, Ecopolitics is determined by multiple centers of power, the immanent conception of power might well prevent us to fully understand and accordingly change the social structures which produce a destructive Anthropocene.

This critical approach was brilliantly constructed by Nikos Poulantzas who was easily giving reason to Foucault's critics about orthodox Marxism, but who was considering the philosopher was willingly making the confusion between this approach and Marxism in general. Indeed, the critics formulated against a static conception of power and a narrow conception of the State as its only core cannot erase the fact that States have been granted with further influence and control on society over the last centuries. As stated by Nikos Poulantzas, "What is truly remarkable is the fact that such discourse, which tends to blot out power by dispersing it among tiny molecular vessels, is enjoying great success at a time when the expansion and weight of the State are assuming proportions never seen before" (Poulantzas, 2000, p. 44). The materiality of States is a strong element Pulantzas opposes to Foucault as an invitation to reconsider the very foundations of power in the economic sphere and the relations between social classes and fractions of those classes.

This perspective allows theorists to introduce some asperities in the conception of power, which opens a door by which it is possible to analyze hierarchical and asymmetrical relations of forces as well as the possibilities to change these configurations. In a nutshell, politics still matters for Ecopolitics and the twist Biopolitics has known invites us to consider the concrete power relations that can avoid a full disaster. As stated by Poulantzas again, "with regard to the dominant class and fractions, the role of law in setting limits expresses the relationship of forces within the power bloc. It becomes concrete above all by delimiting the fields of competence and intervention of the various apparatuses, in which different classes and fractions of this bloc have dominance" (Poulantzas, 2000, p. 92). Those "limits" might be related to various sphere of power which is by principle the State synthetic force. Power relations are of primary importance and though we do not claim here that they should be reduced to class relations under the Anthropocene, the material dimension of the latter strongly invites us to reconsider Poulantzas' debate with Foucault as the main dispute was about the material grounds of power.

Ecopower, as the power to reproduce or to annihilate life conditions, is the most material conceivable form of power. This material form of power invites us, indeed, to analyze the definition of limits determined by law as well as the power relations that produce this definition.

Governing Climate and the Anthropocene by the Limits

There could be an anthropocentric risk in evaluating the possibilities for human action at the edge of the Anthropocene era,[2] which is the clear tendency affirmed by the geoengineering which aims at reducing the transversal dimensions of the crisis we experience to a complex of technical problems (Wallenhorst and Theviot, 2020). The face of (hu)man is reactivated by the Anthropocene which might imply a distinct relation to technological evolution and its relation with economic accumulation, a path explored by the French philosopher Jacques Ellul (Ellul, 2004, 1977).

The paradox of establishing by laws the limits that would make Ecopower go in the direction of perpetuating human life, and life beyond humans, holds in the former twist of Biopolitics. As the latter was aiming at preserving populations, Biopower dynamics built the conditions for Ecopower to become the new and central political stake of our century. It is possible to assume the Anthropocene reproduces the classical problem of human actions' unintended consequences. Still, the stake is unprecedently new, and new forces are coming on stage. Climate, once considered the landscape of *longue durée* (long term approach to history) (Braudel, 1987; Chakrabarty, 2009), disconnected from engineers or economic activities, has come to the forefront of human history: climate change is inducted by material consequences of human material actions.

This new element comes to close a quite short parenthesis of human history, if we follow the interesting insight formulated by Jean-Baptiste Fressoz and Fabien Locher:

> If climate change has been and is a shock for consciences, it is because since the beginning of the 20th century, the industrial civilization and science inculcated two comfortable, but false, ideas. On the one hand, that human action would not disturb climate, and on the other hand that wealthy societies had, for most of it, nothing to fear of its perturbations. Our astonishment in front of the existential crisis of warming is mainly linked to these conforting illusions of both an unwavering and harmless climate. (Fressoz and Locher, 2020, p. 227, my translation)

As these illusions are torn apart by social and political reality, and as they were matched to the emergence of Biopolitics, a new reality of power is emerging, which needs a new conceptualization. Therefore, in this paper, Ecopower is both the extension and related to the discussion of Biopower.

---

[2] Even if we consider Anthropocene begun with the conquest of the Americas or the eighteenth century, it is still possible to conceive humankind is "at the edge" of this new geological time, as a few centuries do not count much on those scales.

The Limits, Ecopower and Poulantzas

The Anthropocene does rehabilitate the idea that compromises by law about new limits determined for human action is the core of politics; more precisely, it is the core of Ecopolitics. Ecopower might push toward the logics of human and vertebrate life perpetuation, if the balance of power decides so.

As we find ourselves in a middle of this huge possible shift, Poulantzas' conception of the State as a producer of laws as limitations of actions within multiple spheres of power is particularly relevant. The extension of Foucault's thought thanks to one thinker that engaged in a detailed discussion with his arguments appears singularly fruitful.

Indeed, class relations matter in the formation of the various balances of power of each sphere of power. The social and political compromises which are going to be made in the years to come are the central piece of Ecopower dynamics. Following Castel, about the necessity to form new moral statements on the current choices made by the most powerful ruling groups (Castel, 2018), it seems quite coherent to produce political theory on the consistency and possibility of new class compromises. These new compromises are a central part of Ecopolitics, but they integrate a fundamentally new dimension: the capitalist class not only produces exploitation, but also bears the main responsibility of the destructive Anthropocene. James O'Connor had conceptualized this new reality as "the second contradiction of capitalism" (O'Connor, 1997). His thesis was that capitalism would not resist the inner contradictions the productive system had unchained by exploiting nature.

It, however, seems that capitalism can resist such a contradiction for a time that might be long enough to severely damage the possibility to reproduce human life on Earth which is a central statement for environmental political theory that cannot be simply dismissed (Keucheyan, 2019). This latter element leads us back to the gray zone opened by the Anthropocene. Though this zone is determined by uncertainty, it calls for political decisiveness, and democrats legitimately wish this decisiveness to be self-determined (Wainwright and Mann, 2018). This contemporary time zone might be indefinitely transitory if Ecopower reveals to favor the perpetuation of life, but it might also be an apocalyptical time which is now a rational, though dreary, hypothesis (Mbembe, 2016). For that, human material actions matter. Politics and the environment matter and their conceptual union refers to what one calls "political ecology" (Deneault, 2019).

Ecopower, then, can be rewritten as the form of power from which humankind will decide if our time will be indefinitely transitory, or if it will abruptly end. In these conditions, Ecopower creates a new regime of historicity like the Anthropocene, except it definitely includes the social and political unlike some technical or messianic conceptions of the Anthropocene.

Conclusion

Biopolitics is a power over life. Michel Foucault pays a lot of attention to the shift from Sovereign politics to Biopolitics. As Biopolitics fails to "foster life," and on the contrary leads to an unprecedented centrality of death, the short-run fate of the species being at stake, a similar shift must be operated now. That is what I have advocated for in this chapter, by defining Ecopower as the new form of power, as the power to perpetuate life conditions of human species and the vertebrate living, or to produce their destruction.

This form of power points toward a new regime of historicity, in which environmental and social justice are deeply entrenched with the very issue of survival. Such a situation implicates, for those who are attached to democratic ideals, a strong insight of lucidity to avoid the trap of a permanent state of emergency. Political theory, under the conditions of the Anthropocene, must deal with two questions without further delay: the power of capital over economic and social dynamics and the possible democratization of those very same processes.

If "Man is Nature becoming conscious of itself," to conclude by Reclus' words, one can hope the recognition and influence of Ecopower relations can be achieved before it is too late.

References

Abélès, M., 2007. "Foucault et la pensée anthropologique." *Revue internationale des sciencessociales* 1, 67–75.
Agamben, G., 1997. *Le pouvoir souverain et la vie nue*. Seuil, Paris.
Anders, G., 2007. *Le temps de la fin*. L'Herne, Paris.
Angus, I., 2018. *Face à l'anthropocène: le capitalisme fossile et la crise du système terrestre*.
AR6 Synthesis Report: Climate Change 2022 — IPCC, 2022. https://www.ipcc.ch/report/sixth-assessment-report-cycle/ (accessed 3.7.22).
Beau, R., Larrère, C. (Eds.), 2018. *Penser l'anthropocène*. Presses de Sciences Po, Paris.
Berque, A., 2014. *La mésologie, pourquoi et pour quoi faire?* Presses universitaires de ParisOuest, Nanterre.
Bonneuil, C., Fressoz, J.-B., 2013. *L'événement anthropocène*. Seuil, Paris.
Braudel, F., 1958. "Histoire et Sciences sociales: la longue durée." *Annales. Histoire Sciences Sociales* 13(4), 725–753. https://doi.org/10.3406/ahess.1958.2781.
Braudel, F., 1987. "Histoire et sciences sociales: la longue durée." *Réseaux* 5, 7–37.
Broswimmer, F.J., 2010. *Une brève histoire de l'extinction en masse des espèces*. Agone, Marseille.
Castel, P.-H., 2018. *Le mal qui vient: essai hâtif sur la fin des temps*. Les éditions du Cerf, Paris.
Césaire, A., 1956. *Lettre d'Aimé Césaire à Maurice Thorez*.
Chakrabarty, D., 2009. "The Climate of History: Four Theses." *Critical Inquiry* 35, 197–222.
Chakrabarty, D., 2017. "The Politics of Climate Change Is More Than the Politics of Capitalism." *Theory, Culture & Society* 34, 25–37.

Charbonnier, P., 2013. "Le rendement et le butin. Regard éclogique sur l'histoire du capitalisme." *Actuel Marx* 1, 92–105.
Deleuze, G., 1990. "Post-scriptum sur la société de contrôle." *L'Autre Journal*.
Deleuze, G., Guattari, F., 1991. *Qu'est-ce que la philosophie?*, Collection "Critique." Editionsde Minuit, Paris.
Deneault, A., 2019. *L'économie de la nature*.
Ellul, J., 1977. *Le système technicien, Liberté de l'esprit*. Calmann-Lévy, Paris.
Ellul,J., 2004. *Exegese des nouveaux lieux communs*. La Table Ronde, Paris.
Engélibert, J.-P., 2019. *Fabuler la fin du monde: la puissance critique des fictions d'apocalypse, L'horizon des possibles*. La Découverte, Paris.
Federau, A., 2017. *Pour une philosophie de l'Anthropocène, 1re éd. ed, L'écologie enquestions*. Presses universitaires de France, Paris.
Foucault, M., 1978. *The History of Sexuality. The Will to Knowledge*, 1st American ed. Pantheon Books, New York.
Foucault,M., 1994. *The Order of Things: An Archaeology of the Human Sciences*. Vintage Books, New York.
Foucault, M., 2001. *Dits et écrits 1954–1975, Gallimard. ed, Dits et écrits*. 1954–1988. Paris.
Foucault, M., 2009. *Security, Territory, Population: Lectures at the Collège de France 1977–1978, Lectures at the Collège de France*. Palgrave Macmillan, Basingstoke.
Fressoz, J.-B., Locher, F., 2020. *Les révoltés du ciel: une histoire du changement climatique (XVe-XXe siècle)*.
Keucheyan, R., 2019. *Les besoins artificiels: comment sortir du consumérisme*. Zones, Paris.
Kuhn, T.S., 1996. *The Structure of Scientific Revolutions*, 3rd ed. University of Chicago Press, Chicago.
Levinson, B., 2010. "Biopolitics in Balance: Esposito's Response to Foucault." *CR: The New Centennial Review* 10, 239–261.
Lewis, S.L., Maslin, M.A., 2018. "L'an 1610 de notre ère. Une date géologiquement et historiquement cohérente pour le début de l'Anthropocène." in: *Penser l'Anthropocène*. La Découverte, Paris, pp. 77–95.
Malm, A., 2017. *L'anthropocène contre l'histoire: le réchauffement climatique à l'ère du capital*.
Mbembe,A., 2016. *Politiques de l'inimitié*. La Découverte, Paris.
Moore, J.W., 2017. "La nature dans les limites du capital (et vice versa)." *Actuel Marx* 1, 24–46.
Morton, T., 2010. *The Ecological Thought*. Harvard University Press, Cambridge.
Motesharrei, S., Rivas, J., Kalnay, E., 2014. "Human and Nature Dynamics (HANDY): Modeling Inequality and Use of Resources in the Collapse or Sustainability of Societies." *Ecological Economics*, 90–102.
O'Connor,J., 1997. *Natural Causes*. The Guilford Press, London.
Petit, V., Guillaume, B., 2016. "Quelle « démocratie écologique » ?" *Raisons Politiques* 64, 49. https://doi.org/10.3917/rai.064.0049.
Poulantzas, N., 2013. *L'Etat, le Pouvoir, le Socialisme*. Les Prairies Ordinaires, Paris.
Poulantzas, N.A., 2000. *State, Power, Socialism*, New ed., Verso classics. Verso Books, London.
Reclus, É., 2015. *L'Homme et la Terre. Livre 1 Les ancêtres*. ENS Éditions, S.L.Schmitt, C., Heller, H., Chamayou, G., 2020. *Du libéralisme autoritaire*. Zones, Paris.

Scott, J.C., 2017. *Against the Grain: A Deep History of the Earliest States, Yale Agrarian Studies*. Yale University Press, New Haven.
secretariat, 2019. *Global Assessment Report on Biodiversity and Ecosystem Services* [WWW Document]. IPBES secretariat. https://www.ipbes.net/global-assessment (accessed 3.7.22).
S.J., Scheffer, M., Winkelmann, R., Schellnhuber, H.J., 2018. "Trajectories of the Earth System in the Anthropocene." *Proceedings of the National Academy of Sciences* 115, 8252–8259. https://doi.org/10.1073/pnas.1810141115.
Thomas, L.-V., 1978. *Mort et pouvoir, Petite bibliothèque Payot*. Payot, Paris.
Wainwright, J., Mann, G., 2018. *Climate Leviathan: A Political Theory of our Planetary Future*. Verso, London; New York.
Wallenhorst, N., Theviot, A., 2020. "Les récits politiques de l'Anthropocène." *Raisons Politiques* 1, 5–34.
Wallerstein, I.M., 2004. *World-Systems Analysis: An Introduction*. Duke University Press, Durham.

Technonaturalism: A Postphenomenological Environ-Mentality

Alexander Stubberfield

Machines become, in technological culture, part of our self-experience and self-expression. They become our familiar counterparts as quasi-others, and they surround us with their presence from which we rarely escape. They become a technological texture of the World and with it they carry a presumption toward totality. In this sense, at every turn, we encounter machines existentially.

—Don Ihde, *Technics and Praxis*[1]

New terrains are always uncertain, but they feature familiar Anthropocenery in which hybrid life-forms, cyborganic ecological systems, and technonatural urban sprawl fuse together new ranges to be managed. Coyotes and falcons may inhabit urban office parks and robotic drones follow waterfowl in their migratory flyways while inventorying foliage densities in the suburban sprawl ingesting wilderness to produce jobs, homes and shops for the human denizens of technonature.

—Timothy W. Luke, *Anthropocene Alerts*[2]

---

[1] Ihde, Don. *Technics and Praxis*. Dordrecht, Holland: D. Reidel Publishing Company, 1979. 15.

[2] Luke, Timothy W. *Anthropocene Alerts: Critical Theory of the Contemporary as Ecocritique*. Candor, NY: Telos Press Publishing. 2019. 215.

A. Stubberfield (✉)
Virginia Polytechnic Institute and State University, Blacksburg, VA, USA
e-mail: havoc123@vt.edu

Introduction: Recording Technonatural Life

The ascendance of the Anthropocene within public and scientific discourse fails to capture the culprits of environmental degradation, danger and decay because the optics embedded within the concept displace responsibility for the emergence of wildness on a global scale from the machinations of artificial persons to all humans, everywhere.[3] I suggest an alternative social scientific imaginary wedding environmentality studies, science and technology studies and world ecology into a localizable toolbox for a new naturalism as a negative to the Anthropocene concept. As many have commented, climate change, global change, the Anthropocene, present problems too big for a single mind to grasp and the changes in the backyard are often ignored for globalistic discourse.[4] Environmental social science needs ways of understanding environmental development in the everyday while enabling multi-methodological and multi-scalar data collection relating that development to the rule of artifice.[5]

Technonaturalism aims at putting "the science" in perspective by recognizing its political dimensions proposing an environmental subjectivity while searching for alternative ways of organizing organic and artificial persons in relation to materiality. Artificial persons are understood as recognized organizations of humans and non-humans into functioning machinery conceived as singular legal agents such as states, private enterprises and non-governmental organizations.[6] This distinction is made analytically to recognize the agency of objects as within notions of personhood without conflating "person" and "human" while allowing reference to extra-human agency within the production of a planetary ecotone. The age we inhabit, if it is to be inhabited, needs a finer approach than implied by "the Anthropocene," if real, local environmental changes are to be understood by the humans most affected by them and if the culprits for environmental damage are to be identified.

---

[3] Luke, Timothy W. "Introduction." In Luke, 2019. 1–17; Moore, Jason W. "The Rise of Cheap Nature." In Moore, Jason W., ed. *Anthropocene or Capitalocene? Nature, History and the Crisis of Capitalism.* Oakland, CA: PM Press, 2016. 78–115.

[4] Moore, Jason W., ed. 2016; White, Damian F., Alan P. Rudy, and Brian J. Gareau. *Environments, Natures and Social Theory: Towards a Critical Hybridity.* New York: Palgrave, 2016.

[5] Baudrillard, Jean. *Simulations.* Foss, Paul, Paul Patton, and Philip Beitchman (trans.). Cambridge, MA: MIT Press: Semiotext(e). 1983; White, Damian F., and Chris Wilbert, eds. *Technonatures: Environments, Technologies, Spaces, and Places in the Twenty-first Century.* Waterloo, Ont.: Wilfrid Laurier University Press, 2009.

[6] This distinction, of course, leaves open the possibility of artificial intelligence understood as persons in the future. For my purposes here, however, AI is machinery and will not be "persons" until granted legal standing. Cutting between organic and artificial is problematic as "person" has historically been shown as a legal invention. Thus, it may be that all "persons" emanate from the artifice of the state and its rule, itself being an artificial person. The distinction, however, is drawn above for analytic purposes cutting between embodied human subjects on the one hand and bodies of humans and non-humans directed toward specific ends and organized primarily through the written word.

Technonaturalism is a philosophical anthropology, and, with the unfortunate quality of being an "-ism" answers calls for a planetary imaginary grounding a critical subjectivity through a new naturalism.[7] However, it is thoroughly postnatural, buying the death of nature thesis through recognizing the profound effects of planetary urbanization, commodity development, ecological disaster, and rule, while venturing a reading of the planetary by eschewing "nature" as a term and adopting "materiality."[8] This recognizes co-productive environmental orders without committing the analyst to a methodological dualism pitting a "nature" against "society," favoring holistic understandings of planetary process as including humans, non-humans and extra-human agencies in step with current hybridity and assemblage theory as well as ecological literature.[9] "Technonaturalism," is borrowed from discussions in critical urban geography tracking the production of *technonatures* and the contributors to that volume are all trying to move beyond Bruno Latour's Actor-Network Theory with different adaptations to read environmental relations.[10] I join them in keeping Latour's hybrid ontology and propose technonaturalism as a scientific anti-science drawn from postphenomenology wedded with environmentality studies to record the production of lifeforms and habitability through geoengineering.[11]

---

[7] Death, Carl, ed. *Critical Environmental Politics.* London: Routledge, 2013.; Gare, Arran. "Naturalism, Theology and Utopia." In Gare, Arran and Wayne Hudson, eds. *For a New Naturalism.* Candor, NY: Telos Press Publishing, 2017. 33–49; Miller, Daniel, ed. *Materiality.* Durham, NC: Duke University Press, 2005; White, Damian F., Alan P. Rudy, and Brian J. Gareau, 2016; White, Damian F., and Chris Wilbert, eds., 2009.

[8] Brenner, Neil, ed. *Implosions/Explosions: Towards a Study of Planetary Urbanization.* Berlin: Jovis Verlag, 2014; Jagtenberg, Tom and David McKie. *Eco-Impacts and the Greening of Postmodernity: New Maps for Communication Studies, Cultural Studies and Sociology.* London: SAGE, 1997; Miller, ed., Ibid; White and Wilbert. "Preface." In White and Wilbert, eds., 2009. vii-viii.

[9] Allenby, Braden R. *Reconstructing Earth: Technology and Environment in the Age of Humans.* Washington, D.C.: Island Press, 2005; Knick, Steven T. and John W. Connelly (eds.) *Greater Sage-grouse: Ecology and Conservation of a Landscape Species and its Habitats.* University of California Press and The Cooper Ornithological Society, 2011; Latour, Bruno. *We Have Never Been Modern.* Translated by Catherine Porter. Cambridge, MA: Harvard University Press, 1991; Sale, Peter F. *Our Dying Planet: An Ecologist's View of the Crisis We Face.* Berkeley: University of California Press, 2011; White and Wilbert. "Introduction." In White and Wilbert, eds., 2009. 1–30.

[10] White and Wilbert, eds., 2009.

[11] Ihde, Don. *Postphenomenology and Technoscience: the Peking University* Lectures. Albany: State University of New York Press, 2009; Luke, Timothy W. *Capitalism, Democracy and Ecology: Departing from Marx.* Chicago: University of Illinois Press, 1999; MacKaye, Benton. *The New Exploration: A Philosophy of Regional Planning, with an Introduction by Lewis Mumford.* Urbana: University of Illinois Press, 1962; Moss, Lenny. "New Naturalism and Critical Theory." In Gare and Hudson, eds., 2017. 78–104; White, Damian F., and Chris Wilbert, eds. 2009.

IHDE IN ENVIRONMENTAL SOCIAL SCIENCE

There have been calls for new understandings of socio-environmental relations through technics echoing earlier moves questioning social scientific optics in pursuit of a new naturalism and efforts have been made unpacking the effects of naturalism within political theory.[12] As some have argued, this necessitates a new naturalism grappling with the biggest problems of our day:[13] "In our view, contemporary social and political theorists need to take account of the need for a new naturalism that is more comprehensive than the naturalism that emerged from the scientific revolution."[14] While attempts have been made, many have rested on Latour's Actor-Network Theory,[15] Haraway's Material Semiotics[16] or Neo-Marxian Metabolic Rift Theory.[17] Despite being an interlocutor of Latour and Haraway's, Don Ihde's thought has not been properly included within conversations in environmental political theory and environmental social science.[18] Technonaturalism employs Ihde's philosophy and the tools he has developed to ground a new naturalism within postphenomenological studies of environmental technics.[19]

Latour's rejection of "capital" within ANT risks losing the effects of corporate activity within the biosphere. Andreas Malm has taken him to task for ignoring this within his theoretical commitments.[20] Malm's Climate Realism, unfortunately, posits an unwarranted dualism through metabolic rift theory, the phenomena of which can be explained through a methodological holism

---

[12] Caraccioli, Mauro J. *Writing the New World: The Politics of Natural History in the Early Spanish Empire*. Gainesville, FL: University of Florida Press, 2021; Luke, Timothy W. *Social Theory and Modernity: Critique, Dissent and Revolution*. Newbury Park, CA: SAGE Publications, 1990; Masco, Joseph. *The Future of Fallout and other Episodes in Radioactive World-Making*. Durham, NC: Duke University Press, 2021; Mitchell, Timothy. *Rule of Experts: Egypt, Technopolitics, Modernity*. Berkeley: University of California Press, 2002.

[13] Pan, David. "Intentionality and Interpretation in Biology and The Humanities." In Gare and Hudson, eds., 2017. 105–123.

[14] Gare and Hudson, eds., 2017. 161.

[15] Latour, Bruno. *Reassembling the Social: An Introduction to Actor Network Theory*. Oxford: Oxford University Press, 2005.

[16] Haraway, Donna J. *Modest_Witness@Second_Millennium.FemaleMan_Meets_OncoMouse: Feminism and Technoscience*. New York: Routledge, 1997.

[17] Foster, John Bellamy. *The Ecological Rift: Capitalism's War on the Earth*. New York: Monthly Review Press, 2010.

[18] Ihde, Don and Evan Selinger, eds. *Chasing Technoscience: Matrix for Materiality*. Bloomington: Indiana University Press, 2003.

[19] Ihde, Don. *Instrumental Realism: The Interface Between Philosophy of Science and Philosophy of Technology*. Bloomington: University of Indiana Press, 1991; Ihde, Don. *Ironic Technics*. USA: Automatic Press, 2008; Ihde, Don. 2009; Ihde, 1979; Ihde, Don. *Technology and the Lifeworld: from Garden to Earth*. Bloomington: University of Indiana Press, 1990.

[20] Malm, Andreas. *The Progress of This Storm: Nature and Society in a Warming World*. London: Verso, 2018.

collapsing the Technology/Nature dyad as Haraway does.[21] Donna Haraway's material semiotic approaches, however, risk reducing environmental relations to textuality thus sucking observers and analysts into the technological nexus of writing and communication.[22] Ihde is an interlocutor of Latour and Haraway and has enjoyed popularity in STS circles for years becoming one of the most celebrated philosophers of the last 50.[23] His writings are toolboxes for analyzing technics that he has labeled as techno-environmental philosophy, yet he has not been properly utilized in environmental social scientific circles.[24] It is possible to unite refinements within environmentality studies with postphenomenology to produce an analytic framework that opens and guides site selection and analysis from embodied explorations of environmental technics at the micro-level to large-scale planetary machinations and climatic changes in environmental governance using Ihde. Importantly, though Ihde is a philosopher, he thinks of postphenomenology less as a systematic approach to the production of understanding, and more as a strategic orientation to understanding technics offering tactics for navigating technologically textured ecosystems.[25]

THE TECHNONATURAL PLANET: PRECIS TO AN ENVIRONMENTAL IMAGINARY

I propose grounding a new naturalism in Ihde's philosophy while extending his thought into a properly recognized Foucauldian register of environmentality and join voices such as Peter-Paul Verbeek's in politicizing postphenomenology to explore a planetary imaginary.[26] Ihde offers a route to evaluating subjective experiences of environment through postphenomenological investigations of planetary technics which takes bodies+technologies as its analytical primitive for understanding environmental relations and the

---

[21] Haraway, 1997.

[22] Eason, Robb. "Hypertext: Rortean Links between Ihde and Haraway." In Ihde, Don and Evan Selinger, eds. *Chasing Technoscience: Matrix for Materiality*. Bloomington: Indiana University Press, 2003. 167–181; Ihde, Don. "Program One: A Phenomenology of Technics." In Ihde, Don. *Technology and the Lifeworld: From Garden to Earth*. Bloomington: Indiana University Press, 1990. 73–123.

[23] Ihde and Selinger, eds., 2003.

[24] Ihde, 1990.

[25] Ihde, Don. *Experimental Phenomenology: An Introduction*. New York: Capricorn books, 1977. 135–153; Ihde, Don, 2009; Ihde, Don, 1990. 162–191; Mathews, Freya. "Invoking the Real: From the Spectacular to the Ontopoetic." In Gare and Hudson, eds., 2017. 144–159.

[26] Feenberg, Andrew. *Technosystem: The Social Life of Reason*. Cambridge, MA: Harvard University press, 2017. 115–134. Verbeek, Peter-Paul. "Politicizing Postphenomenology." In Miller, Glen and Ashley Shew, eds. *Reimagining Philosophy and Technology: Reinventing Ihde*. Springer: Cham, Switzerland, 2020.

production of lifeform through embodiment.[27] Further, Ihde proposes an interrelational ontology strategically oriented to assessing environmental relations and collapsing the human/technology, mind/body, technology/culture and technology/nature dyads recognizing hybridity as an empirical fact.[28] For my purposes, his philosophy is an excellent candidate for investigating the hybrid spaces produced through civilizational expansion recognized by Benton MacKaye as "civilizational wildernesses."[29] It should be possible, therefore, to ground an Ihdean critical environmental subjectivity through the work of C. Wright Mills, Michel Foucault, Timothy W. Luke, Benton MacKaye and Lewis Mumford.

The following augments and extends Ihde's analytic toolkit paralleling his lifeworld analysis with Foucauldian environmentality studies. I begin by recognizing Ihde's deep materialism within his lifeworld analysis as it fits the sociological imagination expounded by Mills that White, Rudy and Gareau use to build a critical subjectivity in more recent work on hybridism and environment.[30] Doing so yields an Ihdean political environmental subjectivity that can then cultivate an investigative community exploring planetary techno-natural ecotones as time-spaces of artificial becoming.[31] This will situate a critical community through common linguistic convention dedicated to understanding environmental relations through technics that can connect and articulate both structural shifts in technological milieu, and embodied experiences through detailed case studies. Further, lifeworld analysis contains a praxis-perception model of techno-environmental subjectivity such that "science" and the "facts" produced by it can be seen as quasi-real thus skirting social constructivism and naïve scientism while offering critiques of technocratic anti-politics.[32]

---

[27] DeLanda, Manuel. "Persons and Networks." In DeLanda, Manuel *A New Philosophy of Society: Assemblage Theory and Social Complexity*. New York: Bloomsbury, 2006. 47–67; Hinchliffe, Steve and Sarah Whatmore. "Living Cities: Toward a Politics of Conviviality." In White and Wilbert, eds., 2009. 105–122; Ihde, Don, 1990. 11- 20; Linder, Fletcher. "Critical Mass: How Built Bodies can forge Environmental Futures." In White and Wilbert, eds., 2009. 149–163; Mike Michael. "The Cellphone-in-the Countryside: On Some Ironic Spatialities of Technonatures." In White and Wilbert, eds., 2009. 85–104.

[28] Ihde, Don. "Technofantasies and Embodiment." In Ihde, Don *Embodied Technics*. Automatic Press, 2010. 1–15; Ihde, 1991; Ihde, Don. *Postphenomenology: Essays in a Postmodern Context*. Evanston: Northwestern University Press, 1993b; Ihde, 1979; Ihde, Don. "Technology and the Lifeworld." In Ihde, 1990. 21–30.

[29] MacKaye, Benton. "The Wilderness of Civilization." In MacKaye, 1962. 16–25.

[30] Mills, C. Wright. *The Sociological Imagination*. New York: Oxford University Press, 1959; White, Rudy and Gareau, 2016.

[31] White and Wilbert, eds., 2009. 12.

[32] Gare, Arran and Wayne Hudson. "The Challenge of a New Naturalism." In Gare and Hudson, eds., 2017. 1–8; Ihde, Don. *Bodies in Technology*. Minneapolis: University of Minnesota Press, 2002. 104–112; Ihde, 1991. ix–xiv.

The Sociological Imagination and the Lifeworld: From Garden to Technoscience

The following expounds C. Wright Mills' sociological imagination to show how Ihde's lifeworld analysis fits Mills' advice to ground a critical and liberatory subjectivity within subjective imaginative capacities regarding a planetary imaginary.[33] I make this move because Mills has enjoyed a resurgence in environmental hybridity and complexity theory, seen best in the recent work of Damian F. White, Alan P. Rudy and Brian J. Gareau as they grapple with the implications of the Anthropocene for social theory in *Environments, Natures, and Social Theory* (2016). Their work highlights the necessity of developing a *socio-ecological imagination* for sorting through the new orientation of humanity-in-the-world in their freshly minted geological epoch, but they selected Donna Haraway as their guide and I depart from them in using Ihde.[34]

I give a brief characterization of Mills' thinking and then move into Ihde's lifeworld analysis as a candidate set of methods for grappling with environments. The discussion forks into governmentality studies in the next section through Timothy W. Luke's ecocritiques of geoengineering to show a bridge from ecocriticism as a genre to environmentality studies, while adding Ihde to elucidate an Ihdean reading of environmentality. Finally, I show how an Ihdean environmentality would include a deep and specific focus on instruments and discuss how this fits into an understanding of geoengineering such that technonature can be explored. If successful, the movements below should yield a critical environmental subjectivity fully equipped to explore technonaturalization and thus the rule of artifice.

Mills critiques the social sciences and attitudes held toward science and engineering during its increasing corporate militaristic bureaucratization in the United States of the late nineteen-fifties.[35] He creates a positive account of what is needed to navigate the informational and dizzying environments of his era terming this account the *Sociological Imagination* (1959). The sociological imagination is an orientation to the world that should be developed by scholars and public intellectuals to translate and evaluate the changing character of their milieux.[36] It aims at uniting the material history and narratives of the society in which subjects find themselves with auto/biographical accounts of their daily life.[37] This unification of larger historical trajectories of society and of the individual within "is the most fruitful form of this self-consciousness."[38]

---

[33] White, Rudy and Gareau, 2016.

[34] Ibid.

[35] Mills, C. Wright. *The Power Elite*. Oxford: Oxford University Press, 1956; Mills, C. Wright. *White collar: The American Middle Classes*. Oxford University Press, 1951.

[36] Mills, 1959. 5.

[37] Ibid., 6–7.

[38] Ibid.

Thus, Mills' sociological imagination is a *Weltanschauung* allowing subjects to elaborate their conditions of existence by connecting them and their place in the world to larger events in history that have helped produce those conditions of existence.

The sociological imagination, in turn, can be broken into two analytic components regarding experience: the personal milieu, and the larger, structural milieu:

> What we experience in various and specific milieux, I have noted, is often caused by structural changes. Accordingly, to understand the changes of many personal milieux we are required to look beyond them. And the number and variety of such structural changes increase as the institutions within which we live become more embracing and more intricately connected with one another. To be aware of the idea of social structure and to use it with sensibility is to be capable of tracing such linkages among a great variety of milieux. To be able to do that is to possess the sociological imagination.[39]

Interestingly, the subjective positionality discussed by Mills is mirrored in Ihde's lifeworld analysis separating, but only analytically, the micro-perceptual positionality of embodied human subjects from larger structurally "given" macro-perceptions conceived as cultural hermeneutics.[40]

Ihde, drawing from Edmund Husserl, Paul Ricœur, Maurice Merleau-Ponty, Martin Heidegger and other phenomenologists, grounds an understanding of being-in-the-world through recognizing how narratives and perceptions overlap and include one another. "The modification I shall make here upon Husserl's example is by distinguishing two senses of *perception*."[41] He moves from both a subjective positionality—constituted by embodiment relations—to a larger hermeneutic positionality that includes narratives about subjects such that it guides meaning-making. This allows Ihde to claim that experience can be divided analytically, and perhaps only analytically, into micro- and macro-perceptions:

> The relation between micro- and macroperception is not one of derivation; rather it is more like that of a figure-to-ground in that microperception occurs within its hermeneutic-cultural context; but all such contexts find their fulfillment only with the range of microperceptual possibility. The implication for the notion of lifeworld will be that this inquiry will be nonfoundationalist with some clear qualifications thus necessarily place upon the idea of accumulation.[42]

---

[39] Ibid., 10–11.

[40] Ihde, Don. *Philosophy of Technology: An Introduction*. New York: Paragon House Publishers, 1993a. 5–30; Ihde, 2009. 41–44.

[41] Ihde, 1990. 29. Emphasis in original.

[42] Ihde, 1990. 29–30.

For my purposes, Ihde's lifeworld analysis offers an analytical primitive in that all subjects must be embodied and are located within fields textured by technologies and their influences. Thus, in assemblage theoretic terms, Ihde has a human+technology primitive for any relational readings of subject-in-the-world. This means that all human communities are technologically textured, varying only in degree by that texturing and display a technics. How specific groups and agents understand and interpret themselves is partially mediated by tacit agreements about technologies and their uses and speaks to their material culture. "Use" here smuggles normativity as larger cultural narratives help incline subjects to use technologies in culturally sedimented ways—thus accumulating and sedimenting knowledge practices through technologies. He goes as far to claim that science is a special instance of a global cultural hermeneutic grounded in Western civilization—maybe—that recursively shapes it through (mis)understandings of "nature"—a reification of materiality—emerging from technologies and the communities that use them:

> This is also why, in the already established metaphor metaphysics of the great machine of nature, the anatomical and technological drawings are, at base, alike. It is the prelude, the necessary condition, for the coming age of technology.
>
> Much, much later this would be recognized by the twentieth-century thinker Martin Heidegger. It was he who noted that technology is not some collection of artifacts but a "way of seeing." And it was Leonardo da Vinci who helped to make that "technological" way of seeing into a gestalt which could be refined and extended, even to the current and still perspectival representations that today capture our fascination.[43]

Scientific praxis, Ihde claims, is a consequence of technological intervention in perceiving materiality and solving problems within it. This is related to his remarks on technological priority over scientific development: "But instrumentation, at least for all modern science, is the material condition; better, it is the *embodiment* by which science *perceives*."[44] Thus, technologies are experiential mediators mirroring Peirce's observations about experimental science, "his [Lavoisier's] way was to carry into his mind into his laboratory, and literally to make of his alembics and cucurbits instruments of thought, giving a new conception of reasoning as something which was done with one's eyes open, in manipulating real things instead of words and fancies."[45] For Ihde, the scientific instrument literally and materially helps create worlds as technoscientists relate to "the real" or "the world" or materiality through the nexus of scientific instrumentaria. "Instruments form the conditions for and are the mediators of

---

[43] Ihde, 1993b. 29–30.

[44] Ihde, 1991. 16. Emphasis in original.

[45] Peirce, Charles Saunders. *Philosophical Writings of Peirce*. Justus Buchler, ed. New York: Dover Publications, 1955. 6.

much, if not all, current scientific knowledge. They are the concrete and material operators within scientific praxis."[46] That is, the entities perceived under conditions of laboratory observation are, more often than not, technologically mediated and are embodied in and by those observational apparatuses. "That is to say that science is necessarily embodied in instrumental technologies."[47] Scientific problems, therefore, are both technologically known, articulated and expressed, as well as housed within larger intersubjective narratives guiding technoscientific experience through instrumental relationships in-use.[48]

This means that scientific problems arise from the subjective positionalities of scientists viewing, for example, anomalous, or previously unexperienced phenomena, with specific knowledge communities guiding their interpretation and articulation: "the same applies to science—in a deep sense, science cannot escape the lifeworld, since it too must make all measurements *perceivable to embodied humans*, although it may do so through the mediations of measuring technologies or instruments."[49] That is, what can be said about phenomena is often and only technologically mediated and thus separates observer from direct experience of scientific objects through technological mediations and is often the only way those phenomena can be experienced. "Instrumentation in the knowledge activities, notably science, is the gradual extension of perception into new realms. The desire is to see, but seeing is seeing through instrumentation."[50] This is Ihde's instrumental realism which he remarks makes worlds and recursively shapes both micro- and macro-perceptions through adjustments in scientific hermeneutics speaking to the administration, construction and understanding of both quotidian and laboratory life. "I contend that contemporary science—in contrast to its ancient forms—is both technologically *embodied* in its necessary instrumentation and also *institutionally embodied* in the social structures of a technological society."[51]

The production of a theoretical postulate, such as an electron, is both the product and process of scientific inquiry and the embedding of a theoretical postulate within a specific knowledge community represents a change in both micro- and macro-perceptions as subjects come to understand their worlds through technologically mediated relations that speak back to the technological texture of the lifeworld: "you read the thermometer, and in the immediacy of your reading you *hermeneutically* know that it is cold. There is an instantaneity to such reading, as it is an already constituted intuition (in phenomenological terms). But you should not fail to note that *perceptually* what you have seen is the dial and the numbers, the thermometer 'text.' And

[46] Ihde, 1991. 45.
[47] Ibid., 55.
[48] Ihde, 2009. 41.
[49] Ihde, 2009. 32. Emphasis in original.
[50] Ihde, 1990. 75.
[51] Ihde, 1991. 63. Emphasis in original.

that text has hermeneutically delivered its 'world' reference, the cold."[52] Technological supervenience within scientific communities shows that "science" is a justificatory framework for figuring out and interpreting instrumental mediations across intersubjective communities of investigators who then feed "facts" back into larger cultural narratives of being-in-the-world thus guiding subjective development and adding to the lifeworld through experimentation and explanation. Thus, borrowing from Merleau-Ponty "one can see that science...could be such a perceived culture. Here is a kind of macroperception which informs and orients bodily perception itself."[53] Bodily perception is influenced because scientific communities discipline investigators in the correct use of instruments thus dictating how bodies relate to both micro-perceptual environments and macro-perceptual narrative through cultural gestalts. This is global technoscientific culture for Ihde that is a complex of institutions, humans and non-humans.

Mills analytically separates subjective experience into *troubles* and *issues*.[54] Troubles can be thought of as arising from personal interactions. They are part of the subject's personal experience and, in Mills' language, their immediate milieu. Troubles are part of the subjective biographical account of the world and concern the everyday and "the social setting that is directly open to his personal experience and to some extent his willful activity."[55] Ihde's microperceptual positionality can account for and include subjective troubles both in the laboratory and everyday life while stereoscopically understanding that "troubles" can change according to "issues." The narrower subjective milieu of experience can include the larger mechanics of meaning-making through the adoption of tools that help subjects sort through experiences. This means, for example, that a personal trouble might be related, if not indexed to the larger field of *issues* and that *issues* arise and speak back to their technological contexts of emergence.

Issues, for Mills, are a larger frame of action and involve overlapping and interpenetrating milieux that transcend the immediate and everyday environments of the subject. Issues have to do with the "organization of many such milieux into the institutions of historical society as a whole, with the ways in which various milieux overlap and interpenetrate to form the larger structure of social and historical life."[56] Ihde has recognized that technoscience is a specific activity currently housed globally in and practiced through institutions. He not only understands the technoscientist as a breed of investigator, but also one that has been enabled and proliferated because of the institutionalization of scientific activity and technological development within artificial

---

[52] Ihde, 1990. 85. Emphasis in original.
[53] Ihde, 1990. 40.
[54] Mills, C. Wright. "Chapter 1: The Promise." In Mills, 1959. 3–24.
[55] Mills, 1959. 8.
[56] Ibid.

persons such as commercial organizations, NGOs and governments.[57] Additionally, he has recognized the amalgamation of specific investigative contexts in producing global technoscientific culture connected to the machinations and demands of global capitalism, war and commercial organization.[58]

This is important because technological development, which indicates new frontiers in technoscientific understanding, is connected to how issues are articulated for whom and by what. That is, every technology is an organization of materials and energy brought to bear on perceived issues arising in the lifeworld. This means that each issue is a product of its context and that those contexts are composed of, among other things, instruments used to detect and articulate those issues. This move allows Ihde to claim that all scientific knowledge is intersubjectively grounded and cannot make claims to an absolute, or value-free "objectivity."

> *Instrumental Realism* tried to reframe the understanding of science in terms of its late modern technological embodiment in instrumentation, its material incarnation. My take upon science was to try to recall that in addition to mathematizing, modeling and formalizing a world, science also *perceives* its worlds, albeit through instruments, and that is where very contemporary science meets the above stated themes of bodies and technofantasies.[59]

All, therefore, "issues" must be evaluated as belonging to technological and scientific contexts that speak back to sets of relationships resulting from technologies-in-use. Conversely, troubles can be understood by connecting personal relationships to technologies through the larger field of technics backtracing them to *issues* and the instruments articulating them.

At stake for Mills then is the trajectory of scientific, technological and thus social development through the social institutions of economics, politics, culture and, therefore, the everyday lives of subjects.[60] The countermeasure to the domination of the elite is his form of self-consciousness.[61] Ihde's lifeworld analysis, with its sophisticated understanding of instruments within micro- and macro-perceptions, can be the countermeasure to elite domination that troubled Mills "Contemporary technoscience in its now technological and

---

[57] Ihde, 1993a. 17.

[58] Braverman, Harry. *Labor and Monopoly Capital: The Degradation of Work in the Twentieth Century.* New York: Monthly Review Press, 1974; Foster, John Bellamy. *The Theory of Monopoly Capitalism: An Elaboration of Marxian Political Economy.* New York: Monthly Review Press, 1986; Ihde, Don. "Stupidity in the Knowledge Society." In Ihde, 2008. 1–17; Mills, 1956. 198–224; Virilio, Paul. *The Information Bomb.* London: Verso, 2000.

[59] Ihde, 2002. xv. Emphasis in original.

[60] Mills, 1959. 16.

[61] Ibid. 15.

corporate form requires a very complex, large and society-supported infrastructure for the production of its "products."[62] In other words, Ihde recognizes the increasing intimacy of military, commercial, non-governmental and governmental institutions and globally intersubjective constitutions of technoscientific communities and understands technologies as containing cultural gestalts that sediment ways of seeing and interacting with "the world."

> [T]his description is what I am calling the third step toward a postphenomenology. It is the step away from generalization about *technology uberhaupt* and a step into their examination of *technologies in their particularities*. It is the step away from a high altitude or transcendental perspective and an appreciation of the multidimensionality of technologies as *material cultures* with a *lifeworld*.[63]

Thus, technoscience begets technoculture witnessed and understood through its artifacts as global culture forming a pluricultural megatechnics.[64]

Repeatedly, environmental issues trouble Ihde and his investigations of technics can be adapted to understand and chart environmental development as part of environmental governance through understanding relationships to "nature" through technological artifacts and their artifacts.[65] "[I]n each set of human-technology relations, the model is that of an interrelational ontology. This style of ontology carries with it a number of implications, including the one that there is a co-constitution of humans and their technologies. Technologies transform our experience of the world and our perceptions and interpretations of our world, and we in turn become transformed in this process."[66] Instruments should be the starting point for analyzing environmental technics subjectively and Ihde has demonstrated numerous examples.[67] As the body is an analytical primitive and must be thought of in relation to technology, this should allow Ihdeans to understand technoscientific production in a similar vein to Haraway including White, Rudy and Gareau's adaptation as including organizational gestalts and narratives within micro-perception.[68] This could

---

[62] Ibid., 60.

[63] Ihde, 2009. 22. Emphasis in original.

[64] Ihde, 1990. 27, 165–191.

[65] Gray, Chris Harbles, Heidi J. Figueroa-Sarriera, and Steven Mentor. "Cyborgology: Constructing the Knowledge of Cybernetic Organisms." In Gray, Chris Harbles, Heidi J. Figueroa-Sarriera, and Steven Mentor, eds. *The Cyborg Handbook*. New York: Routledge, 1995. 1–14; Ihde, 2008; Ihde, 1979; Ihde, 1990; Ihde, 1993a; Ihde, 1993b; White, Rudy and Gareau, 2016.

[66] Ihde, 2009. 44.

[67] Ihde, 2008; Ihde, 1979; Ihde, 1990; Ihde, 1993a; Ihde, 1993b.

[68] Ihde, 2009; Ihde, Don. "The Experience of Technology: Human–Machine Relations." In Ihde, 1979. 3–15; Haraway, 1997; Haraway, Donna J. "A Cyborg Manifesto: Science, Technology, and Socialist-Feminism in the Late Twentieth Century." In Haraway,

rely on semiotic analyses, but must understand subjective milieux as materially conditioned by artificial persons, their actions and narratives, and must be recognized as technonatural organisms composed of humans, non-humans and machinery that leave evidence of their being through artifacts.[69] Importantly, this means that the planetary body contains evidence of artificial persons within its materiality and can speak to the instruments used to produce it. Taken on a case-by-case basis, a material hermeneutics of landscape, for example, is possible when instruments, such as artificial persons, operate through it.

Postphenomenological Environmentality

As a point of departure, Ihde supports and recognizes Haraway's contributions to understanding materiality. He thinks of humans as cyborg subjects and argues that humanity may be a technological animal before anything else—collapsing the human/technology dyad through historically recognizing that technologies predate any communities taken as *homo sapiens*.[70] However, his key difference is how he understands narratives related to scientific inquiry.[71] This is part of his material hermeneutic approaches which he likens to forensic science as it concerns what can be said about materiality[72]: "I take this as an example of a material hermeneutic, in which 'things' are given voices: pollen, grain, metal, and tooth enamel have all 'spoken' in spite of being situated in a context that itself is without proper linguistic phenomena."[73] That is, how an object is seen is both a matter of larger cultural gestalts and subjective orientation to that object: "If one moves into historic periods, it seems possible to parallel textual and linguistic evidence with the material evidence suggested."[74] Key to understanding material hermeneutics is the Latin meaning of *habeas corpus* as the material contains evidence for assisting inference within narrative construction about a state of affairs and scholars such as, William Cronon, have demonstrated material hermeneutics—a landscape hermeneutics—in the humanities beautifully.[75]

Donna J. *Simians, Cyborgs and Women: The Reinvention of Nature*. New York: Routledge, 1991. 149–181.

[69] Haraway, Donna J. "Cyborgs and Symbionta: Living Together in the New World Order." In Gray, Figueroa- Sarriera, and Mentor, eds. 1995, xi–xx; Haraway, 1997; Haraway, 1991.

[70] Ihde, 2008. 31–57; Ihde, 2009. 38.

[71] Ihde, 2002. 89–100.

[72] Ihde, 2009. 68–80.

[73] Ibid., 72.

[74] Ibid., 73.

[75] Cronon, William. *Changes in the Land: Indians, Colonists, and the Ecology of New England*. New York: Hill and Wang, 1983.

Studies like Cronon's[76] speak to Ihde's remarks about how materiality "writes back" through scientific investigation so that "the material hermeneutics reveal the written accounts to be partial and, in some ways, to show phenomena that are in tensions with the written accounts."[77] This is a deep, Peircean, understanding of materiality in that it neither treats scientific investigation as producing objective "truths" nor does he adopt a scientific anti-realism as a constructivist.[78] Instead, materiality in a sense "talks back" to the scientist through their instruments and the scientist interrogates materiality. What can be said about any object under investigation is grounded in larger cultural gestalts reactive to the agency of objects within scientific investigation. Seen in this way, science is less of a text and more of a conversation with materiality in the production of understanding.[79] That is, "science" is, at best, a set of stories about materiality pragmatically verified against its behavior under observation, but those stories can be questioned and interrogated. Importantly, this seats Ihde's brand of phenomenology with American pragmatism and produces a non-foundationalist postphenomenology through technics wholly woke to the myth of the given troubling naturalistic ontologies.[80]

The above can ground a community of representors through the production of common theoretical objects regarding the planetary body. It must be sufficiently critical of structural conditions and connect those conditions to both the extra-human agencies producing the background conditions of planetary life through technological extension and the subjective positions—the troubles—of those within lifeworlds. This community should be plurimethodological[81] and below I show but one route for a technonaturalist through a special case of Ihdean hermeneutic analyses: machine-hermeneutics. This parallels other political adaptations of Ihde's philosophy for political analysis and Peter-Paul Verbeek, for example, has demonstrated applications of Ihde's hermeneutics through an adaptation called *political hermeneutics*.[82] His contribution recognizes Ihde's Deweyan inflections and the deep role

---

[76] Silver, Timothy. *Mount Mitchell and the Black Mountains: An Environmental History of the Highest Peaks in Eastern America*. Chapel Hill: University of North Carolina Press, 2003.

[77] Ihde, 2009. 73.

[78] Peirce, Charles Saunders. "How to Make our Ideas Clear" and "The Approach to Metaphysics." In Buchler, Justus, ed. *The Philosophical Writings of Peirce*. New York: Dover Publications, 1955. 23–41, 310–314; Harney, Maurita. "Peirce and Phenomenological Naturalism." In Gare and Hudson, eds., 2017. 124–143.

[79] Malabou, Catherine. "Epigenesis of the Text: New Paths in Biology and Hermeneutics." In Gare and Hudson, 2017. 66–77.

[80] Macarthur, David. "Liberal Naturalism and Philosophy of the Manifest Image." In Gare and Hudson, eds., 2017. 50–65; Sellars, Wilfrid. "Empiricism and the Philosophy of Mind." In Wilfrid Sellars. *Science, Perception and Reality*. Atascadero, CA: Ridgeview Press, 1963. 127–196.

[81] Feyerabend, Paul. *Against Method*. New York: Verso, 1975. xvii–xxvii.

[82] Verbeek in Miller and Shew, 2020.

technologies play in the construction of democratic publics. His attempt is to open further space in political discourse by recognizing how technologies build publics as well as connecting those technological matters of fact to the circulation of truth and thus the articulation of issues.

Verbeek's Arendt-Ihde contribution is another step in the right direction for the Ihdean and seats Ihde in political theory. However, technonaturalism is an approach to understanding the rule of artifice through its artifacts, including the planetary body. Technonaturalization is the process through which this is accomplished and refers to an instrumentalization of something within an environ such that it becomes governable and displays a potential to govern environmental development as a technology that *naturalizes* disciplinary relations to that instrument.[83] In a Foucauldian register, issues within the topography of milieu are identified and addressed as security and surveillance functions relating to the object as well as space of population understood through specific instruments and their culturally conditioned uses.[84] Specifically, governmentalizing actions are taken to reorganize existing relations within milieux through instruments in an effort to dominate and account for threats to populations with the recognition that threats will never be completely erased.[85] This is accomplished through the deployment and use of instruments within and through milieux forming recursive loops between what is understood through materiality while security and policing actions produce and inscribe it with normativity as truth-effects.[86] Recursive loops within naturalization processes thus are investigable sites for the Ihdean as are relations among things given as "natural."

---

[83] Foucault, Michel. Foucault, Michel. *The Foucault Reader*. Edited by Paul Rabinow. New York: Pantheon Books, 1984. 188–205.

[84] Debrix, Francois. "Cyberterror and Media-Induced Fears: The Production of Emergency Culture." *Strategies*, Vol. 14, No. 1, 2001; Foucault, 1984. 53–55, 66–67, 206–213, 241–243; Foucault, Michel. *The Birth of Biopolitics Lectures at the Collège De France, 1978–1979*. Edited by Michel Senellart. Translated by Graham Burchell, Picador, 2008. 76, 317; Officer, James E. "The Indian Service and its Evolution." In Cadwalader, Sandra L. and Vine Deloria, Jr. *The Aggressions of Civilization: Federal Indian Policy Since the 1880s*. Philadelphia: Temple University Press, 1984. 59–104; Rabinow, Paul. "Introduction." In Foucault, 1984. 20–23.

[85] Debrix, Francois. "Katechontic Sovereignty: Security Politics and the Overcoming of Time." *International Political Sociology* (2015) 9, 143–157; Foucault, Michel. *Security, Territory, Population Lectures at the College De France, 1977–1978*. François Ewald, Alessandro Fontana, and Michel Senellart, eds. Basingstoke: Palgrave Macmillan, 2007. 18, 325–6, 338, 348–49; Deloria, Vine, Jr. "'Congress in Its Wisdom.'" In Cadwalader and Deloria, 1984. 106–130.

[86] Debrix, Francois; and Alexander Barder. "Nothing to Fear but Fear: Governmentality and the Biopolitical Production of Terror." *International Political Sociology* (2009) 3, 398–413. Foucault, Michel. *"Society Must Be Defended": Lectures at the Collège De France, 1975–1976*. Edited by Mauro Bertani and Allesandro Fontana. Translated by David Macey. New York: Picador, 2003. 10–13, 28–29, 34–39; Higgs, Eric. *Nature by Design: People, Natural Process, and Ecological Restoration*. Cambridge, MA: MIT Press, 2003. 46–58.

An Ihdean subjectivity in environmentality studies opens autoethnography of environmental technics grounded in understanding cultural gestalts through instrumental and material hermeneutics and can include embodied first-person investigations of technics.[87] This would allow analysts to practice an inverted nature writing wherein "natural" spaces, such as US Forest Service Wilderness Areas, are seen as technonatural prostheses thus avoiding a romanticized "Nature."[88] Further, attempts at decolonizing travel writing have been made such that postphenomenology can be used to explore environmental technics the world over and could record the production of technonatures through understanding instrumental relationships to naturalized technologies deployed under the aegis of ecological modernization.[89] This would allow a critical embodied assessment of environmental relations to, following the above, artificial persons conceived as technoscientific institutional actors and would move closer to de-colonizing the lifeworld, if those writings and reflections were collected and shared common theoretical commitments. Pluricultures seem an obvious site and might offer another avenue into modern ethnographies as Ihde's praxis-perception model of technics recognizes technological transfer as cultural transfer.[90] Archaeologies of environmental technics, thus, are appropriate for understanding how discursive relations can change with the introduction of new instruments through instrumental hermeneutics.[91] "If we humans 'invent' technologies; then reciprocally, our technologies re-shape our lifeworlds and thus us within these worlds as well."[92]

Doing Technonaturalism

Technonatural case studies could borrow from both Timothy W. Luke and Ihde methodologically recognizing irony as a method for dissimulation and a critical material approach to understanding "environmental communications."[93] Ihdeans, to satisfy Mills, must recognize how technoscientific

---

[87] Ihde, Don. *Embodied Technics*. Automatic Press, 2010; Ihde, Don. *Listening and Voice Phenomenologies of Sound*. Albany: SUNY Press, 2007; Ihde, Don. "Bach to Rock, A Musical Odyssey." In Ihde, 1979. 93–100.

[88] Bowerbank, Sylvia. "Nature Writing as Self-Technology." In Darier, Eric, ed. *Discourses of the Environment*. Oxford: Blackwell Publishers, 1998. 152–178; Luke, Timothy W. "The Dreams of Deep Ecology." In Luke, 2019. 45–75.

[89] Lisle, Debbie. *The Global Politics of Contemporary Travel Writing*. Cambridge, UK: Cambridge University Press, 2006; Rutherford, Paul. "Ecological Modernization and Environmental Risk." In Darier, ed., 1998. 95–118.

[90] Ihde, 1993b. 28–33; Ihde, 2009. 22–23.

[91] Ihde, 2008, 1–16; Ihde, 2009. 61–62; Ley, Lukas. *Building on Borrowed Time: Rising Seas and Failing Infrastructure in Semarang*. Minneapolis: University of Minnesota Press, 2021.

[92] Ihde, 2008. vi.

[93] Ihde, 2008; Luke, Timothy W. "Introduction." In Luke, Timothy W. *Screens of Power: Ideology, Domination and Resistance in the Information Society*. Urbana: University of Illinois Press, 1989. 1–16.

understanding is thread through environmental administration and subjective governance. As Luke remarked concerning technological governmentality, technologies mediate "the central workings of power over people and things for environments in the grip of bureaucracies, markets, and systems," and are central to new modes of governmentality.[94] Technologies, and technological deployments, therefore, are part of the art of state through the co-construction of lifeworlds and technological development may imply co-evolutionary orders of governmental and subjective development through technological governmentality.[95] Marrying Ihdean postphenomenology with Luke's technological and ecocriticism should produce a distributed network of investigators operating at global scale speaking to their local conditions through both embodied and hermeneutic relations to artificial persons.[96]

A postphenomenological understanding of subjects-in-the-world, then, must include an account of everyday subjective governance through instruments and should cultivate resistance to the environing processes of artifice through embodied analysis and larger hermeneutic investigations reminiscent of governmentality studies: "Irony, I claim, has a critical effect, urging multiple perspectives for all examinations of technics…I argue that technologies are *multistable*, that while the materiality of the different artifacts display constraints in relation to human users, monolithic uses are impossible. Thus, prediction is complicated and the proliferation of variable trajectories can always occur."[97] A technonatural subjectivity must recognize artificial persons as productive orders of technoscientific fact and can see how worlds are made through the "beliefs" of artificial persons creating self-repeating loops between subjective micro-perceptions and narratives produced by extra-humans about materiality.[98] Technocracies, for example, cannot predict the development trajectories of their technologies and thus one can dismiss attempts at fixing narratives around technocratic authority related to technonaturalized environmental risk.[99] This would speak to the dominance of artificial persons globally

---

[94] Luke, Timothy W. "Technology." In Death, 2013. 268, 272.

[95] DeLanda, Manuel. *War in the Age of Intelligent Machines*. New York: Zone, 1991.

[96] Luke, 1989. 8.

[97] Ihde, 2008. vi. Emphasis in original.

[98] Baudrillard, 1983. 103–159; Luke, Timothy W. "After One-Dimensionality: Culture and Politics in the Age of Artificial Negativity." In Luke, 1990. 159–181; Higgs, 2003; Luke, Timothy W. "The System of Sustainable Degradation," *Capitalism Nature Socialism*, 17:1, 99–112. 2006. https://doi.org/10.1080/10455750500505556; Mitchell, Timothy. "Egypt at the Exhibition." In Mitchell, Timothy. *Colonizing Egypt*. Berkeley, CA: University of California Press, 1991. 1–33; Mitchell, Timothy. "The Character of Calculability." In Mitchell, 2002. 80–119.

[99] Ihde, 2008. 12–29; Mitchell, Timothy. "Can the Mosquito Speak?" In Mitchell, 2002. 19–53.

through investigations in material culture related to power and would investigate relationships to artificial persons through co-productive orders of material fact as part of machine-hermeneutics.[100]

Luke understands environments as composed of counterbalancing, and competing powers vying for the construction of the real through the production of material fact.[101] This recognizes the shrinking remit of sovereign state power through an understanding of institutional environmental governance and the increasing salience of disciplinary power connected to non-state actors in the construction and administration of "environment."[102] Central to his understanding of environmental governmentality—or environmentality—are extra-human agencies embodied by artificial persons transcending both individual human agency and overlapping and enfolding such that space is intimately glocal within environments constructed by transnational corporate organization.[103] The corporation for Luke, is a sort of machine, or technology of environmental orders that regiment, discipline and support human and non-human populations.[104] Those machines are partially capital, partially labor, partially material, partially symbolic and, as actors, are conceived governmentally as artificial persons—rights-bearing agents that organize, in part, organic persons and direct their energies to the accumulation of capital and the production of operational legitimacy.[105] Fundamentally, those extra-human agencies are the engines of planetary production and have the power to

---

[100] Ihde, 1979. 3–15; Luke, Timothy W. *Museum Politics: Power Plays at the Exhibition*. Minneapolis: University of Minnesota Press, 2002; Luke, Timothy W. "Technique in Marx's Method of Political Economy." In Luke, 1990. 51- 68.

[101] Luke, Timothy W. "The Dangers of Discourse: Polyarchy and Megatechnics as Environmental Forces." In *Capitalism, Democracy and Ecology: Departing from Marx*. Urbana: University of Illinois Press, 1999. 88–117.

[102] Luke, Timothy W. "Simulated Sovereignty, Telematic Territoriality: The Political Economy of Cyberspace." In Featherstone, Mike and Scott Lash, eds. *Spaces of Culture: City, Nation, World*. SAGE: London, 1999. 27–48; Luke, Timothy W. "New World Orders and Neo-World Orders: Power, Politics and Ideology in Infromationalizing Glocalities." In Featherstone, Mike, Scott Lash, and Roland Robertson (eds.). *Global Modernities*. 91–107.

[103] Luke, Timothy W. "Biospheres and Technospheres: Moving from Ecology to Hyperecology with the New Class. In Luke, 1999. 59–87; Luke, Timothy W. "Ecodiscipline and the Post-Cold War Global Economy: Rethinking Environmental Critiques of Geo-Economics." In Luke, 1999. 143–169.

[104] Luke, Timothy W. "The Property Boundaries/Boundary Properties in Technonature Studies: "Inventing the Future." In White and Wilbert, eds., 2009. 193–213.

[105] Luke, Timothy W. "Cyborg Enchantments: Commodity Fetishism and Human/Machine Interactions." *Strategies*: Vol. 13, No. 1, 2000. https://doi.org/10.1080/10402130050007511; Luke, Timothy W. "Liberal Society and Cyborg Subjectivity: The Politics of Environments, Bodies and Nature." *Alternatives: Global, Local, Political*: Vol. 21, No. 1, Jan-Mar. 1996. https://doi.org/10.1177/030437549602100101; "On Environmentality: Geo-Power and Eco-Knowledge in the Discourses of Contemporary Environmentalism." *Cultural Critique*, No. 31, The Politics of Systems and Environments, Part II. (Autumn, 1995), pp. 57–81.

shape the environments and thus the lives of many living in their polyarchic environs through environmentalities.[106] Their environmentalities are enabled and produced by specific assemblages of machines, humans and non-humans adding up to technocratic production of "facts" and their culturally conditioned application to issues occurring within governmentalized localities.

"The Anthropocene" is an example of how issues can be instrumentalized by extra-human agencies becoming a technology of governance.[107] Luke's ecocritique of the Worldwatch Institute, for example, discusses the production and circulation of knowledge about planetary milieux such that Worldwatch is both an instrument and institution of global environmental governance through biopolitical policy recommendations grounded in the use and production of "facts."[108] This echoes work in political ecology recognizing information flows as paramount to the production and administration of order as part of disciplinary power and anchors the art of state in the administration of symbolic flows about lifeworlds constructed through governmentality.[109] Further, Luke's ecocritique of the Worldwatch is a bridge into understanding environmentality thus opening larger structural analyses for the Ihdean.[110]

A thoroughly Ihdean reading of the above would recognize the Worldwatch as producing information that is "real," but is pragmatically understood as incomplete and thus cannot ground claims to total or absolute knowledge. Further, an Ihdean understanding of environmentality would recognize the Worldwatch as attempting intersubjective adjustments to larger cultural gestalts in pursuit of adjusting micro-perceptual sense-making through the Worldwatch as an instrument and connected to or working against other globally massive organizations. This could be analyzed as a special instance in macro-perception through Ihdean machine-relations recognizing the place of the human body within assemblages such as globally massive organizations.[111]

---

[106] Luke, Timothy W. "Environmentality." In *Oxford Handbook of Climate Change and Society*. Dryzek, John S., Richard B. Norgaard, and David Schlosberg, eds. Oxford: Oxford University Press, Jan 2012. https://doi.org/10.1093/oxfordhb/9780199566600.003.0007; Luke, 1995. 57–81; Luke, Timothy W. "On the Politics of the Anthropocene." In. Luke, 2019. 205–229.

[107] Luke, 2019. 1–15.

[108] Luke, Timothy W. "Worldwatching at the Limits of Growth." In *Ecocritique: Contesting the Politics of Nature, Economy and Culture*. Minneapolis: University of Minnesota Press, 1997. 75–94.

[109] Braverman, Irus. "En-listing Life: Red is the Color of Endangered Species Lists." In Gillespie, Catherine and Rosemary-Clare Collard, eds. *Critical Animal Geographies: Politics, Intersections and Hierarchies in a Multispecies World*. New York: Routledge, 2015. 184–202; Luke, 1989; Rutherford, Stephanie. *Governing the Wild: Ecotours of Power*. Minneapolis: University of Minnesota Press, 2011; Whitehead, Mark, Rhys Jones, and Martin Jones. *The Nature of the State: Excavating the Political Ecologies of the Modern State*. Oxford: Oxford University Press, 2007.

[110] Luke, 1995. 57–81.

[111] Luke, 1995. 65.

"We have now, two distinctive types of human-machine relations which stand in contrast: embodiment relations in which something is experienced *through* a machine, and hermeneutic relations in which the machine becomes an 'other' as a focal object of experience."[112]

In the first instance, how humans relate to the machinery of the Worldwatch is through how knowledge and experience are produced within that machine. This again can be broken into two:

> [I]nvariants to note in terms of embodiment relations. First, in embodiment relations we experience an otherness and in this sense the experience through a machine must be described as a partially transparent relation. Secondly and simultaneously, the experience is a transformed experience which has a difference between it and all 'face to face' or 'in the flesh' experience. This transformation contains the possibilities, again co-implied, of both a certain extension and amplification of experience and of a reduction and transformation of experience.[113]

Worldwatch, according to Luke is an eco-disciplinary organization translating knowledge about "the environment" into geo-power as an instrument of planetary monitoring and development "as biological existence was refracted through economic, political, and technological existence, 'the facts of life' passed into fields of control for eco-knowledge and spheres of intervention for geo-power."[114] Thus the "facts" produced by Worldwatch are attempting to discipline others into normatively "correct" behavior concerning the planetary body while actually producing the planetary body through the acts of examination and intellection inherent in the production of scientific knowledge. In this way, Luke is speaking to an Ihdean point about knowing and understanding the planetary through technoscientific institutions such as Worldwatch: "important here is to note that we have moved from experiencing through machines to experiences of machines."[115]

Speaking back to the technological context through which Worldwatch, as a machine, produces information about that context Luke writes:

> In turn, Nature's energies, materials, and sites are redefined by the eco-knowledges of resource managerialism as the source of "goods" for sizable numbers of some people, even though greater material and immaterial "bads" also might be inflicted upon even larger numbers of other people who do not reside in or benefit from the advanced national economies that basically monopolize the use of world resources at a comparative handful of highly developed regional and municipal sites. Many of these eco-knowledge assumptions and

---

[112] Ihde, 1979. 13. Emphasis in original.
[113] Ibid., 10.
[114] Luke, 1995. 67.
[115] Ihde, 1979. 11.

geo-power commitments can be seen at work in the discourses of the Worldwatch Institute as it develops its own unique vision of environmentality for a global resource managerialism.[116]

How that knowledge is produced, what machines are used, who uses them and how they are used are important questions for the Ihdean, especially regarding how "the environment" could be known and articulated. "Through the machine something (presumably) still happens elsewhere, only in this case the engineer does not experience the terminus of the intention which traverses the machine…His primary experiential terminus is with the machine. I shall thus call this relation a *hermeneutic relation*. There is a partial opacity between the machine and the World, and thus the machine is something like a text."[117] In other words, the scale of technoscientific knowledge about an object so big no one perceives it—"the environment"— is known primarily through relating to information produced through globalized machinery. Perhaps, most importantly, Ihde sees this relation as "Human → (Machine-World),"[118] in the production of knowledge about embodiment relations seen as "(Human-Machine) → World."[119]

The technological opacity above can be seen in the reductive focus of resource managerialism to continuing business-as-usual without recognizing the boundaries of facts placed by that technological reduction. This speaks to the disciplinary power of machinery over subjects, as machines, like the Worldwatch, are composed partially of humans oriented to specific problemata with specific instruments for understanding and articulating them. "This is not to say in any case that the machine has intentionality—but it is to point to the source of such pseudo-problems in the structure of the relationships itself. In relations in which machines are focal 'others' all of the ambiguity of other relations becomes a possibility. The machine is capable of anthropomorphization in terms of its 'otherness.'"[120] The machines are, in contemporary technoscience what speaks for materiality through the production of fact "in a world where corporate capital can pretend it brings all good things to life, one must ask how the built and yet to be built environments are shaped to sustain such good things, what ecologies generate which forms of life for whom, and where good things do get brought into being when corporate science and technology get down to business."[121] Fundamentally missing in Ihde is the notion of the artificial person as machinery of social organization that do "act" with intentions seen in processes of geotechnic environing producing lifeforms through adjustments in materiality.

[116] Luke, 1995. 71.
[117] Ihde, 1979. 12. Emphasis in original.
[118] Ibid. 12.
[119] Ibid. 13.
[120] Ibid.
[121] Luke, 1999. 51–52.

Environing for Luke is a process within geoengineering:

> Environmentalized places become sites of supervision, where environmentalists see from above and from without through the enveloping designs of administratively delimited systems. Encircled by enclosures of alarm, environments can be disassembled, recombined, and subjected to the disciplinary designs of expert management. Enveloped in these interpretive frames, environments can be redirected to fulfill the ends of other economic scripts, managerial directives, and administrative writs. Environing, then, engenders "environmentality," which embeds instrumental rationalities in the policing of ecological spaces.[122]

This means that Ihde can recognize both the disciplinary power of "facts" and the disciplinary power of instruments grounded in intersubjective use in the production of "environment" through the materiality of those machines themselves and the perceptual transformations they perform in use. This can include processes related to environing such as: extermination regimes, the weaponization of landscape, the production of technoregions, the instrumentalization of species and the production of environmental subjects. And these processes can be understood as intentional through larger narratives about the objects under administration.

This is, in other words, a process of instrumentalization displaying an instrumentality connected to the domination of machinery within cyborganic subjective hermeneutics. The technonatural imaginary de-anthropomorphizes "the Anthropocene" to recognize processes of planetary colonization by technologies belonging to no one and operating with a *seeming* independence from any one human actor. Anthropologically, this can imply multi-scalar data collection regarding structural positionality within hierarchy as well as subjective orientations to machines. "People in institutions such as bureaucracy appear mostly as the product of sheer density and authority constituted by institutionalized materiality—that is, as subjected to forms, regulations, conventions, and procedures...It is at this institutional level that the general point becomes remarkably clear: that power is, among other things, a property of materiality."[123] This is not to imply a total and complete covering of the Earth by artificial persons, but speaks to the instrumentalism inherent in thought guiding planetary development. All things are potentially instrumentalizable and among the entities materially and financially interested in the circulation of "Nature," or "the Anthropocene," "Purity," "the Given" and other myths are artificial persons pursuing dominance within the lifeworld through the sedimentation of relationships to the Real reflective of and favorable to their actions. New Class managers can and should reflect on their structural positions to machines and commodities and offer technonatural studies.

---

[122] Luke, 1995. 65.
[123] Miller, 2005. 20.

Colonization of the Lifeworld displays the truths of historical materialism and the need for understanding planetary affairs through both an ecosystems of capital approach and a machine ecosystems approach.[124] The ascendance of the machine as a background condition for geotechnic development was already noticed by Marx who, commenting on the spread of global capitalism said, "In machinery, the appropriation of living labor by capital…[that] confronts the worker in a coarsely sensuous form: capital absorbs labour into itself—'as though its body were by love possessed.'"[125] Similarly, Luke grounds his understanding of planetary production through systems of machines operating for the remote extractive interests of urban capital by directing labor thus regimenting environmentalized relations to materiality.[126] However, his approach eschews "Capital" as referring to the Bourgeoise and, instead, finds the independence and autonomy of industrialized megamachinery as resulting from segments of instrumentalized humans ritualizing and reproducing commodity form logics within their lifeworlds as forms of social, political, ethical and environmental consciousness.[127] This is connected to the ascendance of New Class of managerial "experts" endemic to industrialized systems of production and not simply "capitalism."[128]

The pivot above de-territorializes "capital" as a theoretical object finding it instead as a productive force more appropriately understood as an environmental gestalt—as a means of social reproduction within, through and because of artificial persons. Refocusing on commodities displaces the importance of any one actor and recognizes planetary production understood through the commodity nexus in the production of material fact through globe-spanning

---

[124] Mumford, Lewis. *Technics and Civilization*. New York: Harcourt Brace Jovanovich, 1934; Mumford, Lewis. *The City in History: Its Origins, Its Transformations and Its Prospects*. New York: Harcourt, 1961; Mumford, Lewis. *The Myth of the Machine: Technics and Human Development*. New York: Harcourt Brace Jovanovich, 1966; Mumford, Lewis. *The Myth of the Machine: The Pentagon of Power*. New York: Harcourt Brace Jovanovich, 1970.

[125] Marx, Karl. *Grundrisse: Foundations of the Critique of Political Economy*. Martin Nicolaus, trans. New York: Random House, 2005. 703–704.

[126] Luke, Timothy W. "Searching for Alternatives: Postmodern Populism and Ecology." In Luke, 2019. 114–140; Luke, 1999.

[127] Luke, Timothy W. "Community and Ecology." In Luke, 2019. 76–87; Luke, Timothy W. "Departures from Marx: Rethinking Ecologies and Economies." In Luke, 1999. pp. 29–58.

[128] Djilas, Milovan. *The New Class: An Analysis of the Communist System*. London: Thames and Hudson, 1957; Gouldner, Alvin Ward. *The Future of Intellectuals and the Rise of the New Class: A Frame of Reference, Theses, Conjectures, Arguments, and an Historical Perspective on the Role of Intellectuals and Intelligentsia in the International Class Contest of the Modern Era*. London: Macmillan, 1979; Konrád, György, and Szelényi, Iván. *The Intellectuals on the Road to Class Power*. Translated by Andrew Arato and Richard E. Allen. London: Harvester, 1979; Lasch, Christopher. *The Revolt of the Elites and the Betrayal of Democracy*. New York: W. W. Norton & Company, 1996; Luke, Timothy W. *Ideology and Soviet Industrialization*. Westport, CT: Greenwood Press. 1985. 61–72; Luke, Timothy W. "Informationalism and Ecology." In Luke, 2019. 25–44.

circuitry and consumptive networks guided by corporate development logics. "Sustainable development enthusiasts are celebrants of a complete and final transition to a fully processed world which must be subjected to total administration to maximize efficiency and maybe even equity."[129]

Ihde would and does recognize machinery and its products as forming embodied background relations in the lifeworld and defines them as that which "we live and move and engage with [in] an immediate environment, much in the environment is unthematized and taken for granted. And, in any technologically saturated 'world' this background includes innumerable technologies to which we most infrequently attend."[130] This is, as Feenberg in an homage to Ihde, has titled a recent contribution (2017)[131] the *technosystem*[132] which includes, analytically, background, transparency and alterity relations of technologies to bodies within them and is drawn from Lewis Mumford's machine ecosystem analyses of megatechnics.[133] Importantly, relations to technologies vary by degree through transparency relations speaking to the praxis-perception models created through their adoption. In short, many in the most industrialized technoregions are dependent on megatechnics for their daily lives to the point that those relations form ecosystems and enable some forms of life over others.[134] This can literally and materially extinguish lifeforms that do not "fit" into the materiality of planetary megatechnics[135] and assimilate others into "productive" circuitry.[136] As a technology breaks, or simply stands outside quasi-transparencies, it makes itself "other"

---

[129] Luke, 2019. 296.

[130] Ihde, 2009. 43.

[131] Feenberg, 2017.

[132] Ihde, 1990. 3.

[133] Ihde, 1990. 26–27.

[134] Logan, Nneka. "Corporate Personhood and the Corporate Responsibility to Race." *Journal of Business Ethics* (2019) 154: pp. 977–988 https://doi.org/10.1007/s10551-018-3893-3; Logan, Nneka. "Corporate Voice and Ideology: An Alternate Approach to Understanding Public Relations History." *Public Relations Review* 40 (2014): pp. 661–668; Logan, Nneka. "The White Leader Prototype: A Critical Analysis of Race in Public Relations." *Journal of Public Relations Research*, 23:4, 442–457, DOI: 10.1080/1062726X.2011.605974; Mitchell, Timothy. *Carbon Democracy: Political Power in the Age of Oil*. New York: Verso, 2011; Tomlinson, John. *The Culture of Speed: The Coming of Immediacy*. London: SAGE, 2007.

[135] Bending, Tim. *Penan Histories: Contentious Narratives in Upriver Sarawak*. Leiden, Netherlands: KITLV Press, 2006; Brockington, Dan. *Fortress Conservation: The Preservation of the Mkomazi Game reserve, Tanzania*. Bloomington: Indiana University Press, 2002; Sandoz, Mari. *Cheyenne Autumn*. New York: Avon books. 1953.

[136] Joyce, John, Joseph Nevins, and Jill S. Schneiderman. "Commodification, Violence, and the Making of Workers and Ducks at Hudson Valley Foie Gras." In Katherine and Collard, eds. 93–108; Mitchell, Timothy. "The Invention and Reinvention of the Peasant." In Mitchell, 2002. 123–152; Morin, Karen M. "Wildspace: The cage, the Supermax, and the Zoo." In Gillespie and Collard, eds., 2015. 73–92.

for Ihde and displays alterity relations.[137] The Earth, though not *controlled* by technologies, is dominated by them to the point that reading geological history without recognizing the influence of transnational capitalism is wrong-headed from a material perspective.[138] Though not the only geological game in town, an understanding of how planetary megatechnics and thus technonature [mal]functions requires a historically grounded view of the planetary body through the imbrications and variations of capital's development in commodity production and circulation as connected to globe-spanning machinery that forms the background conditions for life, as it can be known currently, on Earth.[139]

Environmentality is keenly useful in criticizing ecological modernization and offers ways into understanding the production of space, subjectivation, normalization, eco-discipline, naturalization and technological rule through artificial persons.[140] Pairing Luke and Ihde connects both to a common influence pre-dating Foucauldian governmentality studies through Lewis Mumford. Mumford's megatechnics connects Ihde and Luke back to approaches in world ecology and this provides an archive for the technonaturalist in understanding planetary technics as resulting from humans+extrahuman agency conceived as geotechnics and geoengineering.

Marrying the above with World Ecology approaches akin to Jason W. Moore's work should allow analytical moves from the positionality of Ihde's R&D philosopher to the technonaturalist.[141] Ihde, Moore and Luke all draw from Lewis Mumford and this common link allows the technonaturalist to see the naturalization and deployment of technologies within environmental development as well as understand sites, such as urbanity, as technologies for technonaturalization that span, following Benton MacKaye—an influence and friend of Mumford's—global networks obliterating urban/rural dyads creating "civilizational wildernesses."[142]

[137] Ihde, 2009. 43.

[138] Luke, 1999; Luke, 2019; Luke, Timothy W. "Marcuse and the Politics of Radical Ecology." In Luke, 1997. 137–152. Luke, 1999; Luke, 2019; Luke, Timothy W. "Marcuse and the Politics of Radical Ecology." In Luke, 1997. 137–152.

[139] Cronon, William. *Nature's Metropolis: Chicago and the Great West*. New York: W.W. Norton Company, 1991; Harvey, David. *Spaces of Global Capitalism: Towards a Theory of Uneven Geographical Development*. New York: Verso, 2006; Mitchell, Timothy. "Principles True in Every Country." In Mitchell, 2002. 54–79; Tsing, Anna Lowenhaupt. *Friction: An Ethnography of Global Connection*. Princeton: Princeton University Press, 2004: White, Rudy and Gareau, 2016. 36–50.

[140] Agrawal, Arun. *Environmentality: Technologies of Government and the Making of Subjects*. Durham, NC: Duke University Press, 2005; Rutherford, 2011.

[141] Ihde, 2002. 103–126.

[142] MacKaye, Benton. "The Wilderness of Civilization." In MacKaye, 1962. 16–25; Mumford, Lewis. "Introduction." In MacKaye, 1962. vii–xxii.

Conclusion: The Technonaturalist and the Wilds of Civilization

Civilizational wildernesses must be explored because of the collision of mechanized being with what falls outside of it: "They were confronted by a wilderness of nature; we are confronted by a wilderness of near infinite complexity; it is one also of monotonous, standardized, mechanized uniformity. And, what is of chief significance, it is a wilderness which '*flows*.' We are invested not merely by a wilderness of civilization, but by an invasion of civilization."[143] The technonaturalist must be aware of the extensive and morphogenic powers of capital in the production of material fact and MacKaye claims that geotechnics and the new exploration are concerned with habitability through the understanding of flows—including commodities—essential to civilizational viability.[144] Those flows will have collection points physically that express relations to environment that are frequently *naturalized*. These hybrid spaces can range from slums to "wilderness" areas, but should aim at unseating megatechnic dominance in the construction and production of precarious life through naturalization.[145] Further, naturalized civilizational prosthetics, such as novel ecosystems and rewilded places, can be explored as technological space through recognizing their landscape hermeneutics and the beings enabled by their instrumentality.[146] As institutions become aware of The Sixth extinction, for example, their governmental reach can include the production of lifeforms perceived as "natural" through rewilding programs.[147]

Technonaturalism is an attempt to read materiality in terms of how civilizational support systems dominate the planet through the gradual absorption and linking of organic processes and histories to the machinations of artificial

---

[143] MacKaye, 1962. 225.

[144] MacKaye, Benton. *From Geography to Geotechnics*. Urbana: University of Illinois Press, 1968. 99, 102–141; Marx, Karl. *Capital: A Critique of Political Economy*. Edited by Friedrich Engels. Vol. II. Moscow: Foreign Languages Publishing House, 1961; Marx, Karl. *The German Ideology: including Theses on Feuerbach and Introduction to the Critique of Political Economy*. Amherst: NY: Prometheus Books, 1998.

[145] Boo, Katherine. *Behind the Beautiful Forevers: Life, Death and Hope in a Mumbai Undercity*. New York: Random House, 2012; Lerner, Steve. *Sacrifice Zones: The Front Lies of Toxic Chemical Exposure in the United States*. Cambridge, MA: The MIT Press. 2010; MacKaye, 1968. 91, 102, 110–173.

[146] Barnard, Timothy P. *Imperial Creatures: Humans and Other Animals in Colonial Singapore, 1819–1942*. Singapore: National University of Singapore Press, 2019; Barnard, Timothy P. *Nature's Colony: Empire, Nation and Environment in the Singapore Botanic Gardens*. Singapore: National University of Singapore Press, 2016; Sheail, John. *Nature's Spectacle: The World's First National Parks and Protected Places*. London: Earthscan, 2010.

[147] Hannibal, Mary Ellen. *The Spine of the Continent: The Race to Save America's Last, Best Wilderness*. Gilford, CT: Lyons Press; 2012; Marris, Emma. *Rambunctious Garden: Saving Nature in a Post-wild World*. New York: Bloomsbury, 2011; Monbiot, George. *Feral: Rewilding the Land, the Sea, and Human Life*. Chicago: University of Chicago Press, 2014.

persons.[148] More specifically, its focus on technology as that which mediates the processes of technonaturalization implies that humanity and human history need not be treated monolithically in coming to grips with the present state of affairs at the planetary scale.[149] As segments of humanity came to dominate their environments, those with techno-power of artificial personhood came to dominate other humans simultaneously linking and collapsing the natural and the social through synthetic assemblages of instruments linked together through communicative networks enabled by the written word.[150] In keeping with Latour, the activities of artificial persons have populated the planetary body with hybrids emanating from purification within the lifeworld producing ironic spatiality and organisms resulting from geotechnics.[151] Naturalization, in the Ihdean sense is when a technology seeps into background relations such that it is taken as a given and the historical record shows artificial persons attempting to naturalize their relations to humans through geotechnic extensions of "nature."[152]

This is a historically accurate way of understanding the production of "the Anthropocene" not as the result of the *Anthropos* but of artificial persons that have shaped and continue to shape lifeworlds of cyborg subjects networked to them globally through technonatural environing. Not only does this allow for critical power in understanding the state of environmental affairs and issues, but more closely resembles a true negative to Anthropocene discourse by de-naturalizing "the environment" through anti-anthropocentric, environmental technic analyses, reflections and records.[153] Those reflections and records, it is hoped, will point to the culprits responsible for technological accidents, such as global environmental degradation. Seen as unexpected through the ambit of machinic certainty, megatechnic malfunctions are the return of *the wild* through technonatural environs—not a return of the *Anthropos*. Ihde resists totalizing narratives about technology and maintains that any human is capable of "unplugging."[154] Planetary background conditions and the forms of life

---

[148] Oosterveer, Peter. "Governing Global Environmental flows: Ecological Modernization in Technonatural Time/Spaces." In White and Wilbert, eds., 2009. 33–60.

[149] Swyngedouw, Erik. "Circulations and Metabolisms: [Hybrid] Natures and [Cyborg] Cities." In White and Wilbert, eds., 2009. 61–84.

[150] Cadwalader, and Deloria, Jr., eds., 1984; Davison, Aidan. "Living Between Nature and Technology: The Suburban Constitution of Environmentalism in Australia." In White and Wilbert, 2009. 167–189; Deloria, Vine, Jr. *Custer Died for Your Sins: An Indian Manifesto*. New York: Macmillan, 1969; Ihde, 1990. 8, 20, 124–160, 190, 191; Luke, 1999; Mumford, Lewis, 1966; Mumford, Lewis, 1970.

[151] Latour, 1993; Luke, 1996; MacKaye, Benton, 1968.

[152] Barnard, Timothy P, Corinne Heng. "A City in a Garden." In Barnard, Timothy P., ed. *Nature Contained: Environmental Histories of Singapore*. Singapore: National University of Singapore Press, 2014. 281–306.

[153] Luke, Timothy W. "Reflections from a Damaged Planet: Adorno as Accompaniment to Environmentalism in the Anthropocene." In Luke, 2019. 287–304.

[154] Ihde, Don. "Of Which Human are We Post?" In Ihde, 2008. 43–57.

they enable beg to differ, but Luke's ecotechnic populism within an Ihdean R&D philosopher offers hope for a new planetary becoming. Importantly, the only way for Ihde that "Technology" could entirely dominate "Humanity" is after "we may already be close to having irreversibly fouled our own earth—just at the moment we have fully inherited it."[155] The technonaturalist, then should "reveal within our innate country, despite the fogs and chaos of cacophonous mechanization *a land in which to live*—a symphonious environment of melody and mystery in which, throughout all ages, we shall 'learn to reawaken and keep ourselves awake, not by mechanical aids,' but by that 'infinite expectation of the dawn' which faces the horizon of an ever-widening vision."[156]

References

Agrawal, Arun. *Environmentality: Technologies of Government and the Making of Subjects*. Durham, NC: Duke University Press, 2005.

Allenby, Braden R. *Reconstructing Earth: Technology and Environment in the Age of Humans*. Washington, D.C.: Island Press, 2005.

Barnard, Timothy P. *Imperial Creatures: Humans and Other Animals in Colonial Singapore, 1819–1942*. Singapore: National University of Singapore Press, 2019.

Barnard, Timothy P. *Nature's Colony: Empire, Nation and Environment in the Singapore Botanic Gardens*. Singapore: National University of Singapore Press, 2016.

Barnard, Timothy P. ed. *Nature Contained: Environmental Histories of Singapore*. Singapore: National University of Singapore Press, 2014.

Baudrillard, Jean. *Simulations*. Foss, Paul, Paul Patton, and Philip Beitchman (trans.). Cambridge, MA: MIT Press: Semiotext(e). 1983.

Bending, Tim. *Penan Histories: Contentious Narratives in Upriver Sarawak*. Leiden, Netherlands: KITLV Press, 2006.

Boo, Katherine. *Behind the Beautiful Forevers: Life, Death and Hope in a Mumbai Undercity*. New York: Random House, 2012.

Braverman, Harry. *Labor and Monopoly Capital: The Degradation of Work in the Twentieth Century*. New York: Monthly Review Press, 1974.

Brenner, Neil, ed. *Implosions/Explosions: Towards a Study of Planetary Urbanization*. Berlin: Jovis Verlag, 2014.

Brockington, Dan. *Fortress Conservation: The Preservation of the Mkomazi Game reserve, Tanzania*. Bloomington: Indiana University Press, 2002.

Cadwalader, Sandra L. and Vine Deloria, Jr., eds. *The Aggressions of Civilization: Federal Indian Policy Since the 1880s*. Philadelphia: Temple University Press, 1984.

Caraccioli, Mauro J. *Writing the New World: The Politics of Natural History in the Early Spanish Empire*. Gainesville, FL: University of Florida Press, 2021.

Cronon, William. *Changes in the Land: Indians, Colonists, and the Ecology of New England*. New York: Hill and Wang, 1983.

---

[155] Ihde, 1990. 163.

[156] MacKaye, 1962. 228. Emphasis in original.

Cronon, William. *Nature's Metropolis: Chicago and the Great West*. New York: W.W. Norton Company, 1991.
Darier, Eric, ed. *Discourses of the Environment*. Oxford: Blackwell Publishers, 1998.
Death, Carl, ed. *Critical Environmental Politics*. London: Routledge, 2013.
Debrix, Francois. "Cyberterror and Media-Induced Fears: The Production of Emergency Culture." *Strategies* (2001), Vol. 14, No. 1: pp. 149–168.
Debrix, Francois. Katechontic Sovereignty: Security Politics and the Overcoming of Time." *International Political Sociology* (2015) 9: pp. 143–157.
Debrix, Francois, and Alexander Barder. "Nothing to Fear but Fear: Governmentality and the Biopolitical Production of Terror." *International Political Sociology* (2009) 3: pp. 398–413.
DeLanda, Manuel *A New Philosophy of Society: Assemblage Theory and Social Complexity*. New York: Bloomsbury, 2006.
DeLanda, Manuel. *War in the Age of Intelligent Machines*. New York: Zone, 1991.
Deloria, Vine, Jr. *Custer Died for Your Sins: An Indian Manifesto*. New York: Macmillan, 1969.
Djilas, Milovan. *The New Class: An Analysis of the Communist System*. London: Thames and Hudson, 1957.
Dryzek, John S., Richard B. Norgaard, and David Schlosberg, eds. *Oxford Handbook of Climate Change and Society*. Oxford: Oxford University Press, Jan 2012. https://doi.org/10.1093/oxfordhb/9780199566600.003.0007
Featherstone, Mike and Scott Lash, eds. *Spaces of Culture: City, Nation, World*. SAGE: London, 1999.
Featherstone, Mike, Scott Lash, and Roland Robertson, eds. *Global Modernities*. SAGE: London, 1995.
Feenberg, Andrew. *Technosystem: The Social Life of Reason*. Cambridge, MA: Harvard University press, 2017.
Feyerabend, Paul. *Against Method*. New York: Verso, 1975.
Foster, John Bellamy. *The Ecological Rift: Capitalism's War on the Earth*. New York: Monthly Review Press, 2010.
Foster, John Bellamy. *The Theory of Monopoly Capitalism: An elaboration of Marxian Political Economy*. New York: Monthly Review Press, 1986.
Foucault, Michel. *Security, Territory, Population Lectures at the College De France, 1977–1978*. François Ewald, Alessandro Fontana, and Michel Senellart, eds. Basingstoke: Palgrave Macmillan, 2007.
Foucault, Michel. *"Society Must Be Defended": Lectures at the College De France, 1975–1976*. Edited by Mauro Bertani and Allesandro Fontana. Translated by David Macey. New York: Picador, 2003.
Foucault, Michel. *The Birth of Biopolitics Lectures at the Collège De France, 1978–1979*. Edited by Michel Senellart. Translated by Graham Burchell, Picador, 2008.
Foucault, Michel. *The Foucault Reader*. Paul Rabinow, ed. New York: Pantheon Books, 1984.
Gare, Arran and Wayne Hudson, eds. *For a New Naturalism*. Candor, NY: Telos Press Publishing, 2017.
Gouldner, Alvin Ward. *The Future of Intellectuals and the Rise of the New Class: A Frame of Reference, Theses, Conjectures, Arguments, and an Historical Perspective on the Role of Intellectuals and Intelligentsia in the International Class Contest of the Modern Era*. London: Macmillan, 1979.

Gillespie, Catherine and Rosemary-Clare Collard, eds. *Critical Animal Geographies: Politics, Intersections and Hierarchies in a Multispecies World*. New York: Routledge, 2015.
Gray, Chris Harbles, Heidi J. Figueroa-Sarriera, and Steven Mentor, eds. *The Cyborg Handbook*. New York: Routledge, 1995.
Hannibal, Mary Ellen. *The Spine of the Continent: The Race to Save America's Last, Best Wilderness*. Gilford, CT: Lyons Press, 2012.
Haraway, Donna J. *Modest_Witness@Second_Millennium. FemaleMan_Meets_OncoMouse: Feminism and Technoscience*. New York: Routledge, 1997.
Haraway, Donna J. *Simians, Cyborgs and Women: The Reinvention of Nature*. New York: Routledge, 1991.
Harvey, David. *Spaces of Global Capitalism: Towards a Theory of Uneven Geographical Development*. New York: Verso, 2006.
Higgs, Eric. *Nature by Design: People, Natural Process, and Ecological Restoration*. Cambridge, MA: MIT Press, 2003.
Ihde, Don. *Bodies in Technology*. Minneapolis: University of Minnesota Press, 2002.
Ihde, Don. *Embodied Technics*. USA: Automatic Press, 2010.
Ihde, Don. *Experimental Phenomenology: An Introduction*. New York: Capricorn books, 1977.
Ihde, Don. *Instrumental Realism: The Interface Between Philosophy of Science and Philosophy of Technology*. Bloomington: University of Indiana Press, 1991.
Ihde, Don. *Ironic Technics*. USA: Automatic Press, 2008.
Ihde, Don. *Listening and Voice Phenomenologies of Sound*. Albany: SUNY Press, 2007.
Ihde, Don. *Philosophy of Technology: An Introduction*. New York: Paragon House Publishers, 1993a.
Ihde, Don. *Postphenomenology: Essays in a Postmodern Context*. Evanston: Northwestern University Press, 1993b.
Ihde, Don. *Postphenomenology and Technoscience: The Peking University Lectures*. Albany: State University of New York Press, 2009.
Ihde, Don. *Technics and Praxis*. Dordrecht, Holland: D. Reidel Publishing Company, 1979.
Ihde, Don. *Technology and the Lifeworld: From Garden to Earth*. Bloomington: Indiana University Press, 1990.
Ihde, Don and Evan Selinger, eds. *Chasing Technoscience: Matrix for Materiality*. Bloomington: Indiana University Press, 2003.
Jagtenberg, Tom and David McKie. *Eco-Impacts and the Greening of Postmodernity: New Maps for Communication Studies, Cultural Studies and Sociology*. London: SAGE, 1997.
Knick, Steven T. and John W. Connelly, eds. *Greater Sage-grouse: Ecology and Conservation of a Landscape Species and its Habitats*. University of California Press and The Cooper Ornithological Society, 2011.
Konrád, György, and Szelényi, Iván. *The Intellectuals on the Road to Class Power*. Translated by Andrew Arato and Richard E. Allen. London: Harvester, 1979.
Lasch, Christopher. *The Revolt of the Elites and the Betrayal of Democracy*. New York: W. W. Norton & Company, 1996.
Latour, Bruno. *Reassembling the Social: An Introduction to Actor Network Theory*. Oxford: Oxford University Press, 2005.

Latour, Bruno. *We Have Never Been Modern*. Translated by Catherine Porter. Cambridge, MA: Harvard University Press, 1991.
Lerner, Steve. *Sacrifice Zones: The Front Lines of Toxic Chemical Exposure in the United States*. Cambridge, MA: The MIT Press. 2010.
Ley, Lukas. *Building on Borrowed Time: Rising Seas and Failing Infrastructure in Semarang*. Minneapolis: University of Minnesota Press, 2021.
Lisle, Debbie. *The Global Politics of Contemporary Travel Writing*. Cambridge, UK: Cambridge University Press, 2006.
Logan, Nneka. "Corporate Personhood and the Corporate Responsibility to Race." *Journal of Business Ethics* (2019) 154: pp. 977–988. https://doi.org/10.1007/s10551-018-3893-3.
Logan, Nneka. "Corporate Voice and Ideology: An Alternate Approach to Understanding Public Relations History." *Public Relations Review* 40 (2014): pp. 661–668.
Logan Nneka. "The White Leader Prototype: A Critical Analysis of Race in Public Relations." *Journal of Public Relations Research*, 23:4 (2011): pp. 442–457. https://doi.org/10.1080/1062726X.2011.605974.
Luke, Timothy W. *Ideology and Soviet Industrialization*. Westport, CT: Greenwood Press. 1985.
Luke, Timothy W. *Screens of Power: Ideology, Domination and Resistance in the Information Society*. Urbana: University of Illinois Press, 1989.
Luke, Timothy W. *Social Theory and Modernity: Critique, Dissent and Revolution*. Newbury Park, CA: SAGE Publications, 1990.
Luke, Timothy W. "On Environmentality: Geo-Power and Eco-Knowledge in the Discourses of Contemporary Environmentalism." *Cultural Critique*, No. 31, The Politics of Systems and Environments, Part II. (Autumn, 1995), pp. 57–81.
Luke, Timothy W. "Liberal Society and Cyborg Subjectivity: The Politics of Environments, Bodies and Nature." *Alternatives: Global, Local, Political*: Vol. 21, No. 1, Jan-Mar. (1996): p. 1–30. https://doi.org/10.1177/030437549602100101.
Luke, Timothy W. *Ecocritique: Contesting the Politics of Nature, Economy and Culture*. Minneapolis: University of Minnesota Press, 1997.
Luke, Timothy W. *Capitalism, Democracy and Ecology: Departing from Marx*. Chicago: University of Illinois Press, 1999.
Luke, Timothy W. "Cyborg Enchantments: Commodity Fetishism and Human/Machine Interactions." *Strategies*: Vol. 13, No. 1, (2000): p. 39–62. https://doi.org/10.1080/10402130050007511.
Luke, Timothy W. *Museum Politics: Power Plays at the Exhibition*. Minneapolis: University of Minnesota Press, 2002.
Luke, Timothy W. *Anthropocene Alerts: Critical Theory of the Contemporary as Ecocritique*. Candor, NY: Telos Press Publishing. 2019.
MacKaye, Benton. *The New Exploration: A Philosophy of Regional Planning, with an Introduction by Lewis Mumford*. Urbana: University of Illinois Press, 1962.
MacKaye, Benton. *From Geography to Geotechnics*. Urbana: University of Illinois Press, 1968.
Malm, Andreas. *The Progress of This Storm: Nature and Society in a Warming World*. London: Verso, 2018.
Marris, Emma. *Rambunctious Garden: Saving Nature in a Post-wild World*. New York: Bloomsbury, 2011.

Marx, Karl. *Capital: A Critique of Political Economy*, Edited by Friedrich Engels. Vol. II. Moscow: Foreign Languages Publishing House, 1961.
Marx, Karl. *The German Ideology: Including Theses on Feuerbach and Introduction to the Critique of Political Economy*. Amherst: NY: Prometheus Books, 1998.
Marx, Karl. *Grundrisse: Foundations of the Critique of Political Economy*. Martin Nicolaus, trans. New York: Random House, 2005.
Masco, Joseph. *The Future of Fallout and other Episodes in Radioactive World-Making*. Durham, NC: Duke University Press, 2021.
Miller, Daniel, ed. *Materiality*. Durham, NC: Duke University Press, 2005.
Miller, Glen and Ashley Shew, eds. *Reimagining Philosophy and Technology: Reinventing Ihde*. Springer: Cham, Switzerland, 2020.
Mills, C. Wright. *The Power Elite*. Oxford: Oxford University Press, 1956.
Mills, C. Wright. *The Sociological Imagination*. New York: Oxford University Press, 1959.
Mills, C. Wright. *White Collar: The American Middle Classes*. Oxford University Press, 1951.
Mitchell, Timothy. *Colonizing Egypt*. Berkeley, CA: University of California Press, 1991.
Mitchell, Timothy. *Rule of Experts: Egypt, Technopolitics, Modernity*. Berkeley: University of California Press. 2002.
Mitchell, Timothy. *Carbon Democracy: Political Power in the Age of Oil*. New York: Verso, 2011.
Monbiot, George. *Feral: Rewilding the Land, the Sea, and Human Life*. Chicago: University of Chicago Press, 2014.
Moore, Jason W, ed. *Anthropocene or Capitalocene? Nature, History and the Crisis of Capitalism*. Oakland, CA: PM Press, 2016.
Mumford, Lewis. *Technics and Civilization*. New York: Harcourt Brace Jovanovich, 1934.
Mumford, Lewis. *The City in History: Its Origins, Its Transformations and Its Prospects*. New York: Harcourt, 1961.
Mumford, Lewis. *The Myth of the Machine: Technics and Human Development*. New York: Harcourt Brace Jovanovich, 1966.
Mumford, Lewis. *The Myth of the Machine: The Pentagon of Power*. New York: Harcourt Brace Jovanovich, 1970.
Peirce, Charles Saunders. *Philosophical Writings of Peirce*. Justus Buchler, ed. New York: Dover Publications, 1955.
Rutherford, Stephanie. *Governing the Wild: Ecotours of Power*. Minneapolis: University of Minnesota Press, 2011.
Sale, Peter F. *Our Dying Planet: An Ecologist's View of the Crisis We Face*. Berkeley: University of California Press, 2011.
Sandoz, Mari. *Cheyenne Autumn*. New York: Avon books, 1953.
Sellars, Wilfrid. *Science, Perception and Reality*. Atascadero, CA: Ridgeview Press, 1963.
Sheail, John. *Nature's Spectacle: The World's First National Parks and Protected Places*. London: Earthscan, 2010.
Silver, Timothy. *Mount Mitchell and the Black Mountains: An Environmental History of the Highest Peaks in Eastern America*. Chapel Hill: University of North Carolina Press, 2003.

Tomlinson, John. *The Culture of Speed: The Coming of Immediacy*. London: SAGE, 2007.
Tsing, Anna Lowenhaupt. *Friction: An Ethnography of Global Connection*. Princeton: Princeton University Press, 2004.
Virilio, Paul. *The Information Bomb*. London: Verso, 2000.
White, Damian F., Alan P. Rudy, and Brian J. Gareau. *Environments, Natures and Social Theory: Towards a Critical Hybridity*. New York: Palgrave, 2016.
White, Damian F., and Chris Wilbert, eds. *Technonatures: Environments, Technologies, Spaces, and Places in the Twenty-first Century*. Waterloo, Ont.: Wilfrid Laurier University Press, 2009.
Whitehead, Mark, Rhys Jones, and Martin Jones. *The Nature of the State: Excavating the Political Ecologies of the Modern State*. Oxford: Oxford University Press, 2007.

Correction to: Contemporary Youth Environmental Activism: Lessons from France and Italy

Paolo Stuppia

Correction to:
Chapter "Contemporary Youth Environmental Activism: Lessons from France and Italy" in: J. J. Kassiola and T. W. Luke (eds.), *The Palgrave Handbook of Environmental Politics and Theory*, Environmental Politics and Theory,
https://doi.org/10.1007/978-3-031-14346-5_23

The original version of the Chapter "Contemporary Youth Environmental Activism: Lessons from France and Italy" was inadvertently published with an incorrect author name "Paola Stuppia", which has been changed to "Paolo Stuppia". The chapter has been updated with the change.

The updated original version of this chapter can be found at
https://doi.org/10.1007/978-3-031-14346-5_23

© The Author(s), under exclusive license to Springer Nature Switzerland AG 2023
J. J. Kassiola and T. W. Luke (eds.), *The Palgrave Handbook of Environmental Politics and Theory*, Environmental Politics and Theory,
https://doi.org/10.1007/978-3-031-14346-5_28

Bibliography

Adams, Robert McCormick and Hans J. Nissen. *The Uruk Countryside: The Natural Setting of Urban Societies*. Chicago: University of Chicago Press, 1972.
Alpers, Paul. *What Is Pastoral*. Chicago: Chicago University Press, 1996.
Azar Gat. *War in Human Civilization*. Oxford: Oxford University Press, 2008.
Benfield, F.K. *People Habitat: 25 Ways to Think about Greener, Healthier Cities*. Washington, DC: Island Press, 2014.
Berry, Wendell. *A Continuous Harmony: Essays Cultural and Agricultural*. New York: Harcourt Brace Jovanovich, 1970.
Berry, Wendell. "Foreword," In *The Vandana Shiva Reader*, Vandana Shiva. Lexington: The University Press of Kentucky, 2014.
Berry, Wendell. "In Distrust of Movements." *The Land Report* 65 (1999): 3–7.
Berry, Wendell. *Sex, Economy, Freedom, & Community*. New York: Pantheon, 1992.
Berry, Wendell. *The Art of the Commonplace: The Agrarian Essays of Wendell Berry*, edited by Norman Wirzba. Berkley: Counterpoint, 2002.
Berry, Wendell. *The Long-Legged House*. New York: Harcourt, 1969.
Berry, Wendell. *The Unsettling of America: Culture and Agriculture*. San Francisco: Sierra Club Books, 1977.
Browne, Erik E. *Mound Sites of the Ancient South: A Guide to the Mississippian Chiefdoms*. Athens, GA: University of Georgia Press, 2013.
Bryant, Levi. "Military Technology and Socio-Political Change in the Ancient Greek City." *The Sociological Review* 38, no. 3 (1990): 484–516.
Buell, Lawrence. *Environmental Imagination*. Cambridge: Harvard University Press, 1995.
Deleuze, Gilles and Felix Guattari. *A Thousand Plateaus: Capitalism and Schizophrenia*. Translated Brian Massumi. Minneapolis: University of Minnesota Press, 1987.
Dunlap, Alexander. "'Agro si, mina NO!' The Tia Maria Copper Mine, State Terrorism, and Social War by Every Means in the Tambo Valley, Peru." *Political Geography* (2019): 10–25.
Fairer, David. "'Where Fuming Trees Refresh the Thirsty Air': The World of Eco-Georgic." *Studies in Eighteenth-Century Culture* 40, no. 1 (2011): 201–218.

Forman, Dave. *Confessions of an Eco-Warrior.* New York: Harmony Books, 1991.
Foster, John Bellamy, Brett Clark, and Richard York. *The Ecological Rift: Capitalism's War on the Earth.* New York: New York University Press, 2010.
Fox, Liam. "Bougainville President Says Panguna Mine Moratorium Remains in Place." *Radio Australia, Australian Broadcasting Company,* February 9, 2021. https://www.abc.net.au/radio-australia/programs/pacificbeat/bougainville-president-says-panguna-mine-off-limits/13134904.
Freyfogle, Eric T., ed. *The New Agrarianism: Land, Culture, and the Community of Life.* Washington, DC: Island Press, 2001.
Graham, A.C. "The 'Nung-chia' 農家 'School of the Tillers' and the Origins of Peasant Utopianism in China." *Bulletin of the School of Oriental and African Studies* 42, no. 1 (1979): 66.
Greear, Jake P. "Decentralized Production and Affective Economies: Theorizing the Ecological Implications of Localism." *Environmental Humanities* 7, no. 1 (2016): 107–127.
Grundy, Kenneth W. *River of Tears: The Rise of the RioTinto Zinc Mining Corporation.* London: Earth Island Press, 1974.
Hanagan, Nora. "From Agrarian Dreams to Democratic Realities: A Deweyan Alternative to Jeffersonian Food Politics." *Political Research Quarterly* 68, no. 1 (2015): 34–45.
Hanson, Victor David. *The Other Greeks: The Family Farm and the Agrarian Roots of Western Civilization.* New York: Free Press, 1995.
Heidegger, Martin. "Why Do I Stay in the Provinces?" In *Heidegger, Philosophical and Political Writings,* edited by Manfred Stassen. New York: Continuum International, 2006.
Hodder, Ian. *The Leopard's Tale.* London: Thames and Hudson, 2006.
Ingram, Darby. "Bougainville President Rejects Panguna Mine Claims." *National Indigenous Times.* February 1, 2021. https://nit.com.au/bougainville-president-rejects-panguna-mine-claims/.
Jones, Daniel. "Panguna Mine Dilemma." Filmed 2008, *Eel Films.* https://www.youtube.com/watch?v=Sv8Q5hH0cys.
Lesur-Dumoulin, Claire, Eric Malézieux, Tamara Ben-Ari, Christian Langlais, and David Makowski. "Lower Average Yields but Similar Yield Variability in Organic Versus Conventional Horticulture: A Meta-analysis." *Agronomy for Sustainable Development* 37, no. 5 (2017): 45.
Lynch, Kevin. *Good City Form.* Boston, MA: MIT Press, 1981.
Major, William. "Other Kinds of Violence: Wendell Berry, Industrialism, and Agrarian Pacifism." *Environmental Humanities* 3, no. 1 (2013): 31.
Martinez-Alier, Joan. *The Environmentalism of the Poor: A Study of Ecological Conflicts and Valuation.* Northampton: Edward Elgar Publishing, 2003.
Marx, Leo. *The Machine in the Garden.* Oxford: Oxford University Press, 1964.
Mcdonald, Joshua. "Will Bougainville Reopen the Panguna Mine?" *The Diplomat,* November 22, 2019. https://thediplomat.com/2019/11/will-bougainville-reopen-the-panguna-mine/.
McGuirk, Rod. "Bougainville Votes for Independence from Papua New Guinea." *The Diplomat,* December 14, 2019. https://thediplomat.com/2019/12/bougainville-votes-for-independence-from-papua-new-guinea/.
McKibben, Bill. *The End of Nature.* New York: Random House, 1991.

Milner, G.R. "Mississippian Period Population Density in a Segment of the Central Mississippi River valley." *American Antiquity* 51, no. 2 (1986): 228.

Montmarquet, J.A. "Philosophical Foundations for Agrarianism." *Agriculture and Human Values* 2, no. 2 (1985): 5.

Morison, James, Rachel Hine, and Jules Pretty. "Survey and Analysis of Labour on Organic Farms in the UK and Republic of Ireland." *International Journal of Agricultural Sustainability* 3, no. 1 (2005): 24–43

Morton, Timothy. *Being Ecological*. Cambridge: The MIT Press, 2018.

Most, Glen W., ed. and trans. *Theogony, Works and Days, Testimonia*. New Haven, CT: Harvard, Loeb Classical Library, 2006.

Mumford, Lewis. *Technics and Civilization*. New York: Harcourt Brace and Co., 1934.

Nelson, Stephanie. "Hesiod, Virgil, and the Georgic Tradition." *The Oxford Handbook of Hesiod*. Oxford: Oxford University Press, 2018: 368.

Nelson, Stephanie. *God and the Land, The Metaphysics of Farming in Hesiod and Vergil*. Oxford: Oxford University Press 2008.

Oelschlaeger, Max. *The Idea of Wilderness: From Prehistory to the Age of Ecology*. New Haven: Yale University Press, 1991.

Pauketat, T.R. *Cahokia: Ancient America's Great City on the Mississippi*. New York: Penguin Random House, 2009.

Pearson, Craig J. "Regenerative, Semi-closed Systems: A Priority for Twenty-first-century Agriculture." *Bioscience* 57, no. 5 (2007): 409–418.

Reagan, Anthony J. "Causes and course of the Bougainville conflict." *The Journal of Pacific History* 33, no. 3 (November 1998): 269–285. https://www.jstor.org/stable/25169410.

Ritchie, Hannah and Max Roser. "Urbanization." OurWorldInData.org (2018). Accessed February 1, 2022, https://ourworldindata.org/urbanization.

Russell, Nerissa. "Spirit Birds at Neolithic Çatalhöyük." *Environmental Archaeology* 24, no. 4 (2019): 377–386.

Scales, Ivan. "Green Consumption, Ecolabelling and Capitalism's Environmental Limits." *Geography Compass* 8, no. 7 (2014): 477–489.

Schneider, Mindi and Philip McMichael. "Deepening, and Repairing, the Metabolic Rift." *The Journal of Peasant Studies* 37, no. 3 (2010): 461–484.

Schwartz, Judith D. "Soil as Carbon Storehouse: New Weapon in Climate Fight?" *Yale Environment 360*. Accessed March 13, 2019, https://e360.yale.edu/features/soil_as_carbon_storehouse_new_weapon_in_climate_fight.

Scott, James C. *Against the Grain: A Deep History of the Earliest States*. New Haven, CT: Yale University Press, 2017.

Scott, James C. *The Art of Not Being Governed*. New Haven: Yale University Press, 2009.

Sharer, Robert J., and Loa P. Traxler. *The Classic Maya*. Stanford, CA: Stanford University Press, 2006.

Sterckx, Roel. "Ideologies of Peasant and Merchant in Warring States China." In *Ideologies of Power and Power of Ideologies in Ancient China*, edited by Yuri Pines, Paul R. Goldin, and Martin Kern, 211–248. Boston: Brill, 2015.

Sweet, Timothy. *American Georgics*. Philadelphia: University of Pennsylvania Press, 2022.

Thompson, Paul B. "Chapter 1: Sustainability and Environmental Philosophy." In *The Agrarian Vision: Sustainability and Environmental Ethics*, 18–41. Lexington: University Press of Kentucky, 2010.

Trigger, Bruce. *Understanding Early Civilizations*. Cambridge: Cambridge University Press, 2003.

Walter, Mariana and Lucrecia Wagner. "Mining struggles in Argentina. The keys of a successful story of mobilization." *The Extractive Industries and Society* 8, no. 4 (2021).

Wesley-Smith, Terence and Eugene Ogan. "Copper, Class, and Crisis: Changing Relations of Production in Bougainville." *The Contemporary Pacific* 4, no. 2 (Fall 1992): 245–267. https://www.jstor.org/stable/23699898.

White, Lynn Jr. "The Historical Roots of Our Ecologic Crisis." *Science* 155, no. 3767 (1967): 1203–1207.

Wittfogel, Karl August. *Oriental Despotism: A Comparative Study of Total Power*. New Haven: Yale, 1957.

Worster, Donald. *Nature's Economy*. Cambridge: Cambridge University Press, 1994.

Yoffee, Norman. *Myths of the Archaic State*. Cambridge: Cambridge University Press, 2005.

Ziser, Michael G. "Walden and the Georgic mode." *Nineteenth-Century Prose* 31, no. 2 (2004): 186–208.

Zomer, R.J., Deborah A. Bossio, Rolf Sommer, and Louis V. Verchot. "Global Sequestration Potential of Increased Organic Carbon in Cropland Soils." *Scientific Reports* 7, no. 1 (November 2017): 1–8.

INDEX

A
"A person is a person through other persons" ("I am because we are"), 504, 522
A Theory of Justice, John Rawls, 29
the absolutist conception of private property, 524
Actor-Network Theory, 309, 450, 669, 670
adaptation learning, 21, 617
adaptation to climate change, 393, 639
advocacy in philanthropy, 379
African eco-humanism, 504, 512, 518, 535
African political thought, 18, 503
agency, 56, 59, 79, 97, 103, 139, 202, 215, 260, 261, 271–273, 275, 276, 278–281, 284, 285, 289–291, 302, 310, 311, 414, 504, 507, 517, 528, 529, 534, 535, 540, 548, 559, 579, 605, 632, 634, 656, 658, 668, 681, 685, 692
agnotology, 329, 332
agonistic theory of democracy, 56
agrarian environmentalism, 13, 14, 149, 151, 153, 165, 168
agrarianism, 14, 149–151, 155–158, 164
agricultural zoning, 458
agri-food networks, 17, 430–437, 439, 440, 446, 450, 453, 456–460

alienation, 164–167, 196, 283, 284, 399, 411–413, 416–418, 524
All We Can Save Project, environmental organization, 215
anamnestic solidarity, 517, 518
anarchy, 540, 554, 556, 557, 559
animal citizens, 13
animal machines, 107
animism, 185, 504, 514, 535
the Anthropocene, 15, 20–23, 70, 91, 92, 97, 99, 103, 108, 121, 123, 168, 186–188, 223–225, 227, 238, 239, 241, 243, 244, 246, 248, 252, 327, 328, 473, 484, 515, 541, 599–603, 605, 606, 608, 610, 613, 614, 621, 627–636, 638–642, 650–653, 655, 656, 658–664, 668, 673, 686, 689, 694
the Anthropocene Epoch Hypothesis, 22, 627–629, 638, 643
the Anthropocene Working Group (AWG), 21, 628, 630, 635–638, 640, 642, 643
Anthropocene starting dates, 21
anthropocentrism, 13, 32, 175, 176, 179–185, 188–191, 217, 226, 351, 468, 482, 535
anthropomorphic policy, 262
anticipatory ecosystem policy, 261
anticipatory regulation, 266
apoliticality, 154

Aristotle, 109, 110
artificial persons, 23, 668, 678, 680, 683–685, 688–690, 692, 694
the Association of Caribbean States (ACS), 547, 550, 552, 555
"autogestion" (self-management), 414, 415, 419, 422, 423
axiology, 56, 349, 387, 392

B
Barak, Nir, 17
Bataille, Georges, 13, 132, 136, 138, 139
Beck, Ulrich, 327, 350
Behrens III, William W., 2, 91
Bendell, Jem, 12, 89, 91–97, 99–103
Berry's eco-agrarianism, 151
Berry, Thomas, 173
Berry, Wendell, 13, 149, 151–154, 158, 165–167
Beyond Nature and Culture, Philippe Descola, 428
biocentrism, 180, 189, 242, 246
biopolitics, 22, 23, 650, 651, 653–660, 662–664
Bookchin, Murray, 38, 47, 398, 582
Bougainville, Island, Papua New Guinea, 14, 147
Bourban, Michel, 9, 11, 12, 67, 76, 78, 79, 81, 83
Bourblanc, Magalie, 15, 16
Brundtland Report, United Nations World Commission on Environment and Development (WCED), "Our Common Future", 353, 354

C
Cadalen, Pierre-Yves, 9, 22, 23, 649
capitalism, 14, 34, 36, 37, 96, 131, 137, 141, 148, 149, 188, 200–207, 209, 211, 216–218, 230, 239, 241, 244, 356, 413, 467, 472, 474, 479, 482, 491, 511, 534, 570, 581, 660, 663, 664, 678, 690, 692
Caribbean environmental system, 558
Caribbean territories, 544, 547, 552, 553
Carson, Rachel, 4, 201, 326, 327, 477
categories of environment regulations, 74, 307
ceaseless flux, 535
ch'i, 180, 181, 185
citadin, 414
citification, 398
citizenship, 11–13, 44, 49–59, 67, 77, 78, 83, 84, 140, 199, 279, 397, 398, 413, 414, 569, 589, 635
city-nature dualism, 387
city-nature hybrid, 17, 392
city-nature relationship, 387, 389, 395
city-zens (citizens in cities), 386, 397, 398
civic ecologism, 387, 397–399
civic humanist republicanism, 44
climate change, 2, 5, 8, 9, 12, 14, 15, 17, 19, 27, 51, 66, 70–73, 75–77, 81, 82, 84, 89, 90, 92–95, 97–99, 101, 102, 129, 187, 196–199, 201–205, 207–210, 216, 217, 224, 241, 260, 268, 305, 328, 367–373, 375, 380, 393, 470, 471, 482, 510, 549, 568, 569, 581, 587, 589, 590, 600, 603, 605–609, 611, 615–618, 620, 633, 638, 639, 650, 659, 660, 663
climate change deniers, 9
climate emergency, 15, 195–199, 208–213, 215–218
climate generation, 570, 575, 589
climate philanthropy, 16, 368–371, 373
cognitive relativism, 187, 308
collapse of society, 96, 581, 611
collapsology, 89
collective action, 80, 82, 159, 214, 273–278, 280, 281, 293, 313, 336, 368, 375, 569, 579, 599, 654, 661
collective goods, 266, 267, 279, 292
collective risk, 267, 281
colonialism, 14, 131, 201, 203, 205, 208, 217, 218, 417, 474, 530, 531, 557
The Communist Manifesto, Karl Marx and Friedrich Engels, 35
community-engaged scholarship, 14, 196, 217
complexification, 613

complex wholeness, 521
conceptualizations, 165, 271, 387, 396, 472, 512, 519, 540, 559, 639, 660, 663
conceptual maps, 92
the Conference of the Parties (COP), United Nations, 568, 569, 571, 577, 591
Confucian Green Theory, 180
consciencism, 510–512
conservation, 39, 89, 95, 247, 282, 283, 351, 388, 438, 475, 477, 508, 543, 545, 555, 556
cosmocentrism, 180
cosmology, 180, 181, 185, 189–191, 261, 282, 391
cosmopolitical perspectives, 474
COVID-19 pandemic, 2, 6, 16, 83, 94, 131, 324, 372, 567, 572, 574, 639
Crews, Chris, 18, 473
critical ecofeminism, 14, 196, 199, 210–212, 214, 216, 217
critical geography, 24
critical realism, 302
critique of feminism, 203
Cronon, William, 680, 681, 692
Crutzen, Paul J., 186–188, 327, 599, 621, 629–631, 638
Crutzen, Paul J. and Eugene F. Stoermer, Global Change Networks
Crutzen, Paul J. "Geology of Mankind", 223

D
D'Alisa, Giacomo, 9, 13, 511
dark green environmental policies, 27, 38
Davies-Venn, Michael, 21, 22, 627
the *Death of Industrial Civilization*, Joel Jay Kassiola, 4
the *Death of Nature*, Caroline Merchant, 183, 337
the *Division of Labor in Society*, Emile Durkheim, 452
decolonial feminism, 206, 207
decoloniality, 204–207
deep adaptation, 12, 89, 91, 92, 96, 97, 99, 100, 102

deep adaptation agenda, 92, 94, 95, 98, 99, 101, 102
deep ecology, 106, 114, 123, 165, 227
degrowth (*la decroissance*) society, 13, 130, 132, 137, 140
Delmas, Corinne, 16
democracy, 12, 29, 31–34, 36, 43–46, 48, 51, 53, 54, 56, 58, 59, 92, 122, 140, 168, 190, 336, 348, 396, 477, 479, 569, 585, 589
democratic citizenship, 12
Descola, Philppe, 309, 428, 450
Dobson, Andrew, 27, 31, 33, 38, 39, 52, 53, 55, 77–79, 82, 241
"Do It Ourselves" politics (DIO politics), 20, 579, 585, 589
Down to Earth, Bruno Latour, 440, 447, 448
Drewes Farm Partnership v. City of Toledo, Ohio, 489, 490
Durkheim, Emile, 29, 135, 300, 452, 454, 455

E
early civilizations, 162
early states, 159–161, 164, 167
Earth-centered politics, 473, 482
Earth in the Balance, Al Gore, 93
Earth jurisprudence, 477, 482, 483
Earth rights, 477
Earth spirits, 507–509, 514, 527–531, 534, 535
Earth Summit, United Nations, 82, 354, 355
eco-agrarianism, 151
eco-anxiety, 12, 66–80, 83, 84
eco-anxiety coping mechanisms, 12, 76, 78
eco-anxiety disorders, 12, 66, 72, 73, 84
ecocentrism, 180, 391
eco-citizenship, 11
Ecocritique: Contesting the Politics of Nature, Timothy W. Luke, 4, 686
ecofeminism, 14, 196, 197, 199–204, 207, 210–214, 216, 217, 337, 338
ecofeminist knowledge production, 210
eco-fiction, 70, 71, 84
eco-humanism, 504, 512, 515, 518, 535

ecological citizenship, 12, 67, 77, 78, 83, 84, 199
ecological disaster, 2, 70, 74, 84, 95, 516, 669
ecological inequality, 336
ecological kinship, 18, 476, 478, 481, 483, 484, 494
ecological overshoot, 89
ecological ownership, 243–245
ecological predicament, 14, 138, 225, 238, 241, 252
ecological risks, 67, 69, 80
ecologism, 11, 27–29, 31–34, 36–41, 82, 241, 351
eco-Marxism, 166
economic growth, 2, 5, 8, 12, 13, 50, 60, 89, 93, 130, 149, 188, 210, 248, 353, 356, 395, 542, 631
ecopower, 22, 23, 651, 657–664
eco-socialism, 33, 36, 37
ecosystem, 2, 18, 19, 39, 67, 68, 74, 75, 77, 81, 96, 114, 119, 120, 122, 123, 152, 176–178, 181, 183, 190, 198, 210, 215, 216, 264, 271, 276, 279, 280, 282–285, 289, 291, 301, 324, 326–328, 338, 373, 377, 393, 394, 399, 459, 473, 477, 479, 481, 484, 490–494, 540–542, 544, 548, 552, 558, 559, 601, 608, 609, 611, 613, 614, 618, 619, 633, 671, 690, 691, 693
ecosystem law, 264, 282
ecosystem policy, 264, 285, 287
efficiency of public policy, 265
embeddedness, 392, 394, 396, 535
endocrine disruptors, 326, 329–333
the Entropy Law and Economic Process, Nicholas Georgescu-Roegen, 249
environmental activism, 19, 468, 478, 578, 590
environmental boundaries, 293, 468, 473, 495, 602, 639
environmental ethics, 11, 13, 14, 83, 149, 156, 174–178, 180, 181, 183–185, 187–192, 262, 263, 337, 388, 392, 394, 396, 504, 517, 535
environmental health, 15, 16, 188, 324, 326, 329, 330, 332, 333, 336–338

environmental health problems, 324, 329, 331–333, 338
environmentalism, 9, 11, 13, 14, 28, 31, 149, 151, 153, 165, 168, 224, 241, 353, 386, 389, 396, 527, 570, 582, 591
environ-mentality, 24, 668, 669, 671–673, 680, 683, 685, 686, 688, 689, 692
environmental justice, 14, 18, 53, 54, 131, 150, 197, 201, 209, 210, 217, 333–338, 370, 386, 395, 399, 424, 468–470, 472–474, 481, 484, 485, 494, 495
environmental kinship, 495
environmental limits, 128
environmental philanthropy, 369
environmental policy analysis, 300, 318
environmental political theory (EPT), 4, 5, 7–12, 22, 44, 45, 47, 50, 51, 75, 174, 175, 183, 188, 189, 191, 217, 346, 469, 503, 664, 670
environmental politics and theory, 2–4, 7, 10, 11, 15–17, 19, 21, 23, 24, 98, 102
environmental risk policy, 260, 261, 265–267, 271–273, 292
ethics, 11, 13, 14, 51, 54, 67, 80, 83, 106, 149, 174–178, 180, 181, 183–185, 187–192, 227, 247, 262, 263, 292, 293, 337, 388, 392, 394, 396, 420, 478, 483, 484, 504, 508, 515, 517, 535, 588, 639
Ethics of the Fathers, Rabbi Tarfon, 5
European Union integration, 357
evidence-based policy, 304
Exclusive Economic Zone (EEZ), 555
expert knowledge, 16, 300, 303–306, 310, 311, 316–319, 558
Extinction Rebellion (XR), environmental organization, 102, 197, 472, 568

F
Feminist Environmental Research Network (FERN), 14, 196, 197, 216, 217
Flipo, Fabrice, 8, 11, 27

food miles, 17, 430, 436
food relocalization, 431–435, 437–440, 446, 447, 450–452, 454, 455, 457–460
food supply chains, 67
forms of power, 410, 417, 421, 654
Foucault, Michel, 22, 23, 136, 650–664, 672, 682
foundations as political actors, 378
foundations committed to supporting the climate, 371
fourth wave of feminism, 14, 199
framing social policy, 352
France, 8, 19, 28, 34, 37, 39, 40, 117, 224, 229, 242, 325, 327, 329, 331, 333, 336, 338, 349, 361, 363, 369, 372, 374–376, 407, 431–433, 436, 437, 440, 460, 570, 573–576, 578, 586, 589–591, 642
French youth environmental activist organizations, 578
Fridays for Future (FFF), ("School Strikes for Climate"), Greta Thunberg, 567, 571

G
gender, 14, 21, 39, 55, 60, 131, 197, 199–205, 207, 215, 216, 324, 337, 395, 575, 577, 578, 587, 589, 637, 640
gender inequality, 337, 338
genealogy, 477
Genesis, 177, 183, 190
the Geological community, 635, 637
Geological Time Scale, 604, 607, 628, 635, 642, 643
geology, 628, 629, 637
Georgescu-Roegen, Nicholas, 130, 249
Ghanaian Chief of Medicine (*BagenabaS*), 505
Gillroy, John Martin, 15, 16
Global Boundary Stratotype Section and Points (GSSP), 635, 636, 641
global environmental politics, 634, 638
Global Goals for Adaptation, 97
global governance, 620, 638, 639

globalization, 176, 183, 434, 435, 440–442, 470, 471, 515, 585, 612, 631
the Global North, 351, 634, 636, 638, 640
the Global South, 201, 202, 204, 540, 541, 634, 636–638, 640
the Golden Spike, 628, 641
the Great Divide, 17, 428–430
the Great Transformation, Karl Polanyi, 442
Global Standard Stratigraphic Age (GSSA), 628, 641
global warming, 1, 5, 8, 15, 68, 92, 94, 153, 168, 470, 568, 569, 571, 573, 577, 585, 590, 591, 599, 633
Gore, Al, 93, 94
governmentality, 314, 673, 684, 685, 692
Gramsci, Antonio, 13, 141, 143, 650
Greear, Jake, 9, 13, 14, 154
green citizenship, 11, 12, 51–59, 77
green gentrification, 394, 396
greenhouse gas emissions, 2, 3, 8, 12, 82, 93, 94, 99, 241, 557, 572, 601, 603, 640
greenhouse gases, 2, 8, 90, 183, 185, 240, 438, 631
greening of democracy, 54
green political economy, 395
Green Political Thought, Andrew Dobson, 27, 33, 38, 241
green virtue, 12, 79, 81, 84
gross domestic product (GDP), 39, 131
growth imaginary, 128, 129, 131

H
harmony with nature, 28, 52, 158, 164, 481–483, 581
Hayek, Friedrich, 30, 50
health and the environment, 215, 324
health risks, 323, 325, 328, 329, 550
Hesiod's ecological agrarianism, 157
"The Historical Roots of Our Ecological Crisis," Lynn White, Jr., 190
Holocene Geological Epoch, 186, 628, 630, 641–643
horizontalism, 137–140

human settlements, 90, 348, 352, 354, 356–358, 606, 607, 616
human supremacist ethics, 175, 176, 181, 184
hybridity, 392, 669, 672, 673
hybrid socio-natural conceptions, 311, 316

I
Ignatov, Anatoli, 9, 18, 19, 505, 528
Ihde, Don, 23, 24, 667, 669–681, 683–685, 687–689, 691, 692, 694, 695
Ihde's lifeworld theory, 673–675
inaction as a psychological defense, 72, 83
inaction as paralysis, 72, 83
indigenous land management, 208
inequality, 23, 35, 46, 55, 113, 131, 138, 333, 335–338, 394, 399, 512
inevitable near-term human extinction ("INTHE"), 92, 97, 103
"infowhelm", 71
instrumental value, 32, 81, 262–264, 266, 270, 280, 281, 284, 289, 292, 293, 507
instrumented expert knowledge, 16, 300, 310, 316–319
Intergovernmental Panel on Climate Change (IPCC), 2, 3, 5, 6, 8, 9, 70, 74, 82, 93, 94, 97, 198, 373, 568, 660
international actors, 19, 542, 546, 559
international agents, 549
International Commission on Stratigraphy (ICS), 628, 635, 636, 638, 641–643
international environmental policy, 539, 540, 547, 555, 559
International Labor Organization's Indigenous and Tribal People's Convention, 18
internationally socially constructed system, 558
International Mother Earth Day, 473, 481, 482, 484
International Philanthropy Commitment on Climate Change, 372
international regimes, 345–348, 360, 362, 558
international relations, 16, 345–351, 355–357, 359, 362, 363, 379, 541, 543, 547, 549, 550
international relations between housing and the environment, 346, 351, 356
intersectionality, 199, 201–204, 207
intrinsic value, 31, 33, 81, 262–267, 272, 273, 276, 279–281, 284–286, 289, 290, 292, 293, 315, 479, 492, 507
Italian youth environmental activist organizations, 573
Italy, 19, 130, 372, 374, 570, 573, 574, 576–578, 580, 584, 589, 591

J
Japanese model of infrastructure planning and policies, 618

K
Kantian conservationism, 276
Kantian morality, 16
Kant, Immanuel, 31, 109, 110, 113, 261, 271–279, 281, 282, 284–289, 291–293, 557
Kassiola, Joel Jay, 4–6, 9, 11, 13, 14, 24, 174, 175, 226, 259
kincentric ecologies, 18
kinship, 18, 473–476, 478–481, 483, 484, 494, 495, 520
kith

L
Lake Erie Bill of Rights (LEBOR), 484, 486–494
Lake Erie's rights of nature, 18, 484, 488, 490, 491, 493, 494
land ethics of Aldo Leopold, 177
Latouche, Serge, 130, 138, 139
Latour, Bruno, 17, 121, 122, 308, 309, 311, 428, 440–450, 453, 457, 459, 669–671, 694
Laudato Si, ("Praise be to you"), Pope Francis, 475

Lawrence, Jennifer L., 9, 11, 14
Lefebvre, Henri, 17, 407–416, 418, 420, 423
Leopold, Aldo, 14, 176–180, 185, 190, 191, 394, 477
Leopold's conception of "land", 177
liberalism, 11, 29–34, 40, 45, 53–55, 210, 468, 524, 650
liberal morality, 45, 47
light green environmental policies, 27
limitless economic growth, 5
limits, 2, 9, 38, 68, 91, 93, 116, 128, 132, 134, 143, 191, 201, 207, 225, 250, 301, 325, 352, 356, 386, 390, 429, 437, 441, 443, 446, 456, 458, 460, 469, 481, 487, 526, 528, 545, 556, 599, 602, 612, 614, 621, 642, 662, 663
the *Limits to Growth*, Meadows et al., 4, 9, 82, 252
Luke, Timothy W., 4, 9, 12, 23, 92, 99, 667–670, 672, 673, 683–692, 694, 695

M
machines, 29, 30, 237, 238, 248, 284, 685–690
the market, 29, 31, 35, 37, 240, 267, 269–271, 273–276, 278–281, 283, 285, 290, 293, 357, 361, 442
the market economy, 46, 242, 356
Marx, Karl, 30, 35, 36, 149, 416, 554, 690, 693
material embeddedness of social life, 300
materialism, 35, 188, 214, 216, 226–228, 235, 309, 310, 512, 534, 672, 690
McKibben, Bill, 168
Meadows, Dennis L., 2, 91, 252
Meadows, Donnella H., 2, 91, 252
mental health and climate change, 72
Merchant, Caroline, 183, 337
milieux (environments), 22, 650
Mills, C. Wright, 23, 672–674, 677, 678, 683
mindfulness, 12, 79–82, 84
mitigation of climate change, 9, 639

mobilization of philanthropy for the climate, 367, 368, 375
modernity, 32, 33, 35, 39, 90, 102, 134, 135, 156, 165, 167, 183, 186, 188, 205, 206, 227, 241, 242, 244, 251, 327, 428, 441, 443, 445, 446, 450, 458
Monier, Anne, 15–17
monism, 17, 200, 387, 392, 396, 399
Morales, Evo, 18, 481, 482
morality, 16, 45, 47, 49, 51, 54, 110, 176, 177, 180, 182, 187–189, 274, 285, 286, 288, 289, 515, 519, 639
Mother Earth rights, 481, 482
Mousie, Joshua L., 17
multispecies justice, 18, 476
Mumford, Lewis, 23, 149, 672, 690–692, 694

N
Nasr, Clemence, 17
natural city, 387, 389–392
natural systems, 262–264, 266, 272, 276, 280–287, 289, 291, 293, 638
nature/human entanglement, 614
nature in cities, 393, 394, 396
nature vs. culture, 23, 52, 53, 226, 237, 387, 389, 428, 430, 450, 476, 515, 521, 526
Navdanya, environmental organization, 214, 215
negritude, 510–514
Nemoz, Sophie, 16
neoclassical economics (NCE), 248
neoliberalism, 139, 203, 206, 210, 470
new naturalism, 23, 668–671
Nkrumah, Kwame, 511–515, 526
non-anthropocentrism, 182, 184, 185, 388, 390
non-Western spirituality, 478
Non-Western thought, 11
normativity, 675, 682

O
oeuvre (a work), 411–414, 416–419
ontology, 23, 24, 309, 411, 419, 424, 473, 512, 513, 522, 669, 672, 679

ontology of the city, 411, 419
orbis hypothesis, 634
Ostrom, Elinor, 224, 544
outermost European territories, 547

P

paradigm, 24, 28, 32, 182, 187, 188, 205, 210, 224, 225, 244, 251, 260, 261, 263, 264, 269–271, 273–293, 301, 303, 331, 359, 374, 389, 391, 467, 471, 473, 475, 481–483, 494, 540, 546, 600, 602, 614, 620, 653, 655, 657
Paris Climate Agreement, 2015, United Nations, 638
Parra-Leylavergne, Andrea, 19
Pattberg, Philippe, 21, 22, 627
phenomenon, 34, 83, 121, 197, 214, 280, 316, 356, 370, 389, 416, 440, 442, 454, 459, 540, 548, 556, 575, 577, 607, 609, 615, 616, 634, 653, 658, 659
philanthropy and politics, 377, 378
Philia, Philanthropy European Association, 372
philosophical policy and legal design (PPLD), 274, 291–293
place-based kinship, 473
planetary boundaries, 21, 22, 68, 183, 599, 605, 606, 620, 641
Polanyi, Karl, 138, 442
the political, 10, 11, 21, 24, 33, 39, 43, 56, 57, 82, 106, 110, 112, 114, 115, 121, 139, 141, 155, 158, 164, 188, 276, 277, 279, 281, 288, 313, 315, 347, 349, 356, 362, 378, 379, 386, 387, 392, 395, 396, 407, 422, 431, 432, 442, 444, 445, 448, 449, 451, 456–458, 476, 484, 487, 508, 528, 531, 533, 545, 550, 600, 607, 610
political analysis, 11, 661, 681
political ecology, 28, 302, 347, 349, 395, 399, 429, 434, 435, 686
political hermeneutics, 681
political ideologies, 11, 12, 20, 29
political moralism, 45

political science, 3, 174–176, 196, 302, 304–306, 348, 352, 369, 370, 544
political violence, 208
the politics of deep adaptation, 12, 13, 91
polycentricity, 658, 660
Pope Francis, 475
positivism, 260, 261, 285, 290
postcolonialism, 199, 504, 512, 528, 535, 634
posthumanism, 199
postphenomenological environmentality, 24, 680
postphenomenology, 669, 671, 679, 681, 683, 684
The *Power Elite*, C. Wright Mills, 673
Poulantzas, Nicos, 23, 32, 658
power, 14, 15, 22, 23, 29, 32–34, 36, 37, 39, 43, 44, 46, 48, 49, 54, 56, 58, 59, 101, 120, 122, 123, 141, 142, 157, 159, 161, 164, 198–200, 202, 205, 207, 211–214, 216–218, 225–227, 229–234, 237–240, 244, 245, 247–249, 251, 252, 264, 283, 290, 305, 306, 308, 310–312, 314, 315, 318, 319, 330, 333, 348–350, 358–360, 362, 363, 370, 377–379, 394–396, 407, 409, 410, 415, 417, 421, 423, 442, 444, 448, 456, 457, 472, 477, 479, 484, 487, 491, 495, 505, 509, 528–534, 544, 621, 634, 635, 640, 650–652, 654–658, 660–664, 685, 686, 688, 689, 693, 694
precarity of life, 101, 103
private goods, 267
private risk, 268
production of knowledge, 211, 214, 300, 305, 688
progressive liberalism, 44
property, 14, 15, 36, 46, 48, 111, 131, 175, 210, 223–231, 233, 234, 237–239, 241, 243–245, 247, 251, 252, 277, 279, 281, 421, 423, 525, 526, 689
property as the fundamental mechanism of social power, 227, 231
property as the fundamental mechanism of the law, 227, 230

protests, 2, 19, 20, 38, 46, 60, 98, 306, 361, 408, 416, 471, 474, 567, 569–574, 576, 578, 579, 582–585, 589–591
public policy instrumentation, 300, 312, 313, 318, 319
the public policy instrumentation framework, 314

R
the radicalness of movements, 582
Randers, Jorgen, 2, 91, 252
Rawls, John, 29
Ray, Emily, 9, 11, 14
relocalized society, 17, 430–432, 455–458, 460
republicanism, 45, 51, 55
republican realism, 44
resilience, 12, 21, 44, 76, 77, 79, 81, 90, 97, 98, 100–102, 201, 215, 243, 604, 611, 617
return to the Earth, 442
return to the soil, 445
rights of nature, 18, 31, 478, 481, 491, 493
Rights of Nature (RON) movement, 18, 468, 472, 476–480, 484, 491, 493, 495
the *Right to the City*, Henri Lefebvre, 17, 407
Rio Tinto Zinc, 147, 148
risk, 16, 38, 67, 69, 72, 97, 139, 242, 259–274, 278, 280, 281, 285, 290–293, 324, 327–332, 335–338, 359, 433, 454, 472, 484, 491, 552, 555, 569, 606, 607, 609, 614, 616, 617, 619, 620, 654, 656, 662, 670, 671, 684
risk assessment, 262, 333, 591
Risk Society, Ulrich Beck, 327
Rockström, Johan *et al*. "Planetary Boundaries", 252
Romano, Onofrio, 9, 13, 136, 137
ruralization, 159–162, 167

S
Saint Francis of Assisi, 177, 178, 190

Samoa Agreement on Small Island Developing States (SIDS), 545
sargassum, 19, 539, 540, 547–556, 558, 559
sargassum management system, 546
Scerri, Andy, 11, 12, 49, 52, 58
Schmaltz, Benoit, 14, 15, 226
Schmitt, Carl, 39
Science and Technology Studies (STS), 16, 299–312, 318, 668, 671
science-society co-constitution, 305
scientific knowledge, 16, 50, 70, 82, 240, 300, 303, 304, 306, 307, 311, 312, 317, 333, 605, 631, 676, 678, 687, 688
Senghor, Léopold Sédar, 512–515, 526
Silent Spring, Rachel Carson, 4, 201
Simon, Turquoise Samantha, 9, 11, 13
Singer, Peter, zoocentrism, 111, 113–115, 177, 182, 189
sixth mass species extinction, 68
small island states, 15
social change, 9, 13, 18, 139, 215, 231, 239, 240, 367, 584
the social construction of power relations
social constructivism, 19, 542, 672
social metabolism, 15, 227, 234, 235, 237–239, 243, 244, 248, 251, 252, 622
social power, 14, 225, 232–234, 237–240, 252
social sciences, 3, 22, 175, 189, 282, 290, 291, 299, 300, 303, 307, 324, 336, 346, 348, 367, 369, 431, 541, 600, 602, 634, 652, 668, 673
societal collapse, 12, 67–69, 89, 90, 92–96, 98–101, 612
the *Social Theory of International Relations*, Alexander Wendt
societal transformation, 6, 17
society, 5, 9, 13–15, 17, 18, 23, 24, 31, 35, 38–40, 50, 96, 100, 101, 103, 116, 127–134, 136, 137, 139–143, 148, 152, 158, 160, 163, 164, 166, 168, 182, 183, 187–190, 200, 205, 211, 212, 214, 227, 229, 231, 235, 237, 238, 246, 247, 249, 270, 274, 281, 292, 293, 305, 307–309, 311, 314, 318, 319, 327, 329, 334, 336,

367, 387, 392, 409, 415, 427, 429–432, 442, 445–458, 460, 495, 509, 514, 517, 539, 544, 569, 577, 581, 582, 585, 587, 609, 611–614, 618, 651, 654, 656, 661, 664, 669, 673, 677
socio-ecological imagination, 673
socio-ecology, 38
the sociology of scientific knowledge, 307
the sovereign, 22, 651, 652, 657, 658
the state, 13, 30, 57, 58, 71, 72, 78, 80, 82, 96, 107, 133, 134, 141, 143, 155, 157, 159, 160, 162–164, 167–169, 203, 207–209, 213, 230, 232, 264, 269–272, 275, 277, 278, 281, 314, 325, 331, 358, 377, 378, 386, 397–399, 410, 415, 420–422, 484, 485, 487, 492, 494, 504, 529, 533, 556, 559, 602, 605, 658, 661–663, 668, 694
the Statutes and Bylaws of the International Quaternary Union of Geological Sciences (IUGS), 628–630, 634–638, 641–643
stewardship, 473, 474, 524, 528
stratigraphy, 641, 643
Stubberfield, Alexander, 9, 23, 24
Stuppia, Paolo, 19, 20, 570, 573, 577, 588
the Subcommission on Quaternary Stratigraphy (SQS), 628, 635, 643
Supreme Court of Ohio, 487, 488
surplus management, 134
sustainability, 10, 12, 22, 53, 76–78, 91, 92, 95–98, 128, 129, 154, 185, 187, 192, 195, 211, 238, 260, 353, 360, 386, 387, 392–394, 396–399, 545, 588, 600, 601, 610, 612, 613, 621
sustainable agriculture, 154
sustainable urbanism, 17, 385–387, 396
systems of denial, 95

T
Tarfon, Rabbi, 5, 6, 10, 24
Taylor, Paul W., 14, 177–180, 185, 190, 191

technology, 23, 31–34, 37, 149, 166, 227, 241–243, 249, 283, 284, 287, 292, 293, 355, 358, 360, 395, 472, 612–615, 621, 672, 675, 678–680, 682, 685, 686, 688, 691, 694
technonaturalism, 23, 668–670, 682, 683, 693
technonatural subjectivity, 684
territory, 15, 29, 106, 116, 197, 207, 208, 317, 431, 440, 442–447, 451–453, 457–459, 529, 534, 547, 550, 556, 609, 616, 618, 619, 652
thermo-industrial civilization, 20, 21, 603, 606, 613, 614, 620
Thunberg, Greta, 6, 19, 143, 201, 209, 568, 571, 577, 585, 588, 590
tindaanas, 505, 527, 529–534
Transgenerational model of environmental ethics, 517

U
ubuntu, 504, 510, 516, 517, 521–523, 535
ukama, 504, 510, 516–521, 523, 535
uncivilization, 12, 89, 92, 103
United Nations Center for Human Settlements (UNCHS), 353
United Nations Conference on Environment and Development (UNCED), 354
United Nations Conference on Housing and Sustainable Urban Development, 345, 356
United Nations Conference on Human Settlements, 345, 352
United Nations Convention for the Protection and Development of the Marine Environment in the Wider Caribbean Region (WCR) (Cartagena Convention), 543
United Nations Convention on International Trade and Endangered Species (CITES), 545
United Nations Convention on the Law of the Seas (UNCLOS), 542, 553, 556

United Nations Declaration on the Rights of Indigenous People (UNDRIP), 207, 473, 475, 476
United Nations Environment Program (UNEP), 198, 351, 542, 548, 550, 551, 558
United Nations Framework Convention on Climate Change (UNFCCC), 82, 373, 471
Universal Declaration of the Rights of Mother Earth, Bolivia, 482
urbanization, 16, 161, 186, 327, 328, 355, 356, 361, 388, 391, 395, 398, 615, 619, 629, 669
urban nature, 393
utopianism, 391

V
verticalism, 137, 139, 140
Vidal, Florian

W
web of life, 32, 475, 608
We Do, environmental organization, 215, 216
Wendt, Alexander, 540–543, 546, 548, 551, 554–558
Western worldview, 14, 508

White, Lynn, Jr., 14, 166, 176, 178–180, 184, 185, 189–191, 475, 482
wilderness, 149, 150, 156, 165, 167, 387, 388, 521, 693
Women's Voices of the Earth (WVE), environmental organization, 215
World Commission on Environment and Development (WCED), 353, 354
world of becoming, 521, 535
worldviews, 5, 8, 11, 14, 19, 23, 45, 78, 176, 178, 180, 181, 183, 185, 190, 191, 203, 205, 206, 212, 224, 239–241, 247, 354, 391, 476, 482, 483, 507, 508, 514, 526, 542

Y
yin/yang, 181, 185
youth environmental activism, 19

Z
zero-infinity problem, 261
Zhang Zai, 180, 181, 185, 191
Zones for Defense (ZADs), environmental organization, 37, 573–575, 577, 578, 582, 584, 585, 589, 591
zoocentrism, 177, 180, 189
Zoopolis, 13, 105, 115, 117, 118